Spectroscopic Properties of Inorganic and Organometallic Compounds

Volume 23

A Specialist Periodical Report

Spectroscopic Properties of Inorganic and Organometallic Compounds

Volume 23

A Review of the Recent Literature Published up to Late 1989

Senior Reporters

G. Davidson, *Department of Chemistry, University of Nottingham*

E. A. V. Ebsworth, F.R.S.E., *Department of Chemistry, University of Edinburgh*

Reporters

J. H. Carpenter, *University of Newcastle upon Tyne*
S. J. Clark, *City University, London*
K. B. Dillon, *University of Durham*
J. D. Donaldson, *City University, London*
S. M. Grimes, *City University, London*
B. E. Mann, *University of Sheffield*
D. W. H. Rankin, *University of Edinburgh*
H. E. Robertson, *University of Edinburgh*
M. J. K. Thomas, *Thames Polytechnic, London*

ROYAL SOCIETY OF CHEMISTRY

ISBN 0-85186-213-6
ISSN 0584-8555

Published by The Royal Society of Chemistry,
Thomas Graham House, The Science Park, Cambridge CB4 4WF

Printed and bound by Bookcraft (Bath) Ltd.

Preface

There are no substantial changes in this year's volume. The topics covered are the same as in recent years, and the Reporters have as always done wonders in covering so much material in so compressed a way. I am sorry that Dr. S. Cradock can no longer contribute the chapter on Microwave Spectroscopy, for he has been a most efficient and reliable colleague, but I am very grateful indeed to Dr. Carpenter for taking over from him so well.

This is the last volume for which I shall act as Senior Reporter; from next year Dr. G. Davidson will be in sole charge. I have overseen the production of fourteen volumes in the series. I believe that these reviews, so well established by my predecessor, Professor N.N. Greenwood, are valuable to the community of inorganic chemists; I certainly use them, and I am delighted to see how they have continued despite uncertainties from one year to another. In the long term, the future of such publications as these will depend on developments in the form of scientific literature and on the impact of developing technology on the retrieval of information; for the present, I only wish that the price could be lower. I have enjoyed working with the Reporters and the staff of the Royal Society of Chemistry, and I wish the series well.

E.A.V. Ebsworth
August 1990

Contents

Conversion Factors

1 kJ mol^{-1}

2.3901×10^{-1}	kcal mol^{-1}
1.0364×10^{-2}	eV atom^{-1}
8.3593×10	cm^{-1}
2.5061×10^{6}	MHz

1 kcal mol^{-1}

4.1840	kJ mol^{-1}
4.3364×10^{-2}	eV atom^{-1}
3.4976×10^{2}	cm^{-1}
1.0486×10^{7}	MHz

1 cm^{-1}

1.1963×10^{-2}	kJ mol^{-1}
2.8592×10^{-3}	kcal mol^{-1}
1.2399×10^{-4}	eV atom^{-1}
2.9979×10^{4}	MHz

1 MHz

3.9903×10^{-7}	kJ mol^{-1}
9.5370×10^{-8}	kcal mol^{-1}
4.1357×10^{-9}	eV atom^{-1}
3.3356×10^{-5}	cm^{-1}

1 eV atom^{-1}

9.6485×10	kJ mol^{-1}
2.3060×10	kcal mol^{-1}
8.0655×10^{3}	cm^{-1}
2.4180×10^{8}	MHz

Moessbauer spectra: $E_{\gamma}(^{57}Fe) = 14.413$ keV

1 mm s^{-1}

4.639×10^{-6}	kJ mol^{-1}
1.109×10^{-6}	kcal mol^{-1}
4.808×10^{-8}	eV atom^{-1}
3.878×10^{-4}	cm^{-1}
1.162×10	MHz

For other Moessbauer nuclides, multiply the above conversion factors by E_{γ}(keV)/14.413.

1
Nuclear Magnetic Resonance Spectroscopy

BY B. E. MANN

1 Introduction

Following the criteria established in earlier volumes, only books and reviews directly relevant to this chapter are included, and the reader who requires a complete list is referred to the Specialist Periodical Reports 'Nuclear Magnetic Resonance',[1] where a complete list of books and reviews is given. Reviews which are of direct relevance to a section of this Report are included in the beginning of that section rather than here. Papers where only ^{1}H, ^{2}H, ^{19}F, and/or ^{31}P n.m.r. spectroscopy is used are only included when they make a non-routine contribution, but complete coverage of relevant papers is still attempted where nuclei other than these are involved. In view of the greater restrictions on space, and the ever growing numbers of publications, many more papers in marginal areas have been omitted. This is especially the case in the sections on solid-state n.m.r. spectroscopy, silicon and phosphorus.

Several relevant reviews have been published, including 'Ligand field interpretation of metal n.m.r. chemical shifts in octahedral d^6 transition metal complexes',[2] 'N.m.r. spectroscopy of the early transition metals',[3] 'Nuclear magnetic resonance of transition metal compounds', which contains ^{95}Mo n.m.r. spectroscopy,[4] 'Metal cluster complexes containing heteroatom-substituted carbene ligands', which contains ^{13}C n.m.r. spectroscopy,[5] 'Metalloproteins',[6] 'N.m.r. spectroscopy', which contains the use of ^{23}Na n.m.r. spectroscopy for the determination of intracellular sodium,[7] and 'Information obtained by liquid crystal n.m.r.', which contains its application to organic tin compounds.[8]

A number of papers have been published which are too broadly based to fit into a later section and are included here. Anomalous scalar couplings of up to 1400 Hz in trihyd-

1 'Nuclear Magnetic Resonance', ed. G.A. Webb (Specialist Periodical Reports), The Royal Society of Chemistry, London, 1989, Vol. 18; 1990, Vol. 19.
2 N. Juranić, *Coord. Chem. Rev.*, 1989, **96**, 253.
3 J. Mason, *Polyhedron*, 1989, **8**, 1657.
4 H. Liu, *Bopuxue Zazhi*, 1989, **6**, 87 (*Chem. Abstr.*, 1989, **111**, 107 433).
5 R.D. Adams, *Chem. Rev.*, 1989, **89**, 1703.
6 G.R. Hanson and G.L. Wilson, *Electron Spin Reson.*, 1989, **11**, 209 (*Chem. Abstr.*, 1989, **111**, 73 131).
7 T. Kutsuzawa, M. Haida, S. Shioya, and H. Yamabayashi, *Kokyu*, 1989, **8**, 2 (*Chem. Abstr.*, 1989, **110**, 131 501).
8 H. Fujiwara, *Seisan to Gijutsu*, 1988, **40**, 10 (*Chem. Abstr.*, 1989, **110**, 224 125).

rides such as $[(\eta^5\text{-}C_5H_5)Ir(AsPh_3)H_3]^+$ have been explained as being due to quantum mechanical exchange.[9] σ-Bond polarities in organometallic compounds have been studied by ^{19}F n.m.r. spectroscopy and exchange equilibria methods.[10] The structures of the hexamethylenecarbodithioates of transition metals[11] and $[M(Se_2C_2S_2CS)_2]^{2-}$, M = Ni, Pt, Zn, Cd, Hg,[12] have been studied by ^{13}C n.m.r. spectroscopy.

2 Stereochemistry

This section is subdivided into ten parts which contain n.m.r. information about Groups IA and IIA and transition-metal complexes presented by Groups according to the Periodic Table. Within each Group, classification is by ligand type.

Complexes of Groups IA and IIA.—The ^7Li quadrupole coupling constants in Li_2, LiMe, $LiCH_2X$ and related compounds have been calculated.[13] A ^{13}C spin-echo technique with gated ^6Li decoupling has been developed for selective observation of different $[RLi]_n$ aggregates.[14] *J* modulation of the ^{13}C signals by $^1J(^{13}C,^6Li)$ has been used to characterise the multiplicity of n.m.r. signals of lithiated carbons in Bu^nLi and PhLi by $^{13}C\{^6Li\}$ SEFT.[15] $[Li(tmeda)CH_2PMe_{2-n}R_n]$ shows $^1J(^{31}P\text{-}^7Li)$ below -70 °C in the ^7Li and ^{31}P n.m.r. spectra. The ^{13}C n.m.r. spectrum was also recorded.[16] ^{77}Se N.m.r. spectroscopy has been used to optimise α-selenoalkyl lithium synthesis.[17] ^6Li and ^{13}C n.m.r. spectroscopy has been used to study the lithiation of some silyl ethers in hydrocarbon solvents by Bu^tLi.[18] Distinct ^6Li, ^7Li, ^{13}C, and ^{31}P n.m.r. signals have been observed in mixtures of $LiBEt_3Ph$, LiBr, and LiPh in $(Me_2N)_3PO$.[19] The *ortho*-metallation of C_6H_5-OMe by Bu^nLi has been shown by ^{13}C n.m.r. spectroscopy to be a tetrameric aggregate. $^6Li,^1H$ HOESY has been used to demonstrate the solvation of Bu^nLi by C_6H_5OMe.[20] ^{13}C n.m.r. spectroscopy has been used to show that 2-lithiobenzoselenophene-tmeda and lithiobenzothiophene-tmeda are dimeric with C_{2v} symmetry.[21] The ^1H, ^7Li, and ^{13}C

9 D.H. Jones, J.A. Labinger, and D.P. Weitekamp, *J. Am. Chem. Soc.*, 1989, **111**, 3087; K.W. Zilm, D.M. Heinekey, J.M. Millar, N.G. Payne, and P. Demou, *J. Am. Chem. Soc.*, 1989, **111**, 3088.
10 D.N. Kravtsov, *Metallorg. Khim.*, 1989, **2**, 157 (*Chem. Abstr.*, 1989, **111**, 78 053).
11 A.K. Singh, B.K. Puri, and R.K. Rawlley, *Indian J. Chem., Sect. A*, 1989, **28A**, 59.
12 R.-M. Olk, A. Röhr, J. Sieler, K. Köhler, R. Kirmse, W. Dietzsche, E. Hoyer, and B. Olk, *Z. Anorg. Allg. Chem.*, 1989, **577**, 206.
13 S. Gerber and H. Huber, *J. Phys. Chem.*, 1989, **93**, 545.
14 R.D. Thomas and D.H. Ellington, *Magn. Reson. Chem.*, 1989, **27**, 628.
15 O. Eppers and H. Günther, *Tetrahedron Lett.*, 1989, **30**, 6155.
16 L.T. Byrne, L.M. Engelhardt, G.E. Jacobsen, W.-P. Leung, R.I. Papasergio, C.L. Raston, B.W. Skelton, P. Twiss, and A.H. White, *J. Chem. Soc., Dalton Trans.*, 1989, 105.
17 A. Krief, W. Dumont, M. Clarembeau, G. Bernard, and E. Badaoui, *Tetrahedron*, 1989, **45**, 2005.
18 T.F. Bates and R.D. Thomas, *J. Organomet. Chem.*, 1989, **359**, 285.
19 H.J. Reich and D.P. Green, *J. Am. Chem. Soc.*, 1989, **111**, 8729.
20 W. Bauer and P.v.R. Schleyer, *J. Am. Chem. Soc.*, 1989, **111**, 7191; S. Harder, J. Boersma, L. Brandsma, G.P.M. van Mier, and J.A. Kanters, *J. Organomet. Chem.*, 1989, **364**, 1.
21 S. Harder, J. Boersma, L. Brandsma, J.A. Kanters, W. Bauer, R. Pi, P.v.R. Schleyer, H. Schoellhorn, and U. Thewalt, *Organometallics*, 1989, **8**, 1688.

n.m.r. spectra of (1) show the presence of several aggregates.[22] ^{13}C n.m.r. data have also been reported for $[PhCH_2Li(tmeda)]$,[23] and $[LiCNMeCH(CH_2NMe_2)_2]_2$.[24]

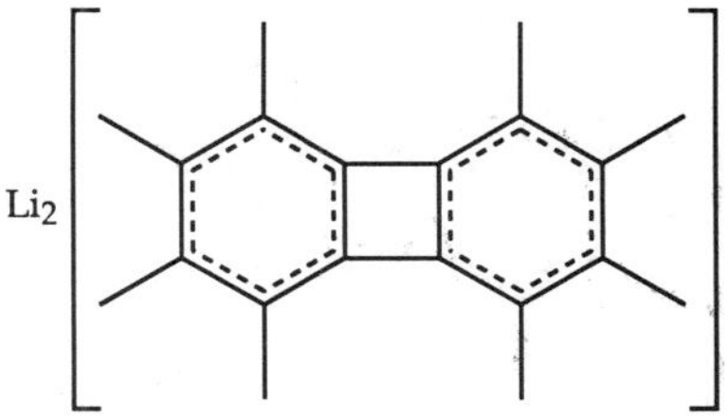

(1)

The ^7Li n.m.r. spectrum of $[FBu^t_2GePLi(C_6H_2Bu^t_3\text{-}2,4,6)]$ shows $^1J(^{31}P\text{-}^7Li)$.[25] The ^6Li-^1H HOESY two dimensional spectrum of $[Li(OCBu^t{=}CH_2)]_4$ has been recorded.[26] The standard chemical potential of Li^+ in 4-butyrolactone has been determined by a ^7Li spin-lattice relaxation method.[27] ^{13}C N.m.r. spectroscopy has been used to perform the conformational analysis of lithium complexes of dicyclohexano-12-crown-4 ethers.[28] In arene solutions, $[Li_2Br_2\{OP(NMe_2)_3\}_3]$ gives a 1:3:3:1 quartet in the ^7Li n.m.r. spectrum due to ^{31}P coupling.[29] Nuclear quadrupole coupling constants of ^2H, ^7Li, and ^{35}Cl have been calculated for HCl and LiCl.[30] N.m.r. data have also been reported for $[H_2C(CH_2)_5NLi]_6$, (^7Li),[31] $[Li(5,12,17\text{-}Me_3\text{-}1,5,9,12,17\text{-}pentaazabicyclo[7.5.5]nonadecane)]^+$, ($^7Li$, ^{13}C),[32] $LiNPr^i_2$, (^6Li, ^{13}C, ^{15}N),[33] $[Pr^nCHCBu^tNHLi(hmpa)_m]_n$, ($^7Li$, ^{13}C),[34] $[Li(pyrrolidide)]_3$, (^6Li, ^7Li),[35] $[R_2BOLi(tmeda)]$, (^7Li, ^{11}B, ^{13}C),[36] $[1,2\text{-}\{Li(thf)PH\}_2C_6H_4]$, ($^7Li$, ^{13}C),[37] and $[Bu^t_2SiFPLi(mesityl)]$, (^7Li, ^{13}C, ^{29}Si).[38]

22 J.W. Bausch, P.S. Gregory, G.A. Olah, G.K.S. Prakash, P.v.R. Schleyer, and G.A. Segel, *J. Am. Chem. Soc.*, 1989, **111**, 3633.

23 W. Zarges, M. Marsch, K. Harms, and G. Boche, *Chem. Ber.*, 1989, **122**, 2303.

24 W. Moene, M. Vos, F.J.J. de Kanter, G.W. Klumpp, and A.L. Spec, *J. Am. Chem. Soc.*, 1989, **111**, 3463.

25 H. Ranaivonjatovo, J. Escudié, C. Couret, J. Satgé, and M. Dräger, *New. J. Chem.*, 1989, **13**, 389.

26 E.M. Arnett, F.J. Fisher, M.A. Nichols, and A.A. Ribeiro, *J. Am. Chem. Soc.*, 1989, **111**, 748.

27 A.I. Mishustin, *Zh. Fiz. Khim.*, 1989, **63**, 828 (*Chem. Abstr.*, 1989, **110**, 220 018).

28 G.W. Buchanan, R.A. Kirby, and K. Bourque, *Can J. Chem.*, 1989, **67**, 449.

29 D. Barr, M.J. Doyle, R.E. Mulvey, P.R. Raithby, D. Reed, R. Snaith, and D.S. Wright, *J. Chem. Soc., Chem. Commun.*, 1989, 318.

30 A. Lal, V. Keshari, and P. Chandra, *J. Mol. Struct.*, 1989, **192**, 263.

31 D. Barr, W. Clegg, S.M. Hodgson, G.R. Lamming, R.E. Mulvey, A.J. Scott, R. Snaith, and D.S. Wright, *Angew. Chem., Int. Ed. Engl.*, 1989, **28**, 1241.

32 A. Bencini, A. Bianchi, A. Borselli, M. Ciampolini, E. Garcia-Espana, P. Dapporto, M. Micheloni, P. Paoli, J.A. Ramirez, and B. Valtancoli, *Inorg. Chem.*, 1989, **28**, 4279.

33 A.S. Galiano-Roth and D.B. Collum, *J. Am. Chem. Soc.*, 1989, **111**, 6772.

34 P.C. Andrews, D.R. Armstrong, M. MacGregor, R.E. Mulvey, and D. Reed, *J. Chem. Soc., Chem. Commun.*, 1989, 1341.

35 D.R. Armstrong, D. Barr, W. Clegg, S.M. Hodgson, R.E. Mulvey, D. Reed, R. Snaith, and D.S. Wright, *J. Am. Chem. Soc.*, 1989, **111**, 4719.

36 G. Beck, P.B. Hitchcock, M.F. Lappert, and I.A. MacKinnon, *J. Chem. Soc., Chem. Commun.*, 1989, 1312.

37 E. Hey, *J. Organomet. Chem.*, 1989, **378**, 375.

38 M. Andrianarison, U. Klingebiel, D. Stalke, and G.M. Sheldrick, *Phosphorus, Sulphur, Silicon*, 1989, **46**, 183.

^{23}Na n.m.r. spectroscopy has been used to discriminate between two possible structures for a piperazino-crown ether derived from adrenalin.[39] A number of biological systems have been studied by ^{23}Na n.m.r. spectroscopy,[40] by ^{2}H, ^{23}Na, ^{35}Cl, and ^{59}Co n.m.r. spectroscopy,[41] and by ^{87}Rb n.m.r. spectroscopy.[42] The nuclear electric hexapole coupling of ^{133}Cs has been determined for Cs^{+} in a nematic lyotropic liquid crystalline solution.[43] ^{133}Cs n.m.r. spectra of oriented CsDNA have been reported.[44]

$^{1}J(^{13}C-^{1}H)$ values for the α-hydrogen nuclei of organomagnesium compounds decrease by 20 Hz due to the presence of magnesium.[45] One and two dimensional ^{1}H and ^{13}C n.m.r. spectroscopy has been used to investigate the solution conformations of $CaCl_2$ and $Ca(NO_3)_2$ complexes of chiral tetramethyl 18-crown-6 macrocycles.[46] ^{13}C n.m.r. data have also been reported for $[(Me_3Si)_2C(MgBr)_2]$,[47] unsaturated organo-Grignard compounds,[48] the Mg complex of (2),[49] $[(Me_3SiC_5H_4)_2Ca(thf)]$, (^{29}Si),[50] $[(Me_5C_5)Ca(\mu\text{-}I)(thf)_2]_2$,[51] a barium complex of a crown ether containing a Ni Schiff base,[52] and barium and lead Schiff base complexes.[53]

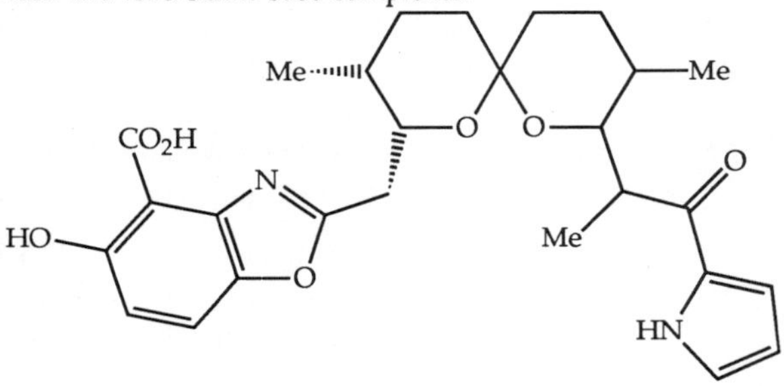

(2)

39 H.D.H. Stöver and C. Detellier, *Tetrahedron Lett.*, 1989, **30**, 4333.
40 L.A. Jelicks and R.K. Gupta, *J. Magn. Reson.*, 1989, **83**, 146; S.J. Kohler, E.K. Smith, and N.H. Kolodny, *J. Magn. Reson.*, 1989, **83**, 423; Y. Boulanger and P. Vinay, *Can. J. Physiol. Pharmacol.*, 1989, **67**, 820 (*Chem. Abstr.*, 1989, **111**, 211 216); D. Burstein and M. Mattingly, *J. Magn. Reson.*, 1989, **83**, 197.
41 L.A. Jelicks and R.K. Gupta, *J. Biol. Chem.*, 1989, **264**, 15230 (*Chem. Abstr.*, 1989, **111**, 190 762).
42 J.L. Allis and R.K. Radda, *J. Magn. Reson.*, 1989, **84**, 372.
43 A.S. Tracey and K. Radley, *Liq. Cryst.*, 1989, **6**, 319 (*Chem. Abstr.*, 1989, **111**, 246 557).
44 L. Einarsson, J. Kowalewski, L. Nordenskiöld, and A. Rupprecht, *J. Magn. Reson.*, 1989, **85**, 228.
45 L.M. Khalilov, A.A. Panasenko, R.R. Muslukhov, A.G. Ibragimov, G.A. Tolstikov, and U.M. Dzhemilev, *Izv. Akad. Nauk SSR, Ser. Khim.*, 1988, 548 (*Chem. Abstr.*, 1989, **110**, 135 282).
46 J.F. Lewison, R.G. Ghirardelli, and R.A. Palmer, *Inorg. Chem.*, 1989, **28**, 3909.
47 M. Hogenbirk, N.J.R. van Eikema Hommes, G. Schat, O.S. Akkerman, F. Bickelhaupt, and G.W. Klumpp, *Tetrahedron Lett.*, 1989, **30**, 6195.
48 R.R. Muslukhov, L.M. Khalilov, A.A. Ibragimov, U.M. Dzhemilev, and A.A. Panasenko, *Metalloorg. Khim.*, 1988, **1**, 680 (*Chem. Abstr.*, 1989, **110**, 95 299).
49 A.M. Albrecht-Gary, S. Blanc-Parasote, D.W. Boyde, G. Dauphin, G. Jeminet, J. Juillard, M. Pridhomme, and C. Tissier, *J. Am. Chem. Soc.*, 1989, **111**, 8598.
50 P. Jutzi, W. Leffers, G. Müller, and B. Huber, *Chem. Ber.*, 1989, **122**, 879.
51 M.J. McCormick, S.C. Sockwell, C.E.H. Davies, T.P. Hanusa, and J.C. Huffman, *Organometallics*, 1989, **8**, 2044.
52 F.C.J.M. van Veggel, M. Bos, S. Harkema, W. Verboom, and D.N. Reinhoudt, *Angew. Chem., Int. Ed. Engl.*, 1989, **28**, 746; F.C.J.M. van Veggel, S. Harkema, M. Bos, W. Verboom, C.J. van Staveren, G.J. Gerritsma, and D.N. Reinhoudt, *Inorg. Chem.*, 1989, **28**, 1133.
53 S.S. Tandon and V. McKee, *J. Chem. Soc., Dalton Trans.*, 1989, 19.

Complexes of Groups IIIA and IVA, the Lanthanides, and Actinides.—The ^{1}H and ^{13}C n.m.r. spectra of complexes such as $[(\eta^5\text{-}C_5H_5)Fe(\eta^5\text{-}C_5H_4YbI)]$ and $[(\eta^5\text{-}C_5H_5)Mo(CO)_3\text{-}YbI]$ have been interpreted in terms of σ- *versus* π-bonding.[54] In $[Y(C{\equiv}CBu^t)_4Li(thf)]$, $^{1}J(^{89}Y\text{-}^{13}C)$ = 96 Hz.[55] The first high resolution ^{171}Yb n.m.r. spectra have been reported for a series of Yb(II) complexes. $^{1}J(^{171}Yb\text{-}^{14}N)$ = 117.6 Hz was observed from $[Yb(NR_2)_2\text{-}(OEt_2)_2]$.[56] The ^{235}U quadrupole spin relaxation mechanisms in UF_6 have been examined.[57] A similar study has also included ^{97}Mo in MoF_6 as well as ^{235}U in UF_6.[58] N.m.r. data have also been reported for $[(\eta^5\text{-}C_5H_5)_2ScOCMe{=}Rh(CO)(\eta^5\text{-}C_5H_5)]$, ($^{13}C$),[59] $[(\eta^5\text{-}C_5Me_5)Y(C_6H_4\text{-}2\text{-}CH_2NMe_2)_2]$, ($^{13}C$),[60] $[(\eta^5\text{-}C_5H_5)Lu(CH_2SiMe_3)_2(thf)_3]$, ($^{13}C$),[61] $[Sc_2\{\eta^8\text{-}C_8H_6(SiMe_3)_2\text{-}1,4\}_2(\mu\text{-}Cl)_2(\mu\text{-}thf)]$, ($^{7}Li$, ^{13}C),[62] $[(\eta^5\text{-}C_5Me_5)La\{CH(SiMe_3)_2\}_2(thf)]$, ($^{13}C$),[63] $[Sm\{CH(SiMe_3)_2\}_3(\mu\text{-}Me)Li(pmdeta)]$, ($^{7}Li$, ^{13}C),[64] $[(\eta^5\text{-}C_5Me_5)LuBu^t_2(thf)]$, ($^{13}C$),[65] [(ring bridged dicyclopentadienyl)MCl], (M = Y, Lu; ^{13}C),[66] $[(\eta^5\text{-}C_4Me_4P)_2Y(\mu\text{-}Cl)_2Li(dme)]$, ($^{7}Li$, ^{13}C, ^{89}Y),[67] $[(\eta^5\text{-}C_5H_5)_3La(\eta^1\text{-}C_5H_5)Na(THF)]$, ($^{13}C$),[68] $[(\eta^8\text{-}C_8H_8)Ln(\eta^5\text{-}C_5Me_5)(thf)]$, (M = Y, La, Pr, Sm, Lu; ^{13}C),[69] $[(\eta^5\text{-}C_5H_5)_2Ce(OBu^t)_2]$, ($^{13}C$),[70] $[(\eta^5\text{-}C_5Me_5)_2Ce(OC_6H_2Bu^t_3\text{-}2,4,6)]$, ($^{13}C$),[71] $[\{(\eta^5\text{-}C_5Me_5)_2Sm\}_2\{\mu,\eta^4\text{-}(PhHC{=}NNCHPh\text{-})_2\}]$, ($^{13}C$),[72] $[(\eta^8\text{-}C_8H_8)Lu\{\eta^5\text{-}C_5(CH_2\text{-}Ph)_5\}]$, ($^{13}C$),[73] $[(\eta^5\text{-}C_5Me_5)_2Yb(thf)_2]$, ($^{13}C$, ^{171}Yb),[74] $[(\eta^8\text{-}C_8H_8)(\eta^5\text{-}C_5Me_5)ThCl(THF)_n]$, ($^{13}C$),[75] $[(\eta^8\text{-}C_8H_8)LaI(thf)_3]$, ($^{13}C$),[76] $Li[(\eta^8\text{-}Bu^nC_8H_7)Lu(tmeda)_2]$, ($^{13}C$),[77] $[Y_3(\mu\text{-}3,5\text{-}Me_2\text{-}$

54 G.Z. Suleimanov, Yu.S. Bogachev, P.V. Petrovskii, E.I. Kazimirchuk, I.L. Zhuravleva, R.Yu. Mekhdiev, T.A. Agdamskii, N.E. Kolobova, and I.P. Beletskaya, *Zh. Org. Khim.*, 1989, **25**, 507 (*Chem. Abstr.*, 1989, **111**, 194 993).

55 W.J. Evans, D.K. Drummond, T.P. Hanusa, and J.M. Olofson, *J. Organomet. Chem.*, 1989, **376**, 311.

56 A.G. Avent, M.A. Edelman, M.F. Lappert, and G.A. Lawless, *J. Am. Chem. Soc.*, 1989, **111**, 3423.

57 I. Ursu, M. Bogdan, F. Balibanu, Z. Gulacsi, M. Gulasci, V. Tosa, and D.E. Demco, *Can. J. Phys.*, 1989, **67**, 52 (*Chem. Abstr.*, 1989, **111**, 69 573).

58 F. Brunet, P. Rigny, and J. Virlet, *Rev. Roum. Phys.*, 1988, **33**, 445 (*Chem. Abstr.*, 1989, **110**, 17 563).

59 M.A. St. Clair, B.D. Santarsiero, and J.E. Bercaw, *Organometallics*, 1989, **8**, 17.

60 M. Booij, N.H. Kiers, H.J. Heeres, and J.H. Teuben, *J. Organomet. Chem.*, 1989, **364**, 79.

61 H. Schumann, J.A. Meese-Marktscheffel, and A. Dietrich, *J. Organomet. Chem.*, 1989, **377**, C5.

62 N.C. Burton, F.G.N. Cloke, P.B. Hitchcock, H.C. de Lemos, and A.A. Sameh, *J. Chem. Soc., Chem. Commun.*, 1989, 1462.

63 H. van der Heijden, C.J. Schaverien, and A.G. Orpen, *Organometallics*, 1989, **8**, 255.

64 P.B. Hitchcock, M.F. Lappert, and R.G. Smith, *J. Chem. Soc., Chem. Commun.*, 1989, 369.

65 H. van der Heijden, P. Pasman, E.J.M. de Boer, C.J. Schaverien, and A.G. Orpen, *Organometallics*, 1989, **8**, 1459.

66 C. Chen, X. Zhong, C. Qian, Z. Xie, and Y. Huang, *Youji Huaxue*, 1988, **8**, 235 (*Chem. Abstr.*, 1989, **110**, 212 986).

67 F. Nief and F. Mathey, *J. Chem. Soc., Chem. Commun.*, 1989, 800.

68 K. Jacob, M. Glanz, K. Tittes, K.-H. Thiele, I. Pavlik, and A. Lyčka, *Z. Anorg. Allg. Chem.*, 1989, **577**, 145.

69 H. Schumann, R.D. Köhn, F.-W. Reier, A. Dietrich, and J. Pickardt, *Organometallics*, 1989, **8**, 1388.

70 W.J. Evans, T.J. Deming, and J.W. Ziller, *Organometallics*, 1989, **8**, 1581.

71 H.J. Heeres, A. Meetsma, J.H. Teuben, and R.D. Rogers, *Organometallics*, 1989, **8**, 2637.

72 W.J. Evans and D.K. Drummond, *J. Am. Chem. Soc.*, 1989, **111**, 3329.

73 H. Schumann, C. Janiak, R.D. Köhn, J. Loebel, and A. Dietrich, *J. Organomet. Chem.*, 1989, **365**, 137.

74 G.B. Deacon, P.B. Hitchcock, S.A. Holmes, M.F. Lappert, P. MacKinnon, and R.H. Newnham, *J. Chem. Soc., Chem. Commun.*, 1989, 935.

75 T.M. Gilbert, R.R. Ryan, and A.P. Sattelberger, *Organometallics*, 1989, **8**, 857.

76 K. Mashima and H. Takaya, *Tetrahedron Lett.*, 1989, **30**, 3697.

77 H. Schumann, P.R. Lee, and J. Loebel, *Chem. Ber.*, 1989, **122**, 1897.

$C_3HN_2)_6(\eta^2$-3,5-$Me_2C_3HN_2)_3(\mu_3$-O)$Na_2\{H(3,5$-$Me_2C_3HN_2)_2\}_2]$, (^{13}C),[78] [La{2-furyl-C(O)-NHN=CMe-2-pyridyl}$_2$Cl]$^{2+}$, (^{13}C),[79] [La{PhC(O)NHCH$_2$C(O)NHN=CMeC$_6$H$_4$OH-2}-$Cl_2(OH_2)_2]^+$, (^{13}C),[80] [$Y_5O(OPr^i)_{13}$], (^{13}C, ^{89}Y),[81] [Y(OSiPh$_3$)$_3$(thf)], (^{13}C),[82] [Y(OC$_6$H$_3$Me$_2$-2,6)$_3$(thf)$_3$], (^{13}C),[83] [La{PhC(O)NHCHC(O)NHNH$_2$}$_2$Cl(OH$_2$)$_2$]$^{2+}$, (^{13}C),[84] La complex of acetone (*N*-benzoyl)glycyl hydrazone, (^{13}C),[85] La and Ce complexes of 2,6-diformyl-4-methylphenol, (^{13}C),[86] [Ce(OBu^t)$_n$(NO$_3$)$_{4-n}$(THF)$_2$], (^{13}C),[87] [Ce(MeOCH$_2$CH$_2$OMe)-(OSiPh$_3$)$_4$], (^{13}C, ^{29}Si),[88] [Ln(S$_2$COEt)$_4$]$^-$, (^{13}C),[89] and [ScCl$_{6-n}$L$_n$]$^{n-3}$, (^{45}Sc).[90]

The large positive $^2J(^1H$-$^1H)$ in [MeTiCl$_3$] is due primarily to the σ-donor and π-acceptor property of the TiCl$_3$ moiety and not to any flattening of the methyl group.[91] Variable temperature ^1H and ^{13}C n.m.r. data of [MMe$_6$]$^{2-}$, M = Zr, Hf, suggest that these complexes adopt trigonal prismatic geometries.[92] The ^{13}C n.m.r. spectrum of (3) has δ -25.1 and $^1J(^{13}C$-$^1H)$ = 113 Hz.[93] The ^1H n.m.r. signal of the CH$_2$CH$_2$ group in [(η^5-C$_5$Me$_5$)$_2$Zr$_2$Cl$_2$(μ-Cl)$_2${CH$_2$CH$_2$Si(SiMe$_3$)$_3$}$_2$] is [AX]$_2$. The ^{13}C and ^{29}Si n.m.r. spectrum was also reported.[94] N.m.r. data have also been reported for [(η^5-C$_5$H$_5$)$_2$-ZrOCH$_2$B(C$_8$H$_{14}$)H]$_2$, (^{13}C),[95] [(η^5-C$_5$Me$_5$)$_2$HfH(Ph$_2$PCO$_2$)], (^{13}C),[96] [(η^5-C$_5$Me$_5$)TiMe-(MeNNCPh$_2$)]$_2$(μ-O), (^{13}C),[97] [(η^5-C$_5$Me$_5$)Ti(CH$_2$SiMe$_3$)$_{3-n}$Cl$_n$], (^{13}C),[98] [μ-{2-(CH$_2$)$_2$C$_6$H$_4$}[(η^5-C$_5$Me$_5$)Ti{2-(CH$_2$)$_2$C$_6$H$_4$}]$_2$], (^{13}C),[99] [Me$_2$Si(η^5-C$_5$H$_4$)$_2$MCH$_2$SiMe$_2$CH$_2$], (M = Ti, Zr; ^{13}C),[100] [(η^5-C$_5$H$_5$)$_2$Ti(μ-CH$_2$)(μ-Cl)Rh(η^4-cod)], (^{13}C),[101] [(η^5-C$_5$H$_5$)$_2$Ti(μ-CH$_2$)(μ-Cl)PtMe(PMe$_2$Ph)], (^{13}C),[102] [(η^5-C$_5$Me$_5$)MeTi(μ-O)(μ-CMe$_2$O)TiMe$_2$(η^5-C$_5$Me$_5$)],

78 H. Schumann, P.R. Lee, and J. Leobel, *Angew. Chem., Int. Ed. Engl.*, 1989, **28**, 1033.
79 B. Singh and P.K. Singh, *Transition Met. Chem. (Weinheim, Ger.)*, 1989, **14**, 411.
80 T.R. Rao and G. Singh, *Transition Met. Chem. (Weinheim, Ger.)*, 1989, **14**, 471.
81 O. Poncelet, W.J. Sartain, L.G. Hubert-Pfalzgraf, K. Folting, and K.G. Caulton, *Inorg. Chem.*, 1989, **28**, 263.
82 M.J. McGeary, P.S. Coan, K. Folting, W.E. Streib, and K.G. Caulton, *Inorg. Chem.*, 1989, **28**, 3284.
83 M.J. Evans, J.M. Olofson, and J.W. Ziller, *Inorg. Chem.*, 1989, **28**, 4309.
84 T.R. Rao, G. Singh, and I.A. Khan, *Transition Met. Chem. (Weinheim, Ger.)*, 1989, **14**, 15.
85 T.R. Rao and G. Singh, *Synth. Inorg. Metal-Org. Chem.*, 1989, **19**, 263.
86 S. Sitran, D. Fregona, and G. Faraglia, *J. Coord. Chem.*, 1988, **18**, 269.
87 W.J. Evans, T.J. Deming, J.M. Olofson, and J.W. Ziller, *Inorg. Chem.*, 1989, **28**, 4027.
88 P.S. Gradeff, K. Yunlu, A. Gleizes, and J. Galy, *Polyhedron*, 1989, **8**, 1001.
89 A.F. Savost'yanova, V.V. Skopenko, M. Ya. Rakhlin, and T.A. Sukhan, *Dokl. Akad. Nauk Ukr. SSR, Ser. B: Geol., Khim. Biol. Nauki*, 1989, 51 (*Chem. Abstr.*, 1989, **111**, 32 633).
90 G.A. Kirakosyan, V.P. Tarasov, and Yu.A. Buslaev, *Magn. Reson. Chem.*, 1989, **27**, 103.
91 R.L. Williamson and M.B. Hall, *ACS Symp. Ser.*, 1989, **394**, 17 (*Chem. Abstr.*, 1989, **111**, 181 144).
92 P.M. Morse and G.S. Girolami, *J. Am. Chem. Soc.*, 1989, **111**, 4114.
93 S.L. Buchwald, E.A. Lucas, and W.M. Davis, *J. Am. Chem. Soc.*, 1989, **111**, 397.
94 J. Arnold, M.P. Engeler, F.H. Elsner, R.H. Heyn, and T.D. Tilley, *Organometallics*, 1989, **8**, 2284.
95 G. Erker, U. Hoffmann, R. Zwettler, and C. Krüger, *J. Organomet. Chem.*, 1989, **367**, C15.
96 GA. Vaughan, G.L. Hillhouse, and A.L. Rheingold, *Organometallics*, 1989, **8**, 1760.
97 R. Serrano, J.C. Flores, P. Royo, M. Mena, M.A. Pellinghelli, and A. Tiripicchio, *Organometallics*, 1989, **8**, 1404.
98 P. Gómez-Sal, M. Mena, F. Palacios, P. Royo, R. Serrano, and S.M. Carreras, *J. Organomet. Chem.*, 1989, **375**, 59.
99 M. Mena, P. Royo, R. Serrano, M.A. Pellinghelli, and A. Tiripicchio, *Organometallics*, 1989, **8**, 476.
100 A. Kabi-Satpathy, C.S. Bajgur, K.P. Reddy, and J.L. Petersen, *J. Organomet. Chem.*, 1989, **364**, 105.
101 P.B. Mackenzie, R.J. Coots, and R.H. Grubbs, *Organometallics*, 1989, **8**, 8.
102 F. Ozawa, J.W. Park, P.B. Mackenzie, W.P. Schaefer, L.M. Henling, and R.H. Grubbs, *J. Am. Chem.*

(^{13}C),[103] $[(\eta^5\text{-}C_5Me_5)_2MN(O)=NCPh=CPh]$, (M = Ti, Zr; ^{13}C),[104] (4), (M = Ti, Zr, Hf; ^{13}C),[105] $[(\eta^5\text{-}C_5H_5)_2Ti(=CHCH=CR^1R^2)(PMe_3)]$, ($^{13}C$),[106] $[(\eta^5\text{-}C_5H_5)_2Ti(C\equiv CSiMe_3)_2]$, ($^{13}C$),[107] $[(\eta^5\text{-}C_5Me_5)_2ZrCH=CHCH_2CH_2]$, ($^{13}C$),[108] $[(\eta^5\text{-}C_5H_5)_2ZrCH=CHCH_2CH_2]$, ($^{13}C$),[109] $[(\eta^5\text{-}C_5Me_5)_2ZrNMeC(=CH_2)SiMe_2CH_2]$, ($^{13}C$),[110] $[(\eta^5\text{-}C_5Me_5)ZrCH=CHC(O)=M(CO)_5]$, (M = Cr, Mo, W; ^{13}C),[111] (5), (^{13}C),[112] $[Zr(OC_6H_3Bu^t{}_2\text{-}2,6)(CH_2Ph)(bipy\text{-}6\text{-}CH_2Ph)]$, ($^{13}C$),[113] $[(\eta^5\text{-}C_5H_4Me)_2ZrBu^iX]$, ($^{13}C$),[114] (6), ($^{13}C$),[115] $[(\eta^5\text{-}C_5H_5)_2Zr(\eta^2\text{-}PhCH=NCH_2Ph)(thf)]$, ($^{13}C$),[116] $[(\eta^5\text{-}C_5H_5)M_2R(=CHPPh_3)]$, (M = Zr, Hf; ^{13}C),[117] $[(\eta^5\text{-}C_5H_4Me)_2ZrR(thf)]^+$, ($^{13}C$),[118] $[(\eta^5\text{-}C_5H_5)(\eta^5\text{-}C_5Me_5)Zr(SiH_2Ph)Cl]$, ($^{13}C$, ^{29}Si),[119] $[(Bu^tO)_3MSi(SiMe_3)_3]$, (M = Zr, Hf; ^{13}C, ^{29}Si),[120] (7), (^{13}C),[121] $[(\eta^5\text{-}C_5H_5)_2ZrCR=CRCH_2S]$, ($^{13}C$),[122] $[(\eta^5\text{-}C_5H_5)Zr(CMe=CMeCMe=CMe)(dmpe)Me]$, ($^{13}C$),[123] $[(\eta^5\text{-}C_5H_5)_2ZrC_6H_4\text{-}2\text{-}C\{W(CO)_5\}=O]$, ($^{13}C$),[124] $[(\eta^5\text{-}C_5H_5)_2Zr(2\text{-}furyl)_2]$, ($^{13}C$),[125] (8), ($^{13}C$),[126] $[(\eta^5\text{-}C_5Me_5)Hf(C\equiv CMe)(Pr^iN=CHCH=NPr^i)]$, ($^{13}C$),[127] (9), ($^{13}C$),[128] $[(\eta^5\text{-}C_5H_5)(\eta^5\text{-}C_5Me_5)Hf(SiHPhSiH_2Ph)Cl]$, ($^{29}Si$),[129] $[Ti(CO)_5(dmpe)]$, (^{13}C),[130] and $[(\eta^5\text{-}C_5H_5)M(CO)_4]^-$, (M = Ti, Zr, Hf; ^{13}C).[131]

Soc., 1989, **111**, 1319.

103 J.C. Flores, M. Mena, P. Royo, and R. Serrano, *J. Chem. Soc., Chem. Commun.*, 1989, 617.

104 G.A. Vaughan, C.D. Sofield, G.L. Hillhouse, and A.L. Rheingold, *J. Am. Chem. Soc*, 1989, **111**, 5491.

105 S.M. Yousaf, M.F. Farona, R.J. Shively, jun., and W.J. Youngs, *J. Organomet. Chem.*, 1989, **363**, 281.

106 P. Binger, P. Müller, R. Benn, and R. Mynott, *Angew. Chem., Int. Ed. Engl.*, 1989, **28**, 610.

107 G.L. Wood, C.B. Knobler, and M.F. Hawthorne, *Inorg. Chem.*, 1989, **28**, 382.

108 R. Beckhaus, K.-H. Thiele, and D. Ströhl, *J. Organomet. Chem.*, 1989, **369**, 43.

109 H.G. Alt and C.E. Denner, *J. Organomet. Chem.*, 1989, **368**, C15.

110 F.J. Berg and J.L. Petersen, *Organometallics*, 1989, **8**, 2461.

111 R. Beckhaus and K.-H. Thiele, *J. Organomet. Chem.*, 1989, **368**, 315.

112 E.-i. Negishi, S.J. Holmes, J.M. Tour, J.A. Miller, F.E. Cederbaum, D.R. Swanson, and T. Takahashi, *J. Am. Chem. Soc.*, 1989, **111**, 3336.

113 L.M. Kobriger, A.K. McMullen, P.E. Fanwick, and I.P. Rothwell, *Polyhedron*, 1989, **8**, 77.

114 Y. Wielstra, S. Gambarotta, A. Meetsma, and A.L. Spek, *Organometallics*, 1989, **8**, 2948.

115 G. Erker, U. Hoffmann, R. Zwettler, P. Betz, and C. Krüger, *Angew. Chem., Int. Ed. Engl.*, 1989, **28**, 630.

116 S.L. Buchwald, B.T. Watson, M.W. Wannamaker, and J.C. Dewan, *J. Am. Chem. Soc.*, 1989, **111**, 4489.

117 G. Erker, U. Korek, R. Schlund, and C. Krüger, *J. Organomet. Chem.*, 1989, **364**, 133.

118 R.F. Jordan, R.E. LaPointe, P.K. Bradley, and N. Baenziger, *Organometallics*, 1989, **8**, 2892.

119 H.-G. Woo and T.D. Tilley, *J. Am. Chem. Soc.*, 1989, **111**, 3757.

120 R.H. Heyn and T.D. Tilley, *Inorg. Chem.*, 1989, **28**, 1768.

121 G. Erker, R. Zwettler, and C. Krüger, *Chem. Ber.*, 1989, **122**, 1377.

122 S.L. Buchwald, R.A. Fisher, and W.M. Davis, *Organometallics*, 1989, **8**, 2082.

123 Y. Wielstra, S. Gambarotta, A. Meetsma, J.L. de Boer, and M.Y. Chiang, *Organometallics*, 1989, **8**, 2696.

124 G. Erker, U. Dorf, R. Lecht, M.T. Ashby, M. Aulbach, R. Schlund, C. Krüger, and R. Mynott, *Organometallics*, 1989, **8**, 2037.

125 G. Erker and R. Petrenz, *J. Chem. Soc., Chem. Commun.*, 1989, 345.

126 B. Hessen, J. Blenkers, J.H. Teuben, G. Helgesson, and S. Jagmer, *Organometallics*, 1989, **8**, 830.

127 B. Hessen, J.E. Bol, J.L. de Boer, A. Meetsma, and J.H. Teuben, *J. Chem. Soc., Chem. Commun.*, 1989, 1276.

128 D.M. Roddick and J.E. Bercaw, *Chem. Ber.*, 1989, **122**, 1579.

129 H.-G. Woo and T.D. Tilley, *J. Am. Chem. Soc.*, 1989, **111**, 8043.

130 J.E. Ellis, *Polyhedron*, 1989, **8**, 1611.

131 S.R. Frerichs and J.E. Ellis, *J. Organomet. Chem.*, 1989, **359**, C41.

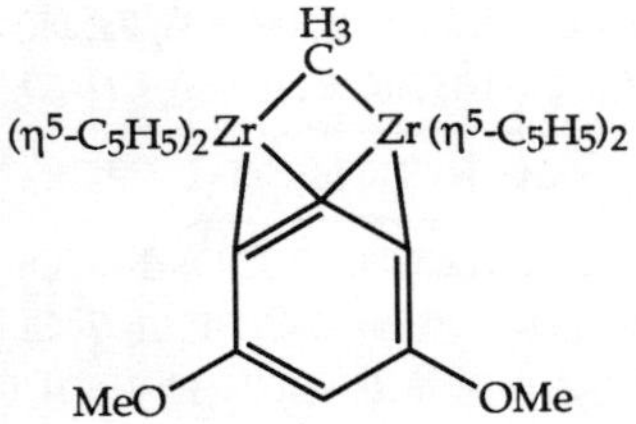

(3)

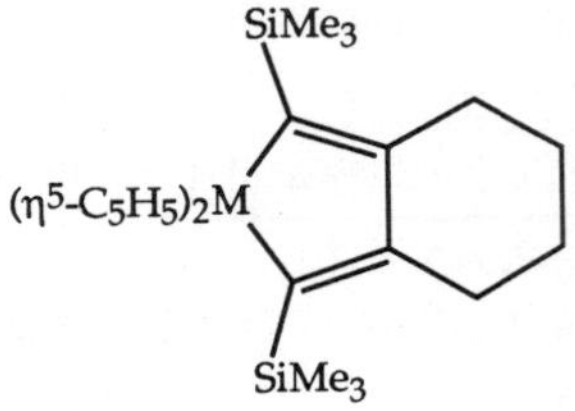

(4)

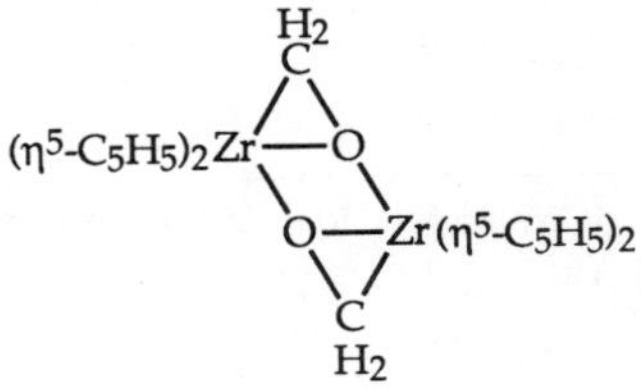

(6)

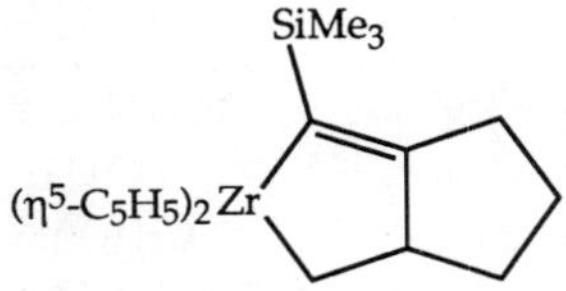

(5)

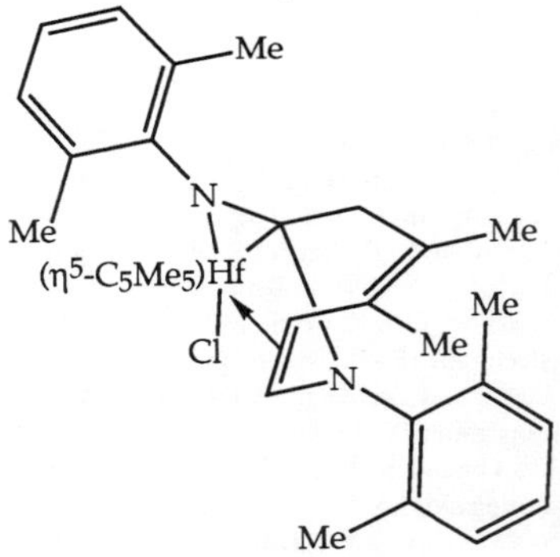

(8)

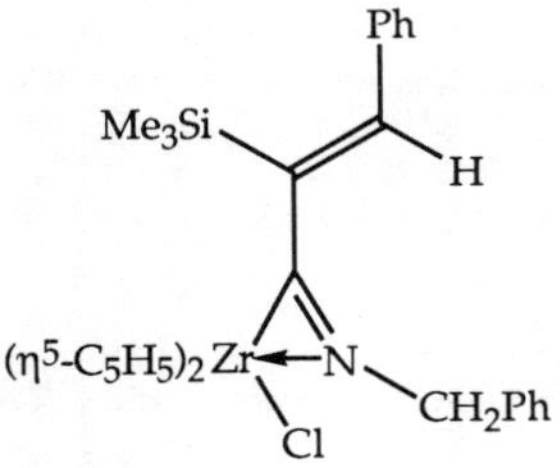

(7)

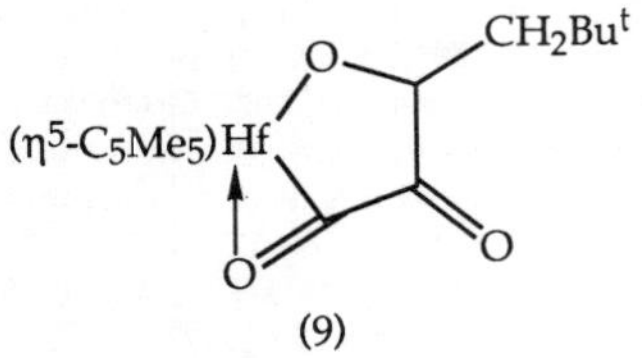

(9)

The indirect $J(^{13}C-^{13}C)$ values have been used to prove the pronounced η^4-diene metal nature for the prone Ti-diene complexes and a substantial participation of bent metallocyclo-3-pentene structure for the supine complexes of $[(\eta^5-C_5Me_5)Ti(\eta^4-C_4H_6)Cl]$.[132] $^1J(^{77}Se-^{77}Se)$ has been determined in enriched samples of $[(\eta^5-C_5H_5)_2TiSe_5]$.[133] ^1H and ^{13}C n.m.r. spectroscopy has been used to determine the

132 H. Yamamoto, H. Yasuda, K. Tatsumi, K. Lee, A. Nakamura, J. Chen, Y. Kai, and N. Kasai, *Organometallics*, 1989, **8**, 105.
133 P. Pekonen, Y. Hiltunen, and R.S. Laitinen, *Acta Chem. Scand.*, 1989, **43**, 414.

solution structures of chiral Ti^{4+} alkoxides.[134] N.m.r. data have also been reported for [(η^5-C_5H_5)$_2$M(η^2-C_2H_4)(PMe_3)], (M = Zr, Hf; ^{13}C),[135] [(η^5-C_5H_4Me)Zr{C(O)CH=CHSiMe$_3$}-Cl], (^{13}C),[136] [(η^5-C_5Me_5)$_2$Zr{η^3-C_3H_4-1-CH_2C(O)O}], (^{13}C),[137] [(η^5-C_5H_5)$_2$Ti{OC(O)-C_6H_4X}$_2$], (^{13}C),[138] [(η^5-C_5H_5)(η^5-C_4Me_4P)$TiCl_2$], (^{13}C),[139] [(η^5-C_5H_5)(η^5-L)$TiCl_2$], {L = (10), R = H; ^{13}C},[140] [(η^5-L)$_2TiCl_2$], {L = (11); ^{13}C},[141] [(η^5-C_5H_5)(η^5-L)MCl_2], {L = (10), R = Me; M = Ti, Zr; ^{13}C},[142] [(η^5-$C_5H_4CHMeCH_2OMe$)MCl_2], (M = Ti, Zr; ^{13}C),[143] [(OC)$_3$Mn(η^5-$C_5H_4CH_2C_5H_4$-η^5)Ti(η^5-C_5H_5)Cl_2], (^{13}C),[144] [{(η^5-C_5Me_5)Ti}$_4O_6$], (^{13}C, ^{17}O),[145] [(η^5-C_5H_5)TiCl(OR)$_2$], (^{13}C),[146] [(η^5-C_5H_5)$_2$Zr(μ-NNCHPh)]$_2$, (^{13}C),[147] [(η^5-C_5H_5)$_2$Zr(μ-PPh_2)(η^1-η^2-OC)Mo(CO)(η^5-C_5H_5)], (^{13}C),[148] [{(η^5-C_5Me_5)$_2$Zr}$_2$(μ-E_2)], (E = S, Se, Te; ^{13}C),[149] [{Me_2Si(η^5-C_5H_4)$_2$}{(η^5-C_5H_5)$ZrCl_2$}(μ-O)], (^{13}C),[150] [(η^5-C_5H_5)$_2$ZrCl($AlCl_4$)], (^{13}C, ^{27}Al),[151] [(η^5-L)$_2ZrCl_2$], {L = (10), R = Me; ^{13}C},[152] [(η^5-C_5H_5)$_2$ZrI]$_2$, (^{13}C),[153] [{(η^5-C_5Me_5)MCl}$_3$O(OH)$_3$Cl], (M = Zr, Hf; ^{13}C),[154] [(η^5-C_5Me_5)M(O)Cl]$_n$, (M = Zr, Hf; ^{13}C),[155] [(η^5-C_5H_5)$ZrCl_3$(OH_2)$_2$(15-crown-5)], (^{13}C),[156] [(η^6-1,3,5-$Bu^t_3C_6H_3$)$_2$M], (M = Ti, Zr, Hf; ^{13}C),[157] [(η^6-C_7H_8)$_2$Zr], (^{13}C),[158] [M(η^6-C_7H_8)(PMe_3)$_2Cl_2$], (M = Zr, Hf; ^{13}C),[159] [$ZrCl_3$-{N($CH_2CH_2NEt_2$)$_2$}], (^{13}C),[160] [Ti(OPr^i)$_3$(acac)], (^{13}C),[161] 2:2 complex between [Ti(OPr^i)$_4$]

134 P.G. Potvin, R. Gau, P.C.C. Kwong, and S. Bianchet, *Can. J. Chem.*, 1989, **67**, 1523.
135 T. Takasashi, M. Tamura, M. Saburi, Y. Uchida, and E.-i. Negishi, *J. Chem. Soc., Chem. Commun.*, 1989, 852.
136 G. Erker, R. Schlund, and C. Krüger, *Organometallics*, 1989, **8**, 2349.
137 H. Yasuda, T. Okamoto, Y. Matsuoka, A. Nakamura, Y. Kai, N. Kanehisa, and N. Kasai, *Organometallics*, 1989, **8**, 1139.
138 Y. Dang, H.J. Geise, R. Dommisse, E. Esmans, and H.O. Desseyn, *Polyhedron*, 1989, **8**, 1844.
139 F. Nief, L. Ricard, and F. Mathey, *Organometallics*, 1989, **8**, 1473.
140 L.A. Paquette, K.J. Moriarty, P. Meunier, B. Gautheron, C. Sornay, R.D. Rogers, and A.L. Rheingold, *Organometallics*, 1989, **8**, 2159.
141 K.J. Moriarty, R.D. Rogers, and L.A. Paquette, *Organometallics*, 1989, **8**, 1512.
142 L.A. Paquette, K.J. Moriarty, J.A. McKinney, and R.D. Rogers, *Organometallics*, 1989, **8**, 1707.
143 Q. Huang, Y. Qian, and Y. Tang, *J. Organomet. Chem.*, 1989, **368**, 277.
144 P. Härter, G. Boguth, E. Herdtweck, and J. Riede, *Angew. Chem., Int. Ed. Engl.*, 1989, **28**, 1008.
145 L.M. Babcock and W.G. Klemperer, *Inorg. Chem.*, 1989, **28**, 2003.
146 M. Riediker, A. Hafner, U. Piantini, G. Rihs, and A. Togni, *Angew. Chem., Int. Ed. Engl.*, 1989, **28**, 498.
147 G.M. Arvanitis, J. Smegal, I. Meier, A.C.C. Wong, J. Schwartz, and D. Van Engen, *Organometallics*, 1989, **8**, 2717.
148 P.Y. Zheng, T.T. Nadasdi, and D.W. Stephan, *Organometallics*, 1989, **8**, 1393.
149 R. Beckhaus and K.-H. Thiele, *Z. Anorg. Allg. Chem.*, 1989, **573**, 195.
150 K.P. Reddy and J.L. Petersen, *Organometallics*, 1989, **8**, 2107.
151 M.V. Gaudet, M.J. Zaworotko, T.S. Cameron, and A. Linden, *J. Organomet. Chem.*, 1989, **367**, 267.
152 L.A. Paquette, K.J. Moriarty, and R.D. Rogers, *Organometallics*, 1989, **8**, 1506.
153 Y. Wielstra, S. Gambarotta, A. Meetsma, and J.L. de Boer, *Organometallics*, 1989, **8**, 250.
154 L.M. Babcock, V.W. Day, and W.G. Klemperer, *Inorg. Chem.*, 1989, **28**, 806.
155 B. Hessen, J. Blenkers, J.H. Teuben, G. Helgesson, and S. Jagner, *Organometallics*, 1989, **8**, 2809.
156 G. Erker, C. Sarter, S. Werner, and C. Krüger, *J. Organomet. Chem.*, 1989, **377**, C55.
157 F.G.N. Cloke, K.A.E. Courtney, A.A. Sameh, and A.C. Swain, *Polyhedron*, 1989, **8**, 1641.
158 M.L.H. Green and N.M. Walker, *J. Chem. Soc., Chem. Commun.*, 1989, 850.
159 M.L.H. Green and N.M. Walker, *J. Chem. Soc., Chem. Commun.*, 1989, 1865.
160 A.R. Wills and P.G. Edwards, *J. Chem. Soc., Dalton Trans.*, 1989, 1253.
161 F. Babonneau, A. Leaustic, and J. Livage, *Mater. Res. Soc. Symp. Proc.*, 1988, **121**, 317 (*Chem. Abstr.*, 1989, **110**, 119 770); A. Leaustic, F. Babonneau, and J. Livage, *Chem. Mater.*, 1989, **1**, 240 (*Chem. Abstr.*, 1989, **110**, 146 597).

and 1,2:5,6-di-*O*-isopropylidene-D-mannitol, (^{13}C),[162] [$Ti_2(\mu$-O)(μ-$O_2CR)_2Cl_4$], (^{17}O),[163] [$Ti_6(\mu_2$-O$)_2(\mu_3$-O$)_2(\mu$-OBu$)_2$(OBu$)_6(O_2$CMe$)_8$], (^{13}C, including solid state),[164] [{*N*-(*E*-but-2-enoyl)bornane-10,2-sultam}$TiCl_4$], (^{13}C),[165] and [$Cl_{4-n}Zr(S_2POCMe_2CMe_2O)_n$], (^{13}C).[166]

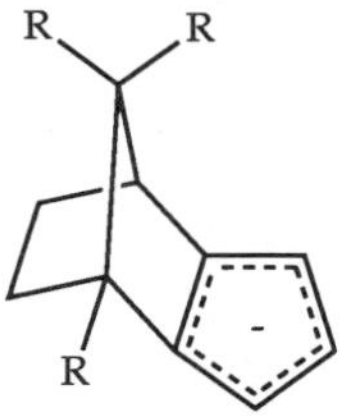

(10)

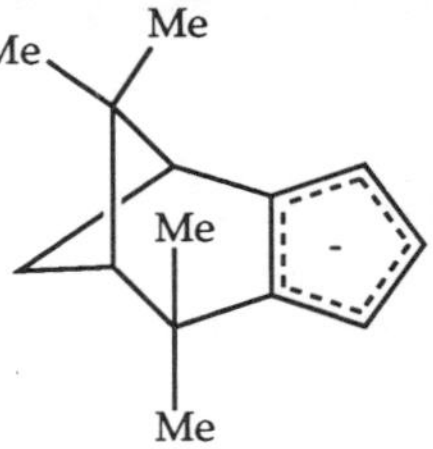

(11)

Complexes of V, Nb, and Ta.——For [V(NO)L]$^-$, L = [N($CH_2CH_2O)_3$]$^{3-}$, the ^{51}V chemical shift correlates with (*d-d* transition energy)$^{-1}$.[167] The dependence of ^{51}V shielding on steric effects of vanadium(V) complexes with OO and ON ligands has been examined. The ^{13}C n.m.r. spectra were also recorded.[168] ^{51}V n.m.r. spectroscopy has been used to study the thermal stability of molybdovanadophosphoric acids.[169] ^{51}V n.m.r. spectroscopy has been used as a probe of vanadium(V) coordination to human apotransferrin.[170] ^{13}C and ^{51}V n.m.r. spectroscopy has been used to study the complexation of oxo ions of V(V), Mo(VI), W(VI), and U(VI) by α-hydroxycarboxylic acids.[171] N.m.r. data have also been reported for [{$C_5Me_3(CH_2)_2$}Ta(H$)_2(PMe_3)_2$], (^{13}C),[172] [(η^5-$C_5H_5)_2$Ta{CH(CN)NCMe$_2CH_2$Bu^t}H], (^{13}C),[173] [(η^5-$C_5H_4R)_2Ta_2(\mu$-X$)_2(B_2H_6)$], (^{11}B, ^{13}C),[174] [(η^5-C_5H_5)V(=CHBu^t)(dmpe)], (^{13}C),[175] [(2,6-Pr$^i_2C_6H_3O)_2$ClTa(CH_2Ph)(CH=CHBu^t)], (^{13}C),[176] [(2,6-Pr$^i_2C_6H_3O)_3$TaCR=CRCHBu^t], (^{13}C),[177] [(η^5-$C_5H_5)_2$TaPh(PMe_3)], (^{13}C),[178] [($Me_3SiO)_3$Ta=C=C=O], (^{13}C),[179] [(η^5-$C_5H_4MR_3$)V(CO$)_4$], (M = Si, Ge, Sn; ^{13}C),[180] [(η^5-$C_5Me_{5-n}H_n$)-

162 P.G. Potvin, R. Gau, P.C.C. Kwong, and S. Bianchet, *Can. J. Chem.*, 1989, **67**, 1523.
163 N.W. Alcock, D.A. Brown, T.F. Illson, S.M. Roe, and M.G.H. Wallbridge, *Polyhedron*, 1989, **8**, 1846.
164 S. Doeuff, Y. Dromzee, F. Taulelle, and C. Sanchez, *Inorg. Chem.*, 1989, **28**, 4439.
165 W. Oppolzer, I. Rodriguez, J. Blagg, and G. Bernardinelli, *Helv. Chim. Acta*, 1989, **72**, 123.
166 J.S. Yadav, R.K. Mehrotra, and G. Srivastava, *Inorg. Chim. Acta*, 1989, **162**, 227.
167 S. Kitagawa, M. Munakata, and M. Ueda, *Inorg. Chem.*, 1989, **164**, 49.
168 C. Weidemann, W. Priebsch, and D. Rehder, *Chem. Ber.*, 1989, **122**, 235.
169 H. Qi, Q. Yan, and X. Fu, *Wuji Huaxue*, 1989, **5**, 1 (*Chem. Abstr.*, 1989, **111**, 246 773).
170 A. Butler and H. Eckert, *J. Am. Chem. Soc.*, 1989, **111**, 2802.
171 V.M.S. Gil, *Pure Appl. Chem.*, 1989, **61**, 841.
172 S.T. Carter, W. Clegg, V.C. Gibson, T.P. Kee, and R.D. Sanner, *Organometallics*, 1989, **8**, 253.
173 J.F. Leboeuf, J.C. Leblanc, and C. Moïse, *J. Organomet. Chem.*, 1989, **364**, C22.
174 C. Ting and L. Messerle, *J. Am. Chem. Soc.*, 1989, **111**, 3449.
175 B. Hessen, A. Meetsma, and J.H. Teuben, *J. Am. Chem. Soc.*, 1989, **111**, 5977.
176 K. Ballard, I.M. Gardiner, and D.E. Wigley, *J. Am. Chem. Soc.*, 1989, **111**, 2159.
177 K.C. Wallace, A.H. Liu, W.M. Davis, and R.R. Schrock, *Organometallics*, 1989, **8**, 644.
178 D.H. Berry and Q. Jiang, *J. Am. Chem. Soc.*, 1989, **111**, 8049.
179 D.R. Neithamer, R.E. LaPointe, R.A. Wheeler, D.S. Richeson, G.D. Van Duyne, and P.T. Wolczanski, *J. Am. Chem. Soc.*, 1989, **111**, 9056.
180 A. Duch, M. Hoch, and D. Rehder, *Chimia*, 1988, **42**, 179 (*Chem. Abstr.*, 1989, **110**, 192 987).

$V(CO)_4]$, (^{51}V),[181] $[(\eta^5\text{-}C_5H_5)Fe(\eta^5\text{-}C_4H_4BPh\text{-}\eta^5)M(CO)_4]$, (M = V, Nb, Ta; ^{11}B, ^{13}C),[182] $[\{\eta^5\text{-}C_5H_{5-n}(SiMe_3)_n\}_2NbCl(CNR)]$, ($^{13}C$),[183] *cis*-$[ClNb(CO)_2(dppe)_2]$, ($^{93}Nb$),[184] $[(\eta^5\text{-}C_5H_5)V(\eta^2\text{-}R^1C{\equiv}CR^2)(PMe_3)_2]$, ($^{13}C$),[185] $[(\eta^5\text{-}C_5H_5)TaCl_2(\eta^2ArC{\equiv}CAr)]$, ($^{13}C$),[186] $[(\eta^7\text{-}C_7H_7)(\eta^4\text{-}C_7H_8)Nb(PMe_3)]$, ($^{13}C$),[187] $[(\eta^5\text{-}C_5Me_5)VX_2O]$, ($^{13}C$, ^{51}V),[188] $[(\eta^5\text{-}C_5Me_5)TaCl_3(OR)]$, ($^{13}C$),[189] $[OVCl_2(NBu^tR)]$, (^{51}V),[190] $[\{(CF_3)_2PN\}_2NVCl_2]$, (^{51}V),[191] $[Nb_2(S_2)_2(NCS)_8]^{4-}$, ($^{14}N$),[192] $[Ta(OC_6H_3Me_2\text{-}2,6)_2(=NBu^t)(Bu^tNCPh{=}CPh_2)]$, ($^{13}C$),[193] $[Ta(NC_6H_3Pr^i{}_2\text{-}2,6)Cl_3L_2]$, ($^{13}C$),[194] $[VO(O_2CBu^t)_3]$, (^{51}V),[195] $[(\mu\text{-}\eta^3\text{-}C_5Me_5O_3)V(O)]_2$, ($^{13}C$, ^{51}V),[196] $[V_5O_{14}]^{3-}$, (^{51}V),[197] $[\{(\eta^5\text{-}C_5Me_5)Rh\}_4(V_6O_{19})]$, ($^{13}C$, ^{51}V),[198] $[\{(\eta^5\text{-}C_5Me_5)M\}_4(V_6O_{19})]$, (M = Rh, Ir; ^{13}C, ^{17}O, ^{51}V),[199] $[(MeCN)(V_{12}O_{32})]^{4-}$, ($^{51}V$),[200] $[PW_{12-n}V_nO_{40}]^{-(3+n)}$, ($^{51}V$),[201] ($^{17}O$, ^{51}V),[202] $[Nb(SO_3F)_{5+n}]^{n-}$, (^{51}V),[203] $[Nb(OC_6H_3Ph\text{-}\eta^4\text{-}C_6H_7)(OC_6H_3Ph_2\text{-}2,6)_2]$, ($^{13}C$),[204] $[M(O_2)_3(glycollate)]^{3-}$, (M = Nb, Ta; ^{13}C),[205] and $[Ta(OC_6H_3PhC_6H_4)_2(OC_6H_3Ph_2\text{-}2,6)]$, ($^{13}C$).[206]

Complexes of Cr, Mo, and W.——The ^{13}C n.m.r. spectrum of $[MoCl(SnCl_3)(CO)_3(NCMe)_2]$ shows several CO signals, indicating the presence of several isomers.[207] N.m.r. data have also been reported for $[MoH_2\{\sigma\text{-}HC{=}CHC(O)OMe\}(dppe)_2]^+$, ($^{13}C$),[208]

181 W.A. Herrmann, E. Herdtweck, and G. Weichselbaumer, *J. Organomet. Chem.*, 1989, **362**, 321.
182 G.E. Herberich, I. Hausmann, and N. Klaff, *Angew. Chem., Int. Ed. Engl.*, 1989, **28**, 319.
183 M. Gömez, J.M. Martinez de Ilarduya, and P. Royo, *J. Organomet. Chem.*, 1989, **369**, 197.
184 C. Felten, J. Richter, W. Priebsch, and D. Rehder, *Chem. Ber.*, 1989, **122**, 1617.
185 B. Hessen and J.H. Teuben, *J. Organomet. Chem.*, 1989, **367**, C18.
186 M.D. Curtis, J. Real, and D. Kwon, *Organometallics*, 1989, **8**, 1644.
187 M.L.H. Green, D. O'Hare, and J.G. Watkins, *J. Chem. Soc., Chem. Commun.*, 1989, 698.
188 M. Herberhold, W. Kremnitz, M. Kuhnlein, M.L. Zeigler, and K. Brunn, *Z. Naturforsch., B*, 1987, **42**, 1520 (*Chem. Abstr.*, 1989, **110**, 154 435).
189 V.C. Gibson and T.P. Kee, *J. Chem. Soc., Chem. Commun.*, 1989, 656.
190 F. Preuss, E. Fuchslocher, E. Leber, and W. Towae, *Z. Naturforsch., B*, 1989, **44**, 271 (*Chem. Abstr.*, 1989, **111**, 32 658).
191 H.W. Roesky, P. Olms, M. Witt, K. Keller, D. Stalke, T. Henkel, and G.M. Sheldrick, *J. Chem. Soc., Chem. Commun.*, 1989, 366.
192 A.V. Mishchenko, V.E. Fedorov, B.A. Kolesov, and M.A. Fedotov, *Koord. Khim.*, 1989, **15**, 200 (*Chem. Abstr.*, 1989, **110**, 224 399).
193 L.R. Chamberlain, B.D. Steffey, I.P. Rothwell, and J.C. Huffman, *Polyhedron*, 1989, **8**, 341.
194 Y-W. Chao, P.A. Weler, and D.E. Wigley, *Inorg. Chem.*, 1989, **28**, 3860.
195 D. Rehder, W. Priebsch, and M.v. Oeynhausen, *Angew. Chem., Int. Ed. Engl.*, 1989, **28**, 1221.
196 F. Bottomley, C.P. Magill, and P.S. White, *J. Am. Chem. Soc.*, 1989, **111**, 3070.
197 V.W. Day, W.G. Klemperer, and O.M. Yaghi, *J. Am. Chem. Soc.*, 1989, **111**, 4518.
198 H.K. Chae, W.G. Klemperer, and V.W. Day, *Inorg. Chem.*, 1989, **28**, 1423.
199 Y. Hayashi, Y. Ozawa, and K. Isobe, *Chem. Lett.*, 1989, 425 (*Chem. Abstr.*, 1989, **111**, 233 153).
200 V.W. Day, W.G. Klemperer, and O.M. Yaghi, *J. Am. Chem. Soc.*, 1989, **111**, 5959.
201 A.D. Kirk, W. Riske, D.K. Lyon, B. Rapko, and R.G. Fink, *Inorg. Chem.*, 1989, **28**, 792.
202 S. Yue, S. Hua, and Z. Qui, *Bopuxue Zazhi*, 1988, **5**, 197 (*Chem. Abstr.*, 1989, **110**, 184 739).
203 W.V. Cicha and F. Aubke, *J. Am. Chem. Soc.*, 1989, **111**, 4328.
204 B.D. Steffey, R.W. Chesnut, J.L. Kerschner, P.J. Pellechia, P.E. Fanwick, and I.P. Rothwell, *J. Am. Chem. Soc.*, 1989, **111**, 378.
205 A.C. Dengel and W.P. Griffiths, *Polyhedron*, 1989, **8**, 1371.
206 B.D. Steffey, L.R. Chamberlain, R.W. Chesnut, D.E. Chebi, P.E. Fanwick, and I.P. Rothwell, *Organometallics*, 1989, **8**, 1419.
207 P.K. Baker and A. Bury, *J. Organomet. Chem.*, 1989, **359**, 189.
208 N. Kashef and R.L. Richards, *J. Organomet. Chem.*, 1989, **365**, 309.

$[MH(O_2CCH{=}CH_2)(\eta^2\text{-}C_2H_4)(PMe_3)_2]_2$, (M = Mo, W; ^{13}C),[209] $[WH_5(PMe_3)][(\eta^5,\eta^5\text{-}Me_4C_5\text{-}CH_2CH_2C_5Me_4)Rh(CO)(PMe_3)]$, ($^{13}C$),[210] $[(\eta^5\text{-}C_5H_5)WH(CO)_2(CNR)]$, ($^{13}C$),[211] $[(\eta^5\text{-}C_5Me_5)WOs_3(CO)_{12}H]$, ($^{13}C$),[212] $[(OC)_5Cr{=}CC(OEt){=}C(OEt)]$, ($^{13}C$),[213] $[(OC)_5Cr{=}CMeNH\text{-}CH_2Ph]$, ($^{13}C$),[214] $[(OC)_5Cr{=}CMeNHCH_2CH{=}CH_2]$, ($^{13}C$),[215] $[(OC)_5Cr{=}CBu^tN{=}CPhO_2CPh]$, ($^{13}C$),[216] $[(OC)_5Cr{=}CR^1N{=}CR^2R^3]$, ($^{13}C$),[217] $[(OC)_5Cr{=}CRN{=}CPhOMe]$, ($^{13}C$),[218] $[(OC)_5\text{-}M{=}CRN{=}CMeO_2CMe]$, (M = Cr, Mo, W; ^{13}C),[219] *cis*-$[(R^1NC)(OC)_4Cr{=}C(NR^2{}_2)Ph]$, ($^{13}C$),[220] $[(OC)_5Cr{=}CPhN{=}CHPh]$, (^{13}C, ^{15}N),[221] (12), (M = Cr, W; ^{13}C),[222] $[(OC)_5Cr{=}C\text{-}(NC_4H_8)N{=}CCl_2]$, ($^{13}C$),[223] (13), (M = Cr, W; ^{13}C),[224] $[(OC)_5M{=}CMeSCH_2CH{=}CH_2]$, (M =

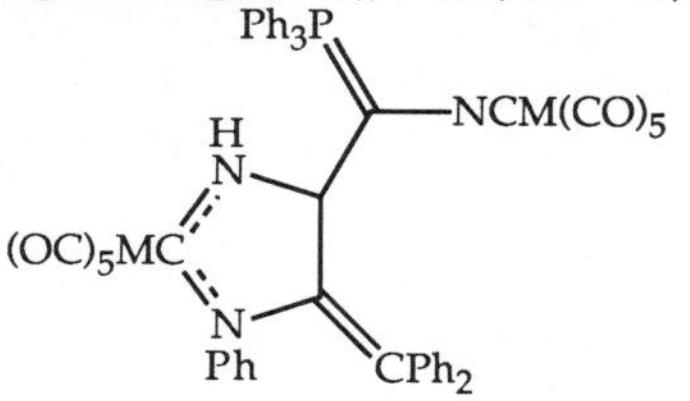

(12)

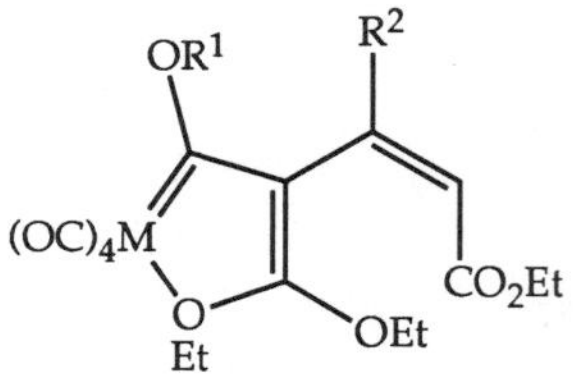

(13)

Cr, W; ^{13}C),[225] $[(OC)_5M{=}CPhSR]$, (M = Cr, W; ^{13}C),[226] $[(OC)_5M{=}CSCH_2CHRCH_2S]$, (M = Cr, W; ^{13}C),[227] $[(OC)_5Cr{=}CXR]$, (^{13}C),[228] $[(\eta^5\text{-}C_5H_5)(OC)M(\mu\text{-}CC_6H_3Me_2\text{-}2,6)(\mu\text{-}CO)Pt\text{-}(PPr^i{}_3)]$, (M = Cr, Mo, W; ^{13}C, ^{195}Pt),[229] $[(\eta^5\text{-}C_5H_5)(OC)_2M(\mu\text{-}ECC_6H_3Me_2\text{-}2,6)Fe(CO)_3]$, (M = Cr, Mo, W; E = S, Se; ^{13}C),[230] $[(\eta^5\text{-}C_5H_4M^1R_3)M^2(CO)_nMe]$, ($M^1$ = Ge, Sn, Pb; M^2 = Fe, n = 2; M^2 = Mo, W, n = 3; ^{13}C),[231] $[(OC)_6Co\{\mu\text{-}Me(OC)_3M(\eta^5\text{-}C_5H_4)C{\equiv}C(\eta^5\text{-}$

209 R. Alvarez, E. Carmona, A. Galindo, E. Gutiérrez, J.M. Marín, A. Monge, M.L. Poveda, C. Ruiz, and J.M. Savariault, *Organometallics*, 1989, **8**, 2430.
210 J.F. Buzinkai and R.R. Schrock, *Inorg. Chem.*, 1989, **28**, 2837.
211 H.G. Alt, H.E. Engelhardt, T. Frister, and R.D. Rogers, *J. Organomet. Chem.*, 1989, **366**, 297.
212 S.-M. Peng, G.-H. Lee, Y. Chi, C.-L. Peng, and L.-S. Hwang, *J. Organomet. Chem.*, 1989, **371**, 197.
213 K.N. Juneau, L.S. Hegedus, and F.W. Roepke, *J. Am. Chem. Soc.*, 1989, **111**, 4762.
214 B. Denise, R. Goumont, A. Parlier, H. Rudler, J.-C. Daran, and J. Vaissermann, *J. Organomet. Chem.*, 1989, **377**, 89.
215 C. Alvarez, A. Parlier, H. Rudler, R. Yefsah, J.C. Daran, and C. Knobler, *Organometallics*, 1989, **8**, 2253.
216 W.D. Wulff, V. Dragisich, J.C. Huffman, R.W. Kaesler, and D.C. Yang, *Organometallics*, 1989, **8**, 2196.
217 K.H. Dötz, D. Grotjahn, and K. Harms, *J. Organomet. Chem.*, 1989, **375**, C47.
218 A. Wienand, H.-U. Reißig, H. Fischer, and J. Hofmann, *Chem. Ber.*, 1989, **122**, 1589.
219 L.S. Hegedus, L.M. Schultze, and J. Montgomery, *Organometallics*, 1989, **8**, 2189.
220 R. Aumann and H. Heinen, *Chem. Ber.*, 1989, **122**, 1139.
221 R. Aumann, S. Althaus, C. Krüger, and P. Betz, *Chem. Ber.*, 1989, **122**, 357.
222 W.P. Fehlhammer, G. Zinner, G. Beck, and J. Fuchs, *J. Organomet. Chem.*, 1989, **379**, 277.
223 W.P. Fehlhammer and G. Beck, *Chem. Ber.*, 1989, **122**, 1901; 1907.
224 F. Camps, J.M. Moretó, S. Ricart, J.M. Viñas, E. Molins, and C. Miravitlles, *J. Chem. Soc., Chem. Commun.*, 1989, 1560.
225 A. Parlier, H. Rudler, and C. Alvarez, *J. Organomet. Chem.*, 1989, **379**, 271.
226 R. Aumann and J. Schröder, *J. Organomet. Chem.*, 1989, **378**, 57.
227 W.P. Fehlhammer and G. Beck, *J. Organomet. Chem.*, 1989, **369**, 105.
228 W.D. Wulff, B.A. Anderson, and A.J. Toole, *J. Am. Chem. Soc.*, 1989, **111**, 5485.
229 T.P. Spaniol and F.G.A. Stone, *Polyhedron*, 1989, **8**, 2271.
230 A.F. Hill, B.A. Nasir, and F.G.A. Stone, *Polyhedron*, 1989, **8**, 179.
231 J. Cervantes, S.P. Vincenti, R.N. Kapoor, and K.H. Pannell, *Organometallics*, 1989, **8**, 744.

$C_5H_4)M(CO)_3Me\}]$, (M = Mo, W; ^{13}C),[232] $[Mo(NBu^t)_2(CH_2Bu^t)_2]$, ($^{13}C$),[233] $[(\eta^5$-$C_5H_5)Mo(CO)_3CH_2R]$, (^{13}C),[234] *trans*-$[(\eta^5$-$C_5H_5)Mo(CO)_2\{P(OPh)_3\}CH_2X]$, ($^{13}C$),[235] $[(\eta^5$-$C_5Me_5)M(NO)_2(CH_2SiMe_3)_2]$, (M = Mo, W; ^{13}C),[236] $[(\eta^5$-$C_5H_5)_2Mo_2(CO)_2(\mu$-$PPh_2)(\mu$-$Ph_2PCH_2)]$, (^{13}C),[237] $[(\eta^5$-$C_5H_5)M(CO)_3C{=}CHCH_2C(O)O]$, (M = Mo, W; ^{13}C),[238] (14), (^{13}C),[239] (15), (^{13}C),[240] $[(\eta^5$-$C_5Me_5)Mo(CO)_2\{P(OR)_3\}CHO]$, ($^{13}C$),[241] $[M(SnCl_3)(CNBu^t)_3$-

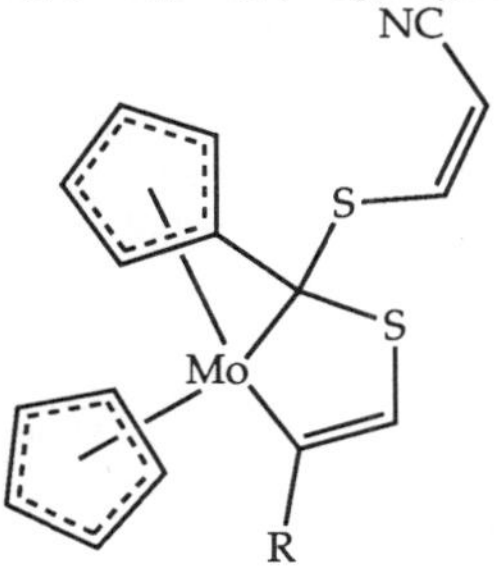

(14)

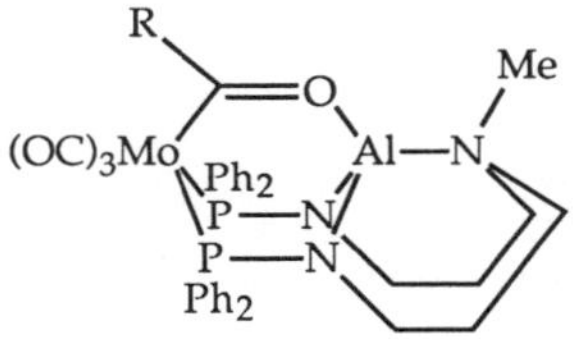

(15)

$(PMe_3)_3][SnCl_3]$, (M = Mo, W; ^{13}C, ^{119}Sn),[242] Mo and W metallocarbenes derived from styrene, (^{13}C),[243] $[HB(Me_2pz)_3Mo(CO)_2CR]$, (^{13}C),[244] $[M(\equiv CC\equiv CBu^t)(O_2CCF_3)(CO)_2L_2]$, (M = Mo, W; ^{13}C),[245] $[Mo_2\{\mu$-$\sigma,\eta^2(4e)C{=}CH_2\}(CO)_4(\eta^5$-$C_5H_5)_2]$, ($^{13}C$),[246] $[(\eta^5$-$C_5H_5)Ni(\mu$-$CR^1{=}CHR^2)(\mu$-$CO)M(\eta^5$-$C_5H_4Me)(CO)(O_2CCF_3)]$, (^{13}C),[247] $[MoFePt(\mu_3$-CC_6H_4Me-$4)(CO)_3(PPr^i_3)(\eta^5$-$C_5H_5)]$, (^{13}C, ^{195}Pt),[248] $[MoCo_2(\mu_3$-CC_6H_4Me-$4)(\mu$-$PPh_2)(CO)_6(\eta^5$-$C_5H_5)]$, (^{13}C),[249] $[Mo_3(\mu$-$CC_6H_3Me_2$-$2,6)_2(CO)_6(\eta^5$-$C_5H_5)_2]$, (^{13}C),[250] $[Mo_2W_3Pt_4(\mu$-CC_6H_4Me-$4)_2(\mu_3$-$CMe)(\mu_3$-CC_6H_4Me-$4)_2(CO)_{10}(\eta^5$-$C_5H_5)_5]$, (^{13}C, ^{195}Pt),[251] $[(\eta^5$-$C_5H_5)(OC)_2M^1(\mu$-

232 C. Lo Sterzo, M.M. Miller, and J.K. Stille, *Organometallics*, 1989, **8**, 2331.
233 G. Schoettel, J. Kress, and J.A. Osborn, *J. Chem. Soc., Chem. Commun.*, 1989, 1062.
234 G.-H. Lee, S.-M. Peng, G.-M. Yang, S.-F. Lush, and R.-S. Liu, *Organometallics*, 1989, **8**, 1106.
235 D.H. Gibson, S.K. Mandal, K. Owens, W.E. Sattich, and J.O. Franco, *Organometallics*, 1989, **8**, 1114.
236 P. Legzdins, E.C. Phillips, and L. Sánchez, *Organometallics*, 1989, **8**, 940.
237 V. Riera, M.A. Ruiz, F. Villafañe, C. Bois, and Y. Jeannin, *J. Organomet. Chem.*, 1989, **375**, C23.
238 A. Wong, R.L. Morgan, sec., J.M. Golder, G.E. Quimbita, and R.V. Pawlick, *Organometallics*, 1989, **8**, 844.
239 F. Conan, J. Sala-Pala, J.E. Guerchais, J. Li, R. Hoffmann, C. Mealli, R. Mercier, and L. Toupet, *Organometallics*, 1989, **8**, 1929.
240 J. Powell, M.R. Gregg, and P.E. Meindl, *Organometallics*, 1989, **8**, 2942.
241 D.H. Gibson, K. Owens, S.K. Mandal, W.E. Sattich, and J.O. Franco, *Organometallics*, 1989, **8**, 498.
242 M.J. Fernandez-Trujillo, M. Jimenez-Tenorio, M.C. Puerta, and D.L. Hughes, *J. Organomet. Chem.*, 1989, **377**, 241.
243 L. Chen, J. Lin, J. Jin, G. Huang, X. Li, H. Chen, and X. Lin, *J. Macromol. Sci., Chem.*, 1989, **A26**, 361 (*Chem. Abstr.*, 1989, **111**, 40 004).
244 D.G. Brower, M. Stoll, and J.L. Templeton, *Organometallics*, 1989, **8**, 2786.
245 I.J. Hart, A.F. Hill, and F.G.A. Stone, *J. Chem. Soc., Dalton Trans.*, 1989, 2261.
246 G.C. Conole, S.F.T. Froom, M. Green, and M. McPartlin, *J. Chem. Soc., Chem. Commun.*, 1989, 92.
247 M.J. Chetcuti, J.C. Gordon, K.A. Green, P.E. Fanwick, and D. Morgenstern, *Organometallics*, 1989, **8**, 1790.
248 A.F. Hill, F. Marken, B.A. Nasir, and F.G.A. Stone, *J. Organomet. Chem.*, 1989, **363**, 311.
249 M.R. Bradford, N.G. Connelly, N.C. Harrison, and J.C. Jeffery, *Organometallics*, 1989, **8**, 1829.
250 S.J. Dossett, A.F. Hill, J.A.K. Howard, B.A. Nasir, T.P. Spaniol, P. Sherwood, and F.G.A. Stone, *J. Chem. Soc., Dalton Trans.*, 1989, 1871.
251 S.J. Davies, J.A.K. Howard, R.J. Musgrove, and F.G.A. Stone, *J. Chem. Soc., Dalton Trans.*, 1989, 2269.

CC$_6$H$_4$Me-4)M^2(μ$_3$-CC$_6$H$_4$Me-4)M^1(η^5-C$_5$H$_5$)(CO)$_2$Pt(η^4-cod)], (M^1 = Mo, W; M^2 = Ni, Pt; ^{13}C, ^{195}Pt),[252] [{(η^4-cod)Pt}$_2$(μ$_3$-CMe)$_3$(μ$_3$-CC$_6$H$_4$Me-4)$_2$W$_3$(η^5-C$_5$H$_5$)$_3$(CO)$_6$Mo$_2$(η^5-C$_5$H$_5$)$_2$(CO)$_4$Pt$_4$], (^{13}C, ^{195}Pt),[253] [(η^5-C$_5$H$_5$)(η^5-C$_5$H$_4$OCHPhPr^i)WMe$_2$], (^{13}C),[254] [Me(OC)$_3$W(η^5-C$_5$H$_4$PPh$_2$)Fe(η^5-C$_5$H$_5$)(COMe)(CO)], (^{13}C),[255] [W$_2$(μ-CAr)Me(CO)$_3$(η^5-C$_5$H$_5$)$_2$], (^{13}C),[256] [(η^5-C$_5$Me$_5$)W(CO)$_3$CH$_2$SPh], (^{13}C),[257] [CH$_2$CH$_2$CH$_2$W(OPr^i)(μ-OPr^i)$_3$W(OPr^i)$_2$(η^2-C$_2$H$_4$)], (^{13}C),[258] [(η^5-C$_5$H$_5$)W(CO)$_3$CF$_3$], (^{13}C, ^{183}W),[259] [(η^5-C$_5$H$_5$)W{C(=CH$_2$)SMe}(CO)$_2$], (^{13}C),[260] [(OC)$_5$W{CPh(OEt)CPh=C=S}]$^-$, (^{13}C),[261] [{WO(η^5-C$_5$H$_4$)Fe(η^5-C$_5$H$_5$)}$_2$O], (^{13}C),[262] [(η^5-C$_5$H$_5$)W{C(CF$_3$)=C(CF$_3$)C(CF$_3$)=C(CF$_3$)}(SR^1)(CNR^2)], (^{13}C),[263] [W(C≡CPh)$_2$(NPh)(η^2-PhC≡CPh)(PMe$_3$)$_2$], (^{13}C),[264] [{η^6-(OC)$_5$WC$_6$H$_5$}Cr(CO)$_3$], (^{13}C),[265] [W(C$_6$H$_4$-2-CH=NCH$_2$CH$_2$NMe$_2$)(CO)$_3$X], (^{183}W),[266] [(η^5-C$_5$H$_5$)W(=CHPh)H(CO)$_2$], (^{13}C),[267] [(η^5-C$_5$H$_5$)(OC)$_2$W=CPhCPh=CHC$_6$H$_4$Me-4]$^+$, (^{13}C),[268] [(OC)$_5$W=CC(=CPhOEt)CH$_2$CH$_2$NMe], (^{13}C),[269] [(*E*)-(OC)$_5$W{C(NMe$_2$)CH=CHC$_5$H$_4$-η^5}Fe-(η^5-C$_5$H$_5$)], (^{13}C),[270] [(OC)$_5$W=C(OMe)CH=CH$_2$], (^{13}C),[271] [EtCMe$_2$OCH$_2$[CMe{C(OMe)=W-(CO)$_5$}CH$_2$]$_3$CMe=CHOMe], (^{13}C),[272] [{(η^5-C$_5$Me$_5$)(OC)$_2$WC(C$_6$H$_4$Me-4)PCBu^t-η^4}Fe(CO)$_3$], (^{13}C),[273] [(Bu^{t}O)$_4$W≡CBu^t], (^{13}C),[274] [W(≡CNEt$_2$)(CNBu^t)$_4$I], (^{13}C),[275] [W(≡CNEt$_2$)-(CNBu^t)$_2$L$_2$I$_2$]$^+$, (^{13}C),[276] [{HC(pz)$_3$}(OC)$_2$W≡CSMe]$^+$, (^{13}C),[277] [WRu(μ-σ,η^2-CH=CH$_2$)-

252 S.J. Davies and F.G.A. Stone, *J. Chem. Soc., Dalton Trans.*, 1989, 785.

253 S.J. Davies, J.A.K. Howard, R.J. Musgrove, and F.G.A. Stone, *Angew. Chem., Int. Ed. Engl.*, 1989, **28**, 624.

254 J.P. McNally and N.J. Cooper, *J. Am. Chem. Soc.*, 1989, **111**, 4500.

255 J.K. Stille, C. Smith, O.P. Anderson, and M.M. Miller, *Organometallics*, 1989, **8**, 1040.

256 I.J. Hart, J.A.K. Howard, R.M. Lowry, T.P. Spaniol, and F.G.A. Stone, *Polyhedron*, 1989, **8**, 2035.

257 H.-J. Kneuper, C. Zimmermann, K. Harms, and G. Boche, *Chem. Ber.*, 1989, **122**, 1043.

258 M.H. Chisholm, J.C. Huffman, and M.J. Hampden-Smith, *J. Am. Chem. Soc.*, 1989, **111**, 5284.

259 D. Naumann and H.G. Varbelow, *J. Fluorine Chem.*, 1988, **41**, 415 (*Chem. Abstr.*, 1989, **111**, 154 010).

260 F.R. Kreissl and N. Ullrich, *Chem. Ber.*, 1989, **122**, 1487.

261 H.G. Raubenheimer, G.J. Kruger, L. Linford, C.F. Marais, R. Otte, J.T.Z. Hattingh, and A. Lombard, *J. Chem. Soc., Dalton Trans.*, 1989, 1565.

262 M. Herberhold and H. Kniesel, *J. Organomet. Chem.*, 1989, **371**, 205.

263 L.J. Canoira, J.L. Davidson, G. Douglas, and K.W. Muir, *J. Organomet. Chem.*, 1989, **362**, 135.

264 G.R. Clark, A.J. Nielson, and C.E.F. Rickard, *J. Chem. Soc., Chem. Commun.*, 1989, 1157.

265 J.A. Heppert, M.E. Thomas-Miller, D.M. Scherubel, F. Takusagawa, M.A. Morgenstern, and M.R. Shaker, *Organometallics*, 1989, **8**, 1199.

266 R. Benn, A. Rufińska, M.A. King, C.E. Osterberg, and T.G. Richmond, *J. Organomet. Chem.*, 1989, **376**, 359.

267 M.J. Winter and S. Woodward, *J. Organomet. Chem.*, 1989, **361**, C18.

268 K.E. Garrett, J.B. Sheridan, D.B. Pourreau, W.C. Feng, G.L. Geoffroy, D.L. Staley, and A.L. Rheingold, *J. Am. Chem. Soc.*, 1989, **111**, 8383.

269 R. Aumann and P. Hinterding, *Chem. Ber.*, 1989, **122**, 365.

270 D.W. Macomber, P. Madhukar, and R.D. Rogers, *Organometallics*, 1989, **8**, 1275.

271 D.W. Macomber, M. Liang, P. Madhukar, and A.G. Verma, *J. Organomet. Chem.*, 1989, **361**, 187.

272 D.W. Macomber and M.-H. Hung, *J. Organomet. Chem.*, 1989, **366**, 147.

273 A.F. Hill, J.A.K. Howard, T.P. Spaniol, F.G.A. Stone, and J. Szameitat, *Angew. Chem., Int. Ed. Engl.*, 1989, **28**, 210.

274 K. Weiss and G. Lössel, *Angew. Chem., Int. Ed. Engl.*, 1989, **28**, 62.

275 A.C. Filippou and E.O. Fischer, *J. Organomet. Chem.*, 1989, **365**, 317.

276 A.C. Filippou, E.O. Fischer, and H.G. Alt, *Z. Naturforsch., B.*, 1988, **43**, 654 (*Chem. Abstr.*, 1989, **110**, 231 806).

$(CO)_4$(*cis*-$Ph_2PCH=CHPPh_2$)(η^5-C_5H_5)], (^{13}C),[278] [(η^5-C_5Me_5)Ni(μ-PhC_2H)(μ-CO)W(CO)-(η^5-C_5H_5)], (^{13}C),[279] [W_3(CMe)(OPr^i)$_9$$(CO)_2$], ($^{13}C$),[280] [($\eta^5$-$C_5H_5$)$WRu_3$$(CO)_9$($\mu$-NPh)-(C≡CPh)], ($^{13}C$),[281] [($\eta^5$-$C_5H_5$)$WOs_3$$(CO)_{11}$(C≡CPh)], ($^{13}C$),[282] [$W_2M_2$($\mu$-CO)($\mu$-CMe)($\mu_3$-CMe)($\mu$-$PPh_2$)$_2$$(CO)_3$($\eta^5$-$C_5H_5$)$_2$], (M = Rh, Ir; ^{13}C),[283] [W_3Rh_2(μ-CO)$_2$(μ-CMe){μ-CMe-C(O)}(μ-PPh_2)$_2$(μ_3-CMe)$(CO)_2$(η^5-C_5H_5)$_3$], (^{13}C),[284] [W_3Rh_2(μ-CO)$_2$(μ-CMe){μ-CMeC(O)}-(μ-PPh_2)$_2$(μ_3-CMe)$(CO)_2$(η^5-C_5H_5)$_3$], (^{13}C),[285] and [W_2RuPt(μ-CC_6H_4Me-4)(μ_3-CC_6H_4Me-4)(CO)(η^5-C_5H_5)$_2$], (^{13}C, ^{195}Pt).[286]

The ^{13}C chemical shift of the ethyne carbon in [WCl_4(η^2-HC≡CH)(OEt_2)] is δ 251 and is therefore carbenoid.[287] The ^{13}C chemical shift has been used to suggest that but-2-yne donates four electrons to the metal in [WI(CO)L(dppm)(η^2-MeC≡CMe)]$^+$.[288] N.m.r. data have also been reported for [(η^6-$C_6H_3R^1_3$)Cr$(CO)_2$($R^2C≡CR^3$)], (^{13}C),[289] [$MoCl_4$(η^2-PhC≡C-Ph)]$_2$, (^{13}C),[290] [(η^5-C_5H_5)$(OC)_2$$M^1$($\mu_2$-$\eta^2$;$\eta^2$-RC≡CR)$M^2$$(CO)_3$L], ($M^1$ = Mo, W; M^2 = Mn, Re; ^{13}C),[291] [(η^5-C_5H_5)$_2$$Mo_2$$(CO)_2$($\mu$-Cl)($\mu$-$Ph_2PCR^1=CR^2$)], ($^{13}C$),[292] [NiMo$(CO)_2$($\mu$-$\eta^2$,$\eta^2$-$CMe_2=C=CH_2$)($\eta^5$-$C_5H_5$)($\eta^5$-$C_5H_4$Me)], ($^{13}C$),[293] [($\eta^5$-$C_5H_5$)$_2$$Mo_2$$(CO)_4${$\mu$-$\eta^1$(*N*),$\eta^2$(*C*,*N*)-4-$MeC_6H_4N=C=CPh_2$}], ($^{13}C$),[294] [HB($Me_2C_3N_2H$)$_3Mo(CO)_2${$\eta^2$-C(O)R}], ($^{13}C$),[295] [{$\eta^3$-$(OC)_5$-W{PPh($CH=CH_2$)}Mo$(CO)_2$($\eta^5$-$C_5H_5$)], ($^{13}C$),[296] [($R_2NCS_2$)$_3$WCPhCPhCPh$CH_2$Ph]$^+$, ($^{13}C$),[297] [W($\mu$-I)Br(CO)(NCMe)($\eta^2$-$R^1C≡CR^2$)]$_2$, ($^{13}C$),[298] [WI$(CO)_2$(dppm)($\eta^2$-MeC≡C-Me)]$^+$, ($^{13}C$),[299] [{HB$(pz)_3$}(OC)($\eta^2$-PhC≡CMe)W(NCMe)]$^+$, ($^{13}C$),[300] [[W($\mu$-I)I(CO){SC-($NMe_2$)$_2$}($\eta^2$-PhC≡CPh)]$_2${SC($NMe_2$)$_2$}], ($^{13}C$),[301] [Na(15-crown-5)][WF_5(η^2-PhC≡CH)], (^{13}C),[302] [(η^5-C_5Me_5)Mo(η^3-$CH_2CHC=CH_2$)$(CO)_2$], (^{13}C),[303] [W{η^3-SC(CF_3)=C(CF_3)H}(η^2-

277 R.A. Doyle and R.J. Angelici, *Organometallics*, 1989, **8**, 2207.
278 S.A. Brew, J.C. Jeffery, and F.G.A. Stone, *Polyhedron*, 1989, **8**, 695.
279 M.J. Chetcuti, B.E. Grant, and P.E. Fanwick, *J. Am. Chem. Soc.*, 1989, **111**, 2743.
280 M.H. Chisholm, K. Folting, J.C. Huffman, J.A. Klang, and W.E. Streib, *Organometallics*, 1989, **8**, 89.
281 Y. Chi, D.-K. Hwang, S.-F. Chen, and L.-K. Liu, *J. Chem. Soc., Chem. Commun.*, 1989, 1540.
282 Y. Chi, G.-H. Lee, S.-M. Peng, and C.-H. Wu, *Organometallics*, 1989, **8**, 1574.
283 S.J. Davies and F.G.A. Stone, *J. Chem. Soc., Dalton Trans.*, 1989, 1865.
284 S.J. Davies, J.A.K. Howard, M.U. Pilotti, and F.G.A. Stone, *J. Chem. Soc., Dalton Trans.*, 1989, 1855.
285 S.J. Davies, J.A.K. Howard, M.U. Pilotti, and F.G.A. Stone, *J. Chem. Soc., Chem. Commun.*, 1989, 190.
286 S.J. Davies, J.A.K. Howard, M.U. Pilotti, and F.G.A. Stone, *J. Chem. Soc., Dalton Trans.*, 1989, 2289.
287 M. Kersting, A. El-Kholi, U. Müller, and K. Dehnicke, *Chem. Ber.*, 1989, **122**, 279.
288 P.K. Baker, K.R. Flower, P.A. Bates, and M.B. Hursthouse, *J. Organomet. Chem.*, 1989, **372**, 263.
289 H.G. Alt, H.E. Engelhardt, R.D. Rogers, and S.T. Abu-Orabi, *J. Organomet. Chem.*, 1989, **378**, 33.
290 I. Paul, K. Dehnicke, and D. Fenske, *Chem. Ber.*, 1989, **122**, 481.
291 H.-J. Müller, K. Polborn, M. Steimann, and W. Beck, *Chem. Ber.*, 1989, **122**, 1901.
292 G. Conole, K.A. Hill, M. McPartlin, M.J. Mays, and M.J. Morris, *J. Chem. Soc., Chem. Commun.*, 1989, 688.
293 M.J. Chetcuti, S.R. McDonald, and N.P. Rath, *Organometallics*, 1989, **8**, 2077.
294 T. Sielisch, W. Ziegler, and U. Behrens, *J. Organomet. Chem.*, 1989, **359**, 315.
295 C.A. Rusik, M.A. Collins, A.S. Gamble, T.L. Tonker, and J.L. Templeton, *J. Am. Chem. Soc.*, 1989, **111**, 2550.
296 C. Hugel-Le Goff, F. Mercier, L. Ricard, and F. Mathey, *J. Organomet. Chem.*, 1989, **363**, 325.
297 S.G. Feng, A.S. Gamble, and J.L. Templeton, *Organometallics*, 1989, **8**, 2024.
298 P.A. Baker, A. Bury, and K.R. Flower, *Polyhedron*, 1989, **8**, 2587.
299 P.K. Baker and K.R. Flower, *Inorg. Chim. Acta*, 1989, **165**, 241.
300 S.G. Feng and J.L. Templeton, *J. Am. Chem. Soc.*, 1989, **111**, 6477.
301 P.K. Baker and E.M. Armstrong, *Polyhedron*, 1989, **8**, 351.
302 P. Neumann, A. El-Kholi, U. Müller, and K. Dehnickep, *Z. Anorg. Allg. Chem.*, 1989, **577**, 185.
303 S.A. Benyunes, M. Green, M. McPartlin, and C.B.M. Nation, *J. Chem. Soc., Chem. Commun.*, 1989, 1887.

$CF_3C{\equiv}CCF_3)(\eta^5\text{-}C_5H_5)]$, ($^{13}C$),[304] $[W(S_2CNEt_2)_2(CO)(\mu\text{-}\eta^2\text{-}Ph_2PC{\equiv}CPPh_2)Co_2(CO)_6(\mu\text{-}CO)_2]$, ($^{13}C$),[305] $[(\eta^5\text{-}C_5H_5)W(\eta^3\text{-}MeCRCO)(PMe_3)(CO)]^+$, ($^{13}C$),[306] $[W_2(OBu^t)_6(\mu\text{-}MeC{\equiv}CMe)\text{-}(CO)]$, ($^{13}C$),[307] $[W_2Cl_4(\mu\text{-}Cl)_2(\mu\text{-}MeC{\equiv}CMe)(THF)_2]$, ($^{13}C$),[308] $[W_4(\mu\text{-}CR)_2(\mu\text{-}MeC{\equiv}CMe)\text{-}(\eta^2\text{-}MeC{\equiv}CMe)_2(OPr^i)_6]$, ($^{13}C$),[309] $[(\eta^5\text{-}C_5H_5)(OC)_2W(\mu\text{-}PhC{=}C{=}CH_2)Fe_2(CO)_6]$, ($^{13}C$),[310] $[WCl_2(CO)(PMePh_2)_2\{C,C^1;\eta^2\text{-}C(O)CPMePh_2\}]$, ($^{13}C$),[311] $[(\eta^5\text{-}C_5H_5)W(\eta^2\text{-}CHMe{=}S)(O_2C\text{-}CF_3)(CO)_2]^+$, ($^{13}C$),[312] $[\{HB(pz)_3\}W(CO)_2\{\eta^2\text{-}C(SMe)SMe\}]^+$, ($^{13}C$),[313] [{HB(1,2,3-benzotriazole)$_3$}Mo(η^3-C_3H_5)(CO)$_2$], (^{13}C),[314] $[(\eta^3\text{-}C_{13}H_9)(\eta^3\text{-}C_3H_5)_3Mo]$, ($^{13}C$),[315] [Mo(*endo-syn*-$\eta^3$-1-$C_3H_4$CHO)(CO)$_2$($\eta^5$-$C_5Me_5$)], ($^{13}C$),[316] [($\eta^5$-$C_5H_5$)Mo($\eta^3$-*anti*-$Ph_3SnCHPh$)(CO)$_2$], ($^{13}C$),[317] $[(\eta^5\text{-}C_5H_5)M\{\eta^3\text{-}1,3\text{-}(2,6\text{-}Me_2C_6H_3N)_2\text{-}2\text{-}R\text{-}C_3\}(CNC_6H_3Me_2\text{-}2,6)]$, (M = Mo, W; ^{13}C),[318] $[(\eta^5\text{-}C_5H_5)W(CO)_2(\eta^3\text{-}C_3H_4R)]$, ($^{13}C$),[319] $[CoW(\mu_3\text{-}PhCCHCR)(\mu\text{-}CO)(CO)_5(\eta^5\text{-}C_5H_5)]$, ($^{13}C$),[320] $[W_2(OBu^t)_6(C_3H_4)(CO)_2]$, ($^{13}C$),[321] $[Cr(CO)_3\{P(OMe)_3\}(\eta^4\text{-}1,5\text{-}cod)]$, ($^{13}C$),[322] $[\{\eta^4\text{-}(EtOPhC)(CH_2)(HOCH_2CH)C\}Cr(CO)_4]$, ($^{13}C$),[323] $[Cr\{\eta^4\text{-}Me_2NBCHCHB(NMe_2)PBu^tP\text{-}Bu^t\}(CO)_4]$, ($^{11}B$, ^{13}C),[324] (16), (^{11}B, ^{13}C),[325] $[Mo(CO)_3(\eta^5\text{-}C_4Ph_4CO)]_2$, ($^{13}C$),[326] [($\eta^5$-$C_5H_5$)

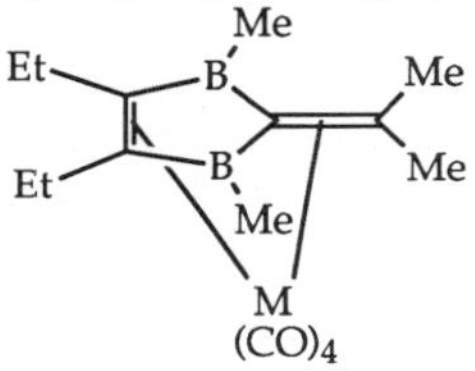

(16)

Mo(η^4-1,4-*anti-anti*-C_4H_4)(NO)], (^{13}C),[327] [(η^4-Me_3Si-bicycloheptadiene)Mo(CO)$_4$],

304 N.M. Agh-Atabay and J.L. Davidson, *J. Chem. Soc., Dalton Trans.*, 1989, 1027.
305 T.M. Nickel, S.Y.W. Yau, and M.J. Went, *J. Chem. Soc., Chem. Commun.*, 1989, 775.
306 N. Ullrich, H. Keller, C. Stegmair, and F.R. Kreissl, *J. Organomet. Chem.*, 1989, **378**, C19.
307 M.H. Chisholm, D. Ho, J.C. Huffman, and N.S. Marchant, *Organometallics*, 1989, **8**, 1626.
308 S.G. Bott, D.L. Clark, M.L.H. Green, and P. Mountford, *J. Chem. Soc., Chem. Commun.*, 1989, 418.
309 M.H. Chisholm, B.W. Eichhorn, and J.C. Huffman, *Organometallics*, 1989, **8**, 80.
310 G.H. Young, A. Wojcicki, M. Calligaris, G. Nardin, and N. Bresciani-Pahor, *J. Am. Chem. Soc.*, 1989, **111**, 6890.
311 A.K. List, G.L. Hillhouse, and A.L. Rheingold, *Organometallics*, 1989, **8**, 2010.
312 F.R. Kreißl, F.X. Müller, D.L. Wilkinson, and G. Müller, *Chem. Ber.*, 1989, **122**, 289.
313 R.A. Doyle and R.J. Angelici, *J. Organomet. Chem.*, 1989, **375**, 73.
314 K.-B. Shiu, F.-M. Shen, S.-L. Wang, and S.-C. Wei, *J. Organomet. Chem.*, 1989, **372**, 251.
315 O. Andell, R. Goddard, S. Holle, P.W. Jolly, C. Krüger, and Y.H. Tsay, *Polyhedron*, 1989, **8**, 203.
316 S.A. Benyunes, M. Green, and M.J. Grimshire, *Organometallics*, 1989, **8**, 2268.
317 M.J. Winter and S. Woodward, *J. Chem. Soc., Chem. Commun.*, 1989, 457.
318 P.K. Baker, G.K. Barker, D.S. Gill, M. Green, A.G. Orpen, I.D. Williams, and A.J. Welch, *J. Chem. Soc., Dalton Trans.*, 1989, 1321.
319 H.G. Alt, H.E. Engelhardt, B. Wrackmeyer, and R.D. Rogers, *J. Organomet. Chem.*, 1989, **379**, 289.
320 G.A. Carriedo, V. Riera, M.G. Sanchez, Y. Jeannin, and M. Philoche-Levisalles, *J. Organomet. Chem.*, 1989, **361**, 197.
321 R.H. Cayton, S.T. Chacon, M.H. Chisholm, M.J. Hampden-Smith, J.C. Huffman, K. Folting, P.D. Ellis, and B.A. Huggins, *Angew. Chem., Int. Ed. Engl.*, 1989, **28**, 1523.
322 D.J. Wink, N.-F. Wang, and B.T. Creagan, *Organometallics*, 1989, **8**, 561.
323 R. Aumann and B. Trentmann, *Chem. Ber.*, 1989, **122**, 1977.
324 M. Drieß, H. Pritzkow, and W. Siebert, *Chem. Ber.*, 1989, **122**, 467.
325 V. Schäfer, H. Pritzkow, and W. Siebert, *Chem. Ber.*, 1989, **122**, 401.
326 R.L. Beddoes, E.S. Cook, and M.J. Morris, *Polyhedron*, 1989, **8**, 1810.
327 N.J. Christensen, A.D. Hunter, and P. Legzdins, *Organometallics*, 1989, **8**, 930.

(^{13}C),[328] and $[Mo_2(dppe)_2(\eta^7\text{-}C_7H_7)_2(\mu\text{-}C_4R_2)]^{2+}$, (^{13}C).[329]

^{1}H, ^{13}C, and ^{17}O n.m.r. spectra of $[(\eta^5\text{-}C_5H_{5\text{-}n}Me_n)M(CO)_3]^-$ and $[(\eta^5\text{-}C_5H_{5\text{-}n}Me_n)M(CO)_2]_2$, M = Cr, Mo, W, indicate that data for the CO groups obey the 'triad' effect. The ^{95}Mo n.m.r. spectra were also recorded.[330] N.m.r. data have also been reported for $[(\eta^6\text{-}Me_4C_3N_2BF_2)Cr(CO)_3]$, (^{11}B, ^{13}C),[331] $[(\eta^5\text{-}C_5H_5)M(CO)_2P_3]$, (M = Cr, Mo, W; ^{13}C),[332] $[(OC)_2(ON)Cr(\eta^5\text{-}C_5H_4)C(O)(\eta^5\text{-}C_5H_4)Fe(\eta^5\text{-}C_5H_5)]$, (^{13}C),[333] $[(OC)_2(ON)Cr(\eta^5\text{-}C_5H_4)CH_2\text{-}(\eta^5\text{-}C_5H_4)Fe(\eta^5\text{-}C_5H_4CH{=}CH_2)]$, (^{13}C),[334] $[(\eta^5{:}\eta^5\text{-}Bu^t{}_4C_{10}H_4)M_2(CO)_6]$, (M = Cr, W; ^{13}C),[335] $[(\eta^5\text{-}C_5H_5)_2Cr_2(CO)_4(\mu\text{-}\eta^2\text{-}P_2)\{Cr(CO)_5\}_2]$, (^{13}C),[336] $[(\eta^5\text{-}C_5H_5)Mo(CNBu^t)_{4\text{-}n}I_n]^{3\text{-}n}$, (^{13}C),[337] $[(\eta^5\text{-}C_5H_4Me)_2Mo_2(\mu\text{-}S)\{\mu\text{-}S(CH_2)_4OH\}(\mu\text{-}SCH_2S)]^+$, (^{13}C),[338] $[Sb\{M(CO)_3(\eta^5\text{-}C_5H_5)\}_3]$, (M = Mo, W; ^{13}C),[339] $[(\eta^5\text{-}C_5H_5)(OC)_2Mo{=}P(C_6H_2Bu^t{}_3\text{-}2,4,6)CH{=}P(C_6H_2Bu^t{}_3\text{-}2,4,6)]$, (^{13}C),[340] $[(\eta^5\text{-}C_5H_5)_2Mo_2(CO)_{4\text{-}n}L_n(\mu\text{-}SR)_2]^{2+}$, (^{13}C),[341] $[(HBpz_3)(OC)_2W(\mu\text{-}CS)Mo(CO)_2(\eta^5\text{-}indenyl)]$, (^{13}C),[342] $[(\eta^5\text{-}C_5H_5)Mo(\eta^2\text{-}AsSCH_2CH_2S)(CO)_2]$, (^{13}C),[343] $[(\eta^5\text{-}C_5H_5)Mo(CO)_2\{\eta^1\text{-}P{=}C\text{-}(SiMe_3)_2\}]$, (^{13}C),[344] $[(\eta^5\text{-}C_5H_5)M(CO)(CNBu^t)_2I]$, (M = Mo, W; ^{13}C),[345] $[(\eta^5\text{-}C_5H_5)Mo\text{-}\{P(OMe)_3\}_2(PPh_2)]$, (^{13}C),[346] $[(\eta^5\text{-}C_5H_5)_2Mo_2\{\mu\text{-}1,2\text{-}(Bu^tP)_2C_6H_4\}Co_2(CO)_5\}]$, (^{13}C),[347] $[Mo_2\text{-}FePt(\mu\text{-}\sigma,\sigma'\sigma''{:}\eta^5\text{-}CC_5H_4)_2(CO)_2(pz_3BH)_2]$, (^{13}C, ^{195}Pt),[348] $[(\eta^5\text{-}C_5H_5)Mo(CO)_2Co_3(CO)_6(\mu\text{-}CO)_3]$, (^{13}C),[349] $[(\eta^5\text{-}C_5H_5)_2W\{C(O)\}_2NMe]$, (^{13}C),[350] $[(\eta^5\text{-}C_5Me_5)Re(CO)(NO){=}CO_2W(\eta^5\text{-}C_5H_5)_2]$, (^{13}C),[351] $[(\eta^5\text{-}C_5H_5)W(CO)_3S(O)NHC_6H_4Me\text{-}2]$, (^{13}C),[352] $[(\eta^5\text{-}C_5H_5)W(CO)_2\text{-}$

328 T.J. Chow and C.-C. Cheng, *J. Organomet. Chem.*, 1989, **368**, 323.
329 R.L. Beddoes, C. Bitcon, A. Ricalton, and M.W. Whiteley, *J. Organomet. Chem.*, 1989, **367**, C21.
330 I.R. Lyatifov, T.Kh. Gasanov, P.V. Petrovski, and A.I. Lutsenko, *J. Organomet. Chem.*, 1989, **361**, 181.
331 N. Kuhn, A. Kuhn, R. Boese, and N. Augart, *J. Chem. Soc., Chem. Commun.*, 1989, 975.
332 L.Y. Goh, C.K. Chu, R.C.S. Wong, and T.W. Hambley, *J. Chem. Soc., Dalton Trans.*, 1989, 1951.
333 Y.-P. Wang, J.-M. Hwu, and S.-L. Wang, *J. Organomet. Chem.*, 1989, **371**, 71.
334 Y.-P. Wang, T.-S. Lin, R.-S. Shyu, J.-M. Hwu, Y. Wang, and M.-C. Cheng, *J. Organomet. Chem.*, 1989, **371**, 57.
335 P. Jutzi and J. Schnittger, *Chem. Ber.*, 1989, **122**, 629.
336 L.Y. Goh, R.C.S. Wong, and T.C.W. Mak, *J. Organomet. Chem.*, 1989, **364**, 363.
337 A.C. Filippou, W. Grünleitner, and E. Herdtweck, *J. Organomet. Chem.*, 1989, **373**, 325.
338 G. Godziela, T. Tonker, and M. Rakowski DuBois, *Organometallics*, 1989, **8**, 2220.
339 A.M. Barr, M.D. Kerlogue, N.C. Norman, P.M. Webster, and L.J. Farrugia, *Polyhedron*, 1989, **8**, 2495.
340 H.H. Karsch, H.-U. Reisacher, B. Huber, G. Müller, K. Jörg, and W. Malisch, *New J. Chem.*, 1989, **13**, 319.
341 M. El Khalifa, M. Guéguen, R. Mercier, F.Y. Pétillon, J.-Y. Saillard, and J. Talarmin, *Organometallics*, 1989, **8**, 140.
342 R.A. Doyle, L.M. Daniels, R.J. Angelici, and F.G.A. Stone, *J. Am. Chem. Soc.*, 1989, **111**, 4995.
343 E.W. Abel, S.R. Allen, and B. Khandelwal, *J. Chem. Soc., Dalton Trans.*, 1989, 885.
344 A.M. Arif, A.H. Cowley, C.M. Nunn, S. Quashie, N.C. Norman, and A.G. Norman, *Organometallics*, 1989, **8**, 1878.
345 A.C. Filippou and W. Grünleitner, *J. Organomet. Chem.*, 1989, **378**, 387.
346 M. Bottrill, M. Green, A.G. Orpen, D.R. Saunders, and I.D. Williams, *J. Chem. Soc., Dalton Trans.*, 1989, 511.
347 E.P. Kyba, M.C. Kerby, R.P. Kashyap, J.A. Mountzouris, and R.E. Davis, *Organometallics*, 1989, **8**, 852.
348 S.J. Davis, A.F. Hill, M.U. Pilotti, and F.G.A. Stone, *Polyhedron*, 1989, **8**, 2265.
349 V.S. Kaganovich, Yu.L. Slovokhotov, A.V. Mironov, Yu.T. Struchkov, and M.I. Rybinskaya, *J. Organomet. Chem.*, 1989, **372**, 339.
350 P. Jernakoff and N.J. Cooper, *J. Am. Chem. Soc.*, 1989, **111**, 7424.
351 R.S. Pilato, G.L. Geoffroy, and A.L. Rheingold, *J. Chem. Soc., Chem. Commun.*, 1989, 1287.
352 W.P. Ehrenreich, M. Herberhold, and A.F. Hill, *J. Organomet. Chem.*, 1989, **371**, 303.

$(PMe_3)]_2$, (^{13}C),[353] $[(\eta^5\text{-}C_5H_5)W(CO)_2S_2CMe]$, ($^{13}C$),[354] $[(\eta^5\text{-}C_5H_5)WI_2(CNBu^t)(CNEt_2)]$, ($^{13}C$),[355] $[(\eta^5\text{-}C_5Me_5)W(O)_2X]$, ($^{13}C$),[356] and $[(\eta^5\text{-}C_5Me_5)W(NO)I]_2$, ($^{13}C$).[357]

^1H, ^{13}C, and ^{19}F n.m.r. spectra of $[\{\eta^6\text{-}1,3\text{-}C_6H_4(CH_2SCH_2)_2C_6H_4\text{-}1,3\}Cr(CO)_3]$ indicate that there is a 60% reduction in ring current on coordination.[358] The ^{13}C n.m.r. spectrum of $[(\eta^6\text{-}C_6H_4Me_2\text{-}1,4)Cr(CO)_3]$ has been discussed based on ring deformation.[359] The conformation of $[(\eta^6\text{-}1\text{-}Cl\text{-}4\text{-}MeC_6H_4)Cr(CO)_3]$ has been studied using ^1H and ^{13}C n.m.r. spectroscopy.[360] Detailed ^1H n.m.r. investigations have been used to identify the site of phenylation of cyclic and acyclic pentadienyl carbanions derived from $[(\eta^6\text{-}FC_6H_5)Cr(CO)_3]$.[361] The ^{19}F n.m.r. chemical shifts for $[(\eta^6\text{-}FC_6H_5)(\eta^6\text{-}C_6H_5X)Cr]$ correlate well with σ_m^+ and σ_m^n substituent constants.[362] N.m.r. data have also been reported for $[(\eta^6\text{-}OC_6H_5)Cr(CO)_3]^-$, ($^{13}C$),[363] $[(\eta^6\text{-arene})Cr(CO)_3]$, ($^{13}C$),[364] (17), (M = Cr, Mo, W; ^{11}B, ^{13}C),[365] $[\{\eta^6\text{-}1,4\text{-}[(\eta^5\text{-}C_5H_5)Fe(CO)_2]_2C_6H_4\}Cr(CO)_3]$, ($^{13}C$),[366] $[(\eta^6,\eta^3\text{-indenyl})\{Cr(CO)_3\}\text{-}\{Rh(\eta^4\text{-cod})\}]$, ($^{13}C$),[367] $[(\eta^6\text{-}4\text{-}Bu^n{}_3SnC_6H_4C_6H_4SnBu^n{}_3\text{-}4)Cr(CO)_3]$, ($^{13}C$),[368] $[(\eta^6\text{-}R^1{}_2\text{-}MCR^2{=}CR^3BR^4NC_6H_5)Cr(CO)_3]$, (M = Si, Sn; ^{11}B, ^{13}C, ^{14}N, ^{29}Si, ^{119}Sn),[369] $[(\eta^6\text{-}MeC_6H_4\text{-}CO_2Me)Cr(CO)_2(\text{neomenthylPPh}_2)]$, ($^{13}C$),[370] $[(\eta^6\text{-dibenzobicyclo}[2.2.2]\text{octadiene})Cr(CO)_3]$, ($^{13}C$),[371] $[(OC)_3Mo(\eta^6\text{-}C_7H_7)]_2$, ($^{13}C$),[372] $[(OC)_3Mo(\mu\text{-}\eta^1{:}\eta^6\text{-}C_7H_7)\text{-}exo\text{-}Re(CO)_5]$, ($^{95}Mo$),[373] and $[M(OC_6H_3\text{-}2\text{-}Ph\text{-}\eta^6\text{-}6\text{-}C_6H_5)XL_2]$, (M = Mo, W; ^{13}C).[374]

353 H.G. Alt, T. Frister, E.E. Trapl, and H.E. Engelhardt, *J. Organomet. Chem.*, 1989, **362**, 125.
354 F.R. Kreissl and N. Ullrich, *J. Organomet. Chem.*, 1989, **361**, C30.
355 A.C. Filippou, *Polyhedron*, 1989, **8**, 1285.
356 H.G. Alt, H.E. Engelhardt, H.I. Hayen, and R.D. Rogers, *J. Organomet. Chem.*, 1989, **366**, 287.
357 F.G. Herring, P. Legzdins, and G.B. Richter-Addo, *Organometallics*, 1989, **8**, 1485.
358 R.H. Mitchell, T.K. Vinod, G.J. Bodwell, and G.W. Bushnell, *J. Org. Chem.*, 1989, **54**, 5871.
359 I. Ono, S. Mita, S. Kondo, and N. Mori, *J. Organomet. Chem.*, 1989, **367**, 81.
360 F. Rose-Munch, E. Rose, A. Semra, and C. Bois, *J. Organomet. Chem.*, 1989, **363**, 103.
361 A. Ceccon, A. Gambaro, F. Gottardi, F. Manoli, and A. Venzo, *J. Organomet. Chem.*, 1989, **363**, 91.
362 V.A. Sergeev, L.P. Yur'eva, A.S. Peregudov, A.Yu. Vasil'kov, I.A. Uralets, N.N. Zaitseva, and D.N. Kravtsov, *Izv. Akad. Nauk SSSR, Ser. Khim.*, 1987, 2576 (*Chem. Abstr.*, 1989, **110**, 24 043).
363 J.A. Heppert, T.J. Boyle, and F. Takusagawa, *Organometallics*, 1989, **8**, 461.
364 K.H. Dötz, A. Tiriliomis, and K. Harms, *J. Chem. Soc., Chem. Commun.*, 1989, 788; M.E. Wright, *J. Organomet. Chem.*, 1989, **376**, 353; J. Suffert and R. Ziessel, *J. Organomet. Chem.*, 1989, **359**, C45; F. Rose-Munch, E. Rose, A. Semra, L. Mignon, J. Garcia-Oricain, and C. Knobler, *J. Organomet. Chem.*, 1989, **363**, 297; F. Rose-Munch, E. Rose, A. Semra, and M. Filoche, *J. Organomet. Chem.*, 1989, **363**, 123; M.S.F. Lie Ken Jie and W.L.K. Lam, *Chem. Phys. Lipids*, 1988, **47**, 209 (*Chem. Abstr.*, 1989, **111**, 194 974); K.H. Dötz, H.-G. Erben, and K. Harms, *J. Chem. Soc., Chem. Commun.*, 1989, 692; B. Ohlsson, C. Ullenius, S. Jagner, C. Grivet, E. Wenger, and E.P. Kündig, *J. Organomet. Chem.*, 1989, **365**, 243; E.P. Kündig, V. Desobry, D.P. Simmons, and E. Wenger, *J. Am. Chem. Soc.*, 1989, **111**, 1804; M. Uemura, T. Minami, K. Hirotsu, and Y. Hayashi, *J. Org. Chem.*, 1989, **54**, 469.
365 S. Allaoud, A. Karim, A. Mortreux, F. Petit, and B. Frange, *J. Organomet. Chem.*, 1989, **379**, 89.
366 A.D. Hunter and J.L. McLernon, *Organometallics*, 1989, **8**, 2679.
367 A. Ceccon, A. Gambaro, S. Santi, G. Valle, and A. Venzo, *J. Chem. Soc., Chem. Commun.*, 1989, 51.
368 M.E. Wright, *Organometallics*, 1989, **8**, 407.
369 R. Köster, G. Seidel, and B. Wrackmeyer, *Chem. Ber.*, 1989, **122**, 1815.
370 G. Jaouen, J.A.S. Howell, M.-C. Tirvengadum, P. McArdle, and D. Cunningham, *J. Organomet. Chem.*, 1989, **370**, 51.
371 T.G. Traylor, M.J. Goldberg, A.R. Miksztal, and C.E. Strouse, *Organometallics*, 1989, **8**, 1101.
372 W. Beck, B. Niemer, and B. Wagner, *Angew. Chem., Int. Ed. Engl.*, 1989, **28**, 1705.
373 H.-J. Müller, U. Nagel, M. Steimann, K. Polborn, and W. Beck, *Chem. Ber.*, 1989, **122**, 1387.
374 J.L. Kerschner, E.M. Torres, P.E. Fanwick, I.P. Rothwell, and J.C. Huffman, *Organometallics*, 1989, **8**, 1424.

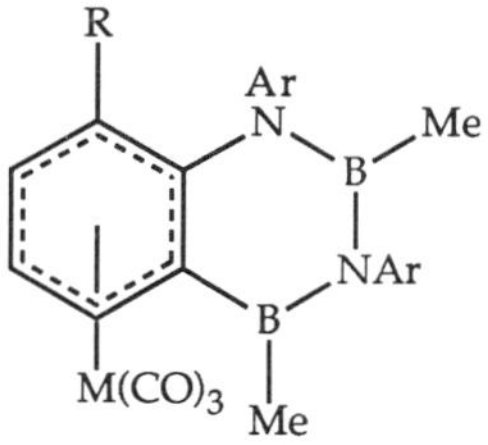
(17)

In contrast to suggestions in the literature, the relation between the ^{13}C parameters for the axial and equatorial CO groups and the σ-donor/π-acceptor properties of the aminophosphine ligand is not straightforward. The ^{15}N resonances of $[(OC)_5Cr\{Bu^t_nP(NH_2)_{3-n}\}]$, n = 1 or 2, and $[(OC)_5CrP(NMe_2)_3]$ are shifted to higher frequencies with respect to those of the free ligands.[375] The ^{95}Mo chemical shift of [Mo(2-phenylazopyridine)$_3$] at δ 1502 is over 1000 p.p.m. more positive than any reported previously for a Mo(O) complex. The ^{13}C and ^{95}Mo n.m.r. spectra of [Mo(2-phenylazopyridine)$_{3-n}$(CO)$_{2n}$] were reported.[376] The ^{53}Cr n.m.r. spectra of *fac*-$[Cr(CO)_3(PF_3)_3]$ and $[Cr(PF_3)_6]$ have been reported and used to show that $[CrO_2ClF]$ is a mixture of $[CrO_2F_2]$ and $[CrO_2Cl_2]$.[377] A good correlation has been found between ^{95}Mo chemical shifts and the Hammett σ constant of the pyridine substituent or the pK_a value of the substituted pyridine for $[Mo(CO)_5(NC_5H_4X)]$.[378] ^{95}Mo n.m.r. spectroscopy has been used to study electronic effects in $Mo(CO)_4$ complexes of aromatic α-diimine chelate ligands.[379] N.m.r. data have also been reported for $[Cr(CO)_5(CNR)]$, (^{13}C),[380] $[(R_3P)(OC)_4OsM(CO)_5]$, (M = Cr, W; ^{13}C),[381] $[\{(OC)_5MPPh_2CH_2CH_2\}_2NP(O)O(CH_2)_3NH]$, (M = Cr, Mo, W; ^{13}C),[382] $[(OC)_5M\{P_3(BNPr^i_2)\}]$, (Cr, W; ^{11}B, ^{13}C),[383] $[(OC)_5MPPr^iNBu^tBNC_5H_6Me_4]$, (M = Cr, W; ^{11}B, ^{13}C, ^{77}Se),[384] $[(OC)_5Cr\{F_3CPFC(NMe_2)PMe_3\}]$, ($^{13}C$),[385] $[(OC)_5MPR^1XNR^2_2]$, (M = Cr, Mo, W; ^{13}C),[386] $[(OC)_5CrECRPhCH_2CMe{=}CMeCH_2]$, (E = S, Se; ^{13}C),[387] $[(OC)_5MSR^1R^2]$, (M = Cr, W; ^{13}C),[388] $[(OC)_4M(\mu\text{-}PPh_2)(\mu\text{-}CO)Pt(\eta^3\text{-cyclooct-4-en-1-yl})]$, (M = Cr, Mo, W;

375 B. Wrackmeyer, K. Schamel, and M. Herberhold, *Z. Naturforsch., B*, 1989, **44**, 55 (*Chem. Abstr.*, 1989, **110**, 224 391).
376 M.N. Ackermann, C.R. Barton, C.J. Deodene, E.M. Specht, S.C. Keill, W.E. Schreiber, and H. Kim, *Inorg. Chem.*, 1989, **28**, 397.
377 M.F.A. Dove, E.M.L. Jones, and R.J. Clark, *Magn. Reson. Chem.*, 1989, **27**, 973.
378 E.C. Alyea, K.J. Fisher, and R.A. Gossage, *Spectrochim. Acta, Part A*, 1989, **45**, 203.
379 E.C. Alyea, J. Malito, S.D. Ernst, W. Kaim, and S.J. Kohlmann, *Polyhedron*, 1989, **8**, 921.
380 W.P. Fehlhammer and G. Beck, *J. Organomet. Chem.*, 1989, **379**, 97.
381 H.B. Davis, F.W.B. Einstein, P.G. Glavina, T. Jones, R.K. Pomeroy, and P. Rushman, *Organometallics*, 1989, **8**, 1030.
382 J. Ellermann, N. Will, and F. Knoch, *J. Organomet. Chem.*, 1989, **366**, 197.
383 G.L. Wood, E.N. Duesler, and R.T. Paine, *Phosphorus Sulphur Silicon*, 1989, **41**, 267.
384 P. Kölle, H. Nöth, and R.T. Paine, *Chem. Ber.*, 1989, **122**, 423.
385 J. Grobe, D.L. Van, and J. Nientiedt, *New J. Chem.*, 1989, **13**, 363.
386 K. Diemert, W. Kuchen, and D. Lorenzen, *J. Organomet. Chem.*, 1989, **378**, 17.
387 H. Fischer, U. Gerbing, and J. Riede, *J. Organomet. Chem.*, 1989, **364**, 155.
388 R. Aumann, J. Schröder, C. Krüger, and R. Goddard, *J. Organomet. Chem.*, 1989, **378**, 185.

^{13}C),[389] [$CH_2pz_2M(CO)_4$], (M = Cr, Mo, W; ^{13}C),[390] [$(OC)_4\overline{CrPBu^tNSNSNH}$], (^{13}C),[391] [$(OC)_4Cr\{(Pr^i_2N)PO\}_5Cr(CO)_3$], (^{13}C),[392] [$OC)_4Cr(Pr^i_2P)_2Te$], (^{13}C, ^{125}Te),[393] [$\{(C_2F_5)_2PCH_2CH_2P(C_2F_5)_2\}M(CO)_4$], (M = Cr, Mo; ^{13}C),[394] [$(OC)_4M(PPh_2CH_2CH_2)_2\overline{N\text{-}P(O)NHCH_2CH_2CH_2O}$], (M = Cr, Mo, W; ^{13}C),[395] (18), (M = Cr, W; ^{13}C),[396] (19), (^{11}B,

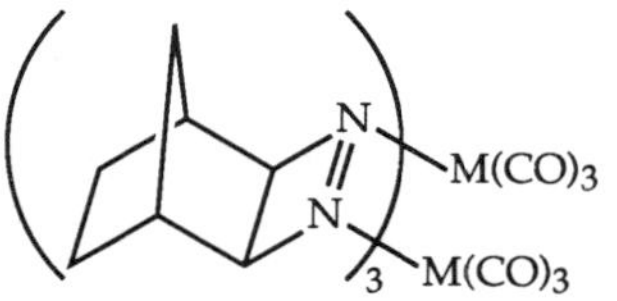

(18)

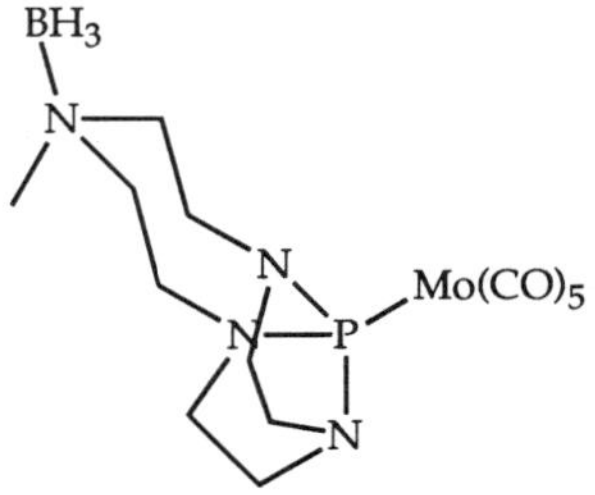

(19)

^{13}C),[397] [$M(CO)_{6-n}(CNR)_n$], (M = Mo, W; ^{13}C, ^{14}N, ^{95}Mo, ^{183}W),[398] [$(OC)_4Mo\{(F_2PNH)_2CO\}$], (^{13}C),[399] [$(OC)_4\overline{MN{=}PPh_2(CH_2)_nE}Ph_2$], (M = Mo, W; E = P, As; ^{29}Si),[400] (20), (E = P, As; ^{13}C),[401] *cis*-[$Mo(CO)_4\{H_2P(CH_2)_3PHCH_2CH{=}CH_2\}$], (^{13}C),[402] *cis*-[$Mo(CO)_4(Ph_2PCH_2NHR)_2$], (^{13}C),[403] *fac*-[$Mo(bipy)(CO)_3\{P(OMe)_3\}$], (^{13}C),[404] *fac*-[$(OC)_3\overline{Mo(Ph_2PCH_2CH_2)_2N}H$], (^{95}Mo),[405] [$M(CO)_3(\mu_3\text{-}OH)]_4^{4-}$, (M = Mo, W; ^{13}C),[406] [$Mo_2(CO)_6\{(Ph_2P)_2O\}_2$], (^{13}C and c.p.m.a.s. ^{31}P),[407] [$M(CO)_3I_2$(1,4,8,11-tetrathiacyclotetradecane)], (M = Mo, W; ^{13}C),[408] [$Mo(CO_3)(CO)\{MeP(CH_2CH_2CH_2PMe_2)_2\}(PMe_3)$], (^{13}C),[409] [$M(S_2COR)_2(CO)(PMe_3)_2$], (M = Mo, W; ^{13}C),[410] [$(\eta^5\text{-}C_5H_5)_2Mo_2(CO)_2(\mu\text{-}PPh_2)_2$], (^{13}C),[411] [$(OC)_5\overline{WPRCH_2C}HPh$], (^{13}C),[412]

389 J. Powell, M.R. Gregg, and J.F. Sawyer, *Inorg. Chem.*, 1989, **28**, 4451.
390 G.G. Lobbia and F. Bonati, *J. Organomet. Chem.*, 1989, **366**, 121.
391 A. Gieren, C. Ruiz-Pérez, T. Hübner, M. Herberhold, K. Schamel, and K. Guldner, *J. Organomet. Chem.*, 1989, **366**, 105.
392 C.M. Banse, M.A. Pray, and M.M. Turnbull, *Polyhedron*, 1989, **8**, 2719.
393 W.-W. Du Mont and R. Hensel, *Z. Anorg. Allg. Chem.*, 1989, **579**, 134.
394 M.F. Ernst and D.M. Roddick, *Inorg. Chem.*, 1989, **28**, 1624.
395 J. Ellermann, M. Moll, and N. Will, *Z. Anorg. Allg. Chem.*, 1989, **574**, 41.
396 M.N. Ackermann, D.E. Adams, J. Pranata, and C.F. Yamauchi, *J. Organomet. Chem.*, 1989, **369**, 55.
397 D.V. Khasnis, M. Lattman, and U. Siriwardane, *J. Chem. Soc., Chem. Commun.*, 1989, 1538.
398 M. Minelli and W.J. Maley, *Inorg. Chem.*, 1989, **28**, 2954.
399 W. Krüger, R. Schmutzler, H.M. Schiebel, and V. Wray, *Polyhedron*, 1989, **8**, 293.
400 K.V. Katti and R.G. Cavell, *Organometallics*, 1989, **8**, 2147.
401 S. Affandi, J.H. Nelson, and J. Fisher, *Inorg. Chem.*, 1989, **28**, 4536.
402 B.N. Diel, P.F. Brandt, R.C. Haltiwanger, M.L.J. Hackney, and A.D. Norman, *Inorg. Chem.*, 1989, **28**, 2811.
403 D.L. Davies, F.I. Healey, J. Howarth, D.R. Russell, and L.J.S. Sherry, *J. Organomet. Chem.*, 1989, **376**, C31.
404 H. Nakazawa, M. Ohta, K. Miyoshi, and H. Yoneda, *Organometallics*, 1989, **8**, 638.
405 J. Ellermann, M. Moll, and N. Will, *J. Organomet. Chem.*, 1989, **378**, 73.
406 J.T. Lin, S.K. Yeh, G.H. Lee, and Y. Wang, *J. Organomet. Chem.*, 1989, **361**, 89.
407 P. Planinić and H. Meider, *Polyhedron*, 1989, **8**, 627.
408 D. Sevdić, M. Ćurić, and Lj. Tušek-Božić, *Polyhedron*, 1989, **8**, 505.
409 L. Dahlenburg and B. Pietsch, *Chem. Ber.*, 1989, **122**, 2085.
410 E. Carmona, L. Contreras, L.J. Sánchez, E. Gútierrez-Puebla, and A. Monge, *J. Chem. Soc., Dalton Trans.*, 1989, 2003.

$[(OC)_5WP(C_6H_2Me_3\text{-}2,4,6)\text{=}CMe_3]$, ($^{13}C$),[413] $[(OC)_5W\overline{P(NHEt)CH_2C}H_2]$, ($^{13}C$),[414] $[(OC)_5W\text{-}$ $\overline{P(NEt_2)CH_2C}HBu^t]$, ($^{13}C$),[415] (21), ($^{29}Si$),[416] $[W(CO)_5(OPh)]^-$, (^{13}C),[417] $[(OC)_5ReWCo_2(CO)_8\text{-}L^1L^2(\mu\text{-}CR)]$, ($^{13}C$),[418] *cis,cis*-$[W(CO)_2(CNR)_2(bipy)]^{2+}$, ($^{13}C$),[419] $[\{HBpz_3(OC)_2W(\mu_2\text{-}CSMe)\}_2Pt]$, ($^{13}C$, ^{195}Pt),[420] $[(Cy_2PCH_2CH_2PPh_2)(OC)_2(ON)WBr]$, ($^{13}C$),[421] $[W(E)Cl_2(CO)(PR_3)_2]$, (^{13}C),[422] and $[W_4(OR)_{12}(CO)_3]$, (^{13}C).[423]

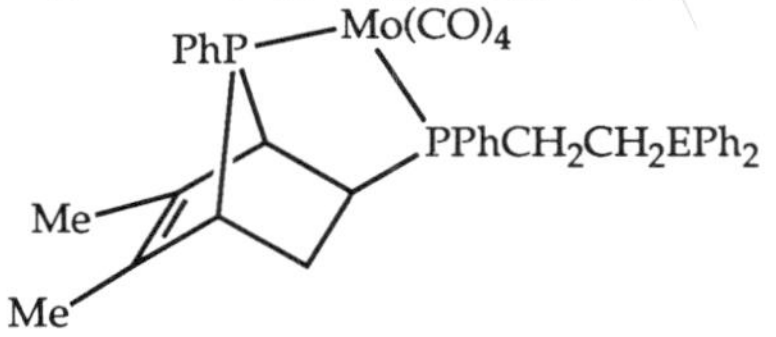

(20)

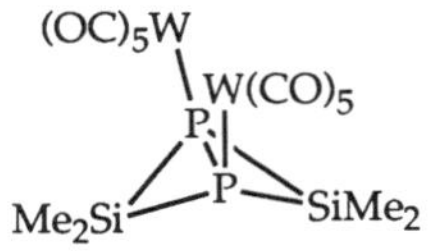

(21)

The ^{31}P n.m.r. spectra of $[MCl_3(NO)(PMe_3)_3]$, M = Mo, W, show well resolved coupling between ^{31}P and ^{14}N. The ^{13}C n.m.r. spectra were also recorded.[424] The diamagnetic anisotropy of the W≡W bond in $[W_2Cl_4(\mu\text{-}Pr^i{}_2PCH_2CH_2CH_2PPr^i{}_2)_2]$ has been estimated as $-3000 \pm 300 \times 10^{-36}$ m^3/molecule.[425] N.m.r. data have also been reported for $[Mo(NMe_2)_3(PBu^t{}_2)]$, (^{13}C),[426] $[Mo_3O_6(OSiMe_2Bu^t)_4(NC_5H_5)_3]$, ($^{13}C$),[427] $[M_2O_4(edta)]^{2-}$, (M = Mo, W; ^{13}C, ^{95}Mo, ^{183}W),[428] $[WOF_4(EtMeNNMeEt)]$, (^{13}C),[429] *trans*-$[Mo(NCH_2)Cl(dppe)_2]$, (^{13}C),[430] $[Mo_2(\mu\text{-}S)_2\{\mu\text{-}S_2P(OEt)_2\}(\equiv NC_6H_4Me\text{-}4)_2(dppe)_2]^+$, ($^{13}C$),[431] $[MoO_2\{SC(NH_2)\text{=}NN\text{=}CH\text{-}2\text{-}furyl\}_2]$, ($^{13}C$),[432] $[Mo(2\text{-}O\text{-}3\text{-}OC(O)CH_2CH_2N\text{=}CH\text{-}naphthyl)(O)_2(OSMe_2)]$, ($^{13}C$),[433] $[Mo(2\text{-}SC_6H_4SCH_2CH_2SC_6H_4S\text{-}2)(NO)_2]$, ($^{13}C$, ^{95}Mo),[434]

411 T. Adatia, M. McPartlin, M.J. Mays, M.J. Morris, and P.R. Raithby, *J. Chem. Soc., Dalton Trans.*, 1989, 1555.
412 A. Marinetti and F. Mathey, *Tethrahedron*, 1989, **45**, 3061.
413 F. Mercier, C. Hugel-Le-Goff, and F. Mathey, *Tetrahedron Lett.*, 1989, **30**, 2397.
414 B. Deschamps, L. Richard, and F. Mathey, *Polyhedron*, 1989, **8**, 2671.
415 F. Mercier, B. Deschamps, and F. Mathey, *J. Am. Chem. Soc.*, 1989, **111**, 9098.
416 M. Driess, A.D. Fanta, D.R. Powell, and R. West, *Angew. Chem., Int. Ed. Engl.*, 1989, **28**, 1038.
417 D.J. Darensbourg, K.M. Sanchez, J.H. Reibenspies, and A.L. Rheingold, *J. Am. Chem. Soc.*, 1989, **111**, 7094.
418 M.R. Bradford, J.C. Jeffery, K.M. Johnson, and S.E. Phillips, *J. Organomet. Chem.*, 1989, **362**, 205.
419 J.A. Connor and J.M.A. Walshe, *Polyhedron*, 1989, **8**, 827.
420 R.A. Doyle, R.J. Angelici, and F.G.A. Stone, *J. Organomet. Chem.*, 1989, **378**, 81.
421 P.V. Bonnesen, C.L. Puckett, R.V. Honeychuck, and W.H. Hersh, *J. Am. Chem. Soc.*, 1989, **111**, 6070.
422 F.-M. Su, J.C. Bryan, S. Jang, and J.M. Mayer, *Polyhedron*, 1989, **8**, 1261.
423 M.H. Chisholm, K. Folting, M.J. Hampden-Smith, and C.E. Hammond, *J. Am. Chem. Soc.*, 1989, **111**, 7283.
424 E. Carmona, E. Gutierrez-Puebla, A. Monge, P.J. Perez, and L.J. Sanchez, *Inorg. Chem.*, 1989, **28**, 2120.
425 M.D. Fryzuk, C.G. Kreiter, and W.S. Sheldrick, *Chem. Ber.*, 1989, **122**, 851.
426 M.H. Chisholm, C.E. Hammond, and J.C. Huffman, *Polyhedron*, 1989, **8**, 1419.
427 G.-S. Kim, D. Huffman, and C.W. DeKock, *Inorg. Chem.*, 1989, **28**, 1279.
428 S. Ikari, Y. Sasaki, A. Nagasawa, C. Kabuto, and T. Ito, *Inorg. Chem.*, 1989, **28**, 1248.
429 S.G. Sakharov, S.A. Zarelua, Yu.V. Kokunov, and Yu.A. Buslaev, *Z. Anorg. Allg. Chem.*, 1989, **577**, 223.
430 D.L. Hughes, M.Y. Mohammed, and C.J. Pickett, *J. Chem. Soc., Chem. Commun.*, 1989, 1399.
431 A.C. Lizano, J.F. Richardson, and M.E. Noble, *Inorg. Chem.*, 1989, **28**, 1451.
432 N. Kanoongo, R. Singh, and J.P. Tandon, *Bull. Chem. Soc. Jpn.*, 1989, **62**, 1385.
433 N. Kanoongo, R. Singh, and J.P. Tandon, *Transition Met. Chem. (Weinheim, Ger.)*, 1989, **14**, 221.

$[W(NBu^t)(NH_2Bu^t)(OC_6H_4Ph\text{-}2)_4]$, ($^{13}C$),[435] $[W_2(OCH_2Bu^t)_6(OCMe_2)(NC_5H_5)]$, ($^{13}C$),[436] and $[W(PF_2Bu^t)_6]$, (^{13}C).[437]

^{13}C and ^{95}Mo n.m.r. spectroscopy has been used to characterise molybdate complexes of aldoses.[438] A molecular orbital localized orbital-local origin method has been used to reproduce ^{95}Mo chemical shift trends in a series of molybdate and thiomolybdate anions.[439] *Ortho*-metallation and hydrogen transport in $[\{(Ph_3P)_2IrH_2\}_3PW_{12}O_{40}]$ have been probed by ^{31}P n.m.r. long-range deuterium isotope effects.[440] The value of ^{17}O n.m.r. spectroscopy to investigate $[MoO_2Cl_2(OH_2)_2]$, $[UO_2(OH_2)_6]^{2+}$, and related compounds has been examined.[441] The nature of peroxomolybdates and peroxotungstates has been investigated using ^{95}Mo and ^{183}W n.m.r. spectroscopy.[442] ^{17}O and ^{183}W n.m.r. studies on an orange complex prepared from acid catalysed hydrolysis of $[K_2WCl_6]$ have shown the presence of $[W_3(\mu_3\text{-}O)(\mu_2\text{-}O)_3(OH_2)_8Cl]^{3+}$ and $[W_3(\mu_3\text{-}O)(\mu_2\text{-}O)_3(OH_2)_9]^{4+}$.[443] N.m.r. data have also been reported for *cis*-$[Mo_2O_5(ND)_2]^{2-}$, (^{13}C),[444] $[MoO_2(OR)_2]$, (^{13}C),[445] *cis*-$[MO_2\{1,2\text{-}(O)_2\text{-}3\text{-}HOC_6H_3\}]^{2-}$, (M = Mo, W; ^{13}C),[446] $[Mo_2O_3\{S_2P(OR)_2\}_4]$, ($^{13}C$),[447] $[Mo_3O_8(OMe)(C_4O_4)_2]^{3-}$, ($^{17}O$),[448] $[M_4(OCH_2R)_{12}]$, (M = Mo, W; ^{13}C),[449] $[Mo_4O_{10}(OMe)_6]^{2-}$, (^{17}O),[450] $[Si_2W_{18}Zr_3O_{71}H_3]^{11-}$, ($^{29}Si$, ^{183}W),[451] $[AlO_4W_{11}O_{35}]^{9-}$, $[Ga(OH)_6Mo_6O_{18}]^{3-}$, ($^{17}O$, ^{27}Al, ^{71}Ga),[452] $[GaMo_{12}O_{40}]^{5-}$, (^{17}O, ^{71}Ga),[453] $[OM(Te_4)_2]^{2-}$, (M = Mo, W; ^{125}Te),[454] $[W_2O_2(\mu\text{-}O)(\mu\text{-}S)(\mu\text{-}edta)]^{2-}$, ($^{13}C$),[455] [(3,3'-$Me_2$-1,1'-bi-2,2'-naphthol)$W_2(OBu^t)_4$], ($^{13}C$),[456] $[WO(O_2)_2(C_2O_4)WO(O_2)_2]^{2-}$, ($^{13}C$),[457] $[H_2W_9O_{37}W_3(OH_2)_3]^{6-}$, ($^{17}O$),[458] $[W_2Se_6]^{2-}$, (^{77}Se),[459] and $[M(WSe_4)_2]^{2-}$, (^{77}Se).[460]

434 D. Sellmann, *J. Organomet. Chem.*, 1989, **372**, 99.
435 J.L. Kerschner, J.S. Yu, P.E. Fanwick, I.P. Rothwell, and J.C. Huffman, *Organometallics*, 1989, **8**, 1414.
436 M.H. Chisholm and J.A. Klang, *J. Am. Chem. Soc.*, 1989, **111**, 2324.
437 W. Krampe, T. Kruck, and K.-F. Tebbe, *Chem. Ber.*, 1989, **122**, 1477.
438 J.F. Verchere and S. Chapelle, *Polyhedron*, 1989, **8**, 333.
439 J.E. Combariza, J.H. Enemark, M. Barfield, and J.C. Facelli, *J. Am. Chem. Soc.*, 1989, **111**, 7619.
440 A.R. Siedle and R.A. Newmark, *J. Am. Chem. Soc.*, 1989, **111**, 2058.
441 J.M. Coddington, R.F. Howe, and M.J. Taylor, *Inorg. Chim. Acta*, 1989, **166**, 13.
442 N.J. Campbell, A.C. Dengel, C.J. Edwards, and W.P. Griffith, *J. Chem. Soc., Dalton Trans.*, 1989, 1203.
443 A. Patel, M.R. McMahon, and D.T. Richens, *Inorg. Chem.*, 1989, **163**, 73.
444 A.M. El-Hendawy, W.P. Griffith, C.A. O'Mahoney, and D.J. Williams, *Polyhedron*, 1989, **8**, 519.
445 G.S. Kim, D. Huffman, and C.W. DeKock, *Inorg. Chem.*, 1989, **28**, 1279.
446 A.M. El-Hendawy and W.P. Griffith, *Inorg. Chim. Acta*, 1989, **160**, 67.
447 R. Ratnani, G. Srivastave, and R.C. Mehrotra, *Inorg. Chim. Acta*, 1989, **161**, 253.
448 Q. Chen, L. Ma, S. Liu, and J. Zubieta, *J. Am. Chem. Soc.*, 1989, **111**, 5944.
449 M.H. Chisholm, K. Folting, C.E. Hammond, M.J. Hampden-Smith, and K.G. Moodley, *J. Am. Chem. Soc.*, 1989, **111**, 5300.
450 H. Kang, S. Liu, S.N. Shaikh, T. Nicholson, and J. Zubieta, *Inorg. Chem.*, 1989, **28**, 920.
451 R.G. Finke, B. Rapko, and T.R. Weakley, *Inorg. Chem.*, 1989, **28**, 1573.
452 M.A. Fedotov and L.P. Kazanskii, *Izv. Akad. Nauk SSSR, Ser. Khim.*, 1988, 2000 (*Chem. Abstr.*, 1989, **110**, 106 952).
453 L.P. Tsiganok, A.B. Vishnikin, and R.I. Maksimovskaya, *Polyhedron*, 1989, **8**, 2739.
454 W.A. Flomer and J.W. Kolis, *Inorg. Chem.*, 1989, **28**, 2513.
455 S. Ikari, Y. Sasaki, and T. Ito, *Inorg. Chem.*, 1989, **28**, 447.
456 J.A. Heppert, S.D. Dietz, T.J. Boyle, and F. Takusagawa, *J. Am. Chem. Soc.*, 1989, **111**, 1503.
457 M. Hashimoto, T. Iwamoto, H. Ichida, Y. Sasaki, and T. Kudo, *Inorg. Chim. Acta*, 1989, **166**, 17.
458 K. Piepgrass and M.T. Pope, *J. Am. Chem. Soc.*, 1989, **111**, 753.
459 Y.J. Lu, M.A. Ansari, and J.A. Ibers, *Inorg. Chem.*, 1989, **28**, 4049.
460 M.A. Ansari, C.N. Chau, C.H. Mahler, and J.A. Ibers, *Inorg. Chem.*, 1989, **28**, 650.

Complexes of Mn, Tc, and Re.——^{55}Mn n.m.r. spectroscopy has been used to probe the structure of various organomanganese(I) complexes. The chemical shift of the metal is indicative of the electron distribution in the complexes and can be correlated with the chemical reactivity.[461] ^{99}Tc n.m.r. spectroscopy has been studied for a variety of technetium compounds and correlations discussed.[462]

Variable temperature ^{1}H T_1 studies of [$ReH_5(PPh_3)_3$] and [$ReH_5(PMe_2Ph)_3$] have been used to determine H—H distances.[463] N.m.r. data have also been reported for [$H_2M(CO)_4$]$^-$, (M = Mn, Re; ^{13}C),[464] [(η^5-C_5H_4Me)$(OC)_2MnH(SnPh_3)$], (^{119}Sn),[465] [(η^5-C_5H_4Me)$Mn(CO)_2H(SiCl_3)$], (^{29}Si),[466] [$ReH_4(PMe_2Ph)_4$]$^+$, (^{13}C),[467] [(η^5-C_5Me_5)(η^5-C_5H_5)ReH], (^{13}C),[468] and [(η^5-C_5Me_5)$_2Re_2H_4$(μ-H)$_2$], (^{13}C).[469]

The rates of demetallation of [$(OC)_4MnC(C_6H_4X$-4)=CHC(O)Me] correlate with the ^{55}Mn chemical shifts. The ^{13}C chemical shifts were also given.[470] The ^{17}O n.m.r. spectrum of [$MeReO_3L$] has been used to investigate complexation by L.[471] The ^{13}C n.m.r. spectrum of [$ReOMe_2Cl(PMe_3)_2$] has been analysed as ABX. The ^{13}C isotope shift is 0.025 p.p.m. and $^{2}J(^{31}P\text{-}^{31}P)$ is 11.4 Hz.[472] N.m.r. data have also been reported for [$(OC)_4$-$MnCHR^1PR^2_2$], (^{13}C),[473] [1,3-C_6H_4\{$Mn(CO)_5$\}$_2$], (^{13}C),[474] [(η^5-C_5H_4Me)$(Ph_3P)(ON)MnC(O)$-$C(O)C_6H_4Me$-4], (^{13}C),[475] [(η^5-C_5H_4Me)$(OC)(R^1{}_3P)Mn{=}C{=}CR^2{}_2$], (^{13}C),[476] [($\mu$-Pb)\{($\eta^5$-$C_5H_5$)$Mn(CO)_2$\}$_2$], (^{13}C),[477] [$MeReO_3$], (^{17}O),[478] [($\eta^5$-$C_5H_5$)Re(NO)($PPh_3$)Me(X)], (^{13}C),[479] [Re(OH)(η^1-C_3H_5)(η^2-MeC≡CMe)$_2$]$^+$, (^{13}C),[480] [(η^5-C_5H_5)Re(NO)(PPh_3)($CH_2CR^1{=}CHR^2$)], (^{13}C),[481] [Re($CH_2CH{=}CHCH{=}CH_2$)$(CO)_3$($Et_2PCH_2CH_2PEt_2$)], (^{13}C),[482] [$Re_2O_5(CH_2Bu^t)_4$],

461 P. DeShong, G.A. Slough, D.R. Sidler, P.J. Rybczynski, W. Von Philipsborn, R.W. Kunz, B.E. Bursten, and T.W. Clayton, jun., *Organometallics*, 1989, **8**, 1381.

462 L.A. O'Connell, R.M. Pearlstein, A. Davison, J.R. Thornback, J.F. Kronauge, and A.G. Jones, *Inorg. Chim. Acta*, 1989, **161**, 39; M. Findeisen, B. Lorenz, and U. Abram, *Z. Chem.*, 1989, **29**, 29 (*Chem. Abstr.*, 1989, **111**, 32 506).

463 F.A. Cotton and R.L. Luck, *J. Am. Chem. Soc.*, 1989, **111**, 5757; *Inorg. Chem.*, 1989, **28**, 6.

464 G.F.P. Warnock, L.C. Moodie, and J.E. Ellis, *J. Am. Chem. Soc.*, 1989, **111**, 2131.

465 U. Schubert, E. Kunz, B. Harkers, J. Willnecker, and J. Meyer, *J. Am. Chem. Soc.*, 1989, **111**, 2572.

466 D.L. Lichtenberger and A. Rai-Chaudhuri, *J. Am. Chem. Soc.*, 1989, **111**, 3583.

467 D.M. Lunder, M.A. Green, W.E. Streib, and K.G. Caulton, *Inorg. Chem.*, 1989, **28**, 4527.

468 R.A. Paciello, P. Kiprof, E. Herdtweck, and W.A. Herrmann, *Inorg. Chem.*, 1989, **28**, 2890.

469 W.A. Herrmann, H.G. Theiler, E. Herdtweck, and P. Kiprof, *J. Organomet. Chem.*, 1989, **367**, 291.

470 P. DeShong, D.R. Sidler, P.J. Rybczynski, A.A. Ogilvie, and W. von Philipsborn, *J. Org. Chem.*, 1989, **54**, 5432.

471 W.A. Herrmann, G. Weichselbaumer, and E. Herdtweck, *J. Organomet. Chem.*, 1989, **372**, 371.

472 D.M. Hoffman and D.A. Wierda, *Polyhedron*, 1989, **8**, 959.

473 E. Lindner, E. Ossig, and M. Darmuth, *J. Organomet. Chem.*, 1989, **379**, 107.

474 A.D. Hunter and A.B. Szigety, *Organometallics*, 1989, 8, 2670.

475 S.L. Bassner, J.B. Sheridan, C. Kelley, and G.L. Geoffroy, *Organometallics*, 1989, **8**, 2121.

476 U. Schubert and J. Grönen, *Chem. Ber.*, 1989, **122**, 1237.

477 W.A. Herrmann, H.-J. Kneuper, and E. Herdtweck, *Chem. Ber.*, 1989, **122**, 445.

478 W.A. Herrmann, J.G. Kuchler, G. Weichselbaumer, E. Herdtweck, and P. Kiprof, *J. Organomet. Chem.*, 1989, **372**, 351.

479 C.H. Winter, W.R. Veal, C.M. Garner, A.M. Arif, and J.A. Gladysz, *J. Am. Chem. Soc.*, 1989, **111**, 4766.

480 E. Spaltenstein, T.K.G. Erikson, S.C. Critchlow, and J.M. Mayer, *J. Am. Chem. Soc.*, 1989, **111**, 617.

481 G.S. Bodner, K. Emerson, R.D. Larsen, and J.A. Gladysz, *Organometallics*, 1989, **8**, 2399.

482 J.R. Bleeke and P.L. Earl, *Organometallics*, 1989, **8**, 2735.

(^{13}C),[483] [(η^5-C_5Me_5)Re(O)(CH_2Bu^t)$_2$], (^{13}C),[484] [(η^5-C_5H_5)(OC)$_2$Re=C(C_6H_4Me-4)NMeCH-Ph], (^{13}C),[485] [2,4,6-$Me_3C_6H_2ReO_3$], (^{13}C, ^{17}O),[486] [(RMe_2P)$_2$(2-MeC_6H_4)$_2$Re(2,3-η^2-1-Me-C_6H_3)]$^+$, (^{13}C),[487] [(η^5-C_5Me_5)Re(NO)(PPh_3)($COCH_2CO$)Re(CO)$_4$(PMe_3)Li(O_3SCF_3)], (^{13}C),[488] [(η^5-C_5H_5)Re(NO)(PPh_3){OC(O)C(CF_3)(OMe)Ph}], (^{13}C),[489] [(η^5-C_5H_5)Re(CO)$_2$-(=CHR)], (^{13}C),[490] *trans*-[ReCl(=C=CHR)(dppe)$_2$], (^{13}C),[491] and [ReCl(CNHBu^t)(CNBu^t)$_2$-(PMe_2Ph)$_2$]$^+$, (^{13}C).[492]

The ^{17}O chemical shift of [(η^5-C_5Me_5)Re(O)(η^2-RC≡CR)] shows the electron deficiency of the terminal oxo group.[493] The ^1H, ^{13}C, and ^{17}O n.m.r. chemical shifts of (η^5-$C_5H_{5-n}Me_n$)M(CO)$_3$], M = Mn, Re, have been discussed to determine trends.[494] The ^{31}P n.m.r. spectrum of [(η^5-C_5Me_5)Fe(P_5){Mn(CO)$_2$(η^5-C_5H_5)}$_4$] has been analysed as AMM'XX'.[495] N.m.r. data have also been reported for [(η^5-C_5H_5)(OC)$_2$Mn(η^2-CH_2=CH-PEt_3)]$^+$, (^{13}C),[496] [(η^5-C_5H_5)Mn(CO)$_2${η^2-HC≡CCH(OR)$_2$}], (^{13}C),[497] [(η^5-C_5H_5)Mn(CO)$_2${η^2-XC≡CCH(OR)$_2$}], (^{13}C),[498] [(η^5-C_5H_5)(OC)$_2$Re(η^2-CH_2=CHMe)], (^{13}C),[499] [ReI(η^2-RC≡CR)$_3$], (^{13}C),[500] [Re(O)(η^2-RC≡CR)$_2$], (^{13}C),[501] [(η^5-C_5Me_5)ReCl$_2${$H_2CC(NC_5H_5)C=CH_2$}], (^{13}C),[502] (22), (^{13}C),[503] [(η^5-CH_2=CHC_5H_3Me)Mn(CO)$_3$], (^{13}C),[504] [(η^5-2-MeO-7-RC_6H_5)Mn(CO)$_2$-(NO)], (^{13}C),[505] [(η^5-arene-R)Mn(CO)$_3$], (^{13}C),[506] [{(η^5-C_5H_4Me)(OC)$_2$Mn}$_2$P($C_6H_2Me_3$-2,4,6)], (^{13}C),[507] [(η^5-C_5Me_5)Mn(CO)($CNCF_3$)L], (^{13}C),[508] [{η^5-$C_5(CH_2Ph)_5$}M(CO)$_3$], (M = Mn, Fe; ^{13}C),[509] (23), (M = Mn, Re; ^{13}C),[510] [{η^5-C_5Cl_4(SR)}Mn(CO)$_3$], (^{13}C),[511] [(η^5-

483 S. Cai, D.M. Hoffman, and D.A. Wierda, *Inorg. Chem.*, 1989, **28**, 3784.
484 J.K. Felixberger, P. Kiprof, E. Herdtweck, W.A. Herrmann, R. Jakobi, and P. Gütlich, *Angew. Chem., Int. Ed. Engl.*, 1989, **28**, 334.
485 B.M. Handwerker, K.E. Garrett, G.L. Geoffroy, and A.L. Rheingold, *J. Am. Chem. Soc.*, 1989, **111**, 368.
486 W.A. Herrmann, M. Ladwig, P. Kiprof, and J. Riede, *J. Organomet. Chem.*, 1989, **371**, C13.
487 J. Arnold, G. Wilkinson, B. Hussain, and M.B. Hursthouse, *Organometallics*, 1989, **8**, 415.
488 J.M. O'Connor, R. Uhrhammer, A.L. Rheingold, and D.L. Staley, *J. Am. Chem. Soc.*, 1989, **111**, 7633.
489 D.M. Dalton and J.A. Gladysz, *J. Organomet. Chem.*, 1989, **370**, C17.
490 C.P. Casey and H. Nagashima, *J. Am. Chem. Soc.*, 1989, **111**, 2353.
491 A.J.L. Pombeiro, S.S.P.R. Almeida, M.F.C.G. Silva, J.C. Jeffrey, and R.L. Richards, *J. Chem. Soc., Dalton Trans.*, 1989, 2381.
492 S. Warner and S.J. Lippard, *Organometallics*, 1989, **8**, 228.
493 W.A. Herrmann, R.A. Fischer, W. Amslinger, and E. Herdtweck, *J. Organomet. Chem.*, 1989, **362**, 333.
494 I.R. Lyatifov, G.M. Jafarov, V.N. Babin, P.V. Petrovskii, and V.D. Zagorevskii, *J. Organomet. Chem.*, 1989, **368**, 223.
495 O.J. Scherer, T. Brück, and G. Wolmershäuser, *Chem. Ber.*, 1989, **122**, 2049.
496 H.G. Alt, H.E. Engelhardt, and R.D. Rogers, *J. Organomet. Chem.*, 1989, **362**, 117.
497 C. Löwe, H.-U. Hund, and H. Berke, *J. Organomet. Chem.*, 1989, **378**, 211.
498 C. Löwe, H.-U. Hund, and H. Berke, *J. Organomet. Chem.*, 1989, **371**, 311.
499 C.P. Casey and E.W. Rutter, jun., *J. Am. Chem. Soc.*, 1989, **111**, 8917.
500 A.B. Manion, T.K.G. Erikson, E. Spaltenstein, and J.M. Mayer, *Organometallics*, 1989, **8**, 1871.
501 E. Spaltenstein, R.R. Conry, S.C. Critchlow, and J.M. Mayer, *J. Am. Chem. Soc.*, 1989, **111**, 8741.
502 R.A. Fischer, R.W. Fischer, W.A. Herrmann, and E. Herdtweck, *Chem. Ber.*, 1989, **122**, 2035.
503 E. Lindner, V. Käss, W. Hiller, and R. Fawzi, *Angew. Chem., Int. Ed. Engl.*, 1989, **28**, 448.
504 H. Nishide, H. Kawakami, Y. Kurimura, and E. Tsuchida, *J. Am. Chem. Soc.*, 1989, **111**, 7175.
505 T.-H. Hyeon and Y.K. Chung, *J. Organomet. Chem.*, 1989, **372**, C12.
506 M.V. Gaudet, A.W. Hanson, P.S. White, and M.J. Zaworotko, *Organometallics*, 1989, **8**, 286.
507 U. Kirchgäßner and U. Schubert, *Chem. Ber.*, 1989, **122**, 1481.
508 D. Lentz and R. Marschall, *Chem. Ber.*, 1989, **122**, 1223.
509 M.D. Rausch, W.-M. Tsai, J.W. Chambers, R.D. Rogers, and H.G. Alt, *Organometallics*, 1989, **8**, 816.
510 T.L. Tisch, T.J. Lynch, and R. Dominguez, *J. Organomet. Chem.*, 1989, **377**, 265.
511 K. Sünkel and D. Steiner, *Chem. Ber.*, 1989, **122**, 609.

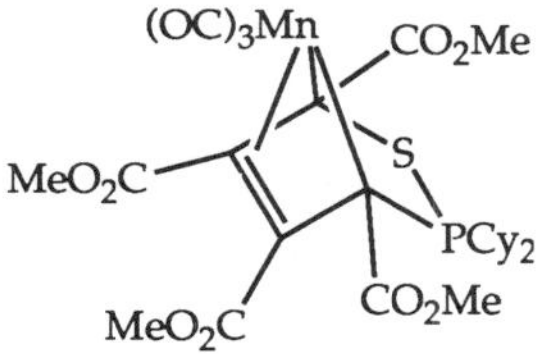

(22)

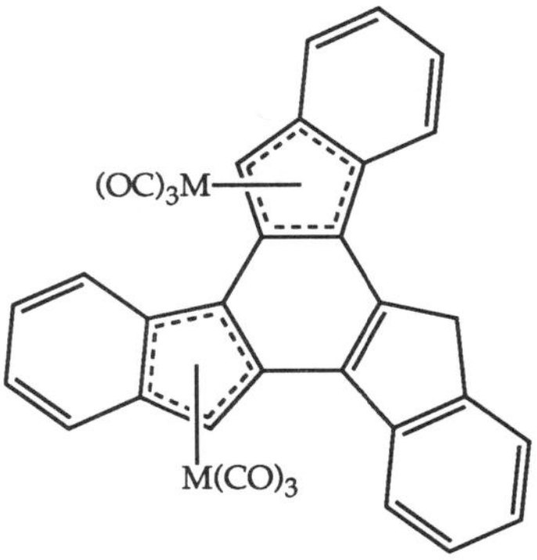

(23)

$C_5Cl_4R)Mn(CO)_3]$, (R = $SnMe_3$, $SiMe_3$, PPh_2, SePh; ^{13}C),[512] $[(\eta^5-C_5Me_5)Re(\eta^5-1,2-Ph_2C_5H_5)Cl]$, ($^{13}C$),[513] $[(\eta^5-C_5H_5)Re(NO)(PPh_3)(OCHRCN)]$, ($^{13}C$),[514] $[\{(\eta^5-C_5H_5)Re(NO)-(PPh_3)\}_2I]^+$, ($^{13}C$),[515] $[(\eta^5-C_5H_5)Re(NO)(PPh_3)(ClCH_2Cl)]^+$, ($^{13}C$),[516] $[(\eta^5-C_5Me_5)ReO(SR)_2]$, ($^{17}O$),[517] $[(\eta^5-C_5Me_5)Re(O)(OSnMe_3)_2]$, ($^{13}C$),[518] $[(\eta^5-C_5Me_5)Re(CO)(PMe_3)(4-N_2C_6H_4-OMe)]$, ($^{13}C$, ^{14}N, ^{15}N),[519] $[(\eta^5-C_5Me_5)Re(CO)_2(SC_4H_4-\eta^4)Fe(CO)_3]$, ($^{13}C$),[520] $[(\eta^5-C_5Me_5)Re-(CO)(PMe_3)X_2]$, ($^{13}C$),[521] $[(\eta^5-C_5Me_5)_2Re_2(OCN)_4(\mu-O)_2]$, ($^{13}C$, ^{17}O),[522] (24), (^{13}C),[523]

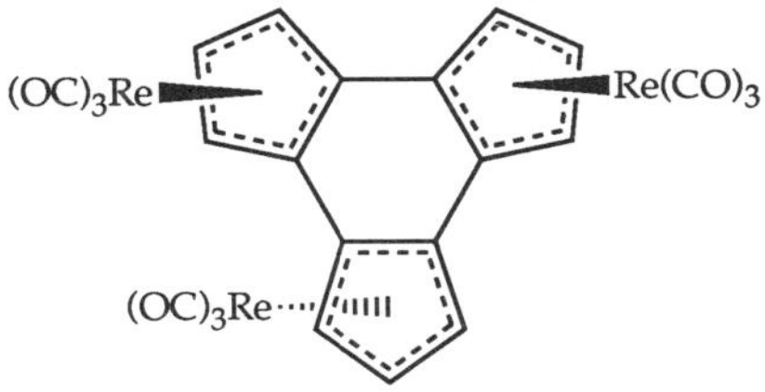

(24)

$[(Bu^tN)_4Re]^-$, (^{13}C),[524] $[1,2-\{Cl(Et_2NCS_2)_2Re{\equiv}N\}_2C_6H_4]$, ($^{13}C$),[525] $[ReN{=}PPh_2CH_2PPh_2-(O)_2(OSiMe_3)_2]$, ($^{13}C$, ^{29}Si),[526] $[Re(OC_6H_3Pr^i{}_2-2,6)_4]$, ($^{13}C$),[527] $[(OC)_4Mn(\eta^2-S{=}PCy_2)]$,

512 K. Sünkel and D. Steiner, *J. Organomet. Chem.*, 1989, **368**, 67.
513 R.A. Fischer and W.A. Herrmann, *J. Organomet. Chem.*, 1989, **377**, 275.
514 C.M. Garner, J.M. Fernández, and J.A. Gladysz, *Tetrahedron Lett.*, 1989, **30**, 3931.
515 C.H. Winter, A.M. Arif, and J.A. Gladysz, *Organometallics*, 1989, **8**, 219.
516 J.M. Fernandez and J.A. Gladysz, *Organometallics*, 1989, **8**, 207.
517 J. Takacs, P. Kiprof, G. Weichselbaumer, and W.A. Herrmann, *Organometallics*, 1989, **8**, 2394.
518 W.A. Herrmann and D.W. Marz, *J. Organomet. Chem.*, 1989, **362**, C5.
519 A.H. Klahn and D. Sutton, *Organometallics*, 1989, **8**, 198.
520 M.-G. Choi and R.J. Angelici, *J. Am. Chem. Soc.*, 1989, **111**, 8753.
521 A.H. Klahn-Olivia, R.D. Singer, J.M. Aramini, and D. Sutton, *Inorg. Chem.*, 1989, **28**, 4217.
522 W.A. Herrmann, K.A. Jung, and E. Herdtweck, *Chem. Ber.*, 1989, **122**, 2041.
523 T.J. Lynch, M.C. Helvenston, A.L. Rheingold, and D.L. Staley, *Organometallics*, 1989, **8**, 1959.
524 A.A. Danopoulos, C.J. Longley, G. Wilkinson, B. Hussain, and M.B. Hursthouse, *Polyhedron*, 1989, **8**, 2657.
525 E.A. Maatta and C. Kim, *Inorg. Chim. Acta*, 1989, **28**, 623.
526 R.G. Cavell and K.V. Katti, *Phosphorus Sulphur Silicon*, 1989, **41**, 43; K.V. Katti and R.G. Cavell, *Inorg. Chem.*, 1989, **28**, 3033.
527 I.M. Gardiner, M.A. Bruck, and D.E. Wigley, *Inorg. Chem.*, 1989, **28**, 1769; I.M. Gardiner, M.A. Bruck,

(^{13}C),[528] [$(OC)_5MnMn(CO)_4O$=$CHCH_2CH_2CH$=CMePh], (^{13}C),[529] [$Tc(CNR^1)_n(CNR^2)_{6-n}$], (^{99}Tc),[530] [$Tc_2(CO)_9(PF_3)$], (^{99}Tc),[531] [$Re(CN)_7$]$^{4-}$, (^{13}C, ^{14}N),[532] [(η^5-C_5H_5)Ru(PMe_3)$_2$-CH=CMeCMe=ORe$(CO)_4$Re$(CO)_5$], (^{13}C),[533] [Re(bipyrimidine)$(CO)_3$(methyl viologen)], (^{13}C),[534] *fac*-[{(η^5-C_5H_5)Fe(η^5-C_5H_4-4-C_5H_4N)}$_2$Re$(CO)_3$Cl], (^{13}C),[535] *fac*-[(bipy)Re$(CO)_3$-($NCBH_3$)], (^{13}C),[536] [Re_2{η^2,η^2-2-$(Ph_2P)_2CHC_5H_4N$}$(CO)_8$], (^{13}C),[537] [TcCl$(PPh_3)_2$-($N_2C_6H_4Br$-4)$_2$], (^{99}Tc),[538] and [Tc(dppe){P$(OMe)_3$}$_4$]$^+$, (^{99}Tc).[539]

Complexes of Fe, Ru, and Os.——A review entitled 'Mixed metal clusters: Structural and reactivity trends' discusses n.m.r. spectra of clusters containing Fe, Ru, Co, and Rh.[540] The magnetic moment of the 3/2-[512] ground state of ^{193}Os has been determined.[541]

The H-H distance in [Fe(η^2-H_2)H$(dppe)_2$]$^+$ has been determined by neutron diffraction and ^1H T_1 measurements.[542] ^1H T_1 measurements have been used to demonstrate that [MH(η^2-H_2){PhP$(OEt)_2$}$_4$]$^+$, M = Fe, Ru, Os, are η^2-H_2 complexes,[543] and to estimate the H—H distance in [(η^2-H_2){$Ph_2P(CH_2)_4PPh_2$}Ru}(μ-Cl)$_3$Ru{$Ph_2P(CH_2)_4PPh_2$}] as 0.86 Å.[544] The ^1H chemical shifts of [$HRuCo_nRh_{3-n}(CO)_{12}$] and [$HRu_2Co_mRh_{3-m}(CO)_{12}$] have been shown to have systematic trends.[545] ^1H n.O.e. measurements have been used to determine the structure of [$Os_3H(CO)_9$(μ_3-η^2:σ:η^2-C_6H_7)].[546] N.m.r. data have also been reported for [$Fe(CO)_2(dppe)H_2$], (^{13}C),[547] [(η^6-C_6H_6)Fe$(PMe_3)_2$H]$^+$, (^{13}C),[548] *cis*-[HFe$(SiPh_3)$-$(CO)_4$], (^{29}Si),[549] [HFe$(SiPh_3)(CO)_2${P$(OEt)_3$}$_2$], (^{29}Si),[550] [(η^5-C_5H_5)FeH$(MMe_3)_2$(CO)], (M = Si, Sn; ^{13}C),[551] [M$(CO)_4$(η^2-B_2H_5)]$^-$, (^{11}B),[552] [$HFe_3(CO)_{11}$]$^-$, (^{13}C),[553] [$HFe_4(CO)_{12}CBH_2$],

P.A. Wexler, and D.E. Wigley, *Inorg. Chem.*, 1989, **28**, 3688.
528 E. Lindner and V. Käss, *Chem. Ber.*, 1989, **122**, 2269.
529 R.M. Bullock, B.J. Rappoli, E.G. Samsel, and A.L. Rheingold, *J. Chem. Soc., Chem. Commun.*, 1989, 261.
530 U. Abram, R. Beyer, R. Muenze, M. Findeisen, and B. Lorenz, *Inorg. Chim. Acta*, 1989, **160**, 139.
531 C.C. Grimm, R.J. Clark, and R. Rosanske, *Inorg. Chim. Acta*, 1989, **159**, 137.
532 T. Taura, *Inorg. Chim. Acta*, 1989, **163**, 131.
533 R.M. Bullock, J.S. Ricci, and D.J. Szalda, *J. Am. Chem. Soc.*, 1989, **111**, 2741.
534 L.N. Winslow, D.P. Rillema, J.H. Welch, and P. Singh, *Inorg. Chem.*, 1989, **28**, 1596.
535 T.M. Miller, K.J. Ahmed, and M.S. Wrighton, *Inorg. Chem.*, 1989, **28**, 2347.
536 J. Guilhem, C. Pascard, J.-M. Lehn, and R. Ziessel, *J. Chem. Soc., Dalton Trans.*, 1989, 1449.
537 B.M. Mattson and L.N. Ito, *Organometallics*, 1989, **8**, 391.
538 T. Nicholson, N. De Vries, A. Davison, and A.G. Jones, *Inorg. Chem.*, 1989, **28**, 3813.
539 U. Abram, R. Beyer, R. Muenze, J. Stach, L. Kaden, B. Lorenz, and M. Findeisen, *Polyhedron*, 1989, **8**, 1201.
540 T.A. Pakkanen, J. Pursiainen, T. Venalainen, and T.T. Pakkanen, *J. Organomet. Chem.*, 1989, **372**, 129.
541 R. Eder, E. Hagn, and E. Zech, *Phys. Rev. C: Nucl. Phys.*, 1989, **40**, 2246 (*Chem. Abstr.*, 1989, **111**, 241 706).
542 J.S. Ricci, T.F. Koetzle, M.T. Bautista, T.M. Hofstede, R.H. Morris, and J.F. Sawyer, *J. Am. Chem. Soc.*, 1989, **111**, 8823.
543 S. Antoniutti, G. Albertin, P. Amendola, and E. Bordignon, *J. Chem. Soc., Chem. Commun.*, 1989, 229.
544 A.M. Joshi and B.R. James, *J. Chem. Soc., Chem. Commun.*, 1989, 1785.
545 J. Pursiainen and T.A. Pakkanen, *Acta Chem. Scand.*, 1989, **43**, 463.
546 M.A. Gallop, B.F.G. Johnson, J. Lewis, and A.H. Wright, *J. Chem. Soc., Dalton Trans.*, 1989, 481.
547 U. Schubert and M. Knorr, *Inorg. Chem.*, 1989, **28**, 1765.
548 L. Howarth and L.-L. Wong, *J. Chem. Soc., Dalton Trans.*, 1989, 1385.
549 G. Bellachioma, G. Cardaci, E. Colomer, R.J.P. Corriu, and A. Vioux, *Inorg. Chem.*, 1989, **28**, 519.
550 M. Knorr and U. Schubert, *J. Organomet. Chem.*, 1989, **365**, 151.
551 M. Akita, T. Oku, and Y. Moro-oka, *J. Chem. Soc., Chem. Commun.*, 1989, 1790.

(^{11}B),[554] [(η^5-C_5Me_5)Ru(μ-PPh_2)(μ-H)(μ-η^2:η^2-C_6H_6)Ru(η^5-C_5Me_5)], (^{13}C),[555] [H(OC)$_3$Ru-(μ-PPh_2)Co(CO)$_3$], (^{13}C),[556] [HRu$_3$(C≡CBu^t)(CO)$_7$(dppe)], (^{13}C),[557] [$H_2BRu_4(CO)_{12}H$], (^{11}B),[558] [HRu$_3$(CO)$_9$(μ_3-COMe)CuPPh$_3$], (^{13}C),[559] [$H_2Ru_3Rh(CO)_{11}(PPh_3)$]$^-$, (^{13}C),[560] [OsH-($\eta^2$-$C_2H_4$)$_2$(PMe$_2$Ph)$_3$]$^+$, (^{13}C),[561] [OsH{P($C_6H_3$-4-Me-2-CH=CH$_2$)($C_6H_4$Me-4)$_2$}{P($C_6H_4$-Me-4)$_3$}$_2$($\eta^2$-$C_2H_4$)]$^+$, (^{13}C),[562] [Os$_3$($\mu$-H)($\mu$-CCH$_2$Pr^i)(CO)$_{10}$], (^{13}C),[563] [HOs$_3(CO)_9$(C≡C-SiMe$_3$)], (^{13}C),[564] [Os$_4$($\mu$-H)(CO)$_{14}$(MR$_3$)], (M = Si, Ge, Sn; ^{13}C),[565] and [Os$_3$Pt(μ-H)$_2$(CO)$_{10}$(PR$_3$)], (^{13}C).[566]

A correlation between δ(^{119}Sn) and the Tolman cone angle for a series of complexes of [(η^5-C_5H_4Me)Fe(CO)L(SnMe$_3$)] has been observed.[567] Two dimensional ^{31}P COSY-90 has been used to determine the relative signs of heteronuclear coupling in [Ru$_2$(CO)$_6$(μ_2-η^2-C≡CPr^i)(μ-PPh$_2$)] and related compounds.[568] ^{199}Hg n.m.r. spectra have been reported for mercury bridged transition metal clusters, *e.g.*, [(μ_3-η^2-C≡CBu^t)(OC)$_9$Ru$_3$(μ-Hg)X]. The chemical shifts cover 1500 p.p.m.[569] Chemical shift anisotropy has been shown to be the dominant factor influencing ^{31}P spin-lattice relaxation in [M$_2$(CO)$_6$(μ_2-η^2-C≡C-Bu^t)(μ-PPh$_2$)], M = Fe, Ru.[570] The determination of δ(^{187}Os), J(^{187}Os), and T_1(^{187}Os) in osmium compounds *via* indirect two dimensional n.m.r. spectroscopy has been reported. The technique was applied to compounds such as [(η^5-C_5H_5)OsL$_2$R].[571] N.m.r. data have also been reported for [(η^5-C_5Me_5)Fe(CO)$_2$Et], (^{13}C),[572] [(η^5-C_5H_5)Fe(CO)$_2$CH$_2$CH$_2$-SMePh]$^+$, (^{13}C),[573] [(η^5-C_5H_5)Fe(CO)$_2$CH$_2$CH$_2$CH(OMe)Ph]$^+$, (^{13}C),[574] [{(η^5-C_5Me_5)Fe-(CO)$_2$CH$_2$}$_2$SR]$^+$, (^{13}C),[575] [{(η^5-C_5Me_5)Fe(CO)$_2$}$_2$(μ-CH$_2$CO)], (^{13}C),[576] [{(η^5-C_5H_5)(OC)$_2$Fe-

552 T.J. Coffy, G. Medford, J. Plotkin, G.J. Long, J.C. Huffman, and S.G. Shore, *Organometallics*, 1989, **8**, 2404.
553 K. Jothimony and S. Vancheezan, *J. Organomet. Chem.*, 1989, **52**, 301.
554 X. Meng, N.P. Rath, and T.P. Fehlner, *J. Am. Chem. Soc.*, 1989, **111**, 3422.
555 H. Omori, H. Suzuki, Y. Take, and Y. Moro-oka, *Organometallics*, 1989, **8**, 2270.
556 S. Guesmi, P.H. Dixneuf, G. Süss-Fink, N.J. Taylor, and A.J. Carty, *Organometallics*, 1989, **8**, 307.
557 E. Sappa, G. Predieri, A. Tiripicchio, and C. Vignali, *J. Organomet. Chem.*, 1989, **378**, 109.
558 F.-E. Hong, T.J. Coffy, D.A. McCarthy, and S.G. Shore, *Inorg. Chem.*, 1989, **28**, 3285.
559 J. Evans, P.M. Stroud, and M. Webster, *Organometallics*, 1989, **8**, 1270.
560 J. Evans, P.M. Stroud, and M. Webster, *J. Chem. Soc., Chem. Commun.*, 1989, 1029.
561 T.J. Johnson, J.C. Huffman, K.G. Caulton, S.A. Jackson, and O. Eisenstein, *Organometallics*, 1989, **8**, 2073.
562 P.J. Desrosiers, L. Cai, and J. Halpern, *J. Am. Chem. Soc.*, 1989, **111**, 8513.
563 M. Green, A.G. Orpen, and C.J. Schaverien, *J. Chem. Soc., Dalton Trans.*, 1989, 1333.
564 B.F.G. Johnson, J. Lewis, M. Monari, D. Braga, and F. Grepioni, *J. Organomet. Chem.*, 1989, **377**, C1.
565 C.Y. Lu, F.W.B. Einstein, V.J. Johnston, and R.K. Pomeroy, *Inorg. Chem.*, 1989, **28**, 4212.
566 P. Ewing and L.J. Farrugia, *Organometallics*, 1989, **8**, 1665.
567 L. Carlton, G. Pattrick, and N.J. Coville, *Inorg. Chim. Acta*, 1989, **160**, 65
568 L.H. Randall, A.A. Cherkas, and A.J. Carty, *Organometallics*, 1989, **8**, 568.
569 S. Hajela, E. Rosenberg, R. Gobetto, L. Milone, and D. Osella, *J. Organomet. Chem.*, 1989, **377**, 85.
570 L.H. Randall and A.J. Carty, *Inorg. Chem.*, 1989, **28**, 1194.
571 R. Benn, E. Joussen, H. Lehmkuhl, F. Lopez Ortiz, and A. Rufińska, *J. Am. Chem. Soc.*, 1989, **111**, 8754.
572 C. Roger, M.-J. Tudoret, V. Guerchais, and C. Lapinte, *J. Organomet. Chem.*, 1989, **365**, 347.
573 C.P. Casey and L.J. Smith, *Organometallics*, 1989, **8**, 2288.
574 M. Brookhart and Y. Liu, *Organometallics*, 1989, **8**, 1569.
575 V. Guerchais, J.-Y. Thépot, and C. Lapinte, *J. Chem. Soc., Chem. Commun.*, 1989, 1554.
576 M. Akita, A. Kondoh, and Y. Moro-oka, *J. Chem. Soc., Dalton Trans.*, 1989, 1083.

$CH_2\}_3CH]$, (^{13}C),[577] $[(OC)_4FeCH_2CH_2C(O)CH_2]$, ($^{13}C$),[578] $[(\eta^5\text{-}C_3H_3BMeNBu^t)Fe(CO)_2R]$, ($^{11}B$, ^{13}C),[579] $[(\eta^5\text{-indenyl})Fe(CO)_2CHMeOEt]$, ($^{13}C$),[580] $[(\eta^5\text{-}C_5H_5)(OC)(PEt_3)FeCH(OMe)\text{-}Ph]$, ($^{13}C$),[581] $[(\eta^5\text{-}C_5H_5)(OC)_2FeCH(OSiHEt_2)Me]$, ($^{13}C$),[582] $[(\eta^5\text{-}C_5H_5)Fe(CO)\{C(SMe)\text{-}(SMe)_2\}]$, ($^{13}C$),[583] *cis*-$[Fe(CO)_4(CF_3)_2]$, ($^{13}C$),[584] $[(OC)_2\{(MeO)_3P\}_2FeX\{C{\equiv}CCH(OR)_2\}]$, ($^{13}C$),[585] $[(\eta^5\text{-}C_5H_5)Fe(CO)_2CS_2Fe(CO)(\eta^5\text{-}C_5H_5)]$, ($^{13}C$),[586] $[(\eta^5\text{-indenyl})Fe(CO)_2\{C(O)Me\}]$, ($^{13}C$),[587] $[(\eta^5\text{-}C_5H_5)Fe(CO)(PR_3)(COMe)]$, ($^{13}C$),[588] $[RC(O)Fe(CO)_4]^-$, (^{13}C),[589] $[MeC(O)Fe\text{-}(CO)_4CO_2Et]$, ($^{13}C$),[590] $[(\eta^5\text{-}C_5H_5)(OC)FeSMeCH{=}C(CN)CO]$, ($^{13}C$),[591] $[(\eta^5\text{-}C_5Me_5)\text{-}Fe(CO)\{C(O)C[C(O)R^1]{=}CR^2P{=}PAr\}]$, ($^{13}C$),[592] $[(OC)_2L^1L^2FeC(O)CH_2C(CO_2Me){=}COMe]$, ($^{13}C$),[593] $[(OC)_2Fe(OC{=}CPh_2)\{P(OR)_3\}_2]$, ($^{13}C$),[594] $[(\eta^5\text{-}C_5H_5)Fe(CO)(EPh_3)(C_6H_4F\text{-}2)]$, (E = P, As, Sb; ^{13}C),[595] $[(\eta^5\text{-}C_5Me_5)Fe({=}CH_2)(dppe)]^+$, ($^{13}C$),[596] $[(\eta^5\text{-}C_5H_5)(OC)\{Ph_2[(S)\text{-}EtMeCH]\text{-}P\}Fe{=}CHPh]^+$, ($^{13}C$),[597] $[L(OC)_3Fe{=}CROEt]$, (^{13}C),[598] *mer*-$[Fe(CO)_3(PPh_3)(HgPh)(SiMePh_2)]$, ($^{199}Hg$),[599] $[(\eta^5\text{-}C_5H_5)Fe(CO)_2Si_2Me_{n+2}Ph_{3-n}]$, ($^{13}C$, ^{29}Si),[600] $[(hmpa)Me_2Si{=}Fe(CO)_4]$, ($^{13}C$, ^{29}Si),[601] $[(\eta^5\text{-}C_5H_5)(OC)_2FeSiMe(SiMe_2)_4SiMe]$, ($^{13}C$, ^{29}Si),[602] (^{29}Si),[603] $[(\eta^5\text{-}C_5H_5)Fe(CO)_2\text{-}SiMe(CH{=}CH_2)Cl]$, ($^{13}C$, ^{29}Si),[604] $[(\eta^5\text{-}C_5H_5)Fe(CO)_2MXR_2]$, (M = Si, Sn, Sb; ^{13}C, ^{29}Si,

577 M. Kobayashi and J.D. Wuest, *Organometallics*, 1989, **8**, 2843.
578 B. Denise, D. Navarre, H. Rudler, and J.C. Daran, *J. Organomet. Chem.*, 1989, **375**, 273.
579 G. Schmid and T. Rohling, *J. Organomet. Chem.*, 1989, **375**, 21.
580 S.A. Levitre, A.R. Cutler, and T.C. Forschner, *Organometallics*, 1989, **8**, 1133.
581 M. Brookhart and R.C. Buck, *J. Am. Chem. Soc.*, 1989, **111**, 559.
582 E.J. Crawford, P.K. Hanna, and A.R. Cutler, *J. Am. Chem. Soc.*, 1989, **111**, 6891.
583 G.N. Glavee, Y. Su, R.A. Jacobson, and R.J. Angelici, *Inorg. Chim. Acta*, 1989, **157**, 73.
584 W. Dukat and D. Naumann, *J. Chem. Soc., Dalton Trans.*, 1989, 739.
585 C. Löwe, H.-U. Hund, and H. Berke, *J. Organomet. Chem.*, 1989, **372**, 295.
586 M.E. Giuseppetti-Dery, B.E. Landrum, J.L. Shipley, and A.R. Cutler, *J. Organomet. Chem.*, 1989, **378**, 421.
587 T.C. Forschner and A.R. Cutler, *J. Organomet. Chem.*, 1989, **361**, C41.
588 M. Brookhart and Y. Liu, *Organometallics*, 1989, **8**, 1572.
589 H. d. Abbayes, J.-C. Clement, P. Laurent, J.-J. Yaouanc, G. Tanguy, and B. Weinberger, *J. Organomet. Chem.*, 1989, **359**, 205.
590 S. Sabo-Etienne, A.-M. Larsonneur, and H. des Abbayes, *J. Chem. Soc., Chem. Commun.*, 1989, 1671.
591 R. Kergoat, M.M. Kubicki, L.C. Gomes de Lima, H. Scordia, J.E. Guerchais, and P. L'Haridon, *J. Organomet. Chem.*, 1989, **367**, 143.
592 L. Weber, M. Frebel, and R. Boese, *New J. Chem.*, 1989, **13**, 303.
593 T.-a. Mitsudo, H. Watanabe, T. Sasaki, Y. Takegami, Y. Watanabe, K. Kafuku, and K. Nakatsu, *Organometallics*, 1989, **8**, 368.
594 R. Birk, H. Berke, H.-U. Hund, G. Huttner, L. Zsolnai, L. Dahlenburg, U. Behrens, and T. Sielisch, *J. Organomet. Chem.*, 1989, **372**, 397.
595 M.A. El-Hinnawi and M.A. Kobeissi, *Inorg. Chim. Acta*, 1989, **166**, 99.
596 C. Roger and C. Lapinte, *J. Chem. Soc., Chem. Commun.*, 1989, 1598.
597 M. Brookhart and R.C. Buck, *J. Organomet. Chem.*, 1989, **370**, 111.
598 S. Lotz, J.L.M. Dillen, and M.M. van Dyk, *J. Organomet. Chem.*, 1989, **371**, 371.
599 E. Kunz and U. Schubert, *Chem. Ber.*, 1989, **122**, 231.
600 K.H. Pannell, J.M. Rozell, jun., and C. Hernandez, *J. Am. Chem. Soc.*, 1989, **111**, 4482.
601 C. Zybill, D.L. Wilkinson, C. Leis, and G. Müller, *Angew. Chem., Int. Ed. Engl.*, 1989, **28**, 203.
602 K.H. Pannell, L.-J. Wang, and J.M. Rozell, *Organometallics*, 1989, **8**, 550.
603 E. Hengge, M. Eibl, and R. Schrank, *J. Organomet. Chem.*, 1989, **369**, C23.
604 N. Auner, J. Grobe, T. Schäfer, B. Krebs, and M. Dartmann, *J. Organomet. Chem.*, 1989, **363**, 7.

^{119}Sn),[605] [(η^5-C_5H_5)Fe(CO)$_2$SiR(N_3)N=PR_3], (^{29}Si),[606] [(OC)$_4$FeGeMe$_2$(CH_2)$_n$GeMe$_2$], (^{13}C),[607] [{(η^5-C_5H_5)(OC)Fe}$_2$(μ-CO)(μ-CHCOR)], (^{13}C),[608] [(η^5-C_5H_5)$_2$Fe$_2$(CO)$_2$(μ-CO)(μ-C=CR^{1}CR^2=CHCN)], (^{13}C),[609] [(η^5-C_5H_5)$_2$Fe$_2$(CO)$_2$(μ-CO){μ-C(SMe)ER}], (E = S, Se; ^{13}C),[610] (25), (^{13}C),[611] [Fe$_2$(CO)$_5$(Pr^{i}N=CHCH=NPr^i)(μ_2,η^2-MeO$_2$CC≡CCO$_2$Me)], (^{13}C),[612] [Fe$_2${μ_2-HC=CH(CO$_2$Me)C(O)}(CO)$_5$(Pr^{i}N=CHCH=NPr^i)], (^{13}C),[613] [(η^5-C_5H_5)$_2$Fe$_2$(μ-CO)-(μ-dppm){μ-C=CHC(CN)=C(CN)$_2$}], (^{13}C),[614] (26), (^{13}C),[615] [(μ-MeOCH=CHC)(μ-Bu^{t}S)-

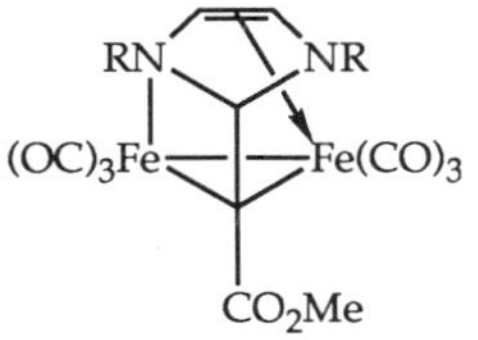

(25)

(26)

Fe$_2$(CO)$_6$], (^{13}C),[616] [(μ-RSC=S)(μ-EtS)Fe$_2$(CO)$_6$], (^{13}C),[617] [(μ-SR^1)(μ-CH$_2$SR^2)Fe$_2$(CO)$_6$], (^{13}C),[618] [(μ-ROC=S)(μ-EtS)Fe$_2$(CO)$_5$L], (^{13}C),[619] [{μ-(F$_3$C)$_2$C=CS}$_2$Fe$_2$(CO)$_6$], (^{13}C),[620] [(μ-R^{1}OC=S)(μ-R^{2}S)Fe$_2$(CO)$_6$], (^{13}C),[621] [Fe$_3$(μ-CO)$_2$(CO)$_6$(RC≡CR)$_2$], (^{13}C),[622] [Fe$_2$Co(CO)$_8$-(PMe$_3$)(CPMe$_3$)]$^-$, (^{13}C),[623] [(η^5-C_5H_5)NiFe$_3$(CO)$_7$(μ-PPh$_2$)(μ_4,η^2-HC≡CPr^i)], (^{13}C),[624] [(η^5-C$_5$Me$_5$)Ru(NO)Me$_2$], (^{13}C),[625] [(η^4-C$_5$Me$_4$CH$_2$)RuMe(PMe$_3$)$_2$]$^+$, (^{13}C),[626] [(η^5-C_5H_5)Ru-(CO)$_2$(CH$_2$)$_n$Ru(CO)$_2$(η^5-C_5H_5)], (n = 5 to 10; ^{13}C),[627] [(OC)$_4$MCH$_2$CH$_2$CH$_2$CH$_2$], (M = Ru, Os; ^{13}C),[628] [M(CH=CHR^1)(PR$^2_3$)$_2$(CO)Cl], (M = Ru, Os; ^{13}C),[629] [Ru(CH=CHR)(PMe$_2$Ph)$_2$-

605 M.F. Lappert, M.J. McGeary, and R.V. Parish, *J. Organomet. Chem.*, 1989, **373**, 107.
606 U. Wachtler, W. Malisch, E. Kolba, and J. Matreux, *J. Organomet. Chem.*, 1989, **363**, C36.
607 J. Barrau, N.B. Hamida, A. Agrebi, and J. Satge, *Organometallics*, 1989, **8**, 1585.
608 C.P. Casey, M. Crocker, P.C. Vosejpka, and A.L. Rheingold, *Organometallics*, 1989, **8**, 278.
609 M. Etienne and J.E. Guerchais, *J. Chem. Soc., Dalton Trans.*, 1989, 2187.
610 N.C. Schroeder, R. Funchess, R.A. Jacobson, and R.J. Angelici, *Organometallics*, 1989, **8**, 521.
611 F. Muller, G. Van Koten, K. Vrieze, D. Heijdenrijk, B.B. Krijnen, and C.H. Stam, *Organometallics*, 1989, **8**, 41.
612 F. Muller, G. van Koten, K. Vrieze, K.A.A. Duineveld, D. Heijdenrijk, A.N.S. Mak, and C.H. Stam, *Organometallics*, 1989, **8**, 1324.
613 F. Muller, I.M. Han, G. van Koten, K. Vrieze, D. Heijdenrijk, J. van Mechelen, and C.H. Stam, *Inorg. Chim. Acta*, 1989, **158**, 99.
614 M. Etienne and L. Toupet, *J. Chem. Soc., Chem. Commun.*, 1989, 1110.
615 C.A. Mirkin, K.-L. Lu, G.L. Geoffroy, A.L. Rheingold, and D.L. Staley, *J. Am. Chem. Soc.*, 1989, **111**, 7279.
616 D. Seyferth, D.P. Ruschke, W.M. Davis, M. Cowie, and A.D. Hunter, *Organometallics*, 1989, **8**, 836.
617 D. Seyferth, G.B. Womack, C.M. Archer, J.P. Fackler, jun., and D.O. Marler, *Organometallics*, 1989, **8**, 443.
618 H.G. Raubenheimer, L. Linford, and A. van A. Lombard, *Organometallics*, 1989, **8**, 2062.
619 H. Patin, A. Darchen, and E.K. Lhadi, *J. Organomet. Chem.*, 1989, **375**, 91.
620 K. Seitz, J. Benecke, and U. Behrens, *J. Organomet. Chem.*, 1989, **371**, 247.
621 A. Darchen, E.K. Lhadi, and H. Patin, *J. Organomet. Chem.*, 1989, **363**, 137.
622 D. Osella, G. Arman, M. Botta, R. Gobetto, F. Laschi, and P. Zanello, *Organometallics*, 1989, **8**, 620.
623 S. Ching and D.F. Shriver, *J. Am. Chem. Soc.*, 1989, **111**, 3243.
624 E. Sappa, D. Belletti, A. Tiripicchio, and M.T. Camellini, *J. Organomet. Chem.*, 1989, **359**, 419.
625 J. Chang, M.D. Seidler, and R.G. Bergman, *J. Am. Chem. Soc.*, 1989, **111**, 3258.
626 D.A. Straus, C. Zhang, and T.D. Tilley, *J. Organomet. Chem.*, 1989, **369**, C13.
627 K.P. Finch, J.R. Moss, and M.L. Niven, *Inorg. Chim. Actsa*, 1989, **166**, 181.
628 E. Lindner, R.-M. Jansen, E.W. Hiller, and R. Fawzi, *Chem. Ber.*, 1989, **122**, 1403.
629 H. Werner, U. Meyer, K. Peters, and H.G. von Schnering, *Chem. Ber.*, 1989, **122**, 2097.

$(CO)_2Cl]$, (^{13}C),[630] $[Ru\{Ph_2PCH(CPh{=}CH)CBu^t{=}O\}(p\text{-cymene})]^+$, ($^{13}C$),[631] $[Ru\{C(CO_2Me){=}C(CO_2Me)C_6H_4X\text{-}4\}(CO)_2(PR_3)_2Cl]$, ($^{13}C$),[632] (27), ($^{13}C$),[633] $[Ru(C{\equiv}CPh)(\eta^3\text{-}PhC_3CHPh)\{PhP(CH_2CH_2CH_2PCy_2)_2\}]$, ($^{13}C$),[634] $[(\eta^6\text{-}C_6H_6)Ru(C_6H_4\text{-}2\text{-}CR{=}NPh)(PMe_3)]^+$, ($^{13}C$),[635] $[(\eta^5\text{-}C_5Me_5)Ru\{C(O)C(CO_2Me){=}CMeP{=}PC_6H_2Bu^t_3\text{-}2,4,6\}(CO)]$, ($^{13}C$),[636] $[(\eta^5\text{-indenyl})Ru({=}C{=}CHPh)(Ph_2PCHMeCHMePPh_2)]$, ($^{13}C$),[637] $[(\eta^6\text{-}C_6Me_6)Ru\{{=}C(OMe)CH_2CR_2OMe\}Cl(PMe_3)]^+$, ($^{13}C$),[638] $[Ru\{C[{=}C(CN)_2]CPh{=}C(CN)_2\}\{P(OMe)_3\}_2(\eta^5\text{-}C_5H_5)]$, ($^{13}C$),[639] $[(OC)_4RuSiF_2CBu^t{=}CHSiF_2]$, ($^{13}C$),[640] $[(\eta^5\text{-}C_5H_5)_2Ru_2(CO)(\mu\text{-}CO)(\mu\text{-}CH_2)(\eta^2\text{-}CF_2{=}CFCF_3)]$, ($^{13}C$),[641] $[(\eta^5\text{-}C_5H_5)_2Ru_2(CO)_2(\mu\text{-}CHMe)(\mu\text{-}CH_2)]$, ($^{13}C$),[642] $[(\eta^5{:}\eta^5\text{-}C_5H_4CH_2C_5H_4)Ru_2(CO)_2(\mu\text{-}CHMe)_2]$, ($^{13}C$),[643] $[(\eta^5\text{-}C_5H_5)_2Ru_2(CO)_2(\mu\text{-}CHMe)(\mu\text{-}CMe_2)]$, ($^{13}C$),[644] $[M_3(CO)_{12}(Cl)(SnCl_3)]$, (M = Ru, Os; ^{13}C),[645] $[Ru_3(\mu\text{-}PPh_2)(\mu\text{-}C{=}CBu^t)(\mu\text{-}CBu^t{=}CHPPh_2)]$, ($^{13}C$),[646] *cis*-$[Os(C_6H_4Me\text{-}2)_2\{CN(2,6\text{-}Me_2C_6H_3)\}_4]$, ($^{13}C$),[647] and $[Os_3(CO)_8\{\eta^1\text{-}C(OMe)Ph\}(\mu_3\text{-}CPh)(\mu_3\text{-}COMe)]$, ($^{13}C$).[648]

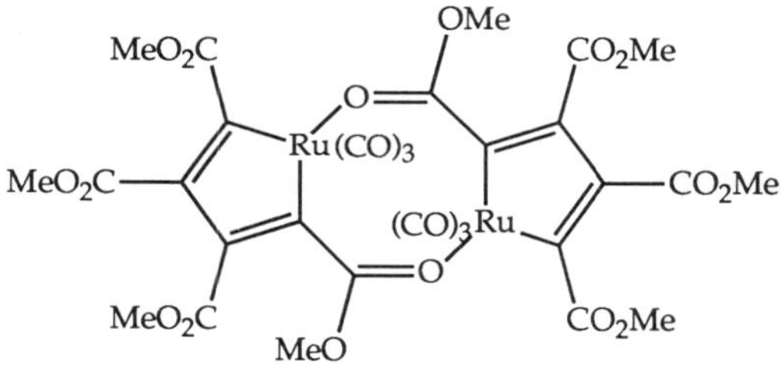

(27)

A new INDO/σ program has been used to calculate ^{13}C chemical shifts for $[(\eta^5\text{-}C_5H_5)Fe(CO)_2(CN)]$.[649] ^1H n.O.e. measurements have been made to determine the preferred conformation of $[(\eta^5\text{-}C_5H_4R)Fe(CO)(PPh_3)I]$.[650] The multiple quantum ^1H n.m.r. spec-

630 J.M. Bray and R.J. Mawby, *J. Chem. Soc., Dalton Trans.*, 1989, 589.
631 B. Guilbert, B. Demerseman, P.H. Dixneuf, and C. Mealli, *J. Chem. Soc., Chem. Commun.*, 1989, 1035.
632 J.R. Crook, B. Chamberlain, and R.J. Mawby, *J. Chem. Soc., Dalton Trans.*, 1989, 465.
633 E. Lindner, R.-M. Jansen, H.A. Mayer, W. Hiller, and R. Fawzi, *Organometallics*, 1989, **8**, 2355.
634 G. Jia, A.L. Rheingold, and D.W. Meek, *Organometallics*, 1989, **8**, 1378.
635 G.C. Martin and J.M. Boncella, *Organometallics*, 1989, **8**, 2968.
636 L. Weber, M. Frebel, and R. Boese, *Chem. Ber.*, 1989, **122**, 2091.
637 F. Morandini, G. Consiglio, A. Sironi, and M. Moret, *J. Organomet. Chem.*, 1989, **370**, 305.
638 H. Le Bozec, K. Ouzzine, and P.H. Dixneuf, *J. Chem. Soc., Chem. Commun.*, 1989, 219.
639 M.I. Bruce, M.P. Cifuentes, M.R. Snow, and E.R.T. Tiekink, *J. Organomet. Chem.*, 1989, **359**, 379.
640 C.Y. Huang and C.S. Liu, *J. Organomet. Chem.*, 1989, **373**, 353.
641 J.A.K. Howard, S.A.R. Knox, N.J. Terrill, and M.I. Yates, *J. Chem. Soc., Chem. Commun.*, 1989, 640.
642 N.M. Doherty, J.A.K. Howard, S.A.R. Knox, N.J. Terrill, and M.I. Yates, *J. Chem. Soc., Chem. Commun.*, 1989, 638.
643 S.A.R. Knox, K.A. Macpherson, A.G. Orpen, and M.C. Rendle, *J. Chem. Soc., Dalton Trans.*, 1989, 1807.
644 R.E. Colborn, D.L. Davies, A.F. Dyke, S.A.R. Knox, K.A. Mead, A.G. Orpen, J.E. Guerchais, and J. Roué, *J. Chem. Soc., Dalton Trans.*, 1989, 1799.
645 A. Becalska, R.K. Pomeroy, and W.A.G. Graham, *Can. J. Chem.*, 1989, **67**, 1236.
646 M. Castiglioni, R. Giordano, and E. Sappa, *J. Organomet. Chem.*, 1989, **362**, 399.
647 J. Arnold, G. Wilkinson, B. Hussain, and M.B. Hursthouse, *Organometallics*, 1989, **8**, 1362.
648 W.-Y. Yeh, S.R. Wilson, and J.R. Shapley, *J. Organomet. Chem.*, 1989, **371**, 257.
649 W. Wu, X. You, and A. Dai, *Sci. Sin., Ser. B (Engl. Ed.)*, 1987, **30**, 1253 (*Chem. Abstr.*, 1989, **110**, 154 481).
650 J. du Toit, D.C. Levendis, J.C.A. Boeyens, M.S. Loonat, L. Carlton, W. Pickl, and N.J. Coville, *J.*

trum of ferrocene in isotropic solution has been reported.[651] ^{13}C and ^{57}Fe n.m.r. spectra of some open, half-open, and closed ferrocenes have been reported. On going from the corresponding closed, *via* the half-open, to open ferrocenes, there is a deshielding of the ^{57}Fe nucleus by about 700 and 1100 p.p.m.[652] The effect of electron-withdrawing and electron-donating substituents on the distribution of electron density has been shown by ^{13}C on $[(\eta^5\text{-}C_5H_5)Fe(\eta^5\text{-}C_5H_4R)]$.[653] The ^{13}C chemical shift of the α-carbon in $[(\eta^5\text{-}C_5H_5)Fe(\eta^5\text{-}C_5H_4C_6H_4R\text{-}2)]$ is influenced by a steric perturbation effect and that of the β-carbon is influenced by the inductive effect.[654] $J\{^{119}Sn\text{-}^{13}C(Et)\}$ has been used to estimate substituent effects in $[(\eta^5\text{-}C_5H_5)Fe(\eta^5\text{-}C_5H_4SnEt_3)]$ and related compounds. The ^{13}C and ^{119}Sn n.m.r. spectra were reported.[655] The interaction between ferrocene and nitrogen or phosphorus has been examined in $[\{(\eta^5\text{-}C_5H_5)Fe(\eta^5\text{-}C_5H_4)\}_{3-n}ER_n]$, E = N, P, using ^{13}C n.m.r. spectroscopy.[656] ^1H COSY and NOESY n.m.r. spectra have been used to assign isomers of $[(\eta^5\text{-}L)_2Fe]$, L = (10).[657] ^1H n.O.e. spectroscopy has been used to assign signals for $[Ru(\eta^5\text{-indenyl})(CO)\{P(OC_6H_4Me\text{-}2)_3\}I]$ and conformations discussed.[658] The ^{13}C chemical shifts of $[(\eta^6\text{-arene})_2Fe]^{2+}$ have been discussed in terms of shielding by the d_{z^2} electrons. The ^{13}C shifts correlate reasonably well with the total charge density at the relevant carbon atom for a variety of sandwich complexes.[659] N.m.r. data have also been reported for $[(\eta^2\text{-}trans\text{-}C_8H_{14})_3Fe(CO)_2]$, ($^{13}C$),[660] $[(\eta^5\text{-}C_5H_5)Fe(CO)_2\{\eta^2\text{-isopropenyl-(-)-menthyl ether}\}]^+$, ($^{13}C$),[661] $[MeO_2C\overline{C{=}C(CO_2Me)CMe(CMe{=}NC_6H_4OMe\text{-}4)_2Fe}(CO)_3]$, ($^{13}C$),[662] $[M_2(CO)_4(Pr^iN{=}CHCH{=}NPr^i)(\mu_2\text{-}R^1C{\equiv}CR^2)]$, (M = Fe, Ru; ^{13}C),[663] $[FeRu(CO)_5(Pr^iN{=}CHCH{=}NPr^i)(\mu_2,\eta^2\text{-}MeO_2CC{\equiv}CCO_2Me)]$, ($^{13}C$),[664] $[M(NH_3)_5(\eta^2\text{-olefin})]^{2+}$, (M = Ru, Os; ^{13}C),[665] $[M(NH_3)_5(\eta^2\text{-1,3-dimethyl uracil})]^{2+}$, (M = Ru, Os; ^{13}C),[666] $[(\eta^5\text{-}C_5H_5)(Me_3P)_2Ru(\eta^2\text{-}HC{\equiv}CH)]^+$, ($^{13}C$),[667] $[Ru_2(CO)_4(\mu_2\text{-}\eta^2\text{-}HNOCR_2)_2]_n$, ($^{13}C$),[668] $[Ru_3(\mu\text{-}CO)_2(CO)_6(RC{\equiv}CR)_2]$, ($^{13}C$),[669] $[Ru_2(CO)_6\{Bu^tN{=}CHCH(NBu^t)CR{=}CR\}]$, ($^{13}C$),[670] $[Os(NH_3)_5\{\eta^2\text{-}$

Organomet. Chem., 1989, **368**, 339.

651 N. Müller, *Monatsh. Chem.*, 1989, **120**, 801.
652 R. Benn, A. Rufińska, M.S. Kralik, and R.D. Ernst, *J. Organomet. Chem.*, 1989, **375**, 115.
653 G.A. Nagy and T. Stefan, *Kem. Kozl.*, 1986, **66**, 351 (*Chem. Abstr.*, 1989, **110**, 154 485).
654 Y. Okada and T. Hayashi, *Nippon Kagaku Kaishi*, 1988, 162 (*Chem. Abstr.*, 1989, **110**, 39 124); Y. Okada and T. Hayashi, *Chem. Express*, 1988, **3**, 227 (*Chem. Abstr.*, 1989, **110**, 213 020).
655 Ch. Krüger, K.-H. Thiele, and M. Dargatz, *Z. Anorg. Allg. Chem.*, 1989, **569**, 97.
656 A. Houlton, P.T. Bishop, R.M.G. Roberts, J. Silver, and M. Herberhold, *J. Organomet. Chem.*, 1989, **364**, 381.
657 V. Bhide, P. Rinaldi, and M.F. Farona, *J. Organomet. Chem.*, 1989, **376**, 91.
658 M.S. Loonat, L. Carlton, J.C.A. Boeyens, and N.J. Coville, *J. Chem. Soc., Dalton Trans.*, 1989, 2407.
659 S. Abdul-Rahman, A. Houlton, R.M.G. Roberts, and J. Silver, *J. Organomet. Chem.*, 1989, **359**, 331.
660 H. Angermund, A.K. Bandyopadhyay, F.-W. Grevels, and F. Mark, *J. Am. Chem. Soc.*, 1989, **111**, 4656.
661 M.K. Begum, K.-H. Chu, T.S. Coolbaugh, M. Rosenblum, and X.-Y. Zhu, *J. Am. Chem. Soc.*, 1989, **111**, 5252.
662 H.-W. Frühauf, F. Seils, and C.H. Stam, *Organometallics*, 1989, **8**, 2338.
663 F. Muller, G. van Koten, K. Vrieze, and D. Heijdenrijk, *Inorg. Chim. Acta*, 1989, **158**, 69.
664 F. Muller, G. van Koten, M.J.A. Kraakman, K. Vrieze, R. Zoet, K.A.A. Duineveld, D. Heijdenrijk, C.H. Stam, and M.C. Zoutberg, *Organometallics*, 1989, **8**, 982.
665 M.G. Elliott, S. Zhang, and R.E. Shepherd, *Inorg. Chem.*, 1989, **28**, 3036.
666 S. Zhang and R.E. Shepherd, *Inorg. Chim. Acta*, 1989, **163**, 237.
667 R.M. Bullock, *J. Chem. Soc., Chem. Commun.*, 1989, 165.
668 F. Neumann, H. Stoeckli-Evans, and G. Süss-Fink, *J. Organomet. Chem.*, 1989, **379**, 139.
669 D. Osella, G. Arman, R. Gobetto, F. Laschi, P. Zanello, S. Ayrton, V. Goodfellow, C.E. Housecroft, and

cis-MeCH=C(OMe)Me}]$^{2+}$, (^{13}C),[671] [(OC)$_8$Os$_2$(μ-HC≡CH)], (^{13}C),[672] [(η^6-C$_6$H$_3$Me$_3$)Fe(η^3-C$_3$H$_5$)(CO)]$^+$, (^{13}C),[673] [(η^3-1,5-Me$_2$-*syn*-C$_5$H$_5$)Fe(CO)$_3$X], (^{13}C),[674] [(η^5-C$_5$H$_5$)Fe(CO)$_2$(η^3-CH$_2$CHCHCO$_2$Me)], (^{13}C),[675] [Ru$_2$(μ-dppm)(CO)$_4${μ$_2$-η^3-HC=CPhC(O)NPh}], (^{13}C),[676] [Ru$_3$(μ-SEt)(μ$_3$-η^3-CCPhCHPh)(CO)$_9$], (^{13}C),[677] [Ru$_2$(CO)(μ-CO){μ-η^1,η^3-CH$_2$C(CH$_2$)$_2$}(η^5-C$_5$H$_5$)$_2$], (^{13}C),[678] [(η^4-diene)Fe(CO)$_3$], (^{13}C),[679] [(η^4-diene)M(CO)$_2$L], (M = Fe, Ru; ^{13}C),[680] [(η^4-1-Me-4-C$_2$H$_3$OC$_4$H$_4$)Fe(CO)$_3$], (^{13}C),[681] [{η^4-1-MeO$_2$C-4-P(O)(OMe)$_2$C$_4$H$_4$}Fe(CO)$_3$], (^{13}C),[682] [(η^4-MeC$_4$H$_5$)Fe(CO)$_3$], (^{13}C),[683] [{η^4-2-PhC(O)C$_4$H$_5$}Fe(CO)$_3$], (^{13}C),[684] [(η^4-2-Et-cyclohexa-1,3-diene)Fe(CO)$_3$], (^{13}C),[685] [(η^4-5-Me-naphthalene)Fe{P(OMe)$_3$}$_3$], (^{13}C),[686] [{η^4-(MeO)MeSiCPh=CHCH=CPh}Fe(CO)$_3$], (^{13}C, ^{29}Si),[687] [(OC)$_2$(PPh$_3$)Fe(η^4-C$_6$H$_7$C$_6$H$_7$-η^4)Fe(CO)$_2$(PPh$_3$)], (^{13}C),[688] [{η^4-(+)-taxodione}Fe(CO)$_3$], (^{13}C),[689] [(η^4-5-R-cyclohexa-1,3-diene)Fe(η^6-C$_6$H$_6$)], (^{13}C),[690] [(η^4-diene)Fe(CO)$_3$], (^{13}C),[691] (28), (^{13}C),[692] [Fe$_2$(R^{1}C=CR^2-CR^3=CR^4)(CO)$_4$(R^{5}N=CHCH=NR^5)], (^{13}C),[693] [Ru(1,4-6-η^4-C$_7$H$_8$)(R^{1}N=CR^{2}CR^2=NR^1)-Cl(NR$^3_2$)], (^{13}C),[694] [(η^5-C$_5$Me$_5$)Ru(μ-η^4-C$_4$H$_4$)(PMe$_3$)Ru(η^5-C$_5$Me$_5$)], (^{13}C),[695] [(η^5-1-R-C$_5$H$_6$)Fe(CO)$_3$]$^+$, (^{13}C),[696] [(η^5-MeBCEtCEtBMeC=CMe$_2$)Fe(CO)$_3$], (^{11}B, ^{13}C),[697] [(η^5-C$_4$H$_4$-

S.M. Owen, *Organometallics*, 1989, **8**, 2689.

670 F. Muller, D.I.P. Dijkhuis, G. van Koten, K. Vrieze, D. Heijdenrijk, M.A. Rotteveel, C.H. Stam, and M.C. Zoutberg, *Organometallics*, 1989, **8**, 992.

671 W.D. Harman, J.C. Dobson, and H. Taube, *J. Am. Chem. Soc.*, 1989, **111**, 3061.

672 M.J. Burns, G.-Y. Kiel, F. Seils, J. Takats, and J. Washington, *J. Am. Chem. Soc.*, 1989, **111**, 6850.

673 M.J. Begley, S.G. Puntambekar, and A.H. Wright, *J. Organomet. Chem.*, 1989, **362**, C11.

674 G.M. Williams and D.E. Rudisill, *Inorg. Chem.*, 1989, **28**, 797.

675 M.-H. Cheng, Y.-J. Wu, S.-L. Wang, and R.-s. Liu, *J. Organomet. Chem.*, 1989, **373**, 119.

676 M. Pizzotti, S. Cenini, C. Crotti, and F. Demartin, *J. Organomet. Chem.*, 1989, **375**, 123.

677 J.W. Ziller, D.K. Bower, D.M. Dalton, J.B. Keister, and M.R. Churchill, *Organometallics*, 1989, **8**, 492.

678 M.J. Fildes, S.A.R. Knox, A.G. Orpen, M.L. Turner, and M.I. Yates, *J. Chem. Soc., Chem. Commun.*, 1989, 1681.

679 J. Rodriguez, P. Brun, J.P. Zahra, and B. Waegell, *Magn. Reson. Chem.*, 1989, **27**, 96.

680 C.M. Adams, A. Hafner, M. Koller, A. Marcuzzi, R. Prewo, I. Solana, B. Vincent, and W. von Philipsborn, *Helv. Chim. Acta*, 1989, **72**, 1658.

681 M. Hastings and G.R. Stephenson, *J. Organomet. Chem.*, 1989, **375**, C27.

682 P. Pinsard, J.-P. Lellouche, J.-P. Beaucourt, L. Toupet, L. Schio, and R. Grée, *J. Organomet. Chem.*, 1989, **371**, 219.

683 D. Kappes, H. Gerlach, P. Zbinden, M. Dobler, W.A. König, R. Krebber, and G. Wenz, *Angew. Chem., Int. Ed. Engl.*, 1989, **28**, 1657.

684 R. Aumann and B. Trentmann, *J. Organomet. Chem.*, 1989, **378**, 171.

685 J. Rodriguez, P. Brun, and B. Waegell, *J. Organomet. Chem.*, 1989, **359**, 343.

686 H. Schaeufele, D. Hu, H. Pritzkow, and U. Zenneck, *Organometallics*, 1989, **8**, 396.

687 F. Carre, R.J.P. Corriu, C. Guerin, B.J.L. Horner, and W.W.C.W.C. Man, *Organometallics*, 1989, **8**, 313.

688 C. Zou, K.J. Ahmed, and M.S. Wrighton, *J. Am. Chem. Soc.*, 1989, **111**, 1133.

689 E. Haslinger and G. Michl, *Liebigs Ann. Chem.*, 1989, 677.

690 D. Mandon and D. Astruc, *Organometallics*, 1989, **8**, 2372.

691 L.A. Paquette, G.A. O'Doherty, B.L. Miller, R.D. Rogers, A.L. Rheingold, and S.L. Geib, *Organometallics*, 1989, **8**, 2167.

692 D. Seyferth, L.L. Anderson, and W.M. Davis, *Organometallics*, 1989, **8**, 1371.

693 F. Muller, I.M. Han, G. van Koten, K. Vrieze, D. Heijdenrijk, R.L. de Jong, and M.C. Zoutberg, *Inorg. Chim. Acta*, 1989, **158**, 81.

694 H. tom Dieck, I. Kleinwächter, E.T.K. Haupt, and D. Bolze-Kuhrt, *J. Organomet. Chem.*, 1989, **365**, 351.

695 H. Omori, H. Suzuki, and Y. Moro-oka, *Organometallics*, 1989, **8**, 1576.

696 W.A. Donaldson and M. Ramaswamy, *Tetrahedron Lett.*, 1989, **30**, 1339.

697 G. Brodt and W. Siebert, *Chem. Ber.*, 1989, **122**, 633.

3Ph)$Fe(CO)_{3-n}L_n$], (^{11}B, ^{13}C),[698] [(η^5-C_5H_5)$M(CO)_2(P{=}NC_6H_2Bu^t_3$-2,4,6)], (M = Fe, Ru; ^{13}C),[699] [(η^5-C_5Me_5)$(OC)_2FeP(OCH{=}CHCH_2){=}PC_6H_2Bu^t_3$-2,4,6], (M = Fe, Ru; ^{13}C),[700] [(η^5-C_5Me_5)$(OC)_2FeAs\{Cr(CO)_5\}AsC_6H_2Bu^t_3$-2,4,6], ($^{13}C$),[701] [($\eta^5$-$C_5H_5$)$(OC)_2M$-$PAs(C_6H_2Bu^t_3$-2,4,6)$PM(CO)_2$($\eta^5$-$C_5H_5$)], (M = Fe, Ru; ^{13}C),[702] [(η^5-C_5Me_5)$Fe(CO)_2AsSP$-$C_6H_2Bu^t_3$-2,4,6], (^{13}C),[703] [(η^5-C_5H_5)$Fe(CO)\{\mu$-$P(OR)_2\}]_2$, (^{13}C),[704] [(η^5-C_5Me_5)$Fe(CO)(PPh_3)$-$AsSPC_6H_2Bu^t_3$-2,4,6], (^{13}C),[705] [{(η^5-C_5H_5)$(Et_3P)(OC)Fe\}_2Te_n$], (^{125}Te),[706] [(η^5-C_5H_5)Fe-(dppm)L]$^+$, (^{13}C),[707] [{η^5-$C_5H_2(SiMe_3)_3\}FeL_3]^+$, ($^{13}C$),[708] [{$\eta^5$-$C_5H_2(SiMe_3)_3\}FeX\{P(OMe)_3\}_2$], ($^{13}C$),[709] [($\eta^5$-$C_5H_5$)Fe($\eta^5$-$C_5H_4R$)], ($^{13}C$),[710] [($\eta^5$-$C_5H_5$)Fe($\eta^5$-$C_5H_4$)$As(N{=}S{=}N)_2As$($\eta^5$-$C_5H_4$)Fe($\eta^5$-$C_5H_5$)], ($^{13}C$),[711] [($\eta^5$-$C_5Me_5$)$Fe(S_2CNMe_2)_2]^+$, ($^{13}C$),[712] [($\eta^5$-$C_5H_5$)Fe-$(S_3C_6H_{12})]^+$, ($^{13}C$),[713] [($\eta^5$-$C_5H_5$)Fe($\eta^5$-$C_6Me_6CH_2Cl$)], ($^{13}C$),[714] [($\eta^5$-$C_5H_5$)Fe{$\eta^5$-$C_6H_5(SO_2$-$C_6H_4Me$-4)CN}], ($^{13}C$),[715] [($\eta^6$,$\eta^6$-arene){Fe($\eta^5$-$C_5H_5$)$_2$}$_2$]$^{2+}$, ($^{13}C$),[716] [{($\eta^5$-$C_5H_5$)Fe($\eta^5$-$C_5H_4$)$CHOHCH{=}CHC_5H_{11}\}Co_2(CO)_6$], ($^{13}C$),[717] [($\eta^5$-$C_5H_5$)Fe($\eta^5$-$C_5H_4C_5H_4$-$\eta^5$)Fe($\eta^6$-$C_6Me_6$)]$^+$, ($^{13}C$),[718] [{($\eta^5$-$C_5H_5$)Fe($\eta^5$-$C_5H_4$)}PMe]$^+$, ($^{13}C$),[719] [($\eta^5$-$C_5H_5$)Fe($\eta^5$-$C_5H_4EH$)], (E = S, Se, Te; ^{13}C),[720] [{(η^5-C_5H_5)Fe(η^5-C_5H_3-1-CH_2NMe_2-2-)}M], (M = Zn, Cd, Hg; ^{13}C),[721] [(η^5-C_5H_5)Fe(η^5-1,3-Ph_2-2-HOC_5H_2)], (^{13}C),[722] [Fe(η^5-$C_5H_4S_3C_5H_3CH_2NMe_2$-η^5)], (^{13}C),[723] [(η^5-C_5H_5)Fe(η^5-C_4Me_4NH)][BF_4], (^{11}B, ^{13}C),[724] [Fe(η^5-$C_4Me_4NBH_3$)$_2$], (^{11}B, ^{13}C),[725] [{Fe(η^5-

698 G.E. Herberich, B. Hessner, and D.P.J. Köffer, *J. Organomet. Chem.*, 1989, **362**, 243.
699 E. Niecke, J. Hein, and M. Nieger, *Organometallics*, 1989, **8**, 2290.
700 L. Weber, M. Frebel, and R. Boese, *Organometallics*, 1989, **8**, 1718.
701 L. Weber, D. Bungardt, A. Müller, and H. Bögge, *Organometallics*, 1989, **8**, 2800.
702 L. Weber, D. Bungardt, and R. Boese, *Z. Anorg. Allg. Chem.*, 1989, **578**, 205.
703 L. Weber and U. Sonnenberg, *Chem. Ber.*, 1989, **122**, 1809.
704 J.T. Spencer, J.A. Spencer, R.A. Jacobson, and J.G. Verkade, *New J. Chem.*, 1989, **13**, 275.
705 L. Weber and U. Sonnenberg, *Chem. Ber.*, 1989, **122**, 1809.
706 M.L. Steigerwald, *Chem. Mater.*, 1989, **1**, 52 (*Chem. Abstr.*, 1989, **110**, 68 438).
707 J. Ruiz, M.-T. Garland, E. Román, and D. Astruc, *J. Organomet. Chem.*, 1989, **377**, 309.
708 J. Okuda, *J. Organomet. Chem.*, 1989, **375**, C13.
709 J. Okuda, *Chem. Ber.*, 1989, **122**, 1259.
710 F. Siglmüller, R. Herrmann, and I. Ugi, *Liebigs Ann.*, 1989, 623.
711 C. Spang, F.T. Edelmann, M. Noltemeyer, and H.W. Roesky, *Chem. Ber.*, 1989, **122**, 1247.
712 M.-H. Desbois and D. Astruc, *Angew. Chem., Int. Ed. Engl.*, 1989, **28**, 460.
713 A.J. Blake, R.D. Crofts, G. Reid, and M. Schröder, *J. Organomet. Chem.*, 1989, **359**, 371.
714 G.E. Herberich and W. Klein, *Chem. Ber.*, 1989, **122**, 2125.
715 R.G. Sutherland, C.-H. Zhang, A. Piórko, and C.C. Lee, *Can. J. Chem.*, 1989, **67**, 137.
716 L.S. Shilovtseva, N.A. Vol'kenau, A.S. Peregudov, P.V. Petrovskii, and D.N. Kravtsov, *Metallorg. Khim.*, 1988, **1**, 441 (*Chem. Abstr.*, 1989, **110**, 213 021).
717 L.L. Troitskaya, V.I. Sokolov, V.I. Bakhmutov, O.A. Reutov, M. Gruselle, C. Cordier, and G. Jaouen, *J. Organomet. Chem.*, 1989, **364**, 195.
718 M.-H. Desbois, D. Astruc, J. Guillin, and F. Varret, *Organometallics*, 1989, **8**, 1848; H. Bönnemann, R. Goddard, J. Grub, R. Mynott, E. Raabe, and S. Wendel, *Organometallics*, 1989, **8**, 1941.
719 H. Schmidbaur, J. Jeong, A. Schier, W. Graf, D.L. Wilkinson, G. Müller, and C. Krüger, *New J. Chem.*, 1989, **13**, 341.
720 M. Herberhold, P. Leitner, C. Dörnhöfer, and J. Ott-Lastic, *J. Organomet. Chem.*, 1989, **377**, 281.
721 C. Krüger, K.-H. Thiele, M. Dargatz, and T. Bartik, *J. Organomet. Chem.*, 1989, **362**, 147.
722 W. Yongskulrote, J.M. Bramlett, C.A. Mike, B. Durham, and N.T. Allison, *Organometallics*, 1989, **8**, 556.
723 I.R. Butler and W.R. Cullen, *Can. J. Chem.*, 1989, **67**, 1851.
724 N. Kuhn, M. Schulten, E. Zauder, N. Augart, and R. Boese, *Chem. Ber.*, 1989, **122**, 1891.
725 N. Kuhn, E.-M. Horn, R. Boese, and N. Augart, *Angew. Chem., Int. Ed. Engl.*, 1989, **28**, 342.

$C_5Me_4N)_2\}_2Ag_2]^{2+}$, ($^{13}C$),[726] [($\eta^5$-$C_5H_5$)Fe($\eta^5$-$C_5H_4COR$)], ($^{13}C$),[727] [$CH_2\{CH_2(\eta^5$-$C_5Me_4)\}_2$-Fe], ($^{13}C$),[728] [{Fe($\eta^5$-$C_5H_4PPh_2)_2\}_2Pt_2(\mu$-$H)_2(\mu$-$CO)_2]^{2+}$, ($^{13}C$),[729] [{Fe($\eta^5$-$C_5H_4SMe)_2$}Pd-($PPh_3$)]$^{2+}$, ($^{13}C$),[730] [{Fe($\eta^5$-$C_5H_4SCH_2$)E}M]$^{2+}$, (M = Pd, Pt; E = S, Se; ^{13}C),[731] [Fe(η^5-C_5H_4-ER)(η^5-C_5H_3-1-ER-2-CHMeNMe_2)], (E = S, Se; ^{13}C),[732] [(η^5-C_5H_5)Fe(η^6-C_8H_8)]$^+$, (^{13}C),[733] [(η^6-$C_6H_3Me_3$-1,3,5)Fe(η^5-$C_6H_3Me_3$-1,3,5)]$^+$, (^{13}C),[734] [(η^6-arene)Fe(η^5-$C_6H_{5-n}Me_nR^1R^2$)]$^+$, (^{13}C),[735] [(η^6-arene)Fe(η^5-C_5H_5)]$^+$, (^{13}C),[736] [(η^6-C_6Et_6)Fe(η^5-C_5H_5)]$^+$, (^{13}C),[737] [{η^6-$C_6(CH_2$-R$)_6$}Fe(η^5-C_5H_5)]$^+$, (^{13}C),[738] [(η^6-C_6Et_5H)Fe(η^5-C_5Me_5)]$^+$, (^{13}C),[739] [(η^5-C_5H_7)Ru(PMe_3)-(PPh_3)(CO)]$^+$, (^{13}C),[740] (29), (^{13}C),[741] [(η^5-C_5H_5)Ru($NC_5H_5)_3$]$^+$, (^{13}C),[742] [(η^5-C_5H_5)Ru(PPh_3)-

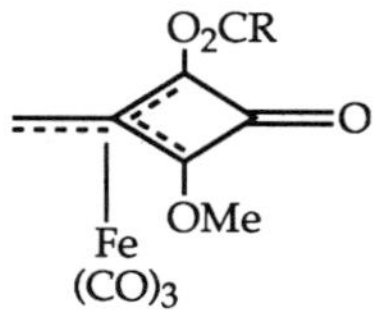

(28)

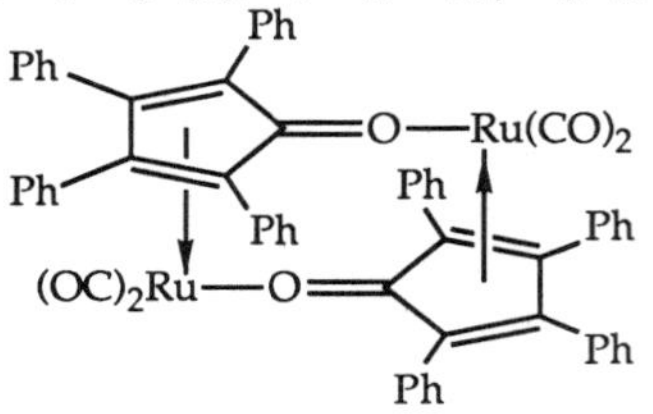

(29)

(CNBu^t)(ICH_2SiMe_3)], (^{13}C),[743] [(η^5-C_5Me_5)Ru(OR)]$_2$, (^{13}C),[744] [(η^5-C_5Me_5)Ru(μ-X)X]$_2$, (^{13}C),[745] [(η^5-C_5H_5)Ru(η^5-C_5R_5)], (^{13}C),[746] [(η^5-C_5Me_5)Ru{η^5-$(CF_3)_4C_5H$}], (^{13}C),[747] [(η^6-arene)Ru(η^5-C_5H_5)], (^{13}C),[748] [(η^6-estrone-3-methyl ester)Ru(η^5-C_5H_5)]$^+$, (^{13}C),[749] [(η^6-

726 N. Kuhn, E.-M. Horn, R. Boese, and D. Bläser, *Chem. Ber.*, 1989, **122**, 2275.

727 P. Carty, J. Grant, and A. Simpson, *Appl. Organomet. Chem.*, 1988, **2**, 277 (*Chem. Abstr.*, 1989, **111**, 134 407).

728 H. Ogino, H. Tobita, H. Habazaki, and M. Shimoi, *J. Chem. Soc., Chem. Commun.*, 1989, 828.

729 A.L. Bandini, G. Banditelli, M.A. Cinellu, G. Sanna, G. Minghetti, F. Demartin, and M. Manassero, *Inorg. Chem.*, 1989, **28**, 404.

730 M. Sato, M. Sekino, M. Katada, and S. Akabori, *J. Organomet. Chem.*, 1989, **377**, 327.

731 M. Sato, H. Asano, K. Suzuki, M. Katada, and S. Akabori, *Bull. Chem. Soc. Jpn.*, 1989, **62**, 3828.

732 C.-K. Lai, A.A. Naiini, and C.H. Brubaker, jun., *Inorg. Chim. Acta*, 1989, **164**, 205; A.A. Naiini, C.-K. Lai, and C.H. Brubaker, jun., *Inorg. Chim. Acta*, 1989, **160**, 241.

733 J. Heck and W. Massa, *J. Organomet. Chem.*, 1989, **376**, C15.

734 D. Mandon and D. Astruc, *J. Organomet. Chem.*, 1989, **369**, 383.

735 M.D. Clerk, K.C. Sturge, M.J. Zaworotko, and P.S. White, *J. Organomet. Chem.*, 1989, **368**, C33.

736 C.C. Lee, C.H. Zhang, A.S. Abd-El-Aziz, A. Piórko, and R.G. Sutherland, *J. Organomet. Chem.*, 1989, **364**, 217; M. Lacoste, H. Rabaa, D. Astruc, A. Le Beuze, J.-Y. Saillard, G. Précigoux, C. Courseille, N. Ardoin, and W. Bowyer, *Organometallics*, 1989, **8**, 2233; A.S. Abd-El-Azis, A. Piórko, C.C. Lee, and R.G. Sutherland, *Can. J. Chem.*, 1989, **67**, 1618.

737 R.H. Dubois, M.J. Zaworotko, and P.S. White, *J. Organomet. Chem.*, 1989, **362**, 155.

738 F. Moulines and D. Astruc, *J. Chem. Soc., Chem. Commun.*, 1989, 614.

739 J.-R. Hamon, J.-Y. Saillard, L. Toupet, and D. Astruc, *J. Chem. Soc., Chem. Commun.*, 1989, 1662; J.-R. Hamon and D. Astruc, *Organometallics*, 1989, **8**, 2243.

740 J.R. Bleeke and D.J. Rauscher, *J. Am. Chem. Soc.*, 1989, **111**, 8973.

741 M.J. Mays, M.J. Morris, P.R. Raithby, Y. Shvo, and D. Czarkie, *Organometallics*, 1989, **8**, 1162.

742 R.H. Fish, H.-S. Kim, and R.H. Fong, *Organometallics*, 1989, **8**, 1375.

743 F.M. Conroy-Lewis, A.D. Redhouse, and S.J. Simpson, *J. Organomet. Chem.*, 1989, **366**, 357.

744 S.D. Loren, B.K. Campion, R.H. Heyn, T.D. Tilley, B.E. Bursten, and K.W. Luth, *J. Am. Chem. Soc.*, 1989, **111**, 4712.

745 U. Koelle and J. Kossakowski, *J. Organomet. Chem.*, 1989, **362**, 383.

746 R.P. Hughes and D.J. Robinson, *Organometallics*, 1989, **8**, 1015.

747 M.J. Burk, A.J. Arduengo, tert., J.C. Calabrese, and R.L. Harlow, *J. Am. Chem. Soc.*, 1989, **111**, 8938.

748 B. Chaudret, X. He, and Y. Huang, *J. Chem. Soc., Chem. Commun.*, 1989, 1844.

749 R.M. Moriarty, Y.-Y. Ku, U.S. Gil, R. Gilardi, R.E. Perrier, and M.J. McGlinchey, *Organometallics*,

coumarin laser dye)Ru(η^5-C_5H_5)]$^+$, (^{13}C),[750] [(η^6-3-AcNHCH$_2$CH$_2$-indole)Ru(η^5-C_5H_5)]$^+$, (^{13}C),[751] [(η^5-C_5Me_5)Os(η^6-C_5Me_4CH$_2$PMe$_2$)Ph]$^+$, (^{13}C),[752] [(η^6-1,3,5-$C_6H_3Me_3$)Ru(η^4-$C_{10}H_8$)], (^{13}C),[753] [(μ-η^5-C_4H_4BPh){Co(η^5-C_5H_5)}{Ru(η^6-C_6Me_6)}]$^+$, (^{11}B),[754] and [(η^6-*p*-cymene)$_3$Ru$_3$S$_2$], (^{13}C).[755]

[Fe(CNCF$_3$)$_5$] is fluxional even at -100 °C.[756] The conformational preferences in [(η^5-C_5H_4Bu^t)Fe(CO)(PPh$_3$)I] have been investigated by ^{1}H n.O.e. measurements and coupling constants. The ^{13}C n.m.r. spectra were also measured.[757] Isotropic ^{1}H and ^{13}C chemical shifts of [Fe(CN)$_4$(en)]$^-$ and [Fe(CN)$_4$(MeNHCH$_2$CH$_2$NHMe)]$^-$ have been analysed.[758] ^{17}O n.m.r. spectroscopy has been used to investigate the binding of C^{17}O to haemoproteins. The ^{17}O nuclear quadrupole coupling constants were determined. ^{13}C n.m.r. spectra were also measured.[759] N.m.r. data have also been reported for [(OC)$_4$Fe-(μ-PCy$_2$)Ir(η^4-C_8H_{12})], (^{13}C),[760] [2-pyridylCH$_2$PPh$_2$Fe(CO)$_4$], (^{13}C),[761] [RSFe(CO)$_4$], (^{13}C),[762] [Fe(CO)$_3$(PPh$_3$)$_2$], (^{13}C),[763] [Fe(CO)$_2$(O$_2$CCH$_2$PPh$_2$)$_2$], (^{13}C),[764] [(OC)(ON)$_2$FeAsBu^t_n-(MMe$_3$)$_{3-n}$], (M = Si, Ge, Sn; ^{13}C, ^{29}Si, ^{59}Co, ^{119}Sn),[765] [(OC)$_4$Fe(μ-PCy$_2$)Rh(η^4-C_8H_{12})], (^{13}C),[766] [(μ-N$_2$C$_4$Bu$^t_3$SiMe$_3$)Fe$_2$(CO)$_6$], (^{13}C),[767] [(Pr^{i}N=CHCH=NPr^i)Fe{μ-Et$_2$NCMe-C(NEt)CMe}(μ-CO)Fe(CO)$_3$], (^{13}C),[768] [Fe$_2$(CO)$_6${Pr^{i}N=CHCHNPr^{i}C(O)CH=CCO$_2$Me}], (^{13}C),[769] [(μ-HNN=CR^{1}CHR^{2}S)Fe$_2$(CO)$_6$], (^{13}C),[770] [Fe$_2$(CO)$_6${μ-PPh$_2$CH=CBu^{t}C(O)}]$^-$, (^{13}C),[771] [(μ-4-MeC$_6$H$_4$C≡CS)(μ-RS)Fe$_2$(CO)$_6$], (^{13}C),[772] [{μ-(EtO$_2$C)(Me$_2$N)C=NCS}(μ-PhCH$_2$S)Fe$_2$(CO)$_6$], (^{13}C),[773] [{μ-OC(CH$_2$PNPr$^i_2$)$_2$}Fe$_2$(CO)$_6$], (^{13}C),[774] [{μ-η^3-SC(SCH=CH$_2$)(μ-

1989, **8**, 960.
750 R.S. Koefod and K.R. Mann, *Inorg. Chem.*, 1989, **28**, 2285.
751 R.M. Moriarty, Y.Y. Ku, and U.S. Gill, *J. Organomet. Chem.*, 1989, **362**, 187.
752 M.I. Rybinskaya, A.Z. Kreindlin, Yu.T. Struchkov, and A.I. Yanovsky, *J. Organomet. Chem.*, 1989, **359**, 233.
753 G.P. Zol'nikova, A.S. Peregudov, Yu.F. Oprunenko, G.M. Babakhina, I.I. Kritskaya, and D.N. Kravtsov, *Metalloorg. Khim.*, 1988, **1**, 79 (*Chem. Abstr.*, 1989, **110**, 231 843).
754 G.E. Herberich, B.J. Dunne, and B. Heßner, *Angew. Chem., Int. Ed. Engl.*, 1989, **28**, 737.
755 J.R. Lockemeyer, T.B. Rauchfuss, and A.L. Rheingold, *J. Am. Chem. Soc.*, 1989, **111**, 5733.
756 D. Lentz, *J. Organomet. Chem.*, 1989, **377**, 305.
757 K.E. Du Plooy, C.F. Marais, L. Carlton, R. Hunter, J.C.A. Boeyens, and N.J. Coville, *Inorg. Chem.*, 1989, **28**, 3855.
758 Y. Kuroda, M. Goto, and T. Sakai, *Bull. Chem. Soc. Jpn.*, 1989, **62**, 3614.
759 H.C. Lee and E. Oldfield, *J. Am. Chem. Soc.*, 1989, **111**, 1584.
760 H.A. Jenkins and S.J. Loeb, *Can. J. Chem.*, 1989, **67**, 1230.
761 H. tom Dieck and G. Hahn, *Z. Anorg. Allg. Chem.*, 1989, **577**, 74.
762 W.-F. Liaw, C. Kim, M.Y. Darensbourg, and A.L. Rheingold, *J. Am. Chem. Soc.*, 1989, **111**, 3591.
763 J.J. Brunet, F.B. Kindela, and D. Neibecker, *J. Organomet. Chem.*, 1989, **368**, 209.
764 J. Takács, L. Markó, P. Kiprof, E. Herdtweck, and W.A. Herrmann, *Polyhedron*, 1989, **8**, 1503.
765 H. Schumann and K.-H. Köhricht, *J. Organomet. Chem.*, 1989, **373**, 307.
766 H.A. Jenkins, S.J. Loeb, and D.W. Stephan, *Inorg. Chem.*, 1989, **28**, 1998.
767 G. Maier and D. Born, *Angew. Chem., Int. Ed. Engl.*, 1989, **28**, 1050.
768 F. Muller, G. Van Koten, M.J.A. Kraakman, K. Vrieze, D. Heijdenrijk, and M.C. Zoutberg, *Organometallics*, 1989, **8**, 1331.
769 F. Muller, G. Van Koten, K. Vrieze, and D. Heijdenrijk, *Organometallics*, 1989, **8**, 33.
770 A.J. Mayr, K.H. Pannell, B. Carrasco-Flores, and F. Cervantes-Lee, *Organometallics*, 1989, **8**, 2961.
771 J. Suades, F. Dahan, and R. Mathieu, *Organometallics*, 1989, **8**, 842.
772 L.Z. Wang, and J. Wang, *Gaodeng Xuexiao Huaxue Xuebao*, 1988, **9**, 915 (*Chem. Abstr.*, 1989, **110**, 127 512).
773 M. Bouzid, J.P. Pradère, P. Palvadeau, J.P. Vénien, and L. Toupet, *J. Organomet. Chem.*, 1989, **369**, 205.
774 L. Weber, E. Luecke, and R. Boese, *Chem. Ber.*, 1989, **122**, 809.

SMe)}$Fe_2(CO)_5L$], (^{13}C),[775] [(μ-η^2-ROCS)$Fe_2L^1L^2(CO)_4$(μ-SMe)], (^{13}C),[776] [{η^4-Fe(CO)$_3$EFe(CO)$_3$PR}Fe(CO)$_3$], (E = Se, Te; ^{13}C, ^{77}Se, ^{125}Te),[777] [$Fe_3(CO)_9(\mu_3\text{-}CCl)_2$], ($^{13}C$),[778] [(μ-$R^1PCR^2$=$CR^2PR^1$)$Fe_3(CO)_9$], ($^{13}C$),[779] [$FeCo_2(CO)_9${$\mu_3$-$CPMe_2CH_2CH_2PMe_2$-$Fe(CO)_4$}], ($^{13}C$),[780] [{μ-(η^5-$C_5H_5$)$CoS_2$}$Fe_2(CO)_6$], ($^{13}C$),[781] [$Fe_4(CO)_{12}Au_2(PMePh_2)_2BH$], ($^{11}B$),[782] [$Fe_2Ru_3(CO)_{14}(\mu_4\text{-}O)]^{2-}$, ($^{17}O$),[783] *trans-trans-trans*-[$RuCl_2(CO)_2(PR_3)_2$], (^{13}C),[784] [$Cl_2(OC)_2Ru${$Ph_2P(CH_2CH_2O)_2Me$}$_2$], (^{13}C),[785] [Ru(2-S-3,5-$Bu^tC_6H_2NHCH_2CH_2NHC_6H_2$-3,5-$Bu^t_2$-2-S)(CO)($PPh_3$)], ($^{13}C$),[786] [(μ-RNCH-2-$C_5H_4N$)$Ru_2(CO)_6$], ($^{13}C$),[787] [$Ru_2(CO)_5$(μ-$PPh_2$)(μ-X)($PPh_3$)], ($^{13}C$),[788] [$Ru_2(dmpm)_2$-$(CO)_5$], ($^{13}C$),[789] [$Ru_2(CO)_4$(η^2-$O_2CR$)$_2$(η^2-diop)], ($^{13}C$),[790] [(3,5-$R^1_2C_3H_2N$)$_2Ru_2(CO)_4$(μ-$O_2CR^2$)$_2$], ($^{13}C$),[791] [$Ru_3${$\mu_2$-NPhC(O)R}$(CO)_{10}]^-$, ($^{13}C$),[792] [$Ru_4(CO)_8$(glutarate)$_2${(-)-diop}$_3$], ($^{13}C$),[793] [$Ru_{18}(C)_2(CO)_{42}Hg_3]^{2-}$, ($^{13}C$),[794] and [$Os_2(O_2CMe)_2(CO)_4(dppm)_2$], ($^{13}C$).[795]

^1H n.m.r. spectroscopy has been used to examine the structure and mobility of cytochrome b_5 in solution.[796] The binding of O_2 to haemoprotein models has been investigated by ^{17}O n.m.r. spectroscopy.[797] Deuteriation has been used to facilitate the analysis SMe)}$Fe_2(CO)_5L$], (^{13}C),[798] [(μ-η^2-ROCS)$Fe_2L^1L^2(CO)_4$(μ-SMe)], (^{13}C),[799] [{η^4-

775 B. Dadamoussa, A. Darchen, P. L'Haridon, C. Larpent, H. Patin, and J.-Y. Thepot, *Organometallics*, 1989, **8**, 564.
776 H. Patin, A. Le Rouzic, E.K. Lhadi, A. Darchen, A. Mousser, and D. Grandjean, *J. Organomet. Chem.*, 1989, **375**, 101.
777 D. Buchholz, G. Huttner, L. Zsolnai, and W. Imhof, *J. Organomet. Chem.*, 1989, **377**, 25.
778 D. Lentz and H. Michael, *Inorg. Chem.*, 1989, **28**, 3397.
779 Th. Fässler, D. Buchholz, G. Huttner, and L. Zsolnai, *J. Organomet. Chem.*, 1989, **369**, 297.
780 S. Ching, M.P. Jensen, M. Sabat, and D.F. Shriver, *Organometallics*, 1989, **8**, 1058.
781 M. Cowie, R.L. DeKock, T.R. Wagenmaker, D. Seyferth, R.S. Henderson, and M.K. Gallagher, *Organometallics*, 1989, **8**, 119.
782 C.E. Housecroft, M.S. Shongwe, and A.L. Rheingold, *Organometallics*, 1989, **8**, 2651.
783 C.K. Schauer, E.J. Voss, M. Sabat, and D.F. Shriver, *J. Am. Chem. Soc.*, 1989, **111**, 7662.
784 R. Vac, J.H. Nelson, E.B. Milosavljevic, and L. Solujic, *Inorg. Chem.*, 1989, **28**, 3831; R. Vac, J.H. Nelson, E.B. Milosavljevic, L. Solujic, and J. Fischer, *Inorg. Chem.*, 1989, **28**, 4132.
785 V.V.S. Reddy, J.E. Whitten, K.A. Redmill, A. Varshney, and G.M. Gray, *J. Organomet. Chem.*, 1989, **372**, 207.
786 D. Sellmann, O. Käppler, and F. Knoch, *J. Organomet. Chem.*, 1989, **367**, 161.
787 F. Muller, G. van Koten, L.H. Polm, K. Vrieze, M.C. Zoutberg, D. Heijenrijk, E. Kragten, and C.H. Stam, *Organometallics*, 1989, **8**, 1340.
788 H. Jungbluth, H. Stöckli-Evans, and G. Süss-Fink, *J. Organomet. Chem.*, 1989, **377**, 339.
789 K.A. Johnson and W.L. Gladfelter, *Organometallics*, 1989, **8**, 2866.
790 G.F. Schmidt and G. Süss-Fink, *J. Organomet. Chem.*, 1989, **362**, 179.
791 F. Neumann and G. Süss-Fink, *J. Organomet. Chem.*, 1989, **367**, 175.
792 S.-H. Han, J.-S. Song, P.D. Macklin, S.T. Nguyen, G.L. Geoffroy, and A.L. Rheingold, *Organometallics*, 1989, **8**, 2127.
793 M. Bianchi, G. Menchi, P. Frediani, F. Piacenti, A. Scrivanti, and U. Matteoli, *J. Mol. Catal.*, 1989, **50**, 277.
794 P.J. Bailey, B.F.G. Johnson, J. Lewis, M. McPartlin, and H.R. Powell, *J. Chem. Soc., Chem. Commun.*, 1989, 1513.
795 E. Singleton, M.O. Albers, and M.M. de V. Steyn, *J. Chem. Soc., Dalton Trans.*, 1989, 2303.
796 N.C. Veitch, D.W. Concar, R.J.P. Williams, and D. Whitford, *FEBS Lett.*, 1988, **238**, 49 (*Chem. Abstr.*, 1989, **110**, 20 038).
797 I.P. Gerothanassis, M. Momenteau, and B. Lock, *J. Am. Chem. Soc.*, 1989, **111**, 7006.
798 B. Dadamoussa, A. Darchen, P. L'Haridon, C. Larpent, H. Patin, and J.-Y. Thepot, *Organometallics*, 1989, **8**, 564.
799 H. Patin, A. Le Rouzic, E.K. Lhadi, A. Darchen, A. Mousser, and D. Grandjean, *J. Organomet. Chem.*,

$Fe(CO)_3EFe(CO)_3PR\}Fe(CO)_3]$, (E = Se, Te; ^{13}C, ^{77}Se, ^{125}Te),[800] $[Fe_3(CO)_9(\mu_3\text{-}CCl)_2]$, ($^{13}C$),[801] $[(\mu\text{-}R^1PCR^2{=}CR^2PR^1)Fe_3(CO)_9]$, ($^{13}C$),[802] $[FeCo_2(CO)_9\{\mu_3\text{-}CPMe_2CH_2CH_2PMe_2\text{-}Fe(CO)_4\}]$, ($^{13}C$),[803] $[\{\mu\text{-}(\eta^5\text{-}C_5H_5)CoS_2\}Fe_2(CO)_6]$, ($^{13}C$),[804] $[Fe_4(CO)_{12}Au_2(PMePh_2)_2BH]$, ($^{11}B$),[805] $[Fe_2Ru_3(CO)_{14}(\mu_4\text{-}O)]^{2-}$, ($^{17}O$),[806] *trans-trans-trans*-$[RuCl_2(CO)_2(PR_3)_2]$, (^{13}C),[807] $[Cl_2(OC)_2Ru\{Ph_2P(CH_2CH_2O)_2Me\}_2]$, ($^{13}C$),[808] $[Ru(2\text{-}S\text{-}3,5\text{-}Bu^tC_6H_2NHCH_2CH_2NHC_6H_2\text{-}3,5\text{-}Bu^t{}_2\text{-}2\text{-}S)(CO)(PPh_3)]$, ($^{13}C$),[809] $[(\mu\text{-}RNCH\text{-}2\text{-}C_5H_4N)Ru_2(CO)_6]$, ($^{13}C$),[810] $[Ru_2(CO)_5(\mu\text{-}PPh_2)(\mu\text{-}X)(PPh_3)]$, ($^{13}C$),[811] $[Ru_2(dmpm)_2\text{-}(CO)_5]$, ($^{13}C$),[812] $[Ru_2(CO)_4(\eta^2\text{-}O_2CR)_2(\eta^2\text{-}diop)]$, ($^{13}C$),[813] $[(3,5\text{-}R^1{}_2C_3H_2N)_2Ru_2(CO)_4(\mu\text{-}O_2CR^2)_2]$, ($^{13}C$),[814] $[Ru_3\{\mu_2\text{-}NPhC(O)R\}(CO)_{10}]^-$, ($^{13}C$),[815] $[Ru_4(CO)_8(glutarate)_2\{(\text{-})\text{-}diop\}_3]$, ($^{13}C$),[816] $[Ru_{18}(C)_2(CO)_{42}Hg_3]^{2-}$, ($^{13}C$),[817] and $[Os_2(O_2CMe)_2(CO)_4(dppm)_2]$, ($^{13}C$).[818]

^1H n.m.r. spectroscopy has been used to examine the structure and mobility of cytochrome b_5 in solution.[819] The binding of O_2 to haemoprotein models has been investigated by ^{17}O n.m.r. spectroscopy.[820] Deuteriation has been used to facilitate the analysis of the ^1H n.m.r. spectrum of $[Ru(2,2\text{-}bipy\text{-}d_8)_2L]^{2+}$, L = (30).[821] The chelation of 5′-dGMP to *trans*-$[RuCl_2(DMSO)_4]$ has been investigated by ^1H n.m.r. spectroscopy.[822] N.m.r. data have also been reported for $[Fe(CN)_4\{(1R,2S)\text{-}cis\text{-}cyclohexanediamine\}]$, ($^{13}C$),[823] $[Fe_2$

1989, **375**, 101.
800 D. Buchholz, G. Huttner, L. Zsolnai, and W. Imhof, *J. Organomet. Chem.*, 1989, **377**, 25.
801 D. Lentz and H. Michael, *Inorg. Chem.*, 1989, **28**, 3397.
802 Th. Fässler, D. Buchholz, G. Huttner, and L. Zsolnai, *J. Organomet. Chem.*, 1989, **369**, 297.
803 S. Ching, M.P. Jensen, M. Sabat, and D.F. Shriver, *Organometallics*, 1989, **8**, 1058.
804 M. Cowie, R.L. DeKock, T.R. Wagenmaker, D. Seyferth, R.S. Henderson, and M.K. Gallagher, *Organometallics*, 1989, **8**, 119.
805 C.E. Housecroft, M.S. Shongwe, and A.L. Rheingold, *Organometallics*, 1989, **8**, 2651.
806 C.K. Schauer, E.J. Voss, M. Sabat, and D.F. Shriver, *J. Am. Chem. Soc.*, 1989, **111**, 7662.
807 R. Vac, J.H. Nelson, E.B. Milosavljevic, and L. Solujic, *Inorg. Chem.*, 1989, **28**, 3831; R. Vac, J.H. Nelson, E.B. Milosavljevic, L. Solujic, and J. Fischer, *Inorg. Chem.*, 1989, **28**, 4132.
808 V.V.S. Reddy, J.E. Whitten, K.A. Redmill, A. Varshney, and G.M. Gray, *J. Organomet. Chem.*, 1989, **372**, 207.
809 D. Sellmann, O. Käppler, and F. Knoch, *J. Organomet. Chem.*, 1989, **367**, 161.
810 F. Muller, G. van Koten, L.H. Polm, K. Vrieze, M.C. Zoutberg, D. Heijenrijk, E. Kragten, and C.H. Stam, *Organometallics*, 1989, **8**, 1340.
811 H. Jungbluth, H. Stöckli-Evans, and G. Süss-Fink, *J. Organomet. Chem.*, 1989, **377**, 339.
812 K.A. Johnson and W.L. Gladfelter, *Organometallics*, 1989, **8**, 2866.
813 G.F. Schmidt and G. Süss-Fink, *J. Organomet. Chem.*, 1989, **362**, 179.
814 F. Neumann and G. Süss-Fink, *J. Organomet. Chem.*, 1989, **367**, 175.
815 S.-H. Han, J.-S. Song, P.D. Macklin, S.T. Nguyen, G.L. Geoffroy, and A.L. Rheingold, *Organometallics*, 1989, **8**, 2127.
816 M. Bianchi, G. Menchi, P. Frediani, F. Piacenti, A. Scrivanti, and U. Matteoli, *J. Mol. Catal.*, 1989, **50**, 277.
817 P.J. Bailey, B.F.G. Johnson, J. Lewis, M. McPartlin, and H.R. Powell, *J. Chem. Soc., Chem. Commun.*, 1989, 1513.
818 E. Singleton, M.O. Albers, and M.M. de V. Steyn, *J. Chem. Soc., Dalton Trans.*, 1989, 2303.
819 N.C. Veitch, D.W. Concar, R.J.P. Williams, and D. Whitford, *FEBS Lett.*, 1988, **238**, 49 (*Chem. Abstr.*, 1989, **110**, 20 038).
820 I.P. Gerothanassis, M. Momenteau, and B. Lock, *J. Am. Chem. Soc.*, 1989, **111**, 7006.
821 S. Chirayil and R.P. Thummel, *Inorg. Chem.*, 1989, **28**, 813.
822 E. Alessio, Y. Xu, S. Cauci, G. Mestroni, F. Quadrifoglio, P. Viglino, and L.G. Marzilli, *J. Am. Chem. Soc.*, 1989, **111**, 7068.
823 Y. Kuroda, N. Tanaka, M. Goto, and T. Sakai, *Inorg. Chem.*, 1989, **28**, 2163.

of the ^{1}H n.m.r. spectrum of $[Ru(2,2\text{-bipy-}d_8)_2L]^{2+}$, L = (30).[824] The chelation of 5′-dGMP to *trans*-$[RuCl_2(DMSO)_4]$ has been investigated by ^{1}H n.m.r. spectroscopy.[825] N.m.r. data have also been reported for $[Fe(CN)_4\{(1R,2S)\text{-}cis\text{-cyclohexanediamine}\}]$, ($^{13}C$),[826] $[Fe_2(S_2O_3)_2(^{15}NO)_4]^{2-}$, ($^{15}N$),[827] $[\{RB(ONC_6H_8NO)_3BR\}Fe]$, ($^{11}B$, ^{13}C),[828] $[\{RB(ONC_8H_{12}NO)_3\text{-}BR\}Fe]$, ($^{13}C$),[829] $[(H_3N)_5RuNCNHC_6H_3Cl_2\text{-}2,6]^{2+}$, ($^{13}C$),[830] $[Ru_2(1,4,7\text{-}Me_3\text{-}1,4,7\text{-triazacyclononane})_2(\mu\text{-}OH)(\mu\text{-}OAc)_2]^{3+}$, ($^{13}C$),[831] $[Ru(bipy)_2(NCS)_2L]$, (^{13}C),[832] *trans*-$[Ru(bipy)_2(NO)Cl]$, (^{13}C),[833] and $[M(SC_4N_2Me_2)_2(PPh_3)_2]$, (M = Ru, Os; ^{13}C).[834]

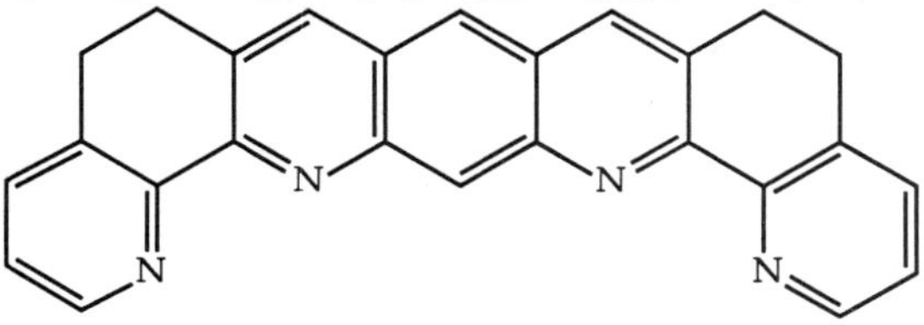

(30)

Complexes of Co, Rh, and Ir.—A review entitled 'Biosynthesis of vitamin B_{12}' has appeared which contains n.m.r. studies.[835]

For $[RhH_2(PPh_3)_2\{NC_5H_4\text{-}2\text{-}C(O)O\}]$, and related compounds, the ^{15}N n.m.r. signal moves to low frequency by 35 to 50 p.p.m. on coordination of the nitrogen. $^{2}J(^{15}N\text{-}^{1}H_{trans})$ = 25 Hz and $^{1}J(^{103}Rh\text{-}^{15}N)$ = 9.1 to 10.1 Hz.[836] Indirect two dimensional $^{31}P,^{103}Rh\{^{1}H\}$ n.m.r. spectroscopy has been used to observe ^{103}Rh in some rhodium complexes.[837] N.m.r. data have also been reported for $[H\{(\eta^5\text{-}C_5H_5)Co\}_4(\mu\text{-}CMe)]$, ($^{13}C$),[838] $[MeC(CH_2PPh_2)_3RhH_2\{Au(AsPh_3)\}_3]^{2+}$, ($^{103}Rh$),[839] $[Rh\{(Ph_2PCH_2CH_2)_3Si\}HCl]$, ($^{13}C$),[840] $[HRh(CH_2CH_2OEt)Cl(PPh_3)_2]$, ($^{13}C$),[841] *fac*-$[IrHEt(InEt_3)(PMe_3)_3]$, ($^{13}C$),[842]

824 S. Chirayil and R.P. Thummel, *Inorg. Chem.*, 1989, **28**, 813.
825 E. Alessio, Y. Xu, S. Cauci, G. Mestroni, F. Quadrifoglio, P. Viglino, and L.G. Marzilli, *J. Am. Chem. Soc.*, 1989, **111**, 7068.
826 Y. Kuroda, N. Tanaka, M. Goto, and T. Sakai, *Inorg. Chem.*, 1989, **28**, 2163.
827 C. Glidewell, R.J. Lambert, M.B. Hursthouse, and M. Motevalli, *J. Chem. Soc., Dalton Trans.*, 1989, 2061.
828 N.A. Voloshin, A.Yu. Nazarenko, E.V. Pol'shin, S.I. Tyukhtenko, and N.A. Kostromina, *Teor. Eksp. Khim.*, 1989, **25**, 322 (*Chem. Abstr.*, 1989, **111**, 224 065).
829 Ya.Z. Voloshin, N.A. Kostromina, A.Yu. Nazarenko, and E.V. Pol'shin, *Ukr. Khim. Zh. (Russ. Ed.)*, 1989, **55**, 7 (*Chem. Abstr.*, 1989, **110**, 204 547).
830 M.L. Naklicki and R.J. Crutchley, *Inorg. Chem.*, 1989, **28**, 4226.
831 P. Neubold, K. Wieghardt, B. Nuber, and J. Weiss, *Inorg. Chem.*, 1989, **28**, 459.
832 R.H. Herber, G. Nan, J.A. Potenza, H.J. Schugar, and A. Bino, *Inorg. Chem.*, 1989, **28**, 938.
833 H. Nagao, H. Nishimura, H. Funato, Y. Ichikawa, F.S. Howell, M. Mukaida, and H. Kakihana, *Inorg. Chem.*, 1989, **28**, 3955.
834 B.G. Olby and S.D. Robinson, *Inorg. Chim. Acta*, 1989, **165**, 153.
835 A.I. Scott, *Lect. Heterocycl. Chem.*, 1987, **9**, 75 (*Chem. Abstr.*, 1989, **110**, 90 662).
836 L. Carlton and M.-P. Belciug, *J. Organomet. Chem.*, 1989, **378**, 469.
837 C.J. Elsevier, J.M. Ernsting, and W.G.J. de Lange, *J. Chem. Soc., Chem. Commun.*, 1989, 585.
838 H. Wadepohl and H. Pritzkow, *Polyhedron*, 1989, **8**, 1939.
839 A. Albinati, F. Demartin, P. Janser, L.F. Rhodes, and L.M. Venanzi, *J. Am. Chem. Soc.*, 1989, **111**, 2115.
840 F.L. Joslin and S.R. Stobart, *J. Chem. Soc., Chem. Commun.*, 1989, 504.
841 M.L. Deem, E.J. Iobst, and M.D. Zink, *Inorg. Chim. Acta*, 1989, **157**, 153.
842 D.L. Thorn and R.L. Harlow, *J. Am. Chem. Soc.*, 1989, **111**, 2575.

$[MeC(pz)_3IrH(CO_2Me)(CO)]^+$, ($^{13}C$),[843] $[IrH(SnCl_3)_5]^{3-}$, (^{119}Sn),[844] and $[Ir(CO)(NC_{13}H_8)(PPh_3)_2]^+$, ($^{13}C$).[845]

^{13}C n.m.r. spectroscopy has been used to derive the electronic parameters in $[CoR(dmgH)_2L]$, R = Me, Et, Pr^i, adamantyl.[846] ^1H n.m.r. spectroscopy has been used to provide evidence for agostic interactions in organocobalt porphyrins.[847] A complete ^1H and ^{13}C n.m.r. assignment of 5′-deoxyadenosylcobinamide has been made using two dimensional n.m.r. spectroscopy.[848] The ^1H n.m.r. spectrum of (31) has a ring current with H_3 at δ 10.91, even though $^{13}C_3$ is at δ 129.8.[849] In $[(\eta^5\text{-}C_5Me_5)(PMe_3)Ir{=}CH_2]$, the

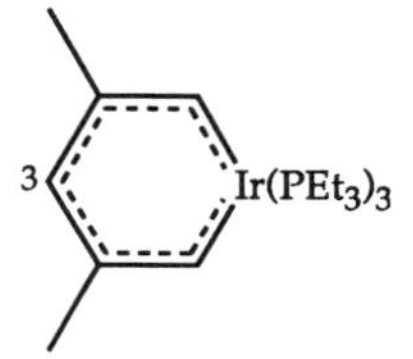

(31)

carbene hydrogen atoms are inequivalent at δ 14.08 and 13.08. The ^{13}C is at δ 189.9, with $^1J(^{13}C\text{-}^1H)$ of 134 and 146 Hz.[850] N.m.r. data have also been reported for $[F_3CCo(dmgH)_2(NC_5H_5)]$, ($^{13}C$),[851] $[R^1R^2C{=}CHCo(dmgH)_2(NC_5H_5)]$, ($^{13}C$),[852] $[PhMeC{=}C{=}C{=}CH\text{-}Co(dmgH)_2(NC_5H_5)]$, ($^{13}C$),[853] $[HC{=}CHCHMeNPr^i{}_2Co(CO)_3]$, ($^{13}C$),[854] $[(OC)_2\text{-}Co(PPh_3)CEt{=}CEtPEt_3{=}S]$, ($^{13}C$),[855] $[Co(C{\equiv}CBu^t)(HC{=}CHBu^t)(MeCN)(PMe_3)_3]^+$, ($^{13}C$),[856] $[\mu\text{-}PhC{\equiv}CSiMe_2Co(CO)_4]$, ($^{13}C$, ^{29}Si),[857] $[PhCCo_3(CO)_8(CHO)]^-$, (^{13}C),[858] *fac*-$[Co\{PhC(O)N\text{-}C(NR_2)E\}_3]$, (E = O, S, Se; ^{59}Co),[859] $[PhCOCo(CO)_3PCy_3]$, (^{13}C),[860] $[(OC)_4CoHgCo_2(CO)_4(\mu\text{-}\eta^2{:}\eta^2\text{-}C_4H_2Ph_2)\{\mu\text{-}HgCo(CO)_4\}]$, ($^{13}C$),[861] $[RhMe(CO)(PMe_3)\{\eta^2\text{-}(Ph_2PCH_2)_3CMe\}]$, ($^{13}C$),[862]

843 M.A. Esteruelas, L.A. Oro, C.M. Claramunt, C. López, J.L. Lavandera, and J. Elguero, *J. Organomet. Chem.*, 1989, **366**, 245.
844 K. Nomura, Y. Saito, and S. Shinoda, *J. Mol. Catal.*, 1989, **50**, 303.
845 F. Neve, M. Ghedini, A. Tiripicchio, and F. Ugozzoli, *Inorg. Chem.*, 1989, **28**, 3084.
846 N.B. Pahor, L. Randaccio, E. Zangrando, and P.A. Marzilli, *J. Chem. Soc., Dalton Trans.*, 1989, 1941.
847 W.A. Al-Akhdar, K.A. Belmore, and M.J. Kendrick, *Inorg. Chim. Acta*, 1989, **165**, 15.
848 T.G. Pagano, P.G. Yohannes, B.P. Hay, J.R. Scott, R.G. Finke, and L.G. Marzilli, *J. Am. Chem. Soc.*, 1989, **111**, 1484.
849 J.R. Bleeke, Y.-F. Xie, W.-J. Peng, and M. Chiang, *J. Am. Chem. Soc.*, 1989, **111**, 4118.
850 D.P. Klein and R.G. Bergmann, *J. Am. Chem. Soc.*, 1989, **111**, 3079.
851 P.J. Toscano, L. Konieczny, S. Liu, and J. Zubieta, *Inorg. Chim. Acta*, 1989, **166**, 163.
852 P.J. Stang and A.K. Datta, *J. Am. Chem. Soc.*, 1989, **111**, 1358.
853 P.J. Stang, A.K. Datta, V. Dixit, and L.G. Wistrand, *Organometallics*, 1989, **8**, 1020.
854 M. Miller and K.M. Nicholas, *J. Organomet. Chem.*, 1989, **362**, C15.
855 E. Lindner, K.E. Frick, M. Stängle, R. Fawzi, and W. Hiller, *Chem. Ber.*, 1989, **122**, 53.
856 N. Habadie, M. Dartiguenave, Y. Dartiguenave, J.F. Britten, and A.L. Beauchamp, *Organometallics*, 1989, **8**, 2564.
857 R.J.P. Corriu, J.J.E. Moreau, and H. Praet, *Organometallics*, 1989, **8**, 2779.
858 D.S. Dumond, S. Hwang, and M.G. Richmond, *Inorg. Chem.*, 1989, **160**, 135.
859 N. Juranić, E. Hoyer, F. Dietze, and L. Beyer, *Inorg. Chim. Acta*, 1989, **162**, 161.
860 N. Kawasaki, K. Masuzoe, F. Ozawa, and A. Yamamoto, *J. Organomet. Chem.*, 1989, **361**, C37.
861 S.J. Tyler and J.M. Burlitch, *J. Organomet. Chem.*, 1989, **361**, 231.
862 G.G. Johnston and M.C. Baird, *J. Chem. Soc., Chem. Commun.*, 1989, 1008.

$[(\eta^5\text{-}C_5Me_5)Rh(CH_2NCN)(PMe_3)]$, ($^{13}C$),[863] $[RhRX\{C_6H_3(CH_2NMe_2)_2\text{-}2,6\}]$, ($^{13}C$, ^{103}Rh),[864] $[(\eta^5\text{-}C_5Me_5)M\{CPh{=}CHC(O)O\}(Me_2SO)]$, (M = Rh, Ir; ^{13}C),[865] $[(Me_3P)_4RhC{\equiv}CR]$, (^{13}C),[866] $[\{(3,5\text{-}Me_2C_6H_3)_4porphyrin\}RhC(O)C(O)Rh\{porphyrin(C_6H_3Me_2\text{-}3,5)\}]$, ($^{13}C$),[867] [(tetramesitylporphyrin)RhC(O)C(O)(tetramesitylporphyrin)], (^{13}C),[868] $[Rh(SnMe_3)(CO)\{(Ph_2P)_3CMe\}]$, ($^{13}C$),[869] $[(\eta^5\text{-}C_5Me_5)_2Rh_2(\mu\text{-}CHMe)(\mu\text{-}CH_2)X_2]$, ($^{13}C$),[870] $[(\eta^5\text{-}C_5Me_5)_2Rh_2(\mu\text{-}CC_6H_4Me\text{-}4)(\mu\text{-}CO)]^+$, ($^{13}C$),[871] $[(\eta^1\text{-}indenyl)IrMe(CNBu^t)_2(PPh_3)_2]^+$, ($^{13}C$),[872] (32), ($^{13}C$),[873] $[IrR^1R^2\{N(SiMe_2CH_2PPh_2)_2\}]$, ($^{13}C$),[874] (33), ($^{13}C$),[875] $[(C_6H_5)Ir(\eta^3\text{-}1\text{-}MeC_3H_4)(PMe_3)_3]^+$, ($^{13}C$),[876] and $[Ir\{C(O)CH_2OCH_2COCl\}(CO)(PPh_3)_2Cl_2]$, ($^{13}C$).[877]

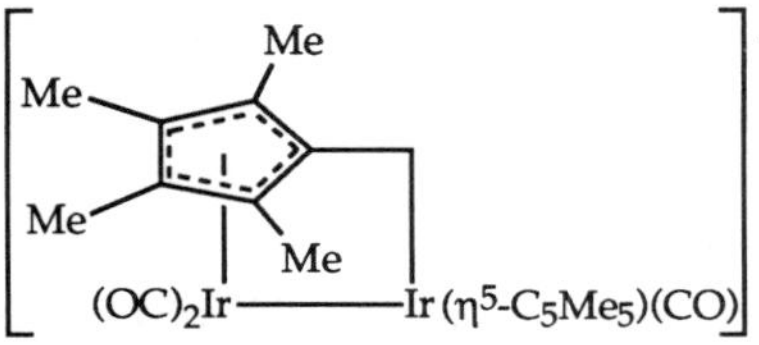

(32)

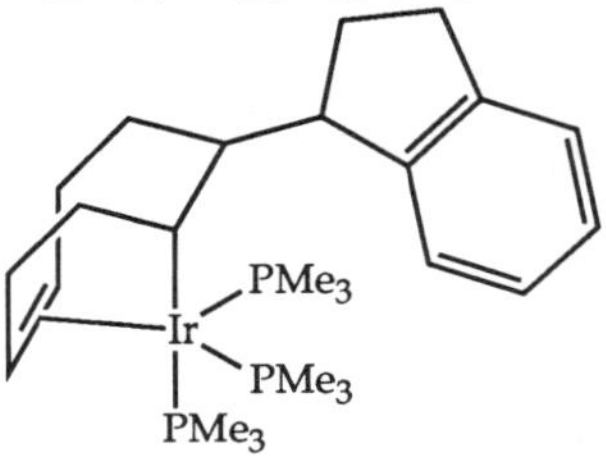

(33)

^1H, ^{13}C, ^{19}F, and ^{31}P n.m.r. spectroscopy has been used to show that $[(\eta^5\text{-}C_5H_5)Rh(PMe_3)(\eta^2\text{-}C_6F_6)]$ is static from 243 to 403 K.[878] The ^{13}C n.m.r. spectrum of $[(\eta^5\text{-}C_5Me_5)Co(2;2,5\text{-}C_7H_9)]^+$ contains an agostic hydrogen atom.[879] ^1H and ^{13}C chemical shifts of the cyclopentadienyl group in $[(\eta^5\text{-}C_5H_5)Co(S_2C_2H_3R)]$ complexes are linearly related to the reversible half wave potential.[880] Attempts to detect a ^{13}C equilibrium isotope effect in $[(\eta^5\text{-}C_5H_5)_2Rh_2(CO)_3]$ have failed.[881] ^{31}P n.m.r. spectroscopy, including COSY, has been

863 M. Steinmetz and H. Werner, *J. Organomet. Chem.*, 1989, **369**, 309.
864 A.A.H. van der Zeijden, G. van Koten, J.M. Ernsting, C.J. Elsevier, B. Krijnen, and C.H. Stam, *J. Chem. Soc., Dalton Trans.*, 1989, 317.
865 G.J. Sunley, P. del C. Menanteau, H. Adams, N.A. Bailey, and P.M. Bailey, *J. Chem. Soc., Dalton Trans.*, 1989, 2415.
866 D. Zargarian, P. Chow, N.J. Taylor, and T.B. Marder, *J. Chem. Soc., Chem. Commun.*, 1989, 540.
867 B.B. Wayland, A.E. Sherry, and V.L. Coffin, *J. Chem. Soc., Chem. Commun.*, 1989, 662.
868 A.E. Sherry and B.B. Wayland, *J. Am. Chem. Soc.*, 1989, **111**, 5010.
869 G.G. Johnston, S.I. Hommeltoft, and M.C. Baird, *Organometallics*, 1989, **8**, 1904.
870 J. Martinez, J.B. Gill, H. Adams, N.A. Bailey, I.M. Saez, and P.M. Maitlis, *Can. J. Chem.*, 1989, **67**, 1698.
871 S.A. Benyunes, P.A. Chaloner, and P.B. Hitchcock, *J. Chem. Soc., Chem. Commun.*, 1989, 1491.
872 A. Habib, R.S. Tanke, E.M. Holt, and R.H. Crabtree, *Organometallics*, 1989, **8**, 1225.
873 F.W.B. Einstein, R.H. Jones, X. Zhang, X. Yan, R. Nagelkerke, and D. Sutton, *J. Chem. Soc., Chem. Commun.*, 1989, 1424.
874 M.D. Fryzuk, P.A. MacNeil, R.L. Massey, and R.G. Ball, *J. Organomet. Chem.*, 1989, **368**, 231.
875 J.S. Merola and R.T. Kacmarcik, *Organometallics*, 1989, **8**, 778.
876 J.S. Merola, *Organometallics*, 1989, **8**, 2975.
877 P.M. Zizelman and J.M. Stryker, *Organometallics*, 1989, **8**, 2075.
878 S.T. Belt, S.B. Duckett, M. Helliwell, and R.N. Perutz, *J. Chem. Soc., Chem. Commun.*, 1989, 928.
879 M.A. Bennett, J.C. Nicholls, A.K.F. Rahman, A.D. Redhouse, J.L. Spencer, and A.C. Willis, *J. Chem. Soc., Chem. Commun.*, 1989, 1328.
880 M. Kajitani, T. Akiyama, A. Sugimori, K. Hirakata, Y. Hoshino, Y. Satsu, G.P. Sato, K. Shimizu, M. Kaise, and C. Nishihara, *J. Electroanal. Chem. Interfacial Electrochem.*, 1988, **251**, 421 (*Chem. Abstr.*, 1989, **110**, 47 189).
881 S. Berger and M. Ochs, *J. Organomet. Chem.*, 1989, **367**, 343.

used to analyse complex spectra such as those from [Rh_2(μ-dppm)$_2$(μ-SO)(CO)$_2$].[882] The ^{11}B n.m.r. spectrum of [$Rh_2Fe_4(CO)_{16}B]^-$ shows $^1J(^{103}Rh-^{11}B)$ = 23 Hz and T_1 of 0.47 s.[883] N.m.r. data have also been reported for [(η^5-C_5Me_5)Co(η^2-C_8H_{12}Se)(η^1-$C_8H_{12}SeN_2$)], (^{13}C),[884] [Co(η^5-$C_5Me_4CH_2CH_2CH$=CH_2-η^2)(CO)], (^{13}C),[885] [Co(η^2-$R^1C≡CR^2$)(PMe$_3$)$_3$]$^+$, (^{13}C),[886] [{$Co_2(CO)_6$(μ-HC≡CCMe$_2$)$_2$NR}, (^{13}C, ^{59}Co),[887] [(PhC≡C)$_2$P{η^2-C≡CPhCo$_2$(CO)$_5$}]$_2$, (^{13}C),[888] [(η^5-C_5Me_5)Rh(η^2-CH_2EPh)(CO)], (E = P, As; ^{13}C),[889] [(Ph$_3$P)$_2$Ir(η^2-C_2H_4)$_2$]$^+$, (^{13}C),[890] [Ir(η^2-MeC≡CMe)(PMe$_2$Ph)$_3$]$^+$, (^{13}C),[891] [{(η^5-C_5H_5)Co}$_2$(η^3:η^3-PhC$_6H_6$Ph)], (^{13}C),[892] [(η^4-C_4H_6)Co(η^5-C_5H_5)], (^{13}C),[893] [(η^5-C_5H_5)Co(η^4-1,3-R^1-2,4-R^2C_4)], (^{13}C),[894] [(η^6-arene)Co(η^4-1-3,6-C_8H_{12})]$^+$, (^{13}C),[895] [(η^5-C_5H_5)Co(η^4-$C_5Ph_4HCO_2Et$)], (^{13}C),[896] [(η^5-C_5H_5)-Co(η^4-$C_5H_5CCl_3$)], (^{13}C),[897] [(η^5-C_5H_5)Co(η^4-$C_5H_5PR_2$=O)], (^{13}C),[898] [($C_3H_3N_2CH_2CH_2N_2$-C_3H_3)Rh(η^4-C_8H_{12})], (^{13}C),[899] [(η^4-C_8H_{12})MCl(1,3-$N_2C_3H_3$R)], (M = Rh, Ir; ^{13}C),[900] [M{Ph$_2$PCH$_2$P(S)Ph$_2$}(η^4-C_8H_{12})], (M = Rh, Ir; ^{13}C),[901] [(η^5-C_5Me_5)Rh(η^4-C_4Me_4S)]$^{2+}$, (^{13}C),[902] [Rh$_2${μ-S(CH$_2$)$_3$NMe$_2$}$_2$(η^4-C_8H_{12})], (^{13}C),[903] [Ir(η^4-C_4H_6){N(SiMe$_2$CH$_2$PPh$_2$)$_2$}], (^{13}C),[904] [{(η^5-C_5Me_5)Co(η^5-C_5Ph_2OEt)}$_2${(CH$_2$)$_4$}$_2$]$^{2+}$, (^{11}B, ^{13}C, ^{59}Co),[905] [(η^5-C_5H_5)Co(η^4-C_4Ph_4BBr)], (^{11}B),[906] [(η^5-CEtCEtBEtNMeSiMe$_2$)ML$_n$], {ML$_n$ = Fe(CO)$_3$, Co(η^5-C_5H_5), Rh(acac); ^{11}B, ^{13}C, ^{14}N, ^{15}N, ^{17}O, ^{29}Si, ^{59}Co, ^{103}Rh},[907] [(η^5-C_5H_5)CoSCPh=CPhSe],

882 A. Neher, O. Heyke, and I.-P. Lorenz, Z. *Anorg. Allg. Chem.*, 1989, **578**, 185.
883 A.K. Bandyopadhyay, R. Khattar, and T.P. Fehlner, *Inorg. Chem.*, 1989, **28**, 4434; R. Khattar, J. Puga, T.P. Fehlner, and A.L. Rheingold, *J. Am. Chem. Soc.*, 1989, **111**, 1877.
884 C.P. Morley, *Organometallics*, 1989, **8**, 800.
885 J. Okuda and K.H. Zimmermann, *Chem. Ber.*, 1989, **122**, 1645.
886 A. Bouayad, M. Dartiguenave, M.-J. Menu, Y. Dartiguenave, F. Bélanger-Gariépy, and A.L. Beauchamp, *Organometallics*, 1989, **8**, 629.
887 L.P. Battaglia, D. Delledonne, M. Nardelli, G. Predieri, G.P. Chiusoli, M. Costa, and C. Pelizzi, *J. Organomet. Chem.*, 1989, **363**, 209.
888 H. Lang and L. Zsolnai, *J. Organomet. Chem.*, 1989, **369**, 131.
889 H. Werner, W. Paul, J. Wolf, M. Steinmetz, R. Zolk, G. Müller, O. Steigelmann, and J. Riede, *Chem. Ber.*, 1989, **122**, 1061.
890 A.R. Siedle and R.A. Newmark, *Organometallics*, 1989, **8**, 1442.
891 G. Marinelli, I. El Idrissi Rachidi, W.E. Streib, O. Eisenstein, and K.G. Caulton, *J. Am. Chem. Soc.*, 1989, **111**, 2346.
892 H. Wadepohl, K. Büchner, and H. Pritzkow, *Organometallics*, 1989, **8**, 2745.
893 R.S. Lokey, N.S. Mills, and A.L. Rheingold, *Organometallics*, 1989, **8**, 1803.
894 Y. Ito, M. Inouye, M. Murakami, and M. Shiro, *J. Organomet. Chem.*, 1989, **359**, C57.
895 H. Bönnemann, R. Goddard, J. Grub, R. Mynott, E. Raabe, and S. Wendel, *Organometallics*, 1989, **8**, 1941.
896 J.M. O'Connor, L. Pu, R. Uhrhammer, J.A. Johnson, and A.L. Rheingold, *J. Am. Chem. Soc.*, 1989, **111**, 1889.
897 N. Kuhn and E. Zauder, *J. Organomet. Chem.*, 1989, **362**, 217.
898 W. Kläui and C.-E. Song, *Inorg. Chem.*, 1989, **28**, 3845.
899 R.M. Claramunt, P. Domiano, J. Elguero, and J.L. Lavandera, *Bull. Soc. Chim. Fr.*, 1989, 472.
900 F. Bonati, L.A. Oro, M.T. Pinillos, C. Tejel, M.C. Apreda, C. Foces-Foces, and F.H. Cano, *J. Organomet. Chem.*, 1989, **369**, 253.
901 J. Browning, K.R. Dixon, and R.W. Hilts, *Organometallics*, 1989, **8**, 552.
902 A.E. Ogilvy, A.E. Skaugset, and T.B. Rauchfuss, *Organometallics*, 1989, **8**, 2739.
903 J.C. Bayón, J. Real, C. Claver, A. Polo, and A. Ruiz, *J. Chem. Soc., Chem. Commun.*, 1989, 1056.
904 M.D. Fryzuk, K. Joshi, and S.J. Rettig, *Polyhedron*, 1989, **8**, 2291.
905 P. Jutzi, U. Siemeling, A. Müller, and H. Bögge, *Organometallics*, 1989, **8**, 1744.
906 F.-E. Hong, C.W. Eigenbrot, and T.P. Fehlner, *J. Am. Chem. Soc.*, 1989, **111**, 949.
907 R. Köster, G. Seidel, B. Wrackmeyer, and D. Schlosser, *Chem. Ber.*, 1989, **122**, 2055.

(^{13}C),[908] [(η^5-C_5H_5)CoI($Ph_2PNHCHMePh$){P(O)(OR^1)R^2}], (^{13}C),[909] [(μ-η^5-$C_5H_4PMe_2$)$_2Rh_2$(CO)$_2$], (^{13}C),[910] [(η^5-C_5H_5)Rh(μ-CO)(μ-2-$Ph_2PC_5H_4N$)Rh(CO)Cl], (^{13}C, ^{103}Rh),[911] [(η^5-C_5H_5)Rh(μ-$MeO_2CC{\equiv}CCO_2Me$)(μ-2-$Ph_2PC_5H_4N$)Rh(CO)(μ-Cl)]$_2$, (^{13}C),[912] [$HBpz_3Rh$(η^5-C_5H_4BPh)], (^{11}B),[913] [(η^5-C_5Me_5)M(phen)Cl]$^+$, (M = Rh, Ir; ^{13}C),[914] [(η^5-C_5H_5)$_3RhM_2$(CO)$_n$], (M = Mo, W, Fe, Ni; ^{13}C),[915] [(η^5-C_5Me_5)IrOC(O)NBu^t], (^{13}C),[916] [($C_8H_{16}N_4P$)Pt(PPh_3)Co(CO)$_4$], (^{13}C),[917] [(ClC)Co_3(CO)$_7$(dppm)], (^{13}C),[918] [Rh(μ-2-OC_5H_4N)(CO)$_2$]$_2$, (^{13}C),[919] *cis*-[Rh(CO)$_2$(NC_5H_5)$_2$]$^+$, (^{13}C),[920] *cis*-[Rh(CO)$_2$(OH_2)X], (^{13}C),[921] [Rh(acac)(CO)$_2$], (^{13}C),[922] [Rh(oxalato)(CO)$_2$]$^-$, (^{13}C),[923] [M_2(CO)$_4${HSCH(CH_2CH_2)$_2$NHMe}]$^{2+}$, (M = Rh, Ir; ^{13}C),[924] [(Me_3P)$_2$Rh(CO)OCHPhCHMeCBu^t=O], (^2H, ^{13}C),[925] [Rh(CO)L(Ph_2P)$_3$CMe]$^+$, (^{13}C),[926] [Si($CH_2CH_2PPh_2$)$_3$Rh(CO)], (^{13}C),[927] [$PdRhCl_3$(CO)(μ-Ph_2PCH_2SMe)$_2$], (^{13}C),[928] [Rh_6(CO)$_{12}${P(OPh)$_3$}$_4$], (^{13}C, ^{103}Rh),[929] [Rh_6C(CO)$_{13}${Au_2(PPh_3)$_2$}], (^{13}C),[930] [Ir_2(CO_3)(CO)$_2$(dmpm)$_2$], (^{13}C),[931] and [Ir_4(CO)$_9$(Ph_2P)$_3$CH], (^{13}C).[932]

In [Co($^{15}NH_3$)$_5$X]$^{2+}$ and *cis*-[CoX_2(en)$_2$]$^+$, the ^{15}N chemical shifts correlate with the Co—N bond strength.[933] The pH dependence of the ^{59}Co T_1 values of [Co(imidazole)$_6$]$^{3+}$ has shown that the linewidths of the ^1H signals are determined by unresolved

908 M. Kajitani, R. Ochiai, K. Dohki, N. Kobayashi, T. Akiyama, and A. Sugimori, *Bull. Chem. Soc. Jpn.*, 1989, **62**, 3266.
909 C. Jablonski, T. Burrow, and P.G. Jones, *J. Organomet. Chem.*, 1989, **370**, 173.
910 M.D. Rausch, W.C. Spink, J.L. Atwood, A.J. Baskar, and S.G. Bott, *Organometallics*, 1989, **8**, 2627; X.D. He, A. Maisonnat, F. Dahan, and R. Poilblanc, *Organometallics*, 1989, **8**, 2618.
911 G. Bruno, S. Lo Schiavo, E. Rotondo, C.G. Arena, and F. Faraone, *Organometallics*, 1989, **8**, 886.
912 E. Rotondo, S. Lo Schiavo, G. Bruno, C.G. Arena, R. Gobetto, and F. Faraone, *Inorg. Chem.*, 1989, **28**, 2944.
913 G.E. Herberich and U. Büschges, *Chem. Ber.*, 1989, **122**, 615.
914 M.-T. Youinou and R. Ziessel, *J. Organomet. Chem.*, 1989, **363**, 197.
915 B. Walther, M. Scheer, H.-C. Böttcher, A. Trunschke, H. Ewald, D. Gutschick, H. Miessner, M. Skupin, and G. Vorbeck, *Inorg. Chim. Acta*, 1989, **156**, 285.
916 D.S. Glueck, F.J. Hollander, and R.G. Bergmann, *J. Am. Chem. Soc.*, 1989, **111**, 2719.
917 D.V. Khasnis, M. Lattman, and U. Siriwardane, *Inorg. Chem.*, 1989, **28**, 2594.
918 S. Aime, M. Botta, R. Gobetto, D. Osella, and L. Milone, *Gazz. Chim. Ital.*, 1987, **117**, 773 (*Chem. Abstr.*, 1989, **110**, 115 064).
919 M.A. Ciriano, B.E. Villarroya, L.A. Oro, M.C. Apreda, C. Foces-Foces, and F.H. Cano, *J. Organomet. Chem.*, 1989, **366**, 377.
920 A.J. Pardey and P.C. Ford, *J. Mol. Catal.*, 1989, **53**, 247.
921 L. Garlaschelli, M. Marchionna, M.C. Iapalucci, and G. Longoni, *J. Organomet. Chem.*, 1989, **378**, 457.
922 A.I. Rubailo, V.P. Selina, and Yu.S. Varshavskii, *Koord. Khim.*, 1988, **14**, 957 (*Chem. Abstr.*, 1989, **110**, 84 412).
923 J. Real, J.C. Bayón, F.J. Lahoz, and J.A. López, *J. Chem. Soc., Chem. Commun.*, 1989, 1889.
924 J.C. Bayón, P. Esteban, J. Real, C. Claver, and A. Ruiz, *J. Chem. Soc., Dalton Trans.*, 1989, 1579.
925 G.A. Slough, R.G. Bergman, and C.H. Heathcock, *J. Am. Chem. Soc.*, 1989, **111**, 938.
926 G.G. Johnston and M.C. Baird, *Organometallics*, 1989, **8**, 1894.
927 D.E. Hendriksen, A.A. Oswald, G.B. Ansell, S. Leta, and R.V. Kastrup, *Organometallics*, 1989, **8**, 1153.
928 G.K. Anderson, *J. Organomet. Chem.*, 1989, **368**, 385.
929 C. Allevi, S. Bordoni, C.P. Clavering, B.T. Heaton, J.A. Iggo, L. Garlaschelli, and C. Seregni, *Organometallics*, 1989, **8**, 385.
930 A. Fumagalli, S. Martinengo, V.G. Albano, D. Braga, and F. Grepioni, *J. Chem. Soc., Dalton Trans.*, 1989, 2343.
931 M.K. Reinking, J. Ni, P.E. Fanwick, and C.P. Kubiak, *J. Am. Chem. Soc.*, 1989, **111**, 6459.
932 J.R. Kennedy, P. Selz, A.L. Rheingold, W.C. Trogler, and F. Basolo, *J. Am. Chem. Soc.*, 1989, **111**, 3615.
933 Y. Nakashima, M. Muto, K. Kawano, Y. Kyogoku, and Y. Yoshikawa, *Bull. Chem. Soc. Jpn.*, 1989, **62**, 2455.

$J(^{59}Co-^1H)$.[934] Addition of tartaric acid to $[Co(en)_3]^{3+}$ results in two ^{59}Co n.m.r. signals due to optical isomers.[935] A broad study of ^{59}Co n.m.r. spectra of Co^{III} porphyrin complexes with axial amine ligands has been reported and the dependence of δ and $\omega_{1/2}$ on ligand substituents was discussed.[936] The ^1H n.m.r. spectrum of cyanocobalamin has been assigned by COSY and NOESY, assisted by ^{13}C n.m.r. spectroscopy and ^{13}C labelling experiments.[937] The *b*, *d*, and *e* analogues of cyanocobalamin as monocarboxylic acids have been identified by ^{15}N n.m.r. spectroscopy.[938] ^1H n.O.e. studies have been performed on Cu_2Co_2 superoxide dismutase.[939] The ^1H, ^{13}C, and ^{31}P n.m.r. spectra of $[Rh\{(S,S)\text{-}Ph_2PCHMeCHMePPh_2\}_2]^+$ have indicated that the six-membered rings have the chair conformation.[940] The ^{31}P n.m.r. spectra of $[RhL(Ph_2PCH_2CH_2)_3P]^+$ have been used to determine the *trans* influence of L.[941] ^{59}Co n.m.r. spectroscopy, including T_1 measurements, has been used to investigate the effect of halide substitution in $[Co(acac)_3]$.[942] N.m.r. data have also been reported for $[(H_3N)_5Co(NHCONHPh)]^{2+}$, ($^{13}C$),[943] $[(C_5H_5N)_3Co_3O(OAc)_5OMe]^+$, ($^{13}C$),[944] $[(en)_2Co(N_3)(C_2O_4)]$, (^{13}C),[945] $[Co(en)_2(SO_3)Cl]$, (^{13}C),[946] *trans*-$[Co(en)_2(\text{Boc-L-valine})Cl]^+$, ($^{13}C$),[947] $[Co(MeNHCH_2CH_2NMeH)_{3-n}(en)_n]^{3+}$, ($^{13}C$),[948] $[\{N(CH_2CH_2NH_2)_3\}\overline{Co(OC_6H_4CHO}\text{-}2)]^{2+}$, ($^{13}C$),[949] *mer*-$[Co(1,4,7,10,13,16\text{-hexaazacyclooctadecane})]^{3+}$, ($^{13}C$),[950] cobalt(III) complexes containing an *N*-glycoside derived from $H_2NCH_2CH_2NH_2$ and an aldose, (^{13}C),[951] $[Co(1,4,7\text{-triazacyclononane})1,4,7\text{-trithiacyclononane})]^{3+}$, ($^{59}Co$),[952] $[N(CH_2CH_2CH_2NH_2)_3Co(O_2CO)]^+$, ($^{13}C$),[953] $[\{H_2NC(CH_2NHCHMeCH_2NHCH_2)_3CNH_2\}Co]^{3+}$, ($^{13}C$),[954] $[\{MeC(CH_2NHCH_2CH_2NHCH_2)_3CCH(NH_3)CO_2H\}Co]^{4+}$, ($^{13}C$),[955] $[\{MeC(CH_2NHCH_2CH_2NH$-

934 G. Navon and R. Panigel, *Inorg. Chem.*, 1989, **28**, 1405.
935 T.Sh. Kapanadze, Yu.V. Kokunov, G.V. Tsintsadze, and Yu.A. Buslaev, *Dokl. Akad. Nauk SSSR*, 1989, **304**, 1139 (*Chem. Abstr.*, 1989, **111**, 89 278).
936 H. Bang, L. Cassidei, H. Danforth, J.O. Edwards, K.I. Hagen, C. Krüger, J. Lachowitz, C.M. Schwab, D.A. Sweigart, and Z. Zhang, *Magn. Reson. Chem.*, 1989, **27**, 1117.
937 K. Kurumaya and M. Kajiwara, *Chem. Pharm. Bull.*, 1989, **37**, 9 (*Chem. Abstr.*, 1989, **110**, 173 453).
938 T.J. DiFeo, R.A. Schiksnis, R.K. Kohli, S.J. Opella, and A. Nath, *Magn. Reson. Chem.*, 1989, **27**, 127.
939 L. Banci, I. Bertini, C. Luchinat, M. Piccioli, A. Scozzafava, and P. Turano, *Inorg. Chem.*, 1989, **28**, 4650.
940 J.Bakos, I. Tóth, G. Szalontai, V. Fülöp, and B. Heil, *J. Organomet. Chem.*, 1989, **371**, 101.
941 J.J. Gambaro, W.H. Hohman, and D.W. Meek, *Inorg. Chem.*, 1989, **28**, 4154.
942 S.C. Busse and E.H. Abbott, *Inorg. Chem.*, 1989, **28**, 488.
943 D.P. Fairlie, W.G. Jackson, and G.M. McLaughlin, *Inorg. Chem.*, 1989, **28**, 1983.
944 C.E. Sumner, jun. and G.R. Steinmetz, *Inorg. Chem.*, 1989, **28**, 4290.
945 M.E. Kastner, D.A. Smith, A.G. Kuzmission, J.N. Cooper, T. Tyree, and M. Yearick, *Inorg. Chim. Acta*, 1989, **158**, 185.
946 W.G. Jackson, A.G. Kuzmission, J.N. Cooper, and J.C. Henry, *Inorg. Chem.*, 1989, **28**, 1553.
947 G.C. Baumann, J.A. Potenza, and S.S. Isied, *Inorg. Chem.*, 1989, **156**, 85.
948 G.H. Searle and F.R. Keene, *Inorg. Chim. Acta*, 1989, **155**, 125.
949 E. Toyota and Y. Yamamoto, *Bull. Chem. Soc. Jpn.*, 1989, **62**, 3817.
950 G.H. Searle, *Bull. Chem. Soc. Jpn.*, 1989, **62**, 4021.
951 K. Ishida, S. Nonoyama, T. Hirano, S. Yano, M. Hidai, and S. Yoshikawa, *J. Am. Chem. Soc.*, 1989, **111**, 1599.
952 H.-J. Küppers, K. Wieghardt, S. Steenken, B. Nuber, and J. Weiss, *Z. Anorg. Allg. Chem.*, 1989, **573**, 43.
953 S.S. Massoud and R.M. Milburn, *Polyhedron*, 1989, **8**, 2389.
954 A.J. Hendry, K.J. Naidoo, and D.A. Thornton, *J. Chem. Soc., Chem. Commun.*, 1989, 998.
955 B. Korybut-Daszkiewicz, R.M. Hartshorn, and A.M. Sargeson, *J. Chem. Soc., Chem. Commun.*, 1989, 1375.

$CH_2)_3P\}Co]^{3+}$, (^{13}C),[956] $[\{MeC(CH_2NHCH_2CH_2NHCH_2)_3As\}Co]^{3+}$, ($^{13}C$),[957] Λ-[Co(α-cyclodextrinato)(1,4,7,10-tetraazacyclododecane)]$^+$, (^{13}C),[958] $[\{MeC(CH_2NHCH_2CH_2NH$-$CH_2)_3CMe\}Co]^{3+}$, (^{13}C),[959] $[Co\{H_2NCH_2CH_2NHCH_2C_4N_2HBrC(O)NHCH_2CH_2C_3N_2H_2\}$-$(OH_2)]^{2+}$, ($^{13}C$),[960] $[\{C_4H_3N_2C(O)NCH_2CH_2C_3H_2N\}_2Co]^{3+}$, ($^{13}C$),[961] [Co{H(ON=CEtCPr=N-$CH_2)CH_2\}Br_2]$, (^{13}C),[962] $[Co\{ON{=}CMeCMe{=}N(CH_2CH_2NH)_2CH_2CH_2N{=}CMeCMe{=}N$-OH}]$^{2+}$, ($^{13}C$),[963] [{H[ON=C(2-furyl)CC(2-furyl)=NO]H}MX(NH_3)], (M = Co; ^{13}C),[964] (M = Co, Rh; ^{13}C),[965] $[Co(dmgH)_2(NC_5H_5)I]$, (^{13}C),[966] [LCo{N^2,$N^{2'}$-propanediylbis(2,3-butane 2-imine-3-oxime)}R]$^+$, (^{13}C),[967] $[Co(tpp)(O_2CH_2CH{=}CH_2)(NC_5H_5)]$, ($^{13}C$),[968] monocarboxylic derivatives of cyanocobalamin, (^{13}C),[969] [Co(carnosinate)(en)]$^+$, (^{13}C),[970] *fac*-$[(H_2NCH_2CH_2O)_3Co]$, (^{59}Co),[971] *trans(Cl,Cl)*, *cis(P,P)*-$[RhCl_2(NH_2CH_2CH_2PMe_2)_2]^+$, ($^{13}C$),[972] [Rh{OCNMeCPhNMeC(O)NCH$_2$(OCH$_2$CH$_2$)$_2$-1-imidazole}Cl$_2$]$^+$, (^{13}C),[973] [Rh$_2${*N*,*N'*-(4-tolyl)$_2$formamidinate}$_2$(O$_2$CCF$_3$)$_2$PPh$_3$], (^{13}C, ^{103}Rh),[974] [Rh$_2${*N*,*N'*-(4-tolyl)$_2$formamidinate}$_3$(O$_3$N)(NC$_5$H$_5$)], (^{13}C, ^{103}Rh),[975] *cis*-[Co(phenylalaninato){(O$_2$C-CH$_2$)$_3$N}]$^+$, (^{13}C),[976] $[Co\{CH_2(NHCHEtCH_2NHCHEt)_2CH_2\}(O_2CCR^1R^2NH_2)]^{2+}$, ($^{13}C$),[977] $[Co\{(O_2CCH_2)_2NCHPhCHPhN(CH_2CO_2)_2\}]^-$, ($^{13}C$),[978] $[Co(S_2CNEt_2)_2\{OC_4H_8NC(S)CH_2$-$C(S)NC_4H_8O\}]^+$, ($^{13}C$),[979] $[RhCl_3(ArR^1C{=}NR^2)_3]$, $[HgCl_2(ArR^1C{=}NR^2)_3]$, ($^{13}C$, ^{199}Hg),[980] $[Ir(NO)(NCMe)(NC_5H_4\text{-}2\text{-}CH{=}NN{=}CH\text{-}2\text{-}C_5H_4N)(PPh_3)_2]^{2+}$, ($^{13}C$),[981] [Rh$_2$(dmpe)$_4$(μ-

956 A. Höhn, R.J. Geue, A.M. Sargeson, and A.C. Willis, *J. Chem. Soc., Chem. Commun.*, 1989, 1644.
957 A. Höhn, R.J. Geue, A.M. Sargeson, and A.C. Willis, *J. Chem. Soc., Chem. Commun.*, 1989, 1648.
958 K. Yamanari, M. Nakamichi, and Y. Shimura, *Inorg. Chem.*, 1989, **28**, 248.
959 P.V. Bernhardt, G.A. Lawrance, and T.W. Hambley, *J. Chem. Soc., Dalton Trans.*, 1989, 1059.
960 S.J. Brown, S.E. Hudson, P.K. Mascharak, and M.M. Olmstead, *J. Am. Chem. Soc.*, 1989, **111**, 6446; S.J. Brown, M.M. Olmstead, and P.K. Mascharak, *Inorg. Chem.*, 1989, **28**, 3720.
961 M. Muetterties, P.K. Mascharak, M.B. Cox, and S.K. Arora, *Inorg. Chim. Acta*, 1989, **160**, 123.
962 Y. Murakami, Y. Hisaeda, S.-D. Fan, and Y. Matsuda, *Bull. Chem. Soc. Jpn.*, 1989, **62**, 2219.
963 M. Mohapatra, V. Chakravortty, and K.C. Dash, *Polyhedron*, 1989, **8**, 1509.
964 A.A.-C. Reádigos, R.C. Vázquez, and J.R.d.B. Arín, *Transition Met. Chem. (Weinheim, Ger.)*, 1989, **14**, 42.
965 M.S. Hussain, H.M. Al-Mohdhar, and A.R. Al-Arfaj, *J. Coord. Chem.*, 1988, **18**, 339.
966 P.J. Toscano, A.L. Seligson, M.T. Curran, A.T. Skrobutt, and D.C. Sonnenberger, *Inorg. Chem.*, 1989, **28**, 166.
967 R. Seeber, W.O. Parker, jun., P.A. Marzilli, and L.G. Marzilli, *Organometallics*, 1989, **8**, 2377.
968 W. Mikolaiski, G. Baum, W. Massa, and R.W. Hoffmann, *J. Organomet. Chem.*, 1989, **376**, 397.
969 H.M. Marques, D.C. Scooby, M. Victor, and K.L. Brown, *Inorg. Chim. Acta*, 1989, **162**, 151.
970 T. Ama, H. Kawaguchi, M. Uchijama, N. Koine, and T. Yasui, *Bull. Chem. Soc. Japan*, 1989, **62**, 3464.
971 A.P. Gulya, Yu.V. Kokunov, S.G. Shova, M.D. Mazus, V.F. Rudik, E.N. Kiryak, and Yu.A. Buslaev, *Dokl. Akad. Nauk SSSR*, 1989, **305**, 627 (*Chem. Abstr.*, 1989, **111**, 205 922).
972 K.P. Simonsen, N. Suzuki, M. Hamada, M. Kojima, S. Ohba, Y. Saito, and J. Fujita, *Bull. Chem. Soc. Jpn.*, 1989, **62**, 3790.
973 F.G.M. Niele, C.F. Martens, and R.J.M. Nolte, *J. Am. Chem. Soc.*, 1989, **111**, 2078.
974 P. Piraino, G. Bruno, G. Tresoldi, S. Lo Schiavo, and F. Nicolo, *Inorg. Chem.*, 1989, **28**, 139.
975 E. Rotondo, B.E. Mann, P. Piraino, and G. Tresoldi, *Inorg. Chem.*, 1989, **28**, 3070.
976 K. Yamanari and M. Nakamichi, *J. Chem. Soc., Chem. Commun.*, 1989, 1723.
977 S. Tsuboyama, S. Miki, T. Chijimatsu, K. Tsuboyama, and T. Sakurai, *J. Chem. Soc., Dalton Trans.*, 1989, 2359.
978 M. Strasak, J. Lucansky, P. Novomesky, and E. Dvorakova, *J. Coord. Chem.*, 1989, **19**, 359.
979 D. Atzei, P. Deplano, and E.F. Trogu, *Transition Met. Chem. (Weinheim, Ger.)*, 1989, **14**, 287.
980 S.S. Al-Showiman, *Orient. J. Chem.*, 1989, **5**, 20 (*Chem. Abstr.*, 1989, **111**, 246 674).
981 M. Ghedini, M. Longeri, F. Neve, A.M.M. Lanfredi, and A. Tiripicchio, *J. Chem. Soc., Dalton Trans.*, 1989, 1217.

dmpe)][C_5H_5]$_2$, (^{13}C),[982] [{H_2C=C(CH_2PPh_2)$_2$}$_2$Rh]$^+$, (^{13}C),[983] [{Ph_2PC(=CH_2)C(=CH_2)-PPh_2}$_2$Rh]$^+$, (^{13}C),[984] [Co(1-ON-2-O-$C_{10}H_6$)$_3$], (^{13}C),[985] [Rh(C_2O_4)$_3$]$^{3-}$, (^{13}C, ^{17}O, ^{103}Rh),[986] [(H_2O)$_4$Ir(μ-OH)$_2$Ir(OH_2)$_4$]$^{4+}$, (^{17}O),[987] [Co(1,4,8,11-tetrathiacyclotetradecane)X_2]$^+$, (^{59}Co),[988] *cis*-[RhCl_2{CH_2($CH_2SCH_2CH_2SCH_2$)$_2CH_2$}]$^+$, (^{13}C),[989] and [Rh{H_2C=CHS(CH_2)$_2$-S(CH_2)$_2$S}(1,4,7-trithiacyclononane)]$^{2+}$, (^{13}C).[990]

Complexes of Ni, Pd, and Pt——A review of X-H correlation, COSY and NOESY as applied to some palladium and platinum complexes has appeared.[991]

^{19}F n.m.r. spectroscopy has been used to study the transmissive ability of *trans*-[Pt-(C_6H_4X)(PPh_3)$_2$(SC_6H_4F-4)][992] and *trans*-[Pt(C_6H_4X)(NC_5H_5)$_2$(SC_6H_4F-4)].[993] N.m.r. data have also been reported for *trans* -[NiHX(PCy_3)$_2$], (^{13}C),[994] [PdH(PPh_3)$_3$]$^+$, (^{13}C),[995] [PtH-(PPh_3)($AuPPh_3$)$_7$]$^{2+}$, (^{195}Pt),[996] *trans*-[PtH(NHPh)(PEt_3)$_2$], (^{13}C, ^{15}N),[997] [Pt_2(H)$_2$(μ-diop)-(diop)$_2$]$^{2+}$, (^{195}Pt),[998] [Ni{OC(CF_3)$_2CH_2PPh_2$}Et(PCy_3)], (^{13}C),[999] [(tmeda)NiCH_2CH_2C(O)O], (^{13}C),[1000] [(η^5-C_5H_5)Ni(PR^1_3)R^2], (^{13}C),[1001] [(η^3-C_3H_5)Ni(η^5-$C_3B_2Me_4R$-η^5)Ni(η^1-C_3H_5)$_2$], (^{11}B, ^{13}C),[1002] [(C_5H_5N)(PMe_3)Ni$C_6H_4CMe_2CH_2$-2], (^{13}C),[1003] [(PMe_3)$_2$Ni$C_6H_4CMe_2CH_2$-2], (^{13}C),[1004] [Ni($CH_2C_6H_4$-2-Me)(PMe_3)$_2$(μ-OH)]$_2$, (^{13}C),[1005] [(dppe)NiCH(SPh)CH_2C(O)N-

982 A.K. Kakkar, N.J. Taylor, and T.B. Marder, *Organometallics*, 1989, **8**, 1765.

983 H. Schmidbaur, C. Paschalidis, O. Steigelmann, and G. Müller, *Chem. Ber.*, 1989, **122**, 1851; *ibid.*, 1825.

984 H. Schmidbaur, K. Dziwok, A. Grohmann, and G. Müller, *Chem. Ber.*, 1989, **122**, 893.

985 S.I. Tyukhtenko, A.T. Pilipenko, N.N. Gerasimchuk, and T.I. Makarenko, *Dokl. Akad. Nauk Ukr. SSR, Ser. B: Geol., Khim. Biol. Nauki*, 1988, 57 (*Chem. Abstr.*, 1989, **111**, 69 783).

986 A.B. Benediktov, A.A. Kazakova, M.A. Fedotov, and A.V. Belyaev, *Koord. Khim.*, 1989, **15**, 822 (*Chem. Abstr.*, 1989, **111**, 208 114).

987 S.E. Castillo-Blum, D.T. Richens, and A.G. Sykes, *Inorg. Chem.*, 1989, **28**, 954.

988 J.J. Jenkinson, W. Levason, R.J. Perry, and M.D. Spicer, *J. Chem. Soc., Dalton Trans.*, 1989, 453.

989 A.J. Blake, G. Reid, and M. Schröder, *J. Chem. Soc., Dalton Trans.*, 1989, 1675.

990 A.J. Blake, A.J. Holder, T.I. Hyde, H.-J. Küppers, M. Schröder, S. Stötzel, and K. Wieghart, *J. Chem. Soc., Chem. Commun.*, 1989, 1600.

991 P.S. Pregosin and C. Aumann, *Pure Appl. Chem.*, 1989, **61**, 1771.

992 S.I. Pombrik, V.M. Pachevskaya, L.S. Golovchenko, A.s. Peregudov, and D.N. Kravtsov, *Metalloorg. Khim.*, 1988, **1**, 150 (*Chem. Abstr.*, 1989, **110**, 154 527).

993 D.N. Kravtsov, S.I. Pombrik, V.M. Pachevskaya, L.S. Golovchenko, A.S. Peregudov, and E.I. Fedin, *Metalloorg. Khim.*, 1988, 1, 468 (*Chem. Abstr.*, 1989, **110**, 57 829).

994 M.Y. Darensbourg, M. Ludwig, and C.G. Riordan, *Inorg. Chem.*, 1989, **28**, 1630.

995 V.N. Zudin, V.D. Chinakov, V.M. Nekipelov, V.A. Rogov, V.A. Likholobov, and Yu.I. Ermakov, *J. Mol. Catal.*, 1989, **52**, 27.

996 R.P.F. Kanters, J.J. Bour, P.P.J. Schlebos, W.P. Bosman, H. Behm, J.J. Steggerda, L.N. Ito, and L.H. Pignolet, *Inorg. Chem.*, 1989, **28**, 2591.

997 R.L. Cowan and W.C. Trogler, *J. Am. Chem. Soc.*, 1989, **111**, 4750.

998 E. Paumard, A. Mortreux, and F. Petit, *J. Chem. Soc., Chem. Commun.*, 1989, 1380.

999 U. Müller, W. Keim, C. Krüger, and P. Betz, *Angew. Chem., Int. Ed. Engl.*, 1989, **28**, 1011.

1000 R. Fischer, B. Nestler, and H. Schütz, *Z. Anorg. Allg. Chem.*, 1989, **577**, 111.

1001 S.I. Black and G.B. Young, *Polyhedron*, 1989, **8**, 585.

1002 G. Brodt, T. Kuhlmann, and W. Siebert, *Chem. Ber.*, 1989, **122**, 829.

1003 E. Carmona, M. Paneque, M.L. Poveda, E. Gutiérrez-Puebla, and A. Monge, *Polyhedron*, 1989, **8**, 1069.

1004 E. Carmona, E. Gutiérrez-Puebla, J.M. Marín, A. Monge, M. Paneque, M.L. Poveda, and C. Ruiz, *J. Am. Chem. Soc.*, 1989, **111**, 2883.

1005 E. Carmona, J.M. Marín, P. Palma, M. Paneque, and M.L. Poveda, *Inorg. Chem.*, 1989, **28**, 1895.

Ph], (^{13}C),[1006] [(η^2-H_2C=$CHCH_2CH_2)_2CHNi(\mu$-$O_2C_6H_4)_2B$], (^{13}C),[1007] [(η^5-C_5H_5)-$NiPBu^t{}_2SC(CO_2Me)$=$C(CO_2Me)$], (^{13}C),[1008] (34), (^{13}C),[1009] [(dppe)$NiCF_2CH_2C(O)NPh$],

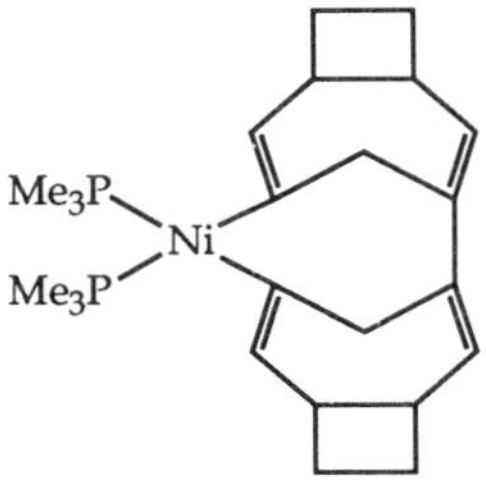

(34)

(^{13}C),[1010] [$(Et_3P)_2NiCPh$=$C(SiMe_3)Si(SiMe_3)C_6H_2Me_3$-2,4,6], ($^{13}C$, ^{29}Si),[1011] [(Bu^tNC)ClNi-C(=NBu^t)C(=NBu^t)CR=NBu^t], (^{13}C),[1012] [$(Me_3P)_2INiC(C{\equiv}CSiMe_3)$=$CSiMe_3C{\equiv}CSiMe_3$], ($^{13}C$),[1013] [($\eta^5$-$C_5H_5$)$NiC_6H_4N$=NPh], ($^{13}C$),[1014] [$(OC)_3Ni$=$CBu(OSiMe_3)$], ($^{13}C$),[1015] [($\eta^6$-$C_6H_3Me_3$-1,3,5)$Ni(SiEt_2Al_2Cl_7)_2$], ($^{13}C$, ^{27}Al, ^{29}Si),[1016] [{$(MeO)_3P$}$_2Ni(SiCl_3)_2(CO)$], (^{13}C),[1017] [(η^5-$C_5H_5)_2Ni_2(\mu$-$Me_3SiC{\equiv}CSiMe_3$)], (^{13}C),[1018] [PdMe(η^3-C_3H_5)L], (^{13}C),[1019] *cis*-[$PdMePh(PEt_2Ph)_2$], (^{13}C),[1020] [(tmeda)$PdCl(CH_2Cl)$], (^{13}C),[1021] [PdCl(μ-C_5H_4N-2-CH_2)-$(PPh_3)_2]_2$, (^{13}C),[1022] [Pd{C(CO_2Me)=C(CO_2Me)CH[C(O)Ph]PPh_2}$_2$], (^{13}C),[1023] [Pd(CR=CRCR=CR-1-2-$C_6H_3XCH_2NMe_2$)Y], (^{13}C),[1024] [PdIPh(tmeda)], (^{13}C),[1025] [(Bu^tS-

1006 H. Hoberg and D. Guhl, *J. Organomet. Chem.*, 1989, **375**, 245.
1007 R. Taube, P. Böhme, and J.-P. Gehrke, *Z. Anorg. Allg. Chem.*, 1989, **578**, 89.
1008 E. Lindner and A. Nothdurft, *Z. Anorg. Allg. Chem.*, 1989, **579**, 200.
1009 D. Bläser, R. Boese, W.A. Brett, P. Rademacher, H. Schwager, A. Stanger, and K.P.C. Vollhardt, *Angew. Chem., Int. Ed. Engl.*, 1989, **28**, 206.
1010 H. Hoberg and D. Guhl, *J. Organomet. Chem.*, 1989, **378**, 279.
1011 J. Ohshita, Y. Isomura, and M. Ishikawa, *Organometallics*, 1989, **8**, 2050.
1012 E. Carmona, J.M. Marín, P. Palma, and M.L. Poveda, *J. Organomet. Chem.*, 1989, **377**, 157.
1013 H.-F. Klein, H. Beck-Hemetsberger, L. Reitzel, B. Rodenhäuser, and G. Cordier, *Chem. Ber.*, 1989, **122**, 43.
1014 G.K. Anderson, R.J. Cross, K.W. Muir, and L. Manojlović-Muir, *J. Organomet. Chem.*, 1989, **362**, 225.
1015 J.L. Simunic and A.R. Pinhas, *Inorg. Chem.*, 1989, **28**, 2400.
1016 S.-B. Choe, H. Kanai, and K.J. Klabunde, *J. Am. Chem. Soc.*, 1989, **111**, 2875.
1017 S.-B. Choe, J.J. Schneider, K.J. Klabunde, L.J. Radonovich, and T.A. Ballantine, *J. Organomet. Chem.*, 1989, **376**, 419.
1018 H. Lehmkuhl, F. Danowski, G. Mehler, J. Popławska, and S. Pasynkiewicz, *J. Organomet. Chem.*, 1989, **363**, 387.
1019 Y. Hayashi, K. Matsumoto, Y. Nakamura, and K. Isobe, *J. Chem. Soc., Dalton Trans.*, 1989, 1519.
1020 F. Ozawa, K. Kurihara, M. Fujimori, T. Hidaka, T. Toyoshima, and A. Yamamoto, *Organometallics*, 1989, **8**, 180.
1021 R. McCrindle, G.J. Arsenault, R. Farwaha, A.J. McAlees, and D.W. Sneddon, *J. Chem. Soc., Dalton Trans.*, 1989, 761.
1022 K. Isobe, Y. Nakamura, and S. Kawaguchi, *Bull. Chem. Soc. Jpn.*, 1989, **62**, 1802.
1023 F. Balegroune, P. Braunstein, T.M.G. Carneiro, D. Grandjean, and D. Matt, *Chem. Ber.*, 1989, 582.
1024 W. Tao, L.J. Silverberg, A.L. Rheingold, and R.F. Heck, *Organometallics*, 1989, **8**, 2550.
1025 W. de Graaf, J. van Wegen, J. Boersma, A.L. Spek, and G. van Koten, *Recl. Trav. Chim. Pays-Bas*, 1989, **108**, 275.

$CH_2Bu^t)PhPd(\mu\text{-}OAc)_2Pd(\mu\text{-}OAc)_2PdPh(Bu^tSCH_2Bu^t)]$, ($^{13}C$),[1026] (35), ($^{13}C$),[1027] (36), (M = Pd, Pt; ^{13}C),[1028] [Pd{3-furyl-2-C(NMe_2)=Se}Cl]$_2$, (^{13}C),[1029] [Pd{$C_6H_3(CH_2SMe)_2$-2,6}Cl], (^{13}C),[1030] [M{CH(CO_2Me)COCHCO_2Me}(CNBu^t)L], (M = Pd, Pt; ^{13}C),[1031] (37), (R = CO_2Me;

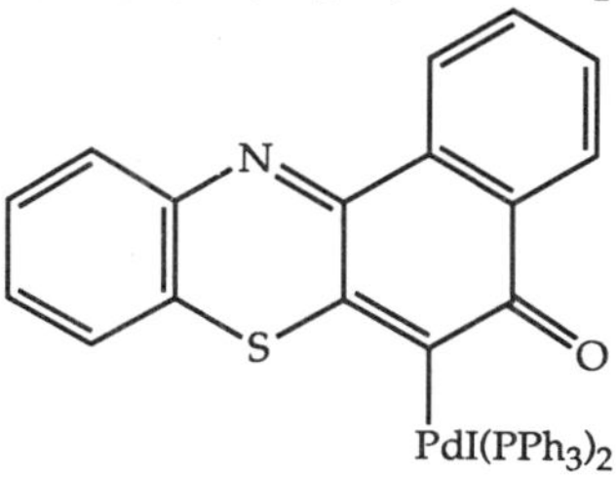

(35)

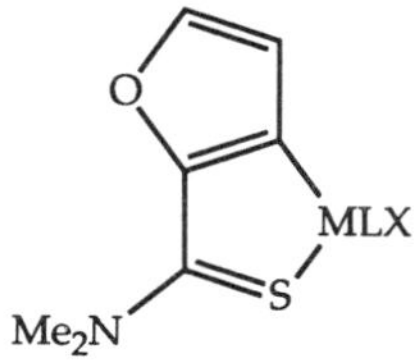

(36)

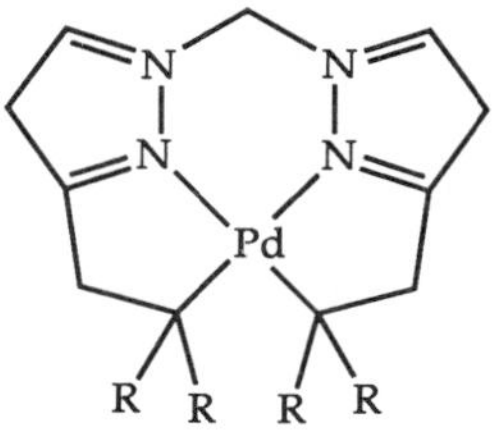

(37)

^{13}C),[1032] [$R^1CH=CR^2CH_2C(O)PdX(PMe_3)_2$], ($^{13}C$),[1033] *trans*-[Pd{C(O)Ph}{C(OH)NR_2}-$(PMe_3)_2]^+$, (^{13}C),[1034] $[N(CH_2CH_2PPh_2)_3MR]^+$, (M = Pd, Pt; R = HgMe, $SnMe_3$; ^{119}Sn, ^{195}Pt, ^{199}Hg),[1035] *cis*-[$(MePPh_2)_2Pd(SiHPh_2)_2$], (^{29}Si),[1036] [$PtMe_3(NO_3)L_2$], (^{13}C),[1037] [Me_2-Pt(C_6H_4-2-CH=$NCH_2CH_2NMe_2$)Br], (^{13}C),[1038] *cis*-[PtMe($CH_2SiMe_2CH_2SiMe_3$)$(PEt_3)_2$], (^{13}C),[1039] *trans*-$[PtEt(CO)(PPh_3)_2]^+$, (^{13}C),[1040] [$Me_2Pt(CH_2CH_2OCO_2)$(bipy)], (^{13}C),[1041] [Pt$(CH_2Bu^t)_2$(bipy)], (^{13}C),[1042] [(η^4-C_8H_{12})Pt$(CH_2SiMe_2Ph)_2$], (^{13}C),[1043] [(η^4-C_7H_8)PtR_2], (^{13}C),[1044] *cis*-[(η^4-C_8H_{12})Pt$(CH_2OH)_2$], (^{13}C),[1045] [Pt(CH_2OAc)(CH_2Cl){C(O)Me}$(PMe_3)_2$Cl], (^{13}C),[1046] [Pt(C_6H_4-2-CH_2)$(PR_3)_2$], (^{13}C),[1047] [Pt{CH(COPh)S(O)$_2$CH(COPh)}$(PPh_3)_2$],

1026 Y. Fuchita, K. Hiraki, Y. Kamogawa, M. Suenaga, K. Tohgoh, and Y. Fujiwara, *Bull. Chem. Soc. Jpn.*, 1989, **62**, 1081.
1027 H. Konishi, C. Matsumura, T. Okano, and J. Kiji, *J. Organomet. Chem.*, 1989, **364**, 245.
1028 M. Nonoyama, *Inorg. Chim. Acta*, 1989, **157**, 9.
1029 M. Nonoyama and K. Nonoyama, *Polyhedron*, 1989, **8**, 2517.
1030 J. Dupont, N. Beydoun, and M. Pfeffer, *J. Chem. Soc., Dalton Trans.*, 1989, 1715.
1031 R.D.W. Kemmitt, P. McKenna, D.R. Russell, and L.J.S. Prouse, *J. Chem. Soc., Dalton Trans.*, 1989, 345.
1032 O. Juanes, J. de Mendoza, and J.C. Rodriguez-Ubis, *J. Organomet. Chem.*, 1989, **363**, 393.
1033 F. Ozawa, T.-i. Son, K. Osakada, and A. Yamamoto, *J. Chem. Soc., Chem. Commun.*, 1989, 1067.
1034 L. Huang, F. Ozawa, K. Osakada, and A. Yamamoto, *Organometallics*, 1989, **8**, 2065.
1035 C.A. Ghilardi, S. Midollini, S. Moneti, A. Orlandini, G. Scapacci, and D. Dakternieks, *J. Chem. Soc., Chem. Commun.*, 1989, 1686.
1036 U. Schubert and C. Müller, *J. Organomet. Chem.*, 1989, **373**, 165.
1037 V. Kjain, *Indian J. Chem., Sect. A*, 1987, **26A**, 1019 (*Chem. Abstr.*, 1989, **110**, 115 076).
1038 C.M. Anderson, R.J. Puddephatt, G. Ferguson, and A.J. Lough, *J. Chem. Soc., Chem. Commun.*, 1989, 1297.
1039 S.K. Thomson and G.B. Young, *Organometallics*, 1989, **8**, 2068.
1040 V.D. Chinakov, G.N. Il'inich, V.N. Zudin, V.A. Likholobov, and V.M. Nekipelov, *J. Organomet. Chem.*, 1989, **366**, 421.
1041 K.-T. Aye, G. Ferguson, A.J. Lough, and R.J. Puddephatt, *Angew. Chem., Int. Ed. Engl.*, 1989, **28**, 767.
1042 V. Christou and G.B. Young, *J. Organomet. Chem.*, 1989, **368**, 391.
1043 B.C. Ankianiec and G.B. Young, *Polyhedron*, 1989, **8**, 57.
1044 R.D. Kelly and G.B. Young, *J. Organomet. Chem.*, 1989, **361**, 123.
1045 J.F. Hoover and J.M. Stryker, *J. Am. Chem. Soc.*, 1989, **111**, 6466.
1046 J.F. Hoover and J.M. Stryker, *Organometallics*, 1989, **8**, 2973.
1047 H.J.R. de Boer, G. Schat, O.S. Akkerman, F. Bickelhaupt, M. de Wit, and C.H. Stam, *Organometallics*, 1989, **8**, 1288.

(^{13}C),[1048] $[\{Ph_2PC(COPh)=C(CO_2Me)CH(CO_2Me)\}_2Pt]$, (^{13}C),[1049] $[Pt\{CH(CO_2Me)_2\}_2(phen)]$, (^{13}C),[1050] $[(\eta^4\text{-}C_7H_8)Pt(1\text{-naphthyl})_2]$, (^{13}C),[1051] $[Pt\{C(C{\equiv}CR)=CMe_2\}(PPh_3)_3]^+$, (^{13}C),[1052] $[Pt(CF_3)_2(\eta^4\text{-}C_7H_8)]$, (^{13}C, ^{195}Pt),[1053] $[PtAr_2(\eta^4\text{-}C_7H_8)]$, (^{13}C),[1054] $[PtCl(C{\equiv}CBu^t)(\mu\text{-dppm})_2Au]^+$, (^{13}C),[1055] $[(Ph_3P)(OC)Pt(AuPPh_3)_6]^{2+}$, (^{13}C),[1056] $[(Cy_2PCH_2CH_2PCy_2)PtCl\{SiR(SiMe_3)_2\}]$, (^{29}Si),[1057] and $[Pt_3(\mu_3\text{-}SnF_3)(\mu_3\text{-}CO)(\mu\text{-dppm})_3]^+$, (^{119}Sn, ^{195}Pt).[1058]

Neighbouring group effects in *trans*-$[PtCl_2(\eta^2\text{-}C_2H_4)L]$ have been examined by ^{1}H, ^{13}C, ^{31}P, and ^{195}Pt n.m.r. spectroscopy.[1059] The ^{13}C chemical shift of the allyl carbon nuclei in $[(\eta^3\text{-}1,1\text{-}Me_2C_3H_3)PdL_2]^+$ has been correlated with reactivity.[1060] N.m.r. data have also been reported for $[(H_3N)Ni(\eta^2\text{-}C_2H_4)_2]$, (^{13}C),[1061] $[(tmeda)Ni(\eta^2\text{-}H_2C=CHCO_2Me)_2]$, (^{13}C),[1062] $[(bipy)Ni(\eta^2\text{-}CH_2=CF_2)]$, (^{13}C),[1063] (38), (^{13}C),[1064] $[Pt(PPh_3)(PR_3)(\eta^2\text{-}C_2H_4)]$,

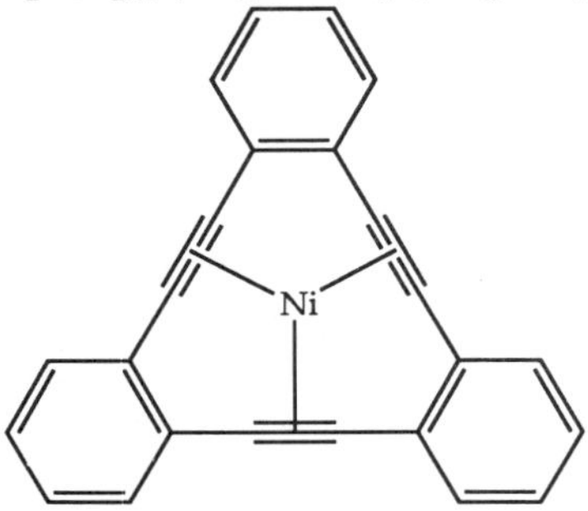

(38)

(^{13}C),[1065] $[(\eta^2\text{-}C_2H_4)PtCl_2(NHCHEtCH_2OCHC_6H_4\text{-}4\text{-}Cl)]$, (^{13}C),[1066] $[(R^1{}_2PCH_2CH_2PR^1{}_2)Pt$-

1048 W. Henderson, R.D.W. Kemmitt, L.J.S. Prouse, and D.R. Russell, *J. Chem. Soc., Dalton Trans.*, 1989, 259.
1049 P. Braunstein, T.M.G. Carneiro, D. Matt, F. Balegroune, and D. Grandjean, *Organometallics*, 1989, **8**, 1737.
1050 G.R. Newkome, K.J. Theriot, F.R. Fronczek, and B. Villar, *Organometallics*, 1989, **8**, 2513.
1051 C. Weisemann, G. Schmidtberg, and H.-A. Brune, *J. Organomet. Chem.*, 1989, **362**, 63.
1052 P.J. Stang and A.K. Datta, *Organometallics*, 1989, **8**, 1024.
1053 T.G. Appleton, R.D. Berry, J.R. Hall, and D.W. Neale, *J. Organomet. Chem.*, 1989, **364**, 249.
1054 C. Weisemann, G. Schmidtberg, and H.-A. Brune, *J. Organomet. Chem.*, 1989, **365**, 403.
1055 L. Manojlović-Muir, A.N. Henderson, I. Treurnicht, and R.J. Puddephatt, *Organometallics*, 1989, **8**, 2055.
1056 L.N. Ito, J.D. Sweet, A.M. Mueting, L.H. Pignolet, M.F.J. Schoondergang, and J.J. Steggerda, *Inorg. Chem.*, 1989, **28**, 3696.
1057 L.S. Chang, M.P. Johnson, and M.J. Fink, *Organometallics*, 1989, **8**, 1369.
1058 G. Douglas, M.C. Jennings, L. Manojlović-Muir, K.W. Muir, and R.J. Puddephatt, *J. Chem. Soc., Chem. Commun.*, 1989, 159.
1059 I.M. Al-Najjar, *Spectrochim. Acta*, 1988, **44A**, 57 (*Chem. Abstr.*, 1989, **110**, 193 049).
1060 B. Åkermark, S. Hansson, B. Krakenberger, K. Zetterberg, and A. Vitagliano, *Chem. Scr.*, 1987, **27**, 525.
1061 W. Kaschube, K.R. Pörschke, W. Bonrath, C. Krüger, and G. Wilke, *Angew. Chem., Int. Ed. Engl.*, 1989, **28**, 772.
1062 W. Kaschube, K.R. Pörschke, K. Seevogel, and C. Krüger, *J. Organomet. Chem.*, 1989, **367**, 233.
1063 H. Hoberg and D. Guhl, *J. Organomet. Chem.*, 1989, **373**, C27.
1064 J.D. Ferrara, A.A. Tanaka, C. Fierro, C.A. Tessier-Youngs, and M.J. Youngs, *Organometallics*, 1989, **8**, 2089.
1065 P.A. Chaloner and G.T.L. Broadwood-Strong, *J. Organomet. Chem.*, 1989, **362**, C21.

(η^2-R^2_2Si=SiR^2_2)], (^{29}Si),[1067] [Pt(η^2-$MeO_2CC{\equiv}CCO_2Me$)(PPh_3)$_2$], (^{13}C),[1068] [Pt_3(μ_3-η^2-HC≡CH){μ_2-η^3-(Ph_2P)$_2$CHCH=CH}(μ-dppm)$_2$]$^+$, (^{195}Pt),[1069] [(η^3-2-MeC_3H_4)Pd(Ph_2POCH_2-$C_4H_7NPPh_2$)]$^+$, (^{13}C),[1070] [(Cy_3P)$_2Pd_2$(*anti*-μ-η^3-CH_2CHCHMe)(μ-SPh)], (^{13}C),[1071] [Pd(η^3-$CH_2CCHCH_2CH_2CHMeCHClCH_2$)Cl]$_2$, ($^{13}C$),[1072] [Pd{$\eta^3$-CHCMeCMe$CH_2$CH(OR)$CH_2$}-Cl]$_2$, ($^{13}C$),[1073] [Pd($\eta^3$-$Me_3SiCR^1CR^2CHC_6H_4NMe_2$-2)Cl]$_2$, ($^{13}C$),[1074] [($\eta^5$-$C_5H_5$)Ni($\eta^4$-$C_4H_5$-BPh)], ($^{11}B$, ^{13}C),[1075] [(η^4-C_8H_{12})$_2Pt_2$(μ-SO_2)(μ-η^2-SO_2)], (^{13}C, ^{195}Pt),[1076] [Zn_4Ni_2(η^5-C_5H_5)$_4$-(η^5-C_5Me_5)$_4$], (^{13}C),[1077] and [Pt_3(μ_3-S)(CE)(μ-dppm)$_3$]$^{2+}$, (E = N^-, O, S; ^{13}C, ^{195}Pt).[1078]

The ^{31}P n.m.r. spectra of [Ni{P(C_6H_4X-4)$_3$}$_4$] show a linear correlation of the coordination shift, pK_a of the tertiary phosphine, and the Hammett sigma function.[1079] ^{31}P n.m.r. spectroscopy has been used to determine the enantiomeric purity of 1-aminoalkylphosphonic acids *via* their diastereomeric Pd(II) complexes.[1080] Platinum amine induced distortions in single and double-stranded oligodeoxyribonucleotides have been investigated by ^1H and ^{31}P n.m.r. spectroscopy.[1081] N.m.r. data have also been reported for [(CH_2=$CHCO_2CH_2CH_2CH_2NC$)Pd(PPh_3)Cl_2], (^{13}C),[1082] [(Ph_3P)$_2$PtN=CPhN=PPh_2N-$SiMe_3$], (^{13}C, ^{29}Si),[1083] [($RNHCH_2CH_2NH_2$)$PtCl_2$], (^{13}C, ^{195}Pt),[1084] [M{NHNHC(S)Ph}$_2$], (M = Ni, Zn; ^{13}C),[1085] [Ni{ON=CR^1CMe=NNHC(SR^2)=NN=NMeCR^1=NO}], (^{13}C),[1086] [M{HN($CH_2CH_2CH_2PMe_2$)$_2$}], (M = Ni, Pd, Pt; ^{13}C),[1087] [M(ECMeCHCMeE)$_2$], (E = O, S, NH; M = Ni, Zn; ^{13}C),[1088] [{CH(CMeN-1,2-C_6H_4NCMe)$_2$CH}Ni], (^{13}C),[1089] (39), (^{13}C),[1090]

1066 A. Albinati, C. Arz, and P.S. Pregosin, *J. Organomet. Chem.*, 1989, **371**, C18.
1067 E.K. Pham and R. West, *J. Am. Chem. Soc.*, 1989, **111**, 7667.
1068 J.A. Davies, C.T. Eagle, D.E. Otis, and U. Venkataraman, *Organometallics*, 1989, **8**, 1080.
1069 M.C. Jennings, L. Manojlović-Muir, and R.J. Puddephatt, *J. Am. Chem. Soc.*, 1989, **111**, 745.
1070 E. Cesarotti, M. Grassi, and L. Prati, *J. Chem. Soc., Dalton Trans.*, 1989, 161.
1071 K. Osakada, T. Chiba, Y. Nakamura, T. Yamamoto, and A. Yamamoto, *Organometallics*, 1989, **8**, 2602.
1072 W.A. Donaldson, J. Wang, V.G. Cepa, and J.D. Suson, *J. Org. Chem.*, 1989, **54**, 6056.
1073 B. Åkermark, B.C. Söderberg, and S.S. Hall, *J. Org. Chem.*, 1989, **54**, 1110.
1074 F. Maassarani, M. Pfeffer, and G. Van Koten, *Organometallics*, 1989, **8**, 871.
1075 G.E. Herberich, I. Hausmann, B. Hessner, and M. Negele, *J. Organomet. Chem.*, 1989, **362**, 259.
1076 D.H. Farrar and R.R. Gukathasan, *J. Chem. Soc., Dalton Trans.*, 1989, 557.
1077 B. Fischer, H. Kleijn, J. Boersma, G. van Koten, and A.L. Spek, *Organometallics*, 1989, **8**, 920.
1078 A.M. Bradford, M.C. Jennings, and R.J. Puddephatt, *Organometallics*, 1989, **8**, 2367.
1079 K.J. Fisher and E.C. Alyea, *Polyhedron*, 1989, **8**, 13.
1080 Z. Głowacki, M. Topolski, E. Matczak-Jon, and M. Hoffmann, *Magn. Reson. Chem.*, 1989, **27**, 922.
1081 T.P. Kline, L.G. Marzilli, D. Live, and G. Zon, *J. Am. Chem. Soc.*, 1989, **111**, 7057.
1082 U. Casellato, B. Corain, M. Zecca, R.A. Michelin, M. Mozzon, and R. Graziani, *Inorg. Chim. Acta*, 1989, **156**, 165.
1083 H.W. Roesky, U. Scholz, and M. Noltemeyer, *Z. Anorg. Allg. Chem.*, 1989, **576**, 255.
1084 B.E. Bowler, K.J. Ahmed, W.I. Sundquist, L.S. Hollis, E.E. Whang, and S.J. Lippard, *J. Am. Chem. Soc.*, 1989, **111**, 1299.
1085 N.K. Singh and U. Sharma, *Synth. React. Inorg. Metal-Org. Chem.*, 1989, **19**, 235.
1086 N.V. Gerbeleu, A.A. Dobrov, and L.A. Nemchinova, *Zh. Neorg. Khim.*, 1988, **33**, 1190 (*Chem. Abstr.*, 1989, **110**, 68 450).
1087 P.G. Edwards, and R.G. Jaouhari, *Polyhedron*, 1989, **8**, 25.
1088 A.M.A. Bennett, G.A. Foulds, and D.A. Thornton, *Polyhedron*, 1989, **8**, 2305.
1089 M. Basato, B. Corain, G. Favero, G. Valle, and J. Eilmes, *Inorg. Chim. Acta*, 1989, **159**, 59.
1090 N.A. Stephenson, H.E. Tweedy, and D.H. Busch, *Inorg. Chem.*, 1989, **28**, 4376.

[Ni{SCHMeC(O)NCH$_2$C(O)O}(CN)]$^-$, (^{13}C),[1091] [{Pd(en)(guanine ring)}$_4$], (^{15}N),[1092] [M(HNCPhNCPhNH$_2$)$_2$], (M = Pd, Pt; ^{13}C),[1093] [M{1-(2′,2″-bipy-6′-ylmethyl)-1,4,7-triazacyclononane}], (M = Pd, Zn; ^{13}C),[1094] [PdL$_2$Cl$_2$], {L = (40); ^{13}C},[1095] [M(2,4-Me$_2$-6-hydroxy-

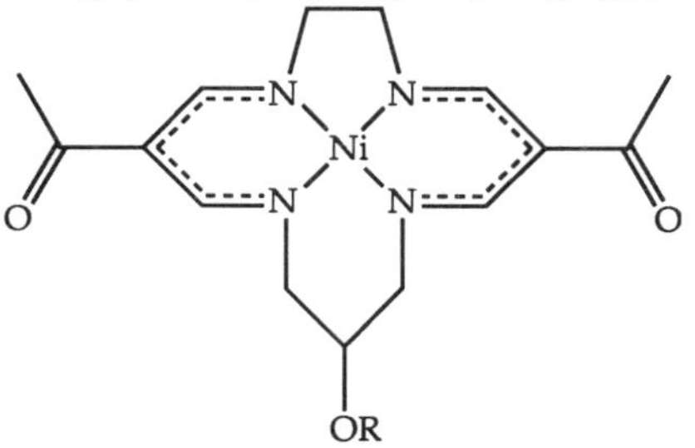

(39)

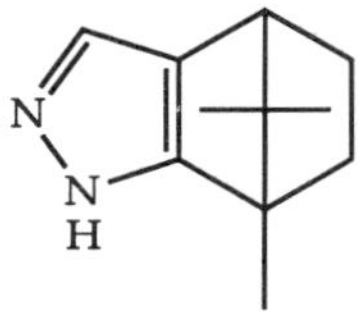

(40)

pyrimidine)$_2$Cl$_2$], (M = Pd, Pt; ^{13}C),[1096] [{1-Bu^{t}C$_6$H$_3$-3,5-[CH$_2$C(CMeNH)CMeCO]$_2$}$_2$Pd$_2$], (^{13}C),[1097] [Pd(NC$_4$H$_6$CO$_2$)(NC$_4$H$_6$)Cl], (^{13}C),[1098] *cis*-[Pt(NH$_3$)$_2$(N-donor)Cl]$^+$, (^{195}Pt),[1099] *trans*-[{PtCl$_2$(NH$_3$)}$_2${H$_2$N(CH$_2$)$_n$NH$_2$}], (^{13}C, ^{195}Pt),[1100] [Pt$_2$(NH$_3$)$_4$(C$_4$H$_6$NO)$_2$]$_2^{4+}$, (^{13}C, ^{195}Pt),[1101] [Pt$_2$(en)$_2$(C$_5$H$_4$NO)$_2$]$^{2+}$, (^{195}Pt),[1102] [Pt{1,2-(H$_2$N)$_2$C$_6$H$_{10}$}(PhPC$_{12}$H$_8$)$_2$]$^{2+}$, (^{13}C, ^{195}Pt),[1103] *cis*-[(RCO$_2$)$_2$Pt{*R*,*R*-1,2-(H$_2$N)$_2$C$_6$H$_{10}$}], (^{195}Pt),[1104] (^{13}C, ^{195}Pt),[1105] [{1,2-C$_6$H$_4$-(NH$_2$)$_2$}PtO$_2$CCHPhO], (^{13}C),[1106] [Pt{NC$_5$H$_4$-2-CH$_2$(CH$_2$CO$_2$H)$_2$}Cl$_2$], (^{195}Pt),[1107] [Pt{(O$_2$C)$_2$-C$_4$H$_6$}(1,3-dioxane-5,5-dimethanamine)], (^{13}C),[1108] [PtCl(PEt$_3$)(cyclamPH)]$^+$, (^{13}C, ^{195}Pt),[1109] [PtCl$_2${NHC(O)NH$_2$}$_2$], (^{195}Pt),[1110] *trans*-[PtCl$_2$L(1,4-diazine)], (^{195}Pt),[1111] [(H$_2$B-pz$_2$)Pt(PEt$_3$)Cl], (^{11}B, ^{13}C),[1112] [Pt(pzH)$_2$Cl$_2$], (^{13}C),[1113] [Pt(bipy)(μ-OH)]$_2^{2+}$, (^{13}C, ^{195}Pt),[1114]

1091 N. Baidya, M.M. Olmstead, and P.K. Mascharak, *Inorg. Chem.*, 1989, **28**, 3426.
1092 K. Uchida, A. Toyama, Y. Tamura, M. Sugimura, F. Mitsumori, Y. Furukawa, H. Takeuchi, and I. Harada, *Inorg. Chem.*, 1989, **28**, 2067.
1093 J. Barker, M. Kilner, M.M. Mahmoud, and S.C. Wallwork, *J. Chem. Soc., Dalton Trans.*, 1989, 837.
1094 N.W. Alcock, F. McLaren, P. Moore, G.A. Pike, and S.M. Roe, *J. Chem. Soc., Chem. Commun.*, 1989, 629.
1095 A.A. Watson, D.A. House, and P.J. Steel, *Polyhedron*, 1989, **8**, 1345.
1096 A.O. Adeyemo, *Inorg. Chim. Acta*, 1989, **159**, 99.
1097 J.R. Bradbury, J.L. Hampton, D.P. Martone, and A.W. Maverick, *Inorg. Chem.*, 1989, **28**, 2392.
1098 B. Wagner, U. Taubald, and W. Beck, *Chem. Ber.*, 1989, **122**, 1031.
1099 L.S. Hollis, A.R. Amundsen, and E.W. Stern, *J. Med. Chem.*, 1989, **32**, 128 (*Chem. Abstr.*, 1989, **110**, 32 871).
1100 N. Farrell and Y. Qu, *Inorg. Chem.*, 1989, **28**, 3416.
1101 K. Matsumoto, H. Miyamae, and H. Moriyama, *Inorg. Chem.*, 1989, **28**, 2959.
1102 T.V. O'Halloran and S.J. Lippard, *Inorg. Chem.*, 1989, **28**, 1289.
1103 T.K. Miyamoto, Y. Matsuura, K. Okude, H. Ichida, and Y. Sasaki, *J. Organomet. Chem.*, 1989, **373**, C8.
1104 A.R. Khokhar, S. Al-Baker, I.H. Krakoff, and R. Perez-Soler, *Cancer Chemother. Pharmacol.*, 1989, **23**, 219 (*Chem. Abstr.*, 1989, **110**, 199 066).
1105 A.R. Khokhar, S. Al-Baker, and G.J. Lumetta, *J. Coord. Chem.*, 1988, **18**, 291.
1106 T.K. Miyamoto, K. Okude, K. Maeda, H. Ichida, Y. Sasaki, and T. Tashiro, *Bull. Chem. Soc. Jpn.*, 1089, **62**, 3239.
1107 J.A. Winter, F.T.A. Lin, and R.E. Shepherd, *Inorg. Chim. Acta*, 1989, **155**, 155.
1108 P. Bitha, S.G. Carvajal, R.V. Citarella, E.F. Delos Santos, F.E. Durr, J.J. Hlvaka, S.A. Lang, jun., H.L. Lindsay, J.P. Thomas, R.E. Wallace, and Y.-i. Lin, *J. Med. Chem.*, 1989, **32**, 2063.
1109 D.E. Berry, J. Browning, G.W. Bushnell, K.R. Dixon, and A. Pidcock, *Can. J. Chem.*, 1989, **67**, 48.
1110 J. Anastassopoulou, P.K. Ganguli, and T. Theophanides, *Inorg. Chim. Acta*, 1989, **159**, 237.
1111 A. Albinati, F. Isaia, W. Kaufmann, C. Sorato, and L.M. Venanzi, *Inorg. Chem.*, 1989, **28**, 112.
1112 D.L. Reger, J.C. Baxter, and L. Lebioda, *Inorg. Chim. Acta*, 1989, **165**, 201.

[$Pt(guanosine)_2Cl_2$], (^{13}C, ^{15}N),[1115] [$Pt\{SC(O)CHMeNH_2\}(PBu^n_3)Cl$], ($^{13}C$),[1116] [$Ni(dppe)Br_2$], ($^{13}C$),[1117] (41), ($^{13}C$),[1118] [$Pd(PPh_3)_2$], ($^{13}C$),[1119] (42), (M = Pd, Pt; ^{13}C),[1120] [Pd(SCMeCH-

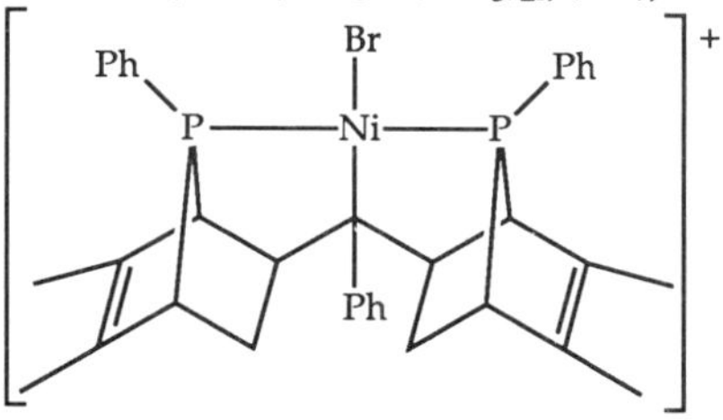

(41)

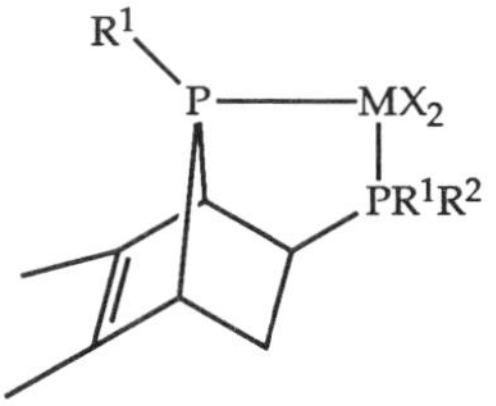

(42)

CMeS)I(PPh_3)], (^{13}C),[1121] [$Au_8Pd(PPh_3)_8$]$^{2+}$, (^{13}C),[1122] [$Pt(PR_3)_2X_2$], (^{195}Pt),[1123] [Pt{Et_2NP-($NMeCONMe)_2P$}$(PPh_3)_2$]$^+$, (^{195}Pt),[1124] [$(Me_2PhP)_2$PtSC{C(O)Me}=CMeO], (^{13}C),[1125] [$Pt(dmpe)_2I_2$]$^{2+}$, (^{195}Pt),[1126] [Pt_2Cl_3\{(*R**)-μ-$Pr^iPCH_2PPr^i_2$\}\{(*R**)-μ-$HPr^iPCH_2PPr^i_2$\}], (^{195}Pt),[1127] [Pt(1,1,4,7,10,10-Ph_6-1,4,7,10-tetraphosphadecane)]$^{2+}$, (^{195}Pt),[1128] [Pt_2(μ-S)(μ-dppm)(η^1-dppm)$_2$], (^{195}Pt),[1129] [Pt_2(μ-dppm)$_2$(μ-$AuPPh_3$)Cl_2]$^+$, (^{195}Pt),[1130] *cis*-[M{Et_2N-C(S)NHC(S)Ph}$_2$], (M = Ni, Pd, Pt; ^{13}C, ^{195}Pt),[1131] [Pd(acac)(acacH)]$^+$, (^{13}C),[1132] 1,3-dithiole-2-thione-4,5-diselenolate bis chelates of Pd, Au, Zn, Cd, and Hg, (^{77}Se),[1133] [M{meso-$PhTe(CH_2)_3TePh$}Br_2], (M = Pd, Pt; ^{125}Te, ^{195}Pt),[1134] [$Pt(Se_4)_3$]$^{2-}$, (^{77}Se),[1135] and [$Pt(S_2CNHCRCO_2H)_2$], (^{13}C).[1136]

1113 M.A. Cinellu, S. Stoccoro, G. Minghetti, A.L. Bandini, G. Banditelli, and B. Bovio, *J. Organomet. Chem.*, 1989, **372**, 311.
1114 S. Wimmer, P. Castan, F.L. Wimmer, and N.P. Johnson, *J. Chem. Soc., Dalton Trans.*, 1989, 403.
1115 G. Barbarella, A. Bertoluzza, M.A. Moreilli, M.R. Tosi, and V. Tugnoli, *Gazz. Chim. Ital.*, 1988, **118**, 637 (*Chem. Abstr.*, 1989, **110**, 154 783).
1116 I. Zahn, K. Polborn, and W. Beck, *Chem. Ber.*, 1989, **122**, 253.
1117 J.A. Rahn, A. Delian, and J.H. Nelson, *Inorg. Chem.*, 1989, **28**, 215.
1118 Lj. Solujić, E.B. Milosačljević, J.H. Nelson, N.W. Alcock, and J. Fischer, *Inorg. Chem.*, 1989, **28**, 3453.
1119 H. Urata, H. Suzuki, Y. Moro-oka, and T. Ikawa, *J. Organomet. Chem.*, 1989, **364**, 217.
1120 J.A. Rahn, M.S. Holt, G.A. Gray, N.W. Alcock, and J.H. Nelson, *Inorg. Chem.*, 1989, **28**, 217.
1121 M.A. Bennett, K.J. Cavell, and K.Y. Chan, *Inorg. Chim. Acta*, 1989, **163**, 153.
1122 L.N. Ito, B.J. Johnson, A.M. Mueting, and L.H. Pignolet, *Inorg. Chem.*, 1989, **28**, 2026.
1123 Z. Yang, X. Han, J. Hu, X. Liu, and Z. Liu, *Bopuxue Zazhi*, 1987, **4**, 317 (*Chem. Abstr.*, 1989, **110**, 24 070).
1124 L. Ernst, P. Look-Herber, R. Schmutzler, and D. Schomburg, *Polyhedron*, 1989, **8**, 2485.
1125 S. Yamazaki, T. Ama, M. Hojo, and T. Ueno, *Bull. Chem. Soc. Jpn.*, 1989, **62**, 4036.
1126 L.R. Hanton, W. Levason, and N.A. Powell, *Inorg. Chim. Acta*, 1989, **160**, 205.
1127 J.C. Huffman, B.R. Lloyd, and W.E. Streib, *Polyhedron*, 1989, **8**, 2169.
1128 P. Brueggeller, *Inorg. Chim. Acta*, 1989, **155**, 45.
1129 N. Hadj-Bagheri and R.J. Puddephatt, *Inorg. Chem.*, 1989, **28**, 2384.
1130 G.J. Arsenault and R.J. Puddephatt, *Can. J. Chem.*, 1989, **67**, 1800.
1131 R. Koehler, R. Richter, H. Tschoe, M. Moll, L. Beyer, and E. Hoyer, *Z. Chem.*, 1988, **28**, 369 (*Chem. Abstr.*, 1989, **110**, 87 384).
1132 I. Guibert, D. Neibecker, and I. Tkatchenko, *J. Chem. Soc., Chem. Commun.*, 1989, 1850.
1133 B. Olk, R.M. Olk, and E. Hoyer, *Z. Chem.*, 1989, **29**, 64 (*Chem. Abstr.*, 1989, **110**, 224 250).
1134 T. Kemmitt, W. Levason, and M. Webster, *Inorg. Chem.*, 1989, **28**, 692.
1135 M.A. Ansari and J.A. Ibers, *Inorg. Chem.*, 1989, **28**, 4068.
1136 J.J. Criado, A. Carrasco, B. Macias, J.M. Salas, M. Medarde, and M. Castillo, *Inorg. Chim. Acta*, 1989, **160**, 37.

Complexes of Cu, Ag, and Au——The nature of both lower and higher order cuprates containing allylic ligands has been investigated using variable temperature ^{13}C n.m.r. experiments.[1137] $^{1}J(^{13}C\text{-}^{7}Li)$ and $^{1}J(^{13}C\text{-}^{107}Ag)$ have been used to establish the structure of $[M_2Li_2(C_6H_4CH_2NMe_2\text{-}2)_4]$, M = Cu, Ag.[1138] COSY has been used to determine the structure of (43).[1139] Solvent effects on ^{19}F chemical shifts in $[(Ph_3P)AuC{\equiv}CC_6H_4F\text{-}4]$ have been examined.[1140] N.m.r. data have also been reported for (44), (^{13}C),[1141] $[(CF_3)_2\text{-}$

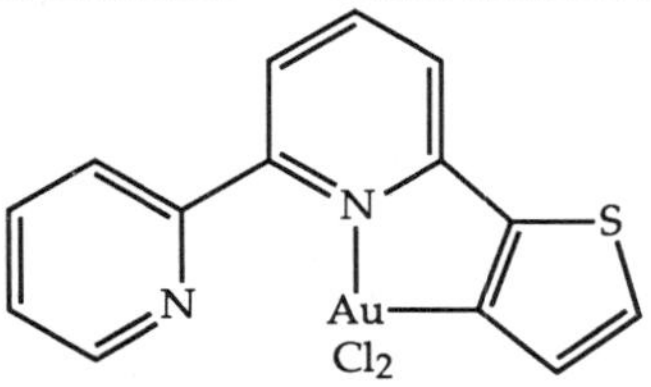

(43)

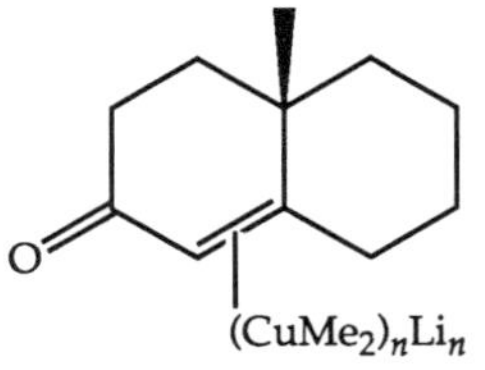

(44)

$Cu(S_2CNEt_2)]$, (^{13}C),[1142] $[2,4,6\text{-}Pr^i{}_3C_6H_2Cu]_4$, ($^{13}C$),[1143] $[(Me_3Si)_2C(BNPr^i{}_2)_2CHAu(PPh_3)]$, ($^{11}B$, ^{13}C),[1144] $[(Ph_3P)AuCHPhSO_2Bu^t]$, (^{13}C),[1145] $[Et_2NC(O)C\{Au(PPh_3)\}CH_2CH_2]$, ($^{13}C$),[1146] $[(Me_3Si)_2C(BNPr^i{}_2)_2C\{Au(PPh_3)\}_2]$, ($^{13}C$),[1147] $[(Ph_3PAu)_5C]^+$, (^{13}C),[1148] $[\{\eta^1\text{-}(PhCH_2)_5C_5\}AuPPh_3]$, ($^{13}C$),[1149] $[Au\{CN(CH_2Ph)CH{=}CHNH\}_2]^+$, ($^{13}C$),[1150] [(1,4,7-triazacyclononane)Cu(η^2-alkene)]$^+$, (^{13}C),[1151] $[Cu_4(\mu\text{-}O_2CC_6H_4\text{-}2\text{-}X)_4(\mu\text{-}RC{\equiv}CR)_2]$, ($^{13}C$),[1152] $[Cu_4(O_2CCF_3)_4(\mu\text{-}EtC{\equiv}CEt)_2]$, ($^{13}C$),[1153] $[C_5H_3N\text{-}2,6\text{-}\{CMe{=}NCH(CO_2Me)CH_2Ph\}_2Cu]^+$, ($^{13}C$),[1154] [Cu(thiamine)Cl], (^{13}C),[1155] $[Ag(NC_5H_5)_4]^+$, (^{13}C),[1156] $[Au(PPh_3)(N\text{-base})]$, ($^{13}C$),[1157] $[(\mu\text{-dppe})(AuSR)_2]$, ($^{13}C$),[1158] $[Cu(SC_6H_4\text{-}2\text{-}SiMe_3)]_{12}$, ($^{13}C$),[1159] [Cu(1,3,6,9,11,14-

1137 B.H. Lipshutz, E.L. Ellsworth, S.H. Dimock, and R.A.J. Smith, *J. Org. Chem.*, 1989, **54**, 4977.
1138 G. van Koten and J.T.B.H. Jastrzebski, *Tetrahedron*, 1989, **45**, 569.
1139 E.C. Constable, R.P.G. Henney, and T.A. Leese, *J. Organomet. Chem.*, 1989, **361**, 277.
1140 A.S. Peregudov, E.G. Perevalova, E.I. Smyslova, and D.N. Kravtsov, *Metalloorg. Khim.*, 1988, **1**, 1402 (*Chem. Abstr.*, 1989, **110**, 164 842).
1141 S.H. Bertz and R.A.J. Smith, *J. Am. Chem. Soc.*, 1989, **111**, 8276.
1142 M.A. Willert-Porada, D.J. Burton, and N.C. Baenziger, *J. Chem. Soc., Chem. Commun.*, 1989, 1633.
1143 D. Nobel, G. van Koten, and A.L. Spek, *Angew. Chem., Int. Ed. Engl.*, 1989, **28**, 208.
1144 G. Karger, P. Hornbach, A. Krämer, H. Pritzkow, and W. Siebert, *Chem. Ber.*, 1989, **122**, 1881.
1145 H.-J. Kneuper, K. Harms, and G. Boche, *J. Organomet. Chem.*, 1989, **364**, 275.
1146 E.G. Perevalova, I.G. Bolesov, Ye.S. Kalyuzhnaya, T.I. Voyevodskaya, L.G. Kuzmina, V.I. Korsunsky, and K.I. Grandberg, *J. Organomet. Chem.*, 1989, **369**, 267.
1147 G. Karger, P. Hornbach, A, Krämer, H. Pritzkow, and W. Siebert, *Chem. Ber.*, 1989, **122**, 1881.
1148 F. Scherbaum, A. Grohmann, G. Müller, and H. Schmidbaur, *Angew. Chem., Int. Ed. Engl.*, 1989, **28**, 463.
1149 H. Schumann, F.H. Görlitz, and A. Dietrich, *Chem. Ber.*, 1989, **122**, 1423.
1150 F. Bonati, A. Burini, B.R. Pietroni, and B. Bovio, *J. Organomet. Chem.*, 1989, **375**, 147.
1151 P. Chaudhuri and K. Oder, *J. Organomet. Chem.*, 1989, **367**, 249.
1152 H.L. Aalten, G. van Koten, E. Riethorst, and C.H. Stam, *Inorg. Chem.*, 1989, **28**, 4140.
1153 D.L. Reger, M.F. Huff, T.A. Wolfe, and R.D. Adams, *Organometallics*, 1989, **8**, 848.
1154 W.L. Kwik and A.W.N. Tay, *Polyhedron*, 1989, **8**, 1321.
1155 A. Adeyemo and R. Oderinde, *Bull. Soc. Chim. Belge*, 1989, **98**, 91.
1156 J. Iqbal, D.W.A. Sharp, and J.M. Winfield, *J. Chem. Soc., Dalton Trans.*, 1989, 461.
1157 E. Colacio, A. Romerosa, J. Ruiz, P. Román, J.M. Gutiérrez-Zorrilla, and M. Martinez-Ripoll, *J. Chem. Soc., Dalton Trans.*, 1989, 2323.
1158 K.T. Garnes, J.R. Heys, S.W. Landvatter, and J. Statler, *J. Organomet. Chem.*, 1989, **378**, 293.
1159 E. Block, M. Gernon, H. Kang, G. Ofori-Okai, and J. Zubieta, *Inorg. Chem.*, 1989, **28**, 1263.

hexathiacyclohexadecane)]$^+$, (^{13}C),[1160] [Cu(S_2CR)], (^{13}C),[1161] [Cu(S_2CAr)]$_4$, (^{13}C),[1162] [Cu$_8$-{S_2C=C(CO_2Bu^t)$_2$}$_5${HS_2C=C(CO_2Bu^t)$_2$}]$^{3-}$, (^{13}C),[1163] [M(SePh)$_n$], (M = Cu, Ag, Zn, Cd, Tl, Sn; ^{77}Se),[1164] [Au(1,4,7-trithiacyclononane)Cl], (^{13}C),[1165] and [($C_2H_4N_2R_2$C=S)Au(SCN)], (^{13}C).[1166]

Complexes of Zn, Cd, and Hg.——^{13}C n.m.r. spectroscopy has been used to investigate the intramolecular interaction between Zn and C≡C in [Zn($CH_2CH_2CH_2CH_2$C≡CH)$_2$].[1167] The structure of [Hg(CF_3)$_2$] dissolved in different nematic phases has been determined by ^{13}C, ^{19}F, and ^{199}Hg n.m.r. spectroscopy. A correlation was observed between the anisotropy of the ^{199}Hg chemical shift and the Hg–C bond length.[1168] Stereospecificity of ^2J(^{199}Hg-^{13}C) has been observed in [XCH=CHHgCl]. An intramolecular Hg←Cl interaction takes place in [*cis*-ClCH=CHHgCl].[1169] N.m.r. data have also been reported for (45),

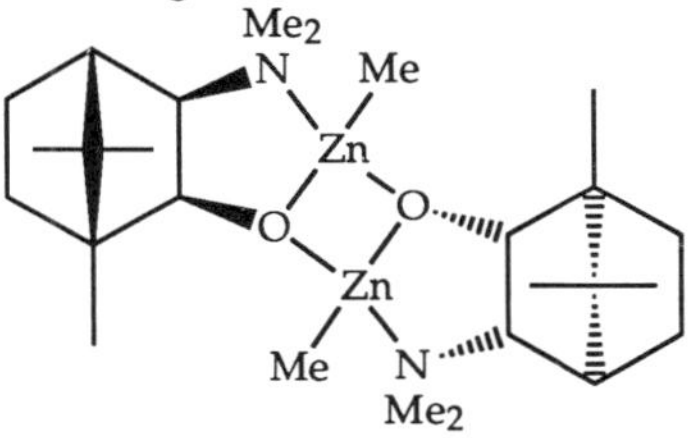

(45)

(^{13}C),[1170] [Zn(indenyl)$_2$], (^{13}C),[1171] [(C_5Me_5)$_2$Zn], (^{13}C),[1172] [Cd(CF_3)$_3$L]$^-$, (^{113}Cd),[1173] MeHg$^+$ complex with cyclopentanone thiosemicarbazone, (^{13}C),[1174] [ArN(CH_2CH_2HgX)$_2$], (^{199}Hg),[1175] [RHgO_2CMe], (^{13}C),[1176] [(η^1-C_5Me_5)HgCl]$_n$, (^{13}C),[1177] (46), (^{13}C),[1178] [4-X-C_6H_4-

1160 B. De Groot and S.J. Loeb, *Inorg. Chem.*, 1989, **28**, 3573.
1161 G. Pellizer, N. Marsich, and A. Camus, *Inorg. Chim. Acta*, 1989, **155**, 167.
1162 A. Camus, N. Marsich, A.M.M. Lanfredi, and F. Ugozzoli, *Inorg. Chim. Acta*, 1989, **161**, 87.
1163 D. Coucouvanis, D. Swenson, N.C. Baenziger, R. Pedelty, M.L. Caffery, and S. Kanodia, *Inorg. Chem.*, 1989, **28**, 2829.
1164 R. Kumar and D.G. Tuck, *Can. J. Chem.*, 1989, **67**, 127.
1165 D. Parker, P.S. Roy, G. Ferguson, and M.M. Hunt, *Inorg. Chim. Acta*, 1989, **155**, 227.
1166 A.A. Isab, *Polyhedron*, 1989, **8**, 2823.
1167 E. Oknińska and K.B. Starowieyski, *J. Organomet. Chem.*, 1989, **376**, 7.
1168 Yu.K. Grishin, A.A. Solomin, and Yu.A. Ustynyuk, *Metalloorg. Khim.*, 1989, **2**, 404 (*Chem. Abstr.*, 1989, **111**, 144 527).
1169 Yu.K. Grishin, D.V. Bazhenov, V.A. Roznyatovskii, M.A. Kazankova, V.R. Kartashov, and Yu.A. Ustynyuk, *Metalloorg. Khim.*, 1988, **1**, 335 (*Chem. Abstr.*, 1989, **110**, 24 028).
1170 M. Kitamura, S. Okada, S. Suga, and R. Noyori, *J. Am. Chem. Soc.*, 1989, **111**, 4028.
1171 B. Fischer, J. Boersma, G. van Koten, W.J.J. Smeets, and A.L. Spek, *Organometallics*, 1989, **8**, 667.
1172 B. Fischer, P. Wijkens, J. Boersma, G. van Koten, W.J.J. Smeets, A.L. Spek, and P.H.M. Budzelaar, *J. Organomet. Chem.*, 1989, **376**, 223.
1173 D. Naumann and W. Tyrra, *J. Organomet. Chem.*, 1989, **368**, 131.
1174 A. Macias, M.C. Rodriguez-Argüelles, M.I. Suárez, A. Sánchez, J.S. Casas, J. Sordo, and U. Englert, *J. Chem. Soc., Dalton Trans.*, 1989, 1787.
1175 G.B. Deacon, B.M. Gatehouse, L.W. Guddat, and S.C. Ney, *J. Organomet. Chem.*, 1989, **375**, C1.
1176 M. Bassetti, B. Floris, and G. Spadafora, *J. Org. Chem.*, 1989, **54**, 5934.
1177 J. Lorberth, T.F. Berlitz, and W. Massa, *Angew. Chem., Int. Ed. Engl.*, 1989, **28**, 611.
1178 J.A. Ganske, R.K. Pandey, M.J. Postich, K.M. Snow, and K.M. Smith, *J. Org. Chem.*, 1989, **54**, 4801.

HgClL], (^{13}C),[1179] [$MeO_2CC_6H_4HgCl$], (^{13}C),[1180] [ArHg(maltolate)], (^{13}C),[1181] (47), (^{13}C),[1182] and [1,8-$(ClHg)_2$-naphthalene], (^{13}C).[1183]

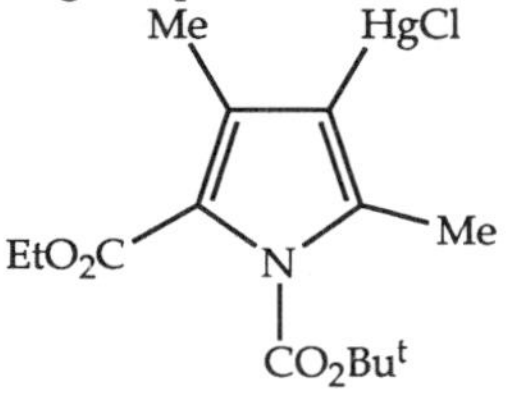

(46)

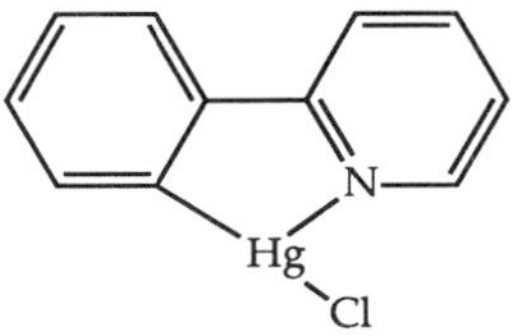

(47)

^{31}P, ^{77}Se, and ^{113}Cd n.m.r. studies have indicated that Cd^{2+} prefers Se or S rather than P on coordination to $Ph_2P(E)CH_2PPh_2$, E = S, Se.[1184] ^{13}C and ^{113}Cd n.m.r. spectroscopy has been used to investigate dithiocarbamate mobilization of Cd from intracellular deposits.[1185] N.m.r. data have also been reported for [ZnL_2], L = (48), (^{13}C),[1186] [Zn{2-thiophenylC(O)NHN=$CMeC_6H_4NH_2$-2}$_2Cl_2$], (^{13}C),[1187] [Zn{2,3,7,8,12,13,17,18-$(HOCH_2CH_2)_8$-porphyrin}], (^{13}C),[1188] [Zn{(4-Me-$C_6H_4)_4$-porphyrin}], (^{13}C),[1189] [Zn{(naphthyl)$_4$-porphyrin}], (^{13}C),[1190] [Zn{[3,4,5-$(MeO)_3$-$C_6H_4]_4$-porphyrin}], (^{13}C),[1191] [(4-Me-1-O-C_6H_2-2,6-CH=$NCH_2)_4Zn_4(O_2CMe)_3OH$], (^{13}C),[1192] ^{113}Cd substituted gene 32 protein, (^{35}Cl, ^{113}Cd),[1193] Cd^{2+} complexes of pentacyclic hexa-amines, (^{13}C, ^{113}Cd),[1194] ^{113}Cd complexes of carboxypeptidase, (^{113}Cd),[1195] ^{113}Cd complexes of parvalbumin, (^{113}Cd),[1196] Cd^{2+} adduct of an 18 residue peptide from HIV-1 nucleic acid binding protein, (^{113}Cd),[1197] Cd^{2+} substituted haem and myoglobin, (^{13}C, ^{113}Cd),[1198] [Cd(*N*-tosyl-β-alaninato-*O*)$_2$-$(OH_2)_4$], (^{113}Cd),[1199] [M(thiobenzoyl hydrazone)L], (M = Zn, Cd; ^{13}C),[1200] [$(HCpz_3)_n$-

1179 S. Bhatia, N.K. Kaushik, and G.S. Sodhi, *Z. Naturforsch., B*, 1988, **43**, 318 (*Chem. Abstr.*, 1989, **110**, 154 430).
1180 M.D. Hylarides, D.S. Wilbur, S.W. Hadley, and A.R. Fritzberg, *J. Organomet. Chem.*, 1989, **367**, 259.
1181 S. Bhatia, N.K. Kaushik, and G.S. Sodhi, *Bull. Chem. Soc. Jpn.*, 1989, **62**, 2693.
1182 E.C. Constable, T.A. Leese, and D.A. Tocker, *J. Chem. Soc., Chem. Commun.*, 1989, 570.
1183 H. Schmidbaur, H.-J. Öller, D.L. Wilkinson, B. Huber, and G. Müller, *Chem. Ber.*, 1989, **122**, 31.
1184 A.M. Bond, R. Colton, J. Ebner, and S.R. Ellis, *Inorg. Chem.*, 1989, **28**, 4509.
1185 R.J. Topping, M.M. Jones, G.R. Gale, and A.B. Smith, *J. Inorg. Biochem.*, 1989, **36**, 115.
1186 M. Bonamico, V. Fares, A. Flamini, P. Imperatori, and N. Poli, *Angew. Chem., Int. Ed. Engl.*, 1989, **28**, 1049.
1187 B. Singh and A.K. Srivastav, *Transition Met. Chem. (Weinhein, Ger.)*, 1989, **14**, 415.
1188 B.A. Gregg, M.A. Fox, and A.J. Bard, *J. Am. Chem. Soc.*, 1989, **111**, 3024.
1189 R. Cosmo, C. Kautz, K. Meerholz, J. Heinze, and K. Müllen, *Angew. Chem., Int. Ed. Engl.*, 1989, **28**, 604.
1190 A. Harriman and J. Davila, *Tetrahedron*, 1989, **45**, 4737.
1191 R.Y. Saleh and D.K. Straub, *Inorg. Chim. Acta*, 1989, **156**, 9.
1192 M. Bell, A.J. Edwards, B.F. Hoskins, E.H. Kachab, and R. Robson, *J. Am. Chem. Soc.*, 1989, **111**, 3603.
1193 D.P. Giedroc, B.A. Johnson, I.M. Armitage, and J.E. Coleman, *Biochemistry*, 1989, **28**, 2410; D.P. Giedroc and J.E. Coleman, *UCLA Symp. Moll. Cell. Biol., New Ser.*, 1989, **98**, 3 (*Chem. Abstr.*, 1989, **110**, 168 494).
1194 H. Strasdeit and S. Pohl, *Z. Naturforsch., B*, 1988, **43**, 1579 (*Chem. Abstr.*, 1989, **111**, 49 248).
1195 P.D. Ellis, *J. Biol. Chem.*, 1989, **264**, 3108.
1196 A.L. Swain and E.L. Amma, *Inorg. Chim. Acta*, 1989, **163**, 5.
1197 T.L. South, B. Kim, and M.F. Summers, *J. Am. Chem. Soc.*, 1989, **111**, 395.
1198 M.A. Kennedy and P.D. Ellis, *J. Am. Chem. Soc.*, 1989, **111**, 3195.
1199 G. Battistuzzi Gavioli, L. Menabue, M. Saladini, M. Sola, A. Bomartini Corradi, and L.P. Battaglia, *J. Chem. Soc., Dalton Trans.*, 1989, 1345.
1200 K.N. Zelenin, L.A. Khorseeva, V.V. Alekseev, O.V. Arapov, Z.M. Matveeve, V.G. Yusupov, M.M.

($HgX_2)_m$], (^{13}C),[1201] Hg^{2+} complexes of methionine and selenamethionine, (^{13}C),[1202] Hg^{2+} complexes of vitamin B_1, (^{13}C),[1203] [(μ-TeR^1)$_6$(HgPR$^2_3$)$_4$]$^{2+}$, (^{125}Te, ^{199}Hg),[1204] [M(D-gluconate)$_2$], (M = Zn, Cd, Hg; ^{13}C),[1205] [Cd(S$_2$CNR^{1}R^2)$_2$], (^{13}C, ^{113}Cd),[1206] [Cd(S$_m$)(S$_n$)]$^{2-}$, (^{113}Cd),[1207] and [Hg(1,4,7-trithiacyclononane)$_2$]$^{2+}$, (^{13}C).[1208]

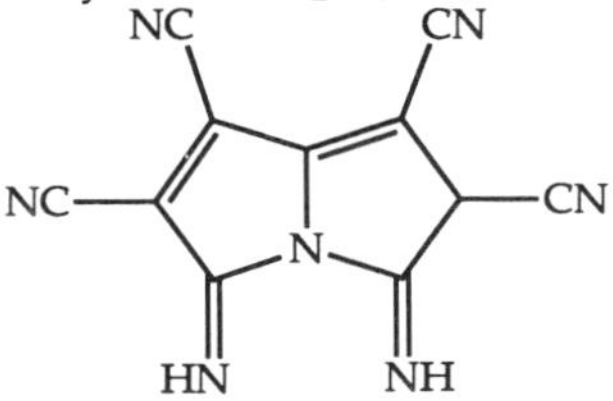

(48)

3 Dynamic Systems

This section is in three main parts: (i) 'Fluxional Molecules', dealing with rate processes involving no molecular change, (ii) 'Equilibria', dealing with the use of n.m.r. spectroscopy to measure the position of equilibria and ligand-exchange reactions, including solvation, and (iii) 'Course of Reactions', dealing with the use of n.m.r. spectroscopy to monitor the course of reactions. Each section is ordered by the Periodic Table.

Ten relevant reviews have appeared: 'Applications of dynamic n.m.r. spectroscopy to the dissociation of chemical bonds in organic, organometallic, and coordination compounds',[1209] 'Rates and mechanisms of complexation reactions of cations with crown ethers and related macrocycles',[1210] 'N.m.r. studies of alkali anions in non-aqueous solvents',[1211] '^{23}Na n.m.r. spectroscopy as a tool for studying cation interactions with biological molecules and cell membranes',[1212] 'Measurement of pH and Na^+ by nuclear magnetic resonance',[1213] 'N.m.r. spectroscopy of $^{23}Na^+$ in intact cells',[1214] 'Biomedical

Karimov, and N.A. Parpiev, *Zh. Obshch. Khim.*, 1989, **59**, 1191 (*Chem. Abstr.*, 1989, **111**, 186 270).
1201 G.G. Lobbia, F. Bonati, A. Cingolani, and D. Leonesi, *Synth. React. Inorg. Metal-Org. Chem.*, 1989, **19**, 827.
1202 A.A. Isab, *Transition Met. Chem. (London)*, 1989, **14**, 235.
1203 A.O. Adeyemo, *Inorg. Chim. Acta*, 1989, **160**, 253.
1204 P.A.W. Dean, V. Manivannan, and J.J. Vittal, *Inorg. Chem.*, 1989, **28**, 2360.
1205 H.A. Tajmir-Riahi, *Can. J. Chem.*, 1989, **67**, 651.
1206 J.S. Casas, A. Sanchez, J. Bravo, S. Garcia-Fontan, E.E. Castellano, and M.M. Jones, *Inorg. Chim. Acta*, 1989, **158**, 119.
1207 R.M.H. Banda, I.G. Dance, T.D. Bailey, D.C. Craig, and M.L. Scudder, *Inorg. Chem.*, 1989, **28**, 1862.
1208 A.J. Blake, A.J. Holder, T.I. Hyde, G. Reed, and M. Schröder, *Polyhedron*, 1989, **8**, 2041.
1209 M. Ōki, *Pure Appl. Chem.*, 1989, **61**, 699.
1210 E.M. Eyring and S. Petrucci, *Report*, 1989, TR-7; Order No. AD-A203436, 50 pp. Avail. NTIS. From *Gov. Rep. Announce. Index (U.S.)*, 1989, **89**, Abstr. No. 929 120 (*Chem. Abstr.*, 1989, **111**, 161 299).
1211 P.P. Edwards, A. Ellaboudy, D.M. Holton, and N.C. Pyper, *Annu. Rep. N.M.R. Spectrosc.*, 1988, **20**, 315.
1212 S.N. Dvoryantsev, *Biol. Membr.*, 1989, **6**, 5 (*Chem. Abstr.*, 1989, **110**, 169 457).
1213 L.S. Okerlund and R.J. Gillies, *Na$^+$/H$^+$ Exch.*, 1988, 21. Ed. by S. Grinstein, CRC: Boca Raton, Fl. (*Chem. Abstr.*, 1989, **110**, 150 656).
1214 R. Gupta and L.A. Jelicks, *Indian J. Chem., Sect. A*, 1988, **27A**, 829 (*Chem. Abstr.*, 1989, **110**, 53 888).

applications of ^{23}Na magnetic resonance imaging and ^{39}K n.m.r. spectroscopy',[1215] 'Structure of calcium ion-binding proteins - X-ray and n.m.r.',[1216] 'Transition metals in living systems - for the development of highly selective and efficient reactions' which contains ^{13}C n.m.r. spectroscopy,[1217] and '(^{1}H, ^{13}C, ^{77}Se, and ^{195}Pt) N.m.r. studies of the dynamic stereochemistry of sulphur and selenium complexes of platinum'.[1218]

Fluxional Molecules.——*Lithium.* The ^{1}H n.m.r. spectrum of [Ph_2PCH_2Li(tmeda)] shows one CH_2 signal at room temperature, which splits into two at low temperatures. The ^{13}C n.m.r. spectrum was also reported.[1219]

Sodium. Dynamic ^{1}H and ^{13}C n.m.r. spectroscopy has been used to investigate conformation flexibility of [Na{C_6H_3-1,3,5-(CH_2NRCH_2-3-C_5H_3N-2-2'-C_5H_3N-3-CH_2NRCH_2)$_3$-1,3,5-C_6H_3}]$^+$.[1220]

Beryllium. The rate of molecular inversion of [$(C_5H_5)_2Be$] has been estimated as $10^{10\pm1}$ s^{-1} at room temperature from partially relaxed spin-spin coupling between ^{13}C and ^{9}Be.[1221]

Calcium, Strontium, Barium, and Lanthanum. Dynamic n.m.r. spectroscopy has been used to investigate intramolecular acetate scrambling in Ca^{2+}, Sr^{2+}, Ba^{2+}, and La^{3+} cyclohexanediaminetetraacetate complexes, and $\Delta H^{\ddagger}$ and $\Delta S^{\ddagger}$ were determined.[1222]

Yttrium. ^{1}H n.m.r. spectroscopy has shown exchange of the diastereotopic NMe groups is $\Delta G^{\ddagger}_{388} = 84 \pm 2$ kJ mol^{-1} for [(η^5-C_5Me_5)Y(2-$C_6H_4CH_2NMe_2$)$_2$]. The ^{13}C n.m.r. spectrum was also reported.[1223]

The Lanthanides. Intramolecular ligand reorganisation processes in [Ln($HBpz_3$)$_2$(acac)] have been investigated.[1224]

Thorium ^{1}H n.m.r. studies of [(η^5-C_5Me_5)$_2$ThCl{(CH_2)(CH_2)PR^1R^2}] have shown two CH_2 groups at low temperature which undergo exchange at room temperature.[1225]

1215 Y. Seo and M. Maeda, *Igaku no Ayumi*, 1989, **148**, 331 (*Chem. Abstr.*, 1989, **111**, 35 800).
1216 M. Ikura and K. Hikichi, *Tanpakushitsu Kakusan Koso*, 1988, 33, 1994 (*Chem. Abstr.*, 1989, **110**, 3841).
1217 K. Araki and S. Shiraishi, *Seisan Kenkyu*, 1989, **41**, 151 (*Chem. Abstr.*, 1989, **111**, 154 230).
1218 K.G. Orrell, *Coord. Chem. Rev.*, 1989, **96**, 1.
1219 G. Fraenkel, W.R. Winchester, and P.G. Williard, *Organometallics*, 1989, **8**, 2308.
1220 F. Ebmeyer and F. Vögtle, *Chem. Ber.*, 1989, **122**, 1725.
1221 K.W. Nugent, J.K. Beattie, and L.D. Field, *J. Phys. Chem.*, 1989, **93**, 5371.
1222 Y. Ba, R.-f. Song, and Z.-w. Qui, *Magn. Reson. Chem.*, 1989, **27**, 916.
1223 M. Booij, N.H. Kiers, A. Meetsma, J.H. Teuben, W.J.J. Smeets, and A.L. Spek, *Organometallics*, 1989, **8**, 2454.
1224 M.A.J. Moss, C.J. Jones, and A.J. Edwards, *J. Chem. Soc., Dalton Trans.*, 1989, 1393.
1225 R.E. Cramer, S. Roth, F. Edelmann, M.A. Bruck, K.C. Cohn, and J.W. Gilje, *Organometallics*, 1989, **8**, 1192.

Titanium, Zirconium, and Hafnium. ^1H and ^{11}B n.m.r. spectroscopy has been used to investigate BH_4 rotation in [Zr{η^5-1,3-$(Me_3Si)_2C_5H_3$}$_2$(η^2-BH_4)Cl]. The ^{13}C n.m.r. spectrum was also recorded.[1226] ^{13}C EXSY has been used to investigate but-1-ene rotation in [(η^5-C_5H_5)$_2$Zr(η^2-EtCH=CH_2)(PMe_3)].[1227] Dynamic ^1H and ^{13}C n.m.r. spectroscopy has been used to study butadiene and cyclopentadienyl rotation in [(η^5-$Bu^nMe_2CC_5H_4$)$_2$Zr-(η^4-C_4H_6)]. The ^{13}C c.p.m.a.s. n.m.r. spectrum was also reported.[1228] ^1H n.m.r. spectroscopy has been used to demonstrate exchange in [(η^5-C_5H_5)$_2$Zr(C≡CR)$_2$Zr(η^5-C_5H_4Me)$_2$] and the ^{13}C n.m.r. spectrum was also recorded.[1229] $\Delta G^\ddagger$ for ring rotation in [{(η^5-C_5H_4-Bu^t)$_2$Zr}$_2$(μ-E^1)(μ-E^2)], E^1, E^2 = O, Se, Te, has been estimated.[1230] The activation energy for ring inversion in [(η^5-$C_5H_4Bu^t$)$_2MS_5$] has been determined[1231] Restricted rotation of the M-aryl bond in [(η^5-C_5Me_4H)$_2MAr_2$], M = Ti, Zr, has been investigated using ^1H and ^{13}C n.m.r. spectroscopy.[1232] The variable temperature ^{31}P n.m.r. spectrum of [M(η^4-C_4H_6)Ph{N($SiMe_2CH_2PMe_2$)$_2$}], M = Zr, Hf, shows an AB spectrum at 210 K which is averaged at higher temperature. The ^{13}C spectrum was also reported.[1233] $\Delta G^\ddagger$ has been determined for enantiomerization of the chiral ring and for hindered rotation of the N-Bu^t group in [(η^5-C_5H_5)$_2$MONRCH$_2$CH=CHCH$_2$NRO], M = Zr, Hf. The ^1H and ^{13}C n.m.r. spectra were reported.[1234] Similar enantiomerization has been observed for [(η^5-C_5H_5)$_2$ZrOC{=M(CO)$_5$}CH$_2$CH=CHCH$_2$CMe$_2$O], M = Cr, Mo, W, by ^1H and ^{13}C n.m.r. spectroscopy.[1235] Ring inversion in [(η^5-C_5H_5)$_2$Ti($Te_2C_6H_4$)] has been studied by ^1H n.m.r. spectroscopy. The ^{77}Se n.m.r. spectrum of the selenium analogue was also recorded.[1236] Variable temperature ^1H n.m.r. spectroscopy has been used to investigate the activation energy for C—N bond rotation as a function of X in [(η^5-C_5H_5)$_2$ZrX(S_2C-NMe_2)].[1237]

Vanadium. The barrier for hindered rotation of the alkyne in [(η^5-C_5Me_5)V(CO)L(η^2-R^1C≡CR^2)] has been determined. The ^{13}C and ^{51}V n.m.r. spectra were also recorded.[1238] The temperature dependence of the ^{13}C n.m.r. spectrum of (49) has been interpreted as

1226 A. Antiñolo, G.S. Bristow, G.K. Campbell, A.W. Duff, P.B. Hitchcock, R.A. Kamarudin, M.F. Lappert, R.J. Norton, N. Sarjudeen, and D.J.W. Winterborn, *Polyhedron*, 1989, **8**, 1601.
1227 P. Binger, P. Müller, R. Benn, A. Rufińska, B. Gabor, C. Krüger, and P. Betz, *Chem. Ber.*, 1989, **122**, 1035.
1228 G. Erker, R. Nolte, C. Krüger, R. Schlund, R. Benn, H. Grondey, and R. Mynott, *J. Organomet. Chem.*, 1989, **364**, 119.
1229 G. Erker, W. Frömberg, R. Benn, R. Mynott, K. Angermund, and C. Krüger, *Organometallics*, 1989, **8**, 911.
1230 G. Erker, R. Nolte, G. Tainturier, and A. Rheingold, *Organometallics*, 1989, **8**, 454.
1231 G. Tainturier, M. Fahim, and B. Gautheron, *J. Organomet. Chem.*, 1989, **362**, 311.
1232 P. Courtot, V. Labed, R. Pichon, and J.Y. Salaun, *J. Organomet. Chem.*, 1989, **359**, C9.
1233 M.D. Fryzuk, T.S. Haddad, and S.J. Rettig, *Organometallics*, 1989, **8**, 1723.
1234 G. Erker and M.G. Humphrey, *J. Organomet. Chem.*, 1989, **378**, 163.
1235 G. Erker, F. Sosna, R. Zwettler, and C. Krüger, *Organometallics*, 1989, **8**, 450.
1236 T.M. Klapötke, *Phosphorus Sulphur Silicon*, 1989, **41**, 105.
1237 D.E. Femec, M.E. Silver, and R.C. Fay, *Inorg. Chem.*, 1989, **28**, 2789.
1238 H.G. Alt, H.E. Engelhardt, A. Razavi, M.D. Rausch, and R.D. Rogers, *Z. Naturforsch., B*, 1988, **43**, 438 (*Chem. Abstr.*, 1989, **110**, 231 783).

restricted cyclo-octatetraene rotation.[1239]

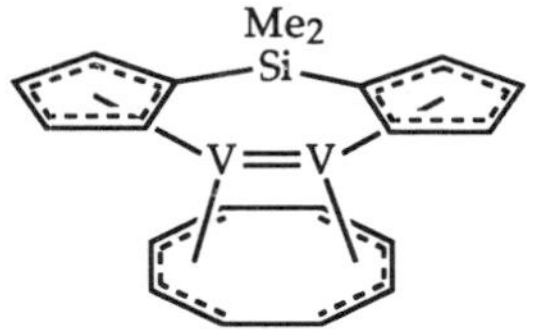

(49)

Niobium and Tantalum. Dynamic ^1H n.m.r. studies of exchange processes in [(η^5-C_5H_5)$_2$M(*exo*-η^2-RCH=CH_2)H], M = Nb, Ta, have led to the proposal for a new mechanism involving an intermediate with agostic bonding.[1240] Quantitative phase sensitive NOESY has been used to show that the observed scrambling of the terminal and bridging hydrogen atoms of (μ-H)$_2BH_2$ ligand of [(η^5-C_5H_5)(η^5-C_5Me_5)Ta(μ-H)$_2BH_2$] proceeds by a predominant mechanism in which there is no direct exchange between the two terminal hydrogen atoms. The ^{11}B n.m.r. spectrum was also recorded.[1241] The variable temperature ^1H n.m.r. spectrum of [Nb(SCH_2CH_2S)$_3$]$^+$ shows one CH_2 signal at room temperature and is [AB]$_2$ at low temperature.[1242]

Chromium, Molybdenum and Tungsten. The dynamic ^1H n.m.r. spectra of [Mo(*cis*-Ph_2PCH=$CHPPh_2$)$_2$(η^2-CH_2=CH_2)$_2$H]$^+$ have been reinterpreted as arising from several fluxional processes. The ^{13}C n.m.r. spectrum was also reported.[1243] The ^{31}P n.m.r. spectrum of [WH_5(PMe_2Ph)$_3$]$^-$ is a singlet at room temperature and AB_2 at -90 °C.[1244]

The ^1H n.m.r. spectrum of (50) shows two interconverting isomers due to inversion at phosphorus with $\Delta G^\ddagger$ = 24 kcal mol^{-1} compared with the usual 35 kcal mol^{-1}.[1245] The barriers for alkyne rotation in [$W_2R^1_2$(η^2-$R^2C{\equiv}CR^3$)$_2$(OPr^i)] and [W_2(C_6H_4Me-4)$_2$(η^2-Me-C≡CMe)$_2$(OBu^t)$_4$] have been determined from ^1H and ^{13}C measurements.[1246] There is restricted rotation about the N-C bond in the ^{13}C n.m.r. spectrum of [M($CHBu^tCH_2CH_2$-COX)($NC_6H_3Pr^i_2$-2,6)(OR)$_2$], M = Mo, W.[1247] [(Ph_2C=NN=)W(CH_2SiMe_3)$_2$(μ-Me_3SiC=)$_2$-W(CH_2SiMe_3)] is fluxional due to a pseudorotational process occurring about the trigonal-bipyramidal tungsten atom.[1248] The variable temperature ^1H and ^{13}C n.m.r. spectrum of [(η^5-C_5H_5)WM_2(CO)$_8$(C≡CPh)], M = Ru, Os, has provided evidence that the

1239 B. Bachmann, F. Hahn, J. Heck, and M. Wuensch, *Organometallics*, 1989, **8**, 2523.
1240 J.E. Bercaw, B.J. Burger, M.L.H. Green, B.D. Santarsiero, A. Sella, M.S. Trimmer, and L.-L. Wong, *J. Chem. Soc., Chem. Commun.*, 1989, 734.
1241 M.L.H. Green and L.-L. Wong, *J. Chem. Soc., Chem. Commun.*, 1989, 571; M.L.H. Green and L.-L. Wong, *J. Chem. Soc., Dalton Trans.*, 1989, 2133.
1242 J.L. Martin and J. Takats, *Can. J. Chem.*, 1989, **67**, 1914.
1243 A.E. Derome, M.L.H. Green, and L.-L. Wong, *New J. Chem.*, 1989, **13**, 747.
1244 D. Alvarez, jun., E.G. Lundquist, J.W. Ziller, W.J. Evans, and K.G. Caulton, *J. Am. Chem. Soc.*, 1989, **111**, 8392.
1245 S.K. Noh, R.A. Heintz, and K.H. Theopold, *Organometallics*, 1989, **8**, 2071.
1246 M.H. Chisholm, B.W. Eichhorn, K. Folting, and J.C. Huffman, *Organometallics*, 1989, **8**, 49; M.H. Chisholm, B.W. Eichhorn, and J.C. Huffman, *Organometallics*, 1989, **8**, 67.
1247 J. Feldman, J.S. Murdzek, W.M. Davis, and R.R. Schrock, *Organometallics*, 1989, **8**, 2260.
1248 M.H. Chisholm, J.A. Heppert, J.C. Huffman, and C.D. Ontiveros, *Organometallics*, 1989, **8**, 976.

acetylide ligand undergoes a 360° rotation on the face of the WRu_2 triangle.[1249] The variable temperature ^{13}C n.m.r. spectrum of $[(\eta^5-C_5H_5)WOs_3(CO)_8(PMePh_2)(\mu-O)(\mu_3-CCH_2C_6H_4Me-4)]$ shows carbonyl exchange.[1250]

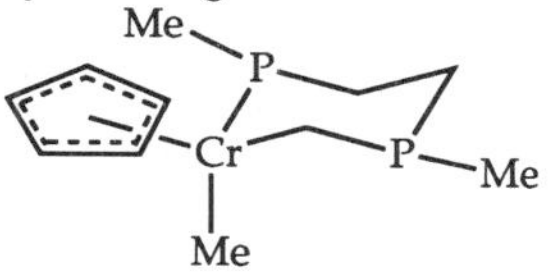

(50)

$\Delta G^{\ddagger}$ for rotation of the P-C bond in $[(\eta^5-C_5H_5)(OC)_2MoCHRPPh_2]$ has been determined by ^1H n.m.r. spectroscopy. The ^{13}C n.m.r. spectrum was also recorded.[1251] Olefin rotation occurs in *mer*-$[(\eta^2\text{-cyclooctene})(\eta^4-C_7H_8)W(CO)_3]$ and *mer*-$[(\eta^2-C_2H_4)(\eta^4-C_7H_8)W(CO)_3]$ and E_a was determined. Unusually large ^{13}C n.m.r. coordination shifts of the olefinic carbon atoms were observed.[1252] The barrier to but-2-yne rotation in $[WI(CO)\{P(OPr^i)_3\}(dppm)(\eta^2-MeC\equiv CMe)]^+$ is 56.9±0.5 kJ mol^{-1}. The ^{13}C n.m.r. spectrum indicates that the alkyne donates four electrons.[1253] Similar studies have been made for $[WI_2(CO)\{P(OPr^i)_3\}_2(\eta^2-MeC\equiv CMe)]$,[1254] and $[WI_2(CO)L_2(\eta^2-RC\equiv CR)]$.[1255] The ^1H and ^{13}C n.m.r. spectra of $[WCo_2(\mu-EtC\equiv CEt)(\mu-C_4Et_4)(CO)_8]$ have been assigned using COSY. A fluxional process produces a mirror plane.[1256]

Stereochemical non-rigidity in $[\{MeN(CH_2N_2C_3HR_2)_3\}M(CO)_2(\eta^3-C_3H_5)]^+$ and $[\{H_2C(N_2C_3HMe_2)_2\}M(CO)_2(\eta^3-C_3H_5)Br]$, M = Cr, Mo, W, has been investigated by ^1H and ^{13}C n.m.r. spectroscopy.[1257] ^1H n.m.r. spectroscopy has been used to investigate allyl rotation in $[(\eta^6-1,3,5-C_6H_3Me_3)Mo(CO)_2(\eta^3-C_3H_5)]^+$, and $[(\eta^5-C_5H_5)M(CO)_2(\eta^3-C_3H_5)]^+$, M = Mn, Re.[1258] Hindered intramolecular carbonyl exchange in (51) has been demonstrated by ^1H and ^{13}C n.m.r. spectroscopy and $\Delta G^{\ddagger}$ determined.[1259]

The variable temperature n.m.r. spectrum of $[(\eta^5-C_5H_7)Mo(CO)_2(dmpe)]$ has been interpreted as dienyl rotation.[1260] The variable temperature ^{13}C n.m.r. spectrum of $[(\eta^5-C_5Me_5)Mo(CO)_3Mo(CO)_2(\eta^7-C_7H_7)]$ shows one ^{13}CO signal at room temperature and three at -80 °C.[1261] Variable temperature ^{19}F n.m.r. spectroscopy shows restricted rotation of the C_6F_5 groups in $[TlMo(SC_6F_5)_2L_2(\eta^5-C_5H_5)]$. The ^{13}C and ^{205}Tl n.m.r. spectra

1249 Y. Chi, B.-J. Liu, G.-H. Lee, and S.-M. Peng, *Polyhedron*, 1989, **8**, 2003.
1250 Y. Chi, J.R. Shapley, M.R. Churchill, and J.C. Fettinger, *J. Organomet. Chem.*, 1989, **372**, 273.
1251 E. Lindner, M. Staengle, W. Hiller, and R. Fawzi, *Chem. Ber.*, 1989, **122**, 823.
1252 F.W. Grevels, J. Jacke, P. Betz, C. Krüger, and Y.H. Tsay, *Organometallics*, 1989, **8**, 293.
1253 P.K. Baker, K.R. Flower, M.G.B. Drew, and G. Forsyth, *J. Chem. Soc., Dalton Trans.*, 1989, 1903.
1254 P.K. Baker, E.M. Armstrong, and M.G.B. Drew, *Inorg. Chem.*, 1989, **28**, 2406.
1255 E.M. Armstrong, P.K. Baker, M.E. Harman, and M.B. Hursthouse, *J. Chem. Soc., Dalton Trans.*, 1989, 295.
1256 I.D. Scott, D.O. Smith, M.J. Went, and L.J. Farrugia, *J. Chem. Soc., Dalton Trans.*, 1989, 1375.
1257 K.B. Shiu, K.S. Liou, C.P. Cheng, B.R. Fang, Y. Wang, G.H. Lee, and W.J. Vong, *Organometallics*, 1989, **8**, 1219; K.B. Shiu and C.J. Chang, *J. Chin. Chem. Soc. (Taipei)*, 1987, **34**, 297 (*Chem. Abstr.*, 1989, **110**, 115 029).
1258 V.V. Krivykh, O.V. Gusev, P.V. Petrovskii, and M.I. Rybinskaya, *J. Organomet. Chem.*, 1989, **366**, 129.
1259 C.G. Kreiter, G. Wendt, and J. Kaub, *Chem. Ber.*, 1989, **122**, 215.
1260 G.H. Lee, S.M. Peng, I.C. Tsung, D. Mu, and R.S. Liu, *Organometallics*, 1989, **8**, 2248.
1261 A. Ricalton and M.W. Whiteley, *J. Organomet. Chem.*, 1989, **361**, 101.

were also reported.[1262] ^{13}CO EXSY has been carried out on $[W_2(CO)_6(\eta^5\text{-}C_5H_5)_2]$ to show that the interconversion of rotamers is rotation or pseudo-rotation. $[WM(CO)_3(PPh_3)_2\text{-}(\eta^5\text{-}C_5H_5)]$, M = Ru, Cu, were also examined.[1263] The temperature dependent ^{31}P n.m.r. spectrum of $[(\eta^5\text{-}C_5Me_5)(OC)_2W(\mu\text{-}\eta^3\text{-}P_3)\{Cr(CO)_5\}_3]$ shows hindered P_3 rotation.[1264]

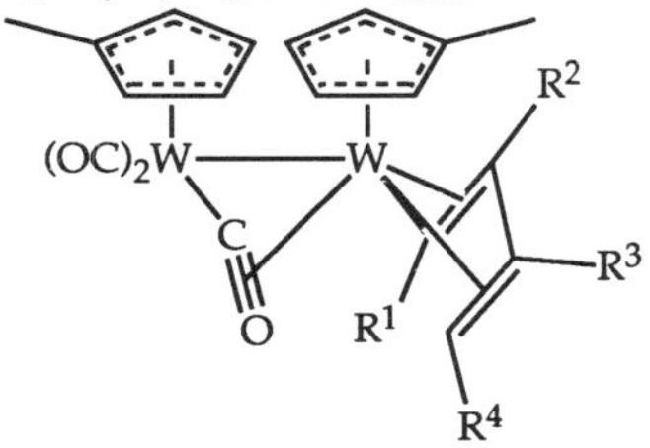

(51)

The ^{13}C n.m.r. spectrum of (52) shows carbonyls in the ratio 2:1 at 142 K and interchange with $\Delta G^{\ddagger}_{142}$ = *ca.* 27 kJ mol^{-1}. It was therefore suggested that the dynamic process in $[(\eta^6\text{-}C_6Et_6)Cr(CO)_3]$ is not restricted rotation.[1265] The activation energy for C(=O)-N bond rotation in $[(\eta^6\text{-}R^1C_6H_4\text{-}2\text{-}NR^2{}_2)Cr(CO)_3]$ has been determined from ^1H and ^{13}C n.m.r. spectroscopy.[1266] The variable temperature ^1H n.m.r. spectrum of $[(\eta^6\text{-}C_7H_8)_2Mo]$ shows restricted rotation and $\Delta G^{\ddagger}$ was determined as 50.7±0.7 kJ mol^{-1}. The ^{13}C n.m.r. spectrum was also recorded.[1267] Variable temperature ^{13}C n.m.r. spectroscopy has been used to show that there is restricted rotation in $[\{\eta^6\text{-}1,3,5\text{-}Et_3\text{-}2,4,6\text{-}(Me_3SiCH_2)_3C_6\}Mo\text{-}(CO)_2(PPh_3)]$.[1268]

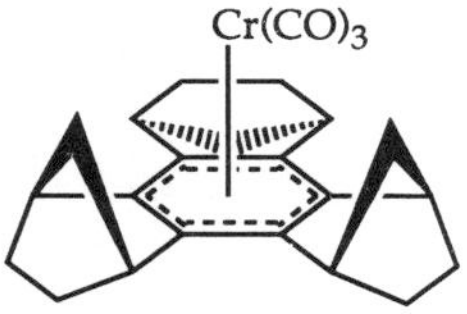

(52)

Pyramidal inversion of S or Se in *cis*-$[M(CO)_4\{(\eta^5\text{-}C_5H_4EMe)_2Fe\}]$, M = Cr, Mo, W, E = S, Se, has been studied by one dimensional and two dimensional EXSY using ^1H, ^{13}C, and ^{77}Se n.m.r. spectroscopy and $\Delta H^{\ddagger}$ and $\Delta S^{\ddagger}$ determined.[1269] Two dimensional ^{31}P EXSY has been used to study isomerization processes in $[Cr(CO)_2(CE)\{P(OMe)_3\}_3]$, E = O, S, Se. The lowest energy pathway is the trigonal-twist.[1270] The sulphur inversion bar-

1262 W.A.W.A. Bakar, J.L. Davidson, W.E. Lindsell, K.J. McCullough, and K.W. Muir, *J. Chem. Soc., Dalton Trans.*, 1989, 991.
1263 W.E. Lindsell and P.J. Tomb, *J. Organomet. Chem.*, 1989, **378**, 245.
1264 O.J. Scherer, J. Schwalb, and G. Wolmershäuser, *New J. Chem.*, 1989, **13**, 399.
1265 P. Kremminger, W. Weissensteiner, C. Kratky, G. Hunter, and R.L. MacKay, *Monatsh Chem.*, 1989, **120**, 1175.
1266 A. Gryff-Keller, P. Szczeciński, and H. Kozieł, *J. Organomet. Chem.*, 1989, **372**, 225.
1267 M.L.H. Green, P.A. Newman, and J.A. Bandy, *J. Chem. Soc., Dalton Trans.*, 1989, 331.
1268 J.A. Chudek, G. Hunter, R.L. Mackay, G. Färber, and W. Weissensteiner, *J. Organomet. Chem.*, 1989, **377**, C69.
1269 E.W. Abel, N.J. Long, K.G. Orrell, A.G. Osborne, V. Šik, P.A. Bates, and M.B. Hursthouse, *J. Organomet. Chem.*, 1989, **367**, 275.
1270 A.A. Ismail, F. Sauriol, and I.S. Butler, *Inorg. Chem.*, 1989, **28**, 1007.

rier for *cis*-$[Mo(CO)_4(RSCH_2CH_2SR)]$, *cis*-$[Mo(CO)_4($*cis*-$Bu^tSCH{=}CHSBu^t)]$ and *cis*-$[Mo(CO)_4(Bu^tSCH_2CH_2SMe)]$ has been determined by variable temperature ^1H EXSY. The ^{95}Mo chemical shifts were rationalised in terms of the relative strengths of the Mo—S bonds.[1271] Variable temperature ^{31}P n.m.r. spectra of *mer-cis*-$[(R_3P)(OC)_3(ON)$-$W(\mu\text{-}F)EF_n]$, EF_n = SbF_5, BF_3, PF_5, show restricted rotation of $[EF_{n+1}]^-$ and $\Delta G^\ddagger$ was determined.[1272]

$\Delta G^\ddagger$ has been determined for M—NR_2 bond rotation in $[Mo(NO)(NR_2)$(2,3,8,9-dibenzo-1,4,7,10-tetrathiadecane)].[1273] Exchange between π-bound and free OAr groups in $[W(OC_6H_3\text{-}2\text{-}Ph\text{-}\eta^6\text{-}6\text{-}C_6H_5)(OC_6H_3Ph\text{-}2,6)(PMePh_2)]$ has been observed by ^1H and ^{13}C n.m.r. spectroscopy.[1274] ^1H EXSY has been used to show that $[W_4(OPr^i)_{12}]$ undergoes a dynamic process in which the W—W double and single bonds migrate around the W_4 ring. This motion of the metal atoms is coupled with a correlated rotation about the W—O bonds of the OPr^i ligands attached to the wingtip W atoms so that proximal and distal W—OPr^i group exchange occurs.[1275] ^{77}Se n.m.r. spectroscopy has been used to study dynamic processes in $[W_2Se_2\{Se_2C_2(CO_2Me)_2\}_4]^{2-}$.[1276]

Manganese. $[Mn(\eta^5\text{-}C_5Me_5)(\eta^6\text{-}C_8H_8)]$ has been shown by ^1H and ^{13}C n.m.r. spectroscopy to be fluxional with the cyclo-octatetraene undergoing 1,3- or random shifts.[1277]

Rhenium. ^1H n.m.r. spectroscopy has been used to show that the hydrides of $[ReH_7$-$(Ph_2PC_5H_4\text{-}\eta^5)_2Fe]$ are fluxional giving one signal at room temperature and three at low temperature.[1278] Variable temperature ^1H n.m.r. spectra of $[ReH_7(pz_2CH_2)]$ on D substitution coupled with T_1 measurements show that the structure is classical. It is also fluxional.[1279] The ^1H n.m.r. spectrum of $[Re_2H_8(dppe)_2]$ shows hydrogen exchange between the bridge and terminal positions.[1280] A similar study has been performed on $[Re_2H_8(PPh_3)_4]$.[1281] The ^1H n.m.r. spectrum of $[Re_2H_5(PMe_3)_6]^{2+}$ shows exchange of hydrides between bridge and terminal positions.[1282] The variable temperature ^1H and ^{31}P n.m.r. spectra of $[ReCl_2(\eta^2\text{-}H_2)(PMe_2Ph)_4]$ and $[ReH_3(PMePh_2)_4]$ have been reported.[1283] ^1H n.m.r. spectroscopy has been used to investigate the exchange of hydrides in $[(\eta^5\text{-}C_5H_5)ReH_2(PPh_3)(NO)]^+$, and $\Delta G^\ddagger_{295}$ = 13 kcal mol^{-1}.[1284]

1271 E.W. Abel, D.E. Budgen, I. Moss, K.G. Orrell, and V. Šik, *J. Organomet. Chem.*, 1989, **362**, 105.
1272 R.V. Honeychuck and W.H. Hersh, *J. Am. Chem. Soc.*, 1989, **111**, 6056; *Inorg. Chem.*, 1989, **28**, 2869.
1273 D. Sellmann, G. Poehlmann, F. Knoch, and M. Moll, *Z. Naturforsch., B*, 1989, **44**, 312 (*Chem. Abstr.*, 1989, **111**, 49 289).
1274 J.L. Kerschner, P.E. Fanwick, I.P. Rothwell, and J.C. Huffman, *Organometallics*, 1989, **8**, 1431.
1275 M.H. Chisholm, D.L. Clark, and M.J. Hampden-Smith, *J. Am. Chem. Soc.*, 1989, **111**, 574.
1276 M.A. Ansari, C.H. Mahler, and J.A. Ibers, *Inorg. Chem.*, 1989, **28**, 2669.
1277 R.C. Hemond, R.P. Hughes, and A.L. Rheingold, *Organometallics*, 1989, **8**, 1261.
1278 X.L. Luo and R.H. Crabtree, *Inorg. Chem.*, 1989, **28**, 3775.
1279 D.G. Hamilton, X.L. Luo, and R.H. Crabtree, *Inorg. Chem.*, 1989, **28**, 3198.
1280 P.E. Fanwick, D.R. Root, and R.A. Walton, *Inorg. Chem.*, 1989, **28**, 395.
1281 F.A. Cotton and R.L. Luck, *Inorg. Chem.*, 1989, **28**, 4522.
1282 D.R. Root and R.A. Walton, *Inorg. Chem.*, 1989, **28**, 2503.
1283 F.A. Cotton and R.L. Luck, *Inorg. Chem.*, 1989, **28**, 2181.
1284 M.S. Chinn, D.M. Heinekey, N.G. Payne, and C.D. Sofield, *Organometallics*, 1989, **8**, 1824.

The variable temperature ^{1}H and ^{13}C n.m.r. spectra of [(η^5-$C_5H_4CCl_3$)Re(NO)(PPh_3)-($PPhCl_2$)] has been reported.[1285] Carbonyl scrambling in $[Re_7C(CO)_{21}HgY]^{2-}$ has been investigated by ^{13}C n.m.r. spectroscopy.[1286]

Iron and Ruthenium. The ^{1}H n.m.r. spectrum of [$RuH_4(Cy_2PCH_2CH_2CH_2)_2PPh$] shows exchange between η^2-H_2 and $(H)_2$.[1287] The variable temperature ^{1}H-{^{31}P} n.m.r. spectrum of [(η^5-C_5Me_5)RuH_3(PPr^i_3)] changes from a singlet at room temperature to AB_2 at low temperature.[1288] Similar behaviour is observed for [(η^5-C_5Me_5)RuH_3(PCy_3)CuCl].[1289] ^{1}H n.m.r. spectroscopy has been used to study exchange in [HRu(η^5-C_5H_5)(η^5-2,4-$Me_2C_5H_5$)]$^+$ between the hydride and CH_2 protons. The ^{13}C n.m.r. spectrum was also recorded.[1290] [Ru(η^5-2,4-$Me_2C_5H_5$)$_2$H]$^+$ is fluxional showing three separate dynamic processes. The lowest two are shown in Scheme 1 and the highest involves exchange between the hydride and the CH_2 groups. The ^{1}H and ^{13}C n.m.r. spectra were recorded.[1291] Variable temperature ^{1}H and ^{31}P n.m.r. studies of [Au_2Ru_4(μ_3-H)(μ-H){μ-Ph_2P-$(CH_2)_nPPh_2$}$(CO)_{12}$] have demonstrated dynamic behaviour involving the Au atom site exchange.[1292] ^{13}C n.m.r. spectroscopy has been used to investigate CO scrambling in $[Ru_{10}C(CO)_{24}]^{2-}$ and $[HRu_{10}C(CO)_{24}]^-$.[1293]

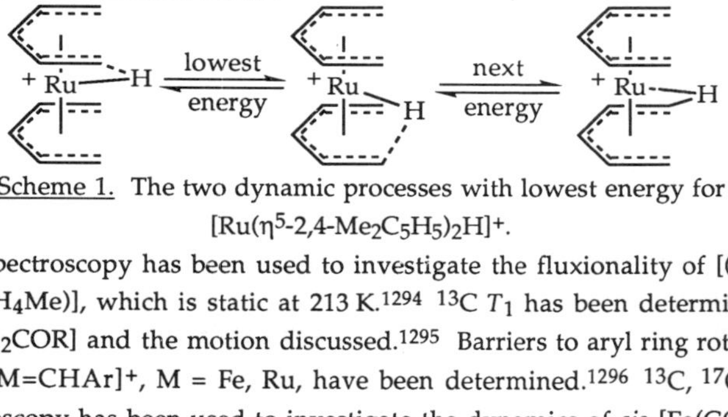

Scheme 1. The two dynamic processes with lowest energy for [Ru(η^5-2,4-$Me_2C_5H_5$)$_2$H]$^+$.

^{1}H n.m.r. spectroscopy has been used to investigate the fluxionality of [(η^5-C_5H_5)Fe-$(CO)_2$(η^1-C_5H_4Me)], which is static at 213 K.[1294] ^{13}C T_1 has been determined for [(η^5-C_5H_5)Fe$(CO)_2$COR] and the motion discussed.[1295] Barriers to aryl ring rotation in [(η^5-C_5H_5)(OC)LM=CHAr]$^+$, M = Fe, Ru, have been determined.[1296] ^{13}C, ^{17}O, and ^{119}Sn n.m.r. spectroscopy has been used to investigate the dynamics of *cis*-[Fe$(CO)_4$$(SnMe_3)_2$]

1285 W.E. Behro, A.M. Arif, and J.A. Gladysz, *Inorg. Chem.*, 1989, **28**, 3837.
1286 T.J. Henly and J.R. Shapley, *Organometallics*, 1989, **8**, 2729.
1287 G. Jia and D.W. Meek, *J. Am. Chem. Soc.*, 1989, **111**, 757.
1288 T. Arliguie, C. Border, B. Chaudret, J. Devillers, and R. Poilblanc, *Organometallics*, 1989, **8**, 1308.
1289 B. Chaudret, G. Commenges, F. Jalon, and A. Otero, *J. Chem. Soc., Chem. Commun.*, 1989, 210.
1290 R. Gleiter, I. Hyla-Kryspin, M.L. Ziegler, G. Sergeson, J.C. Green, L. Stahl, and R.D. Ernst, *Organometallics*, 1989, **8**, 298.
1291 D.N. Cox and R. Roulet, *J. Chem. Soc., Chem. Commun.*, 1989, 175.
1292 S.S.D. Brown, I.D. Salter, A.J. Dent, G.F.M. Kitchen, A.G. Orpen, P.A. Bates, and M.B. Hursthouse, *J. Chem. Soc., Dalton Trans.*, 1989, 1227; C.P. Blaxill, S.S.D. Brown, J.C. Frankland, I.D. Salter, and V. Šik, *J. Chem. Soc., Dalton Trans.*, 1989, 2039.
1293 P.J. Bailey, B.F.G. Johnson, J. Lewis, M. McPartlin, and H.R. Powell, *J. Organomet. Chem.*, 1989, **377**, C17.
1294 M.E. Wright, J.F. Hoover, R.S. Glass, and V.W. Day, *J. Organomet. Chem.*, 1989, **364**, 373.
1295 L.E. Nance, H.J. Will, D.B. MacQueen, J.M. Garrison, and J.M. Nicklaw, *Magn. Reson. Chem.*, 1989, **27**, 895.
1296 M. Brookhart, W.B. Studabaker, M.B. Humphrey, and G.R. Husk, *Organometallics*, 1989, **8**, 132.

and *fac*-[$Fe(CO)_3(CS)(SnMe_3)_2$].[1297]

Benzyl rotation in [(η^5-C_5H_5)$Fe(CO)_2$(η^3-MeOCHPh)] has been investigated.[1298] Restricted rotation of the cyclopentadienyl group in [(η^5-C_5H_4Me)Fe(CO)LI] has been attributed to ring distortion and/or interaction with the ring methyl group.[1299] The variable temperature ^1H and ^{13}C n.m.r. spectra of [(μ,η^2-$CHCCMe_2$)(μ-Bu^{t}S)$Fe_2(CO)_6$] show methyl group exchange.[1300] The variable temperature ^{13}C n.m.r. spectra of [(η^2-PhCH=CH)$Fe_2(CO)_7$],[1301] [Ru_3Pt(μ_4-η^2-C=CHBu^t)$(CO)_9$(dppe)],[1302] [(μ-C_7H_7)$M^1(CO)_3M^2$(CO)(PPh_3)], M^1 = Fe, Ru, M^2 = Rh, Ir,[1303] and [(η^4-$Me_2CCHCHCHCOMe$)$Fe(CO)_3$][1304] show CO exchange. The barrier to ring rotation in [Fe{η^5-1,3-$(Me_3Si)_2C_5H_3$}$_2$] has been determined as 11 kcal mol^{-1} from ^1H and ^{13}C n.m.r. spectroscopy.[1305] Phosphorus inversion in [(η^5-C_5H_5){1,2-$C_6H_4(PMePh)_2$}Fe(PMePh)] has been determined.[1306] Hindered ring rotation in [{η^5-Bu^t$(Me_3Si)_2C_5H_2$}$_2$Fe] has been observed in the ^1H and ^{13}C n.m.r. spectra.[1307] The variable temperature ^1H n.m.r. spectra of some cyclic ferrocene derivatives have been measured.[1308] Dynamic ^1H n.m.r. spectra of [Os_3{μ_3-CHC(C_5H_4-η^5)Fe(η^5-C_5H_5)}(μ-CO)$(CO)_9$] show restricted rotation of the substituted cyclopentadienyl.[1309]

The fluxionality of [$Fe_3(CO)_{11}(CNCF_3)$] has been investigated by ^{13}C n.m.r. spectroscopy.[1310] The variable temperature ^{13}C n.m.r. spectra of [$Fe_2Co(CO)_8${$P(OPh)_3$}(CCO)]$^-$ show one carbonyl n.m.r. signal at -60 °C and four at -120 °C.[1311] The mechanism of CO scrambling in [$Fe_3(CO)_{12-n}${$P(OR)_3$}$_n$] has been redetermined from ^{13}C n.m.r. spectra.[1312] Variable temperature ^{13}C n.m.r. spectroscopy has been used to show CO exchange in [Ru_2(μ-dppm)(μ-CO)$(CO)_6$].[1313] Variable temperature ^{13}C and ^{31}P n.m.r. spectra of a mixture of diastereomers of [$Ru_3(CO)_9$(μ_3-η^2-C≡CBu^t)(μ_4-Hg)$Ru_3(CO)_8$(μ_3-η^2-C≡CBu^t)($PMePhCH_2Ph$)] show racemisation which was interpreted as a rearrangement of the metal skeleton.[1314]

The fluxionality of [RuX$(Cy_2PCH_2CH_2PCy_2)_2$] has been studied by ^{31}P n.m.r. spectro-

1297 W. Petz, B. Wrackmeyer, and W. Storch, *Chem. Ber.*, 1989, **122**, 2261.
1298 M. Brookhart, R.C. Buck, and E. Danielson, tert., *J. Am. Chem. Soc.*, 1989, **111**, 567.
1299 P. Johnston, L. Denner, C.F. Marais, J.C.A. Boeyens, and N.J. Coville, *J. Crystallogr. Spectrosc. Res.*, 1988, **18**, 403 (*Chem. Abstr.*, 1989, **110**, 16 357).
1300 D. Seyferth, G.B. Womack, C.M. Archer, and J.C. Dewan, *Organometallics*, 1989, **8**, 430.
1301 A. Ishihara, T.-A. Mitsudo, and Y. Watanabe, *J. Organomet. Chem.*, 1989, **368**, 199.
1302 P. Ewing and L.J. Farrugia, *Organometallics*, 1989, **8**, 1246.
1303 S.T. Astley and J. Takats, *J. Organomet. Chem.*, 1989, **363**, 167.
1304 H. Kalchhauser, *Monatsh. Chem.*, 1989, **120**, 809.
1305 J. Okuda and E. Herdtweck, *J. Organomet. Chem.*, 1989, **373**, 99.
1306 G.T. Crisp, G. Salem, S.B. Wild, and F.S. Stephens, *Organometallics*, 1989, **8**, 2360.
1307 J. Okuda, *Chem. Ber.*, 1989, **122**, 1075.
1308 P.D. Beer, A.D. Keefe, and M.G.B. Drew, *J. Organomet. Chem.*, 1989, **378**, 437.
1309 K.I. Hardcastle, A.J. Deeming, D. Nuel, and N.I. Powell, *J. Organomet. Chem.*, 1989, **375**, 217.
1310 D. Lentz, *Z. Naturforsch., B*, 1987, **42**, 839 (*Chem. Abstr.*, 1989, **110**, 24 054).
1311 S. Ching, M. Sabat, and D.F. Shriver, *Organometallics*, 1989, **8**, 1047.
1312 H. Adams, N.A. Bailey, G.W. Bentley, and B.E. Mann, *J. Chem. Soc., Dalton Trans.*, 1989, 1831; S. Aime, M. Botta, O. Gambino, R. Gobetto, and D. Osella, *J. Chem. Soc., Dalton Trans.*, 1989, 1277.
1313 G.-Y. Kiel and J. Takats, *Organometallics*, 1989, **8**, 839.
1314 S. Hajela, B.M. Novak, and E. Rosenberg, *Organometallics*, 1989, **8**, 468.

scopy.[1315] ^{31}P n.m.r. spectroscopy shows hindered rotation of the monodentate $(Ph_2PCH_2CH_2)_2PPh$ in $[RuCl_2\{(Ph_2PCH_2CH_2)_2PPh\}_2]$.[1316] Variable temperature ^{1}H and ^{13}C n.m.r. spectra of $[(2,3,4,5\text{-}Me_4thiophene)RuCl_2(PPh_3)]$ have demonstrated hindered rotation about the Ru—P bond.[1317]

Osmium. The ^{1}H n.m.r. spectrum of $[Os_3H_2(CH{=}COEt)(CO)_9]$ shows two hydride signals at -68.5 °C and one at room temperature.[1318] The fluxional behaviour of $[Os_3Pt(\mu\text{-}H)_2(CO)_9(PCy_3)(CNCy)]$ and $[Os_3Pt(\mu\text{-}H)_2(\mu\text{-}CH_2)(CO)_9(PCy_3)(CNCy)]$ has been studied using ^{13}C EXSY. In $[Os_3Pt(\mu\text{-}H)_2(\mu\text{-}CH_2)(CO)_9(PCy_3)(CNCy)]$, three exchange processes were observed.[1319] Carbonyl exchange in $[Os_3X(CO)_{11}]^-$ has been examined by ^{13}C n.m.r. spectroscopy.[1320] The ^{13}C n.m.r. spectrum of $[Os_3(CO)_{10}(\mu\text{-}Br)(CH{=}CHPh)]$ shows six CO signals at room temperature and ten at 168K.[1321] The variable temperature ^{1}H and ^{13}C n.m.r. spectra of $[Os_3(CO)_9(\mu_3\text{-}\eta^2\text{-}EtC{\equiv}CEt)PR_3]$ and $[Os_3(CO)_9(\mu_3\text{-}\eta^2\text{-}MeC{\equiv}CMe)\{P(OMe)_3\}]$ show two exchange processes.[1322] Carbonyl scrambling in $[OsRh_4(CO)_9(\mu_2\text{-}CO)_6]^{2-}$ has been studied by ^{13}C n.m.r. spectroscopy.[1323]

Cobalt. ^{1}H and ^{13}C n.m.r. spectroscopy has been used to study dynamic processes in $[(\eta^5\text{-}C_5Me_5)LCo(CH_2CH_2\text{-}\mu\text{-}H)]^+$ and three dynamic processes were observed.[1324] Restricted rotation of the C_7H_7 ring in $[(\mu\text{-}\eta^3{:}\eta^4\text{-}C_7H_7)Co_2(CO)_3(\eta^5\text{-}C_5Me_5)(\mu\text{-}H)]$ has been observed by ^{1}H and ^{13}C n.m.r. spectroscopy.[1325] The fluxionality of $[Co(\eta^4\text{-}1,3\text{-}C_7H_8)(PMe_3)_3]^+$ and $[Co(\eta^4\text{-}1,3\text{-}C_8H_8)(PMe_3)_3]^+$ has been studied.[1326] Variable temperature ^{1}H n.m.r. spectroscopy has been used to investigate restricted ring rotation in $[Co\{\eta^5\text{-}C_5H_{5-n}(SiMe_3)_n\}_2]^+$. The ^{13}C n.m.r. spectrum was also reported.[1327]

The ^{13}C T_1 values of the protonated carbon atoms in $[Co_3(CO)_9(\mu_3\text{-}CPh)]$ have been measured and used to calculate tumbling times and activation energies.[1328] Interconversion of diastereoisomers of $[Co_3(CO)_7(Ph_2AsCH_2CH_2PPh_2)CCO_2(menthyl)]$ has been investigated by variable temperature ^{13}C and ^{31}P n.m.r. spectroscopy.[1329] Variable temperature ^{13}C n.m.r. spectra of $[Co_4(CO)_8(\mu_4\text{-}PPh)_2(dmpe)]$ have revealed

1315 A. Mezzetti, A. Del Zotto, and P. Rigo, *J. Chem. Soc., Dalton Trans.*, 1989, 1045.
1316 R.R. Guimerans, E.C. Hernandez, M.M. Olmstead, and A.L. Balch, *Inorg. Chim. Acta*, 1989, **165**, 45.
1317 J.R. Lockemeyer, T.B. Rauchfuss, A.L. Rheingold, and S.R. Wilson, *J. Am. Chem. Soc.*, 1989, **111**, 8828.
1318 E. Boyar, A.J. Deeming, M.S.B. Felix, S.E. Kabir, T. Adatia, R. Bhusate, M. McPartlin, and H.R. Powell, *J. Chem. Soc., Dalton Trans.*, 1989, 5.
1319 L.J. Farrugia, *Organometallics*, 1989, **8**, 2410.
1320 J.L. Zuffa, S.J. Kivi, and W.L. Gladfelter, *Inorg. Chem.*, 1989, **28**, 1888.
1321 Y. Chi, B.-F. Chen, S.-L. Wang, R.-K. Chiang, and L.-S. Hwang, *J. Organomet. Chem.*, 1989, **377**, C59.
1322 E. Rosenberg, J. Bracker-Novak, R.W. Gellert, S. Aime, R. Gobetto, and D. Osella, *J. Organomet. Chem.*, 1989, **365**, 163.
1323 A. Fumagalli, L. Garlaschelli, and R. Della Pergola, *J. Organomet. Chem.*, 1989, **362**, 197.
1324 M. Brookhart, D.M. Lincoln, A.F. Volpe, jun., and G.F. Schmidt, *Organometallics*, 1989, **8**, 1212.
1325 H. Wadepohl, W. Galm, and H. Pritzkow, *Angew. Chem., Int. Ed. Engl.*, 1989, **28**, 345.
1326 L.C.A. de Carvalho, M. Dartiguenave, and Y. Dartiguenave, *J. Organomet. Chem.*, 1989, **367**, 187.
1327 J. Okuda, *J. Organomet. Chem.*, 1989, **367**, C1.
1328 S.P. Wang, A.F.T. Chen, M.G. Richmond, and M. Schwartz, *J. Organomet. Chem.*, 1989, **371**, 81.
1329 D.T. Clark, K.A. Sutin, and M.J. McGlinchey, *Organometallics*, 1989, **8**, 155.

two fluxional processes which exchange CO groups.[1330] ^1H EXSY has been applied to ligand movements in $[Co([18]\text{-crown-6})(MeOH)_3]^{2+}$.[1331]

Rhodium and Iridium. The ^{13}C n.m.r. spectrum of $[HIr_4(CO)_{11}]^-$ shows fluxionality, but is static at -95 °C.[1332] Three separate exchange processes have been observed for $[\{Pr^i{}_2P(CH_2)_nPPr^i{}_2\}_2Rh_2(H)(\mu\text{-}H)_3]$ in the ^1H and ^{31}P n.m.r. spectra; intermolecular exchange with free H_2, and two intramolecular processes exchanging the four hydride ligands.[1333] ^{31}P n.m.r. spectroscopy has been used to investigate exchange in $[\{Pr^i{}_2P(CH_2)_2PPr^i{}_2\}_2Ir_2(H)(\mu\text{-}H)_3]$.[1334] The variable temperature ^1H n.m.r. spectrum of $[Ir_2(H)_4Cu(\mu\text{-}Cl)Cl(CO)_2\{(\mu\text{-}Ph_2PCH_2)_2AsPh\}]^+$ shows a dynamic process. The ^{13}C n.m.r. spectrum was also recorded.[1335] ^1H n.m.r. spectroscopy has been used to investigate exchange in $[(Ph_3P)M(\mu\text{-}H)IrH_2(PPh_3)_3]$, M = Ag, Au. Two dimensional ^{109}Ag n.m.r. spectra were also recorded.[1336]

The ^1H and ^{31}P n.m.r. spectra of *cis-mer*-$[RhMe_2Cl(PMePh_2)_3]$ show two dynamic processes; hindered rotation about the Rh—P and P—C bonds and dissociation of the unique phosphorus *trans* to one methyl group.[1337] The fluxionality of $[(Me_3P)_nRhMe]$, n = 3 or 4, has been studied by ^{13}C and ^{31}P n.m.r. spectroscopy.[1338] Stereospecific exchange of the (octaethylporphyrin)Rh groups in $[(\text{octaethylporphyrin})RhCH_2CHMeRh(\text{octaethylporphyrin})]$ has been studied.[1339] Inversion at phosphorus in $[N(SiMe_2CH_2PPh_2)_2IrMe(PR_2)]$ has been investigated by ^1H and ^{31}P n.m.r. spectroscopy.[1340] Fluxionality of $[N(CH_2CH_2PPh_2)_3Rh(\eta^2\text{-}HC{\equiv}CCO_2Et)]^+$ has been studied by ^{31}P n.m.r. spectroscopy.[1341]

$\Delta G^\ddagger$ values for ethene rotation in $[\eta^5\text{-}1,2,3\text{-}Me_3\text{-indenyl})Rh(\eta^2\text{-}CH_2{=}CH_2)_2]$[1342] and $[HBpz_3Ir(\eta^2\text{-}CH_2{=}CH_2)_2]$[1343] have been determined. The ^{13}C n.m.r. spectra were also recorded. The fluxionality of $[\{Rh(\mu\text{-}2\text{-}SC_5H_4N)(\eta^4\text{-}C_8H_{12})\}_2]$ has been investigated by ^1H and ^{13}C n.m.r. spectroscopy.[1344] Variable temperature ^1H n.m.r. spectroscopy, including magnetisation transfer, has been used to investigate facile intramolecular

1330 C.L. Schulman, M.G. Richmond, W.H. Watson, and A. Nagl, *J. Organomet. Chem.*, 1989, **368**, 367.
1331 F.L. Dickert and H.U. Meißner, *Ber. Bunsenges. Phys. Chem.*, 1989, **93**, 1450.
1332 D.M. Vandenberg, T. Chin-Choy, and P.C. Ford, *J. Organomet. Chem.*, 1989, **366**, 257.
1333 M.D. Fryzuk, W.E. Piers, F.W.B. Einstein, and T. Jones, *Can. J. Chem.*, 1989, **67**, 883.
1334 H. El-Amouri and A. Bouayad, *Polyhedron*, 1989, **8**, 1945.
1335 A.L. Balch, B.J. Davis, F. Neve, and M.M. Olmstead, *Organometallics*, 1989, **8**, 1000.
1336 A. Albinati, C. Anklin, P. Janser, H. Lehner, D. Matt, P.S. Pregosin, and L.M. Venanzi, *Inorg. Chem.*, 1989, **28**, 1105.
1337 E.G. Thaler, E.G. Lundquist, and K.G. Caulton, *Polyhedron*, 1989, **8**, 2689.
1338 R.T. Price, R.A. Andersen, and E.L. Muetterties, *J. Organomet. Chem.*, 1989, **376**, 407.
1339 B.B. Wayland, Y. Feng, and S. Ba, *Organometallics*, 1989, **8**, 1438.
1340 M.D. Fryzuk and K. Joshi, *Organometallics*, 1989, **8**, 722.
1341 C. Bianchini, D. Masi, A. Meli, M. Peruzzini, J.A. Ramirez, A. Vacca, and F. Zanobini, *Organometallics*, 1989, **8**, 2179.
1342 A.K. Karrar, N.J. Taylor, J.C. Calabrese, W.A. Nugent, D.C. Roe, E.A. Connaway, and T.B. Marder, *J. Chem. Soc., Chem. Commun.*, 1989, 990.
1343 R.S. Tanke and R.H. Crabtree, *Inorg. Chem.*, 1989, **28**, 3444.
1344 M.A. Ciriano, F. Viguri, J.J. Perez-Torrente, F.J. Lahoz, L.A. Oro, A. Tiripicchio, and M. Tiripicchio-Camellini, *J. Chem. Soc., Dalton Trans.*, 1989, 25.

bridging ligand exchange and M—M bond cleavage in $[M(\eta^4\text{-}C_8H_{12})(\mu\text{-}2\text{-}OC_5H_4N)]_2$, M = Rh, Ir.[1345] The two conformers of $[\{(\eta^5\text{-}C_5H_5)Rh(CNBu^t)\}_2(\mu\text{-}CNBu^t)]$ exchange with $\Delta G^{\ddagger} = 45.7 \pm 1.6$ kJ mol^{-1} and there is bridge↔terminal isonitrile exchange with $\Delta G^{\ddagger} = 61.8 \pm 0.8$ kJ mol^{-1}. The ^{13}C and ^{29}Si n.m.r. spectra of $[(\eta^5\text{-}C_5H_5)Rh(\eta^2\text{-}CH_2{=}CHSiMe_3)_2]$ were also recorded.[1346]

The ^{31}P n.m.r. spectrum of $[Ir_2(CNR)_4(dmpm)_2]$ is $[AB]_2$ at -60 °C and broad at room temperature.[1347] Two dynamic processes for carbonyl scrambling have been observed by magnetisation transfer in the ^{13}C n.m.r. spectrum of $[Ir_4(CO)_{11}(PEt_3)]$.[1348] Variable temperature ^{13}C and ^{195}Pt n.m.r. spectroscopy has been used to investigate carbonyl exchange in $[PtIr_4(CO)_{14}]^{2-}$.[1349]

The variable temperature ^{13}C n.m.r. spectrum of $[N(CH_2CH_2N{=}CHC_6H_4\text{-}2\text{-}O)_3Rh]$ shows fluxionality due to the Bailar twist with $\Delta G^{\ddagger}_{218} = 46$ kJ mol^{-1}.[1350] The fluxional behaviour of $[MeC(CH_2PPh_2)_3Rh(S_2CNR)]^+$ has been studied by ^1H and ^{31}P n.m.r. spectroscopy.[1351] ^1H, ^{13}C, and ^{31}P n.m.r. spectroscopy has been used to show restricted rotation in *trans*-$[IrCl_2(PMe_2Ph)_4]^+$, and three rotamers were observed.[1352]

Nickel. ^{13}C and ^{31}P n.m.r. spectroscopy has been used to investigate the fluxionality of $[Ni(\eta^1\text{-}CH_2Ph)Cl(PMe_3)_2]$ and $[Ni(\eta^3\text{-}CH_2Ph)Cl(PMe_3)]$.[1353] The fluxionality of $[(\eta^2\text{-naphthalene})\{Pr^i_2P(CH_2)_nPPr^i_2\}Ni]$ has been studied in the solid state and in solution by c.p.m.a.s. and two dimensional ^{13}C n.m.r. spectroscopy.[1354] The ^1H and ^{13}C n.m.r. spectra of $[(Et_3P)Ni(\eta^2\text{-}HC{\equiv}CH)_2]$ show restricted rotation of the ethyne, with $\Delta G^{\ddagger} = 46$ kJ mol^{-1}.[1355] The fluxionality of $[Ni(\eta^3\text{-}CH_2Ph)(PPh_3)_2]^+$ has been studied by ^1H, ^{13}C, and ^{31}P n.m.r. spectroscopy.[1356]

^{31}P n.m.r. spectroscopy has been used to show that $[(NC)_2Ni\overline{PPh(OCMe_2CH_2P}Ph_2)_2]$ is fluxional, and this has been interpreted as an equilibrium with a four coordinate intermediate.[1357] The mechanism of fluxionality of (53) has been investigated using ^1H and ^{13}C n.m.r. spectroscopy.[1358] Tetrahedral inversion of (54) has been investigated.[1359]

1345 G.S. Rodman and K.R. Mann, *J. Organomet. Chem.*, 1989, **378**, 255.
1346 S.T. Belt, S.B. Duckett, D.M. Haddleton, and R.N. Perutz, *Organometallics*, 1989, **8**, 748.
1347 J. Wu, P.E. Fanwick, and C.P. Kubiak, *J. Am. Chem. Soc.*, 1989, **111**, 7812.
1348 B.E. Mann, B.T. Pickup, and A.K. Smith, *J. Chem. Soc., Dalton Trans.*, 1989, 889.
1349 A. Fumagalli, R. Della Pergola, F. Bonacina, L. Garlaschelli, M. Moret, and A. Sironi, *J. Am. Chem. Soc.*, 1989, **111**, 165.
1350 G. Hunter and N. Kilcullen, *J. Chem. Soc., Dalton Trans.*, 1989, 2115.
1351 C. Bianchini, A. Meli, F. Laschi, F. Vizza, and P. Zanello, *Inorg. Chem.*, 1989, **28**, 227.
1352 A.J. Deeming, S. Doherty, J.E. Marshall, and N.I. Powell, *J. Chem. Soc., Chem. Commun.*, 1989, 1351.
1353 E. Carmona, M. Paneque, and M.L. Poveda, *Polyhedron*, 1989, 8, 285.
1354 R. Benn, R. Mynott, I. Topalovic, and F. Scott, *Organometallics*, 1989, **8**, 2299.
1355 K.-R. Pörschke, *J. Am. Chem. Soc.*, 1989, **111**, 5691.
1356 J.R. Ascenso, M.A.A.F. de C.T. Carrondo, A.R. Dias, P.T. Gomes, M.F.M. Piedade, C.C. Romão, A. Revillon, and I. Tkatchenko, *Polyhedron*, 1989, **8**, 2449.
1357 A.W.G. Platt and P.G. Pringle, *Inorg. Chim. Acta*, 1989, **160**, 33.
1358 J.H. Cameron and S. Graham, *J. Chem. Soc., Dalton Trans.*, 1989, 1599.
1359 A.I. Uraev, A.L. Nivorozhkin, A.S. Frenkel, A.S. Antsishkina, M.A. Porai-Koshits, L.E. Konstantinovsky, G.K.-I. Magomedov, and A.D. Garnovsky, *J. Organomet. Chem.*, 1989, **368**, 303.

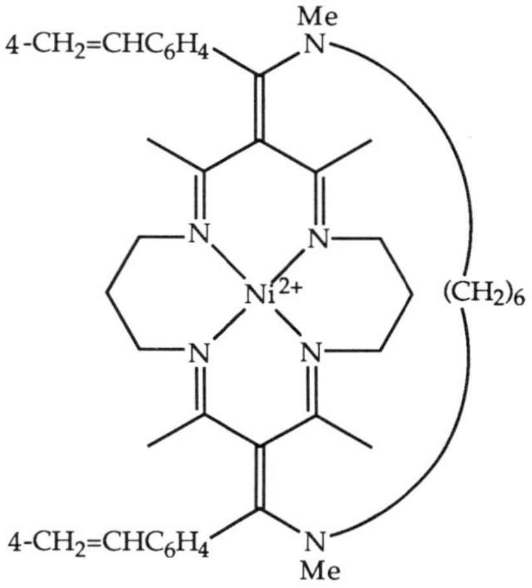

(53)

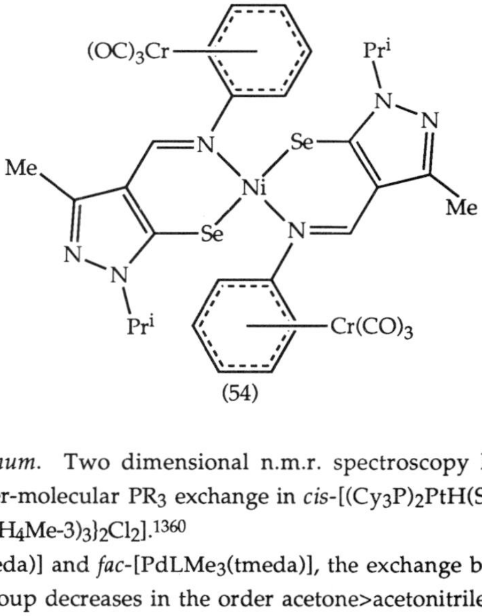

(54)

Palladium and Platinum. Two dimensional n.m.r. spectroscopy has been applied to study intra- and inter-molecular PR_3 exchange in *cis*-[$(Cy_3P)_2PtH(SiR_3)$], [$(Cy_3P)_2Pt(\eta^2$-$C_2H_4)$], and [$Hg\{P(C_6H_4Me$-$3)_3\}_2Cl_2$].[1360]

In *fac*-[$PdXMe_3$(tmeda)] and *fac*-[$PdLMe_3$(tmeda)], the exchange between the axial and equatorial methyl group decreases in the order acetone>acetonitrile>I^->Br^-. The ^1H and ^{13}C n.m.r. spectra were measured.[1361] Exchange in $PtMe_3X$ complexes of $(MeS)_2CHCH$-$(SMe)_2$ has been studied by ^1H and ^{195}Pt n.m.r. spectroscopy.[1362] Variable temperature ^1H n.m.r. studies of [$(PtMe_2SMe)_2(PtMe_3X)_2$] and [$(PtMe_2SMe)_3(PtMe_3Cl)$] show that methyl scrambling occurs with $\Delta G^{\ddagger}$ in the range 61 to 72 kJ mol^{-1}. The ^{13}C n.m.r. spec-

1360 M.J. Hampden-Smith and H. Ruegger, *Magn. Reson. Chem.*, 1989, **27**, 1107.

1361 W. De Graaf, J. Boersma, W.J.J. Smeets, A.L. Spek, and G. Van Koten, *Organometallics*, 1989, **8**, 2907.

1362 E.W. Abel, T.P.J. Coston, K.M. Higgins, K.G. Orrell, V. Šik, and T.S. Cameron, *J. Chem. Soc., Dalton Trans.*, 1989, 701; E.W. Abel, T.P.J. Coston, K.G. Orrell, and V. Šik, *J. Chem. Soc., Dalton Trans.*, 1989, 711.

trum was also recorded.[1363] Variable temperature ^{1}H and ^{195}Pt n.m.r. spectroscopy has been used to examine S and Se inversion in $[PtXMe_3(\eta^5\text{-}MeEC_5H_4)_2Fe]$, E = S, Se. $\Delta G^{\ddagger}$ was determined for sulphur inversion and ring inversion.[1364] The variable temperature ^{31}P n.m.r. spectrum of $[(\eta^5\text{-}C_5H_5)(OC)_3MoPtMe(PPh_3)_2]$ shows two signals at 195 °C and one at 265 °C.[1365] The ^{1}H and ^{13}C n.m.r. spectra of $[Fe(\eta^5\text{-}C_5H_4PPh_2)_2\text{-}Pt(CH_2SiMe_2CH{=}CH_2)]$ show two signals for the cyclopentadienyl at room temperature and four at low temperature due to restricted rotation about the Fe-C_5H_5 bond.[1366] Variable temperature ^{1}H n.m.r. spectroscopy has been used to demonstrate two disctinct rearrangements in $[Pt(3,4\text{-diacetyl-}2,4\text{-haxadiene-}2,5\text{-diol})(SEt_2)_2]$.[1367] ^{31}P EXSY has been applied to exchange in $[Pt(SnCl_3)(COEt)\{Ph_2P(CH_2)_4PPh_2\}]$.[1368] ^{1}H n.m.r. spectra of two isomers of $[Pt(\eta^3\text{-}2\text{-}MeC_3H_4)(SnCl_3)(\eta^2\text{-}PhCH{=}CH_2)]$ have been interpreted using ^{1}H-$\{^{195}Pt\}$ inverse correlation, *J*-resolved, COSY-90, and NOESY. The two isomers are involved in slow exchange.[1369] The solution behaviour of $[M(\eta^3\text{-}2\text{-}MeC_3H_4)(SnCl_3)(CO)]$, M = Pd, Pt, has been investigated by ^{1}H and ^{13}C n.m.r. spectroscopy. The compound is dynamic by CO dissociation. The ^{119}Sn and ^{195}Pt n.m.r. spectra were also recorded.[1370]

The dynamic ^{1}H n.m.r. spectra, including EXSY, of $[MX_2E\{(CH_2)_nCR{=}CR_2\}_2]$, M = Pd, Pt, show exchange of the double bonds and $\Delta G^{\ddagger}$ was determined.[1371] The ^{1}H n.m.r. spectrum of $[(\eta^3\text{-}C_3H_5)Pd(CNMe)_2]^+$ is A_4X at 68 °C but AMM'XX' at -43 °C, and $\Delta G^{\ddagger}$ was determined. The ^{13}C n.m.r. spectrum was also recorded.[1372] ^{1}H and ^{31}P DANTE and NOESY have been used to investigate isomerisation in $[(\eta^3\text{-}2\text{-}MeC_3H_4)\text{-}PdPh_2PNC_4H_7PPh_2]^+$.[1373] According to ^{1}H and ^{13}C n.m.r. spectroscopy, the *syn*- and *anti*-protons of $[Pt\{\eta^3\text{-}CH_2C(O)CH_2\}(PPh_3)_2]$ exchange with $\Delta G^{\ddagger} = 9.2$ kcal mol^{-1}.[1374]

Variable temperature ^{31}P n.m.r. spectroscopy shows 1,2-shifts in *trans*-$[M(P_3C_2Bu^t_2)\text{-}(PEt_3)_2]$, M = Pd, Pt.[1375] The dynamic behaviour of $[(diop)_2M]$, M = Pd, Pt, has been reinvestigated and somewhat different conclusions from those presented by Brown and Chaloner were drawn.[1376] ^{1}H n.m.r. spectroscopy has been used to investigate the isomerism and fluxional behaviour of $[PdCl_2L(Me_2SNR)]$.[1377] Exchange is observed in (55)

1363 E.W. Abel, K.G. Orrell, and D. Stephenson, *J. Organomet. Chem.*, 1989, **373**, 401.
1364 E.W. Abel, N.J. Long, K.G. Orrell, A.G. Osborne, and V. Šik, *J. Organomet. Chem.*, 1989, **378**, 473.
1365 M. Ferrer, O. Rossell, M. Seco, and P. Braunstein, *J. Chem. Soc., Dalton Trans.*, 1989, 379.
1366 R.D. Kelly and G.B. Young, *Polyhedron*, 1989, **8**, 433.
1367 S. Yamazaki and T. Ama, *Bull. Chem. Soc. Jpn.*, 1989, **62**, 931.
1368 C. Ammann, P.S. Pregosin, and A. Scrivanti, *Inorg. Chim. Acta*, 1989, **155**, 217.
1369 C. Ammann, P.S. Pregosin, H. Rüegger, M. Grassi, and A. Musco, *Magn. Reson. Chem.*, 1989, **27**, 355.
1370 M. Grassi, S.V. Meille, A. Musco, R. Pontellini, and A. Sironi, *J. Chem. Soc., Dalton Trans.*, 1989, 615.
1371 E.W. Abel, D.G. Evans, J.R. Koe, V. Šik, M.B. Hursthouse, and P.A. Bates, *J. Chem. Soc., Dalton Trans.*, 1989, 2315.
1372 F.R. Lemke and C.P. Kubiak, *J. Organomet. Chem.*, 1989, **373**, 391.
1373 E. Cesarotti, M. Grassi, L. Prati, and F. Demartin, *J. Organomet. Chem.*, 1989, **370**, 407.
1374 J. Fawcett, W. Henderson, M.D. Jones, R.D.W. Kemmitt, D.R. Russell, B. Lam, S.K. Kang, and T.A. Albright, *Organometallics*, 1989, **8**, 1991.
1375 P.B. Hitchcock, J.F. Nixon, and G.J.D. Sillett, *New J. Chem.*, 1989, **13**, 353.
1376 E.M. Campi, P.S. Elmes, W.R. Jackson, R.J. Thomson, and J.A. Weigold, *J. Organomet. Chem.*, 1989, **371**, 393.
1377 J.L. Davidson, P.N. Preston, S.A.R. Spankie, G. Douglas, and K.W. Muir, *J. Chem. Soc., Dalton Trans.*, 1989, 497.

due to ring inversion with concurrent inversion at sulphur. The ^{13}C n.m.r. spectrum was also recorded.[1378] ^1H n.m.r. spectroscopy has been used to investigate the dimerisation of $[Pt(amino\ acid)(OSMe_2)(OH_2)]^+$ and the rates and activation parameters were determined.[1379] The dynamics of Pt—P restricted rotation in *trans*-$[PtCl_2(PMeBu^t{}_2)_2]$ has been investigated using dynamic ^{31}P n.m.r. spectroscopy.[1380] The ^1H and ^{195}Pt n.m.r. spectra of *cis*-$[PtCl_2\{[(\mu\text{-}Ph_2PCH_2)_2PPh]PtCl_2\}_2]$ show restricted rotation about the Pt—P bond with $\Delta G^‡$ of *ca.* 48 kJ mol^{-1}.[1381] ^1H n.m.r. spectroscopy has been used to determine the mechanism of sulphur inversion in some thioether complexes such as $[(PPh_3)_2\overline{PtO_2CCH_2S}Pr^i]$.[1382]

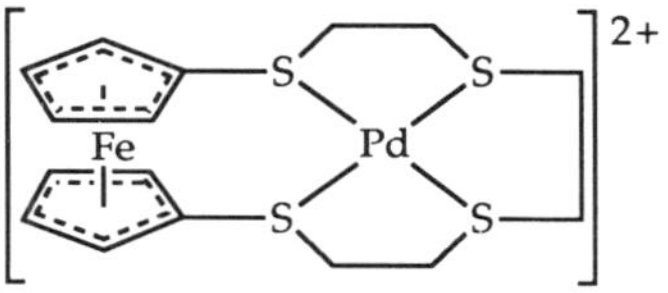

(55)

Copper. The ^1H n.m.r. spectrum of $[Cu_4Br_2\{C_6H_3(CH_2NMe_2)_2\text{-}2,6\}]$ shows dynamic behaviour.[1383] The pyrazole in $[Cu(pzH)_2(CO)Cl]$ is fluxional according to ^1H and ^{13}C n.m.r. spectroscopy.[1384] The ^1H and ^{13}C n.m.r. spectra of $[Cu(SC_6H_4CHMeNMe_2\text{-}2)]_3$ are temperature dependent.[1385]

Cadmium. Dynamic n.m.r. spectra have been observed for [Cd{glycol(β-aminoethyl ether)$_2$-*N,N'*-tetraacetate}]$^{2-}$.[1386]

Mercury. The activation energy of the inversion of $[HgL_2]$, L = (56), has been determined as 54.3 kJ mol^{-1}. Ligand exchange was also studied.[1387]

Boron. Fluxionality in $[B_3H_6X]$ has been investigated by ^{11}B and ^{13}C n.m.r. spectroscopy.[1388] In *closo*-$[2,2\text{-}(PhMe_2P)_2\text{-}1,2\text{-}SePtB_{10}H_{10}]$, the $Pt(PMe_2Ph)_2$ moiety rotates with

1378 S. Sato and S. Akabori, *Bull. Chem. Soc. Jpn.*, 1989, **62**, 3492.
1379 L.E. Erickson, I.B. Burgeson, E. Eidsmoe, and R.G. Larsen, *Inorg. Chem.*, 1989, **28**, 1315.
1380 L.A. Luck, W.L. Elcesser, J.L. Hubbard, and C.H. Bushweller, *Magn. Reson. Chem.*, 1989, **27**, 488.
1381 K.A. Asker, K.R. Seddon, and M.C. Grossel, *New J. Chem.*, 1989, **13**, 587.
1382 S. Toyota, Y. Yamada, M. Kaneyoshi, and M. Ōki, *Bull. Chem. Soc. Jpn.*, 1989, **62**, 1509.
1383 E. Wehman, G. Van Koten, C.J.M. Erkamp, D.M. Knotter, J.T.B.H. Jastrzebski, and C.H. Stam, *Organometallics*, 1989, **8**, 94.
1384 M.A. Angaroni, G.A. Ardizzoia, T. Beringhelli, G. D'Alfonso, G. La Monica, N. Masciocchi, and M. Moret, *J. Organomet. Chem.*, 1989, **363**, 409.
1385 D.M. Knotter, G. van Koten, H.L. van Maanen, D.M. Grove, and A.L. Spek, *Angew. Chem., Int. Ed. Engl.*, 1989, **28**, 341.
1386 Y. Bu, R. Song, and Z. Qiu, *Kexue Tongbao (Foreign Lang. Ed.)*, 1988, **33**, 1087 (*Chem. Abstr.*, 1989, **110**, 87 200).
1387 A.L. Nivorozhkin, E.V. Sukholenko, L.E. Nivorozhkin, N.I. Borisenko, V.I. Minkin, Yu.K. Grishin, O.A. D'yachenko, T.G. Takhirov, and D.B. Tagiev, *Polyhedron*, 1989, **8**, 569.
1388 H. Binder, B. Brellochs, B. Frei, A. Simon, and B. Hettich, *Chem. Ber.*, 1989, **122**, 1049.

$\Delta G^{\ddagger}$ = *ca*. 58 kJ mol^{-1}. The ^{11}B n.m.r. spectrum was recorded.[1389] Restricted rotation about the B—N bond in [1-(η^5-C_5Me_5)-4-Et_2N-*isocloso*-1-$RhB_{10}H_9$] has been observed with $\Delta G^{\ddagger}_{272}$ = *ca*. 54 kJ mol^{-1}.[1390] The concentration dependence of the rotational motion of $[BPh_4]^-$ and $[AsPh_4]^+$ has been determined from ^{13}C spin-lattice relaxation times.[1391] The fluxionality of $[(RBO)_3N_2H_4]$ has been investigated by ^{11}B n.m.r. spectroscopy.[1392]

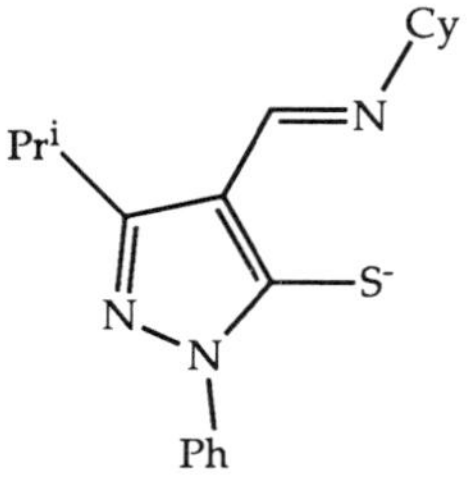

(56)

Aluminium. The variable temperature ^1H n.m.r. spectrum of [Me_2Al{OCHPhCH-(NHMe)Me}] shows two methyl groups at 22 °C and one at 70 °C. $\Delta G^{\ddagger}$ was determined and the ^{13}C n.m.r. spectrum was also measured.[1393]

Indium. The ^{13}C n.m.r. spectrum of [In($C_6H_2Me_3$-2,4,6)$_3$] shows two *ortho*-methyl signals, one with agostic interactions, and $^1J(^{13}C-^1H)$ = 102.0 Hz.[1394]

Thallium. The fluxional behaviour of [Tl($C_5H_4NO_2$)] and [K(C_5H_4CHO)] has been investigated by ^1H and ^{13}C n.m.r. spectroscopy.[1395]

Silicon, Germanium, Tin, and Lead. Restricted rotation of the mesityl group in [(η^5-C_5Me_5)$_2$Zr{SiH($C_6H_2Me_3$-2,4,6)$_2$}Me] is stopped at -80 °C with $\Delta G^{\ddagger}$ = 10.4 ± 0.2 kcal mol^{-1}. The ^{13}C and ^{29}Si n.m.r. spectra were recorded.[1396]

^1H n.m.r. spectroscopy has been used to investigate the steric barrier for rotation about the CH_2—Ph bond in $PhCH_2MMe_3$, M = C, Si, Ge, Sn, Pb.[1397] The rapid metallotropic rearrangement in trimethylstannyldihydropentalene has been investigated using ^1H

1389 Faridoon, O.N. Dhubhghaill, T.R. Spalding, G. Ferguson, B. Kaitner, X.L.R. Fontaine, and J.D. Kennedy, *J. Chem. Soc., Dalton Trans.*, 1989, 1657.
1390 E.J. Ditzel, X.L.R. Fontaine, N.N. Greenwood, J.D. Kennedy, and M. Thornton-Pett, *J. Chem. Soc., Chem. Commun.*, 1989, 1115.
1391 Y. Masuda and H. Yokoyama, *Chem. Lett.*, 1988, 1769 (*Chem. Abstr.*, 1989, **111**, 174 259).
1392 M.K. Das, J.F. Mariategui, and K. Niedenzu, *Report*, 1987, UK/DC-TR-16; Order No. AD-A182688, 11pp. Avail. NTIS. From *Gov. Rep. Announce. Index (U.S.)*, 1987, **87**, Abstr. No. 748 265 (*Chem. Abstr.*, 1989, **110**, 95 304).
1393 M.L. Sierra, V.S.J. de Mel, and J.P. Oliver, *Organometallics*, 1989, **8**, 2486.
1394 J.T. Leman and A.R. Bacon, *Organometallics*, 1989, **8**, 2214.
1395 M. Arthurs, P. Gross, G. Kubal, L. Paniwnyk, and E. Curzon, *J. Organomet. Chem.*, 1989, **366**, 223.
1396 D.M. Roddick, R.H. Heyn, and T.D. Tilley, *Organometallics*, 1989, **8**, 324.
1397 T. Schaefer, G.H. Penner, C.S. Takeuchi, and C. Beaulieu, *Can. J. Chem.*, 1989, **67**, 1283.

and ^{13}C n.m.r. spectroscopy. The rearrangements are *via* [1,5] sigmatropic shifts.[1398] The activation energy for inversion of (57) has been determined from ^{13}C n.m.r. measurements.[1399] Racemisation of [$\overline{Si(1\text{-naphthyl-8-}N}Me_2)_2R^1R^2$] has been investigated and the ^{29}Si n.m.r. spectrum recorded.[1400] Hindered rotation about the Si—C bonds in (58) has been studied.[1401]

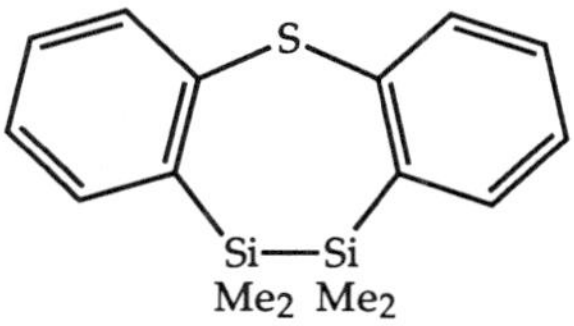

(57)

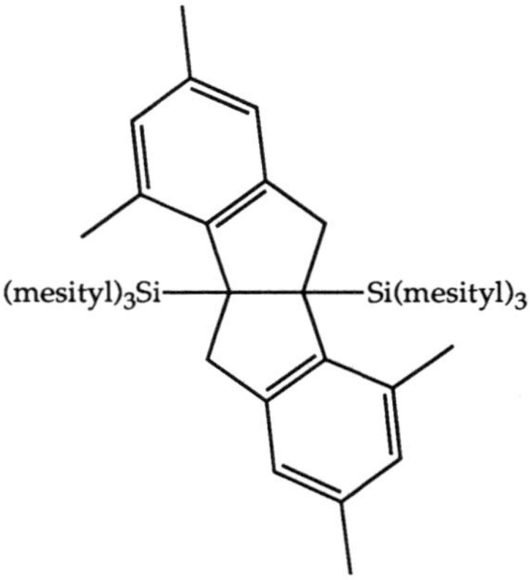

(58)

^{1}H n.m.r. spectroscopy has been used to study restricted rotation in $Me_2NC(S)SN(SiMe_3)_2$. The ^{13}C, ^{15}N, and ^{29}Si n.m.r. spectra were also recorded.[1402] The $\Delta G^\ddagger$ for exchange of the silicon between acetyl groups in [$\{MeC(O)\}_2NCH_2SiMe_2X$] has been determined from ^{1}H, ^{13}C, and ^{29}Si n.m.r. spectroscopy.[1403] The Berry pseudorotation in $[R_3SiF_2]^-$ has been studied by ^{19}F and ^{29}Si n.m.r. spectroscopy.[1404] Some compounds of the type [$SnCl_{3-n}R_n(pz_3BH)$] are fluxional according to ^{13}C and ^{119}Sn n.m.r. spectroscopy.[1405]

Ring inversion barriers have been determined for 4,5;7,8-dibenzo- and dinaphtho-1,3,2-dioxasilocins.[1406] Local side-group dynamics of poly(methylphenylsiloxane) have been studied by ^{2}H n.m.r. spectroscopy.[1407] Substituent contributions to the exchange barriers of $[R^1R^2SiF_3]^-$ have been measured by ^{19}F and ^{13}C n.m.r. spectroscopy.[1408] Berry

1398 A.K. Shestakova, V.A. Chertkov, and Yu.A. Ustynyuk, *Metalloorg. Khim.*, 1988, **1**, 132.
1399 V.A. Polyakov, O.G. Rodin, and V.F. Traven, *Zh. Obshch. Khim.*, 1988, **58**, 1840 (*Chem. Abstr.*, 1989, **111**, 97 332).
1400 C. Breliere, R.J.P. Corriu, G. Royo, and J. Zwecker, *Organometallics*, 1989, **8**, 1834.
1401 M. Weidenbruch, Y. Pan, K. Peters, and H.G. von Schnering, *Chem. Ber.*, 1989, **122**, 1483.
1402 G. Schubert and G. Gattow, *Z. Anorg. Allg. Chem.*, 1989, **573**, 75.
1403 I.D. Kalikhman, O.B. Bannikova, L.I. Belousova, B.A. Gostevskii, E. Liepinsh, O.A. Vyazankina, N.S. Vyazankin, and V.A. Pestunovich, *Metallorg. Khim.*, 1988, **1**, 683 (*Chem. Abstr.*, 1989, **111**, 57 840).
1404 S.E. Johnson, J.S. Payne, R.O. Day, J.M. Holmes, and R.R. Holmes, *Inorg. Chem.*, 1989, **28**, 3190.
1405 G.G. Lobbia, F. Bonati, P. Cecchi, A. Cingolani, and A. Lorenzotti, *J. Organomet. Chem.*, 1989, **378**, 139.
1406 B.A. Arbuzov, V.V. Klochkov, L.V. Egorova, K.A. Il'yasov, and R.P. Arshinova, *Dokl. Akad. Nauk SSSR*, 1989, **305**, 102 (*Chem. Abstr.*, 1989, **111**, 134 265).
1407 G. Meier, F. Fujara, and W. Petry, *Macromolecules*, 1989, **22**, 4421 (*Chem. Abstr.*, 1989, **111**, 234 009).
1408 R. Damrauer, B. O'Connell, S.E. Danahey, and R. Simon, *Organometallics*, 1989, **8**, 1167.

pseudorotation in $[(2,4,6\text{-}Me_3C_6H_2)_2SiF_3]^-$ has been determined by ^{19}F n.m.r. spectroscopy.[1409] The activation energy for ring inversion in $Me_2Ge(SeCH_2)_2CMe_2$ has been determined as 32.4 kJ mol^{-1} from ^{1}H, ^{13}C, and ^{77}Se n.m.r. spectroscopy.[1410] ^{1}H, ^{13}C, and ^{119}Sn n.m.r. spectroscopy has been used to study Berry pseudorotation at the tin in [(8-Me_2N-naphthyl-1-$)_2SnW(CO)_5$].[1411]

The barrier to pseudorotation in $[RSi(O_2CCH_2O)_2]^-$ has been determined and the ^{1}H, ^{13}C, and ^{29}Si n.m.r. spectra recorded.[1412] The activation energy for Si—N bond breaking in $[Si(C_6H_4CHMeNMe_2)(O_2C_6H_4)_2]^-$ has been determined.[1413] Pseudorotation in $[2,4,6\text{-}Me_3C_6H_2SiF_4]^-$ has been investigated by ^{19}F and ^{29}Si n.m.r. spectroscopy.[1414]

Nitrogen. Degenerate and pseudodegenerate 1,3-nitrogen shifts in $PhCN_5S_3$ have been studied using ^{15}N n.m.r. spectroscopy.[1415]

Phosphorus. The dynamic properties of liquid crystalline polyphosphazene films have been studied by n.m.r. spectroscopy.[1416]

Sulphur. The rotational barrier for $(F_5S)_2CHCF_2Cl$ is hindered with $\Delta G^{\ddagger} = 61.7$ kJ mol^{-1} according to ^{19}F n.m.r. measurements.[1417]

Selenium. Ring inversion in $[1,3\text{-}C_6H_4(CH_2SeCH_2)C_6H_4\text{-}1,3]$ and $[2,6\text{-}C_5NH_4(CH_2SeCH_2)C_5NH_4\text{-}2,6]$ has been studied by ^{1}H, ^{13}C, and ^{77}Se n.m.r. spectroscopy.[1418] Ring inversion in some selenium coronands has also been studied by ^{1}H, ^{13}C, and ^{77}Se n.m.r. spectroscopy.[1419]

Tellurium. Exchange of the non-equivalent fluorine atoms in *cis*-$[F_2TePh_3Cl]$ has been investigated using ^{19}F and ^{125}Te n.m.r. spectroscopy.[1420]

1409 S.E. Johnson, J.A. Deiters, R.O. Day, and R.R. Holmes, *J. Am. Chem. Soc.*, 1989, **111**, 3250.
1410 S. Tomoda, M. Shimoda, M. Sanami, Y. Takeuch, and Y. Iitaka, *J. Chem. Soc., Chem. Commun.*, 1989, 1304.
1411 J.T.B.H. Jastrebski, P.A. van der Schaff, J. Boersma, G. van Koten, D. Heijdenrijk, K. Goubitz, and D.J.A. de Ridder, *J. Organomet. Chem.*, 1989, **367**, 55.
1412 E. Liepins, G. Ancens, N.P. Erchak, and E. Lukevics, *Zh. Obshch. Khim.*, 1988, **58**, 384 (*Chem. Abstr.*, 1989, **110**, 23 952).
1413 F. Carré, G. Cerveau, C. Chuit, R.J.P. Corriu, and C. Réyé, *Angew. Chem., Int. Ed. Engl.*, 1989, **28**, 489.
1414 S.E. Johnson, R.O. Day, and R.R. Holmes, *Inorg. Chem.*, 1989, **28**, 3182.
1415 K.T. Bestari, R.T. Boere, and R.T. Oakley, *J. Am. Chem. Soc.*, 1989, **111**, 1579.
1416 I.B. Sokol'skaya, V.V. Kireev, and Yu.V. Zelenev, *Vysokomol. Soedin., Ser. B*, 1988, **30**, 795 (*Chem. Abstr.*, 1989, **110**, 39 615).
1417 R. Gerhardt and K. Seppelt, *Chem. Ber.*, 1989, **122**, 463.
1418 M. Hojjatie, S. Muralidharan, and H. Freiser, *Tetrahedron*, 1989, **45**, 1611; S. Muralidharan, M. Hojjatie, M. Firestone, and H. Freiser, *J. Org. Chem.*, 1989, **54**, 393.
1419 R.J. Batchelor, F.W.B. Einstein, I.D. Gay, J.H. Gu, B.D. Johnston, and B.M. Pinto, *J. Am. Chem. Soc.*, 1989, **111**, 6582.
1420 A.F. Janzen and M. Jang, *Can. J. Chem.*, 1989, **67**, 71.

Equilibria.—*Solvation Studies of Ions. Group IA.* ^{17}O n.m.r. spectroscopy has been used to study ion hydration in acetone and acetonitrile.[1421] The self diffusion of water molecules in aqueous electrolyte solutions has been studied.[1422] The relative ^{1}H T_1 and self-diffusion measurements have been performed for DMSO solutions of NaBr, KBr, NaI, KI, RbI, and CsI.[1423] ^{1}H n.m.r. spectroscopy has been used to investigate the solvation of Na^+, K^+, Rb^+, and Cs^+ by $(Me_2N)_2CO$.[1424] Interionic interactions of Li^+, Na^+, K^+, Cs^+, and Tl^+ with counterions in *N*-methylformamide have been investigated by ^{7}Li, ^{23}Na, ^{39}K, ^{133}Cs, and ^{205}Tl n.m.r. spectroscopy.[1425]

^{1}H n.m.r. spectroscopy has been used to investigate MeCN exchange on Li^+.[1426] The preferential solvation of Li^+ cations in aqueous MeCN, acetone, THF, DMF, DMSO, or HMPA has been studied by using ^{7}Li quadrupole relaxation.[1427] Ion solvation and association behaviour has been studied for solutions of $LiClO_4$ and $LiNO_3$ in DMF, MeOH, and EtOH by multinuclear n.m.r. spectroscopy.[1428] The standard Gibbs free energy of transfer of Li^+ from $POCl_3$ to $SOCl_2$ has been determined by a ^{7}Li quadrupole relaxation method.[1429]

^{23}Na n.m.r. spectroscopy has been used to investigate the preferential solvation of Na^+ in mixtures of *N*-methylformamide with water and some non-aqueous solvents,[1430] and NaI in ethylene glycol-MeCN and propylene glycol-MeCN.[1431] The dynamic interactions in dilute KCl and NaCl aqueous solutions have been studied by ^{23}Na, ^{35}Cl, and ^{39}K n.m.r. spectroscopy.[1432] The ^{1}H T_1 values of $HCON(CD_3)_2$ and $DCON(CD_3)_2$ caused by magnetic dipole-dipole interaction with ^{127}I and ^{23}Na in the ions I^- and Na^+ respectively, have been measured.[1433] The concentration and temperature dependence of water and F^- in aqueous solutions of KF and CsF has been studied by n.m.r. spectroscopy.[1434]

Magnesium. The nuclear magnetic relaxation rates of $^{35}Cl^-$ and $^{25}Mg^{2+}$ at infinite dilution in water have been reported with increased accuracy.[1435]

Yttrium. ^{1}H, ^{35}Cl, and ^{89}Y n.m.r. spectroscopy has been used to investigate complex

1421 K. Miura, *J. Magn. Reson.*, 1989, **81**, 552.
1422 V.M. Zagainov, V.A. Sevryugin, S.I. Alekseeva, and M.I. Emel'yanov, *Zh. Fiz. Khim.*, 1989, **63**, 1071 (*Chem. Abstr.*, 1989, **111**, 65 105).
1423 A. Sacco, M. Carbonara, and M. Holz, *J. Chem. Soc., Faraday Trans. 1*, 1989, **85**, 1257.
1424 L.A. Ciolino and B.J. Barker, *Canad. J. Chem.*, 1989, **67**, 2206.
1425 L.L. Soong, G.E. Leroi, and A.I. Popov, *J. Solution Chem.*, 1989, **18**, 561.
1426 R.G. Keil, D.W. Johnson, M.A. Fryling, and J.F. O'Brien, *Inorg. Chem.*, 1989, **28**, 2764.
1427 A.I. Mishustin, *Zh. Fiz. Khim.*, 1988, **62**, 3201 (*Chem. Abstr.*, 1989, **110**, 142 424).
1428 D.W. James, R.E. Mayes, W.H. Leong, I.M. Jamie, and G. Zhen, *Faraday Discuss. Chem. Soc.*, 1988, **85**, 269.
1429 A.I. Mishustin, V.A. Mozalevskaya, and V.P. Ponkratov, *Zh. Fiz. Khim.*, 1989, **63**, 1345 (*Chem. Abstr.*, 1989, **111**, 84 969).
1430 H.-J. Chang, L.-L. Soong, G.E. Leroi, and A.I. Popov, *J. Solution Chem.*, 1989, **18**, 759.
1431 A.K. Covington and M. Dunn, *J. Chem. Soc., Faraday Trans. 1*, 1989, **85**, 2835.
1432 R.K. Mazitov, J.M. Enikeev, and I.F. Ismaev, *Khim. Fiz.*, 1989, **8**, 230 (*Chem. Abstr.*, 1989, **110**, 161 215).
1433 C.K. Finter and H.G. Hertz, *J. Chim. Phys. Phys.-Chim. Biol.*, 1988, **85**, 589.
1434 M.I. Emel'yanov, V.M. Zagainov, and E.A. Nikifirov, *Khim. Fiz.*, 1989, **8**, 419 (*Chem. Abstr.*, 1989, **111**, 141 017).
1435 R.P.W.J. Struis, J. De Bleijser, and J.C. Leyte, *J. Phys. Chem.*, 1989, **93**, 7932; 7943.

formation of $[Y(NO_3)_3]$ and $[Y(ClO_4)_3]$ in aqueous solvent mixtures.[1436]

Lanthanum. The hydration of La^{3+} has been carried out for solutions of $La(ClO_4)_3$ and $La(NO_3)_3$ in aqueous mixtures with acetone-d_6 and Freon-12 using ^1H, ^{35}Cl, and ^{139}La n.m.r. spectroscopy.[1437] Water exchange on $[Ln(OH_2)_6]^{3+}$, Ln = Tb, Dy, Ho, Er, Tm, has been investigated by high pressure ^{17}O n.m.r. spectroscopy.[1438] Solvent exchange on Gd^{3+} has been studied by ^1H n.m.r. spectroscopy.[1439] The hydration of the Gd^{3+} complex of the monopropylamide derivative of 1,4,7,10-tetraazacyclododecane-*N,N',N'',N'''*-tetraacetic acid has been determined from ^1H T_1 measurements.[1440] The solvation of Dy^{3+} in H_2O-MeCN has been investigated using ^1H T_1 measurements.[1441]

Chromium. N.m.r. relaxation has been used to study the preferential solvation of $[Cr(acac)_3]$ with $HCBr_3$ and a solvation number of 2 was found.[1442]

Molybdenum. Acid dissociation and water exchange on $[Mo_3O_4(OH_2)_9]^{4+}$ have been studied using ^{17}O n.m.r. spectroscopy.[1443]

Manganese. The solvation of $[Mn(NH_2Et)_2L]^{2+}$, L = (59), has been investigated by ^1H n.m.r. spectroscopy.[1444]

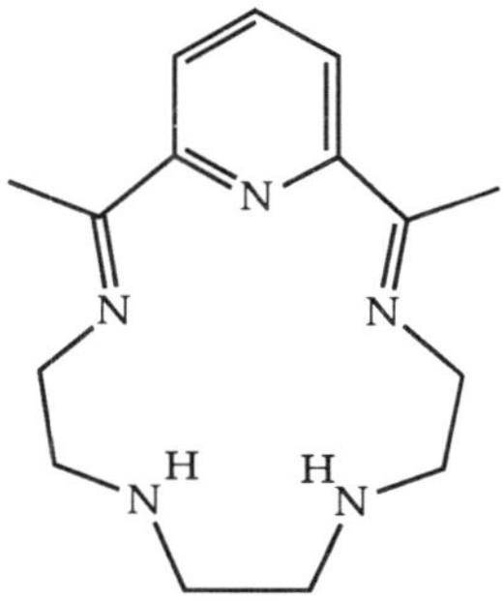

(59)

Cobalt. The $^1H/^2H$ fractionation factors of coordinated water in $[Co(OH_2)_6]^{2+}$ and Co^{II} substituted carbonic anhydrase have been determined from ^1H relaxation rates of the water protons.[1445]

Nickel. Variable temperature and pressure ^{14}N n.m.r. studies of RCN exchange on Ni^{II}

1436 A. Fratiello, V. Kubo-Anderson, T. Bolinger, C. Cordero, B. DeMerit, T. Flores, D. Matejka, and R. Perrigan, *J. Magn. Reson.*, 1989, **83**, 358.

1437 A. Fratiello, V. Kubo-Anderson, T. Bolinger, C. Cordero, B. DeMerit, T. Flores, and R. Perrigan, *J. Solution Chem.*, 1989, **18**, 313.

1438 C. Cossy, L. Helm, and A.E. Merbach, *Inorg. Chem.*, 1989, **28**, 2699.

1439 I.A. Veselov, V.G. Shtyrlin, and A.V. Zakharov, *Koord. Khim.*, 1989, **15**, 567 (*Chem. Abstr.*, 1989, **111**, 13 112).

1440 A.D. Sherry, R.D. Brown, tert., C.F.G.C. Geraldes, S.H. Koenig, K.T. Kuan, and M. Spiller, *Inorg. Chem.*, 1989, **28**, 620.

1441 S.G. Vul'fson, A.E. Nepryakhin, F.V. Devyatov, Yu.I. Sal'nikov, and A.N. Vereshchagin, *Dokl. Akad. Nauk SSSR*, 1989, **305**, 121 (*Chem. Abstr.*, 1989, **110**, 219 979).

1442 P.L. Wang and L.P. Hwang, *J. Chem. Soc., Faraday Trans. 1*, 1989, **85**, 2335.

1443 D.T. Richens, L. Helm, P.A. Pittet, A.E. Merbach, F. Nicolo, and G. Chapuis, *Inorg. Chem.*, 1989, **28**, 1394.

1444 B.K. Wagnon and S.C. Jackels, *Inorg. Chem.*, 1989, **28**, 1923.

1445 J.W. Kassebaum and D.N. Silverman, *J. Am. Chem. Soc.*, 1989, **111**, 2691.

have been reported and $\Delta V^{\ddagger}$, $\Delta H^{\ddagger}$, and $\Delta S^{\ddagger}$ determined.[1446] Variable temperature and pressure ^{1}H n.m.r. spectroscopy has been used to investigate the rates and activation parameters for MeCN or DMF exchange on nickel cyclic amine complexes.[1447]

Palladium. The kinetics of water exchange on $[Pd\{(R_2NCH_2CH_2)_2NR\}(OH_2)]^{2+}$ by thiourea have been determined by ^{1}H, ^{13}C, and ^{17}O n.m.r. spectroscopy and the activation parameters, including $\Delta V^{\ddagger}$, determined.[1448]

Platinum. Water exchange of *trans*-$[PtCl_2(OH_2)_2]$ and $[Pt(OH_2)_4]^{2+}$ has been studied by ^{195}Pt n.m.r. spectroscopy using an ^{18}O tracing technique.[1449] DMSO exchange on *cis*-$[Pt(DMSO)_4]^{2+}$ has been studied by ^{1}H n.m.r. spectroscopy and activation parameters, including $\Delta V^{\ddagger}$, have been determined.[1450]

Zinc. The relaxation rates of ^{2}H and ^{17}O have been obtained in a series of $ZnCl_2$ solutions and used to study the structure and dynamics of the water.[1451]

Aluminium. ^{27}Al m.a.s. n.m.r. spectroscopy has been shown to be superior to conventional ^{27}Al n.m.r. spectroscopy in resolving species present in the $AlCl_3$-H_2O-DMF system.[1452] The solvation behaviour of Al^{3+} cations in HMPT-water mixtures has been investigated by ^{27}Al n.m.r. spectroscopy.[1453] The effect of DMF addition on alumina sol-gel formation has been investigated by ^{27}Al n.m.r. spectroscopy.[1454] ^{27}Al n.m.r. spectroscopy has been used to study the solvation of anhydrous $AlCl_3$ in various Lewis base solutions.[1455] N.m.r. spectroscopy has been used to detect water in $SOCl_2$, $AlCl_3$-$SOCl_2$, and $LiAlCl_4$-$SOCl_2$ systems for Li batteries.[1456]

Gallium. ^{71}Ga n.m.r. spectroscopy has been used to study aqueous Ga^{3+} and its hydrolysis.[1457]

Silicon, Germanium, and Tin. Equilibrium constants and enthalpies of complexation of $CHCl_3$ with Me_3MOOR, M = Si, Ge, Sn, have been determined by n.m.r. spectroscopy.[1458]

Nitrogen. The ^{14}N and ^{17}O T_1 values have been determined for MNO_3 in aqueous solution.[1459]

Ionic Equilibria. ^{1}H T_1 has been measured in micellar solution as a function of added

1446 M. Ishii, S. Funahashi, K. Ishihara, and M. Tanaka, *Bull. Chem. Soc. Jpn.*, 1989, **62**, 1852.
1447 L. Fielding and P. Moore, *J. Chem. Soc., Dalton Trans.*, 1989, 873.
1448 J. Berger, M. Kotowski, R. van Eldik, U. Frey, L. Helm, and A.E. Merbach, *Inorg. Chem.*, 1989, **28**, 3759.
1449 O. Groening and L.I. Elding, *Inorg. Chem.*, 1989, **28**, 3366.
1450 Y. Ducommun, L. Helm, A.E. Merbach, B. Hellquist, and L.I. Elding, *Inorg. Chem.*, 1989, **28**, 377.
1451 J.R.C. Van der Maarel, *J. Magn. Reson.*, 1989, **81**, 92.
1452 E. Hallas, H.-H. Emons, M. Siedler, and B. Thomas, *Magn. Reson. Chem.*, 1989, **27**, 307.
1453 F. Birkeneder, H.-H. Emons, and K. Pollmer, *Z. Anorg. Allg. Chem.*, 1989, **575**, 121.
1454 L.F. Nazar, L.C. Klein, and D. Napier, *Mater. Res. Soc. Symp. Proc.*, 1988, **121**, 133 (*Chem. Abstr.*, 1989, **110**, 119 932).
1455 X. Mao and G. Xu, *Huaxue Wuli Xuebao*, 1988, **1**, 347 (*Chem. Abstr.*, 1989, **110**, 237 770).
1456 H. Cunmao, M. Ke, and Y. Wenzhi, *J. Power Sources*, 1989, **26**, 415 (*Chem. Abstr.*, 1989, **111**, 118 126).
1457 J.W. Akitt and D. Kettle, *Magn. Reson. Chem.*, 1989, **27**, 377.
1458 N.V. Kochnev, N.V. Yablokova, Yu.A. Kurskii, and Yu.A. Aleksandrov, *Zh. Obshch. Khim.*, 1989, **59**, 384 (*Chem. Abstr.*, 1989, **110**, 231 768).
1459 A. Adachi, H. Kiyoyama, M. Nakahara, Y. Masuda, H. Yamatera, A. Shimizu, and Y. Taniguchi, *J. Chem. Phys.*, 1989, **90**, 392.

Cr^{3+}, Co^{2+}, Ni^{2+}, and Cu^{2+} salts.[1460] Equilibrium studies of 8-quinolyl phosphate, 1-naphthyl phosphate, and 8-quinolyl methyl phosphate with Cu^{2+}, Zn^{2+}, Ni^{2+}, Co^{2+}, and Mn^{2+} have been performed using n.m.r. spectroscopy.[1461] Complex formation between nitrilotris(methylenephosphonate) with Mn^{2+}, Fe^{2+}, Co^{2+}, Ni^{2+}, Cu^{2+}, Zn^{2+}, and Cd^{2+} has been studied by ^{31}P n.m.r. spectroscopy.[1462] The formation of Cu^{2+}-(Ni^{2+}, VO^{2+}, Co^{2+}) heterocitrate complexes has been studied using n.m.r. relaxation.[1463] Complex formation equilibria of ethylenediaminetetrapropionic acid with some divalent transition metal ions have been investigated using ^{13}C n.m.r. spectroscopy.[1464]

Group Ia. Exchange between monomer and dimer of PhLi has been studied by ^{7}Li and ^{13}C n.m.r. spectroscopy. When HMPA is added the Li^{+} is solvated and coupling between ^{7}Li and ^{31}P is observed.[1465] A monomer-trimer equilibrium has been observed for [2,6-$(Me_2N)C_6H_4Li$] by ^{13}C n.m.r. spectroscopy.[1466] A ^{13}C T_1 n.m.r. study of ion pairing and aggregation of [Li(indenide)] has been reported.[1467] Ion pairing in lithiated propynes has been investigated by ^{13}C n.m.r. spectroscopy.[1468]

^{7}Li and ^{23}Na n.m.r. spectroscopy has been used to investigate methylamine assisted solubilisation of Li and Na metals.[1469] The selective encapsulation of Li^{+} by 5,12,17-Me_3-1,5,9,12,17-pentaazabicyclo[7.5.5]nonadecane has been studied by ^{7}Li and ^{13}C n.m.r. spectroscopy, and ΔH° and ΔS° determined.[1470] Lithium complexation of jasplakinolide has been investigated by ^{1}H and ^{13}C n.m.r. spectroscopy.[1471] ^{23}Na n.m.r. spectroscopy has been used to study the interaction between NaO_3SCF_3 and poly(bismethoxyethoxyethoxide phosphazene) and local motion was discussed.[1472]

An n.m.r. probe head for diffusion measurements in molten salts has been described and applied to ^{23}Na in molten $NaNO_3$.[1473] ^{7}Li, ^{23}Na, ^{39}K, ^{87}Rb, and ^{133}Cs n.m.r. spectroscopy has been used to investigate ion pairing in alkaline silicate solution.[1474]

^{7}Li T_1 studies of the complexation of 12-crown-4 with $LiClO_4$ in non aqueous solvents

1460 M.J. Hey and C.I. Raven, *J. Chem. Soc., Faraday Trans. 1*, 1989, **85**, 1743.
1461 M.M. Armstrong, U. Kramer, P.W. Linder, R.G. Torrington, and T.A. Modro, *J. Coord. Chem.*, 1989, **20**, 81.
1462 K. Sawada, T. Araki, T. Suzuki, and K. Doi, *Inorg. Chem.*, 1989, **28**, 2687.
1463 N.E. Zhuravlev and Yu.I. Sal'nikov, *Izv. Vyssh. Uchebn. Zaved. Khim. Khim. Tekhnol.*, 1989, **32**, 25 (*Chem. Abstr.*, 1989, **111**, 65 069).
1464 L.H.J. Lajunen, P. Kokkonen, H. Knuutti, and J. Jokisaari, *Acta Chem. Scand.*, 1989, **43**, 2.
1465 H.J. Reich, D.P. Green, and N.H. Phillips, *J. Am. Chem. Soc.*, 1989, **111**, 3444.
1466 S. Harder, J. Boersma, L. Brandsma, J.A. Kanters, W. Bauer, and P.v.R. Schleyer, *Organometallics*, 1989, **8**, 1696.
1467 T. Lejon and U. Edlund, *Acta Chem. Scand.*, 1989, **43**, 275.
1468 P.I. Dem'yanov, I.M. Shtyrkov, D.P. Krut'ko, V.S. Petrosyan, and O.A. Reutov, *Metalloorg. Khim.*, 1988, **1**, 236 (*Chem. Abstr.*, 1989, **111**, 97 316).
1469 M.K. Faber, O. Fussa-Rydel, J.B. Skowyra, L.E.H. McMills, and J.L. Dye, *J. Am. Chem. Soc.*, 1989, **111**, 5957 (*Chem. Abstr.*, 1989, **111**, 49 235).
1470 A. Bencini, A. Bianchi, M. Ciampolini, E. Garcia-Espana, P. Dapporto, M. Micheloni, P. Paoli, J.A. Ramirez, and B. Valtancoli, *J. Chem. Soc., Chem. Commun.*, 1989, 701.
1471 W. Inman, P. Crews, and R. McDowell, *J. Org. Chem.*, 1989, **54**, 2523.
1472 S.G. Greenbaum, K.J. Adamic, Y.S. Pak, M.C. Wintersgill, and J.J. Fontanella, *Solid State Ionics 1987*, (Pub. 1988), **28-30**, 1042 (*Chem. Abstr.*, 1989, **110**, 145 563).
1473 C. Herdlicka, J. Richter, and M.D. Zeidler, *Z. Naturforsch., A*, 1988, **43**, 1075 (*Chem. Abstr.*, 1989, **110**, 141 906).
1474 A.V. McCormick, A.T. Bell, and C.J. Radke, *J. Phys. Chem.*, 1989, **93**, 1733.

have been reported.[1475] ^7Li and ^{23}Na n.m.r. spectroscopy has been used to study intermolecular interactions in the alkali metal anion-radical-salt-crown ether-MeCN system.[1476] ^1H and ^{13}C n.m.r. spectroscopy has been used to investigate conformational changes in some crown ethers on Li^+, Na^+, and K^+ complexation.[1477] The kinetics of dissociation of $[(15\text{-crown-}5)Na]^+$ have been determined by ^{23}Na n.m.r. spectroscopy. The dominant mechanism is associative.[1478] The complexing stabilities of 18-crown-6 with Na^+ and K^+ in methanol have been determined from ^{13}C T_1 measurements.[1479] ^{23}Na and ^{133}Cs n.m.r. spectroscopy has been used to study the system Na^+/Cs^+-DMF-benzo-18-crown-6.[1480] The role of the anion in the dissociation kinetics of [Na(dibenzo-24-crown-8)]$^+$ in CH_3NO_2 has been studied using ^{23}Na n.m.r. spectroscopy.[1481] The interaction of Na^+ with $CH_2\{1,2\text{-naphthyl-}(OCH_2CH_2)_n\}_2O$ has been investigated by ^{23}Na n.m.r. spectroscopy.[1482] ^1H and ^{13}C n.m.r. spectroscopy has been used to investigate the binding of Li^+, Na^+, K^+, and $[NH_4]^+$ to polyquinone crown ethers.[1483] ^{13}C n.m.r. spectroscopy has been used to study the complexation of K^+ by $[M\{\eta^5\text{-}C_5H_4CONHC_6H_3\text{-}3\text{-}(OCH_2CH_2)_4\text{-}4\text{-}O\}_2]$, M = Fe, Ru.[1484]

^1H n.m.r. spectroscopy has been used to study alkali cation-cyclogentiotetraose peracetate complexation.[1485] ^{23}Na n.m.r. spectroscopy has been used to study competitive binding of Li^+, K^+, Rb^+, and Cs^+ to polyelectrolytes in mixed counterion systems.[1486] The self-diffusion constant of $^7Li^+$ in poly(acrylic acid) solution containing a small amount of $MnCl_2$ was determined by pulsed field gradient ^7Li n.m.r. spectroscopy and by field dependent nuclear magnetic relaxation.[1487] The effect of mixing of LiO_3SCF_3 and NaI at high and equal concentrations in a polymer based polyethylene oxide has been investigated by ^7Li n.m.r. spectroscopy.[1488] Pulsed field gradient ^7Li n.m.r. spectroscopy has been used to study Li^+ diffusion in polyethylene oxide-LiO_3SCF_3.[1489] The preferential solvation of Li^+ in $LiBF_4$-$MeOCH_2CH_2OMe$-propylene carbonate has been

1475 A.I. Mishustin, N.B. Generalova, and A.Yu. Tsivadze, *Zh. Neorg. Khim.*, 1989, **34**, 1421 (*Chem. Abstr.*, 1989, **111**, 203 021).
1476 V.S. Kuts, O.V. Zhalko-Titarenko, L.F. Tsekhmistrenko, and V.S. Aleksandrova, *Teor. Eksp. Khim.*, 1989, **25**, 173 (*Chem. Abstr.*, 1989, **111**, 65 106).
1477 E. Kleinpeter, S. Stoss, M. Gäbler, and W. Schroth, *Magn. Reson. Chem.*, 1989, **27**, 676.
1478 K.M. Briere and C. Detellier, *New J. Chem.*, 1989, **13**, 145.
1479 C. Erk, *J. Mol. Liq.*, 1989, **40**, 143 (*Chem. Abstr.*, 1989, **110**, 200 082).
1480 J. Tao, Y. Feng, W. Deng, Y. Ning, and G. Xu, *Qinghua Daxue Xuebao, Ziran Kexueban*, 1989, **27**, 70 (*Chem. Abstr.*, 1989, **110**, 127 337).
1481 H.D.H. Stover and C. Detellier, *J. Phys. Chem.*, 1989, **93**, 3174.
1482 J.C. Lockhart, M.B. McDonnell, W. Clegg, M.N.S. Hill, and M. Todd, *J. Chem. Soc., Dalton Trans.*, 1989, 203.
1483 G. Mehta, K.S. Rao, N. Krishnamurthy, V. Srinivas, and D. Balasubramanian, *Tetrahedron*, 1989, **45**, 2743.
1484 P.D. Beer, H. Sikanyika, A.M.Z. Slawin, and D.J. Williams, *Polyhedron*, 1989, **8**, 879.
1485 G. Bonas and M. Vignon, *Recl. Trav. Chim. Pays-Bas*, 1989, **108**, 161.
1486 C. Qian, M.R. Asdjodi, H.G. Spencer, and G.B. Savitsky, *Macromolecules*, 1989, **22**, 995 (*Chem. Abstr.*, 1989, **110**, 96 264).
1487 R.H. Tromp, J. De Bleijser, and J.C. Leyte, *J. Phys. Chem.*, 1989, 93, 2626.
1488 D.P. Tunstall, A.S. Tomlin, F.M. Gray, J.R. MacCullum, and C.A. Vincent, *J. Phys.: Condens. Matter*, 1989, **1**, 4035 (*Chem. Abstr.*, 1989, **111**, 164 961).
1489 J. Tsuchiya, *Mater. Res. Soc. Symp. Proc.*, 1989, **135**, 357 (*Chem. Abstr.*, 1989, **111**, 124 516).

studied by ^{7}Li relaxation measurements.[1490]

The solvation and counterion binding in a non-aqueous surfactant lamellar phase has been investigated by ^{23}Na n.m.r. spectroscopy.[1491] ^{23}Na n.m.r. spectroscopy has been used to study amorphous poly(ethylene oxide) containing various sodium salts.[1492] ^{1}H n.m.r. spectroscopy has been used to evaluate water-lignosulphonate systems.[1493] ^{1}H and ^{23}Na n.m.r. relaxation time measurements have been used to study the interaction between water and sodium cellulose sulphate.[1494] ^{23}Na and ^{31}P n.m.r. spectroscopy has been used to investigate sonicated cardiolipin sodium salt in aqueous medium.[1495]

The noninvasive measurement of tissue lithium in man by ^{7}Li n.m.r. spectroscopy has been reported.[1496] Li^{+} transport across human erythrocyte membranes has been measured by ^{7}Li n.m.r. spectroscopy.[1497] The effects of membrane surface charge in Li^{+}-membrane interactions and on Li^{+} transport mediated by monensin have been studied by ^{7}Li n.m.r. spectroscopy.[1498] Li^{+} transport in isolated rat hepatocytes has been studied by ^{7}Li n.m.r. spectroscopy.[1499]

^{23}Na n.m.r. imaging in short times has been studied.[1500] The interaction between Bobbyx mori silk fibroin and metal ions has been investigated by ^{23}Na and ^{27}Al n.m.r. spectroscopy.[1501] The transport of Li^{+}, Na^{+}, and K^{+} through phospholipid bilayers mediated by the antibiotic M139603 has been investigated by ^{7}Li, ^{23}Na, and ^{39}K n.m.r. spectroscopy.[1502] ^{23}Na, ^{31}P, and ^{87}Rb n.m.r. spectroscopy has been used to study perfused rat kidney.[1503] The use of ^{23}Na n.m.r. spectroscopy in the analysis of the renal

1490 A.I. Mishustin, V.N. Plakhotnik, V.N. Dam'e, N.F. Tovmash, E.A. Sapranova, and G.A. Baronskii, *Tr. Inst. - Mosk. Khim.-Tekhnol. Inst. im. D.I. Mendeleeva*, 1987, **148**, 109 (*Chem. Abstr.*, 1989, **110**, 200 124).

1491 A.J.I. Ward, G. Rong, and S.E. Friberg, *Colloids Surf.*, 1989, **38**, 285 (*Chem. Abstr.*, 1989, **111**, 219 878).

1492 M.C. Wintersgill, J.J. Fontanella, Y.S. Pak, S.G. Greenbaum, A. Al-Mudaris, and A.V. Chadwick, *Polymer*, 1989, **30**, 1123 (*Chem. Abstr.*, 1989, **111**, 79 015); S.G. Greenbaum, Y.S. Pak, M.C. Wintersgill, and J.J. Fontanella, *Solid State Ionics*, 1988, **31**, 241 (*Chem. Abstr.*, 1989, **110**, 145 628).

1493 H. Hatakeyama, S. Hirose, and T. Hatakeyama, *ACS Symp. Ser.*, 1989, **397**, 274 (*Chem. Abstr.*, 1989, **111**, 156 161).

1494 H. Hatakeyama, H. Iwata, and T. Hatakeyama, *Wood Cellul.*, 1987, 39. Eds. J.F. Kennedy, G.O. Phillips, and P.A. Williams, Horwood, Chichester (*Chem. Abstr.*, 1989, **110**, 194 867).

1495 A. Lai, M. Casu, C. Meloni, and U. Muscatello, *Biochem. Biophys. Res. Commun.*, 1989, **161**, 979 (*Chem. Abstr.*, 1989, **111**, 73 361).

1496 P.F. Renshaw, S. Wicklund, and J.S. Leigh, jun., *Lithium: Inorg. Pharmacol. Psychiatr. Use, Proc. Br. Lithium Congr., 2nd 1987*, (Pub. 1988), 277 (*Chem. Abstr.*, 1989, **110**, 147 067).

1497 M.T. Espanol, R. Ramasamy, and D. Mota de Freitas, *Prog. Clin. Biol. Res.*, 1989, **292**, 33 (*Chem. Abstr.*, 1989, **111**, 32 993).

1498 F.G. Riddell, S. Arumugam, and B.G. Cox, *Lithium: Inorg. Pharmacol. Psychiatr. Use, Proc. Br. Lithium Congr., 2nd*, (Pub. 1988), 293 (*Chem. Abstr.*, 1989, **110**, 147 166).

1499 G.M.H. Thomas and R. Olufunwa, *Lithium: Inorg. Pharmacol. Psychiatr. Use, Proc. Br. Lithium Congr., 2nd 1987* , (Pub. 1988), 289 (*Chem. Abstr.*, 1989, **110**, 147 165).

1500 S.J. Kohler, N.H. Kolodny, D.J. D'Amico, S. Balasubramaniam, P. Mainardi, and E.S. Gragoudas, *J. Magn. Reson.*, 1989, **82**, 505.

1501 T. Asakura, M. Demura, and M. Tsutsumi, *Makromol. Chem., Rapid Commun.*, 1988, **9**, 835 (*Chem. Abstr.*, 1989, **110**, 4024).

1502 F.G. Riddell and S. Arumugam, *Biochim. Biophys. Acta*, 1989, **984**, 6 (*Chem. Abstr.*, 1989, **111**, 169 501).

1503 J.L. Allis, Z.H. Endre, and G.K. Radda, *Biochem. Soc. Trans.*, 1989, **17**, 236 (*Chem. Abstr.*, 1989, **110**, 111 142).

function has been reported.[1504] Clinical applications of ^{23}Na n.m.r. spectroscopy for measurement of red cell Na^+ concentrations have been discussed.[1505] The androgen sensitivity of rat prostate carcinoma has been studied by ^{31}P n.m.r. spectroscopy and by ^{1}H and ^{23}Na magnetic resonance imaging.[1506] Dextran-magnetitie has been proposed as a contrast agent for ^{23}Na imaging.[1507]

The complexing of phenylenedioxydiacetamides with K^+ and Ca^{2+} has been investigated using ^{1}H and ^{13}C n.m.r. spectroscopy.[1508] ^{39}K n.m.r. spectroscopy has been applied to potassium transport in mung bean root tip.[1509] The direct measurement of K^+ movement in perfused mandibular salivary gland stimulated with acetylcholine has been carried out by ^{39}K n.m.r. spectroscopy.[1510] ^{87}Rb n.m.r. spectroscopy has been used to study K^+ fluxes in human erythrocytes,[1511] has been proposed as a novel method of measuring cation flux in intact kidney,[1512] and has been used to study the perfused rat heart.[1513] The ionic interactions in solutions of mixed lecithin/diacetylphosphate vesicles has been studied by ^{133}Cs n.m.r. spectroscopy.[1514] ^{43}Ca and ^{133}Cs n.m.r. spectroscopy has been used to study ion interactions with the gramicidin A transmembrane channel.[1515]

Group IIA. Magnesium interactions with Na_2ATP and Li_4ATP have been investigated by ^{31}P n.m.r. spectroscopy.[1516] ^{31}P and ^{113}Cd n.m.r. spectroscopy has been used to study Ca^{2+} binding to trypsin, chymotrypsin, and subtilisin.[1517] The interaction of Ca^{2+} with DNA, as probed by ^{43}Ca n.m.r. spectroscopy, is not influenced by terminal phosphate groups at ends and nicks.[1518] ^{25}Mg and ^{113}Cd n.m.r. spectroscopy has been used to study the divalent metal binding site of NADP-dependent isocitrate dehydrogenase from pig

1504 R.S. Balaban, S.D. Wolff, and C. Moonen, *U.S. Pat. Appl.* US 160 993 15 Jul, 1988, Appl. 26 Feb, 1988; 17 pp. Avail. NTIS. Order No, PAT-APPL-7-160 993 (*Chem. Abstr.*, 1989, **110**, 208 595).

1505 S. Kojima, M. Kanashiro, F. Hayashi, K. Yoshida, H. Abe, M. Imanishi, M. Kawamura, Y. Kawano, T. Ashida, *et al.*, *Scand. J. Clin. Lab. Invest.*, 1989, **49**, 489 (*Chem. Abstr.*, 1989, **111**, 228 337).

1506 D.B. Vigneron, H. Hricak, T.L. James, P.B. Jajodia, L. Nunes, and P. Narayan, *Magn. Reson. Med.*, 1989, **11**, 152 (*Chem. Abstr.*, 1989, **111**, 149 621).

1507 R.M. Summers, P.M. Joseph, P.F. Renshaw, and H.L. Kundel, *Magn. Reson. Med.*, 1988, **8**, 427 (*Chem. Abstr.*, 1989, **110**, 110 912).

1508 W.O. Lin, M.C.B.V. De Souza, and H.G. Alt, *Quim Nova*, 1988, **11**, 393 (*Chem. Abstr.*, 1989, **111**, 186 242).

1509 Y. Yazaki, K. Maki, T. Sato, E. Ohta, and M. Sakata, *Plant Cell Physiol.*, 1988, **29**, 1417 (*Chem. Abstr.*, 1989, **110**, 72 560).

1510 M. Murakami, E. Suzuki, S. Miyamoto, Y. Seo, and H. Watari, *Pfluegers Arch.*, 1989, **414**, 385 (*Chem. Abstr.*, 1989, **111**, 190 714).

1511 J.L. Allis, R.M. Dixon, A.M. Till, and G.K. Radda, *J. Magn. Reson.*, 1989, **85**, 529.

1512 Z.H. Endre, J.L. Allis, P.J. Ratcliffe, and G.K. Radda, *Kidney Int.*, 1989, **35**, 1249 (*Chem. Abstr.*, 1989, **111**, 111 843).

1513 J.L. Allis, C.D. Snaith, A.M.L. Seymour, and G.K. Radda, FEBS Lett., 1989, **242**, 215 (*Chem. Abstr.*, 1989, **110**, 72 052).

1514 R. Rymden, M. Jansson, K. Edwards, M. Almgren, and W. Brown, *J. Colloid Interface Sci.*, 1989, **128**, 477 (*Chem. Abstr.*, 1989, **110**, 179 914).

1515 D.W. Urry, N. Jing, T.L. Trapane, C.H. Luan, and M. Waller, *Curr. Top. Membr. Transp.*, 1988, **33**, 51 (*Chem. Abstr.*, 1989, **110**, 188 735).

1516 M.S. Hughes, S. Partridge, G. Marr, and N.J. Birch, *Magnesium Res.*, 1988, **1**, 35 (*Chem. Abstr.*, 1989, **110**, 90 755).

1517 F. Adebodun and F. Jordan, *Biochemistry*, 1989, **28**, 7524 (*Chem. Abstr.*, 1989, **111**, 129 725).

1518 W.H. Braunlin, L. Nordenskioeld, and T. Drakenberg, *Biopolymers*, 1989, **28**, 1339 (*Chem. Abstr.*, 1989, **111**, 111 150).

heart.[1519] The kinetics of Ca^{2+} binding to calbindin mutants have been determined by ^{43}Ca n.m.r. spectroscopy.[1520] ^{31}P n.m.r. spectroscopy has been used to study the binding of Ca^{2+} to didodecylphosphate vesicles during Ca^{2+}-induced aggregation and fusion.[1521] The thermodynamic parameters for the binding of divalent cations to gramicidin A incorporated into a lipid environment have been determined by ^{205}Tl n.m.r. spectroscopy.[1522] The dissolution mechanism of Bobbyx mori silk fibroin in a calcium chloride aqueous solution has been studied using ^{1}H, ^{13}C, and ^{43}Ca n.m.r. spectroscopy.[1523]

Scandium. The interaction of aquated $[Sc(OH)(OH_2)_5]^{2+}$ with serine and small peptides has been investigated by ^{45}Sc n.m.r. spectroscopy.[1524]

Lanthanides. ^{1}H n.m.r. spectroscopy has been used to study the rate of dissociation of [La(4,7,13,16,21-pentaoxa-1,10-diazabicyclo[8.8.5]tricosane)]$^{3+}$.[1525] Variable pressure ^{139}La n.m.r. spectroscopy has been used to study the kinetics of La^{3+} complex formation with 2,6-dicarboxy-4-hydroxypyridine.[1526] ^{139}La n.m.r. spectroscopy has been used to investigate La^{3+} complexing by 18-crown-6. It was found that Cl^- also coordinates. The equilibrium constant for formation and equilibrium between the two coordination sites were determined.[1527]

The interaction of L-cysteine with lanthanide ions has been studied by ^{1}H and ^{13}C n.m.r. spectroscopy.[1528] Complexation between $Ln(NO_3)_3$, Ln = La, Ce, and $[Zr\{O_2P(OC_6H_{12}\text{-}2\text{-}Et)_2\}_6]^{2-}$ has been investigated by ^{31}P n.m.r. spectroscopy.[1529]

The multistage equilibrium in the $[Eu(fod)_3]$-18-crown-6 system has been studied by ^{1}H n.m.r. spectroscopy.[1530] The delivery of Gd^{3+}-$HO_2CCH_2N\{CH_2CH_2N(CH_2CO_2H)_2\}_2$ to hepatic cells has been studied by ^{31}P n.m.r. spectroscopy.[1531] ^{13}C and ^{17}O n.m.r. spectroscopy has been used to investigate the interaction of Er^{3+} with various acetic acid derivatives.[1532] The rate of electron transfer between $[(Bu^t[8]\text{-annulene})_2Yb]^{2-}$ and $[(Bu^t[8]\text{-annulene})_2Yb]^-$ has been measured by ^{1}H n.m.r. spectroscopy.[1533]

1519 R.S. Ehrlich and R.F. Colman, *Biochemistry*, 1989, **28**, 2058 (*Chem. Abstr.*, 1989, **110**, 110 753).
1520 S. Forsén, S. Linse, E. Thulin, B. Lindegaard, S.R. Martin, P.M. Bayley, P. Brodin, and T. Grundstroem, *Eur. J. Biochem.*, 1988, **177**, 47 (*Chem. Abstr.*, 1989, **110**, 3175).
1521 L.A.M. Rupert, D. Hoekstra, and J.B.F.N. Engberts, *J. Colloid Interface Sci.*, 1989, **130**, 271 (*Chem. Abstr.*, 1989, **111**, 103 175).
1522 J.F. Hinton, J.Q. Fernandez, D.C. Shungu, and F.S. Millett, *Biophys. J.*, 1989, **55**, 327 (*Chem. Abstr.*, 1989, **110**, 188 081).
1523 T. Asakura and N. Nakayama, *Sen'i Gakkaishi*, 1989, **45**, 252 (*Chem. Abstr.*, 1989, **111**, 155 732).
1524 D. Rehder and K. Hink, *Inorg. Chim. Acta*, 1989, **158**, 265.
1525 R.A. Torres and P.A. Baisden, *Inorg. Chem.*, 1989, **28**, 2807.
1526 Y. Ducommun, L. Helm, G. Laurenczy, and A.E. Merbach, *Inorg. Chim. Acta*, 1989, **158**, 3.
1527 Z. Chen, H. Dettman, and C. Detellier, *Polyhedron*, 1989, **8**, 2029.
1528 J. Ren, F. Pei, and W. Wang, *Bopuxue Zazhi*, 1988, **5**, 349 (*Chem. Abstr.*, 1989, **111**, 154 333).
1529 I.S. Pronin, A.A. Vashman, and N.A. Shorokhov, *Zh. Neorg. Khim.*, 1988, **33**, 2775 (*Chem. Abstr.*, 1989, **110**, 106 975).
1530 L. Lu, V.V. Yastrebov, and I.A. Silin-Bekchurin, *Zh. Obshch. Khim.*, 1989, **59**, 1848 (*Chem. Abstr.*, 1989, **111**, 203 034).
1531 J. Vion-Dury, S. Masson, J.M. Devoisselle, M. Sciaky, F. Desmoulin, S. Confort-Gouny, D. Coustaut, and P.J. Cozzone, *J. Pharmacol. Exp. Ther.*, 1989, **250**, 1113 (*Chem. Abstr.*, 1989, **111**, 201 521).
1532 D.F. Mullica, G.A. Wilson, and C.K.C. Lok, *Inorg. Chim. Acta*, 1989, **163**, 139.
1533 D.C. Eisenberg, *Report*, 1988, *LBL-25834*; Order No. DE89000555, 121 pp. Avail. NTIS. From *Energy Res. Abstr.*, 1988, **13**, Abstr. No. 57103 (*Chem. Abstr.*, 1989, **111**, 181 804); D.C. Eisenberg, S.A. Kinsley, and A. Streitwieser, *J. Am. Chem. Soc.*, 1989, **111**, 5769.

Titanium. The interaction of TiF_4 and TaF_5 with MeCN has been investigated by ^{19}F n.m.r. spectroscopy.[1534] The conditions and reactions of formation of Ti(IV) peroxofluoro complexes have been studied in H_2O_2 solutions by ^{19}F n.m.r. spectroscopy.[1535]
Zirconium. A ^{31}P n.m.r. study of the $[Zr\{O_2P(OC_6H_{12}\text{-}2\text{-}Et)_2\}_4\{HO_2P(OC_6H_{12}\text{-}2\text{-}Et)_2\}_2]$-$HNO_3$-$H_2O$-$C_{10}H_{22}$ has established the formation of $[Zr\{O_2P(OC_6H_{12}\text{-}2\text{-}Et)_2\}_{4-n}(NO_3)_n$-$\{HO_2P(OC_6H_{12}\text{-}2\text{-}Et)_2\}_2]$ complexes in the organic phase.[1536] Precipitate formation in the Zr^{4+}-$[(BuO)_2PO_2]^-$-$[NO_3]^-$ system has been investigated by ^{13}C, ^{14}N, and ^{31}P n.m.r. spectroscopy.[1537]
Vanadium. ^{51}V n.m.r. spectroscopy has been used to investigate the interactions of vanadate with hydroxypyridines and pyridinecarboxylates,[1538] $N(CH_2CH_2OH)_3$,[1539] H_2O_2,[1540] and citrate.[1541] ^{17}O, ^{51}V, and ^{95}Mo n.m.r. spectroscopy, including ^{51}V COSY, has been used to carry out speciation when $[MoO_4]^{2-}$ is added to $[V_{10}O_{28}]^{6-}$ and various $[Mo_nV_{10-n}O_{28}]^{-6+n}$ species identified.[1542] ^{13}C, ^{31}P, and ^{51}V n.m.r. spectroscopy has been used to study the interaction of $[VO_4]^{3-}$ with AMP, ADP, and ATP.[1543] The interaction of $[VO_4]^{3-}$ with monosaccharides and nucleosides has been studied by ^{13}C and ^{51}V n.m.r. spectroscopy.[1544] ^{51}V n.m.r. spectroscopy has been used to investigate complexation of $[VO_4]^{3-}$ by desferrioxamine,[1545] and ribonuclease-T_1.[1546]
Niobium and Tantalum. Variable temperature ^1H n.m.r. spectroscopy has been used to show that $[(\eta^5\text{-}C_5H_5)_2Nb(\mu\text{-}SPh)_2Mo(CO)_4]^+$ is a dynamic mixture of *syn* and *anti* isomers.[1547] The interconversion of two geometric isomers of [M(norbornane-*exo*-2,3-dithiolate)$_3$]$^-$, M = Nb, Ta, has been investigated by ^1H and ^{13}C n.m.r. spectroscopy.[1548]
Chromium. The rate of electron self-exchange in $[Cr(CNC_6H_3Pr^i\text{-}2,6)_6]/[Cr(CNC_6H_3Pr^i\text{-}2,6)_6]^+$ has been measured by ^1H n.m.r. spectroscopy.[1549] ^1H spin echo has been used to examine the interaction of Cr(VI) with glutathione in intact erythrocytes.[1550] Protein site recognition by chromium(III) complexes has been probed through n.O.e.

1534 J. Pauli, W. Storek, and L. Riesel, *Z. Chem.*, 1988, **28**, 226 (*Chem. Abstr.*, 1989, **110**, 50 109).
1535 B.N. Chernyshov, N.A. Didenko, and N.G. Bakeeva, *Koord. Khim.*, 1989, **15**, 917 (*Chem. Abstr.*, 1989, **111**, 246 690).
1536 A.A. Vashman, I.S. Pronin, M.M. Borunov, V.N. Shesterikov, I.V. Mukhin, and S.V. Galatkionov, *Zh. Neorg. Khim.*, 1988, **33**, 2896 (*Chem. Abstr.*, 1989, **110**, 45 909).
1537 N. Uetake, *J. Nucl. Sci. Technol.*, 1989, **26**, 329 (*Chem. Abstr.*, 1989, **110**, 220 955).
1538 B. Galeffi and A.S. Tracey, *Inorg. Chem.*, 1989, **28**, 1726.
1539 D.C. Crans, R.L. Bunch, and L.A. Theisen, *J. Am. Chem. Soc.*, 1989, **111**, 7597.
1540 N.J. Campbell, A.C. Dengel, and W.P. Griffith, *Polyhedron*, 1989, **8**, 1379.
1541 P.M. Ehde, I. Andersson, and L. Pettersson, *Acta Chem. Scand.*, 1989, **43**, 136.
1542 O.W. Howarth, L. Pettersson, and I. Andersson, *J. Chem. Soc., Dalton Trans.*, 1989, 1915.
1543 C.F.G.C. Geraldes and M.M.C.A. Castro, *J. Inorg. Biochem.*, 1989, **37**, 213.
1544 C.F.G.C. Geraldes and M.M.C.A. Castro, *J. Inorg. Biochem.*, 1989, **35**, 79.
1545 A. Butler, S.M. Parsons, S.K. Yamagata, and R.I. de la Rosa, *Inorg. Chim. Acta*, 1989, **163**, 1.
1546 D. Rehder, H. Holst, R. Quaas, W. Hinrichs, U. Hahn, and W. Saenger, *J. Inorg. Biochem.*, 1989, **37**, 141.
1547 M.Y. Darensbourg, R. Silva, J. Reibenspies, and C.K. Prout, *Organometallics*, 1989, **8**, 1315.
1548 K. Tatsumi, I. Matsubara, Y. Inoue, A. Nakamura, K. Miki, and N. Kasai, *J. Am. Chem. Soc.*, 1989, **111**, 7766.
1549 K.A. Anderson and S. Wherland, *Inorg. Chem.*, 1989, **28**, 601.
1550 P. O'Brien and Z. Ozolins, *Inorg. Chim. Acta*, 1989, **155**, 25.

quenching.[1551]

Molybdenum. The complexation of $[CeMo_{12}O_{42}]^{8-}$ by Zn^{2+} ions has been studied by ^{17}O, ^{67}Zn, and ^{95}Mo n.m.r. spectroscopy.[1552]

Tungsten. A ^{1}H n.m.r. study of $H_3PW_{12}O_{40}$ has shown that solvent composition not only affects the rate of H^+ exchange but also affects the nature of the heteropoly acid interaction with the organic solvents.[1553]

Manganese. Mn^{2+} induced ^{1}H relaxation of $[Me_4N]^+$ in aqueous $[Me_4N]$ carrageenate solutions has been studied.[1554] ^{1}H and ^{2}H n.m.r. spectroscopy has been used to investigate the L-dopamine-melanin-Mn^{2+} complex in water and ^{27}Al n.m.r. spectroscopy has been used to investigate Al^{3+} binding.[1555] ^{1}H relaxation by Mn^{2+} has been used to probe the active centre of alkaline phosphatase.[1556]

Iron. The temperature, pressure, and electrolyte dependence of $[(\eta^5\text{-}C_5H_5)_2Fe]/[(\eta^5\text{-}C_5H_5)_2Fe]^+$ and its derivatives has been determined.[1557] ^{13}CN exchange in $[Fe(CN)_5(NO)]^{2-}$ catalysed by haemoglobin has been studied by ^{13}C n.m.r. spectroscopy.[1558] The activation parameters for the $\delta\leftrightarrow\lambda$ interconversion of the 5-membered diamine chelate rings in $[Fe(CN)_4(1,2\text{-diamine})]^-$ have been determined by line shape analysis.[1559] Ligand exchange and redox kinetics of $[Fe(phen)_3]^{2/3+}$ have been studied.[1560] ^{1}H n.m.r. spectroscopy has been used to investigate the dissociation rates and formation constants for imidazoles coordinated to low spin complexes of tetraphenylporphyrin.[1561]

Ruthenium. Pressure and temperature effects on the $[(\eta^5\text{-}C_5H_5)_2Ru]/[(\eta^5\text{-}C_5H_5)_2RuBr]^+$ have been studied by ^{1}H n.m.r. s pectroscopy.[1562]

Cobalt. The self exchange rates for the $[Co(norbornyl)_4]^-/[Co(norbornyl)_4]/[Co(norbornyl)_4]^+$ system have been measured by ^{1}H and ^{59}Co n.m.r. spectroscopy. The rate of interconversion of isomers of $[Co(norbornyl)_4]^+$ has been determined by ^{59}Co NOESY.[1563] Solvent dynamical effects for electron exchange in $[(\eta^5\text{-}C_5H_5)_2Co]/[(\eta^5\text{-}C_5H_5)_2Co]^+$ have been determined.[1564] ^{1}H T_1 measurements have been used to investigate complex

1551 J.R. Bocarsly and J.K. Barton, *Inorg. Chem.*, 1989, **28**, 4189.
1552 M.A. Petrukhin, M.A. Fedotov, I.V. Tat'yanina, and E.A. Torchenkova, *Zh. Neorg. Khim.*, 1989, **34**, 849 (*Chem. Abstr.*, 1989, **111**, 186 231).
1553 Sh.A. Gafarov and V.F. Chuvaev, *Azerb. Khim. Zh.*, 1988, 121 (*Chem. Abstr.*, 1989, **111**, 141 715).
1554 B.J. Kvam and H. Grasdalen, *Macromolecules*, 1989, **22**, 3919 (*Chem. Abstr.*, 1989, **111**, 174 552).
1555 S. Aime, M. Botta, and I. Camurati, *J. Inorg. Biochem.*, 1989, **36**, 1.
1556 C. Schulz, I. Bertini, M.S. Viezzoli, R.D. Brown, tert., S.H. Koenig, and J.E. Coleman, *Inorg. Chem.*, 1989, **28**, 1490.
1557 K. Kirchner, S.Q. Dang, M. Stebler, H.W. Dodgen, S. Wherland, and J.P. Hunt, *Inorg. Chem.*, 1989, **28**, 3604; G.E. McManis, R.M. Nielson, A. Gochev, and M.J. Weaver, *J. Am. Chem. Soc.*, 1989, **111**, 5533; R.M. Nielson, G.E. McManis, and M.J. Weaver, *J. Phys. Chem.*, 1989, **93**, 4703; R.M. Nielson, G.E. McManis, L.K. Safford, and M.J. Weaver, *J. Phys. Chem.*, 1989, **93**, 2152; R.M. Nielson, M.N. Golovin, G.E. McManis, and M.J. Weaver, *Report*, 1987, **TR-69**; Order No. AD-A188779, 22 pp. Avail. NTIS. From *Gov. Rep. Announce. Index (U.S.)*, 1988, **88**, Abstr. No. 827 067 (*Chem. Abstr.*, 1989, **111**, 97 470).
1558 P.R. Shafer, D.E. Wilcox, H. Kruszyna, R. Kruszyna, and R.P. Smith, *Toxicol. Appl. Pharmacol.*, 1989, **99**, 1 (*Chem. Abstr.*, 1989, **111**, 70 246).
1559 Y. Kuroda, N. Tanaka, M. Goto, and T. Sakai, *Inorg. Chem.*, 1989, **28**, 997.
1560 S. Schmid, K. Kirchner, and V.N. Sapunov, *Inorg. Chem.*, 1989, **28**, 4167.
1561 M. Nakamura, *Inorg. Chim. Acta*, 1989, **161**, 73.
1562 K. Kirchner, H.W. Dodgen, S. Wherland, and J.P. Hunt, *Inorg. Chem.*, 1989, **28**, 604.
1563 E.K. Byrne and K.H. Theopold, *J. Am. Chem. Soc.*, 1989, **111**, 3887.
1564 R.M. Nielson, G.E. McManis, M.N. Golovin, and M.J. Weaver, *Report*, 1988, **TR-77**; Order No. AD-

formation between $[Co(en)_3]^{3+}$ and $[Co(edta)]^-$.[1565] ^{31}P n.m.r. spectroscopy has been used to determine the equilibrium constants for $(MeO)_2PO_2H$ binding to $[N(CH_2CH_2CH_2NH_2)_3Co(OH_2)_2]^{3+}$ and related compounds.[1566] The ionization of pyridine bases and their reactions with Co^{II} and Ni^{II} acetates in anhydrous acetic acid have been studied by ^{13}C n.m.r. spectroscopy.[1567] ^1H n.m.r. spectroscopy has been used to study the complexation of $Co(ClO_4)_2$, $Ni(ClO_4)_2$, or $Cu(ClO_4)_2$ with 4-amino-1,2,4-triazole.[1568] The interaction of Co^{2+} or Ni^{2+} with cetyltrimethyl ammonium bromide has been investigated by ^{13}C n.m.r. spectroscopy.[1569]

Rhodium. Hapticity changes and the rate of electron exchange in the $[(\eta^5\text{-}C_5Me_5)(\eta^6\text{-}C_6Me_6)Rh]^{2+}/[(\eta^5\text{-}C_5Me_5)(\eta^6\text{-}C_6Me_6)Rh]^+/[(\eta^5\text{-}C_5Me_5)(\eta^6\text{-}C_6Me_6)Rh]$ system have been investigated by ^1H line broadening.[1570] Electron exchange in the $[Rh_2(O_2CCH_3)_4(NCMe)_2]/[Rh_2(O_2CCH_3)_4(NCMe)_2]^+$ system with Ni tetraaza macrocycles and Fe and Ru tris(polypyridine) complexes has been investigated by ^1H n.m.r. spectroscopy.[1571]

Nickel, Palladium, and Platinum. Dynamic n.m.r. spectroscopy has been used to study intermolecular ligand exchange in $[AuMe_2Cl(guanosine)]$ and $[EtPt(\eta^4\text{-}C_8H_{12})Pt(cytidine)]^+$.[1572] Equilibrium constants for the coordination of olefins to [PtClR(phen)] have been studied by ^1H n.m.r. spectroscopy.[1573] Low temperature ^1H n.m.r. spectroscopy has been used to show an equilibrium between $[M(\eta^3\text{-}2\text{-}MeC_3H_4)(acac)]$, M = Pd, Pt, and PPh_3 to give $[M(\eta^3\text{-}2\text{-}MeC_3H_4)(PR_3)_2]^+[acac]^-$.[1574]

The acid-base equilibria for $[Pt(^{15}NH_3)_3(OH_2)]^{2+}$ have been investigated by ^{15}N and ^{195}Pt n.m.r. spectroscopy.[1575] ^1H, ^{13}C, ^{15}N, and ^{195}Pt n.m.r. spectroscopy has been used to study equilibria between *cis*-$[Pt(NH_3)_2(OH_2)_2]^{2+}$ and amino acids[1576] and *N*-acetylglycine.[1577] ^1H and ^{31}P n.m.r. spectroscopy has been used to investigate the interaction of [M{(4-*N*-methylpyridinium)$_4$porphyrin}], M = Ni, Pd, Zn, with oligodeoxyribonucleosides.[1578] The transition to the glassy state for $NiCl_2$ and $AlCl_3$ solutions has been monitored by ^1H n.m.r. spectroscopy.[1579]

Copper. ^1H and ^{13}C n.m.r. spectroscopy has been used to investigate the equilibria and

A192101, 55 pp. Avail. NTIS. From *Gov. Rep. Announce. Index (U.S.)*, 1988, **88**, Abstr. No. 840 885 (*Chem. Abstr.*, 1989, **111**, 97 479).
1565 R.A. Marusak and A.G. Lappin, *J. Phys. Chem.*, 1989, **93**, 6856.
1566 J. Chin, M. Banaszczyk, V. Jubian, and X. Zou, *J. Am. Chem. Soc.*, 1989, **111**, 186.
1567 K. Azuma, N. Nakasuka, and M. Tanaka, *Inorg. Chem.*, 1989, **28**, 3667.
1568 V.A. Kalibabchuk, L.V. Glushchenko, O.N. Miroshnikov, A.D. Garnovskii, and I.V. Romanova, *Zh. Neorg. Khim.*, 1989, **34**, 2315 (*Chem. Abstr.*, 1989, **111**, 220 248).
1569 J.B. Nagy, *Colloids Surf.*, 1989, **35**, 201 (*Chem. Abstr.*, 1989, **111**, 141 157).
1570 R.M. Nielson and M.J. Weaver, *Organometallics*, 1989, **8**, 1636.
1571 M.A.S. Aquino, D.A. Foucher, and D.H. Macartney, *Inorg. Chem.*, 1989, **28**, 3357.
1572 S. Komiya, Y. Mizuno, and T. Shibuya, *Chem. Lett.*, 1988, 367 (*Chem. Abstr.*, 1989, **110**, 213 031).
1573 M.E. Cucciolito, V. De Felice, A. Panunzi, and A. Vitagliano, *Organometallics*, 1989, **8**, 1180.
1574 H. Kurosawa, K. Ishii, Y. Kawasaki, and S. Murai, *Organometallics*, 1989, **8**, 1756.
1575 T.G. Appleton, J.R. Hall, S.F. Ralph, and C.S.M. Thompson, *Inorg. Chem.*, 1989, **28**, 1989.
1576 T.G. Appleton, J.W. Connor, J.R. Hall, and P.D. Prenzler, *Inorg. Chem.*, 1989, **28**, 2030.
1577 T.G. Appleton, J.R. Hall, and P.D. Prenzler, *Inorg. Chem.*, 1989, **28**, 815.
1578 J.A. Strickland, L.G. Marzilli, W.D. Wilson, and G. Zon, *Inorg. Chim. Acta*, 1989, **28**, 4191.
1579 V.A. Finkel'stein, I.D. Isaev, and A.G. Lundin, *Dokl. Akad. Nauk SSSR*, 1988, **299**, 877 (*Chem. Abstr.*, 1989, **110**, 14 220).

reactions of $[Me_2CuLi]$ or $[Ph_2CuLi]$ with 1-MeO_2CCH=CHC_6H_4-2-CHMeOMe.[1580] The reaction and equilibria of $[PhMe_2SiLi]$ with CuCN[1581] and CuBr[1582] have been studied by ^1H, ^7Li, ^{13}C, and ^{29}Si n.m.r. spectroscopy.

The rate of the $[Cu(2,9\text{-}Me_2\text{-}1,10\text{-phenanthroline})_2]^+$/$[Cu(2,9\text{-}Me_2\text{-}1,10\text{-phenanthroline})_2]^{2+}$ electron-transfer has been measured by ^1H n.m.r. spectroscopy.[1583] The kinetics of the electron self-exchange for the Cu(I)/Cu(II) couple of the pentacoordinate complex of 2,6-[1-{(2-imidazol-4-ylethyl)imino}ethyl]$_2$pyridine have been studied by ^1H n.m.r. spectroscopy.[1584] Hydrolytic reactions of Cu(II) bipyridine complexes have been investigated by ^1H n.m.r. spectroscopy.[1585] Variable temperature ^1H and ^{31}P n.m.r. spectroscopy has been used to show that MeCN and L = 2-$Ph_2PC_5H_4N$ dissociation and rapid ligand exchange take place in solution in $[Cu_2(2\text{-}Ph_2PC_5H_4N)_3(NCMe)_n]^{2+}$.[1586] Complexation, stability, and lability of Cu(II) complexes of glycylglycine have been studied.[1587] The binding of F^- to copper zinc superoxide dismutase has been investigated by ^{19}F n.m.r. spectroscopy.[1588]

Silver. Ag(I)-polyamine systems in DMSO have been investigated by ^1H n.m.r. spectroscopy.[1589] ^{13}C n.m.r. spectroscopy has been used to investigate the complexation of Ag^+ by monensin.[1590] The complexation of Ag(I) with polyfunctional organic sulphides has also been investigated.[1591]

Gold. ^{31}P n.m.r. spectroscopy has been used to investigate equilibria between $[AuBr_2]^-$ and tertiary phosphines.[1592] ^{13}C n.m.r. spectroscopy has been used to study reactions of aurothiomalate with selenopropionate.[1593]

Zinc. The pH dependence of Zn(II) complexes of 4,7-diazadecane-1,10-diamine, 4,8-diazaundecane-1,11-diamine, and tetraethylenepentamine has been investigated.[1594] ^{31}P n.m.r. spectroscopy has been used to investigate the equilibrium between Zn^{2+}, Al^{3+} and Ga^{3+} complexes of Schiff bases with $H_2OCCH_2CH(NH_2)PO_3H_2$ and 5′-deoxypyridoxal.[1595] Zn(II) complexes of 1-hydroxyethylidene-1,1-diphosphonic acid have been studied by ^{31}P n.m.r. spectroscopy.[1596] ^1H n.m.r. spectroscopy has been used to carry out

1580 B. Christenson, T. Olsson, and C. Ullenius, *Tetrahedron*, 1989, **45**, 523.
1581 S. Sharma and A.C. Oehlschlager, *Tetrahedron*, 1989, **45**, 557.
1582 S. Sharma and A.C. Oehlschlager, *J. Org. Chem.*, 1989, **54**, 5383.
1583 H. Doine, Y. Yano, and T.W. Swaddle, *Inorg. Chem.*, 1989, **28**, 2319.
1584 J.A. Goodwin, L.J. Wilson, D.M. Stanbury, and R.A. Scott, *Inorg. Chem.*, 1989, **28**, 42.
1585 I. Fabian, *Inorg. Chem.*, 1989, **28**, 3805.
1586 E. Lastra, M.P. Gamasa, J. Gimeno, M. Lanfranchi, and A. Tiripicchio, *J. Chem. Soc., Dalton Trans.*, 1989, 1499.
1587 V.G. Shtyrlin, E.L. Gogolashvili, and A.V. Zakharov, *J. Chem. Soc., Dalton Trans.*, 1989, 1293.
1588 L. Banci, I. Bertini, C. Luchinat, A. Scozzafava, and P. Turano, *Inorg. Chem.*, 1989, **28**, 2377.
1589 A. Cassol, P. Di Bernardo, P. Zanonato, R. Portanova, M. Tolazzi, G. Tomat, V. Cucinotta, and D. Sciotto, *J. Chem. Soc., Faraday Trans. 1*, 1989, **85**, 2445.
1590 H. Tsukube and H. Sohmiya, *Inorg. Chim. Acta*, 1989, **166**, 27.
1591 V.N. Maistrenko, L.D. Sokolova, and Yu.I. Murinov, *Zh. Neorg. Khim.*, 1989, **34**, 2257 (*Chem. Abstr.*, 1989, **111**, 241 345).
1592 W.E. Hill, M.Q. Islam, T.R. Webb, and C.A. McAuliffe, *Inorg. Chim. Acta*, 1989, **157**, 215.
1593 A.A. Isab and A.P. Arnold, *J. Coord. Chem.*, 1989, **20**, 95.
1594 D.N. Hague and A.D. Moreton, *J. Chem. Soc., Dalton Trans.*, 1989, 1171.
1595 A.E. Martell and B. Szpoganicz, *Inorg. Chem.*, 1989, **28**, 4199.
1596 R. Zhuang and Q. Li, *Wuji Huaxue*, 1989, **5**, 68 (*Chem. Abstr.*, 1989, **111**, 203 013).

speciation of $ZnCl_2/KCl$ solutions.[1597]

Cadmium. The binding of pyridine and benzimidazole to Cd(II) complexes of 'expanded porphyrin' has been investigated by ^1H and ^{13}C n.m.r. spectroscopy.[1598] ^1H and ^{31}P n.m.r. spectroscopy has been used to investigate speciation between Cd^{2+} and Pb^{2+} and $\{(H_2O_3PCH_2)_2NCH_2CH_2\}_2NCH_2PO_3H_2$.[1599] ^{13}C n.m.r. spectroscopy has been used to investigate complex formation of Cd(II) with guanosine.[1600] ^{113}Cd n.m.r. spectroscopy has been used to provide evidence for cooperative interaction between amino- and carboxyl-terminal domains of calmodulin.[1601]

Mercury. ^{81}Br n.m.r. spectroscopy has been used to detect formation of bromide-organomercurial-protein complexes.[1602]

Boron. Sodium peroxoborate formation has been studied by ^{11}B n.m.r. spectroscopy.[1603] ^{11}B and ^{19}F n.m.r. spectroscopy has also been used.[1604] Complex formation between borate and diols has been studied by ^{11}B and ^{13}C n.m.r. spectroscopy.[1605] The pH dependence of the stability of boric acid and boric esters of amino acids and amino diols has been studied by ^1H and ^{11}B n.m.r. spectroscopy.[1606] ^1H n.m.r. spectroscopy has been used to study the reaction between boric acid and some polyethylene glycols.[1607] The kinetics of F^- substitution by $[OH]^-$ or $[MeO]^-$ in $[BF_3(OMe)]^-$ have been investigated by ^{19}F n.m.r. spectroscopy.[1608]

Aluminium and Gallium. Equilibrium constants for

$$2[AlH_n(OR)_{4-n}]^- \rightleftharpoons [AlH_{n+1}(OR)_{3-n}]^- + [AlH_{n-1}(OR)_{5-n}]^-$$

have been determined by ^{27}Al n.m.r. spectroscopy.[1609] Complexes of Al^{3+} with amino-polycarboxylic acids have been investigated by ^{27}Al n.m.r. spectroscopy.[1610] ^{27}Al n.m.r. spectroscopy has been used to investigate the complexation of Al^{3+} with propionic and acetic acid.[1611] The influence of acetate, oxalate, and citrate anions on the hydrolytic reactions of aluminium hydroxides has been investigated using ^{27}Al n.m.r. spectro-

1597 A.D. Zdunek and J.R. Selman, *Electrochim. Acta*, 1989, **34**, 1461 (*Chem. Abstr.*, 1989, **111**, 242 431).
1598 J.L. Sessler, T. Murai, and V. Lynch, *Inorg. Chem.*, 1989, **28**, 1333.
1599 R.D. Gillard, P.D. Newman, and J.D. Collins, *Polyhedron*, 1989, **8**, 2077.
1600 S.T. Trisak and B.M. Rode, *Inorg. Chim. Acta*, 1989, **160**, 249.
1601 M. Ikura, N. Hasegawa, S. Aimoto, M. Yazawa, K. Yagi, and K. Hikichi, *Biochem. Biophys. Res. Commun.*, 1989, **161**, 1233 (*Chem. Abstr.*, 1989, **111**, 149 072).
1602 J.J. Pesek and F. Abhanan, *Inorg. Chem.*, 1989, **28**, 485.
1603 I.E. Il'yin, N.B. Kondrikov, O.V. Brovkina, and B.N. Chernyshov, *Latv. PSR Zinat. Akad. Vestis, Kim. Ser.*, 1989, 275 (*Chem. Abstr.*, 1989, **111**, 86 136).
1604 B.N. Chernyshov and O.V. Brovkina, *Latv. PSR Zinat. Akad. Vestis, Kim. Ser.*, 1988, 422 (*Chem. Abstr.*, 1989, **110**, 64 446).
1605 J.M. Coddington and M.J. Taylor, *J. Coord. Chem.*, 1989, **20**, 27.
1606 J. van Haveren, J.A. Peters, J.G. Batelaan, A.P.G. Kieboom, and H. van Bekkum, *Recl. Trav. Chim. Pays-Bas*, 1989, **108**, 179.
1607 A. Kamars and E.M. Shvarts, *Latv. PSR Zinat. Akad. Vestis, Kim. Ser.*, 1989, 243 (*Chem. Abstr.*, 1989, **111**, 154 479).
1608 L.V. Kharchenko, V.N. Plakhotnik, N.G. Parkhomenko, V.D. Kopanev, and V.V. Sokolov, *Kinet. Katal.*, 1988, **29**, 955 (*Chem. Abstr.*, 1989, **110**, 122 330).
1609 S. Heřmánek, O. Křížž, J. Fusek, Z. Černý, and B. Čásenský, *J. Chem. Soc., Perkin Trans.*, 2, 1989, 987.
1610 R.K. Iyer, S.B. Karweer, and V.K. Jain, *Magn. Reson. Chem.*, 1989, **27**, 328.
1611 E. Markland, L.O. Öhman, and S. Sjöeberg, *Acta Chem. Scand.*, 1989, **43**, 641.

scopy.[1612] The complexation of Al^{3+} with *N*-substituted 3-hydroxy-4-pyridones has been studied using ^{27}Al n.m.r. spectroscopy.[1613] ^{27}Al and ^{31}P n.m.r. spectroscopy has been used to study complexation of Al^{3+} with phosphate.[1614] Equilibria in the H^+-Al^{3+}-$[MoO_4]^{2-}$ system have been investigated using ^{27}Al n.m.r. spectroscopy.[1615] ^{27}Al n.m.r. spectroscopy has been used to investigate ferron-hydroxo-polynuclear aluminium interactions.[1616] ^{13}C n.m.r. spectroscopy has been used to characterise Al^{3+} and Ga^{3+} complexes of desferrioxamine B and activation parameters for isomerisation were determined.[1617] The composition and structure of aluminium mixed-ligand complexes in aqueous and aqueous peroxide solutions have been determined by ^1H, ^{19}F, and ^{27}Al n.m.r. spectroscopy.[1618] Temperature effects on the ^{27}Al n.m.r. spectra of polymeric aluminium hydrolysis species have been investigated by ^{27}Al n.m.r. spectroscopy.[1619] Aluminosilicate species have been detected in aqueous solution by ^{27}Al and ^{29}Si n.m.r. spectroscopy.[1620] The structure and properties of aluminosilicate solutions and gels have been investigated by n.m.r. spectroscopy.[1621] ^{23}Na, ^{27}Al, and ^{31}P n.m.r. spectroscopy has been used to investigate the reaction of $NaAlCl_4$ with $POCl_3$.[1622] Chemical exchange between different polyatomic species in butylpyridinium chloride-$AlCl_3$ melts has been studied by ^{27}Al n.m.r. spectroscopy.[1623] Donor-acceptor properties of chloroaluminate melts with $Et_3P{=}O$ have been investigated by ^{31}P n.m.r. spectroscopy.[1624] ^1H, ^{13}C, ^{17}O, and ^{27}Al n.m.r. spectroscopy has been used to investigate complex formation between RC(O)Br and $AlBr_3$.[1625]

Thallium. The formation, equilibrium properties, and decomposition in aqueous solution of $[RTl]^{2+}$ containing $RC(O)CH_2$ groups have been investigated by ^{205}Tl n.m.r. spectroscopy.[1626] The kinetics of ligand exchange in the Tl^{3+}/Cl^- have been studied by ^{205}Tl n.m.r. spectroscopy.[1627] ^{205}Tl n.m.r. spectroscopy has been used to determine the thermodynamic parameters for binding of monovalent cations to gramicidins A and C.[1628]

Silicon. ^{29}Si n.m.r. spectroscopy has been used to determine the structure of mixtures

1612 M.K. Wang, K.K. Lim, and K.H. Houng, *Chung-kwo Nung Yeh Hua Hsueh Hui Chih*, 1988, **26**, 245 (*Chem. Abstr.*, 1989, **110**, 38 238).
1613 D.J. Clevette, W.O. Nelson, A. Nordin, C. Orvig, and S. Sjoeberg, *Inorg. Chem.*, 1989, **28**, 2079.
1614 V.A. Urikh and I.N. Zadubrovskaya, *Izv. Akad. Nauk Kaz. SSR, Ser. Khim.*, 1989, 16 (*Chem. Abstr.*, 1989, **111**, 84 937); M.A. Wilson, P.J. Collin, and J.W. Akitt, *Anal. Chem.*, 1989, **61**, 1253.
1615 L.O. Oehman, *Inorg. Chem.*, 1989, **28**, 3629.
1616 P.M. Bertsch, M.A. Anderson, and W.J. Layton, *Magn. Reson. Chem.*, 1989, **27**, 283.
1617 B. Borgias, A.D. Hugi, and K.N. Raymond, *Inorg. Chem.*, 1989, **28**, 3538.
1618 B.N. Chernyshov and V.V. Kon'shin, *Koord. Khim.*, 1988, **14**, 1066 (*Chem. Abstr.*, 1989, **110**, 50 113).
1619 J.J. Fitzgerald, L.E. Johnson, and J.S. Frye, *J. Magn. Reson.*, 1989, **84**, 121.
1620 S.D. Kinrade and T.W. Swaddle, *Inorg. Chem.*, 1989, **28**, 1952.
1621 G. Harvey and L.S.D. Glasser, *ACS Symp. Ser.*, 1989, **398**, 49 (*Chem. Abstr.*, 1989, **111**, 181 988).
1622 F. Birkeneder, R.W. Berg, H.A. Hjuler, and N.J. Bjerrum, *Z. Anorg. Allg. Chem.*, 1989, **573**, 170.
1623 K. Ichikawa, T. Jin, and T. Matsumoto, *J. Chem. Soc., Faraday Trans. 1*, 1989, **85**, 175.
1624 T.A. Zawodzinski, jun., and R.A. Osteryoung, *Inorg. Chem.*, 1989, **28**, 1710.
1625 M. Vol'pin, I. Akhrem, and A. Orlinkov, *New J. Chem.*, 1989, **13**, 771.
1626 B. Györi, A. Sánchez, and J. Glaser, *J. Organomet. Chem.*, 1989, **361**, 1.
1627 I. Banyai and J. Glaser, *J. Am. Chem. Soc.*, 1989, **111**, 3186.
1628 J.F. Hinton, J.Q. Fernandez, D.C. Shungu, W.L. Whaley, R.E. Koeppe, sec., and F.S. Millett, *Biophys. J.*, 1988, **54**, 527 (*Chem. Abstr.*, 1989, **110**, 3096).

of $[EtN(CH_2CH_2OH)_3]_4[SiO_4]$ and $[Et_3NCH_2CH_2OH]_4[SiO_4]$.[1629] The influence of alkali metal cations on silicon exchange and ^{29}Si T_1 and T_2 has been investigated in alkaline silicate solutions.[1630] The effect of formamide[1631], fluoride,[1632] and HCl[1633] on silica sol-gel processes has been studied using ^{29}Si n.m.r. spectroscopy. The interactions between phosphate and silicate ions in aqueous solution have been investigated by ^{31}P n.m.r. spectroscopy.[1634]

Germanium. ^{14}N, including phase sensitive NOESY, and ^{73}Ge n.m.r. spectroscopy has been used to study the reaction products and chemical exchange processes in the system $GeCl_4$/KSCN/acetone-d_6.[1635]

Tin. Complexation of halide to $[XSn\{(CH_2)_n\}_3SnX]$ has been investigated by ^{13}C and ^{119}Sn n.m.r. spectroscopy.[1636]

Lead. The effects of pH on aqueous solutions of Pb(II) ethanoate and perchlorate have been studied using ^{207}Pb n.m.r. spectroscopy.[1637]

Nitrogen. The structure of aqueous solutions of Bu_4NBr has been studied by ^{13}C, ^{14}N, ^{17}O, and ^{79}Br n.m.r. spectroscopy.[1638]

Phosphorus. N.m.r. spectroscopy has been used to determine the pK_a values for H_3PO_2, H_3PO_3, and H_3PO_4.[1639]

Arsenic. The kinetics of the equilibrium

$$[AsF_5OH]^- + HF \rightleftharpoons [AsF_6]^- + H_2O$$

have been determined by ^1H and ^{19}F n.m.r. spectroscopy.[1640] The molecular dynamics and decomposition mechanism of the poly(4-phenylene)-$[AsF_6]^-$ system have been studied by ^1H and ^{19}F n.m.r. spectroscopy.[1641]

Tellurium. The exchange reactions of Te(II) and Te(IV) dithiocarbomate complexes have been investigated by ^{125}Te n.m.r. spectroscopy.[1642]

1629 F. Schlenkrich, O. Rademacher, and H. Scheler, *Z. Chem.*, 1989, **29**, 72 (*Chem. Abstr.*, 1989, **110**, 220 020).

1630 A.V. McCormick, A.T. Bell, and C.J. Radke, *J. Phys. Chem.*, 1989, **93**, 1737; A.V. McCormick, A.T. Bell, and C.J. Radke, *Mater. Res. Soc. Symp. Proc.*, 1988, **121**, 67 (*Chem. Abstr.*, 1989, **110**, 83 183).

1631 A.H. Boonstra, T.N.M. Bernards, and J.J.T. Smits, *J. Non-Cryst. Solids*, 1989, **109**, 141 (*Chem. Abstr.*, 1989, **111**, 119 595).

1632 R. Winter, J.B. Chan, R. Frattini, and J. Jonas, *J. Non-Cryst. Solids*, 1988, **105**, 214 (*Chem. Abstr.*, 1989, **110**, 43 414).

1633 A.H. Boonstra and T.N.M. Bernards, *J. Non-Cryst. Solids*, 1989, **108**, 249 (*Chem. Abstr.*, 1989, **111**, 27 640).

1634 V.V. Malyarenko, N.F. Chuiko, V.Yu. Tretinnik, and S.V. Pakhovchishin, *Ukr. Khim. Zh. (Russ. Ed.)*, 1988, **54**, 686 (*Chem. Abstr.*, 1989, **110**, 14 279).

1635 Ē. Kupče, E. Upena, M. Trušule, and E. Lukevics, *Polyhedron*, 1989, **8**, 2641.

1636 M.T. Blanda, J.H. Horner, and M. Newcomb, *J. Org. Chem.*, 1989, **54**, 4626.

1637 S.A. Campbell, R. Lowry, R.D. Newmark, and L.M. Peter, *Chem. Phys. Lett.*, 1989, **155**, 89 (*Chem. Abstr.*, 1989, **110**, 161 218).

1638 I.A. Koppel, V.A. Eiber, J.B. Koppel, and T.J. Pehk, *Org. React. (Tartu)*, 1988, **25**, 58 (*Chem. Abstr.*, 1989, **111**, 84 960).

1639 J.W. Larson and M. Pippin, *Polyhedron*, 1989, **8**, 527.

1640 V.N. Plakhotnik, N.F. Towmash, and B.N. Chernyshov, *Koord. Khim.*, 1989, **15**, 344 (*Chem. Abstr.*, 1989, **110**, 219 831).

1641 V.G. Shteinberg, B.A. Shumm, A.F. Zueva, O.N. Efimov, and L.N. Erofeev, *Synth. Met.*, 1989, **29**, E7 (*Chem. Abstr.*, 1989, **111**, 16 516).

1642 A.M. Bond, D. Dakternieks, R. Di Giacomo, and A.F. Hollenkamp, *Inorg. Chem.*, 1989, **28**, 1510.

Chlorine. Cl^- binding to polyammonium receptor molecules has been investigated by ^{35}Cl n.m.r. spectroscopy.[1643] Proton mobility in liquid and frozen $HClO_4.5.5H_2O$ has been studied by n.m.r. spectroscopy.[1644]

Equilibria among Uncharged Species. Lithium. ^1H, ^6Li, and ^{13}C n.m.r. spectroscopy has been used to study dynamic exchange, rotational behaviour, and aggregation of 2-methylallyllithium.[1645]

Europium. The chemical equilibrium and structures of complexes of 18-crown-6 with $[Eu(fod)_3]$ have been determined by ^1H n.m.r. spectroscopy.[1646]

Uranium. $\Delta G^‡$ for Bu^tNC exchange on $[U(OC_6H_3Bu^t{}_2\text{-}2,6)_3]$ has been determined.[1647]

Titanium. ^{31}P n.m.r. spectroscopy has been used to investigate the equilibrium constant for PR_3 exchange in $[(\eta^5\text{-}C_5H_5)_2Ti(=CH_2)(PR_3)]$ and the kinetics were determined. The ^{13}C n.m.r. spectrum was also determined.[1648]

Zirconium and Hafnium. The activation energy for

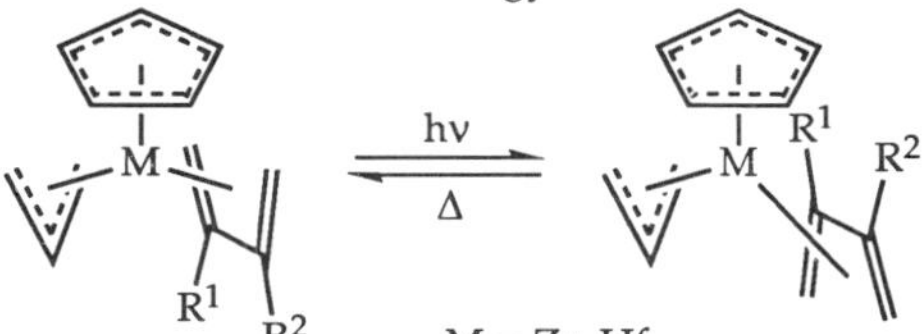

M = Zr, Hf

has been determined by ^1H and ^{13}C n.m.r. spectroscopy.[1649]

Chromium. ^1H and ^{13}C n.m.r. spectroscopy has been used to determine $\Delta G^‡$, $\Delta H^‡$, and $\Delta S^‡$ for the $\eta^6 \leftrightarrow \eta^5$ rearrangement of (60).[1650] Haptotropic rearrangements between (61), R^1= Me, R^2 = H, and (61), R^1= H, R^2 = Me, have been studied by ^1H n.m.r. spectroscopy.[1651]

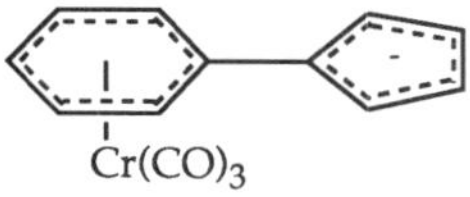

(60)

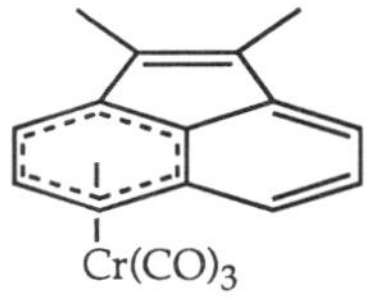

(61)

1643 M.W. Hosseini, J.-P. Kintzinger, J.-M. Lehn, and A. Zahidi, *Helv. Chim. Acta*, 1989, **72**, 1078.
1644 T.H. Huang, R.A. Davis, U. Frese, and U. Stimming, *Report*, 1988, TR-1; Order No. AD-A198981, 15 pp. Avail. NTIS. From *Gov. Rep. Announce. Index (U.S.)*, 1989, 89, Abstr. No. 903 260 (*Chem. Abstr.*, 1989, **111**, 65 120).
1645 G. Fraenkel and W.R. Winchester, *J. Am. Chem. Soc.*, 1989, **111**, 3794.
1646 L. Lu, Q. Song, X. Yang, X. Wang, and V.V. Yastrebov, *Wuji Huaxue*, 1988, **4**, 104 (*Chem. Abstr.*, 1989, **111**, 84 917).
1647 W.G. Van der Sluys and A.P. Sattelberger, *Inorg. Chem.*, 1989, **28**, 2496.
1648 J.D. Meinhart, E.V. Anslyn, and R.H. Grubbs, *Organometallics*, 1989, **8**, 583.
1649 C. Sontag, H. Berke, C. Sarter, and G. Erker, *Helv. Chim. Acta*, 1989, **72**, 1676.
1650 A.Ceccon, A. Gambaro, F. Gottardi, S. Santi, A. Venzo, and V. Luccini, *J. Organomet. Chem.*, 1989, **379**, 67.
1651 Yu.F. Oprunenko, S.G. Malyugina, O.Ya. Babushkina, Yu.A. Ustynyuk, P.N. Nesterenko, N.A. Ustynyuk, D.N. Kravtsov, and V.A. Piven, *Metallorg. Khim.*, 1988, **1**, 384 (*Chem. Abstr.*, 1989, **111**, 7553).

Molybdenum. *Cis-trans* isomerism of [(η^5-C_5Me_5)Mo$(CO)_2$(CHO)(PMe_2Ph)] has been investigated by ^{1}H n.m.r. spectroscopy.[1652] The equilibrium

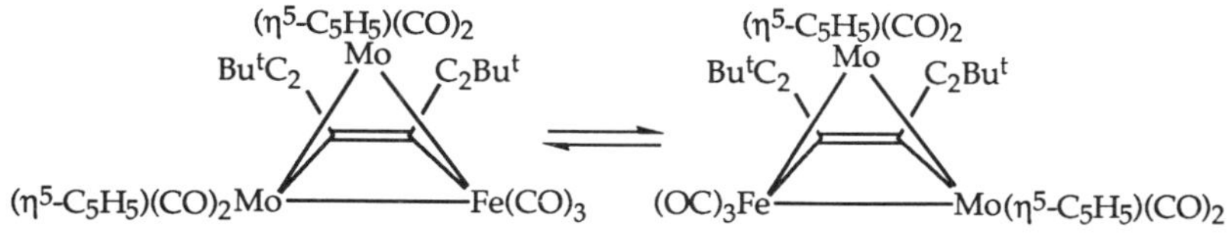

has been demonstrated by ^{13}C n.m.r. spectroscopy.[1653]

Tungsten. The ^{1}H n.m.r. spectrum of [(η^5-C_5H_5)$WOs_3(CO)_{11}$(μ-H)$_3$] is dynamic due to interconverting isomers.[1654] The interconversion of trigonal-bipyramid and square-pyramidal [(2,6-$Pr^i_2C_6H_3$)N=W{$(CH_2)_2$CHBu^t}($OCMe_2CF_3$)$_2$] has been investigated by ^{1}H and ^{13}C n.m.r. spectroscopy.[1655] The equilibrium constant for

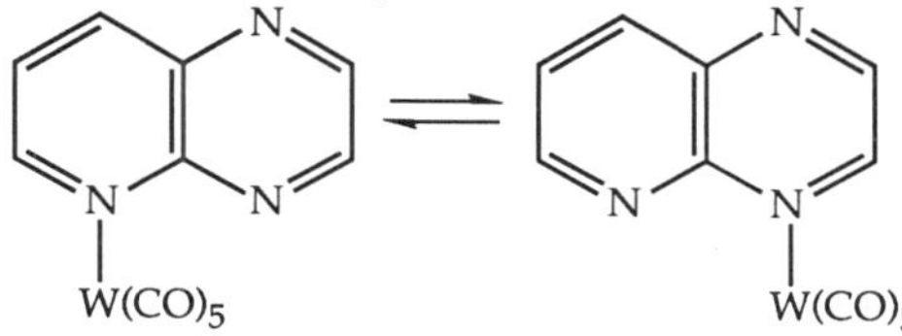

has been determined from ^{1}H n.m.r. measurements.[1656] ^{1}H, ^{13}C, and ^{77}Se n.m.r. spectroscopy has been used to show that [(OC)$_5$W(Se=CHC_6H_4Me-4)] is fluxional due to the tungsten moving between η^1 and η^2 coordination.[1657] ^{31}P n.m.r. spectroscopy has been used to show that there is a fast equilibrium between [W_2Cl_4(μ-Cl)$_2$(PEt_3)$_4$] and [W_2Cl_3(μ-Cl)$_3$(PEt_3)$_3$] and PEt_3.[1658]

Rhenium. The equilibrium between *triangulo*-[Re_3(μ-OPr^i)$_3$(OPr^i)$_6$] and *triangulo*-[Re_3(μ-OPr^i)$_3$H(OPr^i)$_5$] has been investigated by ^{1}H n.m.r. spectroscopy.[1659]

Iron. The rate constant for interconversion of (62) and (63) has been determined using magnetisation transfer and line shape analysis.[1660] Complex formation between SO_2 and [(η^5-C_5H_5)Fe(η^5-C_5H_4CHMeNMe$_2$)] has been investigated by ^{1}H and ^{13}C n.m.r. spectroscopy.[1661] ^{1}H and ^{13}C n.m.r. spectroscopy has suggested that there is a rapid equilibrium between [(η^5-C_5H_5)Fe{η^2-MeSC(SMe)$_2$}(CO)] and [(η^5-C_5H_5)Fe{=C(SMe)$_2$}(SMe)(CO)],[1662] and between ketene derivatives of [(η^5-C_5R_5)Fe$(CO)_2$]$_2$.[1663] ^{1}H n.m.r. spectroscopy has been used to study the kinetics and mechanisms of the isomerization reac-

1652 A. Asdar, C. Lapinte, and L. Toupet, *Organometallics*, 1989, **8**, 2708.
1653 S.J. Etches, I.J. Hart, and F.G.A. Stone, *J. Chem. Soc., Dalton Trans.*, 1989, 2281.
1654 Y. Chi, C.-Y. Cheng, and S.-L. Wang, *J. Organomet. Chem.*, 1989, **378**, 45.
1655 J. Feldman, W.M. Davis, and R.R. Schrock, *Organometallics*, 1989, **8**, 2266.
1656 C. Bessenbacher and W. Kaim, *J. Organomet. Chem.*, 1989, **369**, 83.
1657 H. Fischer, S. Zeuner, U. Gerbing, J. Riede, and C.G. Kreiter, *J. Organomet. Chem.*, 1989, **377**, 105.
1658 S.T. Chacon, M.H. Chisholm, W.E. Streib, and W.V.D. Sluys, *Inorg. Chem.*, 1989, **28**, 5.
1659 D.M. Hoffman, D. Lappas, and D.A. Wierda, *J. Am. Chem. Soc.*, 1989, **111**, 1531 (*Chem. Abstr.*, 1989, **110**, 127 562).
1660 Z. Goldschmidt and H.E. Gottlieb, *J. Organomet. Chem.*, 1989, **361**, 207.
1661 S. Brownstein, N.F. Han, E. Gabe, and F. Lee, *Can. J. Chem.*, 1989, **67**, 93.
1662 G.N. Glavee and R.J. Angelici, *J. Am. Chem. Soc.*, 1989, **111**, 3598.
1663 M. Akita, T. Kawahara, M. Terada, N. Kakinuma, and Y. Moro-oka, *Organometallics*, 1989, **8**, 687.

tions of [(μ-$PhCH_2S$)(μ-RS)$Fe_2(CO)_6$],[1664] and the influence of subunit assembly on the low-spin $\rightleftharpoons$ high-spin equilibrium of met-azido haemoglobin A.[1665]

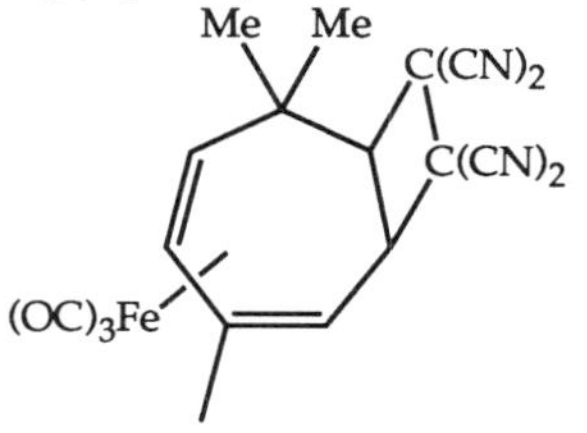

(62)

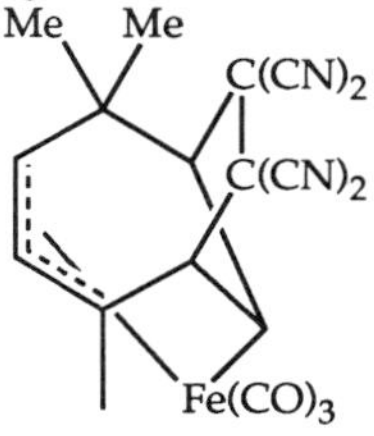

(63)

Ruthenium. Variable temperature ^1H and ^{31}P n.m.r. spectroscopy has been used to demonstrate the equilibrium between [$Ru(H)_2(O_2CCF_3)_2(PCy_3)_2$] and [$Ru(\eta^2-H_2)(O_2CCF_3)_2(PCy_3)_2$].[1666] The *fac*↔*mer* isomerisation of some tris-β-dieketonate Ru(III) complexes has been investigated by ^1H n.m.r. spectroscopy.[1667]

Cobalt. The thermodynamics of the equilibrium between carbonyl bridged and non-bridged isomers of [$Co_2(CO)_4(\mu-R_2PCH_2PR_2)_2$]4 have been determined from ^1H, ^{13}C, and ^{31}P n.m.r. spectroscopy.[1668]

Rhodium. ^1H and ^{13}C n.m.r. spectroscopy has been used to demonstrate an equilibrium of [(octaethylporphyrin)RhH] with CO to give [(octaethylporphyrin)RhCHO].[1669] High pressure n.m.r. spectroscopy has been used to study H_2 induced $P(C_6H_4SO_3Na\text{-}3)$ dissociation from [$HRh(CO)P(C_6H_4SO_3Na\text{-}3)_3$].[1670] ^1H and ^{13}C n.m.r. spectroscopy has been used to demonstrate an equilibrium between [$Rh_2(\mu-H)(\mu-C_4H_5)(Pr^i{}_2PCH_2CH_2PPr^i{}_2)_2$] and [$Rh_2(\mu-C_4H_6)(Pr^i{}_2PCH_2CH_2PPr^i{}_2)_2$].[1671]

Nickel. The aggregation of nickel complexes of substituted salen has been studied by ^1H n.m.r. spectroscopy.[1672]

Palladium and Platinum. Ligand exchange rates for *cis*-[$PtPh_2(SMe_2)_2$] and *cis*-[$PtPh_2(OSMe_2)_2$] have been determined as a function of pressure.[1673] ^{13}C n.m.r. spectroscopy has been used to investigate the equilibrium between [$PtPh_2(SBu^n{}_2)_2$] and [$Pt_2Ph_4(SBu^n{}_2)_2$] with the liberation of $SBu^n{}_2$.[1674] ^{31}P n.m.r. spectroscopy has been used to

1664 Q. Zhang, C. Zhu, L. Song, R. Chen, X. Lu, and G. Li, *Huaxue Xuebao*, 1988, **46**, 617 (*Chem. Abstr.*, 1989, **110**, 213 025).
1665 Y. Yamamoto and G.N. La Mar, *Biochim. Biophys. Acta*, 1989, **996**, 187 (*Chem. Abstr.*, 1989, **111**, 129 241).
1666 T. Arliguie and B. Chaudret, *J. Chem. Soc., Chem. Commun.*, 1989, 155.
1667 K. Kobayashi, Y. Satsu, Y. Hoshino, S. Arai, K. Hatakeyama, A. Endo, K. Shimizu, and G.P. Satô, *Bull. Chem. Soc. Jpn.*, 1989, **62**, 3252.
1668 D.J. Elliot, H.A. Mirza, R.J. Puddephatt, D.G. Holah, A.N. Hughes, R.H. Hill, and W. Xia, *Inorg. Chem.*, 1989, **28**, 3282 (*Chem. Abstr.*, 1989, **111**, 121 812).
1669 B.B. Wayland, K.J. Balkus, jun., and M.D. Farnos, *Organometallics*, 1989, **8**, 950.
1670 I.T. Horvath, R.V. Kastrup, A.A. Oswald, and E.J. Mozeleski, *Catal. Lett.*, 1989, **2**, 85 (*Chem. Abstr.*, 1989, **110**, 215 149).
1671 M.D. Fryzuk, W.E. Piers, S.J. Rettig, F.W.B. Einstein, T. Jones, and T.A. Albright, *J. Am. Chem. Soc.*, 1989, **111**, 5709.
1672 K. Miyamura, K. Satoh, and Y. Gohshi, *Bull. Chem. Soc. Jpn.*, 1989, **62**, 45; K. Miyamura, K. Satoh, and Y. Gohshi, *J. Coord. Chem.*, 1988, **18**, 149.
1673 U. Frey, L. Helm, A.E. Merbach, and R. Romeo, *J. Am. Chem. Soc.*, 1989, **111**, 8161.
1674 G. Alibrandi, D. Minniti, L.M. Scolaro, and R. Romeo, *Inorg. Chem.*, 1989, **28**, 1939.

investigate PPh_3 exchange in [Pt(PPh_3)$_3$] which is catalysed by PPh_3.[1675] An equilibrium between [N($CH_2CH_2PPh_2$)$_3$Pt(PPh_3)] and [N($CH_2CH_2PPh_2$)$_3$Pt] and PPh_3 has been investigated by ^{31}P n.m.r. spectroscopy.[1676] ^{31}P and ^{195}Pt n.m.r. spectroscopy has been used to investigate anion redistribution between [$MX_2(PR_3)_2$] and [$MY_2(PR_3)_2$] to give [MXY-(PR_3)$_2$], M = Pd, Pt.[1677] The kinetics of acetate exchange on [Pt_4(μ-O_2CCH_3)$_4$(μ-O_2C-CCl_3)$_4$] have been determined using ^1H magnetisation transfer.[1678]

Copper. The structure and the solution behaviour of [MeCu(PR_3)$_n$] have been determined by variable temperature ^1H and ^{31}P n.m.r. spectroscopy.[1679]

Gold. Equilibria involving PR_3 on AuBr have been investigated by ^{31}P n.m.r. spectroscopy.[1680] ^{31}P n.m.r. spectroscopy has been used to investigate the binding of [AuCl(PEt_3)] to bovine serum albumin.[1681]

Zinc. ^1H n.m.r. spectroscopy has been used to examine ligand exchange on a trimeric zinc porphyrin.[1682]

Cadmium and Mercury. Ligand exchange on [M(S_2COPr^i)$_2$] and [M{$S_2P(OPr^i)_2$}$_2$], M = Cd, Hg, has been studied by ^{113}Cd and ^{199}Hg n.m.r. spectroscopy.[1683]

Aluminium and Indium. The lability of complexes of $EtMe_2M$, M = Al, In, has been investigated using ^1H n.m.r. spectroscopy.[1684] ^1H n.m.r. spectroscopy has been used to investigate exchange between isomers of [Al_2Me_4(μ-NC_5H_4S-2)$_2$].[1685] ^1H n.m.r. spectroscopy has been used to investigate complexation reactions of aluminium alkyls with electron donors.[1686] Molecular association in gels derived from [$(BuO)_2AlOSi(OEt)_3$] has been studied by ^{27}Al and ^{29}Si n.m.r. spectroscopy.[1687]

Silicon. ^1H and ^{13}C n.m.r. spectroscopy has been used to determine the axial/equatorial preference of SiH_3 and SiH_2Me on cyclohexane.[1688] Redistribution reactions of [$SiF_3SiFHBr$] and [SiF_2HSiF_2X] have been studied by ^{19}F n.m.r. spectroscopy.[1689] The degenerate *E,Z* isomerisation of [R_2NP=C($SiMe_3$)$_2$] has been observed by ^{13}C n.m.r. spectroscopy.[1690] Isomerisation of [Me_3SiN=$PPh_2CH_2PPh_2$] has been studied by ^{29}Si and

1675 V.V. Grushin, I.S. Akhrem, and M.E. Vol'pin, *J. Organomet. Chem.*, 1989, **371**, 403.
1676 C.A. Ghilardi, S. Midollini, S. Moneti, A. Orlandini, and J.A. Ramirez, *J. Chem. Soc., Chem. Commun.*, 1989, 304.
1677 J.A. Rahn, M.S. Holt, and J.H. Nelson, *Polyhedron*, 1989, **8**, 897.
1678 T. Yamaguchi, Y. Sasaki, A. Nagasawa, T. Ito, N. Koga, and K. Morokuma, *Inorg. Chem.*, 1989, **28**, 4311.
1679 P.S. Coan, K. Folting, J.C. Huffman, and K.G. Caulton, *Organometallics*, 1989, **8**, 2724.
1680 W.E. Hill, M.Q. Islam, T.R. Webb, and C.A. McAuliffe, *Inorg. Chim. Acta*, 1989, **157**, 215.
1681 A.A. Isab, A.L. Hormann, D.T. Hill, D.E. Griswold, M.J. DiMartino, and C.F. Shaw, tert., *Inorg. Chem.*, 1989, **28**, 1321.
1682 H.L. Anderson and J.K.M. Sanders, *J. Chem. Soc., Chem. Commun.*, 1989, 1714.
1683 B.F. Abrahams, G. Winter, and D. Dakternieks, *Inorg. Chim. Acta*, 1989, **162**, 211.
1684 D.C. Bradley, H. Chudzynska, and D.M. Frigo, *Chemtronics*, 1988, **3**, 159 (*Chem. Abstr.*, 1989, **111**, 194 828).
1685 R. Kumar, V.S.J. de Mel, and J.P. Oliver, *Organometallics*, 1989, **8**, 2488.
1686 T.M. Ushakova, I.N. Meshkova, and M.A. Markevich, *Metalloorg. Khim.*, 1988, **1**, 929 (*Chem. Abstr.*, 1989, **111**, 194 829).
1687 J.C. Pouxviel and J.P. Boilot, *J. Mater. Sci.*, 1989, **24**, 321 (*Chem. Abstr.*, 1989, **110**, 158 978).
1688 K.G. Penman, W. Kitching, and W. Adcock, *J. Org. Chem.*, 1989, **54**, 5390.
1689 J.J. D'Errico and K.G. Sharp, *Inorg. Chem.*, 1989, **28**, 2886.
1690 M.I. Povolotskii and V.V. Negrebetskii, *Zh. Obshch. Khim.*, 1988, **58**, 231 (*Chem. Abstr.*, 1989, **110**, 23 993).

^{31}P n.m.r. spectroscopy.[1691] ^{13}C and ^{31}P n.m.r. spectroscopy has been used to study a tautomeric mixture of [$(Me_3Si)_2PC(OSiMe_3)=NPr^i$] and [$(Me_3Si)_2PC(O)NPr^iSiMe_3$].[1692]
Tin. ^{13}C c.p.m.a.s and solution n.m.r. spectroscopy has been used to study axial/equatorial equilibria in 3-Ph_3Sn-cyclohexene and 3-Ph_3Sn-cycloheptene.[1693] Ligand coordination to [$(Me_2ClSnCH_2CH_2)_2P(O)Ph$] has been studied by ^1H, ^{13}C, ^{31}P, and ^{119}Sn n.m.r. spectroscopy.[1694] Equilibrium constants for the formation of 1:1 adducts between $MeSnCl_3$, Bu^nSnCl_3, and Ph_3SnCl_3 with MeCN, $Me_2NC(O)Me$, Ph_3PO, and Et_3PO have been determined by ^{13}C, ^{31}P, and ^{119}Sn n.m.r. spectroscopy.[1695]
Selenium. The conformational behaviour of *cis*- and *trans*-2-(PhSe)-1,3-diselenanes has been investigated by ^{13}C and ^{77}Se n.m.r. spectroscopy.[1696] The kinetics of symmetric selenol/diselenide exchange reactions involving selenocysteamine/selenocystamine have been studied by ^1H and ^{13}C n.m.r. spectroscopy.[1697]

Course of Reactions.—*Scandium.* The reaction of $Sc(NO_3)_3$ with [$UMo_{18}O_{42}$] has been followed by ^{17}O, ^{45}Sc, and ^{95}Mo n.m.r. spectroscopy.[1698] The exchange of [$Th(acac)_4$] with deuterated acetylacetone has been monitored by ^1H n.m.r. spectroscopy.[1699]
Titanium and Zirconium. [$(\eta^5-C_5H_4Bu^t)_2Zr(CH_2PPh_2)_2$] reduction has been monitored by ^1H n.m.r. spectroscopy.[1700] ^1H n.m.r. spectroscopy has been used to monitor the rate of reaction of [$(\eta^5-C_5Me_5)_2ZrPh(OH)$] and an isotope effect of 4.3 ± 0.1 was found.[1701] ^1H n.m.r. spectroscopy has been used to follow the decomposition of [$(\eta^5-C_5H_5)_2Zr(CH_2O$-$CH_2Ph)Cl$] to give [$(\eta^5-C_5H_5)_2Zr(OCH_2CH_2Ph)Cl$]. The ^{13}C n.m.r. spectrum was also reported.[1702] The kinetics of the formation of silica-titania-zirconia gels from the alkoxides have been followed by ^{29}Si n.m.r. spectroscopy.[1703]
Vanadium. The reaction of $[PW_{11}O_{39}]^{7-}$ with $[VO_3]^-$ or $[HV_{10}O_{28}]^{5-}$ has been studied by ^{31}P and ^{51}V n.m.r. spectroscopy.[1704]
Tantalum. ^1H and ^{13}C n.m.r. spectroscopy has been used to monitor the reaction of

1691 K.V. Katti and R.G. Cavell, *Inorg. Chem.*, 1989, **28**, 413.
1692 R. Appel and M. Poppe, *Angew. Chem., Int. Ed. Engl.*, 1989, **28**, 53.
1693 W. Kitching, K.G. Penman, G. Valle, G. Tagliavini, and P. Ganis, *Organometallics*, 1989, **8**, 785.
1694 M. Dargatz, H. Hartung, E. Kleinpeter, B. Rensch, D. Schollmeyer, and H. Weichmann, *J. Org. Chem.*, 1989, **361**, 43.
1695 J.N. Spencer, B.G. Enders, A. Grushow, S.P. Kneizys, W.L. Nachlis, D. Mokrynka, S.M. Coley, J.C. Otter, and C.H. Yoder, *J. Org. Chem.*, 1989, **362**, 53.
1696 B.M. Pinto, B.D. Johnston, and R. Nagelkerke, *Heterocycles*, 1989, **28**, 389 (*Chem. Abstr.*, 1989, **111**, 77 399).
1697 J.C. Pleasants, W. Guo, and D.L. Rabenstein, *J. Am. Chem. Soc.*, 1989, **111**, 6553.
1698 E.P. Samokhvalova, M.A. Fedotov, I.V. Tat'yanina, and E.A. Torchenkova, *Zh. Neorg. Khim.*, 1989, **34**, 660 (*Chem. Abstr.*, 1989, **110**, 241 576).
1699 N. Fujiwara, T. Nakagawa, H. Tomiyasu, and H. Fukutomi, *Bull. Chem. Soc. Jpn.*, 1989, **62**, 2087.
1700 R. Choukroun, D. Gervais, and C.-E. Rifaï, *J. Organomet. Chem.*, 1989, **368**, C11.
1701 M. Segi, T. Koyama, Y. Takata, T. Nakajima, and S. Suga, *J. Am. Chem. Soc.*, 1989, **111**, 8749.
1702 S.L. Buchwald, R.B. Nielsen, and J.C. Dewan, *Organometallics*, 1989, **8**, 1593.
1703 W. Beier, A.A. Goektas, G.H. Frischat, C. Wies, K. Meise-Gresch, and W. Mueller-Warmuth, *Phys. Chem. Glasses*, 1989, **30**, 69 (*Chem. Abstr.*, 1989, **110**, 197 806).
1704 A.K. Il'yasova, K.Zh. Serikpaeva, and R.A. Geskina, *Koord. Khim.*, 1989, **15**, 649 (*Chem. Abstr.*, 1989, **111**, 166 117).

$[(\eta^5\text{-}C_5H_5)_2TaH_3]$ with $RO_2CC{\equiv}CCO_2R$ or $PhC{\equiv}CPh$.[1705]

Chromium. The reaction between *cis*-azobenzene and $[Me(MeO)C{=}Cr(CO)_5]$ has been followed by ^1H and ^{13}C n.m.r. spectroscopy.[1706] ^1H n.m.r. spectroscopy has been used to follow the exchange of aliphatic amines with $[Cr(CO)_6]$.[1707] ^1H and ^{13}C n.m.r. spectroscopy has been used to follow the reaction of $[(\eta^5\text{-}C_5H_5)_2Cr_2(CO)_4Se]$ with S_8.[1708]

Molybdenum and Tungsten. ^1H and ^{13}C n.m.r. spectroscopy has been used to monitor hydrogen exchange in, and methane elimination, from $[(\eta^5\text{-}C_5H_5)_2W(H)CH_3]$.[1709] ^1H n.m.r. spectroscopy has been used to monitor alkane elimination from $[(\eta^5\text{-}C_5Me_5)_2\text{-}WRH]$ and to investigate the ^2H effect.[1710] The oxygen exchange rate for $[(\eta^5\text{-}C_5Me_5)_2\text{-}W{=}O]$ with $H_2{}^{17}O$ has been measured by ^{17}O n.m.r. spectroscopy. The ^{13}C n.m.r. spectrum was also recorded.[1711] ^{31}P n.m.r. spectroscopy has been used to show that $[M(CO)_5\text{-}(PPh_2H)]$ and *cis*-$[M(CO)_4(PPh_2H)_2]$, M = Mo, W, are precursors to phosphido bridged derivatives.[1712] Variable temperature ^{13}C n.m.r. spectroscopy has been used to show that in $[Mo(CO)_3(\text{diglyme})]$, the diglyme is displaced by acetone, methanol, and ethanol.[1713] The reaction between 1,2-$[W_2(NMe_2)_4(PCy_2)_2]$ and $^{13}CO_2$ has been investigated by ^{13}C n.m.r. spectroscopy.[1714] The equilibration of *syn*- and *gauche*-isomers of $[Mo_2Cl_7\text{-}(PMe_3)_2]^-$ has been monitored by ^1H n.m.r. spectroscopy.[1715] The reaction of $[MoO_2\{$R-substituted-2-(salicylideneamino)phenol$\}_2(\text{solv})_2]$ with $(R_f)_2SO$ or PR_3 has been monitored by ^{19}F and ^{31}P n.m.r. spectroscopy.[1716]

Manganese. Nucleophilic addition of H^- and $[CH(CO_2Me)_2]^-$ to $[(1\text{-}5\text{-}\eta^5\text{-}6\text{-}RC_6H_6)Mn\text{-}(CO)(NO)(PMe_3)]^+$ has been monitored by ^1H n.m.r. spectroscopy.[1717]

Rhenium. The decomposition of *trans*-$[(\eta^5\text{-}C_5H_5)ReMe(COMe)(CO)_2]$ has been monitored by ^1H and ^{13}C n.m.r. spectroscopy to give acetone, and a transient "$[(\eta^5\text{-}C_5H_5)Re\text{-}(CO)_2]$".[1718] The rate of linkage isomerism of $[(\eta^5\text{-}C_5Me_5)Re(CO)_2(^{15}N{\equiv}^{14}N)]$ to give $[(\eta^5\text{-}C_5Me_5)Re(CO)_2(^{14}N{\equiv}^{15}N)]$ has been monitored by ^{15}N n.m.r. spectroscopy.[1719] ^1H n.m.r. spectroscopy has been used to follow the reaction of $[ReCl_2(CNPr^i{}_2)(PMePh_2)_2]^+$ and Pr^iNC.[1720]

Iron and Ruthenium. The radical pair mechanism in hydrometalation and stoichio-

1705 G.E. Herberich and I. Savvopoulos, *J. Organomet. Chem.*, 1989, **362**, 345.
1706 L.S. Hegedus and B.R. Lundmark, *J. Am. Chem. Soc.*, 1989, **111**, 9194.
1707 T.J. Burkey, *Polyhedron*, 1989, **8**, 2681.
1708 L.Y. Goh, *J. Chem. Soc., Dalton Trans.*, 1989, 431.
1709 R.M. Bullock, C.E.L. Headford, K.M. Hennessy, S.E. Kegley, and J.R. Norton, *J. Am. Chem. Soc.*, 1989, **111**, 3897.
1710 G. Parkin and J.E. Bercaw, *Organometallics*, 1989, **8**, 1172.
1711 G. Parkin and J.E. Bercaw, *J. Am. Chem. Soc.*, 1989, **111**, 391.
1712 R.K. Keiter, E.A. Keiter, K.N. Mittelberg, J.S. Martin, V.M. Meyers, and J.G. Wang, *Organometallics*, 1989, **8**, 1399.
1713 G.W. Wagner and B.E. Hanson, *Polyhedron*, 1989, **8**, 2925.
1714 W.E. Behro, M.H. Chisholm, J.D. Martin, J.C. Huffman, K. Folting, and W.E. Streib, *J. Am. Chem. Soc.*, 1989, **111**, 8149.
1715 F.A. Cotton and R.L. Luck, *Inorg. Chem.*, 1989, **28**, 182.
1716 J.A. Craig, E.W. Harlan, B.S. Snyder, M.A. Whitener, and R.H. Holm, *Inorg. Chem.*, 1989, **28**, 2082.
1717 R.D. Pike, W.J. Ryan, G.B. Carpenter, and D.A. Sweigart, *J. Am. Chem. Soc.*, 1989, **111**, 8535.
1718 K.I. Goldberg and R.G. Bergman, *J. Am. Chem. Soc.*, 1989, **111**, 1285.
1719 A. Cusanelli and D. Sutton, *J. Chem. Soc., Chem. Commun.*, 1989, 1719.
1720 S. Warner and S.J. Lippard, *Inorg. Chem.*, 1989, **28**, 3008.

metric hydrogenation reactions of $[(\eta^5\text{-}C_5H_5)Fe(CO)_2H]$ with conjugated dienes has been monitored by ^1H and ^{13}C n.m.r. spectroscopy.[1721] ^1H n.m.r. spectroscopy has been used to monitor the reaction of $[HRu_3(CO)_{11}]^-$ with PPh_3. The ^{13}C n.m.r. spectra were also recorded.[1722] The relative rates of reaction of $[(\eta^5\text{-}C_5H_5)Fe(CO)_2CH_2Br]$ with PR_3 have been determined by ^1H and ^{13}C n.m.r. spectroscopy.[1723] ^{13}C n.m.r. spectroscopy has been used to follow the photolysis of $[(\eta^5\text{-}C_5H_5)Fe\{C(O)(\overline{CPhCH_2CH_2CH_2})\}(CO)_2]$.[1724] Low temperature ^1H n.m.r. studies of the reactions of $[\{(4\text{-}MeC_6H_4)_4porphyrin\}FeR]$ with O_2 have shown that $[\{(4\text{-}MeC_6H_4)_4porphyrin\}FeO_2R]$ is initially formed, and then decomposes.[1725] ^1H n.m.r. spectroscopy has been used to monitor the reaction of $[(\eta^4\text{-}C_7H_7D)\text{-}Fe(CO)_3]$ with t.c.n.e.[1726] The kinetics of the 1,3-rearrangement in $[(\eta^4\text{-heptafulvene})Fe\text{-}(CO)_3]$ have been determined.[1727] The reactions of $[(\eta^3\text{-}C_3H_5)Fe(CO)_3]^-$ with carbon electrophiles have been monitored by ^1H and ^{13}C n.m.r. spectroscopy.[1728] The reaction of $[(\eta^6\text{-}C_6Me_6)Ru(\eta^6\text{-anthracene})]^{2+}$ with MeCN has been monitored by ^1H n.m.r. spectroscopy.[1729] The PMe_3 exchange kinetics have been determined for $[(\eta^5\text{-}C_5Me_5)Ru\text{-}(PMe_3)_2X]$ by adding $P(CD_3)_3$.[1730] The kinetics and mechanism of PR_3 substitution for CO in $[Fe_2Co(CO)_9(CCO)]^-$ have been studied by ^1H and ^{13}C n.m.r. spectroscopy.[1731] The selective *trans* addition of D_2 to the coordinated $R^1C{=}CR^2CR^2{=}NR^1$ ligand in $[FeRu\text{-}(CO)_6(R^1C{=}CR^2CR^2{=}NR^1)]$ has been monitored by ^1H n.m.r. spectroscopy under high pressure.[1732] ^{13}C n.m.r. spectroscopy has been used to show $[Fe(^{13}CN)_5(NO)]^{2-}$ is converted in blood to give $[S^{13}CN]^-$.[1733] The aquation kinetics of $[Ru_2OCl_{10}]^{4-}$ have been determined by ^1H n.m.r. spectroscopy.[1734]

Osmium. The rate of ligand exchange of $[Os(PMe_3)_5]$ with $P(CD_3)_3$ has been determined by ^1H, ^{13}C, and ^{31}P n.m.r. spectroscopy.[1735]

Cobalt. ^1H and ^{13}C n.m.r. spectroscopy has been used to monitor ligand shift between $[(1\text{-}4\text{-}\eta^4\text{-}H_2C{=}CHCH{=}CHCH{=}CD_2)Co(\eta^5\text{-}C_5H_5)]$ and $[(3\text{-}6\text{-}\eta^4\text{-}H_2C{=}CHCH{=}CHCH{=}CD_2)Co\text{-}(\eta^5\text{-}C_5H_5)]$.[1736] The reactions of $[Co(NH_3)_5(OH_2)]^{3+}$ and related compounds with $[SO_3]^{2-}$

1721 T.A. Shackleton and M.C. Baird, *Organometallics*, 1989, **8**, 2225.
1722 H. Jungbluth, G. Süss-Fink, M.A. Pellinghelli, and A. Tiripicchio, *Organometallics*, 1989, **8**, 925.
1723 G.C.A. Bellinger, H.B. Friedrich, and J.R. Moss, *J. Organomet. Chem.*, 1989, **366**, 175.
1724 R.L. Trace and W.M. Jones, *J. Organomet. Chem.*, 1989, **376**, 103.
1725 R.D. Arasasingham, A.L. Balch, and L. Latos-Grazynski, *Stud. Org. Chem. (Amsterdam)*, 1988, **33**, 417 (*Chem. Abstr.*, 1989, **110**, 173 409).
1726 Z. Goldschmidt, E. Genizi, and H.E. Gottlieb, *J. Organomet. Chem.*, 1989, **368**, 351.
1727 Z. Goldschmidt, D. Hezroni, H.E. Gottlieb, and S. Antebi, *J. Organomet. Chem.*, 1989, **373**, 235.
1728 M. Brookhart, J. Yoon, and S.K. Noh, *J. Am. Chem. Soc.*, 1989, **111**, 4117.
1729 D.A. Freedman and K.R. Mann, *Inorg. Chem.*, 1989, **28**, 3926.
1730 H.E. Bryndza, P.J. Domaille, R.A. Paciello, and J.E. Bercaw, *Organometallics*, 1989, **8**, 379.
1731 S. Ching and D.F. Shriver, *J. Am. Chem. Soc.*, 1989, **111**, 3238.
1732 R. Zoet, C.J. Elsevier, G. Van Koten, P. Versloot, K. Vrieze, M. Van Wijnkoop, C.A. Duineveld, K. Goubitz, D. Heijdenrijk, and C.H. Stam, *Organometallics*, 1989, 8, 23.
1733 A.R. Butler, C. Glidewell, and A.E. Waddon, *Polyhedron*, 1989, 8, 2627.
1734 N.G. Maksimov, O.K. Shpigunova, T.T. Babanakova, N.M. Sinitsyn, E.D. Korniets, V.E. Volkov, and I.D. Danilov, *Zh. Neorg. Khim.*, 1989, **34**, 419 (*Chem. Abstr.*, 1989, **110**, 199 915).
1735 S.P. Ermer, R.S. Shinomoto, M.A. Deming, and T.C. Flood, *Organometallics*, 1989, **8**, 1377.
1736 J.A. King, jun. and K.P.C. Vollhardt, *J. Organomet. Chem.*, 1989, 369, 245.

have been studied by ^{59}Co n.m.r. spectroscopy.[1737] The rates of base catalysed proton exchange and inversion in $[Co\{(OCMe)_2CX\}_2(tmeda)]^+$ have been determined by ^1H n.m.r. spectroscopy.[1738] Hydrolytic cleavage of AMP with $[Co\{(H_2NCH_2CH_2CH_2)_3N\}(OH_2)_2]^{3+}$ has been monitored by ^{31}P n.m.r. spectroscopy.[1739] ^{13}C n.m.r. spectroscopy has been used to investigate the hydrolysis of MeO_2CCH_3 catalysed by $[Co\{(H_2NCH_2CH_2CH_2)_3N\}(OH_2)(OH)]^{2+}$.[1740] The hydrolysis of *trans*-$[Co(5,7,7,12,14,14\text{-}Me_6\text{-}1,4,8,11\text{-tetraazacyclotetradeca-4,11-diene})Cl_2]^+$ has been monitored by ^1H n.m.r. spectroscopy.[1741] ^1H and ^{13}C n.m.r. spectroscopy has been used to investigate the mechanism of vitamin B_{12} biosynthesis.[1742]

Rhodium. The homogeneous hydrogenation of PhC≡CH catalysed by $[RhCl(PPh_3)_3]$ has been investigated using *para*-hydrogen induced polarisation.[1743] The hydrogenation catalyst species present in solution on treating $[Rh(\eta^4\text{-}C_7H_8)(diphosphine)]^+$ with hydrogen has been determined by ^{31}P n.m.r. spectroscopy.[1744] Deuteriation of the 2,2′-bipyridine ligands in $[Rh(bipy)_3]^{3+}$ has been monitored by ^1H and ^{13}C n.m.r. spectroscopy.[1745] ^{13}C n.m.r. spectroscopy has been used to investigate the mechanism of decarbonylation and dehydration of ketose sugars catalysed by $[RhCl(PPh_3)_3]$, $[Rh(CO)Cl(PPh_3)_2]$, and $[Rh(Ph_2PCH_2CH_2CH_2PPh_2)_2]^+$.[1746]

Iridium. ^1H n.m.r. spectroscopy has been used to investigate H/D exchange between olefins and C_6D_6 catalysed by $[IrH_5(PPr^i_3)_2]$.[1747] The protonation of $[\{(\eta^5\text{-}C_5Me_5)Ir(CO)\}_2(\mu\text{-}CH_2)]$ has been monitored by ^1H and ^{13}C n.m.r. spectroscopy.[1748] ^{31}P n.m.r. spectroscopy has been used to determine the kinetics of hydrolysis of *cis*-$[(en)_2Ir(OH_2)\{OP(O)(OEt)(OC_6H_4NO_2)\}]^{2+}$ as a function of pH.[1749] ^1H and ^{31}P n.m.r. signal enhancement from *para*-hydrogen addition catalysed by [IrBr(CO)(dppe)] and related compounds has been reported.[1750] ^{15}N n.m.r. spectroscopy has been used to study the nitration of $[IrCl_6]^{2-}$.[1751]

Nickel. ^1H, ^{13}C, and ^{31}P n.m.r. spectroscopy has been used to follow the transformation of $[Ni(CPh{=}CHCOR)Cl(PMe_3)_2]$ to $[\{\eta^2\text{-}(Me_3P)CPh{=}CHC(O)R\}NiCl(PMe_3)]$.[1752] ^{13}C n.m.r. spectroscopy has been used to investigate the reaction of $[(\eta^3\text{-}1,3\text{-}Me_2C_3H_3)NiMe(CO)_2]$

1737 T.Sh. Kapanadze, M.A. Elerdashvili, V.P. Tarasov, G.V. Tsintsadze, and Yu.A. Buslaev, *Koord. Khim.*, 1989, **15**, 248 (*Chem. Abstr.*, 1989, **111**, 89 270).
1738 G. Ma, M. Kojima, and J. Fujita, *Bull. Chem. Soc. Jpn.*, 1989, **62**, 2547.
1739 J. Chin and M. Banaszczyk, *J. Am. Chem. Soc.*, 1989, **111**, 4103.
1740 J. Chin and M. Banaszczyk, *J. Am. Chem. Soc.*, 1989, **111**, 2724.
1741 C.L. Schwarz and J.F. Endicott, *Inorg. Chem.*, 1989, **28**, 4011.
1742 A.I. Scott, *Pure Appl. Chem.*, 1989, **61**, 501.
1743 R.U. Kirss and R. Eisenberg, *J. Organomet. Chem.*, 1989, **359**, C22.
1744 J. Bakos, I. Toth, G. Szalontai, and B. Heil, *Kem. Kozl.*, 1986, 66, 153 (*Chem. Abstr.*, 1989, **110**, 173 430).
1745 E.C. Constable, *Polyhedron*, 1989, **8**, 83.
1746 M.A. Andrews, *Organometallics*, 1989, **8**, 2703.
1747 J.W. Faller and C.J. Smart, *Organometallics*, 1989, **8**, 602.
1748 D.M. Heinekey, S.T. Michel, and G.K. Schulte, *Organometallics*, 1989, **8**, 1241.
1749 P. Hendry and A.M. Sargeson, *J. Am. Chem. Soc.*, 1989, **111**, 2521.
1750 T.C. Eisenschmid, J. McDonald, R. Eisenberg, and R.G. Lawler, *J. Am. Chem. Soc.*, 1989, **111**, 7267.
1751 A.B. Venediktov, M.A. Fedotov, S.V. Korenev, and A.V. Belyaev, *Koord. Khim.*, 1989, **15**, 556.
1752 E. Carmona, E. Gutiérrez-Puebla, A. Monge, J.M. Marín, M. Paneque, and M.L. Poveda, *Organometallics*, 1989, **8**, 967.

with CO at low temperature.[1753] Oligomerisation of styrene catalysed by $[(\eta^3\text{-}2\text{-Me-}C_3H_4)Ni(\eta^4\text{-}C_8H_{12})]^+$ has been monitored by ^1H and ^{13}C n.m.r. spectroscopy.[1754] The oxidation of $[Ni(S_2PBu_2)_2]$ by Bu^tOOH has been followed by ^1H n.m.r. spectroscopy.[1755]

Palladium and Platinum. ^1H, ^{13}C, and ^{195}Pt n.m.r. spectroscopy has been used to monitor the transformation of *trans*-$[PtH(SC_4H_6)(PEt_3)_2]^+$ to $[Pt(\eta^2\text{-}\overline{S{=}CHCH_2CH_2CH_2})\text{-}(PEt_3)_2]^+$.[1756] ^1H n.m.r. spectroscopy has been used to investigate the oxidative addition of MeI to $[Me_2Pd(NC_5H_4\text{-}2\text{-})_2CHMe]$.[1757] The reaction of $[PdIMe_3(bipy)]$ with $[PtMe_2\text{-}(bipy)]$ has been monitored by ^1H n.m.r. spectroscopy.[1758] ^1H and ^{13}C n.m.r. spectroscopy has been used to follow the *ortho*-metallation of $[L_2Pt(CH_2CMe_2Ph)_2]$ and a large deuterium isotope effect was found.[1759] The oxidative addition of alkylvinyl triflates to $[Pt(PPh_3)_4]$ and $[(\eta^2\text{-}C_2H_4)Pt(PPh_3)]$ has been monitored by ^1H, ^{13}C, ^{19}F, ^{31}P, and ^{195}Pt n.m.r. spectroscopy.[1760] ^1H and ^{13}C n.m.r. spectroscopy has been used to monitor the thermal decomposition of *trans*-$[Pt(CMe{=}CH_2)(C{\equiv}CBu^t)(PPh_3)_2]$.[1761] The loss of Bu^n from $[(2\text{-}Bu^nTeC_6H_4CH{=}NCH_2)_2CH_2]$ on coordination to $PtCl_2$ has been monitored by ^{125}Te and ^{195}Pt n.m.r. spectroscopy.[1762] ^1H n.m.r. spectroscopy has been used to demonstrate polarisation when $PhC{\equiv}CH$ is hydrogenated in the presence of $[Pd_2(dppm)_3]$ using *para*-hydrogen.[1763] The reaction of $[Pt(S_2CNEt_2)_2]$ with dppe and related ligands has been studied by ^{31}P, ^{77}Se, and ^{195}Pt n.m.r. spectroscopy.[1764] ^{31}P n.m.r. spectroscopy has been used to monitor the reaction of $[PtCl_2(dppm)]$ with KOH to give *cis*- and *trans*-$[Pt_2\{PPh_2(O)\}_2(\mu\text{-}OH)_2(PMePh_2)_2]$,[1765] and *mer*→*fac* isomerism of $[Pt(S_4N_4)Cl_2(PMe_2Ph)]$.[1766]

Gold. ^1H n.m.r. spectroscopy has been used to monitor the thermal decomposition of $[Me_2Au(acac)_2]$.[1767] ^{31}P n.m.r. spectroscopy has been used to investigate the reaction of $[AuBr_2]^-$ with $PHPh_2$ at 180 °C. The ^{31}P m.a.s. n.m.r. spectrum was also recorded.[1768]

Mercury. Mercuriation of acetone has been monitored by ^1H and ^{199}Hg n.m.r.

1753 T. Bartik, I. Gerdes, P. Heimbach, and H.-G. Schulte, *J. Organomet. Chem.*, 1989, **367**, 359.
1754 J.R. Ascenso, A.R. Dias, P.T. Gomes, C.C. Romao, Pham Quang Tho, D. Neibecker, and I. Tkatchenko, *Macromolecules*, 1989, **22**, 998 (*Chem. Abstr.*, 1989, **110**, 95 841).
1755 S. Al-Malaika, M. Coker, and G. Scott, *Polym. Degrad. Stab.*, 1988, **22**, 147 (*Chem. Abstr.*, 1989, **110**, 96 438).
1756 G.N. Glavee, L.M. Daniels, and R.J. Angelici, *Organometallics*, 1989, **8**, 1856.
1757 P.K. Byers, A.J. Canty, R.T. Honeyman, and A.A. Watson, *J. Organomet. Chem.*, 1989, **363**, C22.
1758 K.-T. Aye, A.J. Canty, M. Crespo, R.J. Puddephatt, J.D. Scott, and A.A. Watson, *Organometallics*, 1989, **8**, 1518.
1759 D.C. Griffiths and G.B. Young, *Organometallics*, 1989, **8**, 875.
1760 P.J. Stang, M.H. Kowalski, M.D. Schiavelli, and D. Longford, *J. Am. Chem. Soc.*, 1989, **111**, 3347.
1761 P.J. Stang and M.H. Kowalski, *J. Am. Chem. Soc.*, 1989, **111**, 3356.
1762 N.I. Al-Salim and W.R. McWhinnie, *Polyhedron*, 1989, **8**, 2769.
1763 R.U. Kirss and R. Eisenberg, *Inorg. Chem.*, 1989, **28**, 3372.
1764 R. Colton and J. Ebner, *Inorg. Chem.*, 1989, **28**, 1559.
1765 P. Bergamini, S. Sostero, O. Traverso, T.J. Kemp, and P.G. Pringle, *J. Chem. Soc., Dalton Trans.*, 1989, 2017.
1766 P.F. Kelly and J.D. Woollins, *Polyhedron*, 1989, **8**, 2907.
1767 R.B. Klassen and T.H. Baum, *Organometallics*, 1989, **8**, 2477.
1768 D.B. Dyson, R.V. Parish, C.A. McAuliffe, R.G. Pritchard, R. Fields, and B. Beagley, *J. Chem. Soc., Dalton Trans.*, 1989, 907.

spectroscopy.[1769]

Boron. ^{11}B n.m.r. spectroscopy has been used to investigate the pyrolysis of $[B_3H_8]^-$.[1770] ^{11}B n.m.r. spectroscopy has been used to investigate the thermal decomposition of $[B_9H_{13}]^{2-}$.[1771] The fluoridation of $[BCl_3]$ by $CFCl_3$ has been monitored by ^{11}B n.m.r. spectroscopy.[1772] Redistribution reactions of four coordinate haloboron cations from R_3N adducts of mixed boron trihalides have been studied by ^{11}B and ^{19}F n.m.r. spectroscopy.[1773] ^{11}B n.m.r. spectroscopy has been used to investigate reactions of $[B(OH)_4]^-$ or $[B(OH)_3]$ with H_2O_2 as a function of pH and a number of peroxoborate species were identified.[1774] The decomposition of peroxyborates has been studied by n.m.r. spectroscopy.[1775]

Aluminium. ^1H and ^{13}C CIDNP has been observed for the reaction of $[R_3Al]$ with PhC(O)O_2C(O)Ph.[1776] The mechanism of reaction of $[(EtO)_2AlEt]$ and $[XeF_2]$ has been studied by ^1H and ^{19}F CIDNP.[1777] The prolonged hydrolysis of Al^{3+} with aluminium metal has been monitored by ^1H and ^{27}Al n.m.r. spectroscopy.[1778] The hydrolysis of $[Al(OPr^i)_3]$ has been followed by ^{27}Al n.m.r. spectroscopy.[1779] ^{27}Al n.m.r. spectroscopy has been used to investigate the effect of Al^{3+} on the polymerisation of silicic acid in aqueous solution and on the deposition of silica.[1780] The liquid phase of aluminosilicates at different crystallisation lives has been investigated by ^{27}Al and ^{29}Si n.m.r. spectroscopy.[1781]

Silicon. ^1H, ^{13}C, and ^{29}Si n.m.r. spectroscopy has been used to monitor the pyrolysis of 1-allyl-1-silacyclopent-3-enes.[1782] The dilution effect on the crystallisation kinetics of dimethylsiloxane in toluene has been studied by using ^1H n.m.r. spectroscopy.[1783] The kinetics of thermal decomposition of $[Bu_3SiO_3H]$ have been determined.[1784] The initial stages of $[Si(OMe)_4]$ sol-gel formation have been studied by ^{29}Si n.m.r. spectroscopy.[1785]

1769 F.A. Johnson and W.D. Perry, *Organometallics*, 1989, **8**, 2646.

1770 B. Bonnetot, B. Frange, H. Mongeot, S. El Shamy, A. Ouassas, and C. R'Kha, *Bull. Soc. Chim. Fr.*, 1989, 632.

1771 T.D. Getman, J.A. Krause, P.M. Niedenzu, and S.G. Shore, *Inorg. Chem.*, 1989, **28**, 1507.

1772 D.A. Saulys, J. Castillo, and J.A. Morrison, *Inorg. Chem.*, 1989, **28**, 1619.

1773 M.J. Farquharson and J.S. Hartman, *Can. J. Chem.*, 1989, **67**, 1711.

1774 J. Flanagan, W.P. Griffith, R.D. Powell, and A.P. West, *J. Chem. Soc., Dalton Trans.*, 1989, 1651.

1775 B.N. Chernyshov, A.Yu. Vaseva, O.V. Brovkina, and E.G. Ippolitov, *Dokl. Akad. Nauk SSSR*, 1989, **306**, 1140 (*Chem. Abstr.*, 1989, **111**, 161 373).

1776 R.A. Sadykov and N.M. Shishlov, *J. Organomet. Chem.*, 1989, **369**, 1.

1777 R.A. Sadykov, V.N. Yakovlev, R.G. Bulgakov, V.P. Kazakov, G.A. Tolstikov, and A.I. Kruppa, *J. Organomet. Chem.*, 1989, **377**, 1.

1778 J.W. Akitt, J.M. Elders, X.L.R. Fontaine, and A.K. Kundu, *J. Chem. Soc., Dalton Trans.*, 1989, 1889.

1779 K. Maeda, F. Mizukami, M. Watanabe, N. Arai, S. Niwa, M. Toba, and K. Shimizu, *Shokubai*, 1989, **31**, 132 (*Chem. Abstr.*, 1989, **111**, 13 017).

1780 T. Yokoyama, Y. Takahashi, C. Yamanaka, and T. Tarutani, *Geothermics*, 1989, **18**, 321 (*Chem. Abstr.*, 1989, **110**, 234 566).

1781 E. Popovici, *Rev. Chim. (Bucharest)*, 1989, **40**, 115 (*Chem. Abstr.*, 1989, **111**, 68 149).

1782 J.-P. Béteille, M.P. Clarke, I.M.T. Davidson, and J. Dubac, *Organometallics*, 1989, **8**, 1292.

1783 G. Feio, G. Buntinx, and J.P. Cohen-Addad, *J. Polym. Sci., Part B: Polym. Phys.*, 1989, **27**, 1 (*Chem. Abstr.*, 1989, **110**, 136 061).

1784 B.I. Tarunin, V.N. Tarunina, and Yu.A. Kurskii, *Zh. Obshch. Khim.*, 1988, **58**, 1060 (*Chem. Abstr.*, 1989, **110**, 114 921).

1785 R.A. Assink and B.D. Kay, *Mater. Res. Soc. Symp. Proc.*, 1988, **121**, 25 (*Chem. Abstr.*, 1989, **110**, 119

^1H n.m.r. spectroscopy has been used to monitor the hydrolysis of $[Si(OEt)_4]$.[1786] A similar study has used ^{29}Si n.m.r. spectroscopy.[1787] ^{29}Si n.m.r. spectroscopy has been used to study the hydrolysis of $[Si(OEt)_4]$ in the presence of $[Ti(OEt)_4]$.[1788] The hydrolysis of $[Si(OR)_4]$ in the presence of $[AlCl_3]$ has also been studied.[1789] The oligomerisation of silicates has been studied by ^{29}Si n.m.r. spectroscopy.[1790] ^{13}C and ^{29}Si n.m.r. spectroscopy has been used to follow the hydrolysis of $[RSiCl_3]$.[1791]

Tin. The reaction of $[Ph_3C][BF_4]$ with deuterium labelled ethyl derivatives of tin has been studied by ^1H and ^{13}C n.m.r. spectroscopy.[1792] The deposition of tin from alkaline solutions on hydrolysis has been studied by ^{119}Sn n.m.r. spectroscopy.[1793] The mechanism of $[SnCl_4]$ promoted additions of allylstannanes to aldehydes has been studied using ^{13}C and ^{119}Sn n.m.r. spectroscopy.[1794] Two dimensional ^{119}Sn NOESY n.m.r. spectroscopy has been used to study the hydrolysis of $[SnCl_6]^{2-}$.[1795]

Phosphorus. ^{31}P CIDNP has been observed in the oxidation of $P(OPh)_3$ to $P(O)(OPh)_3$ by $Me_2C(OH)O_3H$.[1796]

Arsenic. The kinetics of the formation of $[AsF_6]^-$ from $[AsF_5(OH)]^-$ and HF have been studied using ^1H and ^{19}F n.m.r. spectroscopy.[1797]

Selenium. ^{77}Se n.m.r. spectroscopy has been used to investigate the disproportionation of $[Se_2Cl_2]$ and $[Se_2Br_2]$.[1798]

Tellurium. The reaction of $[TeF_5]^-$ with amino alcohols has been followed by ^{19}F and ^{125}Te n.m.r. spectroscopy.[1799] ^{19}F and ^{125}Te n.m.r. spectroscopy has been used to investigate the hydrolysis of $[TeClF_5]$.[1800]

925).
1786 V.B. Lazarev, S.G. Ambratsumyan, V.I. Privalov, G.P. Panasyuk, A.I. Mishustin, and A.G. Panasyuk, *Vysokochist. Veshchestva*, 1989, 43 (*Chem. Abstr.*, 1989, **110**, 158 998).
1787 J.C. Pouxviel and J.P. Boilot, *Mater. Res. Soc. Symp. Proc.*, 1988, **121**, 37 (*Chem. Abstr.*, 1989, **110**, 119 926); L.W. Kelts and N.J. Armstrong, *J. Mater. Res.*, 1989, **4**, 423 (*Chem. Abstr.*, 1989, **110**, 159 008).
1788 E.N. Boulos, K.R. Carduner, R.O. Carter, tert., and M.F. Best, *Mater. Res. Soc. Symp. Proc.*, 1988, **121**, 531 (*Chem. Abstr.*, 1989, **110**, 119 949).
1789 J.M. Boulton, K. Jones, and H.G. Emblem, *J. Mater. Sci.*, 1989, **24**, 979 (*Chem. Abstr.*, 1989, **110**, 178 171).
1790 C.T.G. Knight, *Zeolites*, 1989, **9**, 448 (*Chem. Abstr.*, 1989, **111**, 241 409).
1791 F.J. Feher, D.A. Newman, and J.F. Walzer, *J. Am. Chem. Soc.*, 1989, **111**, 1741.
1792 E.V. Uglova, V.V. Dashkova, D.V. Bazhenov, Yu.K. Grishin, and O.A. Reutov, *Metalloorg. Khim.*, 1988, **1**, 903 (*Chem. Abstr.*, 1988, **111**, 78 212).
1793 A. Molenaar and J.W.G. De Bakker, *J. Electrochem. Soc.*, 1989, **136**, 378 (*Chem. Abstr.*, 1989, **110**, 139 656).
1794 G.E. Keck, M.B. Andrus, and S. Castellino, *J. Am. Chem. Soc.*, 1989, **111**, 8136.
1795 X.-A. Mao, X.-Z. You, and A.-B. Dai, *Magn. Reson. Chem.*, 1989, **27**, 836.
1796 E.I. Gol'dfarb, V.V. Shereshovets, R.K. Yanbaev, O.A. Cherkasova, V.D. Komissarov, N.A. Mukmeneva, and G.A. Tolstikov, *Izv. Akad. Nauk SSSR, Ser. Khim.*, 1988, 493 (*Chem. Abstr.*, 1989, **110**, 23 997).
1797 V.N. Plakhotnik, N.F. Towmash, and B.N. Chernyshov, *Koord. Khim.*, 1989, **15**, 344 (*Chem. Abstr.*, 1989, **110**, 219 831).
1798 M. Lamoureux and J. Milne, *Can. J. Chem.*, 1989, **67**, 1936.
1799 Yu.V. Kokunov, V.M. Afanas'ev, M.P. Gustyakova, and Yu.A. Buslaev, *Dokl. Akad. Nauk SSSR*, 1988, **303**, 637 (*Chem. Abstr.*, 1989, **110**, 127 515).
1800 L.J. Lawlor, A. Martin, M.P. Murchie, J. Passmore, and J.C.P. Sanders, *Can. J. Chem.*, 1989, **67**, 1501.

4 Paramagnetic Complexes

In this section, compounds of *d*-block transition elements will be considered first and then those of the lanthanide and actinide elements. Papers concerning the use of paramagnetic complexes as 'shift reagents' are usually omitted.

Six reviews have appeared: 'Nuclear spin relaxation in paramagnetic solution',[1801] 'An introduction to lanthanide shift reagents',[1802] 'Lanthanide chelate complexes as supplementary agents in n.m.r. spectroscopy',[1803] 'The nature of the lanthanide shift reagent-substrate complex',[1804] 'Interaction of monosaccharides with metal ions - n.m.r. analysis of lanthanide complexes',[1805] and 'Binuclear lanthanide(III)-silver(I) n.m.r. shift reagents'.[1806]

Dynamic nuclear polarisation of ^{7}Li n.m.r. signals has been observed.[1807]

The Transition Metals.——^{1}H and ^{2}H n.m.r. spectra of [(η^5-$C_5H_5)_2$M], M = V, Cr, Mn, Co, Ni, have been reported. For all but M = Co, linewidth ratios of 36 to 37 were found in agreement with theory, but [(η^5-$C_5H_5)_2$Co] showed an additional coupling contribution. Primary isotope shifts of 2.33, 1.24, and 1.45 p.p.m. were found for [(η^5-$C_5D_5)_2$-Mn]/[(η^5-$C_5H_5)_2$Mn], [(η^5-$C_5D_5)_2$Co]/[(η^5-$C_5H_5)_2$Co], and [(η^5-$C_5D_5)_2$Ni]/[(η^5-$C_5H_5)_2$Ni] respectively.[1808] Suspensions of sulphonated polystyrene ion exchange resin impregnated with paramagnetic ions have been shown to substantially enhance the T_1^{-1} of water.[1809]

Chromium. ^{2}H n.m.r. spectra have been correlated with torsion angles for some acetate chelate rings for polyamine-*N*-carboxylatochromate(III) complexes. A Karplus like relationship was found.[1810] N.m.r. data have also been reported for [(η^5-C_5Me_5)$CrMe_2$-(PMe_3)], (^{2}H),[1811] [(η^5-C_5Me_5)CrMeCl]$_2$, (^{1}H),[1812] [$CrR_2Cl\{(Me_2PCH_2)_3CMe\}$], (^{1}H),[1813] [($\eta^5$-$C_5H_4R$)Cr($\mu$,$\eta^5$-$As_5$)Cr($\eta^5$-$C_5H_4R$)], (^{1}H),[1814] [($\eta^5$-$C_5Me_5$)$CrX_2$]$_2$, (^{1}H),[1815] and Cr(III) acetates, (^{1}H).[1816]

1801 Q. Lin, *Ziran Zazhi*, 1989, **12**, 177 (*Chem. Abstr.*, 1989, **111**, 125 247).
1802 T.C. Morrill, *Methods Stereochem. Anal.*, 1986, **5**, 1 (*Chem. Abstr.*, 1989, **111**, 56 604).
1803 K. Peters, K. Doerffel, Z. Ksandr, and R. Radeglia, *Wiss. Z. Tech. Hochsch. "Carl Schorlemmer" Leuna-Merseburg*, 1989, **31**, 172 (*Chem. Abstr.*, 1989, **111**, 107 420).
1804 D.J. Raber, *Methods Stereochem. Anal.*, 1986, **5**, 55 (*Chem. Abstr.*, 1989, **110**, 106 879).
1805 K. Izumi, *Kagaku Kogyo*, 1988, **39**, 918 (*Chem. Abstr.*, 1989, **110**, 87 263).
1806 T.J. Wenzel, *Methods Stereochem. Anal.*, 1986, **5**, 151 (*Chem. Abstr.*, 1989, **110**, 127 144).
1807 D.J. Singel, H. Seidel, R.D. Kendrick, and C.S. Yannoni, *J. Magn. Reson.*, 1989, **81**, 145.
1808 N. Hebendanz, F.H. Köhler, F. Scherbaum, and B. Schlesinger, *Magn. Reson. Chem.*, 1989, **27**, 798.
1809 J.H. Braybrook and L.D. Hall, *J. Chem. Soc., Chem. Commun.*, 1989, 396.
1810 S. Kaizaki and M. Hayashi, *J. Chem. Soc., Dalton Trans.*, 1989, 1947.
1811 A. Grohmann, F.H. Köhler, G. Müller, and H. Zeh, *Chem. Ber.*, 1989, **122**, 897.
1812 D.S. Richeson, J.F. Mitchell, and K.H. Theopold, *Organometallics*, 1989, **8**, 2570.
1813 E.G. Thaler, K. Folting, J.C. Huffman, and K.G. Caulton, *J. Organomet. Chem.*, 1989, **376**, 343.
1814 O.J. Scherer, W. Wiedemann, and G. Wolmershäuser, *J. Organomet. Chem.*, 1989, **361**, C11.
1815 F.H. Köhler, J. Lachmann, G. Müller, H. Zeh, H. Brunner, J. Pfauntsch, and J. Wachter, *J. Organomet. Chem.*, 1989, **365**, C15.
1816 J.E. Tackett, *Appl. Spectrosc.*, 1989, **43**, 490 (*Chem. Abstr.*, 1989, **110**, 239 381).

Molybdenum and Tungsten. The ^{17}O n.m.r. spectra of paramagnetic heteropoly blue anions such as $[PM_{12}O_{40}]^{3-}$, M = Mo, W, have been discussed.[1817] ^{183}W n.m.r. isotropic shifts have been measured for α-$[CoO_4W_{12}O_{36}]^{6-}$.[1818] N.m.r. data have also been reported for $[W(NC_5H_4\text{-}2\text{-}CH{=}NR)Br_4]$, ($^1H$),[1819] $[WCl_2(OR)_2(PMe_2Ph)_2]$, (^1H),[1820] $[W(OC_6H_3\text{-}Cl_2\text{-}2,6)_4Cl(NC_5H_5)]$, ($^1H$, ^{13}C),[1821] $[(PhS)_2FeS_2MS_{2-n}O_n]^{2-}$, (M = Mo, W; ^1H),[1822] $[MoFe_3S_4\{1,3,5\text{-}(4,6\text{-}Me_2\text{-}3\text{-}SC_6H_2S)_3\text{-}2,4,6\text{-}(4\text{-}MeC_6H_4S)_3C_6\}(dmpe)(SEt)]^-$, ($^1H$),[1823] $[Mo_2Fe_7S_8(SR)_{12}]$, (^1H),[1824] and $[Mo_2Fe_7S_8(\mu\text{-}SEt)_6(cysteinate)_6]^{3-}$, ($^1H$).[1825]

Manganese. Mn(II) *N,N'*-dipyridoxylethylenediamine-*N,N'*-dicetate 5,5'-bisphosphate has been proposed for magnetic resonance image enhancement.[1826] The effect of D_2O on ^7Li and ^{133}Cs T_1 values in the presence of $MnCl_2$ or $CoCl_2$ has been determined.[1827] N.m.r. data have also been reported for $[Mn_2\{4\text{-}MeC_6H_2(O\text{-}1)\text{-}2,6\text{-}[CH_2N(CH_2\text{-}2\text{-}C_5H_4N)_2]_2\}(\mu\text{-}O_2CCH_3)_2]^+$, ($^1H$),[1828] $[Mn\{Mn(salicylhydroximate)\}_4(O_2CCH_3)_2(dmf)_6]$, ($^1H$),[1829] $[Mn_2(N,N'\text{-}disalicylidene\text{-}2\text{-}hydroxypropanediamine)_2]^+$, ($^1H$),[1830] $[Mn\{ethylene(2\text{-}HOC_6H_4\text{-}glycine)_2\}]^-$, ($^1H$),[1831] and $[Mn(S_2C_6H_4)_2]^-$, (^1H).[1832]

Rhenium. The population of the paramagnetic excited state for $[Re(octaethylporphyrin)]_2$ has been estimated from ^1H n.m.r. spectroscopy.[1833] N.m.r. data have also been reported for $[Re_2Fe_7S_8(SEt)_{12}]^{2-}$, ($^1H$).[1834]

Iron. Electron delocalisation into ligands has been studied for some Fe(II) and Fe(III) complexes of imidazole ligands.[1835] The ^1H n.m.r. spectra of a variety of low- and high-spin Fe(III) complexes of pyropheophorbide *a* methyl ester have been determined.[1836]

1817 K. Piepgrass, J.N. Barrows, and M.T. Pope, *J. Chem. Soc., Chem. Commun.*, 1989, 10.
1818 R. Acerate, N. Casañ-Pastor, J. Bas-Serra, and L.C.W. Baker, *J. Am. Chem. Soc.*, 1989, **111**, 6049.
1819 G.-E. Matsubayashi, T. Ohara, and T. Tanaka, *Inorg. Chim. Acta*, 1989, **161**, 67.
1820 J.L. Kerschner, P.E. Fanwick, I.P. Rothwell, and J.C. Huffman, *Inorg. Chem.*, 1989, **28**, 780.
1821 R.M. Kolodziej, R.R. Schrock, and J.C. Dewan, *Inorg. Chem.*, 1989, **28**, 1243.
1822 D. Wu, Z. Gou, and H. Liu, *Bopuxue Zazhi*, 1989, **6**, 111 (*Chem. Abstr.*, 1989, **111**, 129 704).
1823 S. Ciurli and R.H. Holm, *Inorg. Chem.*, 1989, **28**, 1685.
1824 B. Kang, H. Liu, J. Cai, L. Huang, Q. Liu, D. Wu, L. Weng, and J. Lu, *Transition Met. Chem. (Weinheim, Ger.)*, 1989, **14**, 427.
1825 D.J. Evans, G.J. Leigh, and J.B. Parra-Soto, *Polyhedron*, 1989, **8**, 1865.
1826 S.M. Rocklage, W.P. Cacheris, S.C Quay, F.E. Hahn, and K.N. Raymond, *Inorg. Chem.*, 1989, **28**, 477.
1827 I.E. Ismaev, R.K. Mazitov, A.V. Il'yasov, and A.F. Borina, *Izv. Akad. Nauk SSSR, Ser. Khim.*, 1988, 2428 (*Chem. Abstr.*, 1989, **110**, 146 432).
1828 H. Diril, H.-R. Chang, M.J. Nilges, X. Zhang, J.A. Potenza, H.J. Schugar, S.S. Isied, and D.N. Hendrickson, *J. Am. Chem. Soc.*, 1989, **111**, 5102.
1829 M.S. Lah and V.L. Pecoraro, *J. Am. Chem. Soc.*, 1989, **111**, 7258.
1830 J.A. Bonadies, M.J. Maroney, and V.L. Pecoraro, *Inorg. Chem.*, 1989, **28**, 2044.
1831 X. Li and V.L. Pecoraro, *Inorg. Chem.*, 1989, **28**, 3403.
1832 K. Greiwe, B. Krebs, and G. Henkel, *Inorg. Chem.*, 1989, **28**, 3713.
1833 J.P. Collman, J.M. Garner, and L.K. Woo, *J. Am. Chem. Soc.*, 1989, **111**, 8141.
1834 S. Ciurli, M.J. Carney, R.H. Holm, and G.C. Papaefthymiou, *Inorg. Chem.*, 1989, **28**, 2696.
1835 F.J. Wu and D.M. Kurtz, jun., *J. Am. Chem. Soc.*, 1989, **111**, 6563.
1836 S. Licoccia, M.J. Chatfield, G.N. La Mar, K.M. Smith, K.E. Mansfield, and R.R. Anderson, *J. Am. Chem. Soc.*, 1989, **111**, 6087.

^{15}N n.m.r. resonances of imidazole bound to Fe(III) porphyrin have been located.[1837] The spin equilibrium in Fe(III) octaethylporphyrin complexes has been studied.[1838] F^- coordination to Fe(II) tetraphenylporphyrin has been investigated.[1839] The metal centred pseudo-contact shift contribution in [Fe(protoporphyrin)$(CN)_2$] has been determined from ^1H and ^{13}C n.m.r. measurements.[1840] ^1H n.m.r. spectroscopy has been used to study Fe(II) haem in micellar aqueous solution and isotropic shifts determined.[1841] The coordination environment of protonated forms of Fe(III) catechoylamides has been probed by ^2H n.m.r. spectroscopy.[1842] N.m.r. data have also been reported for [Fe(porphyrin)CHR_2], (^1H, ^2H),[1843] $[\{\eta^5\text{-}(PhCH_2)_2C_2B_4H_4\}Fe(\mu,\eta^5\text{-}Et_2MeC_3B_2Et)Co(\eta^5\text{-}C_5H_5)]^-$, ($^1H$, ^{11}B),[1844] $[(\eta^5\text{-}C_5Me_4H)_2Fe]$, ($^1H$),[1845] [Fe(2,4,4,9,11,11-Me_6-1,5,8,12-tetraazacyclotetradeca-1,8-diene)(S_2O_5)], (^1H),[1846] [{(tetraphenylporphyrin)Fe}$_2$O], (^1H),[1847] $[Fe(tetraphenylporphyrin)(CNR)_2]^+$, ($^1H$),[1848] [Fe(tetraphenylporphyrin)$OTeF_5$], (^1H, ^{19}F),[1849] [Fe(tetramesitylporphyrin)O_2Bu^t], (^1H),[1850] myoglobin containing etiohaemin, (^1H),[1851] $[Fe_3O(O_2CCH_3)_6(NC_5H_5)_3]$, ($^2H$),[1852] Fe(III) complex with peptidoglycan, (^1H, ^{13}C),[1853] [Fe(15-crown-5)Cl_2][$FeCl_4$], (^{13}C),[1854] $[Fe_2\{2,6\text{-}[(2\text{-}C_5H_4NCH_2)_2N]_2\text{-}4\text{-}MeC_6H_2O\}(O_2CR)_2]^{2+}$, ($^1H$),[1855] and $[Fe_4S_4(S\text{-}2\text{-}C_6H_4SBu^t)_4]^{3-}$, ($^1H$).[1856]

Ruthenium. The temperature variation of the magnetic susceptibility of $[Ru(bipy)_2(NC_5H_5)O]^{2+}$ has been determined from ^1H n.m.r. measurements.[1857] N.m.r. data have

1837 Y. Yamamoto, N. Nanai, Y. Inoue, and R. Chûjô, *J. Chem. Soc., Chem. Commun.*, 1989, 1419.
1838 L.B. Dugad, V.R. Marathe, and S. Mitra, *Proc. Indian Natl. Sci. Acad., Part A*, 1988, **54**, 769 (*Chem. Abstr.*, 1989, **111**, 125 504).
1839 B.S. Yu and H.M. Goff, *J. Am. Chem. Soc.*, 1989, **111**, 6558.
1840 Y. Yamamoto, N. Nanai, Y. Inoue, and R. Chûjô, *Bull. Chem. Soc. Jpn.*, 1989, **62**, 1771.
1841 O.K. Medhi, S. Mazumdar, and S. Mitra, *Inorg. Chem.*, 1989, **28**, 3243.
1842 M.E. Cass, T.M. Garrett, and K.N. Raymond, *J. Am. Chem. Soc.*, 1989, **111**, 1677.
1843 R.D. Arasasingham, A.L. Balch, C.R. Cornman, and L. Latos-Grazynski, *J. Am. Chem. Soc.*, 1989, **111**, 4357.
1844 M.D. Attwood, K.K. Fonda, R.N. Grimes, G. Brodt, D. Hu, U. Zenneck, and W. Siebert, *Organometallics*, 1989, **8**, 1300.
1845 J.S. Miller, D.T. Glatzhofer, D.M. O'Hare, W.M. Reiff, A. Chakraborty, and A.J. Epstein, *Inorg. Chem.*, 1989, **28**, 2930.
1846 M.S. Reynolds and R.H. Holm, *Inorg. Chim. Acta*, 1989, **155**, 113.
1847 A. Nanthakumar and H.M. Goff, *Inorg. Chem.*, 1989, **28**, 4559.
1848 G. Simonneaux, F. Hindre, and M. Le Plouzennec, *Inorg. Chem.*, 1989, **28**, 823.
1849 P.J. Kellett, M.J. Pawlik, L.F. Taylor, R.G. Thompson, M.A. Levstik, O.P. Anderson, and S.H. Strauss, *Inorg. Chem.*, 1989, **28**, 440.
1850 R.D. Arasasingham, C.R. Cornman, and A.L. Balch, *J. Am. Chem. Soc.*, 1989, **111**, 7800.
1851 S. Neya, N. Funasaki, and K. Imai, *Biochim. Biophys. Acta*, 1989, **996**, 226 (*Chem. Abstr.*, 1989, **111**, 149 079).
1852 Y. Kaneko, M. Nakano, M. Sorai, Ho G. Jang, and D.N. Hendrickson, *Inorg. Chem.*, 1989, **28**, 1067.
1853 M. Tonković, O. Hadžija, B. Ladešić, B Klaić, and S. Musić, *Inorg. Chim. Acta*, 1989, **161**, 81.
1854 K.B. Yatsimirskii, E.V. Rybak-Akimova, and G.G. Talanova, *Dokl. Akad. Nauk Ukr. SSR, Ser. B*, 1989, 50 (*Chem. Abstr.*, 1989, **111**, 246 678).
1855 A.S. Borovik, V. Papaefthymiou, L.F. Taylor, O.P. Anderson, and L. Que, jun., *J. Am. Chem. Soc.*, 1989, **111**, 6183.
1856 M.J. Carney, G.C. Papaefthymiou, R.B. Frankel, and R.H. Holm, *Inorg. Chem.*, 1989, **28**, 1497.
1857 J.C. Dobson, J.H. Helms, P. Doppelt, B.P. Sullivan, W.E. Hatfield, and T.J. Meyer, *Inorg. Chem.*, 1989, **28**, 2200.

also been reported for [Ru(octaethylporphyrin)O(ROH)], (^{1}H),[1858] and [Ru(tfac)$_3$], (^{1}H, ^{19}F).[1859]

Cobalt. The electronic structure of substituted cobaltocenes has been deduced from ^{1}H and ^{13}C chemical shifts.[1860] A theory for the electron-nucleus interaction in the Co_4S_{11} cluster in Co(II) reconstituted metallothioneins has been developed to explain ^{1}H chemical shifts.[1861] ^{13}C isotropic shifts have been recorded for a number of $[CoX_4]^{2-}$ salts containing various quaternary ammonium cations.[1862] N.m.r. data have also been reported for [{(η^5-C_5H_5)Co}$_3$(CO)$_2$], (^{1}H),[1863] [Co{OC[CMe$_2$NC(O)CMe$_2$NCMe$_2$]$_2$CO}]$^-$, (^{1}H),[1864] [Co{HB(3-Pr^{i}pz)$_2$(5-Pr^{i}pz)}$_2$], (^{1}H),[1865] Co(II) substituted carbonic anhydrase, (^{1}H),[1866] and [M(OSR$_2$)(OH$_2$)$_5$]$^{2+}$, (M = Co, Ni; ^{1}H).[1867]

Rhodium. The temperature dependence of the ^{1}H n.m.r. signals from [(η^6-C_6Me_6)Rh-(η^5-C_5Me_5)] is consistent with a doublet ground state.[1868]

Nickel. Spin delocalisation into the alkyl chains of [Ni(bipy)(S$_2$COR)$_2$] has been measured by ^{1}H and ^{13}C n.m.r. spectroscopy.[1869] The combined use of ^{1}H and ^{13}C T_1 and contact shift measurements has been used to investigate the conformational and dynamic properties of nickel acetate and propionate complexes.[1870]

Copper. Selective line broadening of paramagnetic Cu^{2+} on ^{13}C linewidths of nitrogen compounds has been studied.[1871] ^{1}H n.m.r. spectroscopy has been used to study dicopper(II) complexes with binucleating ligands containing imidazoles.[1872] N.m.r. data have also been reported for [Cu(triazatetrabenzoporphyrinato)I], (^{1}H, ^{13}C),[1873] Cu^{2+} complexes with DL-asparagine, (^{1}H, ^{13}C),[1874] Cu^{2+} substituted transferrin, (^{1}H),[1875] and

1858 W.H. Leung and C.M. Che, *J. Am. Chem. Soc.*, 1989, **111**, 8812.
1859 N.B. Morozova, V.N. Mit'kin, I.K. Igumenov, S.V. Zemskov, and O.G. Potapova, *Koord. Khim.*, 1989, **15**, 110 (*Chem. Abstr.*, 1989, **110**, 184 830).
1860 H. Eicher and F.H. Köhler, *Chem. Phys.*, 1988, **128**, 297 (*Chem. Abstr.*, 1989, **111**, 233 134).
1861 I. Bertini, C. Luchinat, L. Messori, and M. Vasak, *J. Am. Chem. Soc.*, 1989, **111**, 7300 (*Chem. Abstr.*, 1989, **111**, 125 509).
1862 N. Chen, P.J. Witton, C.E. Holloway, and I.M. Walker, *J. Coord. Chem.*, 1988, **19**, 113.
1863 C.E. Barnes, J.A. Orvis, D.L. Staley, A.L. Rheingold, and D.C. Johnson, *J. Am. Chem. Soc.*, 1989, **111**, 4992.
1864 T.J. Collins and E.S. Uffelman, *Angew. Chem., Int. Ed. Engl.*, 1989, **28**, 1501.
1865 S. Trofimenko, J.C. Calabrese, P.J. Domaille, and J.S. Thompson, *Inorg. Chem.*, 1989, **28**, 1091.
1866 L. Banci, I. Bertini, C. Luchinat, R. Monnanni, and J. Moratal Mascarell, *Gazz. Chim. Ital.*, 1989, **119**, 23 (*Chem. Abstr.*, 1989, **110**, 169 284).
1867 Sh.A. Markaryan, *Zh. Strukt. Khim.*, 1988, **29**, 70 (*Chem. Abstr.*, 1989, **110**, 68 346).
1868 J. Merkert, R.M. Nielson, M.J. Weaver, and W.E. Geiger, *J. Am. Chem. Soc.*, 1989, **111**, 7084.
1869 S.K. Sheppard, G.R. Eaton, and S.S. Eaton, *Inorg. Chem.*, 1989, **28**, 4496.
1870 M. Casu, A. Lai, C. Meloni, and G. Saba, *J. Mol. Liq.*, 1989, **40**, 25 (*Chem. Abstr.*, 1989, **110**, 127 335).
1871 G. Szalontai, *Monatsh. Chem.*, 1989, **120**, 1049.
1872 M. Maekawa, S. Kitagawa, M. Munakata, and H. Masuda, *Inorg. Chem.*, 1989, **28**, 1904.
1873 K. Liou, M.Y. Ogawa, T.P. Newcomb, G. Quirion, M. Lee, M. Poirier, W.P. Halperin, B.M. Hoffman, and J.A. Ibers, *Inorg. Chem.*, 1989, **28**, 3889.
1874 L. Lomozik and A. Wojciechowska, *Polyhedron*, 1989, **8**, 1.
1875 I. Bertini, C. Luchinat, R.D. Brown, tert., and S.H. Koenig, *J. Am. Chem. Soc.*, 1989, **111**, 3532.

[$Cu(OCH_2CH_2OCH_2CH_2OMe)_2$], ($^1H$).[1876]

Complexes of the Lanthanides and Actinides.——*Lanthanides.* The ^{17}O and ^{31}P pseudocontact shifts of [$Ln(P_3O_{10})_2(OH_2)]^{7-}$ are of opposite signs.[1877] The isotropic shifts of [Ln_2(octaethylporphyrin)$_3$], Ln = Ce, Pr, Nd, Sm, and Eu, have been used to deduce structural information on congestion.[1878] A correlation has been sought between the hypersensitive pseudoquadrupole transitions and lanthanide induced ^1H, ^{13}C, and ^{17}O shifts by Nd^{3+}.[1879] Some Gd Schiff base complexes have been assessed as contrast agents for magnetic resonance imaging.[1880] ^{19}F magnetic resonance imaging with Gd-DTPA has been studied.[1881] A comparison of paramagnetic shifts induced in ^7Li and ^{23}Na by various Dy^{3+} and Tm^{3+} shift reagents has been reported.[1882] The use of shift reagents in ^{23}Na imaging has been reported.[1883] The paramagnetic shifts in the histidine ^1H n.m.r. spectrum have been determined as functions of $CaCl_2$ concentration in the presence of Yb(III).[1884] N.m.r. data have also been reported for lanthanide hydroborate complexes,[1885] [$(\eta^5-C_5Me_5)_2CeOC_6H_3Bu^t_2$-2,6], ($^1H$, ^{13}C),[1886] [$(\eta^5-C_5H_5)_2Ce(NO_3)_2Na(thf)_2$], ($^1H$, ^{13}C),[1887] [Pr_2(tetraphenylimidodiphosphinato)$_4$(3-phenylpropionato)$_2$], (^1H),[1888] [$(\eta^5-C_5Me_5)_2Eu(H_2)$],[1889] [$(\eta^5-C_5Me_5)_2Sm(\eta^3-C_3H_5)$], ($^1H$),[1890] [$(\eta^5-C_5Me_5)_2Sm(OC_5H_{10})$], ($^{13}C$),[1891] [$(\eta^5-C_5Me_5)_2Sm(P{=}CBu^t)]_2$, ($^1H$),[1892] and [$(\eta^5-C_5Me_5)_2YbC_6F_5$], ($^1H$).[1893]

Actinides. ^1H n.m.r. spectroscopy has been used to investigate the stereochemistry of the insertion of cyclic ketones into a U—C bond in [$\{(Me_3Si)_2N\}_2\overline{UCH_2SiMe_2N}$-$SiMe_3$].[1894] N.m.r. data have also been reported for [$(\eta^6$-1,3,5-$Me_3C_6H_3)U(BH_4)_3$],

1876 S.C. Goel, K.S. Kramer, P.C. Gibbons, and W.E. Buhro, *Inorg. Chem.*, 1989, **28**, 3619.
1877 J.A. Peters, A. Sinnema, A.P.G. Kieboom, and H. van Bekkum, *Inorg. Chim. Acta*, 1989, **160**, 7.
1878 J.W. Buchler, M. Kihn-Botulinski, J. Loeffler, and M. Wicholas, *Inorg. Chem.*, 1989, **28**, 3770.
1879 D.F. Mullica, G.A. Wilson, and C.K.C. Lok, *Inorg. Chim. Acta*, 1989, **156**, 159.
1880 P.H. Smith, J.R. Brainard, D.E. Morris, G.D. Jarvinen, and R.R. Ryan, *J. Am. Chem. Soc.*, 1989, **111**, 7437.
1881 A.V. Ratner, S. Quay, H.H. Muller, B.B. Simpson, R. Hurd, and S.W. Young, *Invest. Radiol.*, 1989, **24**, 224 (*Chem. Abstr.*, 1989, **110**, 188 557).
1882 R. Ramasamy, M.C. Espanol, K.M. Long, D.M. de Freitas, and C.F.G.C. Geraldes, *Inorg. Chim. Acta*, 1989, **163**, 41.
1883 J.A. Balschi, S.J. Kohler, J.A. Bittl, C.S. Springer, jun., and J.S. Ingwall, *J. Magn. Reson.*, 1989, **83**, 138.
1884 V.D. Buiklinskii, V.T. Panyushkin, S.L. Bel'skaya, and G.V. Moseiko, *Koord. Khim.*, 1989, **15**, 1000 (*Chem. Abstr.*, 1989, **111**, 181 973).
1885 Z. Ding, T. Wu, L. Liu, and L. Zhang, *Wuji Huaxue*, 1988, **4**, 37 (*Chem. Abstr.*, 1989, **110**, 87 376).
1886 H.J. Heeres, J.H. Teuben, and R.D. Rogers, *J. Organomet. Chem.*, 1989, **364**, 87.
1887 P.S. Gradeff, K. Yunlu, T.J. Deeming, J.M. Olofson, J.W. Ziller, and W.J. Evans, *Inorg. Chem.*, 1989, **28**, 2600.
1888 C. Alvarez, L. Barkaoui, N. Goasdoue, J.C. Daran, N. Platzer, H. Rudler, and J. Vaissermann, *J. Chem. Soc., Chem. Commun.*, 1989, 1507.
1889 S.P. Nolan and T.J. Marks, *J. Am. Chem. Soc.*, 1989, **111**, 8538.
1890 S.P. Nolan, D. Stern, and T.J. Marks, *J. Am. Chem. Soc.*, 1989, **111**, 7844.
1891 W.J. Evans and T.A. Ulibarri, *Polyhedron*, 1989, **8**, 1007.
1892 A. Recknagel, D. Stalke, H.W. Roesky, and F.T. Edelmann, *Angew. Chem., Int. Ed. Engl.*, 1989, **28**, 445.
1893 C.J. Burns and R.A. Andersen, *J. Chem. Soc., Chem. Commun.*, 1989, 136.
1894 A. Dormond, A. Elbouadili, and C. Moïse, *J. Organomet. Chem.*, 1989, **369**, 171.

(^1H),[1895] [(η^5-2,4-$Me_2C_5H_5)_2U(BH_4)_2$], (^1H),[1896] [(η^5-6,6-$Me_2C_6H_5)_2U(BH_4)_2]^-$, ($^1H$),[1897] [Li(tmeda)$_2$][$Li_2$(tmeda)$_2$($\mu$-$\eta^5$,$\eta^5$-$MeC_5H_4$)][($\eta^5$-$C_5H_4Me)_6U_2(\mu$-Me)]$_2$, ($^1H$),[1898] [($\eta^5$-$C_5H_5)_3U(SiPh_3)$], ($^1H$),[1899] [($\eta^5$-$C_5H_5)_3UCl]^-$, ($^1H$),[1900] [($\eta^5$-$C_5H_5)_3UX]^-$, ($^1H$),[1901] [($\eta^5$-$C_5Me_5)_2UCl_2(HNPPh_3)$], ($^1H$),[1902] [U{1,1,3-$Me_3$-2-($CH_2SiMe_2NSiMe_2$)-2-O}{N($SiMe_3)_2$}$_2$], ($^1H$),[1903] [U{($Me_3SiN)_2PPh_2$}$_2Cl_2$], ($^1H$),[1904] [U($OC_6H_3Bu^t_2$-2,6)$_4$], ($^1H$),[1905] [($Bu^t_3CO)_2UCl_2(thf)_2$], ($^1H$),[1906] and [$UI_3(MeOCH_2CH_2OMe)_2$], ($^1H$).[1907]

5 Solid-state N.M.R. Spectroscopy

This section consists of three main parts: 'Motion in Solids', 'Structure of Solids', and 'Molecules Sorbed Onto Solids'.

A book entitled 'Nuclear magnetic resonance spectroscopy of water in heterogeneous systems',[1908] and a number of reviews have appeared: 'Characterisation of materials by ^{13}C and ^{29}Si n.m.r. spectroscopy',[1909] 'Nuclear magnetic resonance techniques for the study of mechanisms of diffusion in solids',[1910] 'Magnetic resonance of functional ceramics',[1911] 'N.m.r. relaxation mechanisms and linewidths in insulators below 1K',[1912] 'Displacive order-disorder crossover in perovskite and antifluorite crystals undergoing rotational phase transitions',[1913] '^{17}O nuclear magnetic resonance spectroscopic study of high T_c superconductors',[1914] 'Nuclear magnetic resonance in high T_c superconductors',[1915] 'N.m.r. in high T_c oxide superconductors',[1916] 'N.m.r.

1895 D. Baudry, M. Bulot, P. Charpin, M. Ephritikhine, M. Lance, M. Nierlich, and J. Vigner, *J. Organomet. Chem.*, 1989, **371**, 155.
1896 D. Baudry, E. Bulot, P. Charpin, M. Ephritikhine, M. Lance, M. Nierlich, and J. Vigner, *J. Organomet. Chem.*, 1989, **371**, 163.
1897 D. Baudry, E. Bulot, and M. Ephritikhine, *J. Chem. Soc., Chem. Commun.*, 1989, 1316.
1898 S.D. Stults, R.A. Andersen, and A. Zalkin, *J. Am. Chem. Soc.*, 1989, **111**, 4507.
1899 M. Porchia, N. Brianese, U. Casellato, F. Ossola, G. Rossetto, P. Zanella, and R. Graziani, *J. Chem. Soc., Dalton Trans.*, 1989, 677.
1900 J.-F. Le Maréchal, C. Villiers, P. Charpin, M. Lance, M. Nierlich, J. Vigner, and M. Ephritikhine, *J. Chem. Soc., Chem. Commun.*, 1989, 308.
1901 J.F. Le Maréchal, C. Villiers, P. Charpin, M. Nierlich, M. Lance, J. Vigner, and M. Ephritikhine, *J. Organomet. Chem.*, 1989, **379**, 259.
1902 R.E. Cramer, S. Roth, and J.W. Gilje, *Organometallics*, 1989, **8**, 2327.
1903 A. Dormond, C. Moïse, A. Elbouadili, and H. Bitar, *J. Organomet. Chem.*, 1989, **371**, 175.
1904 A. Recknagel, M. Witt, and F.T. Edelmann, *J. Organomet. Chem.*, 1989, **371**, C40.
1905 W.G. Van der Sluys, A.P. Sattelberger, W.E. Streib, and J.C. Huffman, *Polyhedron*, 1989, **8**, 1247.
1906 C. Baudin and M. Ephritikhine, *J. Organomet. Chem.*, 1989, **364**, C1.
1907 D.L. Clark, A.P. Sattelberger, S.G. Bott, and R.N. Vrtis, *Inorg. Chem.*, 1989, **28**, 1771.
1908 V.V. Mank and N.N. Lebovka, 'Nuclear Magnetic Resonance Spectroscopy of Water in Heterogeneous Systems' (Naukova Dunka: Kiev, USSR), 1988, 201 pp. (Russ) (*Chem. Abstr.*, 1989, **111**, 49 168).
1909 M. Imanari and K. Hioka, *Zairyo Kagaku*, 1989, **25**, 252 (*Chem. Abstr.*, 1989, **111**, 16 375).
1910 A.V. Chadwick, *Int. Rev. Phys. Chem.*, 1988, **7**, 251 (*Chem. Abstr.*, 1989, **110**, 141 706).
1911 W. Windsch, G. Voelkel, B. Milsch, and J. Haase, *Silikattechnik*, 1989, **40**, 234 (*Chem. Abstr.*, 1989, **111**, 238 124).
1912 O. Gonen and J.S. Waugh, *Physica A (Amsterdam)*, 1989, **156**, 219 (*Chem. Abstr.*, 1989, **111**, 145 323).
1913 R.L. Armstrong, *Prog. Nucl. Magn. Reson. Spectrosc.*, 1988, **21**, 151.
1914 E. Oldfield, C. Coretsopoulos, S. Yang, L. Reven, H.C. Lee, J. Shore, O.H. Han, E. Ramli, and D. Hinks, *Phys. Rev. B: Condens. Matter*, 1989, **40**, 6832 (*Chem. Abstr.*, 1989, **111**, 246 304).
1915 M. Mehring, *IBM J. Res. Dev.*, 1989, **33**, 342 (*Chem. Abstr.*, 1989, **111**, 145 330).
1916 Y. Kitaoka, K. Ishida, K. Fujiwara, Y. Kohori, K. Asayama, H. Katayama-Yoshida, Y. Okabe, and

study of magnetism and superconductivity in high T_c oxides', which includes ^{17}O, ^{63}Cu, and ^{139}La n.m.r. spectra,[1917] 'Magnetism and superconductivity by n.m.r. study',[1918] 'Study of superconduction by n.m.r.',[1919] 'Microscopic study of high-transition temperature superconductive oxides by n.m.r. methods',[1920] 'Nuclear magnetic and nuclear quadrupole resonance studies in high temperature superconductors',[1921] 'Nuclear resonance studies of $YBa_2Cu_3O_{7-\delta}$',[1922] 'Charge differentiation of inequivalent copper sites in $YBa_2Cu_3O_{7-\delta}$ investigated by n.q.r. and n.m.r.',[1923] 'Nuclear resonance studies in $YBa_2Cu_3O_{7-\delta}$ and $Tl_2Ba_2Ca_2Cu_3O_{10+x}$',[1924] '^{29}Si and ^{31}P n.m.r. investigations of the structures of NASICON compounds',[1925] 'Magnetic resonance and electronic spectroscopy of $M_2P_2S_6$ intercalation compounds',[1926] 'Spin-1 behaviour of systems of dipolar coupled pairs of spin $\frac{1}{2}$ nuclei',[1927] '^{31}P n.m.r. investigations of batch reactions and recrystallisation phenomena of fluoroaluminate glasses',[1928] 'Recent advances in the study of fast ionically conducting glasses using n.m.r. techniques',[1929] 'Structural characterization of non-oxide chalcogenide glasses using solid state n.m.r.',[1930] 'Highly resolved proton magnetic resonance studies on the acidity of zeolites and related catalysts',[1931] 'Applications of n.m.r. spectroscopy to the study of zeolite synthesis',[1932] 'N.m.r. spectroscopy. A unique technique for zeolites',[1933] 'Solid state n.m.r. studies of zeolite catalysts',[1934] 'Brønsted and Lewis acidity of zeolites and related catalysts studied by n.m.r. spectroscopy',[1935] 'Magnetic resonance as a tool for graphite intercalation compound studies',[1936] '^{29}Si n.m.r. determinations of silicon-oxygen-silicon bond angle distributions in silica',[1937] '^{29}Si n.m.r. as a structural probe for silicate glasses',[1938]

T. Takahashi, *Springer Ser. Mater. Sci.*, 1989, **11**, 148 (*Chem. Abstr.*, 1989, **111**, 144 642).
1917 Y. Kitaoka, K. Ishida, K. Fujiwara, K. Asayama, H. Katayama-Yoshida, Y. Okabe, and T. Takahashi, *IBM J. Res. Dev.*, 1989, **33**, 277 (*Chem. Abstr.*, 1989, **111**, 124 468).
1918 K. Asayama, Y. Kitaoka, and Y. Kohori, *J. Phys. Colloq.*, 1988, (C8, Proc. Int. Congr. Magn., Pt 3), C8-2125/C8-2129 (*Chem. Abstr.*, 1989, **111**, 124 441).
1919 K. Asayama, *Seisan to Gijutsu*, 1987, **39**, 45 (*Chem. Abstr.*, 1989, **110**, 240 731).
1920 H. Yasuoka, *Funtai oyobi Funmatsu Yakin*, 1988, **35**, 457 (*Chem. Abstr.*, 1989, **110**, 80 935).
1921 L. Mihaly, *Phys. Scr., T*, 1989, **T25**, 80 (*Chem. Abstr.*, 1989, **110**, 241 323).
1922 R.E. Walstedt and W.W. Warren, jun., *Springer Ser. Mater. Sci.*, 1989, **11**, 137 (*Chem. Abstr.*, 1989, **111**, 144 641).
1923 H. Yasuoka, *Springer Ser. Mater. Sci.*, 1989, **11**, 156 (*Chem. Abstr.*, 1989, **111**, 144 643).
1924 R. Vijayaraghavan, A.K. Rajarajan, and L.C. Gupta, *Stud. High Temp. Supercond.*, 1989, **2**, 79 (*Chem. Abstr.*, 1989, **111**, 206 251).
1925 C. Jaeger, G. Scheler, S. Barth, and A. Feltz, *Exp. Tech. Phys.*, 1988, **36**, 339.
1926 E. Lifshitz and A.H. Francis, *NATO ASI Ser., Ser. B*, 1987, **172**, 425 (*Chem. Abstr.*, 1989, **110**, 181 721).
1927 A. Keller, *Adv. Magn. Reson.*, 1988, **12**, 183 (*Chem. Abstr.*, 1989, **110**, 87 212).
1928 C. Jaeger and D. Ehrt, *Exp. Tech. Phys.*, 1988, **36**, 349 (*Chem. Abstr.*, 1989, **110**, 100 169).
1929 S.W. Martin, *Mater. Chem. Phys.*, 1989, **23**, 225 (*Chem. Abstr.*, 1989, **111**, 164 800).
1930 H. Eckert, *Angew. Chem., Int. Ed. Engl.*, 1989, **28**, 1723.
1931 H. Ernst, *Z. Phys. Chem. (Leipzig)*, 1988, **269**, 1073 (*Chem. Abstr.*, 1989, **110**, 122 264).
1932 A.T. Bell, *ACS Symp. Ser.*, 1989, **398**, 66 (*Chem. Abstr.*, 1989, **111**, 181 265).
1933 B. Norden and U. Edlund, *Kem. Tidskr.*, 1989, **101**, 129 (*Chem. Abstr.*, 1989, **111**, 246 292).
1934 J. Klinowski, *Colloids Surf.*, 1989, **36**, 133 (*Chem. Abstr.*, 1989, **111**, 121 634).
1935 H. Pfeifer, *Colloids Surf.*, 1989, **36**, 169 (*Chem. Abstr.*, 1989, **111**, 121 636).
1936 J. Conard, *NATO ASI Ser., Ser. B*, 1987, **172**, 357 (*Chem. Abstr.*, 1989, **111**, 16 371).
1937 R.F. Pettifer, R. Dupree, I. Farnan, and U. Sternberg, *J. Non-Cryst. Solids*, 1988, **106**, 408 (*Chem. Abstr.*, 1989, **110**, 101 997).
1938 R. Dupree, *Exp. Tech. Phys.*, 1988, **36**, 315 (*Chem. Abstr.*, 1989, **110**, 100 166).

'Ionically conductive glasses based on SiS_2',[1939] 'Nuclear magnetic resonance studies of simple molecules on metal surfaces',[1940] 'The main limitations for an application of n.m.r. pulsed field gradient technique of self-diffusion studies in zeolites',[1941] 'Molecular motion in zeolite sorbends, studied by various n.m.r. methods',[1942] 'High-resolution nuclear magnetic resonance spectroscopy of adsorbed molecules',[1943] and '^{195}Pt n.m.r. studies of supported catalysts'.[1944]

Motion in Solids.——The theory of spin-lattice relaxation of diffusing light nuclei in glasses has been published.[1945]

Li^+ migration in laponite has been studied using ^7Li, ^{27}Al, and ^{29}Si n.m.r. spectroscopy.[1946] ^7Li n.m.r. spectroscopy has been used to study ionic conduction in composite solid electrolytes such as LiI.[1947] Ion dynamics in $Li_{2-x}Na_xSO_4$ have been studied by ^7Li and ^{23}Na n.m.r. spectroscopy.[1948] ^7Li T_1 has been measured to study Li^+ diffusion in B_2S_3-Li_2S-LiX glasses.[1949] ^2H n.m.r. spectroscopy has been used to study D_2O dynamics in glassy and liquid $LiCl.H_2O$.[1950] T_1 of ^7Li and ^{127}I in LiI has been used to investigate Li^+ motion.[1951]

^{23}Na n.m.r. spectroscopy has been used to demonstrate motional narrowing due to Na^+ motion in cation conducting polymer electrolytes.[1952] The reorientation of water in sodium phosphoenolpyruvate[1953] and the disodium salt of glucose 1-phosphate[1954] has been studied by ^1H, ^{23}Na, and ^{31}P n.m.r. spectroscopy. The dislocation jump distance during creep of NaCl single crystals has been studied by ^{23}Na n.m.r. spectroscopy.[1955]

1939 J.H. Kennedy, *Mater. Chem. Phys.*, 1989, **23**, 29 (*Chem. Abstr.*, 1989, **111**, 206 292).
1940 C.P. Slichter, *AIP Conf. Proc.*, 1988, **180**, 25 (*Chem. Abstr.*, 1989, **111**, 202 415).
1941 C. Foerste, W. Heink, J. Kaerger, H. Pfeifer, N.N. Feokistova, and S.P. Zhdanov, *Zeolites*, 1989, **9**, 299 (*Chem. Abstr.*, 1989, **111**, 121 479).
1942 H. Lechert and W.D. Basler, *J. Phys. Chem. Solids*, 1989, **50**, 497 (*Chem. Abstr.*, 1989, **111**, 84 522).
1943 W. Meiler and T. Wutscherk, *Isotopenpraxis*, 1989, **25**, 41 (*Chem. Abstr.*, 1989, **111**, 207 734).
1944 J.P. Bucher, J. Buttet, J.J. Van der Klink, M. Graetzel, E. Newson, and T.B. Truong, *Colloids Surf.*, 1989, **36**, 155 (*Chem. Abstr.*, 1989, **111**, 121 635).
1945 W. Schirmacher and A. Schirmer, *Solid State Ionics*, 1987, **28-30 (Pt. 1)**. 134 (*Chem. Abstr.*, 1989, **110**, 127 301).
1946 V. Luca, C.M. Cardile, and R.H. Meinhold, *Clay Miner.*, 1989, **24**, 115 (*Chem. Abstr.*, 1989, **111**, 81 501).
1947 T. Asai, C.H. Hu, H. Yumoto, and S. Kawai, *Mem. Inst. Sci. Ind. Res., Osaka Univ.*, 1989, **46**, 71 (*Chem. Abstr.*, 1989, **111**, 144 709).
1948 K.D. Junke, M. Mali, J. Roos, D. Brinkmann, A. Lunden, and B. Graneli, *Solid State Ionics, 1987* (Pub. 1988), **28-30(Pt. 2)**, 1287 (*Chem. Abstr.*, 1989, **110**, 184 619).
1949 M. Grüne, H. Meierkord, W. Müller-Warmuth, P. zum Hebel, B. Krebs, and M. Wulff, *Ber. Bunsenges. Phys. Chem.*, 1989, **93**, 1313.
1950 A. Hagemeyer, O. Kanert, G. Balzer-Joellenbeck, and H. Jain, *Phys. Rev. B: Condens. Matter*, 1989, **39**, 15 (*Chem. Abstr.*, 1989, **110**, 160 688).
1951 M. Mali, J. Roos, D. Brinkmann, J.B. Phipps, and P.M. Skarstad, *Solid State Ionics*, 1987, (Pub. 1988), **28-30 (Pt. 2)**, 1089 (*Chem. Abstr.*, 1989, **110**, 106 828).
1952 H. Liu, Y. Okamoto, T. Skotheim, Y.S. Pak, S.G. Greenbaum, and K.J. Adamic, *Mater. Res. Soc. Symp. Proc.*, 1989, **135**, 337 (*Chem. Abstr.*, 1989, **111**, 175 068).
1953 E.C. Reynhardt, *Chem. Phys. Lett.*, 1989, **159**, 287.
1954 E.C. Reynhardt and S. Froneman, *Mol. Phys.*, 1989, **66**, 921.
1955 K.L. Murty, D. Begert, R. Munter, and O. Kanert, *Appl. Phys. Lett.*, 1989, **55**, 628 (*Chem. Abstr.*, 1989,

^{35}Cl T_1 measurements have been used to study motion in $[Rb_{1-x}(NH_4)_x]_2[ZnCl_4]$.[1956] The diffusion and ordering of ions in RbF and CsF have been studied by ^{19}F, ^{87}Rb, and ^{133}Cs n.m.r. spectroscopy.[1957]

^{19}F n.m.r. spectroscopy has been used to investigate ion transport in $Sr_{0.75}La_{0.25}F_{2.25}$ and $Sr_{0.75}La_{0.25}F_{2.25}$.[1958] ^{27}Al n.m.r. spectroscopy has been used to study Ba^{2+} motion in barium aluminium priderite.[1959]

Hydrogen diffusion in $YBa_2Cu_3O_{6.9}H_x$ has been studied using ^{1}H T_1 and $T_{1\rho}$ measurements.[1960]

The diffusion processes in crystalline and vitreous $Li_{0.75}Th_{0.25}F_{1.75}$ and Li_3ThF_7 have been studied using ^{7}Li and ^{19}F n.m.r. spectroscopy.[1961] Ionic motion in LaF_3 superionic conductor has been studied by ^{19}F n.m.r. spectroscopy with homonuclear decoupling.[1962]

Muon motion in $TiH_{1.99}$ has been investigated by ^{1}H n.m.r. spectroscopy.[1963] ^{1}H n.m.r. spectroscopy has been used to study hydride diffusion in Ti_3SbH_x.[1964] Defect motion and cation diffusion in a single crystal of TiO_{2-x} has been studied using ^{49}Ti T_1 and $T_{1\rho}$ measurements.[1965] ^{1}H nuclear spin relaxation caused by diffusion in HfV_2H_x has been studied.[1966]

^{2}H n.m.r. spectroscopy has been used to study deuterium migration in vanadium, niobium and tantalum deuterides.[1967] Pulsed field gradient measurements have been applied to diffusion measurements in $TaH_{0.73}$.[1968] ^{1}H, ^{2}H, and ^{51}V n.m.r. spectroscopy has been used to study low-frequency local motion of hydrogen or deuterium atoms in TaV_2.[1969] ^{7}Li n.m.r. spectroscopy has been used to determine the activation energy for

111, 205 849).

1956 F. Milia, G. Papavassiliou, and E. Giannakopoulos, *Phys. Rev. B: Condens. Matter*, 1989, **39**, 12349 (*Chem. Abstr.*, 1989, **111**, 165 860).

1957 M. Upschulte and H. Richtering, *Z. Phys. Chem. (Munich)*, 1989, **161**, 35 (*Chem. Abstr.*, 1989, **111**, 121 288).

1958 A.N. Matsulev, V.M. Buznik, A.I. Livshits, P.P. Fedorov, and B.P. Sobolev, *Fiz. Tverd. Tela (Leningrad)*, 1988, **30**, 3554 (*Chem. Abstr.*, 1989, **110**, 145 722).

1959 Y. Onoda, M. Watanabe, Y. Fujiki, Y. Kudo, T. Erata, S. Yoshikado, T. Ohachi, and I. Taniguchi, *Solid State Ionics, 1987*, (Pub. 1988), **28-30(Pt. 1)**, 179 (*Chem. Abstr.*, 1989, **110**, 105 966).

1960 K. Morimoto, W. Ye, S. Orimo, T. Takabatake, H. Fujii, and T. Hihara, *Solid State Commun.*, 1989, **71**, 291 (*Chem. Abstr.*, 1989, **111**, 223 945).

1961 S.H. Pulcinelli and J. Senegas, *Eur. J. Solid State Inorg. Chem.*, 1988, **25**, 249 (*Chem. Abstr.*, 1989, **110**, 183 872); J. Senegas and S.H. Pulcinelli, *J. Fluorine Chem.*, 1989, **42**, 31 (*Chem. Abstr.*, 1989, **110**, 126 450).

1962 A.F. Privalov, H.M. Vieth, and I.V. Murin, *J. Phys. Chem. Solids*, 1989, **50**, 395 (*Chem. Abstr.*, 1989, **111**, 125 520).

1963 J.R. Kempton, K.G. Petzinger, W.J. Kossler, H.E. Schone, and C.E. Stronach, *Phys. Rev. B; Condens. Matter*, 1989, **40**, 59 (*Chem. Abstr.*, 1989, **111**, 145 545).

1964 N. Rajalakshmi, K.V.S.R. Rao, and A. Weiss, *Ber. Bunsenges. Phys. Chem.*, 1989, **93**, 157.

1965 H. Kolem and O. Kanert, *Z. Metallkd.*, 1989, **80**, 227 (*Chem. Abstr.*, 1989, **111**, 16 537).

1966 D. Ding, H. Chen. Z. Guo, Q. Jin, and D. Zhou, *Wuli Xuebao*, 1988, **37**, 1859 (*Chem. Abstr.*, 1989, **110**, 146 453).

1967 N. Salibi, B. Ting, D. Cornell, and R.E. Norberg, *Phys. Rec. B: Condens. Matter*, 1988, **38**, 4416 (*Chem. Abstr.*, 1989, **110**, 49 979).

1968 R.M. Cotts, M.J.R. Hoch, T. Sun, and J.T. Markert, *J. Magn. Reson.*, 1989, **83**, 252.

1969 A.V. Skripov, M. Yu. Belgaev, S.V. Rychkova, and A.P. Stepanov, *J. Phys.: Condens. Matter*, 1989, **1**, 2121 (*Chem. Abstr.*, 1989, **111**, 32 514).

Li motion in $Li_{1+x}V_3O_8$.[1970] Lithium diffusion in $Li_xV_2O_5$ with 5% P_2O_5 has been studied by means of ^7Li n.m.r. spectroscopy.[1971] Ammonium motion in NH_4TaWO_6, ND_4TaWO_6, and NH_4NbWO_6 has been studied using ^1H and ^2H n.m.r. spectroscopy.[1972] ^{23}Na T_1 has been used to measure Na^+ diffusion in $Ba_2NaNb_5O_{15}$.[1973] Anisotropic dynamics of $Cs_x(CH_2F_2)_y(NbS_2)$ have been investigated using ^1H and ^{19}F n.m.r. spectroscopy.[1974]

Exchange dynamics of ^{23}Na in $NaMo_4O_6$ have been determined.[1975]

Rotation of the pyridine in $[Mn_3O(O_2CCD_3)_6(NC_5H_5)_3](NC_5H_5)$ has been studied by ^2H n.m.r. spectroscopy.[1976]

The reorientational motion of cyclopentadienyl ligands in the ^{13}C c.p.m.a.s. n.m.r. spectrum of *cis*- and *trans*-$[(\eta^5\text{-}C_5H_5)_2Fe_2(CO)_4]$ has been analysed using atom-atom pairwise potential energy calculations.[1977] Molecular reorientation in $[Fe_3(CO)_{12}]$ and $[Co_4(CO)_{12}]$ has been examined using potential energy methods and the results compared with ^{13}C m.a.s. n.m.r. spectroscopy.[1978] The effect of high hydrostatic pressure on the ionic motion in $[NH_4]_4[Fe(CN)_6].1.5H_2O$ has been studied by ^1H n.m.r. spectroscopy.[1979] ^1H T_1, $T_{1\rho}$, T_2 measurements have been used to investigate hydrogen motion in $[NH_4]_4[Fe(CN)_6].1.5H_2O$.[1980] The factors influencing the rate of intramolecular electron transfer in the solid state have been studied for $[Fe_3O(O_2CCH_3)_6(NC_5H_4Me\text{-}4)_3]$ using ^2H n.m.r. spectroscopy.[1981]

^1H n.m.r. spectroscopy has been used to investigate hydrogen motion in solid [PPN]-$[HRu_{10}C(CO)_{24}]$.[1982] ^1H T_1 measurements for $[Ni(NH_3)_6][NO_3]_2$ have been reported.[1983] The reorientation of PtF_6 has been studied by ^{19}F n.m.r. spectroscopy.[1984] Ag diffusion

1970 G. Wang, Q. Cao, M. Li, Q. Deng, L. Li, D. Guan, and Q. Li, *Wuli Xuebao*, 1988, **37**, 1217 (*Chem. Abstr.*, 1989, **110**, 50 077).

1971 T. Asai, S. Sugimoto, S. Kawai, S. Okada, and J. Yamaki, *Mater. Res. Bull.*, 1989, **24**, 75 (*Chem. Abstr.*, 1989, **111**, 121 265).

1972 H. Möller, W. Müller-Warmuth, M. Kuntz, and G. Tomandl, *Ber. Bunsenges. Phys. Chem.*, 1989, **93**, 140.

1973 J. Dolinsek, R. Blinc, and J. Schneck, *Solid State Commun.*, 1989, **70**, 1077 (*Chem. Abstr.*, 1989, **111**, 145 599).

1974 H. Möller, W. Müller-Warmuth, E. Wein, Z.T. Lalowicz, and R. Schoellhorn, *J. Magn. Reson.*, 1989, **83**, 1.

1975 P.J. Chu and B.C. Gerstein, *J. Chem. Phys.*, 1989, **90**, 3713.

1976 H.G. Jang, J.B. Vincent, M. Nakano, J.C. Huffman, G. Christou, M. Sorai, R.J. Wittebort, and D.N. Hendrickson, *J. Am. Chem. Soc.*, 1989, **111**, 7778.

1977 D. Braga, C. Gradella, and F. Grepioni, *J. Chem. Soc., Dalton Trans.*, 1989, 1721.

1978 D. Braga and F. Grepioni, *Polyhedron*, 1989, **8**, 2237.

1979 D.R. Balasubramanyan and B.V. Bhat, *J. Phys.: Condens. Matter*, 1989, **1**, 1495 (*Chem. Abstr.*, 1989, **110**, 223 428).

1980 G. Mangamma and S.V. Bhat, *Solid State Ionics 1988*, (Pub. 1989), **35**, 123 (*Chem. Abstr.*, 1989, **111**, 223 948).

1981 H.G. Jang, S.J. Geib, Y. Kaneko, M. Nakano, M. Sorai, A.L. Rheingold, B. Montez, and D.N. Hendrickson, *J. Am. Chem. Soc.*, 1989, **111**, 173 (*Chem. Abstr.*, 1989, **110**, 32 913).

1982 P.J. Bailey and M.J. Duer, *J. Chem. Soc., Chem. Commun.*, 1989, 1139.

1983 J. Czaplicki, N. Weiden, and A. Weiss, *Physica B (Amsterdam)*, 1988, **154**, 93 (*Chem. Abstr.*, 1989, **110**, 204 396).

1984 E.P. Zeer, O.V. Falaleev, Yu.N. Ivanov, G.F. Lybzikov, A.K. Tsvetnikov, and A.A. Uminskii, *Khim. Fiz.*, 1989, **8**, 1067 (*Chem. Abstr.*, 1989, **111**, 144 393).

in $(AgI)_x(Ag_2S.GeS_2)_{1-x}$ has been investigated using ^{109}Ag T_1 and T_2 measurements.[1985]

The mobility of water in $Zn_3(PO_4)_2.4H_2O$ has been studied using ^1H n.m.r. spectroscopy.[1986] ^1H n.m.r. spectroscopy has been used to study self-diffusion and reorientation of $[MeNH_3]^+$ ions in $[MeNH_3]_2[ZnCl_4]$,[1987] and $[NH_4]^+$ in $[NH_4]_2[ZnCl_4]$.[1988] ^1H n.m.r. spectroscopy has been used to investigate proton reorientation and tunnelling in $Cd(NH_3)_6Cl_2$.[1989]

The mobility of Li^+ in Li_2S-B_2S_3 and Li_2S-P_2S_5 has been studied by ^7Li, ^{11}B, and ^{31}P n.m.r. spectroscopy.[1990] Phase transitions and ionic motions in $[NH_4]_{3-n}K_n[AlF_6]$ have been investigated by ^1H, ^{19}F, and ^{27}Al n.m.r. spectroscopy.[1991]

The application of n.m.r. spectroscopy to study diffusion anisotropy in polycrystalline samples has been examined and applied to ZSM-5 crystallites.[1992] N.m.r. measurements have been applied to intra-crystalline diffusion in zeolites.[1993] High resolution solid state ^{13}C n.m.r. spectroscopy has been used to study the dynamic behaviour of $[Me_4N]^+$ ions trapped in zeolites.[1994] ^1H n.m.r. spectroscopy has been used to study the motion of water molecules in $Li_2Al_2Si_4O_{12}.2H_2O$.[1995] Site exchange dynamics in hydrated CsNa-A zeolite have been investigated by ^{133}Cs n.m.r. spectroscopy.[1996] Water diffusion in $Na_8Al_6Si_6O_{24}(OH)_2.4.6H_2O$ and $Na_8Al_6Ge_6O_{24}(OH)_2.6.8H_2O$ has been studied using ^1H T_1 measurements.[1997]

T_1 and second moment measurements for ^1H and ^{19}F nuclei have been used to study reorientational motion and self-diffusion of ions in $[NH_4]_3[InF_6]$ and $[NH_4]_3[GaF_6]$ crystals.[1998] Diffusion and short-range order in $Pb_{1-x}In_xF_{2+x}$ superionic conductor have been investigated by ^{19}F n.m.r. spectroscopy.[1999]

1985 J. Roos, D. Brinkmann, M. Mali, A. Pradel, and M. Ribes, *Solid State Ionics 1987*, (Pub. 1988), **28-30 (Pt. 1)**, 710 (*Chem. Abstr.*, 1989, **110**, 127 302).

1986 A.B. Yaroslavtsev, Z.N. Prozorovskaya, and V.F. Chuvaev, *Zh. Neorg. Khim.*, 1989, **34**, 2036 (*Chem. Abstr.*, 1989, **111**, 141 070).

1987 H. Ishida, T. Iwachido, N. Hayama, R. Ikeda, M. Terashima, and D. Nakamura, *Z. Naturforsch., A*, 1989, **44**, 741 (*Chem. Abstr.*, 1989, **111**, 244 735).

1988 A.H. Vuorimaki, M. Punkkinen, and E.E. Ylinen, *Chem. Phys. Lett.*, 1989, **161**, 561 (*Chem. Abstr.*, 1989, **111**, 246 556).

1989 J. Czaplicki, N. Weiden, and A. Weiss, *Phys. Status Solidi A*, 1989, **113**, 163 (*Chem. Abstr.*, 1989, **111**, 186 147).

1990 Z. Zhang, J.H. Kennedy, J. Thompson, S. Anderson, D.A. Lathrop, and H. Eckert, *Appl. Phys. A*, 1989, **A49**, 41 (*Chem. Abstr.*, 1989, **111**, 100 276).

1991 K. Hirokawa and Y. Furukawa, *J. Phys. Chem. Solids*, 1988, **49**, 1047 (*Chem. Abstr.*, 1989, **110**, 105 445).

1992 B. Zibrowius, J. Caro, and J. Kaerger, *Z. Phys. Chem. (Leipzig)*, 1988, **269**, 1101 (*Chem. Abstr.*, 1989, **110**, 146 440).

1993 J. Kaerger and D.M. Ruthven, *Zeolites*, 1989, **9**, 267 (*Chem. Abstr.*, 1989, **111**, 121 361).

1994 S. Hayashi, K. Suzuki, and K. Hayamizu, *J. Chem. Soc., Faraday Trans. 1*, 1989, **85**, 2973.

1995 K. Larsson, J. Tegenfeldt, and A. Kvick, *J. Phys. Chem. Solids*, 1989, **50**, 107 (*Chem. Abstr.*, 1989, **110**, 241 422).

1996 M.K. Ahn and L.E. Iton, *J. Phys. Chem.*, 1989, **93**, 4924; T. Tokuhiro, M. Mattingly, L.E. Iton, and M.K. Ahn, *J. Phys. Chem.*, 1989, **93**, 5584.

1997 V.A. Detinich, Yu.G. Kriger, and V.Yu. Galitskii, *Kristallografiya*, 1988, **33**, 1538 (*Chem. Abstr.*, 1989, **110**, 68 349).

1998 A. Sasaki, Y. Furukawa, and D. Nakamura, *Ber. Bunsenges. Phys. Chem.*, 1989, **93**, 1142.

1999 J.M. Reau, J. Senegas, J.P. Laval, and B. Frit, *Solid State Ionics*, 1988, **31**, 147 (*Chem. Abstr.*, 1989, **110**, 126 296).

Water diffusion in perfluoro-sulphonic acid cation exchange membranes has been studied.[2000] Molecular motion in diethyl siloxane has been studied by ^{29}Si n.m.r. spectroscopy.[2001] Motion in the crystalline and plastic-crystalline phases of cyclododecamethylhexasilane has been analysed by c.p.m.a.s. ^{13}C and ^{29}Si n.m.r. spectroscopy.[2002] ^{29}Si n.m.r. spectroscopy has been used to study the effects of temperature and composition on silicate glass structure and dynamics.[2003]

^{19}F n.m.r. spectroscopy has been used to investigate XF_6 group reorientation in polycrystals.[2004] ^{1}H T_1 measurements have been used to investigate coupled methyl quantum rotation in Me_2SnCl_2.[2005] Proton conductivity in $HSbO_3.nH_2O$ and $SnO_2.nH_2O$ has been investigated by ^{1}H T_1, $T_{1\rho}$, and T_2 measurements.[2006] The dynamics of hydrogen atoms in $Sn(HPO_4)_2.nH_2O$ have been studied.[2007] The ^{1}H second moment and T_1 have been determined for $[C_5H_5NH]_2[MCl_6]$, M = Te, Sn, and $[C_5H_5N]_2[SnBr_6]$ in order to investigate phase transitions and ionic motions.[2008] The temperature dependence of the ^{35}Cl T_1 has been reported for $[C_5H_5NH]_2[SnCl_6]$ and the activation energies for ion reorientation determined.[2009] ^{19}F n.m.r. spectroscopy has been used to study fluorine motion in $Pb_{1-x}Bi_xO_xF_{2-x}$.[2010] Molecular rotations and phase transitions in $CsPbBr_3$ have been studied using ^{81}Br n.m.r. spectroscopy.[2011]

Motion in Me_4NSCN has been investigated by ^{1}H n.m.r. spectroscopy.[2012] ^{1}H and ^{2}H n.m.r. spectroscopy has been used to investigate molecular motion and phase transitions in $[NH_4][NO_3]$.[2013] Self-diffusion and overall rotation of $[MeNH_3]^+$ in $[MeNH_3]Br$ have been studied.[2014] ^{35}Cl T_1 and ^{1}H second moment measurements have been used

2000 V.I. Volkov, I.A. Nesterov, S.F. Timashev, V.I. Sundukov, and L.V. Kropotov, *Khim. Fiz.*, 1989, **8**, 128 (*Chem. Abstr.*, 1989, **110**, 142 097).

2001 V.M. Litvinov, A.K. Whittaker, A. Hagemeyer, and H.W. Spiess, *Colloid Polym. Sci.*, 1989, **267**, 681 (*Chem. Abstr.*, 1989, **111**, 175 032).

2002 M. Moeller, D. Oelfin, and B. Wunderlich, *Mol. Cryst. Liq. Cryst.*, 1989, **173**, 101 (*Chem. Abstr.*, 1989, **111**, 244 913).

2003 J.F. Stebbins, *J. Non-Cryst. Solids*, 1988, **106**, 359 (*Chem. Abstr.*, 1989, **110**, 102 202).

2004 L.G. Falaleeva, O.V. Falaleev, and E.P. Zeer, *Yader. Magnit. Rezonans i Dinam. Spinov. Sistem, Krasnoyarsk*, 1988, 26. From *Ref. Zh., Fiz. (A-Sh.)*, 1988, Abstr. No. 12N1953 (*Chem. Abstr.*, 1989, **111**, 16 545).

2005 L. v. Wüllen and W. Müller-Warmuth, *Ber. Bunsenges. Phys. Chem.*, 1989, **93**, 1298.

2006 C. Forano, J.P. Besse, J.P. Battut, J. Dupuis, and A. Hajjimohamad, *Solid State Ionics*, 1989, **34**, 7 (*Chem. Abstr.*, 1989, **111**, 124 624).

2007 A.B. Yaroslavtsev, Z.N. Prozorovskaya, V.F. Chuvaev, and V.I. Spitsyn, *Dokl. Akad. Nauk SSSR*, 1988, **301**, 153 (*Chem. Abstr.*, 1989, **111**, 202 147).

2008 Y. Tai, T. Asaji, R. Ikeda, and D. Nakamura, *Z. Naturforsch., A*, 1989, **44**, 300 (*Chem. Abstr.*, 1989, **111**, 48 544).

2009 Y. Tai, A. Ishikawa, K. Horiuchi, T. Asaji, and D. Nakamura, *Z. Naturforsch., A*, 1988, **43**, 1002 (*Chem. Abstr.*, 1989, **110**, 127 296).

2010 P. Laborde, G. Villeneuve, J.M. Reau, and J.L. Soubeyroux, *Solid State Ionics*, 1987, **28-30** **(Pt. 1)**, 56((*Chem. Abstr.*, 1989, **110**, 87 216).

2011 V.V. Petrov, A.V. Losev, A.V. Bogdanova, A.A. Kryuchin, M.I. Dashkevich, and V.G. Pitsyuga, *Fiz. Tverd. Tela (Leningrad)*, 1989, **31**, 309 (*Chem. Abstr.*, 1989, **111**, 106 974).

2012 T. Tanabe, R. Ikeda, and D. Nakamura, *Phys. Status Solidi A*, 1989, **114**, K143 (*Chem. Abstr.*, 1989, **111**, 222 479).

2013 R. Grosescu, U. Haeberlen, and M. Lupu, *Rev. Roum. Phys.*, 1988, **33**, 637 (*Chem. Abstr.*, 1989, **110**, 1 564).

2014 M. Tansho, D. Nakamura, and R. Ikeda, *Z. Naturforsch., A*, 1989, **44**, 738 (*Chem. Abstr.*, 1989, **111**, 22 476).

to study motion in $[C_5H_5NH][SbCl_4]$.[2015] Molecular motion in $[SbMe_4]X$ has been determined from ^1H and ^{19}F n.m.r. spectroscopy.[2016] Ionic dynamics of $[Me_4N][SbF_6]$ have been studied by ^1H and ^{19}F n.m.r. spectroscopy.[2017]

Molecular motion in liquid and solid H_2S and H_2Se has been investigated by ^2H and ^{33}S n.m.r. spectroscopy.[2018] Motion in $[TeF_5O][C_{14}H_{19}N_2]$ has been investigated by ^{19}F n.m.r. spectroscopy.[2019]

The dynamics of $[H_3O][ClO_4]$ have been investigated by ^1H n.m.r. spectroscopy.[2020] Methyl reorientation in 1,4-didecyl-1,4-diazoniabicyclo[2.2.2]octane dibromide has been investigated by ^{79}Br n.m.r. spectroscopy.[2021]

Structure of Solids.——Shift references for high-resolution solid state ^{11}B, ^{13}C, ^{23}Na, ^{27}Al, ^{29}Si, and ^{31}P n.m.r. spectra have been suggested.[2022] ^{29}Si n.m.r. spectra of silicides of Ti, Zr, Nb, Ta, Cr, Mo, W, Rh, Ni, Pd, Pt, and Cu have been reported.[2023] ^{31}P n.m.r. spectra of MgP_4, Mg_3P_2, ZrP, and MPS_3, M = Fe, Ni, Mn, Cd, have been measured.[2024] ^{17}O n.m.r. spectroscopy has been used to show that ^{17}O labelled forms of Tl_2O_3, Bi_2O_3, Y_2O_3, SiO_2, ZSM-5, and BaO_2 are readily prepared from the parent oxides by relatively mild calcination in the presence of $^{17}O_2$ gas.[2025]

The anisotropy of the ^{13}C n.m.r. shielding in Li_2C_2, Na_2C_2, Ca_2C, Sr_2C_2, and Ba_2C_2 has been measured, and anisotropies of 180 to 235 p.p.m. were found.[2026] Resolution improvement of solid state two dimensional ^{23}Na nutation n.m.r. spectra by m.a.s. has been reported.[2027] ^{23}Na and ^{39}K m.a.s. n.m.r. spectroscopy has been used to confirm the presence of Na^- in K^+(1,4,7,10,13,16--Me_6-1,4,7,10,13,16-hexaazacyclen)Na^-.[2028]

High resolution solid state ^{31}P n.m.r. spectra have been measured for alkali phosphates.[2029] ^{19}F and ^{31}P m.a.s. n.m.r. spectroscopy has been used to determine the chemical shift tensors in $K_3Na(PO_3F)_2$.[2030] ^2H n.m.r. spectroscopy has been used to

2015 T. Okuda, Y. Aihara, N. Tanaka, K. Yamada, and S. Ichiba, *J. Chem. Soc., Dalton Trans.*, 1989, 631.
2016 G. Burbach, S.-Q. Dou, and A. Weiss, *Ber. Bunsenges. Phys. Chem.*, 1989, **93**, 1302.
2017 S. Sato, M. Kondo, R. Ikeda, and D. Nakamura, *Ber. Bunsenges. Phys. Chem.*, 1989, **93**, 450.
2018 M.J. Collins, C.I. Ratcliffe, and J.A. Ripmeester, *J. Phys. Chem.*, 1989, **93**, 7495.
2019 P.J. Kellett, O.P. Anderson, S.H. Strauss, and K.D. Abney, *Can. J. Chem.*, 1989, **67**, 2023.
2020 M.H. Herzog-Cance, T.M. Pham, and A. Potier, *J. Mol. Struct.*, 1989, **196**, 291 (*Chem. Abstr.*, 1989, **111**, 69 581).
2021 H. Nakayama, T. Eguchi, N. Nakamura, H. Chihara, T. Nogami, K. Imamura, and Y. Shirota, *Bull. Chem. Soc. Jpn.*, 1989, **62**, 399.
2022 S. Hayashi and K. Hayamizu, *Bull. Chem. Soc. Jpn.*, 1989, **62**, 2429.
2023 T.M. Duncan and D.M. Hamilton, *J. Mater. Res.*, 1988, **3**, 943 (*Chem. Abstr.*, 1989, **110**, 17 560).
2024 R.A. Nissan and T.A. Vanderah, *Report*, 1988, Order No. AD-A199981, 35 pp. Avail. NTIS. From *Gov. Rep. Announce. Index (U.S.)*, 1989, 89, Abstr. No. 908 513 (*Chem. Abstr.*, 1989, **111**, 145 595); R.A. Nissan and T.A. Vanderah, *J. Phys. Chem. Solids*, 1989, **50**, 347 (*Chem. Abstr.*, 1989, **111**, 125 519); S. Torre and J. Ziolo, *Phys. Rev. B: Condens. Matter*, 1989, **39**, 8915 (*Chem. Abstr.*, 1989, **111**, 107 846).
2025 S. Yang, K.D. Park, and E. Oldfield, *J. Am. Chem. Soc.*, 1989, **111**, 7278.
2026 T.M. Duncan, *Inorg. Chem.*, 1989, **28**, 2663.
2027 G. Engelhardt, J.C. Buhl, J. Felsche, and H. Foerster, *Chem. Phys. Lett.*, 1988, **153**, 332 (*Chem. Abstr.*, 1989, **110**, 146 436).
2028 M.E. Kuchenmeister and J.L. Dye, *J. Am. Chem. Soc.*, 1989, **111**, 935.
2029 S. Hayashi and K. Hayamizu, *Bull. Chem. Soc. Jpn.*, 1989, **62**, 3061.
2030 A.-R. Grimmer and J. Neels, *Z. Anorg. Allg. Chem.*, 1989, **576**, 117.

investigate phase transitions in RbD_2PO_4.[2031] ^{133}Cs n.m.r. spectroscopy has been used to study one dimensional fluctuations in $CsHPO_4$ at high pressure.[2032] Line shape in the ^{31}P m.a.s. n.m.r. spectra of $Na_4P_2O_7$, $Na_4P_2O_7.10H_2O$, $Ag_4P_2O_7$, and $(KPO_3)_n$ depends on spin rate.[2033] ^1H, ^{23}Na, and ^{31}P relaxation measurements have been applied to molecular dynamics of the polycrystalline sodium salt of AMP.[2034]

Central line spin-locking phenomenon of half-integer quadrupole nuclei in powder samples has been investigated using ^{23}Na n.m.r. spectroscopy for NaOH and $NaNO_2$.[2035] Spin-relaxation in LiO_2CCH_3 has been reinvestigated.[2036] ^{19}F n.m.r. spectroscopy has been used to investigate hydrogen bonding in $KF.(HO_2CCH_2CH_2CO_2H)$.[2037]

The ^1H, ^7Li, and ^{133}Cs n.m.r. spectra of single crystals of $CsLiH_2(SeO_3)_2$ have been reported.[2038] ^{39}K n.m.r. evidence for the phason gap in K_2SeO_3 has been reported.[2039] High resolution ^{77}Se n.m.r. spectra have been measured for $KHSeO_3$ and $RbHSeO_3$.[2040]

^6Li, ^7Li, ^{29}Si, and ^{31}P n.m.r. spectroscopy has been used to identify microphase separation in glasses of the systems Li_2S-SiS_2, Li_2S-SiS_2-P_2S_5, and Li_2S-P_2S_5-B_2S_3.[2041] A ^7Li n.m.r. study of Li_2S-SiS_2 glasses has been reported.[2042] The local structure of Li_2S-SiS_2-LiI glasses has been investigated by ^6Li, ^7Li, and ^{29}Si n.m.r. spectroscopy.[2043]

Anion substituting impurities in LiF crystals have been investigated by ^7Li and ^{19}F relaxation.[2044] The structural phase transition in $RbCaF_3$ has been investigated by ^{87}Rb T_1 measurements.[2045] Acoustic n.m.r. spectroscopy has been studied for ^{23}Na in NaCl and ^{127}I in CsI.[2046] N.m.r. imaging methods have been used to demonstrate spatial

2031 O. Jarh, J. Dolinsek, B. Topic, B. Zeks, R. Blinc, V.H. Schmidt, L.A. Shuvalov, and A. Baranov, *Phys. Rev. B: Condens. Matter*, 1989, **39**, 2004 (*Chem. Abstr.*, 1989, **110**, 184 647).

2032 P.J. Schuele and V.H. Schmidt, *Phys. Rev. B: Condens. Matter*, 1989, **39**, 2549 (*Chem. Abstr.*, 1989, **110**, 146 037).

2033 S. Hayashi and K. Hayamizu, *Chem. Phys. Lett.*, 1989, **161**, 158 (*Chem. Abstr.*, 1989, **111**, 223 951).

2034 E.C. Reynhardt, S. Froneman, E.R. Andrew, and E. Szczesniak, *J. Magn. Reson.*, 1989, **84**, 110.

2035 G. Li and X. Wu, *Wuli Xuebao*, 1989, **38**, 338 (*Chem. Abstr.*, 1989, **111**, 69 577).

2036 E. Crits, L. Van Gerven, L.P. Ingman, E. Ylinen, M. Punkkinen, and S. Emid, *Physica B (Amsterdam)*, 1988, **154**, 97 (*Chem. Abstr.*, 1989, **110**, 184 625).

2037 E.A. Moore, M. Mortimer, N.F. Peirson, and G. Oates, *Chem. Phys. Lett.*, 1988, **151**, 213 (*Chem. Abstr.*, 1989, **110**, 17 581).

2038 I.S. Vinogradova, V.I. Ponomarev, and S.I. Daryuga, *Kristallografiya*, 1989, **34**, 364 (*Chem. Abstr.*, 1989, **110**, 223 065).

2039 B. Topic, U. Haeberlen, and R. Blinc, *Phys. Rev. B: Condens. Matter*, 1989, **40**, 799 (*Chem. Abstr.*, 1989, **111**, 165 986).

2040 I.S. Vinogradova, A.A. Sukhovskii, and F.F. Khizbullin, *J. Solid State Chem.*, 1989, **78**, 209 (*Chem. Abstr.*, 1989, **110**, 224 234).

2041 H. Eckert, Z. Zhang, and J.H. Kennedy, *Mater. Res. Soc. Symp. Proc.*, 1989, **135**, 259 (*Chem. Abstr.*, 1989, **111**, 107 855).

2042 A. Pradel, M. Ribes, and M. Maurin, *Solid State Ionics 1987*, (Pub. 1988), **28-30(Pt. 1)**, 762 (*Chem. Abstr.*, 1989, **110**, 127 304).

2043 H. Eckert, Z. Zhang, and J.H. Kennedy, *J. Non-Cryst. Solids*, 1989, **107**, 271 (*Chem. Abstr.*, 1989, **111**, 160 812).

2044 L.I. Bryukvina, V.M. Khulugorov, A.G. Shneider, A.N. Gil'manov, and F.F. Gubaidullin, *Ukr. Fiz. Zh. (Russ. Ed.)*, 1988, **33**, 1544 (*Chem. Abstr.*, 1989, **110**, 16 197).

2045 A. Trokiner and H. Zanni-Theveneau, *J. Phys. C: Solid State Phys.*, 1988, **21**, 4913 (*Chem. Abstr.*, 1989, **110**, 48 827).

2046 A.A. Kuleshov, V.M. Mikushev, A.L. Stolypko, and E.V. Charnaya, *Akust. Zh.*, 1989, **35**, 473 (*Chem. Abstr.*, 1989, **111**, 125 530).

variation of the ^{23}Na signal intensity in NaCl.[2047] The ^{35}Cl and ^{37}Cl solid state n.m.r. chemical shifts of NaCl, KCl, RbCl, CsCl, and NH_4Cl are highly correlated with the interionic separation and with the electronegativity of the cation.[2048] The theory of spin dynamics in KBr has been examined.[2049]

^{1}H and ^{19}F n.m.r. investigations have provided evidence of a distortion of octahedra in $CsCaF_{3-x}H_x$.[2050] The ^{1}H n.m.r. line shape of polycrystalline $Ca(OH)_2$ has been reported.[2051] ^{11}B n.m.r. spectroscopy has been used to study the structure of BaO-B_2O_3-Al_2O_3 glasses.[2052] The effects of anisotropic molecular vibrations of $Ca(O_2CH)_2$ have been studied by ^{13}C n.m.r. spectroscopy.[2053] The ^{27}Al n.m.r. spectra of $CaLnAlO_4$, Ln = La, Pr, Eu, and Y, have been determined.[2054] $Ba_2ScCuO_{4.5}$, $Ba_3Sc_4Cu_3O_{12}$, and $Ba_2Sc_2O_5$ have been investigated by ^{45}Sc n.m.r. spectroscopy.[2055] The ^{205}Tl Knight shift has been measured in $TlCa_3BaCu_3O_x$,[2056] and $Tl_2Ba_2CaCu_2O_8$.[2057] ^{29}Si n.m.r. spectroscopy has been used to study vitreous and sol-gel precursors of $CaTiSiO_5$ precursors.[2058] Dynamic angle spinning and double rotation n.m.r. spectroscopy has been applied to ^{17}O in SiO_2 and $CaMgSi_2O_6$.[2059] The ^{29}Si n.m.r. spectrum of $Ca_4(SiO_4)(SO_4)Cl_2$ has been reported.[2060] The ^{1}H T_1 and T_2 values for $CaSO_4$ have been measured.[2061] ^{1}H relaxation and ^{27}Al n.m.r. spectra have been used to study the hydration of $3CaO.3Al_2O_3.CaSO_4$.[2062] The phase diagram of $Ca_{3-x}Zn_x(PO_4)_2$ has been investigated by ^{31}P n.m.r. spectroscopy.[2063] ^{19}F CRAMPS has been applied to the analysis of $CaF_2/Ca_5(PO_4)_3F_x(OH)_{1-x}$

2047 B.H. Suits and J.L. Lutz, *J. Appl. Phys.*, 1989, **65**, 3728 (*Chem. Abstr.*, 1989, **111**, 49 155).
2048 T.L. Weeding and W.S. Veeman, *J. Chem. Soc., Chem. Commun.*, 1989, 946.
2049 B.C. Sanctuary and T.K. Halstead, *Bull. Magn. Reson.*, 1989, **11**, 31 (*Chem. Abstr.*, 1989, **111**, 207 721).
2050 M. Pezat, J. Senegas, G. Villeneuve, H.H. Park, and A. Tressaud, *J. Solid State Chem.*, 1988, **77**, 387 (*Chem. Abstr.*, 1989, **110**, 164 836).
2051 O.P. Kukhlevskii and O.V. Falaleev, *Yader. Magnit. Rezonans i Dinam. Spinov. Sistem, Krasnoyarsk*, 1988, 110. From *Ref. Zh., Fiz. (A-Zh.)*, 1988, Abstr. No. 12N1959 (*Chem. Abstr.*, 1989, **111**, 16 542).
2052 D.J. Cha, Y.J. Kim, S.J. Moon, J.I. Kye, H.T. Kim, H.J. Lee, and S.J. Chung, *Sae Mulli*, 1988, **28**, 593 (*Chem. Abstr.*, 1989, **110**, 80 977).
2053 T. Nakai, J. Ashida, and T. Terao, *Mol. Phys.*, 1989, **67**, 839.
2054 L.S. Vorotilova and L.V. Dmitrieva, *Kristallografiya*, 1989, **34**, 1028 (*Chem. Abstr.*, 1989, **111**, 145 600).
2055 A.A. Gippius, V.P. Denisov, V.V. Moshchalkov, Yu.M. Petrusevich, O.P. Revokatov, E.V. Antipov, A.L. Kharlanov, L.M. Kovba, and L.N. Lykova, *Zh. Eksp. Teor. Fiz.*, 1989, **95**, 2125 (*Chem. Abstr.*, 1989, **111**, 207 945).
2056 A. Poddar, P. Mandal, K.G. Ray, A.N. Das, B. Ghosh, P. Choudhury, and S.K. Lahiri, *Physica C (Amsterdam)*, 1989, **159**, 226 (*Chem. Abstr.*, 1989, **111**, 186 145).
2057 F. Hentsch, N. Winzek, M. Mehring, H. Mattausch, and A. Simon, *Physica C (Amsterdam)*, 1989, **158**, 137.
2058 J.S. Hartman, R.L. Millard, and E.R. Vance, *J. Non-Cryst. Solids*, 1989, **108**, 49 (*Chem. Abstr.*, 1989, **111**, 27 478).
2059 B.F. Chmelka, K.T. Mueller, A. Pines, J. Stebbins, Y. Wu, and J.W. Zwanziger, *Nature (London)*, 1989, **339**, 42 (*Chem. Abstr.*, 1989, **111**, 165 966).
2060 F. Von Lampe, *Guisuanyan Xuebao*, 1988, **16**, 477 (*Chem. Abstr.*, 1989, **111**, 83 077).
2061 P. Kuhns, O. Gonen, and J.S. Waugh, *J. Magn. Reson.*, 1989, **82**, 231.
2062 T.V. Kuznetsova, A.B. Kudryavtsev, and A.V. Pyatkova, *Zh. Prikl. Khim. (Leningrad)*, 1989, **62**, 552 (*Chem. Abstr.*, 1989, **110**, 198 214).
2063 R.J.B. Jakeman, A.K. Cheetham, N.J. Clayden, and C.M. Dobson, *J. Solid State Chem.*, 1989, **78**, 23 (*Chem. Abstr.*, 1989, **110**, 161 065).

mixtures.[2064] The structure of $M(PO_3)_2$-CaF_2-AlF_3, M = Ba, Sr, has been determined from ^{19}F T_1 measurements.[2065] Solid state reactions in the $Sr(PO_3)_2$-$4AlF_3$ system have been investigated.[2066]

The ^{19}F n.m.r. spectrum of CaF_2 has been reported.[2067] ^{27}Al m.a.s. n.m.r. spectroscopy has been used to demonstrate a coordination number of six for Al in AlF_3-BaF_2-CaF_2 glasses.[2068] A ^{19}F n.m.r. study of a BaF_2-LaF_3-ZnF_4 glass has revealed the presence of three fluoride ion sites.[2069] The ^{19}F T_1 of $SrSnF_4$ has been measured.[2070]

The ^{13}C m.a.s. n.m.r. spectrum of $[La_2\{4\text{-Me-1-OC}_6H_2\text{-2,6-(CH=NCH}_2CH_2OCH_2CH_2O\text{-}CH_2CH_2N\text{=CH)}_2\text{-2,6-C}_6H_2\text{-1-O-4-Me}\}]^{4+}$ has been reported.[2071] The hyperfine splitting of ^{165}Ho in $Ho(OH)_3$ has been studied using spin-echo n.m.r. spectroscopy.[2072] The ^{27}Al n.m.r. spectrum of $HoAlO_3$ has been reported.[2073] ^{165}Ho n.m.r. spectroscopy has been used to investigate hyperfine enhanced magnetism of $Cs_2NaHoCl_6$.[2074] ^{7}Li n.m.r. spectroscopy has been used to study $La_2Li_{0.50}M_{0.50}O_4$, M = Co, Ni, Cu, and to determine the structural distortion of LiO_6 octahedra.[2075] The effect of magnetic fields on a single crystal of $ErCrO_3$ has been studied using ^{53}Cr n.m.r. spectroscopy.[2076] The photo-induced change in the ^{57}Fe n.m.r. spectrum of $Y_3Fe_5O_{12}$ has been reported.[2077] The spin-echo ^{57}Fe n.m.r. spectrum of the Fe^{3+} sites in $Y_3Fe_5O_{12}$ and $Gd_3Fe_5O_{12}$ has been studied.[2078]

The field dependence of the hyperfine splitting of Ho^{3+} in $Y(EtOSO_3)_3$ has been studied.[2079] ^{27}Al, ^{29}Si, and ^{139}La n.m.r. spectroscopy has been used to study the La-Si-Al-O-N system.[2080] ^{29}Si m.a.s. n.m.r. spectroscopy has been used to study $La_5Si_3O_{12}N$

2064 K.A. Smith and D.P. Burum, *J. Magn. Reson.*, 1989, **84**, 85.
2065 M. Dubiel and D. Ehrt, *Rostocker Phys. Manuskr.*, 1988, **12**, 35 (*Chem. Abstr.*, 1989, **111**, 200 190).
2066 C. Jaeger and D. Ehrt, *Z. Phys. Chem. (Munich)*, 1989, **162**, 109 (*Chem. Abstr.*, 1989, **111**, 145 778).
2067 N.A. Sergeev, A.V. Sapiga, and D.S. Ryabushkin, *Pis'ma Zh. Eksp. Teor. Fiz.*, 1989, **49**, 323 (*Chem. Abstr.*, 1989, **110**, 224 254).
2068 I. Yasui, H. Hagihara, and Y. Arai, *Mater. Sci. Forum*, 1988, **32**, 173 (*Chem. Abstr.*, 1989, **110**, 178 081).
2069 D.R. MacFarlane, J.O. Browne, T.J. Bastow, and G.W. West, *J. Non-Cryst. Solids*, 1989, **108**, 289 (*Chem. Abstr.*, 1989, **111**, 121 329).
2070 J.E. Callanan, R. Shaviv, E.F. Westrum, jun., and R.D. Weir, *J. Solid State Chem.*, 1989, **81**, 51 (*Chem. Abstr.*, 1989, **111**, 103 395).
2071 I.A. Kahwa, S. Folkes, D.J. Williams, S.V. Ley, C.A. O'Mahoney, and G.L. McPherson, *J. Chem. Soc., Chem. Commun.*, 1989, 1531.
2072 D.St.P. Bunbury, C. Carboni, and M.A.H. McCausland, *J. Phys.: Condens. Matter*, 1989, **1**, 1309 (*Chem. Abstr.*, 1989, **110**, 224 248).
2073 L.S. Vorotilova, L.V. Dmitrieva, V.S. Kasperovich, and R.M. Rakhmankulov, *Fiz. Tverd. Tela (Leningrad)*, 1989, **31**, 298 (*Chem. Abstr.*, 1989, **110**, 224 253).
2074 Y. Masuda and H. Suzuki, *J. Low Temp. Phys.*, 1989, **75**, 159 (*Chem. Abstr.*, 1989, **111**, 185 981).
2075 G. Villeneuve, T. Rojo, G. Demazeau, and P. Hagenmuller, *Mater. Res. Bull.*, 1988, **23**, 1787 (*Chem. Abstr.*, 1989, **110**, 127 306).
2076 N.M. Kovtun, A.S. Moskvin, G.A. Troitskii, and V.M. Khmara, *Fiz. Tverd. Tela (Leningrad)*, 1988, **30**, 3629 (*Chem. Abstr.*, 1989, **110**, 127 340).
2077 S.V. Seregin, R.A. Doroshenko, V.A. Timofeeva, and R.S. Fakhretdinova, *Pis'ma Eksp. Teor. Fiz.*, 1989, **50**, 130 (*Chem. Abstr.*, 1989, **111**, 186 163).
2078 V.D. Doroshev, S.F. Ivanov, A.N. Molchanov, and M.M. Savosta, *Fiz. Tverd. Tela (Leningrad)*, 1989, **31**, 252 (*Chem. Abstr.*, 1989, **110**, 204 431).
2079 C. Carboni, R.L. Cone, Z.P. Han, and M.A.H. McCausland, *J. Phys. Colloq.*, 1988, C8-843 (*Chem. Abstr.*, 1989, **111**, 107 848).
2080 R. Dupree, M.H. Lewis, and M.E. Smith, *J. Am. Chem. Soc.*, 1989, **111**, 5125.

and related systems.[2081] The spin reorientation transitions have been studied in polycrystalline $Y_3Fe_{5-x}Si_xO_{12}$ by ^{57}Fe n.m.r. spectroscopy.[2082] ^{119}Sn n.m.r. spectra have been measured for $Ln_2Sn_2O_7$, Ln = La, Pr, Nd, Sm, Eu, Tm, Yb, Lu, and Y.[2083]

The ^{19}F and ^{141}Pr n.m.r. spectra of Pr^{3+} in LaF_3 have been reported.[2084] The reaction of $Al(PO_3)_3$ with AlF_3 and of $La(PO_3)_3$ with LaF_3 has been investigated by ^{31}P n.m.r. spectroscopy.[2085] The magnetic phases of $Eu_{1-x}Tm_xSe$ have been investigated by ^{153}Eu n.m.r. spectroscopy.[2086]

The zero-field ^{63}Cu and ^{65}Cu n.m.r. spectra of $La_2CuO_{4-\delta}$ have been observed.[2087] The T_1 of ^{139}La n.m.r. spectrum of $La_{2-x}Sr_xCuO_4$ has been measured.[2088] An anomalous ^{63}Cu n.m.r. signal with extremely small e^2qQ has been observed in $Nd_{1.85}Ce_{0.15}CuO_{4-\delta}$.[2089] A comparison of the solid-state ^{27}Al n.m.r. spectra of $Sr_2YCu_{3-x}Al_xO_y$ and the barium analogue suggests a different aluminium environment in these two systems.[2090] The ^{63}Cu and ^{65}Cu n.m.r. spectra of $Bi_2Sr_2YCu_2O_{8+\delta}$ have been reported.[2091]

Based on recent n.m.r. measurements performed on $YB_2Cu_3O_7$, an illustrative model for the formation of hole pairs in $LnBa_2Cu_3O_7$ has been presented.[2092] ^{63}Cu and ^{65}Cu n.m.r. spectra have been recorded for $La_4BaCu_5O_{13+x}$.[2093] ^{17}O and ^{63}Cu n.m.r. spectra of $YBa_2Cu_3O_{7-\delta}$ have been determined.[2094] The spin dynamics in $YBa_2Cu_3O_{7-\delta}$ have been investigated using ^{63}Cu, ^{65}Cu, and ^{89}Y n.m.r. spectroscopy.[2095] The spin lattice relaxation times of ^{63}Cu and ^{89}Y in $YBa_2Cu_3O_{7-\delta}$ and ^{9}Be in UBe_{13} are quite similar.[2096] $YBa_2Cu_3O_{7-\delta}$ has been investigated by ^{17}O,[2097] ^{63}Cu, ^{65}Cu,[2098] and ^{89}Y n.m.r. spectro-

[2081] R.K. Harris, M.J. Leach, and D.P. Thompson, *Chem. Mater.*, 1989, **1**, 336 (*Chem. Abstr.*, 1989, **110**, 217 880).

[2082] V.A. Borodin, V.D. Doroshev, E.N. Kuznetsov, V.A. Ruban, M.M. Savosta, and T.N. Tarasenko, *Fiz. Tverd. Tela (Leningrad)*, 1989, **31**, 297 (*Chem. Abstr.*, 1989, **110**, 224 197).

[2083] C.P. Grey, C.M. Dobson, A.K. Cheetham, and R.J.B. Jakeman, *J. Am. Chem. Soc.*, 1989, **111**, 505.

[2084] M. Lukac, F.W. Otto, and E.L. Hahn, *Phys. Rev. A: Gen. Phys.*, 1989, **39**, 1123 (*Chem. Abstr.*, 1989, **110**, 127 344).

[2085] D. Ehrt and C. Jaeger, *Z. Phys. Chem. (Munich)*, 1989, **162**, 97 (*Chem. Abstr.*, 1989, **111**, 145 777).

[2086] K. Hiraoka, N. Fujiya, K. Kojima, and T. Hihara, *J. Phys. Colloq.*, 1988, C8-887 (*Chem. Abstr.*, 1989, **111**, 145 382).

[2087] T. Tsuda, T. Shimizu, H. Yasuoka, K. Kishio, and K. Kitazawa, *J. Phys. Soc. Jpn.*, 1988, **57**, 2908 (*Chem. Abstr.*, 1989, **110**, 17 570).

[2088] T. Kobayashi, S. Wada, Y. Kitaoka, and K. Asayama, *J. Phys. Soc. Jpn.*, 1989, **58**, 2662 (*Chem. Abstr.*, 1989, **111**, 207 047).

[2089] G.Q. Zheng, Y. Kitaoka, Y. Oda, and K. Asayama, *J. Phys. Soc. Jpn.*, 1989, **58**, 1910 (*Chem. Abstr.*, 1989, **111**, 186 133).

[2090] S.A. Sunshine, L.F. Schneemeyer, T. Siegrist, D.C. Douglass, J.V. Waszczak, R.J. Cava, E.M. Gyorgy, and D.W. Murphy, *Chem. Mater.*, 1989, **1**, 331 (*Chem. Abstr.*, 1989, **110**, 240 981).

[2091] R. De Renzi, G. Guidi, P. Carretta, G. Calestani, and S.J.F. Cox, *Phys. Lett. A*, 1989, **135**, 132 (*Chem. Abstr.*, 1989, **110**, 184 653).

[2092] V. Mueller, *Solid State Commun.*, 1988, **67**, 805 (*Chem. Abstr.*, 1989, **110**, 16 906).

[2093] T. Imai, H. Yasuoka, T. Shimizu, Y. Ueda, and K. Kosuge, *J. Phys. Soc. Jpn.*, 1989, **58**, 1528 (*Chem. Abstr.*, 1989, **111**, 145 572).

[2094] E. Ioon, I. Heinma, V. Miidel, S. Vija, and E. Lippmaa, *Pis'ma Zh. Eksp. Teor. Fiz.*, 1989, **50**, 296 (*Chem. Abstr.*, 1989, **111**, 223 958); K. Ishida, Y. Kitaoka, K. Asayama, H. Katayama-Yoshida, Y. Okabe, and T. Takahashi, *J. Phys. Soc. Jpn.*, 1988, **57**, 2897 (*Chem. Abstr.*, 1989, **110**, 17 568).

[2095] W.W. Warren, jun., R.E. Walstedt, G.F. Brennert, R.J. Cava, R. Tycko, R.F. Bell, and G. Dabbagh, *Phys. Rev. Lett.*, 1989, **62**, 1193 (*Chem. Abstr.*, 1989, **110**, 224 235).

[2096] C. Tien and I.M. Jiang, *Phys. Rev. B: Condens. Matter*, 1989, **40**, 229 (*Chem. Abstr.*, 1989, **111**, 165 985).

[2097] P. Wzietek, D. Koengeter, P. Auban, D. Jerome, J.M. Bassat, J.P. Coutures, B. Dubois, and P. Odier,

scopy.[2099] ^{203}Tl and ^{205}Tl n.m.r. spectroscopy has been used to investigate $Tl_2Ba_2Ca_2$-Cu_3O_{10}.[2100] ^{1}H n.m.r. signals have been obtained from hydrogen doped YBa_2Cu_3-$O_{7-\delta}$.[2101] ^{11}B n.m.r. spectroscopy has been used to analyse boron incorporation into

Europhys. Lett., 1989, **8**, 363 (*Chem. Abstr.*, 1989, **110**, 204 415); C. Coretsopoulos, H.C. Lee, E. Ramli, L. Reven, T.B. Rauchfuss, and E. Oldfield, *Phys. Rev. B: Condens. Matter*, 1989, **39**, 781 (*Chem. Abstr.*, 1989, **110**, 127 318); M. Horvatic, Y. Berthier, P. Butaud, Y. Kitaoka, P. Segransan, C. Berthier, H. Katayama-Yoshida, Y. Okabe, and T. Takahashi, *Physica C (Amsterdam)*, 1989, **159**, 689 (*Chem. Abstr.*, 1989, **111**, 186 160); H. Bleier, P. Bernier, D. Jerome, J.M. Bassat, J.P. Coutures, B. Dubois, and P. Odier, *J. Phys. (Paris)*, 1988, **49**, 1825 (*Chem. Abstr.*, 1989, **110**, 106 808).

2098 C.H. Pennington, D.J. Durand, C.P. Slichter, J.P. Rice, E.D. Bukowski, and D.M. Ginsberg, *Phys. Rev. B: Condens. Matter*, 1989, **39**, 2902 (*Chem. Abstr.*, 1989, **110**, 204 406); A.V. Miller, A.M. Vainrub, I. Heinmaa, A.V. Reinkhold, E. Lippmaa, and I.F. Shchegolev, *Pis'ma Zh. Eksp. Teor. Fiz.*, 1989, **49**, 211 (*Chem. Abstr.*, 1989, **110**, 204 429); M. Horvatic, P. Segransan, C. Berthier, Y. Berthier, P. Butaud, J.Y. Henry, M. Couach, and J.P. Chaminade, *Phys. Rev. B: Condens. Matter*, 1989, **39**, 7332 (*Chem. Abstr.*, 1989, **111**, 16 532); M. Takigawa, P.C. Hammel, R.H. Heffner, and Z. Fisk, *Phys. Rev. B: Condens. Matter*, 1989, **39**, 7371 (*Chem. Abstr.*, 1989, **111**, 16 533); F. Mila and T.M. Rice, *Physica C (Amsterdam)*, 1989, **157**, 561 (*Chem. Abstr.*, 1989, **111**, 16 535); P. Mendels and H. Alloul, *Physica C (Amsterdam)*, 1988, **156**, 355 (*Chem. Abstr.*, 1989, **110**, 49 999); R.A. Brand, C. Sauer, H. Luetgemeier, B. Rupp, and W. Zinn, *Physica C (Amsterdam)*, 1988, **156**, 539 (*Chem. Abstr.*, 1989, **110**, 106 323); M.V. Eremin, I.E. Graboi, A.V. Egorov, V.V. Naletov, M.S. Tagirov, and M.A. Teplov, *Pis'ma Zh. Eksp. Teor. Fiz.*, 1989, **49**, 446 (*Chem. Abstr.*, 1989, **111**, 125 506); H. Yasuoka, T. Shimizu, T. Imai, S. Sasaki, Y. Ueda, and K. Kosuge, *Hyperfine Interact. 1988*, (Pub. 1989), **49**, 167 (*Chem. Abstr.*, 1989, **111**, 165 799); H. Riesemeier, J. Pattloch, K. Luders, and V. Muller, *Solid State Commun.*, 1988, **68**, 251 (*Chem. Abstr.*, 1989, **110**, 32 764); C.H. Pennington, D.J. Durand, C.P. Slichter, J.P. Rice, E.D. Bukowski, and D.M. Ginsberg, *Phys. Rev. B: Condens. Matter*, 1989, **39**, 274 (*Chem. Abstr.*, 1989, **110**, 127 316); M. Takigawa, P.C. Hammel, R.H. Heffner, Z. Fisk, J.L. Smith, and R.B. Schwarz, *Phys. Rev. B: Condens. Matter*, 1989, **39**, 300 (*Chem. Abstr.*, 1989, **110**, 127 317); G. Guidi, C. Bucci, P. Carretta, R. De Renzi, R. Tedeschi, C. Vignali, and F. Licci, *Solid State Commun.*, 1988, **68**, 759 (*Chem. Abstr.*, 1989, **110**, 127 311); W.W. Warren, jun., R.E. Walstedt, G.F. Brennert, R.J. Cava, B. Batlogg, and L.W. Rupp, *Phys. Rev. B: Condens. Matter*, 1989, **39**, 831 (*Chem. Abstr.*, 1989, **110**, 127 319); H. Zimmermann, M. Mali, D. Brinkmann, J. Karpinski, E. Kaldis, and S. Rusiecki, *Physica C (Amsterdam)*, 1989, **159**, 681 (*Chem. Abstr.*, 1989, **111**, 186 159); R.E. Walstedt, W.W. Warren, jun., R.F. Bell, and G.P. Espinosa, *Phys. Rev. B: Condens. Matter*, 1989, **40**, 2572 (*Chem. Abstr.*, 1989, **111**, 207 974); R.E. Walstedt, W.W. Warren, jun., R.F. Bell, G.F. Brennert, G.P. Espinosa, R.J. Cava, L.F. Schneemeyer, and J.V. Waszczak, *Phys. Rev. B: Condens. Matter*, 1988, **38**, 9299 (*Chem. Abstr.*, 1989, **110**, 68 333); R.E. Walstedt, W.W. Warren, jun., R. Tycko, R.F. Bell, G.F. Brennert, R.J. Cava, L. Schneemeyer, and J. Waszczak, *Phys. Rev. B: Condens. Matter*, 1988, **38**, 9303 (*Chem. Abstr.*, 1989, **110**, 68 334); A.V. Bondar, S.M. Ryabchenko, Yu.V. Fedotov, and A.A. Motuz, *Pis'ma Zh. Eksp. Teor. Fiz.*, 1989, **50**, 133 (*Chem. Abstr.*, 1989, **111**, 145 014); H. Yasuoka, *Oyo Butsuri*, 1989, **58**, 739 (*Chem. Abstr.*, 1989, **111**, 246 525); M. Matsumura, H. Yamagata, Y. Yamada, K. Ishida, Y. Kitaoka, K. Asayama, H. Takagi, H. Iwabuchi, and S. Uchida, *J. Phys. Soc. Jpn.*, 1989, **58**, 805 (*Chem. Abstr.*, 1989, **111**, 32 519).

2099 N. Bontemps, D. Fournier, A.C. Boccara, P. Monod, H. Alloul, J. Arabski, and G. Deutscher, *J. Phys. (Paris)*, 1989, **50**, 2895 (*Chem. Abstr.*, 1989, **111**, 221 200); H.B. Brom, G.J. Kramer, J. Van den Berg, B. Pronk, J.C. Jol, and R. Hulstman, *Synth. Met.* 1988, (Pub. 1989), **29**, F553 (*Chem. Abstr.*, 1989, **111**, 89 207); R. Dupree, D.M. Paul, M.E. Smith, and G. Balakrishnan, *Phys. Rev. Lett.*, 1989, **63**, 688 (*Chem. Abstr.*, 1989, **111**, 206 938); H. Alloul, T. Ohno, and P. Mendels, *Phys. Rev. Lett.*, 1989, **63**, 1700 (*Chem. Abstr.*, 1989, **111**, 246 554); H. Alloul, *Phys. Rev. Lett.*, 1989, **63**, 689 (*Chem. Abstr.*, 1989, **111**, 206 939); F.J. Adrian, *Phys. Rev. Lett.*, 1989, **63**, 690 (*Chem. Abstr.*, 1989, **111**, 206 940); F.J. Adrian, *Phys. Rev. Lett.*, 1988, **61**, 2148 (*Chem. Abstr.*, 1989, **110**, 29 236).

2100 A.K. Rajarajan, K.V. Gopalakrishnan, R. Vijayaraghavan, and L.C. Gupta, *Solid State Commun.*, 1989, **69**, 213 (*Chem. Abstr.*, 1989, **110**, 145 903); K. Fujiwara, Y. Kitaoka, K. Asayama, H. Katayama-Yoshida, Y. Okabe, and T. Takahashi, *J. Phys. Soc. Jpn.*, 1988, **57**, 2893 (*Chem. Abstr.*, 1989, **110**, 32 749); M. Lee, Y.Q. Song, W.P. Halperin, L.M. Tonge, T.J. Marks, H.O. Marcy, and C.R. Kannewurf, *Phys. Rev. B: Condens. Matter*, 1989, **40**, 817 (*Chem. Abstr.*, 1989, **111**, 165 987).

2101 S.D. Goren, C. Korn, V. Volterra, M. Schaefer, H. Riesemeier, E. Roessler, H. Stenschke, H.M. Vieth, and K. Lueders, *Solid State Commun.*, 1989, **70**, 279 (*Chem. Abstr.*, 1989, **111**, 145 559); H. Niki, T. Suzuki, S. Tomiyoshi, H. Hentona, M. Omori, T. Kajitani, T. Kamiyama, and R. Igei, *Solid State Commun.*, 1989, **69**, 547 (*Chem. Abstr.*, 1989, **110**, 241 420); H. Nishihara, N. Nishida, T. Takabatake, K. Kishio, A. Ohtomo, K. Hayashi, M. Ishikawa, Y. Nakazawa, K. Koga, et al., *Jpn. J. Appl. Phys., Part 1*, 1988, **27**, 1652 (*Chem. Abstr.*, 1989, **110**, 17 566).

$YBa_2Cu_3O_{7-\delta}$.[2102] The $YBa_2Cu_3O_{7-\delta}$ precursor sol has been examined by ^{13}C n.m.r. spectroscopy.[2103] ^{19}F n.m.r. spectroscopy has been used to investigate the introduction of F^- into $YBa_2Cu_3O_{7-\delta}$.[2104]

The ^7Li n.m.r. spectrum of Li_3ThF_7 has been correlated with the crystal structure.[2105]

^1H T_1 and T_2 have been measured for $Ti_4Fe_2O_{2.4}H_x$.[2106] ^1H n.m.r. spectroscopy has been used to identify hydrogen reduced titania.[2107] ^7Li n.m.r. experiments have been performed on $Li_{1+x}Ti_{2-x}O_4$ to investigate the microscopic properties of the electron spins.[2108] Solid state ^{17}O n.m.r. spectroscopy has been used to determine the molecular structure of hydrous sodium titanate.[2109] ^{47}Ti, ^{49}Ti, and ^{137}Ba n.m.r. spectroscopy has been used to study $BaTiO_3$.[2110] Variable temperature ^1H n.m.r. experiments have been performed on hydrogen reduced TiO_2.[2111] ^{29}Si m.a.s. n.m.r. spectra have been reported for $[H_2NEt_2]_2[Ti_4O_2(OH)_2(OSiPh_2O)_6]$ and related compounds.[2112] ^1H n.m.r. spectroscopy has been used to investigate $M(SO_4)_2.4H_2O$, M = Ti, V, Fe.[2113] A Si-Ti-C-O fibre has been studied by m.a.s. n.m.r. spectroscopy.[2114] $Cu_{1-x}Ti_2S_4$ has been characterised by ^{65}Cu n.m.r. spectroscopy.[2115] $Li_xCu_{0.07}Ti_2S_4$ has been investigated by ^7Li n.m.r. spectroscopy.[2116] The intercalation of Li^+ into $TiSe_2$ and $ZrSe_2$ has been studied by ^{77}Se n.m.r. spectroscopy.[2117]

The electronic structures of V_2HfH_x and $V_2Hf_{0.5}Zr_{0.5}H_x$ have been investigated by n.m.r. spectroscopy.[2118] The c.p.m.a.s. ^{13}C n.m.r. spectrum of $[(\eta^5\text{-}C_5Me_5)_2ZrC_6H_4BPh_3]$

2102 H. Dwelk, D. Freude, R. Herrmann, H. Pfeifer, and N. Pruss, *Phys. Status Solidi A*, 1989, **113**, 159 (*Chem. Abstr.*, 1989, **111**, 223 126).

2103 S. Kramer, G. Moore, G. Kordas, P.A. Keifer, and C.T.G. Knight, *Mater. Res. Soc. Symp. Proc.*, 1988, **121**, 643 (*Chem. Abstr.*, 1989, **110**, 119 781).

2104 S.N. Mishra, D. Riegel, L. Buermann, K.D. Gross, T. Kornrumpf, M. Luszik-Bhadra, H. General, H. Schmidt, and E. Zirngiebl, *Europhys. Lett.*, 1988, **7**, 165 (*Chem. Abstr.*, 1989, **110**, 16 910); J.A. Stuart, P.K. Davies, and T.P. Feist, *Solid State Ionics*, 1989, **32-33**, 1100 (*Chem. Abstr.*, 1989, **111** 144 906).

2105 S.H. Pulcinelli and J. Senegas, *J. Fluorine Chem.*, 1989, **42**, 41 (*Chem. Abstr.*, 1989, **110**, 105 518).

2106 M. Stioui, A. Resnik, M. Reshotko, A. Grayevsky, D. Shaltiel, N. Kaplan, and B. Rupp, *J. Less-Common Met.*, 1988, **141**, 177 (*Chem. Abstr.*, 1989, **110**, 32 660).

2107 P. Jonsen, *Catal. Lett.*, 1989, **2**, 345 (*Chem. Abstr.*, 1989, **111**, 222 458).

2108 M. Itoh, Y. Hasegawa, H. Yasuoka, Y. Ueda, and K. Kosuge, *Physica C (Amsterdam)*, 1989, **157**, 65 (*Chem. Abstr.*, 1989, **110**, 183 996).

2109 B.C. Bunker, C.H.F. Peden, D.R. Tallant, S.L. Martinez, and G.L. Turner, *Mater. Res. Soc. Symp. Proc.*, 1988, **121**, 105 (*Chem. Abstr.*, 1989, **110**, 142 071).

2110 T.J. Bastow, *J. Phys.: Condens. Matter*, 1989, **1**, 4985 (*Chem. Abstr.*, 1989, **111**, 207 190).

2111 P. Jonsen, *Colloids Surf.*, 1989, **36**, 127 (*Chem. Abstr.*, 1989, **111**, 121 705).

2112 N.M. Rutherford, *Report*, 1987, **LBL-23992**; Order No. DE88002081, 231 pp. Avail. NTIS. From *Energy Res. Abstr.*, 1988, **13**, Abstr. No. 6186 (*Chem. Abstr.*, 1989, **110**, 50 131).

2113 A.B. Yaroslavtsev, T.A. Karpukhina, V.F. Chuvaev, and Z.N. Prozorovskaya, *Koord. Khim.*, 1989, **15**, 911 (*Chem. Abstr.*, 1989, **111**, 246 651).

2114 V.S.R. Murthy, M.H. Lewis, M.E. Smith, and R. Dupree, *Mater. Lett.*, 1989, **8**, 263 (*Chem. Abstr.*, 1989, **111**, 200 217).

2115 A.C.W.P. James, J.B. Goodenough, N.J. Clayden, and P.M. Banks, *Mater. Res. Bull.*, 1989, **24**, 143 (*Chem. Abstr.*, 1989, **111**, 142 839).

2116 A.C.W.P. James, J.B. Goodenough, and N.J. Clayden, *J. Solid State Chem.*, 1988, **77**, 356 (*Chem. Abstr.*, 1989, **110**, 164 930).

2117 Y. Chabre and P. Deniad, *NATO ASI Ser., Ser. B*, 1987, **172**, 395 (*Chem. Abstr.*, 1989, **111**, 16 515).

2118 W. Daumer, H.R. Khan, and K. Lueders, *Phys. Rev. B: Condens. Matter*, 1988, **38**, 4427 (*Chem. Abstr.*, 1989, **110**, 16 887).

has been reported.[2119] The ^{7}Li and ^{23}Na n.m.r. spectra of $NaLiZrSi_6O_{15}$ have been measured.[2120] $Na_{1+x}Zr_2(PO_4)_{3-x}(SiO_4)_x$ has been studied by ^{23}Na and ^{31}P n.m.r. spectroscopy.[2121] The ^{7}Li and ^{77}Se n.m.r. spectrum of Li_xZrSe_2 and the ^{31}P n.m.r. spectrum of Li_xNiPS_3 have been reported.[2122]

^{51}V n.m.r. spectroscopy has been used to study δ-LiV_2O_5.[2123] Pure amorphous V_2O_5 has been characterised by ^{51}V n.m.r. spectroscopy.[2124] The composition of the active component of supported vanadium-magnesium catalysts has been determined by ^{51}V n.m.r. spectroscopy.[2125] Vanadium-molybdenum oxide has been characterised by ^{51}V and ^{95}Mo n.m.r. spectroscopy.[2126] The ^{51}V Knight shift has been determined for β-$Pb_xV_2O_5$.[2127] A two dimensional n.m.r. technique has been used to determine the quadrupole splitting and the line shape of half-integral nuclei, and applied to ^{27}Al in a zeolite and ^{51}V in NH_4VO_3.[2128] The relaxation time of ^{51}V in $BiVO_4$ has been determined.[2129] A comparative study of $V_6O_{13.27}$ and V_6O_{13} has been made using ^{51}V n.m.r. spectroscopy.[2130] The state and roles of H_2O in $V_{12-x}Mo_xO_{30+y},nH_2O$ have been studied by ^{1}H n.m.r. spectroscopy.[2131] The relaxation of ^{205}Tl in $Tl_xV_5S_8$ and $Tl_xV_6S_8$ has been studied.[2132]

^{1}H n.m.r. spectroscopy has been used to study $Ta_{1-x}Nb_xH_y$.[2133] ^{13}C c.p.m.a.s. n.m.r. spectra have been measured for $[(\eta^5\text{-}C_5Me_5)Cl_3Ta\{\eta^2\text{-}OC(NC_5H_3Me_2\text{-}2,6)SiMe_3\}]$ and related compounds.[2134]

The ^{77}Se n.m.r. spectra of $CdCr_2Se_4$ and $Cd_{0.94}In_{0.06}Cr_2Se_4$ have been determined.[2135] Cation distribution in $CdCr_2Se_4$ doped with indium has been determined using ^{53}Cr n.m.r. spectroscopy.[2136] ^{35}Cl and ^{37}Cl n.m.r. measurements have been used to clarify

2119 G.G. Hlatky, H.W. Turner, and R.R. Eckman, *J. Am. Chem. Soc.*, 1989, **111**, 2728.

2120 S. Ghose, D. Brinkmann, and J. Roos, *Phys. Chem. Miner.*, 1989, **16**, 720 (*Chem. Abstr.*, 1989, **111**, 198 640).

2121 C. Jaeger, S. Barth, and A. Feltz, *Chem. Phys. Lett.*, 1989, **154**, 45 (*Chem. Abstr.*, 1989, **110**, 146 439); C. Jaeger, S. Barth, and A. Feltz, *Chem. Phys. Lett.*, 1988, **150**, 503 (*Chem. Abstr.*, 1989, **110**, 106 804).

2122 Y. Chabre, *Mater. Res. Soc. Symp. Proc.*, 1989, **135**, 265 (*Chem. Abstr.*, 1989, **111**, 203 859).

2123 M. Bose and A. Basu, *J. Solid State Chem.*, 1989, **81**, 30 (*Chem. Abstr.*, 1989, **111**, 165 983).

2124 M. Nabavi, C. Sanchez, F. Taulelle, J. Livage, and A. De Guibert, *Solid State Ionics*, 1987, **28-30**, 1183 (*Chem. Abstr.*, 1989, **110**, 103 773).

2125 O.B. Lapina, A.V. Simakov, V.M. Mastikhin, and S.A. Veniaminov, *J. Mol. Catal.*, 1989, **50**, 55.

2126 N.N. Chumachenko, D.V. Tarasova, G.N. Kustova, R.I. Maksimovskaya, and I.P. Olen'kova, *Izv. Akad. Nauk SSSR, Neorg. Mater.*, 1988, **24**, 1679 (*Chem. Abstr.*, 1989, **110**, 68 427).

2127 S. Uji and H. Nagasawa, *J. Phys. Soc. Jpn.*, 1988, **57**, 2791 (*Chem. Abstr.*, 1989, **111**, 16 501).

2128 A. Samoson, *Exp. Tech. Phys.*, 1988, **36**, 273 (*Chem. Abstr.*, 1989, **110**, 87 201).

2129 A.R. Lim, S.H. Choh, and M.S. Jang, *Sae Mulli*, 1988, **28**, 611 (*Chem. Abstr.*, 1989, **110**, 125 921).

2130 M. Bose and A. Basu, *J. Solid State Chem.*, 1989, **81**, 147 (*Chem. Abstr.*, 1989, **111**, 223 814).

2131 T.A. Denisova, G.S. Zakharova, V.L. Volkov, and R.N. Pletnev, *Koord. Khim.*, 1989, **15**, 221 (*Chem. Abstr.*, 1989, **111**, 28 998).

2132 H. Nishihara, T. Ohtani, and S. Onoue, *Europhys. Lett.*, 1989, **8**, 189 (*Chem. Abstr.*, 1989, **110**, 164 841).

2133 K.H. Richter and A. Weiss, *J. Less-Common Met.*, 1988, **142**, 301 (*Chem. Abstr.*, 1989, **110**, 49 981).

2134 J. Arnold, T.D. Tilley, A.L. Rheingold, S.J. Geib, and A.M. Arif, *J. Am. Chem. Soc.*, 1989, **111**, 149.

2135 A.S. Moskvin, A.A. Shemyakov, and V.K. Prokopenko, *Fiz. Tverd. Tela (Leningrad)*, 1989, **31**, 293 (*Chem. Abstr.*, 1989, **111**, 186 150); G.N. Abelyashev, V.N. Berzhanskii, N.A. Sergeev, and Yu.V. Fedotov, *Pis'ma Zh. Eksp. Teor. Fiz.*, 1988, **48**, 619 (*Chem. Abstr.*, 1989, **110**, 68 353).

2136 T.G. Aminov, V.T. Kalinnikov, V.K. Prokopenko, G.G. Shabunina, and A.A. Shemyakov, *Izv. Akad. Nauk SSSR*, 1988, **24**, 1918 (*Chem. Abstr.*, 1989, **110**, 106 829); G.N. Abelyashev, V.N. Berzhanskii,

the covalency between Cr^{2+} and Cl^-.[2137]

The ^{31}P m.a.s. n.m.r. spectrum of [Mo(CO)$_5$(cyclenphosphorane)] has been used to show that the phosphorane is an open tautomer. The solution ^{13}C n.m.r. spectrum was also reported.[2138] The ^{13}C c.p.m.a.s. n.m.r. spectrum of [(triaza cyclic ligand)Mo(CO)$_3$] has been recorded.[2139] [(Bu^{t}NC)$_4$Mo(SBu^t)$_2$] has been investigated by ^{13}C c.p.m.a.s. n.m.r. spectroscopy.[2140] $Rb_{0.3}MoO_3$ has been studied using ^{87}Rb n.m.r. spectroscopy.[2141] Low temperature structural transformations in $BaMo_6S_8$ have been investigated using ^{95}Mo T_1 measurements.[2142]

Evidence from ^1H, ^{13}C, and ^{31}P n.m.r. spectroscopy has been used to show that EtOH in $H_3PW_{12}O_{40}$ is present as Et^+ and $[EtOH_2]^+$.[2143] The ^{27}Al m.a.s. n.m.r. spectrum of $H_5[AlO_4W_{12}O_{36}].6H_2O$ has been reported.[2144] The dehydration of $[Al(OH_2)][PW_{12}O_{40}].4H_2O$ has been followed by ^{27}Al and ^{31}P n.m.r. spectroscopy.[2145]

Variable temperature ^{13}C m.a.s. n.m.r. spectra have been reported for $KFe_2Mn(CO)_{12}$.[2146] ^{55}Mn and ^{59}Co n.m.r. spectra have been measured for $NiMnO_3$ and $CoMnO_3$.[2147] ^{55}Mn, ^{63}Cu, and ^{65}Cu n.m.r. spectroscopy has been used to study $CaCu_{3-x}Mn_xMn_4O_{12}$.[2148] ^{54}Mn n.m.r. spectra have been determined for $Mn(O_2CCH_3)_2.4H_2O$.[2149] ^1H n.m.r. spectroscopy has been used to investigate phase transitions of $CsMn_{1-x}Co_xCl_3.2H_2O$,[2150] and $Mn_{1-x}Co_xCl_2.2H_2O$.[2151]

The isotropic ^{13}C chemical shifts for axial and equatorial CO in [Fe(CO)$_5$] have been ob-

N.A. Sergeev, and Yu.V. Fedotov, *Ukr. Fiz. Zh. (Russ. Ed.)*, 1989, **34**, 935 (*Chem. Abstr.*, 1989, **111**, 107 863).

2137 H. Kubo, T. Hamasaki, and M. Tanimoto, *J. Phys. Soc. Jpn.*, 1988, **57**, 4379 (*Chem. Abstr.*, 1989, **110**, 146 445).

2138 D.V. Khasnis, M. Lattman, U. Siriwardane, and S.K. Chopra, *J. Am. Chem. Soc.*, 1989, **111**, 3103).

2139 S.P. van Kouwenberg, E.H. Wong, G.R. Weisman, E.J. Gabe, F.L. Lee, and P. Jackson, *Polyhedron*, 1989, **8**, 2333.

2140 S. Yao, Y. Shi, S.Lu, J. Hu, and J. Qiu, *Bopuxue Zazhi*, 1988, **5**, 335 (*Chem. Abstr.*, 1989, **111**, 107 819).

2141 P. Butaud, A. Janossy, P. Segransan, and C. Berthier, *Synth. Met. 1988*, (Pub. 1989), **29**, F499 (*Chem. Abstr.*, 1989, **111**, 48 750).

2142 K.N. Mikhalev, B.A. Aleksashin, S.V. Verkhovskii, and E.Z. Kurmaev, *Fiz. Met. Metalloved.*, 1988, **66**, 871 (*Chem. Abstr.*, 1989, **110**, 68 259).

2143 K.Y. Lee, Y. Kanda, N. Mizuno, T. Okuhara, M. Misono, S. Nakata, and S. Asaoka, *Chem. Lett.*, 1988, 1175 (*Chem. Abstr.*, 1989, **110**, 172 570).

2144 H.J. Lunk, M. Salmen, H. Weiner, C. Baeker, W. Wilde, and D. Müller, *Z. Anorg. Allg. Chem.*, 1989, **568**, 171.

2145 A.R. Siedle, T.E. Wood, M.L. Brostrom, D.C. Koskenmaki, B. Montez, and E. Oldfield, *J. Am. Chem. Soc.*, 1989, **111**, 1665.

2146 G.W. Wagner, A.A. Chen, and M.A. Vannice, *Catal. Lett.*, 1989, **2**, 413 (*Chem. Abstr.*, 1989, **111**, 220 058).

2147 I.O. Troyanchuk, A.A. Shemyakov, and V.K. Prokopenko, *Phys. Status Solidi A*, 1989, **113**, K107 (*Chem. Abstr.*, 1989, **111**, 186 019).

2148 I.O. Troyanchuk, L.A. Bashkirov, A.A. Shemyakov, and V.K. Prokopenko, *Phys. Status Solidi A*, 1988, **109**, K59 (*Chem. Abstr.*, 1989, **110**, 87 197).

2149 M. Le Gros, A. Kotlicki, and B.G. Turrell, *Hyperfine Interact.*, 1988, **43**, 311 (*Chem. Abstr.*, 1989, **110**, 184 659); M. Le Gros, A. Kotlicki, and B.G. Turrell, *Hyperfine Interact.*, 1988, (Pub. 1989), **51**, 111 (*Chem. Abstr.*, 1989, **111**, 165 990).

2150 H. Kubo, T. Hamasaki, N. Uryu, H. Nishihara, and H. Yasuoka, *Fukuoka Kogyo Daigaku Erekutoronikusu Kenkyusho Shoho*, 1986, **3**, 21 (*Chem. Abstr.*, 1989, **110**, 87 192).

2151 H. Kubo, T. Hamasaki, H. Deguchi, and K. Takeda, *J. Phys., Colloq.*, 1988, (C8, Proc. Int. Congr. Magn., 1988, Pt. 2), C8-831 (*Chem. Abstr.*, 1989, **111**, 145 560).

served for solid [$Fe(CO)_5$] at -50 °C.[2152] ^{13}C enriched [$Fe_3(CO)_{12}$], [$Ru_3(CO)_{12}$], [$Os_3(CO)_{12}$], [$Rh_6(CO)_{16}$], and [$Ir_4(CO)_{12}$] have been examined by ^{13}Cm.a.s. n.m.r. spectroscopy. Isotropic chemical shifts and principal components of the chemical shift tensors were determined.[2153] A similar ^{13}C n.m.r. study has been performed on [$Fe_3(CO)_{12}$] and [$Fe_3(CO)_8(PhC_2Ph)_2$].[2154] ^{13}C n.m.r. spectra have been used to study iodine doped [(η^4-diene)$Fe(CO)_3$].[2155] C.p.m.a.s. ^{31}P n.m.r. spectra have been measured for [$Fe_2(CO)_6$(μ-X)(μ-PPh_2)]. A correlation between chemical shift and bond angle was found.[2156] The ^{13}C shielding tensor has been determined for [$Fe_2(CO)_6(C_4Et_4)$].[2157]

The state of carbon in iron-carbon martensite has been investigated by ^{13}C n.m.r. spectroscopy.[2158] ^1H n.m.r. studies at high hydrostatic pressures and compressibilities have shown that $(NH_4)_4Fe(CN)_6.1.5H_2O$ undergoes a phase transition.[2159] ^{13}C T_1 measurements have been used to investigate iron(II) spin-crossover in [$Fe(Phen)_2(N^{13}CS)_2$].[2160]

The ^7Li n.m.r. spectrum of $Li_{0.5}Fe_{2.5}O_4$ has been determined.[2161] The ^{57}Fe n.m.r. spectra of $MFe_{12}O_{19}$, M = Ba, Sr, Pb,[2162] M = Ba,[2163] and $NiFe_2O_4$[2164] have been measured. Induced hyperfine interactions on nuclei of oxide ions in ferrite spinels have been investigated by ^{17}O n.m.r. spectroscopy.[2165] ^{57}Fe n.m.r. spectroscopy has been used to study $Cd_xCu_{1-x}Fe_2O_4$.[2166] ^{57}Fe signal decay in $FeBO_3$ has been studied.[2167] The n.m.r. spectrum of iron silicates has been determined.[2168] ^{57}Fe n.m.r. spectra of the domains and domain walls of $PbFe_{12}O_{19}$ have been reported.[2169]

2152 B.E. Hanson, *J. Am. Chem. Soc.*, 1989, **111**, 6442.
2153 T.H. Walter, L. Reven, and E. Oldfield, *J. Phys. Chem.*, 1989, **93**, 1320.
2154 G.E. Hawkes, K.D. Sales, L.Y. Lian, and R. Gobetto, *Proc. R. Soc. London, A*, 1989, **424**, 93 (*Chem. Abstr.*, 1989, **111**, 207 951).
2155 S. Miyanaga, H. Yasuda, H. Sakai, and A. Nakamura, *Chem. Mater.*, 1989, **1**, 384 (*Chem. Abstr.*, 1989, **111**, 88 189).
2156 A.J. Carty, C.A. Fyfe, M. Lettinga, S. Johnson, and L.H. Randall, *Inorg. Chem.*, 1989, **28**, 4120.
2157 S. Aime, M. Botta, G.E. Hawkes, K.D. Sales, and L.-Y. Lian, *J. Organomet. Chem.*, 1989, **368**, 331.
2158 F.E. Fujita and S. Nasu, *Mater. Sci. Forum*, 1987, (Pub. 1989), **37**, 235 (*Chem. Abstr.*, 1989, **110**, 216 965).
2159 D.R. Balasubramanyam, S.V. Bhat, M. Mohan, and A.K. Singh, *Solid State Ionics 1987*, (Pub. 1988), **28**, 664 (*Chem. Abstr.*, 1989, **110**, 87 217).
2160 B.R. McGarvey, *J. Magn. Reson.*, 1989, **82**, 253.
2161 A.A. Shemyakov, A.M. Kotel'va, V.A. Klochan, and A.V. Kunevich, *Fiz. Tverd. Tela (Leningrad)*, 1988, **30**, 3140 (*Chem. Abstr.*, 1989, **110**, 17 590).
2162 H. Stepankova, J. Englich, P. Novak, B. Sedlak, and M. Pfeffer, *Hyperfine Interact. 1988*, (Pub. 1989), **50**, 639 (*Chem. Abstr.*, 1989, **111**, 207 885).
2163 S.P. Kuntsevich and A.A. Bezlepkin, *Fiz. Tverd. Tela (Leningrad)*, 1989, **31**, 261 (*Chem. Abstr.*, 1989, **110**, 224 252); A.A. Bezlepkin, S.P. Kuntsevich, and Yu.A. Popkov, *Fiz. Nizk. Temp. (Kiev)*, 1989, **15**, 875 (*Chem. Abstr.*, 1989, **111**, 246 563).
2164 V.B. Tyutyunnik, V.S. Abramov, V.P. Pashchenko, and V.I. Arkharov, *Dokl. Akad. Nauk SSSR*, 1989, **306**, 1118 (*Chem. Abstr.*, 1989, **111**, 89 169).
2165 N.M. Kovtun, A.M. Kotel'va, A.S. Moskvin, and A.A. Shemyakov, *Zh. Eksp. Teor. Fiz.*, 1989, **95**, 1459 (*Chem. Abstr.*, 1989, **111**, 125 516).
2166 J. Hankiewicz, Z. Pajak, and J. Radomski, *Acta Phys. Pol., A*, 1988, (Pub. 1989), **A76**, 63 (*Chem. Abstr.*, 1989, **111**, 246 454).
2167 A.V. Ivanov, V.R. Korneev, A.P. Paugurt, and I.V. Pleshakov, *Pis'ma Zh. Tekh. Fiz.*, 1988, **14**, 2049 (*Chem. Abstr.*, 1989, **110**, 106 833); Kh.G. Bogdanova, V.A. Golenishchev-Kutuzov, L.I. Medvedev, M.I. Kurkin, and E.A. Turov, *Zh. Eksp. Teor. Fiz.*, 1989, **95**, 613 (*Chem. Abstr.*, 1989, **110**, 204 419).
2168 H. Nagata, T. Takeguchi, S. Iwamoto, H. Matsuda, M. Inoue, and T. Inui, *Shokubai*, 1989, **31**, 369 (*Chem. Abstr.*, 1989, **111**, 240 739).
2169 S.P. Kuntsevich and A.A. Bezlepkin, *Ukr. Fiz. Zh. (Russ. Ed.)*, 1989, **34**, 126 (*Chem. Abstr.*, 1989, **110**, 184 626).

The ^{19}F n.m.r. spectrum of $K_2Co_xFe_{1-x}F_4$ has been reported.[2170] A drastic increase of the T_1 of ^{133}Cs in $CsFeCl_3$ has been observed at 4.2 K.[2171] Nuclear relaxation times of ^{31}P have been measured for $[Au_{55}(PPh_3)_{12}Cl_6]$ and $[Ru_{55}(PBu^t_3)_{12}Cl_{20}]$.[2172] Relaxation of ^{35}Cl in K_2OsCl_6 has been determined.[2173]

The ^{13}C m.a.s. n.m.r. spectrum of $[Co_4(CO)_{12}]$ has been determined.[2174] The ^1H T_1 of $[Co(NH_3)_6]X_2$ has been measured.[2175] The ^{13}C m.a.s. n.m.r. spectra of some Co(III) complexes with quadridentate ligands[2176] and $[Bu^n{}_4N]_2$[Co(5,6-dihydro-1,4-dithiin-2,3-dithiolato)$_2$][2177] have been measured. ^1H n.m.r. spectroscopy has been used to study the spin states of the Mn^{2+} impurity in $CoCl_2.2H_2O$.[2178] ^{133}Cs n.m.r. spectra of $CsCoCl_3$ have been determined to investigate the magnetic phase transitions.[2179] The ^{19}F n.m.r. spectrum of M_3RhF_6, M = K, Rb, Cs, has been studied.[2180]

^3He n.m.r. spectroscopy has been used to study ageing effects in palladium tritide.[2181] Five separate ^{13}C resonances are observed for the methylcyclopentadienyl ring carbon atoms in $[(\eta^5\text{-}MeC_5H_4)PtMe_3]$.[2182] The ^{13}C c.p.m.a.s. n.m.r. spectrum of [Ni(5,7,12,14,19,-21,26,28-$^{13}C_8$)phthalocyanin] has been recorded.[2183] The ^{15}N shielding tensors of *cis*-$[Pt(NH_3)_2(SCN)_2]$ show the anisotropy of the heavy atom shift.[2184] Solid state ^{13}C and ^{31}P n.m.r. spectra have been used to show distortions of the nominally square-planar complexes $[M(PR_3)_2X_2]$, M = Pd, Pt.[2185] The solid-state ^{13}C n.m.r. spectra of [tetramethyl-tetraselenafulvene]$_2[Ni\{Se_2C_2(CF_3)_2\}_2]$[2186] and $[M_2(S_2CMe)_4]$[2187] have been recorded.

The ^1H T_1 values of [bis(ethylenedithiolo)tetrathiafulvalene]$_2Cu(NCS)_2$ have been

2170 W.A.H.M. Vlak, M.J. Van Dort, A.F.M. Arts, and H.W. De Wijn, *Phys. Rec. B: Condens. Matter*, 1988, **38**, 11 659 (*Chem. Abstr.*, 1989, **110**, 127 297).

2171 M. Chiba, Y. Ajiro, K. Adachi, and T. Morimoto, *J. Phys. Soc. Jpn.*, 1988, **57**, 3178 (*Chem. Abstr.*, 1989, **110**, 17 572).

2172 M.P.J. Van Staveren, H.B. Brom, L.J. De Jongh, and G. Schmid, *Z. Phys. D: At., Mol. Clusters*, 1989, **12**, 451 (*Chem. Abstr.*, 1989, **111**, 165 982).

2173 M.A. Singh and R.L. Armstrong, *J. Magn. Reson.*, 1989, **84**, 448.

2174 S. Aime, M. Botta, R. Gobetto, and B.E. Hanson, *Inorg. Chem.*, 1989, **28**, 1196.

2175 J. Czaplicki, N. Weiden, and A. Weiss, *Physica B (Amsterdam)*, 1989, **159**, 214 (*Chem. Abstr.*, 1989, **111**, 207 978).

2176 Y. Yamamoto, E. Toyota, and S. Shimokawa, *Bull. Chem. Soc. Jpn.*, 1989, **62**, 2717.

2177 J.H. Welch, R.D. Bereman, P. Singh, and C. Moreland, *Inorg. Chim. Acta*, 1989, **158**, 17.

2178 H. Kubo, T. Hamasaki, and K. Takeda, *J. Phys. Soc. Jpn.*, 1988, **57**, 3568 (*Chem. Abstr.*, 1989, **110**, 68 338).

2179 H. Kikuchi, Y. Ajiro, and T. Goto, *J. Phys. Soc. Jpn.*, 1989, **58**, 43 (*Chem. Abstr.*, 1989, **110**, 146 449); H. Kikuchi and Y. Ajiro, *J. Phys. Soc. Jpn.*, 1989, **58**, 692 (*Chem. Abstr.*, 1989, **110**, 224 239); Y. Ajiro, H. Kikuchi, T. Okita, M. Chiba, K. Adachi, M. Mekata, and T. Goto, *J. Phys. Soc. Jpn.*, 1989, **58**, 390 (*Chem. Abstr.*, 1989, **110**, 224 238).

2180 M.L. Afanas'ev, O.V. Falaleev, V.V. Men'shikov, and E.P. Zeer, *Dokl. Akad. Nauk SSSR*, 1988, **303**, 1396 (*Chem. Abstr.*, 1989, **110**, 127 338).

2181 G.C. Abell and A. Attalla, *Fusion Technol.*, 1988, **14**, 643 (*Chem. Abstr.*, 1989, **111**, 69 570).

2182 Z. Xue, M.J. Strouse, D.K. Shuh, C.B. Knobler, H.D. Kaesz, R.F. Hicks, and R.S. Williams, *J. Am. Chem. Soc.*, 1989, **111**, 8779.

2183 A.G.M. Barrett, W.E. Broderick, B.M. Hoffman, and C.S. Velazquez, *J. Org. Chem.*, 1989, **54**, 3233.

2184 R.A. Santos, W.-J. Chien, G.S. Harbison, J.D. McCurry, and J.E. Roberts, *J. Magn. Reson.*, 1989, **84**, 357.

2185 J.A. Rahn, D.J. O'Donnell, A.R. Palmer, and J.H. Nelson, *Inorg. Chem.*, 1989, **28**, 2631.

2186 A.K. Whittaker, P.C. Stein, P. Bernier, W.B. Heuer, and B.M. Hoffman, *J. Phys. Chem.*, 1989, **93**, 3038.

2187 S. Kinoshita, H. Wakita, and M. Yamashita, *J. Chem. Soc., Dalton Trans.*, 1989, 2457.

measured.[2188] The ^{13}C Knight shifts were also measured.[2189] Quadrupolar effects in the ^{31}P and ^{119}Sn c.p.m.a.s. n.m.r. spectra of $[(Ph_3P)_2Cu(NO_3)]$ and $[Ph_3SnCl]$ have been used to derive $^1J(^{63}Cu-^{31}P)$ and $^1J(^{119}Sn-^{13}C)$.[2190] The ^{31}P c.p.m.a.s. n.m.r. spectra of $[(Ph_3P)_3(CuCl)_2]$,[2191] $[Cu\{P[C_6H_2(OMe)_3-2,4,6]_3\}Cl]$,[2192] and $[PPh_3Me][(Ph_3P)_2CuI_2]$[2193] have been reported. ^{65}Cu and ^{133}Cs n.m.r. spectroscopy has been used to investigate alkali metal promoted copper catalysts.[2194] ^1H and ^{13}C n.m.r. spectroscopy has been used to investigate Cu/Zn/Al oxide methanol synthesis catalyst.[2195] ^{17}O, ^{63}Cu, and ^{65}Cu T_1 values have been determined for superconducting copper oxides.[2196] The ^1H n.m.r. spectrum of $CuSO_4.5H_2O$ has been investigated.[2197] ^{31}P n.m.r. spectra have been measured for $AgI-Ag_2O-P_2O_5$ glasses.[2198] Spin rate dependent chemical shifts have been observed for ^{63}Cu, ^{65}Cu, ^{79}Br, and ^{127}I in CuCl, CuBr, and CuI.[2199] ^1H T_1 has been determined for $[(TMTSF)_3Cu_2Cl_4]$ and $[(TMTSF)CuBr_2]$.[2200]

Single crystals of a zinc aluminophosphate molecular sieve have been examined by n.m.r. spectroscopy.[2201] The ^{31}P m.a.s. n.m.r. spectra of $ZnSiP_2$, $ZnGeP_2$, and $ZnSnP_2$ have been recorded.[2202] The ^{19}F n.m.r. spectrum of $KZnF_3$ shows three non-equivalent sites.[2203] N.m.r. spectroscopy has been used to study Mn^{2+} doped $(NH_4)_2ZnBr_4$.[2204]

The ^{113}Cd c.p.m.a.s. n.m.r. spectrum of a Cd^{2+} complex of 1,4,7,10,13,16-hexaazacyclooctadecane has been reported.[2205] Samples of a $Li_2O-P_2O_5-CdO$ glass have been studied using ^{31}P n.m.r. spectroscopy.[2206] Solid state ^{113}Cd n.m.r. spectroscopy has been used to

2188 T. Takahashi, T. Tokiwa, K. Kanoda, H. Urayama, H. Yamochi, and G. Saito, *Synth. Met.*, 1988, **27**, A319 (*Chem. Abstr.*, 1989, **110**, 184 627).

2189 D. Schweitzer, K. Polychroniadis, T. Klutz, H.J. Keller, I. Hennig, I. Heinen, U. Haeberlen, E. Gogu, and S. Gartner, *Synth. Met.*, 1988, **27**, A465 (*Chem. Abstr.*, 1989, **110**, 145 937).

2190 A.C. Olivieri, *J. Magn. Reson.*, 1989, **81**, 201.

2191 C.F. Albert, P.C. Healy, J.D. Kildea, C.L. Raston, B.W. Skelton, and A.H. White, *Inorg. Chem.*, 1989, **28**, 1300.

2192 G.A. Bowmaker, J.D. Cotton, P.C. Healy, J.D. Kildea, S.B. Silong, B.W. Skelton, and A.H. White, *Inorg. Chem.*, 1989, **28**, 1462.

2193 G.A. Bowmaker, A. Camus, P.C. Healy, B.W. Skelton, and A.H. White, *Inorg. Chem.*, 1989, **28**, 3883.

2194 P.J. Chu, B.C. Gerstein, G.R. Sheffer, and T.S. King, *J. Catal.*, 1989, **115**, 194 (*Chem. Abstr.*, 1989, **110**, 102 491).

2195 P.R. Dennison, K.J. Packer, and M.S. Spencer, *J. Chem. Soc., Faraday Trans. 1*, 1989, **85**, 3537.

2196 Y. Kitaoka, K. Ishida, K. Fujiwara, T. Kondo, K. Asayama, H. Katayama-Yoshida, Y. Okabe, and T. Takahashi, *J. Magn. Magn. Mater.*, 1988, **76**, 527 (*Chem. Abstr.*, 1989, **110**, 164 848).

2197 W.G. Bos, *Report*, 1986, Order No. PB88-244850, 105 pp. Avail. NTIS. From *Gov. Rep. Announce. Index (U.S.)*, 1988, **88**, Abstr. No. 860 134 (*Chem. Abstr.*, 1989, **111**, 107 841); R.J. Wittebort, *J. Magn. Reson.*, 1989, **83**, 626.

2198 S. Hayashi and K. Hayamizu, *J. Solid State Chem.*, 1989, **80**, 195 (*Chem. Abstr.*, 1989, **111**, 181 223).

2199 S. Hayashi and K. Hayamizu, *Chem. Lett.*, 1989, 1419 (*Chem. Abstr.*, 1989, **111**, 186 156).

2200 M. Inoue, M.B. Inoue, C. Cruz-Vazquez, M. Mizuno, T. Asaji, L.S. Prabhumirashi, and D. Nakamura, *Synth. Met.*, 1988, **27**, B151 (*Chem. Abstr.*, 1989, **110**, 241 409).

2201 S. Han, C. Yang, Y. Guo, and H. Yuan, *Jilin Daxue Ziran Kexue Xuebao*, 1989, 104 (*Chem. Abstr.*, 1989, **111**, 87 650).

2202 T.A. Vanderah and R.A. Nissan, *J. Phys. Chem. Solids*, 1988, **49**, 1335.

2203 M. Engelsberg, R.E. De Souza, and P.H.V. Pereira, *Phys. Rev. B: Condens. Matter*, 1989, **40**, 106 (*Chem. Abstr.*, 1989, **111**, 145 596).

2204 P.M. Rao and K.V. Reddy, *Cryst. Res. Technol.*, 1989, **24**, 893 (*Chem. Abstr.*, 1989, **111**, 246 511).

2205 P.S. Marchetti, S. Bank, T.W. Bell, M.A. Kennedy, and P.D. Ellis, *J. Am. Chem. Soc.*, 1989, **111**, 2063.

2206 Z. Li, D. Jiang, Q. Jin, and D. Ding, *Wuli Xuebao*, 1989, **38**, 326 (*Chem. Abstr.*, 1989, **111**, 69 576).

study cadmium speciation in environmentally contaminated sediments.[2207]

A c.p.m.a.s. ^{31}P COSY n.m.r. study of $[Hg(PCy_3)_2(ClO_4)_2]$ has been reported.[2208] ^{31}P c.p.m.a.s. n.m.r. spectroscopy has been used to investigate $[(R_3P)_2HgX_2]$.[2209] ^1H T_1 depends on magnetic field for {bis(ethylenedithio)tetrathiafulvene}$_4Hg_{2.89}Br_8$.[2210]

The m.a.s. ^{11}B n.m.r. spectra of several hydroborates have been recorded.[2211] Quadrupolar and dipolar interactions in polycrystalline $Cs_2B_6H_6$ have been measured using ^1H, ^{11}B, and ^{133}Cs n.m.r. spectroscopy.[2212] The change in structure of B_2O_3 glass between the vitreous and molten states has been investigated by ^{11}B n.m.r. spectroscopy.[2213] In $Li_2O.3B_2O_3$ glass, the T_1 of ^8Li and ^{12}B has been studied using β-radiation detected n.m.r. relaxation.[2214] The relationship between structure and acidity in non-stoichiometric aluminium borate catalysts has been studied by ^{11}B n.m.r. spectroscopy[2215] and by ^{11}B and ^{27}Al n.m.r. spectroscopy.[2216] Lithium borosilicate glasses have been studied by n.m.r. spectroscopy.[2217] The low temperature reflow of borophosphosilicate glasses doped with arsenic has been examined by n.m.r. spectroscopy.[2218] ^{11}B and ^{31}P n.m.r. spectroscopy has been used to study the reaction of boron monophosphate with $Na_4P_2O_7$.[2219]

The application of total suppression of spinning sidebands to m.a.s. ^{27}Al n.m.r. spectra has been reported.[2220] The m.a.s. n.m.r. spectra of satellite transitions of ^{23}Na and ^{27}Al nuclei in solids have been examined.[2221] ^{27}Al n.m.r. spectra have been reported for AlN.[2222]

The direct detection of the ^{27}Al resonance with a SQUID spectrometer has been reported.[2223] ^{27}Al n.m.r. spectroscopy has been used to study mixed alkali effects in β-alum-

2207 S. Bank, J.F. Bank, P.S. Marchetti, and P.D. Ellis, *J. Environ. Qual.*, 1989, **18**, 25 (*Chem. Abstr.*, 1989, **110**, 179 147).

2208 T. Allman, *J. Magn. Reson.*, 1989, **83**, 637.

2209 E.C. Alyea, K.J. Fisher, and S. Johnson, *Can. J. Chem.*, 1989, **67**, 1319.

2210 A.V. Skripov, A.P. Stepanov, V.A. Merzhanov, R.N. Lyubovskaya, and R.B. Lyubovskii, *Pis'ma Zh. Eksp. Teor. Fiz.*, 1989, **49**, 229 (*Chem. Abstr.*, 1989, **111**, 165 965).

2211 H. Mongeot, A. Aboukhassib, B. Bonnetot, and F. Lefebvre, *J. Chim. Phys. Phys.-Chim. Biol.*, 1988, **85**, 979.

2212 V.I. Privalov, V.P. Tarasov, M.A. Meladze, D.M. Vinitskii, K.A. Solntsev, and N.T. Kuznetsov, *Zh. Neorg. Khim.*, 1989, **34**, 1121 (*Chem. Abstr.*, 1989, **111**, 89 213).

2213 S. Xu, L. Pan, F. Tian, and X. Wu, *Wuli Xuebao*, 1988, **37**, 1866 (*Chem. Abstr.*, 1989, **110**, 197 766).

2214 A. Schirmer, P. Heitjans, H. Ackermann, B. Bader, P. Freilaender, and H.J. Stoeckmann, *Solid State Ionics*, 1987, (Pub. 1988), **28-30(Pt. 1)**, 717 (*Chem. Abstr.*, 1989, **110**, 127 303).

2215 K.P. Peil, L.G. Galya, and G. Marcelin, *Proc. - Int. Congr. Catal., 9th*, 1988, **4**, 1712 (*Chem. Abstr.*, 1989, **111**, 121 677).

2216 K.P. Peil, L.G. Galya, and G. Marcelin, *J. Catal.*, 1989, **115**, 441 (*Chem. Abstr.*, 1989, **110**, 102 511).

2217 J. Zhong, X. Wu, M.L. Liu, and P.J. Bray, *J. Non-Cryst. Solids*, 1988, **107**, 81 (*Chem. Abstr.*, 1989, **111**, 121 330).

2218 K. Kobayashi, *Glass Technol.*, 1989, **30**, 110 (*Chem. Abstr.*, 1989, **111**, 62 555).

2219 V.A. Sinyaev, K.T. Sarsenbaev, S.S. Temirova, and V.Zh. Ushanov, *Zh. Neorg. Khim.*, 1989, **34**, 2019 (*Chem. Abstr.*, 1989, **111**, 246 776).

2220 K.R. Carduner, *J. Magn. Reson.*, 1989, **81**, 312.

2221 H.J. Jakobsen, J. Skibsted, H. Bildsøe, and N.C. Nielsen, *J. Magn. Reson.*, 1989, **85**, 173.

2222 J. Haase, D. Freude, T. Froehlich, G. Himpel, F. Kerbe, E. Lippmaa, H. Pfeifer, P. Sarv, H. Schaefer, and B. Seiffert, *Chem. Phys. Lett.*, 1989, **156**, 328 (*Chem. Abstr.*, 1989, **111**, 32 520).

2223 J. Chang, C. Connor, E.L. Hahn, H. Huber, and A. Pines, *J. Magn. Reson.*, 1989, **82**, 387.

ina.[2224] ^{27}Al n.m.r. cross polarisation has been used to examine surface aluminium atoms in aluminas.[2225] The thermolysis of $[Al(OH_2)_6]Cl_3$ has been monitored by ^{27}Al m.a.s. n.m.r. spectroscopy.[2226] M.a.s. ^{17}O n.m.r. spectra have been obtained from ^{17}O labelled α-Al_2O_3, AlO(OH), and θ-Al_2O_3.[2227] ^{27}Al n.m.r. spectroscopy has been used to study aluminium colloids.[2228] Fine alumina powder from $Al(OBu^s)_3$ hydrolysis has been characterised by ^{27}Al n.m.r. spectroscopy.[2229] Optical Raman heterodyne detection of ^{27}Al superhyperfine spectra in ruby has been reported.[2230] The semi-empirical theory of nuclear shielding has been applied to ^{27}Al n.m.r. spectra, especially in glasses.[2231] M.a.s. ^{27}Al n.m.r. spectra have been observed for amorphous anodic oxide films on alumina.[2232] ^{27}Al n.m.r. spectroscopy has been used to investigate the mechanism of mineral tanning.[2233] ^{23}Na T_1 measurements have been made on alkali aluminogermanate glasses.[2234] The ^{27}Al n.m.r. spectra of solid M_3AlF_6 show the effects of condensation of AlF_6 octahedra.[2235]

For InP, the conduction electron density at the ^{31}P and ^{115}In nuclei has been deduced.[2236] The phase transition in $TlGaSe_2$ has been studied by ^{71}Ga n.m.r. spectroscopy.[2237] Indirect exchange nuclear interactions in $TlInSe_2$, TlSe, and $TlInTe_2$ have been studied by ^{203}Tl and ^{205}Tl n.m.r. spectroscopy.[2238] Chemical shift anisotropies of ^{205}Tl and ^{207}Pb in $TlPbI_3$ have been shown to be determined by the distortions of the MI_n coordination polyhedra.[2239]

^{11}B n.m.r. spectroscopy has been used to study the structure of Na_2O-B_2O_3-Al_2O_3-SiO_2

2224 D.E. Hintenlang, E.J. Holupka, P.J. Bray, and S.G. Greenbaum, *Solid State Ionics*, 1989, **34**, 207 (*Chem. Abstr.*, 1989, **111**, 186 131).

2225 H.D. Morris and P.D. Ellis, *J. Am. Chem. Soc.*, 1989, **111**, 6045.

2226 G. Blumenthal, G. Wegner, D. Müller, A. Samoson, and G. Kranz, *Z. Anorg. Allg. Chem.*, 1989, **576**, 43.

2227 T.H. Walter and E. Oldfield, *J. Phys. Chem.*, 1989, **93**, 6744.

2228 J.Y. Bottero, D. Tchoubar, J.M. Cases, J.J. Fripiat, and F. Fiessinger, *Process Technol. Proc.*, 1988, **7**, 459 (*Chem. Abstr.*, 1989, **110**, 63 431).

2229 C.E. Kim, H.W. Lee, H.J. Jung, G.D. Kim, and N.G. Park, *Yoop Hakhoechi*, 1988, **25**, 585 (*Chem. Abstr.*, 1989, **111**, 62 734).

2230 A. Szabo, T. Muramoto, and R. Kaarli, *Opt. Lett.*, 1988, **13**, 1075 (*Chem. Abstr.*, 1989, **110**, 68 348).

2231 E. Hallas and U. Sternberg, *Mol. Phys.*, 1989, **68**, 315.

2232 K. Kobayashi, K. Shimizu, and Y. Kurima, *Hyomen Gijutsu*, 1989, **40**, 486 (*Chem. Abstr.*, 1989, **111**, 16 534); I. Farnan, R. Dupree, Y. Jeong, G.E. Thompson, G.C. Wood, and A.J. Forty, *Thin Solid Films*, 1989, **173**, 209 (*Chem. Abstr.*, 1989, **111**, 204 016); I. Farnan, R. Dupree, A.J. Forty, Y.S. Jeong, G.E. Thompson, and G.C. Wood, *Philos. Mag. Lett.*, 1989, **59**, 189 (*Chem. Abstr.*, 1989, **111**, 64 466).

2233 A.D. Covington, R.A. Hancock, and I.A. Ioannidis, *J. Soc. Leather Technol. Chem.*, 1989, **73**, 1 (*Chem. Abstr.*, 1989, **111**, 59 911).

2234 K.L. Ngai, J.N. Mundy, H. Jain, O. Kanert, and G. Balzer-Jollenbeck, *Phys. Rev. B: Condens. Matter*, 1989, **39**, 6169 (*Chem. Abstr.*, 1989, **111**, 31 822).

2235 D. Müller and U. Bentrup, *Z. Anorg. Allg. Chem.*, 1989, **575**, 17.

2236 B. Gotschy, G. Denninger, H. Obloh, W. Wilkening, and J. Schneider, *Solid State Commun.*, 1989, **71**, 629 (*Chem. Abstr.*, 1989, **111**, 246 536); B. Clerjaud, F. Gendron, H. Obloh, J. Schneider, and W. Wilkening, *Phys. Rev. B: Condens. Matter*, 1989, **40**, 2042 (*Chem. Abstr.*, 1989, **111**, 246 483).

2237 N.T. Mamedov, E.S. Krupnikov, and A.M. Panich, *Fiz. Tverd. Tela (Leningrad)*, 1989, **31**, 290 (*Chem. Abstr.*, 1989, **110**, 125 957).

2238 N.K. Moroz, N.T. Mamedov, and S.P. Gabuda, *Fiz. Tverd. Tela (Leningrad)*, 1989, **31**, 250 (*Chem. Abstr.*, 1989, **111**, 145 590).

2239 A.B. Vassilikou-Dova and G. Lehmann, *Z. Naturforsch., A.*, 1988, **43**, 1072 (*Chem. Abstr.*, 1989, **110**, 146 443).

glasses.[2240] Glasses of the $NaAlSi_3O_8$-$NaBSi_3O_8$ have been investigated using ^{11}B, ^{23}Na, ^{27}Al, and ^{29}Si n.m.r. spectroscopy.[2241] The lack of local structure in oxide glasses quenched during flow has been demonstrated by ^{11}B, ^{27}Al, and ^{29}Si n.m.r. spectroscopy.[2242] N.m.r. spectra have been used to assist the structural interpretation of the low temperature reflow of borosilicate glasses doped with BF_3.[2243] $Na_8[AlSiO_4]_6[B(OH)_4]_2$ has been studied by ^{11}B and ^{29}Si n.m.r. spectroscopy.[2244] ^{29}Si n.m.r. spectroscopy has been used to investigate the structure of amorphous silica-alumina supports.[2245]

Chemically stabilised β-cristobalite has been studied by n.m.r. spectroscopy.[2246] Water in minerals of Al_2SiO_5 has been investigated by ^1H and ^{29}Si n.m.r. spectroscopy.[2247] ^{29}Si n.m.r. spectra have been recorded for sodium aluminosilicate gel.[2248] The homogeneity of alumina-silica glass has been investigated by ^{27}Al n.m.r. spectroscopy.[2249] Cluster analysis of aluminosilicate glasses by ^{29}Si n.m.r. spectroscopy has been reported.[2250] Physical flaws in alumina reinforced with SiC fibres have been detected by n.m.r. imaging.[2251]

Some clay minerals have been characterised by ^{27}Al and ^{29}Si n.m.r. spectroscopy.[2252] The mechanisms of palygorskite and sepiolite alteration have been deduced from ^{27}Al and ^{29}Si n.m.r. spectroscopy.[2253] ^{29}Si m.a.s. n.m.r. spectroscopy has been used to study the hydration and compressive strength of cement paste.[2254] N.m.r. spectroscopy has been used to study HCl treatment of clinoptilolite.[2255] ^{29}Si n.m.r. spectroscopy has been used to study the synthesis of silicon and manganese substituted hydroxyapatites.[2256] The dispersion of charge deficits in a tetrahedral sheet of phyllosilicates has been

2240 Y.J. Kim, S.J. Moon, M.J. Gang, S.Y. Kim, S.K. Hong, S.J. Chung, and D.J. Cha, *Sae Mulli*, 1988, **28**, 599 (*Chem. Abstr.*, 1989, **110**, 80 978).
2241 K.L. Geisinger, R. Oestrike, A. Navrotsky, G.L. Turner, and R.J. Kirkpatrick, *Geochim. Cosmochim. Acta*, 1988, **52**, 2405 (*Chem. Abstr.*, 1989, **110**, 61 324).
2242 J.F. Stebbins, D.R. Spearing, and I. Farnan, *J. Non-Cryst. Solids*, 1989, **110**, 1 (*Chem. Abstr.*, 1989, **111**, 240 449).
2243 K. Kobayashi, *Glass Technol.*, 1988, **29**, 253 (*Chem. Abstr.*, 1989, **110**, 119 708).
2244 J.C. Buhl, G. Engelhardt, and J. Felsche, *Zeolites*, 1989, **9**, 40 (*Chem. Abstr.*, 1989, **110**, 164 928).
2245 M. McMillan, J.S. Brinen, J.D. Carruthers, and G.L. Haller, *Colloids Surf.*, 1989, **38**, 133 (*Chem. Abstr.*, 1989, **111**, 202 759).
2246 A.J. Perrotta, D.K. Grubbs, E.S. Martin, N.R. Dando, H.A. McKinstry, and C.Y. Huang, *J. Am. Ceram. Soc.*, 1989, **72**, 441 (*Chem. Abstr.*, 1989, **110**, 178 266).
2247 A.M. Kalinichenko, A.I. Katalenets, V.Ya. Proshko, and L.F. Pasal'skaya, *Geokhimiya*, 1989, 1024 (*Chem. Abstr.*, 1989, **111**, 137 676).
2248 D.M. Ginter, A.T. Bell, and C.J. Radke, *J. Magn. Reson.*, 1989, **81**, 217 (*Chem. Abstr.*, 1989, **111**, 89 202).
2249 M. Yamane, S. Inoue, A. Yasumori, and M. Iwasaki, *Adv. Fusion Glass, Proc. Int. Conf., 1st*, 1988, **51**, 1. Ed. by D.F. Bickford, Am. Ceram. Soc.: Westerville, Ohio (*Chem. Abstr.*, 1989, **111**, 218 229).
2250 M. Nofz, K. Forkel, F.G. Wihsmann, and H.G. Bartel, *Phys. Chem. Glasses*, 1989, **30**, 46 (*Chem. Abstr.*, 1989, **110**, 197 801).
2251 S. Karunanithy and S. Mooibroek, *J. Mater. Sci.*, 1989, **24**, 3686 (*Chem. Abstr.*, 1989, **111**, 218 319).
2252 D.E. Woessner, *Am. Mineral.*, 1989, **74**, 203 (*Chem. Abstr.*, 1989, **110**, 118 453).
2253 S. Komarneni, *Clays Clay Miner.*, 1989, **37**, 469 (*Chem. Abstr.*, 1989, **111**, 157 535).
2254 G. Parry-Jones, A.J. Al-Tayyib, S.U. Al-Dulaijan, and A.I. Al-Mana, *Cem. Concr. Res.*, 1989, **19**, 228 (*Chem. Abstr.*, 1989, **110**, 198 138).
2255 M.V. Sychev, V.V. Goncharuk, N.G. Vasil'ev, V.V. Myalkovskii, and L.M. Oleinik, *Kinet. Katal.*, 1989, **30**, 410 (*Chem. Abstr.*, 1989, **111**, 64 818).
2256 M. Miyake, K. Mochizuki, Y. Nagayama, T. Suzuki, and H. Kobayashi, *Gypsum Lime*, 1988, **217**, 397 (*Chem. Abstr.*, 1989, **110**, 230 604).

investigated by ^{29}Si n.m.r. spectroscopy.[2257]

Angle-dependent n.m.r. spectroscopy has been used to study the lattice sites of water of hydration in natrolite.[2258] Correlations between ^{29}Si chemical shifts and structure parameters of sodalites have been observed.[2259] The ^{23}Na n.m.r. spectra of sodalites have been measured.[2260] Silicon and aluminium ordering in leucite has been investigated by high resolution ^{27}Al n.m.r. spectroscopy.[2261] ^{23}Na, ^{27}Al, and ^{29}Si n.m.r. spectra have been determined for alkali feldspar.[2262] ^{27}Al and ^{29}Si n.m.r. spectroscopy has been used to study quaternary ammonium cation effects on crystallisation of zeolites of the offretite-erionite family.[2263] ^{27}Al and ^{29}Si n.m.r. spectroscopy has been used to investigate hydrothermal isomorphous substitution of aluminium in faujasite frameworks.[2264] Defects in dealuminated faujasite have been characterised by ^{29}Si n.m.r. spectroscopy.[2265] Allophane and imogolite have been studied by ^{27}Al and ^{29}Si n.m.r. spectroscopy.[2266] Structural and thermal transformations of imogolite have been studied by ^{27}Al and ^{29}Si n.m.r. spectroscopy.[2267] Disordering and surface hydration of diopside have been investigated by ^{29}Si n.m.r. spectroscopy.[2268] ^{27}Al and ^{29}Si n.m.r. spectroscopy has been used to investigate kaolinite dehydroxylation.[2269] The synthesis and sintering of corderite have been investigated by ^{29}Si n.m.r. spectroscopy.[2270] The ^{29}Si n.m.r. spectrum of $Mg_2[Al_2BeSi_6O_{18}]$ indicates that all the silicon sites are located within the 6-membered tetrahedral rings.[2271] Structural changes of zeolites in mechanochemical processes have been studied by ^{27}Al and ^{29}Si n.m.r. spectroscopy.[2272] Activated layered silicates have been investigated by ^{27}Al and ^{29}Si n.m.r. spectro-

2257 C.P. Herrero, J. Sanz, and J.M. Serratosa, *J. Phys. Chem.*, 1989, **93**, 4311.

2258 V.A. Detinich, B.N. Grechushnikov, and V.Yu. Galitskii, *Kristallografiya*, 1989, **34**, 494 (*Chem. Abstr.*, 1989, **110**, 183 495).

2259 G. Engelhardt, S. Luger, J.C. Buhl, and J. Felsche, *Zeolites*, 1989, **9**, 182 (*Chem. Abstr.*, 1989, **111**, 48 566).

2260 R. Janssen, R.E.H. Breuer, E. De Boer, and G. Geismar, *Zeolites*, 1989, **9**, 59 (*Chem. Abstr.*, 1989, **110**, 102 355).

2261 B.L. Phillips, R.J. Kirkpatrick, and A. Putnis, *Phys. Chem. Miner.*, 1989, **16**, 591 (*Chem. Abstr.*, 1989, **111**, 118 280).

2262 B.L. Phillips, R.J. Kirkpatrick, and G.L. Hovis, *Phys. Chem. Miner.*, 1988, **16**, 262 (*Chem. Abstr.*, 1989, **111**, 165 964).

2263 M.L. Occelli, G.P. Ritz, P.S. Iyer, R.D. Walker, and B.C. Gerstein, *Zeolites*, 1989, **9**, 104 (*Chem. Abstr.*, 1989, **110**, 203 122).

2264 H. Hamdan, B. Sulikowski, and J. Klinowski, *J. Phys. Chem.*, 1989, **93**, 350.

2265 G.J. Ray, A.G. Nerheim, and J.A. Donohue, *Zeolites*, 1988, **8**, 458 (*Chem. Abstr.*, 1989, **110**, 29 372).

2266 H. Shimizu, T. Watanabe, T. Henmi, A. Masuda, and H. Saito, *Geochem. J.*, 1988, **22**, 23 (*Chem. Abstr.*, 1989, **110**, 11 152).

2267 K.J.D. MacKenzie, M.E. Bowden, I.W.M. Brown, and R.H. Meinhold, *Clays Clay Miner.*, 1989, **37**, 317 (*Chem. Abstr.*, 1989, **111**, 81 465).

2268 J.A. Peck, I. Farnan, and J.F. Stebbins, *Geochim. Cosmochim. Acta*, 1988, **52**, 3017.

2269 J.F. Lambert, W.S. Millman, and J.J. Fripiat, *J. Am. Chem. Soc.*, 1989, **111**, 3517; R.C.T. Slade and T.W. Davies, *Colloids Surf.*, 1989, **36**, 119 (*Chem. Abstr.*, 1989, **111**, 62 746).

2270 J.C. Broudic, S. Vilminot, and J.C. Bernier, *Mater. Sci. Eng., A*, 1988, **A109**, 243 (*Chem. Abstr.*, 1989, **111**, 11 441).

2271 A. Hoelscher and W. Schreyer, *Eur. J. Mineral.*, 1989, **1**, 21 (*Chem. Abstr.*, 1989, **111**, 100 454).

2272 A.T. Pilipenko, B.Yu. Kornilovich, A.K. Zapol'skii, N.G. Vasil'ev, and E.A. Pashchenko, *Dokl. Akad. Nauk Ukr. SSR, Ser. B: Geol., Khim. Biol. Nauki*, 1987, 52 (*Chem. Abstr.*, 1989, **110**, 141 998).

scopy.[2273] The very large pore molecular sieve MCM-9 has been characterised by multinuclear n.m.r. spectroscopy.[2274]

Bridging hydroxyl groups in zeolites have been determined.[2275] The effect of water vapour on amorphous aluminosilicates has been investigated by n.m.r. spectroscopy.[2276] M.a.s. ^{17}O n.m.r. spectra have been determined for ^{17}O labelled oxides and zeolites.[2277] The ^{27}Al T_1 has been measured for zeolites.[2278] The dealumination of zeolites using aqueous NH_4BF_4 has been investigated by ^{27}Al and ^{29}Si n.m.r. spectroscopy.[2279] Two dimensional ^{29}Si m.a.s. n.m.r. spectra, including COSY, have been used to determine the ^{29}Si-^{29}Si connectivity in some zeolites.[2280] Alkyl amines and ammonium salts trapped in zeolites have been determined by ^{1}H n.m.r. spectroscopy.[2281] Multinuclear n.m.r. spectroscopy has been used to investigate coking in zeolites.[2282]

Zeolite D_ρ shallow bed calcined in steam has been characterised by n.m.r. spectroscopy.[2283] M.a.s. ^{13}C and ^{29}Si n.m.r. spectra have been reported for highly siliceous MFI-type zeolites.[2284] ^{27}Al n.m.r. spectra of Cu^{2+} doped zeolites A and X show modulation.[2285] The growth of large zeolite crystals has been examined by ^{13}C n.m.r. spectroscopy.[2286] Phase transitions in zeolite Na-A have been investigated by ^{23}Na, ^{27}Al, and ^{29}Si n.m.r. spectroscopy.[2287]

^{11}B and ^{29}Si n.m.r. spectroscopy has been used to investigate isomorphous substitution of boron in mordenite and zeolite Y.[2288] The degree of dealumination of

2273 B.Yu. Kornilovich and A.T. Pilipenko, *Izv. Akad. Nauk SSSR, Neorg. Mater.*, 1989, **25**, 817 (*Chem. Abstr.*, 1989, **111**, 160 960); Z. Deng, J.F.H. Lambert, and J.J. Fripiat, *Chem. Mater.*, 1989, **1**, 640 (*Chem. Abstr.*, 1989, **111**, 246 650).

2274 E.G. Derouane, L. Maistriau, Z. Gabelica, A. Tuel, J.B. Nagy, and R. Von Ballmoos, *Appl. Catal.*, 1989, **51**, L13 (*Chem. Abstr.*, 1989, **111**, 208 041).

2275 A.G. Pel'menshchikov, E.A. Paukshtis, V.G. Stepanov, V.I. Pavlov, E.N. Yurchenko, K.G. Ione, G.M. Zhidomirov, and S. Beran, *J. Phys. Chem.*, 1989, **93**, 6725 (*Chem. Abstr.*, 1989, **111**, 121 418).

2276 S.E. Spiridonov, M.L. Pavlov, and S.N. Khadzhiev, *Kinet. Katal.*, 1989, **30**, 756 (*Chem. Abstr.*, 1989, **111**, 241 008).

2277 S. Yang, K.D. Park, and E. Oldfield, *J. Am. Chem. Soc.*, 1989, **111**, 7278.

2278 J. Haase, H. Pfeifer, W. Oehme, and J. Klinowski, *Chem. Phys. Lett.*, 1988, **150**, 189 (*Chem. Abstr.*, 1989, **110**, 17 573).

2279 X. Liu and R. Xu, *J. Chem. Soc., Chem. Commun.*, 1989, 1837.

2280 C.A. Fyfe, H. Gies, and Y. Feng, *J. Chem. Soc., Chem. Commun.*, 1989, 1240.

2281 J.M. Higel, L. Delmotte, J.L. Guth, and R. Wey, *Analusis*, 1989, **17**, 146 (*Chem. Abstr.*, 1989, **111**, 224 362).

2282 T.H. Fleisch, G.W. Zajac, B.L. Meyers, G.J. Ray, and J.T. Miller, *Proc. - Int. Congr. Catal., 9th*, 1988, **1**, 483 (*Chem. Abstr.*, 1989, **111**, 64 834).

2283 R.X. Fischer, W.H. Baur, R.D. Shannon, R.H. Staley, L. Abrams, A.J. Vega, and J.D. Jorgensen, *Acta Crystallogr., Sect. B: Struct. Sci.*, 1988, **B44**, 321 (*Chem. Abstr.*, 1989, **110**, 50 082).

2284 J.M. Chezeau, L. Delmotte, J.L. Guth, and M. Soulard, *Zeolites*, 1989, **9**, 78 (*Chem. Abstr.*, 1989, **110**, 127 314).

2285 D. Goldfarb and L. Kevan, *J. Magn. Reson.*, 1989, **82**, 270.

2286 A. Sacco, jun., A. Dixon, R. Thompson, G. Scott, and J. Dltr, *NASA Contract. Rep.*, 1988, **NASA-CR-183262**, NAS1.26:183262, 5 pp. Avail. NTIS. From *Sci. Tech. Aerosp. Rep.*, 1989, **27**, Abstr. No. N89-14100 (*Chem. Abstr.*, 1989, **111**, 48 403).

2287 R. Janssen, G.A.H. Tijink, W.S. Veeman, T.L.M. Maesen, and J.F. Van Lent, *J. Phys. Chem.*, 1989, **93**, 899.

2288 T.R. Gaffney, R. Pierantozzi, and M.R. Seger, *ACS Symp. Ser.*, 1989, **398**, 374 (*Chem. Abstr.*, 1989, **111**, 241 013).

mordenites has been determined by ^{27}Al m.a.s. n.m.r. spectroscopy.[2289] The partial destruction of ultrastable zeolite Y has been investigated by m.a.s. n.m.r. spectroscopy.[2290] Y zeolites have been characterised using ^{23}Na n.m.r. spectroscopy.[2291] The dealumination of faujasite type zeolites with $SiCl_4$ has been monitored by ^{27}Al and ^{29}Si n.m.r. spectroscopy.[2292] Ethylene polymerisation over PdY zeolites has been studied by n.m.r. spectroscopy.[2293] The dealumination of zeolite Y has been studied by n.m.r. spectroscopy.[2294] Realumination of dealuminated zeolite Y has been monitored by ^{27}Al and ^{29}Si n.m.r. spectroscopy.[2295] ^{27}Al quadrupole nutation n.m.r. spectroscopy has revealed the presence of four kinds of aluminium in dealuminated zeolite Y.[2296] Chemical reactions during alkaline treatment of dealuminated Y zeolites have been investigated by m.a.s. ^{29}Si n.m.r. spectroscopy.[2297] Framework vacancies and non-framework species in dealuminated Y type zeolites have been determined by ^{29}Si n.m.r. spectroscopy.[2298] ^{13}C n.m.r. spectroscopy has been used to study carbonaceous deposits on zeolite catalysts.[2299] The nature and location of framework species in zeolite Y dealuminated with $SiCl_4$ have been determined by ^{27}Al and ^{29}Si n.m.r. spectroscopy.[2300] The hydrothermal transformation of natural clinoptilolite to zeolites Y and P_1 has been investigated by ^{29}Si n.m.r. spectroscopy.[2301] Small clusters of GaP prepared within the pores of zeolite Y have been identified by ^{31}P n.m.r. spectroscopy.[2302] ^{29}Si n.m.r. spectra of metal substituted Y zeolites have been reported.[2303] Some zeolites have been characterised by ^{23}Na n.m.r. spectroscopy.[2304] The structural evolution of dealuminated Y zeolites during various chemical treatments has been studied by ^{27}Al and ^{29}Si n.m.r. spectroscopy.[2305]

2289 F. Goovaerts, E.F. Vansant, P. De Hulsters, and J. Gelan, *J. Chem. Soc., Faraday Trans., 1*, 1989, **85**, 3675; 3685.

2290 A. Yoshida and Y. Adachi, *Zeolites*, 1989, **9**, 111 (*Chem. Abstr.*, 1989, **110**, 204 638).

2291 L.B. Welsh and S.L. Lambert, *Prepr. - Am. Chem. Soc., Div. Pet. Chem.*, 1988, **33**, 631 (*Chem. Abstr.*, 1989, **110**, 17 586).

2292 B. Sulikowski, G. Borbely, H.K. Beyer, H.G. Karge, and I.W. Mishin, *J. Phys. Chem.*, 1989, **93**, 3240.

2293 X. Bai, G. Wang, and L. Zheng, *Cuihua Xuebao*, 1989, **10**, 42 (*Chem. Abstr.*, 1989, **111**, 121 663).

2294 S. Nakata, S. Asaoka, and K. Segawa, *Nippon Kagaku Kaishi*, 1989, 795 (*Chem. Abstr.*, 1989, **111**, 121 779); J.M. Cruz, A. Corma, and V. Fornes, *Appl. Catal.*, 1989, **50**, 287 (*Chem. Abstr.*, 1989, **111**, 121 702).

2295 H. Hamdan and J. Klinowski, *ACS Symp. Ser.*, 1989, **398**, 448 (*Chem. Abstr.*, 1989, **111**, 181 771).

2296 H. Hamdan and J. Klinowski, *ACS Symp. Ser.*, 1989, **398**, 465 (*Chem. Abstr.*, 1989, **111**, 161 315); H. Hamdan and J. Klinowski, *J. Chem. Soc., Chem. Commun.*, 1989, 240; H. Hamdan and J. Klinowski, *Chem. Phys. Lett.*, 1989, **158**, 447 (*Chem. Abstr.*, 1989, **111**, 141 092).

2297 W. Lutz, U. Lohse, and B. Fahlke, *Cryst. Res. Technol.*, 1988, **23**, 925 (*Chem. Abstr.*, 1989, **110**, 32 824).

2298 J. Jeanjean, L. Aouali, D. Delafosse, and A. Dereigne, *J. Chem. Soc., Faraday Trans. 1*, 1989, **85**, 2771.

2299 B.R. Richardson and J.F. Haw, *Anal. Chem.*, 1989, **61**, 1821.

2300 T.A. Vereshchagina, S.D. Kirik, S.A. Dubkova, A.I. Boronin, and A.G. Anshits, *Izv. Akad. Nauk SSSR, Ser. Khim.*, 1989, 1729 (*Chem. Abstr.*, 1989, **111**, 240 646).

2301 C. De las Pozas, D. Diaz Quintanilla, J. Perez-Pariente, R. Roque-Malherbe, and M. Magi, *Zeolites*, 1989, **9**, 33 (*Chem. Abstr.*, 1989, **110**, 138 782).

2302 J.E. MacDougall, H. Eckert, G.D. Stucky, N. Herron, Y. Wang, K. Moller, T. Bein, and D. Cox, *J. Am. Chem. Soc.*, 1989, **111**, 8006.

2303 K.J. Chao and J.Y. Chern, *J. Phys. Chem.*, 1989, **93**, 1401.

2304 S. Nakata, S. Asaoka, and H. Takahashi, *Nippon Kagaku Kaishi*, 1989, 393 (*Chem. Abstr.*, 1989, **110**, 176 046).

2305 L. Aouali, J. Jeanjean, A. Dereigne, P. Tougne, and D. Delafosse, *Zeolites*, 1988, **8**, 517 (*Chem. Abstr.*, 1989, **110**, 14 101).

Solid state n.m.r. spectroscopy has been used to investigate hexane cracking with dealuminated HZSM-5.[2306] The potential of two dimensional ^{29}Si solid state n.m.r. spectroscopy has been used to establish the three dimensional Si-O-Si lattice connectivities of ZSM-39.[2307] The shape selectivity in zeolite ZSM-5 has been investigated by n.m.r. spectroscopy.[2308] Low temperature plasma calcination of zeolite NH_4, Na-Y has been investigated by ^{27}Al n.m.r. spectroscopy.[2309] Acidic and hydrophobic properties of mordenite and ZSM-5 have been studied by ^{27}Al and ^{29}Si n.m.r. spectroscopy.[2310] The crystalline and amorphous phases of ZSM-5 have been characterised by ^{7}Li and ^{23}Na n.m.r. spectroscopy.[2311] The ^{29}Si n.m.r. spectrum of Fe-ZSM-5 has been reported.[2312] The differences in acidity between high-silica zeolites have been characterised by n.m.r. spectroscopy.[2313] The effect of boron on ZSM-5 shape selectivity has been studied by n.m.r. spectroscopy.[2314] The lattice structure of ZSM-11 has been investigated by ^{29}Si n.m.r. spectroscopy.[2315]

^{27}Al n.m.r. spectroscopy has been used to investigate the synthesis of $AlPO_4$-5.[2316] ^{27}Al and ^{31}P n.m.r. spectroscopy has been used to study the aluminophosphate molecular sieve VPI-5.[2317] The synthesis of $AlPO_4$ has been monitored by ^{27}Al and ^{31}P n.m.r. spectroscopy.[2318] Links between ^{31}P n.m.r. spectra of aluminium phosphate minerals and their crystallographic structures have been examined.[2319] The preparation of chabazite type silicoaluminophosphates has been studied by ^{29}Si n.m.r. spectroscopy.[2320] Aluminophosphate surfaces have been characterised by ^{29}Si n.m.r. spectroscopy.[2321] The reaction mechanism of $AlPO_4$-5 with NaF has been studied by ^{31}P n.m.r. spectro-

2306 E. Brunner, H. Ernst, D. Freude, M. Hunger, C.B. Krause, D. Prager, W. Reschetilowski, W. Schwieger, and K.H. Bergk, *Zeolites*, 1989, **9**, 282 (*Chem. Abstr.*, 1989, **111**, 121 707).

2307 C.A. Fyfe, H. Gies, and Y. Feng, *J. Am. Chem. Soc.*, 1989, **111**, 7702.

2308 M.W. Anderson and J. Klinowski, *Nature (London)*, 1989, **339**, 200.

2309 T.L.M. Maesen, B. Sulikowski, H. Van Bekkum, H.W. Kouwenhoven, and J. Klinowski, *Appl. Catal.*, 1989, **48**, 373 (*Chem. Abstr.*, 1989, **110**, 219 792).

2310 K. Segawa, M. Sakaguchi, S. Nakata, and S. Asaoka, *Nippon Kagaku Kaishi*, 1989, 528 (*Chem. Abstr.*, 1989, **110**, 199 870).

2311 J.B. Nagy, P. Bodart, H. Collette, C. Fernandez, Z. Gabelica, A. Nastro, and R. Aiello, *J. Chem. Soc., Faraday Trans. 1*, 1989, **85**, 2749.

2312 S. Qiu, W. Pang, and C. Zhang, *Wuji Huaxue*, 1988, **4**, 11 (*Chem. Abstr.*, 1989, **111**, 69 658).

2313 K. Aomura and M. Nitta, *Kenkyu Hobun - Asahikawa Kogyo Koto Senmon Gakko*, 1989, **26**, 1 (*Chem. Abstr.*, 1989, **111**, 121 654).

2314 M.B. Sayed, A. Auroux, and J.C. Vedrine, *J. Catal.*, 1989, **116**, 1 (*Chem. Abstr.*, 1989, **110**, 233 589).

2315 C.A. Fyfe, H. Gies, G.T. Kokotailo, C. Pasztor, H. Strobl, and D.E. Cox, *J. Am. Chem. Soc.*, 1989, **111**, 2470.

2316 E. Jahn, D. Mueller, W. Wieker, and J. Richter-Mendau, *Zeolites*, 1989, **9**, 177 (*Chem. Abstr.*, 1989, **111**, 69 668).

2317 M.E. Davis, C. Montes, P.E. Hathaway, J.P. Arhancet, D.L. Hasha, and J.M. Garces, *J. Am. Chem. Soc.*, 1989, **111**, 3919.

2318 L. Coury, F. Babonneau, M. Henry, and J. Livage, *C.R. Acad. Sci., Ser. 2*, 1989, **309**, 799 (*Chem. Abstr.*, 1989, **111**, 246 657).

2319 W.F. Bleam, P.E. Pfeffer, and J.S. Frye, *Phys. Chem. Miner.*, 1989, **16**, 455 (*Chem. Abstr.*, 1989, **111**, 26 357).

2320 T. Tsurita and M. Otake, *Nippon Kagaku Kaishi*, 1989, 332 (*Chem. Abstr.*, 1989, **110**, 195 685).

2321 S.M. Wharry, S.J. Martin, and M.P. McDaniel, *J. Catal.*, 1989, **115**, 463 (*Chem. Abstr.*, 1989, **110**, 121 998).

scopy.[2322] Methanol conversion on aluminophosphates has been studied by n.m.r. spectroscopy.[2323] The ordering in the framework of a magnesium aluminophosphate molecular sieve has been studied by ^{27}Al and ^{31}P n.m.r. spectroscopy.[2324] The ^{13}C n.m.r. spectrum of Ga-ferrierite has been measured.[2325] The ^{71}Ga n.m.r. spectrum of MFI-gallosilicates shows the removal of tetrahedral gallium and the depositing of octahedral gallium in the pores on steam treatment.[2326] ^{27}Al n.m.r. spectroscopy shows that aluminium is in tetrahedral sites in calcium fluoroaluminate glasses.[2327] High resolution ^{27}Al n.m.r. spectra have been observed for AlF_3 particles.[2328]

The ^{13}C tensors of C_6H_6 and $Si(CH_3)_4$ have been determined.[2329] ^1H and ^{13}C n.m.r. spectra have been reported for $C_8KH_{0.61}$.[2330] Discommensuration observed through ^{133}Cs n.m.r. spectroscopy in stage II caesium-graphite intercalation compounds has been studied.[2331] ^{19}F n.m.r. spectroscopy has shown a reversible phase transition in $C_nVF_{5.8\text{-}6.0}$.[2332] The structure of $C_2F.0.22(H_3C)_2CO$ and $C_2F.0.22(D_3C)_2CO$ has been studied by ^1H, ^2H, and ^{13}C n.m.r. spectroscopy.[2333] The nature of the fluorine-carbon bond in $Cs_2F.0.13BrF_3$, $C_2F.xClF_3$, $C_2F.0.1Me_2CO$, and $C_2F.0.1MeCN$ has been discussed on the basis of ^{19}F n.m.r. chemical shifts.[2334] The ^{19}F n.m.r. spectrum of $C_{12.4}ClF_{5.6}$ has been reported.[2335] The ^{13}C n.m.r. spectrum of an oriented pyrographite sphere has been reported.[2336]

^{29}Si shielding tensors have been obtained in some solid organosilicon compounds.[2337] Solid state n.m.r. spectra of $(Bu^n{}_2Si)_n$ and $(Pe^n{}_2Si)_n$ have shown that their backbone conformation is a 7/3 helix.[2338] ^{13}C, ^{29}Si, and ^{31}P c.p.m.a.s. n.m.r. spectra of

2322 J. Hou and C. Zhai, *Bopuxue Zazhi*, 1988, **5**, 379 (*Chem. Abstr.*, 1989, **110**, 180 151).
2323 O.V. Kikhtyanin, V.M. Mastikhin, and K.G. Ione, *Appl. Catal.*, 1988, **42**, 1 (*Chem. Abstr.*, 1989, **110**, 172 612).
2324 P.J. Barrie and J. Klinowski, *J. Phys. Chem.*, 1989, **93**, 5972; J. Wang, K. Wang, S. Xiang, H. Li, and L. Hou, *Shiyou Xuebao, Shiyou Jiagong*, 1988, **4**, 22 (*Chem. Abstr.*, 1989, **111**, 121 470).
2325 B. Sulikowski and J. Klinowski, *J. Chem. Soc., Chem. Commun.*, 1989, 1289.
2326 C.R. Bayense, J.H.C. van Hooff, A.P.M. Kentgens, J.W. de Haan, and L.J.M. van de Ven, *J. Chem. Soc., Chem. Commun.*, 1989, 1292.
2327 J.E. Shelby, C.M. Shaw, and M.S. Spess, *J. Appl. Phys.*, 1989, **66**, 1149 (*Chem. Abstr.*, 1989, **111**, 101 706).
2328 K. Kimura and N. Satoh, *Chem. Lett.*, 1989, 271 (*Chem. Abstr.*, 1989, **110**, 219 660).
2329 K. Takegoshi, *Bunko Kenkyu*, 1988, **37**, 287 (*Chem. Abstr.*, 1989, **111**, 207 939).
2330 T. Saito, K. Nomura, K. Mizoguchi, K. Mizuno, K. Kume, and H. Suematsu, *J. Phys. Soc. Jpn.*, 1989, **58**, 269 (*Chem. Abstr.*, 1989, **110**, 164 813).
2331 M. Malki, M. Goldmann, and H. Estrade-Szwarckopf, *Phase Transitions*, 1989, **14**, 217 (*Chem. Abstr.*, 1989, **110**, 183 406).
2332 T. Nakajima, T. Matsui, M. Motoyama, and Y. Mizutani, *Carbon*, 1988, **26**, 831 (*Chem. Abstr.*, 1989, **110**, 50 080).
2333 A.M. Panich, A.V. Sabylinskii, S.P. Gabuda, A.M. Danilenko, and A.S. Nazarov, *Zh. Strukt. Khim.*, 1989, **30**, 66 (*Chem. Abstr.*, 1989, **110**, 223 027).
2334 A.M. Panich, *Zh. Strukt. Khim.*, 1989, **30**, 170 (*Chem. Abstr.*, 1989, **111**, 223 957).
2335 Yu.I. Nikonorov, O.V. Falaleev, O.P. Kukhlevskii, and E.P. Zeer, U.S.S.R. SU 1,422,119 (Cl. G01N24/08), 07 Sep 1988, Appl. 4,139,888, 04 Aug 1986. From *Otkrytiya, Izobret.*, 1988, **(33)**, 177 (*Chem. Abstr.*, 1989, **110**, 146 485).
2336 P. Lauginie, H. Estrade-Szwarckopf, B. Rousseau, and J. Conard, *C.R. Acad. Sci., Ser. 2*, 1988, **307**, 1693 (*Chem. Abstr.*, 1989, **110**, 146 444).
2337 R.K. Harris, T.N. Pritchard, and E.G. Smith, *J. Chem. Soc., Faraday Trans. 1*, 1989, **85**, 1853.
2338 F.C. Schilling, A.J. Lovinger, J.M. Zeigler, D.D. Davis, and F.A. Bovey, *Macromolecules*, 1989, **22**, 3055 (*Chem. Abstr.*, 1989, **111**, 40 201); F.C. Schilling, F.A. Bovey, D.D. Davis, A.J. Lovinger, R.B.

$[X_nMe_{3-n}Si(CH_2)_mPPhD]$ have been reported.[2339]

Carbon rich α-Si-C:H films prepared from SiH_4/CH_4 have been investigated by solid state ^{13}C n.m.r. spectroscopy.[2340] The conversion of polycarbosilane to SiC has been studied by ^{13}C and ^{29}Si n.m.r. spectroscopy.[2341]

Chirally substituted Si—H containing polysiloxanes, immobilised on small particle silica gel, has been studied by ^{29}Si n.m.r. spectroscopy.[2342] The formation of polysiloxane chains on glass has been studied by ^{29}Si n.m.r. spectroscopy.[2343] The ^{13}C and ^{29}Si n.m.r. spectra of $[O_2SiOSiMe_2CH_2CH_2SiMe_2C_6H_4]_n$ have been reported.[2344] The curing mechanism of polycarbosilane fibre has been studied by ^{13}C and ^{29}Si n.m.r. spectroscopy.[2345] A novel glycol-silicone polymer has been characterised by ^{29}Si n.m.r. spectroscopy.[2346] ^{13}C and ^{29}Si n.m.r. studies of the molecular structure and dynamics of $[Et_2SiO]_n$ polymorphs have been reported.[2347] ^{29}Si n.m.r. spectroscopy has been applied to a 1:1 composite of poly(vinyl alcohol)/tetraethylsiloxane gel.[2348] ^{29}Si n.m.r. spectroscopy has been used to characterise polysiloxanes with $(CH_2)_nPPh_2$ side chains.[2349] ^{13}C and ^{29}Si n.m.r. spectroscopy has been used to investigate glass-filled polymer composites.[2350] The curing and decomposition of methyl silicone resin have been studied by ^1H, ^{13}C, and ^{29}Si n.m.r. spectroscopy.[2351] The deactivation of fused silica capillary columns with polymethylhydrosiloxanes has been investigated by ^{29}Si n.m.r. spectroscopy.[2352] Polymeric organosilicon gels, prepared from $Me_2Si(OEt)_2$ and $Si(OEt)_4$, have been studied using ^{29}Si n.m.r. spectroscopy.[2353] Polysilicic acid ester gels have been

Macgregor, jun., C.A. Walsh, and J.M. Zeigler, *Macromolecules*, 1989, **22**, 4645 (*Chem. Abstr.*, 1989, **111**, 234 045).

2339 E. Lindner, A. Bader, E. Glaser, and P. Wegner, *J. Mol. Catal.*, 1989, **56**, 86.

2340 K. Yamamoto, Y. Ichikawa, N. Fukada, T. Nakayama, and Y. Tawada, *Thin Solid Films*, 1989, **173**, 253 (*Chem. Abstr.*, 1989, **111**, 245 353).

2341 T. Taki, M. Inui, K. Okamura, and M. Sato, *J. Mater. Sci. Lett.*, 1989, **8**, 918 (*Chem. Abstr.*, 1989, **111**, 139 198); J. Lipowitz, G.E. LeGrow, T.F. Lim, and N. Langley, *Ceram. Eng. Sci. Proc.*, 1988, **9**, 931 (*Chem. Abstr.*, 1988, **110**, 100 257); J. Lipowitz, G. LeGrow, T. Lim, and N. Langley, *Ceram. Trans.*, 1989, **2**, 421 (*Chem. Abstr.*, 1989, **111**, 158 895); G.W. Wagner, B.K. Na, and M.A. Vannice, *J. Phys. Chem.*, 1989, **93**, 5061.

2342 F.J. Ruffing, J.A. Lux, W. Roeder, and G. Schomburg, *Chromatographia*, 1988, **26**, 19 (*Chem. Abstr.*, 1989, **111**, 208 459).

2343 M. Hetem, G. Rutten, L. Van de Ven, J. De Haan, and C. Cramers, *HRC CC, J. High Resolut. Chromatogr. Chromatogr. Commun.*, 1988, **11**, 510 (*Chem. Abstr.*, 1989, **110**, 87 706).

2344 P.A. Agaskar, *J. Am. Chem. Soc.*, 1989, **111**, 6858.

2345 T. Taki, K. Ohamura, and M. Sato, *J. Mater. Sci.*, 1989, **24**, 1263 (*Chem. Abstr.*, 1989, **111**, 24 835).

2346 M.A. Brook, C.H. Kremers, T. Sebastian, and W. Yu, *J. Polym. Sci., Part C: Polym. Lett.*, 1989, **27**, 229 (*Chem. Abstr.*, 1989, **111**, 58 441).

2347 G. Koegler, A. Hasenhindl, and M. Moeller, *Macromolecules*, 1989, **22**, 4190 (*Chem. Abstr.*, 1989, **111**, 215 175).

2348 N. Zumbulyadis and J.M. O'Reilly, *J. Magn. Reson.*, 1989, **82**, 613.

2349 R.V. Parish, D. Habibi, and V. Mohammadi, *J. Organomet. Chem.*, 1989, **369**, 17.

2350 T.L. Weeding, W.S. Veeman, L.W. Jenneskens, H.A. Gaur, H.E.C. Schuurs, and W.G.B. Huysmans, *Macromolecules*, 1989, **22**, 706 (*Chem. Abstr.*, 1989, **110**, 96 435).

2351 O. Schneider, *Thermochim. Acta*, 1988, **134**, 269 (*Chem. Abstr.*, 1989, **110**, 76 649).

2352 M. Hetem, G. Rutten, B. Vermeer, J. Rijks, L. Van de Ven, J. De Haan, and C. Cramers, *J. Chromatogr.*, 1989, **477**, 3 (*Chem. Abstr.*, 1989, **111**, 202 548).

2353 F. Babonneau, K. Thorne, and J.D. Mackenzie, *Chem. Mater.*, 1989, **1**, 554 (*Chem. Abstr.*, 1989, **111**, 134 857).

analysed by ^{29}Si n.m.r. spectroscopy.[2354] ^{29}Si n.m.r. spectroscopy has been used to study grain boundaries in Si_3N_4 powders in the presence of Y_2O_3 and Al_2O_3 sintering aids.[2355] ^{15}N c.p.m.a.s. n.m.r. spectroscopy has been used to study the structure and reactions of phthalocyaninatopolysiloxanes.[2356]

^{29}Si n.m.r. spectroscopy has been used to investigate calcium silicate hydrates.[2357] The ^{29}Si shielding tensor in quartz has been determined.[2358] Neutron irradiated quartz and amorphous silica have been investigated by ^{29}Si n.m.r. spectroscopy.[2359] Pore structure in porous silica glass has been analysed by n.m.r. spectroscopy.[2360] The effects of the reaction conditions on the hydrolysis and condensation rate of $Si(OMe)_4$ have been investigated by ^{1}H and ^{29}Si n.m.r. spectroscopy.[2361] Silicon oxide filled membranes have been characterised by ^{29}Si n.m.r. spectroscopy.[2362] High resolution ^{29}Si n.m.r. spectroscopy has been used to study decadodecasil.[2363] Stishovite, a dense phase of SiO_2, has been studied by ^{29}Si n.m.r. spectroscopy.[2364] ^{2}H n.m.r. spectroscopy has been used to investigate alkyl modified silica.[2365] Proton environments and hydrogen-bonding in hydrous silicate glasses have been studied using ^{1}H n.m.r. spectroscopy.[2366] Quantitative ^{1}H, ^{2}H, and ^{3}H n.m.r. studies of water in silicate glasses have been reported.[2367] Short-range ordering in $KHSi_2O_5$ has been studied by ^{1}H and ^{29}Si n.m.r. spectroscopy.[2368] By use of ^{29}Si n.m.r. spectroscopy, small amounts of 5- and 6-coordinate silicon have been observed in a $K_2Si_4O_9$ glass.[2369] ^{29}Si n.m.r. spectroscopy has been used to study the effect of finely divided SiO_2 on the hydration of C_3S.[2370] M.a.s. ^{29}Si n.m.r. spectroscopy has been used to characterise a silicogermanate

2354 D.M. Millar, *Report*, 1987, DOE/ER/O1198-T18; Order No. DE88014627, 169 pp. Avail. NTIS. From *Energy Res. Abstr.*, 1988, **13**, Abstr. No. 49188 (*Chem. Abstr.*, 1989, **110**, 236 124).

2355 K.R. Carduner, R.O. Carter, tert., M.J. Rokosz, C. Peters, G.M. Crosbie, and E.D. Stiles, *Chem. Mater.*, 1989, **1**, 302 (*Chem. Abstr.*, 1989, **110**, 217 780).

2356 B. Wehrle, H.-H. Limbach, T. Zipplies, and M. Hanack, *Angew. Chem., Int. Ed. Engl.*, 1989, **28**, 1743.

2357 M. Grutzeck, A. Benesi, and B. Fanning, *J. Am. Ceram. Soc.*, 1989, **72**, 665 (*Chem. Abstr.*, 1989, **111**, 11 695).

2358 D.R. Spearing and J.F. Stebbins, *Am. Mineral.*, 1989, **74**, 956 (*Chem. Abstr.*, 1989, **111**, 81 491).

2359 S.L. Chan, L.F. Gladden, and S.R. Elliott, *J. Non-Cryst. Solids*, 1988, **106**, 413 (*Chem. Abstr.*, 1989, **110**, 106 824).

2360 F. D'Orazio, J.C. Tarczon, W.P. Halperin, K. Eguchi, and T. Mizusaki, *J. Appl. Phys.*, 1989, **65**, 742 (*Chem. Abstr.*, 1989, **110**, 102 432).

2361 M. Spinu, C. Arnold, and J.E. McGrath, *Polym. Prepr. (Am. Chem. Soc., Div. Polym. Chem.)*, 1989, **30**, 125 (*Chem. Abstr.*, 1989, **111**, 200 264).

2362 K.A. Mauritz, R.F. Storey, and C.K. Jones, *ACS Symp. Ser.*, 1989, **395**, 401 (*Chem. Abstr.*, 1989, **111**, 233 699).

2363 N.C.M. Alma-Zeestraten, J. Dorrepaal, J. Keijsper, and H. Gies, *Zeolites*, 1989, **9**, 81 (*Chem. Abstr.*, 1989, **110**, 127 315).

2364 J.F. McHone, R.A. Nieman, C.F. Lewis, and A.M. Yates, *Science (Washington, D.C., 1883-)*, 1989, **243**, 1182.

2365 M. Gangoda, R.K. Gilpin, and J. Figueirinhas, *J. Phys. Chem.*, 1989, **93**, 4815.

2366 S.C. Kohn, R. Dupree, and M.E. Smith, *Nature (London)*, 1989, **337**, 539.

2367 H. Eckert, J.P. Yesinowski, and E.M. Stolper, *Solid State Ionics*, 1988, **32-33**, 298 (*Chem. Abstr.*, 1989, **111**, 238 151).

2368 Z.Q. Deng, J.F. Lambert, and J.J. Fripiat, *Chem. Mater.*, 1989, **1**, 375 (*Chem. Abstr.*, 1989, **110**, 241 543).

2369 J.F. Stebbins and P. McMillan, *Am. Mineral.*, 1989, **74**, 965 (*Chem. Abstr.*, 1989, **111**, 157 667).

2370 C.M. Dobson, D.G.C. Goberdhan, J.D.F. Ramsay, and S.A. Rodger, *J. Mater. Sci.*, 1988, **23**, 4108 (*Chem. Abstr.*, 1989, **110**, 62 609).

zeolite.[2371] The ^{29}Si n.m.r. spectrum of a silica-antimonic acid polymer has been reported.[2372]

The ^{119}Sn n.m.r. spectrum of $[(Bu^n{}_3Sn)_2(SO_4)]$ has been recorded.[2373] Multinuclear ^{13}C, ^{77}Se, ^{119}Sn, and ^{125}Te n.m.r. spectroscopy has been applied to $[Me_2SnE]_3$, E = S, Se, Te. When E = S, there are two inequivalent ^{119}Sn nuclei and three inequivalent ^{13}C nuclei.[2374] ^{31}P and ^{119}Sn n.m.r. spectroscopy has been used to determine the tin coordination in $[Me_2Sn(glycylmethionate)]$ and related compounds.[2375] The ^{119}Sn and ^{207}Pb n.m.r. spectra of $[(Cy_3Sn)_2S]$, $[R_2SnE]_3$, E = S, Se, and $[(Ph_3Pb)_2S]$ have been determined. The resolution was adequate to determine coupling constants.[2376]

The ^{117}Sn and ^{119}Sn n.m.r. spectra of 1,10-dichloro-1,10-distannabicyclo[8.8.8]hexacosane[2377] and 2,2-$Bu^t{}_2$-1,3,2-dioxastannolane[2378] have been reported. Dissociation of HPO_4 in $Sn(HPO_4)_2.2H_2O$ has been investigated by ^{31}P n.m.r. spectroscopy.[2379] High-resolution ^{119}Sn n.m.r. spectra of ternary tin oxides have been reported.[2380] $^1J(^{119}Sn\text{-}^{35}Cl)$ and $^1J(^{119}Sn\text{-}^{37}Cl)$ have been determined from the ^{119}Sn n.m.r. spectrum of $[SnCl_2(acac)_2]$.[2381] An autodyne detector based on FET circuits has been described for ^{119}Sn in $Pb_{0.8}Sn_{0.2}Te$.[2382] The structure of $[N_2H_5][Sn_3F_7]$ has been studied by n.m.r. spectroscopy.[2383] Solid state n.m.r. spectroscopy has been used to demonstrate the contamination of $Pb_{1-x}Sn_xF_2$ by Cu^{2+}.[2384]

^{31}P n.m.r. spectroscopy has been used to study the ferroelectric phase transition in $PbHPO_4$.[2385] The ^1H and ^{35}Cl n.m.r. spectra of $PbClO_3$ have been reported.[2386] The ^{19}F n.m.r. spectrum of $BiPbO_2F$ has been measured.[2387] Experimentally observed trends in the ^{207}Pb chemical shifts of PbX_2, X = F, Cl, Br, I, have been analysed by using electronic structures.[2388] The phase transition in $[NH_4]_2[PbCl_6]$ has been investigated by ^1H n.m.r.

2371 Z. Gabelica and J.-L. Guth, *Angew. Chem., Int. Ed. Engl.*, 1989, **28**, 81.
2372 M. Kone, P. Courtin, and J. Lemerle, *Polyhedron*, 1989, **8**, 2577.
2373 K.C. Molloy, K. Quill, D. Cunningham, P. McArdle, and T. Higgins, *J. Chem. Soc., Dalton Trans.*, 1989, 267.
2374 I.D. Gay, C.H.W. Jones, and R.D. Sharma, *J. Magn. Reson.*, 1989, **84**, 501.
2375 F. Huber, A.K. Saxena, R.K. Harris, and A. Sebald, *J. Organomet. Chem.*, 1989, **368**, 31.
2376 R.K. Harris and A. Sebald, *Magn. Reson. Chem.*, 1989, **27**, 81.
2377 M. Newcomb, J.H. Horner, M.T. Blanda, and P.J. Squattrito, *J. Am. Chem. Soc.*, 1989, **111**, 6294.
2378 P.A. Bates, M.B. Hursthouse, A.G. Davies, and S.D. Salter, *J. Organomet. Chem.*, 1989, **363**, 45.
2379 A.B. Yaroslavtsev, Z.N. Prozorovskaya, V.F. Chuvaev, V.V. Parshutkin, and G.G. Shifanova, *Zh. Neorg. Khim.*, 1989, **34**, 1188 (*Chem. Abstr.*, 1989, **111**, 49 234).
2380 N.J. Clayden, C.M. Dobson, and A. Fern, *J. Chem. Soc., Dalton Trans.*, 1989, 843.
2381 D.C. Apperley, B. Haiping, and R.K. Harris, *Mol. Phys.*, 1989, **68**, 1277.
2382 A.G. Khandozhko, E.I. Slyn'ko, and I.P. Chernysh, *Prib. Tekh. Eksp.*, 1988, 110 (*Chem. Abstr.*, 1989, **110**, 17 585).
2383 W. Granier and S. Vilminot, *Eur. J. Solid State Inorg. Chem.*, 1988, **25**, 53 (*Chem. Abstr.*, 1989, **111**, 244 770).
2384 G. Denes, *J. Solid State Chem.*, 1988, **77**, 54 (*Chem. Abstr.*, 1989, **110**, 50 217).
2385 F. Ermark, B. Topic, U. Haeberlen, and R. Blinc, *J. Phys.: Condens. Matter*, 1989, **1**, 5489 (*Chem. Abstr.*, 1989, **111**, 207 191).
2386 P. Kamasa, F.I. Toth, J. Kasprazak, and J. Pietrzak, *Nauchn. Appar.*, 1988, **3**, 31 (*Chem. Abstr.*, 1989, **111**, 89 209).
2387 C. Follet-Houttemane, J. Canonne, J.C. Boivin, J.C. Champaraud-Mesjard, D. Mercurio, B. Frit, and G. Roult, *Solid State Ionics*, 1987, (Pub. 1988), **28**, 458 (*Chem. Abstr.*, 1989, **111**, 15 793).
2388 M. Nizam, M. Allavena, Y. Bouteiller, B.H. Suits, and D. White, *J. Magn. Reson.*, 1989, **82**, 441.

spectroscopy.[2389]

The temperature dependence of $T_{1\rho}$ of ^{2}H has been measured for $[NH_{4-n}D_n][ClO_4]$.[2390] The nuclear quadrupole interaction tensor of ^{14}N and ^{35}Cl and the diamagnetic shielding tensor of ^{35}Cl in $[NH_4][ClO_4]$ have been determined.[2391] The ^{31}P n.m.r. spectrum of $[Ph_4P]X$, X = Cl, Br, I, shows two signals due to the presence of two crystallographic sites.[2392] $[Bu^n{}_4N][IO_4]$ and $[Bu^n{}_4P][IO_4]$ show ^{127}I n.m.r. spectra with quadrupole coupling constants of 2 to 8 MHz.[2393] The ^{31}P chemical shift tensors of PR_3[2394] and $[Ph_3PI]X$[2395] in the solid state have been determined. Me_2PhPO has been proposed as a compound for setting the magic angle in the m.a.s. ^{31}P n.m.r. spectrum.[2396]

$^1J(^{31}P$-$^{14}N)$ has been determined from the ^{31}P n.m.r. spectra of some monophosphazenes.[2397] The ^{13}C and ^{31}P n.m.r. spectra of a series of alternating copolymers containing a organo-λ^5-phosphazene backbone have been reported.[2398]

The nature of the ^{31}P anisotropic chemical shift interaction has been examined. Linear correlations between the principal values of the ^{31}P chemical shift tensor, the P—O bond lengths, and P—O—P bond angles have been established.[2399]

The crystalline→plastic phase transition of P_4S_3 has been observed using ^{31}P n.m.r. spectroscopy.[2400] The multiplicity of ^{31}P resonances of P_4S_n has been correlated with the structural inequivalences[2401] and local environments.[2402] ^{31}P n.m.r. spectra have been determined for P_4Se_n.[2403] The ^{13}C n.m.r. spectra of $[M(S_2COR)_3]$, M = As, Sb, Bi, have been determined.[2404]

The pressure dependence of the spin density wave properties in $[TMTSF]_2[PF_6]$ has been investigated by ^{1}H n.m.r. spectroscopy.[2405] Locally resolved isotropic and anisotropic ^{13}C Knight shifts in $[fluoranthenyl]_2[MF_6]$, M = P, As, have been

2389 C. Dimitropoulos and J. Pelzl, *Z. Naturforsch., A*, 1989, **44**, 109 (*Chem. Abstr.*, 1989, **110**, 222 970).
2390 K. Morimoto and T. Hihara, *J. Phys. Soc. Jpn.*, 1989, **58**, 2272 (*Chem. Abstr.*, 1989, **111**, 246 518).
2391 T.J. Bastow and S.N. Stuart, *J. Phys.: Condens. Matter*, 1989, **1**, 4649 (*Chem. Abstr.*, 1989, **111**, 207 976).
2392 E.E. Schweizer, C.J. Baldacchini, and A.L. Rheingold, *Acta Crystallogr., Sect. C: Cryst. Struct. Commun.*, 1989, **C45**, 1236 (*Chem. Abstr.*, 1989, **111**, 205 973).
2393 P.K. Burkert and M. Grommelt, *Z. Naturforsch., B*, 1988, **43**, 1381 (*Chem. Abstr.*, 1989, **110**, 50 017).
2394 G.H. Penner and R.E. Wasylishen, *Can. J. Chem.*, 1989, **67**, 1909.
2395 K.B. Dillon and J. Lincoln, *Polyhedron*, 1989, **8**, 1445.
2396 R.C. Crosby and J.F. Haw, *J. Magn. Reson.*, 1989, **82**, 367.
2397 W.P. Power, R.E. Wasylishen, and R.D. Curtis, *Can. J. Chem.*, 1989, **67**, 454.
2398 M. Pomerantz and M.W. Victor, *Macromolecules*, 1989, **22**, 3511 (*Chem. Abstr.*, 1989, **111**, 134 847).
2399 S. Un and M.P. Klein, *J. Am. Chem. Soc.*, 1989, **111**, 5119.
2400 T. Bjorholm and H.J. Jakobsen, *J. Magn. Reson.*, 1989, **84**, 204.
2401 R.K. Harris, P.J. Wilkes, P.T. Wood, and J.D. Woolins, *J. Chem. Soc., Dalton Trans.*, 1989, 809.
2402 H. Eckert, C.S. Liang, and G.D. Stucky, *Report*, 1988, **TR-16**; Order No. AD-A196557, 29 pp. Avail. NTIS. From *Gov. Rep. Announce. Index (U.S.)*, 1988, **88**, Abstr. No. 857 640 (*Chem. Abstr.*, 1989, **111**, 107 847); H. Eckert, C.S. Liang, and G.D. Stucky, *J. Phys. Chem.*, 1989, **93**, 452.
2403 D. Lathrop and H. Eckert, *J. Non-Cryst. Solids*, 1988, **106**, 417 (*Chem. Abstr.*, 1989, **110**, 87 214); D. Lathrop and H. Eckert, *J. Phys. Chem.*, 1989, **93**, 7895; D. Lathrop and H. Eckert, *J. Am. Chem. Soc.*, 1989, **111**, 3536.
2404 A.M. Hounslow, S.F. Lincoln, and E.R.T. Tiekink, *J. Chem. Soc., Dalton Trans.*, 1989, 233.
2405 T. Takahashi, H. Kawamura, T. Ohyama, Y. Maniwa, K. Murata, and G. Saito, *J. Phys. Soc. Jpn.*, 1989, **58**, 703 (*Chem. Abstr.*, 1989, **110**, 184 661); T. Takahashi, *Synth. Met.*, 1988, **27**, B397 (*Chem. Abstr.*, 1989, **110**, 204 413).

determined.[2406] ^{31}P n.m.r. spectroscopy has been used to study PCl_5.[2407]

^{13}C n.m.r. spectra have been reported for $[TMTSF]_2X$ salts.[2408] ^{1}H n.m.r. spectroscopy has been used to investigate hydrogen bonding in polycrystalline trihydrogen selenites.[2409] ^{39}K and ^{77}Se n.m.r. spectroscopy has been used to study the paraelectric-to-incommensurate phase transition of H_2SeO_4.[2410] Phase transitions in NH_4HSeO_4 have been investigated by n.m.r. spectroscopy.[2411] ^{2}H and ^{77}Se n.m.r. spectroscopy has been used to confirm the existence of the superstructure in the ferroelectric phase of NH_4HSeO_4.[2412]

Molecules Sorbed Onto Solids.——This section is divided into two subsections: 'Water sorbed Onto Solids' and 'Atoms and Other Molecules Sorbed Onto Solids'.

Water Sorbed Onto Solids.——The dynamics and ordering of intercalated water in layered metal hydroxides from $Mg(NO_3)_2$-$Al(NO)_3$ have been investigated by ^{1}H n.m.r. spectroscopy.[2413] The spin-lattice relaxation of water in cement gels has been determined.[2414] Pulsed n.m.r. spectroscopy has been used to determine bound and free water on Al_2O_3.[2415] ^{1}H, ^{2}H, ^{13}C, ^{17}O, and ^{23}Na n.m.r. spectroscopy has been used to investigate Na^+ and water reorientation at the interface of a clay.[2416] Intracrystalline self-diffusion of water and short-chain-length paraffins in A-type zeolites has been studied.[2417] The use of deuterated molecules for the determination of self-diffusion coefficients at two-component adsorption on zeolites has been investigated.[2418] ^{1}H n.m.r. spectroscopy has been used to study the hydroxyl structure on the surface of HY zeolite.[2419] The motion of adsorbed water on synthetic porous alunite has been studied by ^{1}H n.m.r. spectro-

2406 D. Koengeter and M. Mehring, *Phys. Rev. B: Condens. Matter*, 1989, **39**, 6361 (*Chem. Abstr.*, 1989, **111**, 49 149).
2407 R.K. Harris and A. Root, *Mol. Phys.*, 1989, **66**, 993.
2408 P.C. Stein and P. Bernier, *Synth. Met.*, 1988, **27**, B77 (*Chem. Abstr.*, 1989, **110**, 204 412); J.M. Delrieu, M. Roger, C. Coulon, R. Laversanne, and E. Dupart, *Synth. Met.*, 1988, **27**, B35 (*Chem. Abstr.*, 1989, **110**, 204 410); F. Creuzet, C. Bourbonnais, P. Wzietek, H. Nelisse, L.G. Caron, D. Jerome, and K. Bechgaard, *Synth. Met.*, 1988, **27**, B65 (*Chem. Abstr.*, 1989, **110**, 204 411).
2409 H. Rosenberger, G. Scheler, and Yu.N. Moskvich, *Magn. Reson. Chem.*, 1989, **27**, 50.
2410 B. Topic, A. Von Kienlin, A. Goelzhaeuser, U. Haeberlen, and R. Blinc, *Phys. Rev. B: Condens. Matter*, 1988, **38**, 8625 (*Chem. Abstr.*, 1989, **110**, 49 543).
2411 I.P. Aleksandrova, P. Colomban, F. Denoyer, N. Le Calve, A. Novak, B. Pasquier, and A. Rozycki, *Phys. Status Solidi A*, 1989, **114**, 531 (*Chem. Abstr.*, 1989, **111**, 245 879).
2412 A.A. Sukhovskii, I.P. Aleksandrova, and Yu.N. Moskvich, *Phys. Status Solidi A*, 1989, **111**, 477 (*Chem. Abstr.*, 1989, **111**, 107 824).
2413 G. Marcelin, N.J. Stockhausen, J.F.M. Post, and A. Schultz, *J. Phys. Chem.*, 1989, **93**, 4646.
2414 R. Blinc, J. Dolinsek, G. Lahajnar, A. Sepe, I. Zupancic, S. Zumer, F. Milia, and M.M. Pintar, *Z. Naturforsch., A*, 1988, **43**, 1026 (*Chem. Abstr.*, 1989, **110**, 127 329).
2415 Anon. (UK). *Res. Discl.*, 1988, **295**, 887 (*Chem. Abstr.*, 1989, **110**, 49 624).
2416 J. Grandjean and P. Laszlo, *J. Magn. Reson.*, 1989, **83**, 128.
2417 J. Kaerger, H. Pfeifer, M. Rosemann, N.N. Feokistova, and S.P. Zdanov, *Zeolites*, 1989, **9**, 247 (*Chem. Abstr.*, 1989, **111**, 45 722).
2418 C. Foerste, A. Germanus, J. Kaerger, G. Moebius, M. Buelow, S.P. Zdanov, and N.N. Feokistova, *Isotopenpraxis*, 1989, **25**, 48 (*Chem. Abstr.*, 1989, **111**, 220 006).
2419 Z. Xue, L. Dai, Y. Wang, and L. Zhu, *Kexue Tongbao (Foreign Lang. Ed.)*, 1988, **33**, 1952 (*Chem. Abstr.*, 1989, **111**, 121 414).

scopy.[2420] An n.m.r. study of the state and diffusion mobility of water in ion-exchange membranes of MK-40 has been reported.[2421] Molecular dynamics simulation of water near silica surfaces has been compared with n.m.r. results.[2422] ^2H n.m.r. spectroscopy has been used to study the state of the Triton X-100 layer sorbed on silica gel.[2423] Pulsed field gradients have been used to study the self-diffusion process of water molecules in porous glass.[2424] The water content of salic minerals has been determined by ^1H n.m.r. spectroscopy.[2425] The adsorption of water molecules on $TlMX_2$ semiconductors has been studied.[2426]

Atoms and Other Molecules Sorbed Onto Solids.——Adsorbed hydrogen on SiO_2 supported ruthenium catalysts has been studied by ^1H n.m.r. spectroscopy.[2427] The effects of H_2 adsorption on Rh/TiO_2 have been investigated by ^1H n.m.r. spectroscopy.[2428] N.m.r. spectroscopy has been used to study H_2 and D_2 adsorbed on palladium supported on SiO_2.[2429] Hydrogen adsorbed on Pt/Al_2O_3 on SiO_2 has been investigated by ^1H n.m.r. spectroscopy.[2430] The ^1H lineshape of hydrogen in hydrogenated amorphous silicon and silicon nitride alloy films has been reported.[2431]

The adsorption of Rb^+ on γ-alumina has been followed by ^{87}Al n.m.r. spectroscopy.[2432] The effect of coating with $Ca_2P_2O_7$, $Ba(PO_3)_2$, and BPO_4 on the oxidation of a corundum-graphite refractory material has been studied by n.m.r. spectroscopy.[2433] ^1H n.m.r. spectroscopy has been used to investigate the reaction of anchored Zr(IV) hydride complexes with C_2H_4.[2434] The characterisation of V_2O_5 supported on Al_2O_3 has been achieved by ^{51}V n.m.r. spectroscopy.[2435] A series of SiO_2 supported V_2O_5 have been

2420 S. Ozeki, Y. Masuda, and H. Sano, *J. Phys. Chem.*, 1989, **93**, 7226.

2421 V.I. Volkov, G.K. Saldadze, R.I. Tagirova, L.V. Kropotov, V.G. Khutsishvili, and N.N. Shapet'ko, *Zh. Fiz. Khim.*, 1989, **63**, 1005 (*Chem. Abstr.*, 1989, **111**, 141 337).

2422 H.S. Kim, S.H. Lee, and H. Pak, *Chayon Kwahak Taehak Nonmunjip (Soul Taehakkyo)*, 1988, **13**, 43 (*Chem. Abstr.*, 1989, **111**, 64 729).

2423 B.N. Nikolaev, A.M. Shlyakov, N.A. Klimenko, and T.V. Karmazina, *Dokl. Akad. Nauk Ukr. SSR, Ser. B: Geol., Khim. Biol. Nauki*, 1988, 54 (*Chem. Abstr.*, 1989, **111**, 161 131).

2424 K. Fukuda, T. Kasuga, T. Mizusaki, A. Hirai, and K. Eguchi, *J. Phys. Soc. Jpn.*, 1989, **58**, 1662 (*Chem. Abstr.*, 1989, **111**, 64 787).

2425 A. Valassis and L.F. Pasal'skaya, *Dokl. Akad. Nauk Ukr. SSR, Ser. B: Geol., Khim. Biol. Nauki*, 1987, 6 (*Chem. Abstr.*, 1989, **110**, 234 771).

2426 S.G. Abdullaeva, N.T. Mamedov, and F.A. Mustafaev, *Dokl. Akad. Nauk Az. SSR*, 1988, **44**, 11 (*Chem. Abstr.*, 1989, **111**, 103 165).

2427 X. Wu, B.C. Gerstein, and T.S. King, *J. Catal.*, 1989, **118**, 238 (*Chem. Abstr.*, 1989, **111**, 121 725).

2428 G. Munuera, A.R. Gonzalez-Elipe, A. Munoz, A. Fernandez, J. Soria, J. Conesa, and J. Sanz, *Sens. Actuators*, 1989, **18**, 337 (*Chem. Abstr.*, 1989, **111**, 206 460).

2429 A.A. Chen, A.J. Benesi, and M.A. Vannice, *J. Catal.*, 1989, **119**, 14 (*Chem. Abstr.*, 1989, **111**, 241 031).

2430 D.H. Lenz, W.C. Conner, jun., and J.P. Fraissard, *J. Catal.*, 1989, **117**, 281 (*Chem. Abstr.*, 1989, **111**, 13 049).

2431 M. Kumeda, S. Nakanishi, and T. Shimizu, *Solid State Commun.*, 1988, **67**, 585 (*Chem. Abstr.*, 1989, **110**, 17 559).

2432 J.T. Cheng and P.D. Ellis, *J. Phys. Chem.*, 1989, **93**, 2549.

2433 S.A. Suvorov, E.V. Chaikun, S.L. Korobkin, and G.S. Sokolova, *Ogneupory*, 1988, 24 (*Chem. Abstr.*, 1989, **110**, 12 324).

2434 S.A. Vasnetsov, A.V. Nosov, V.M. Mastikhin, and V.A. Zakharov, *J. Mol. Catal.*, 1989, **53**, 37.

2435 K.V.R. Chary, V.V. Rao, and V.M. Mastikhin, *J. Chem. Soc., Chem. Commun.*, 1989, 202; H. Eckert and I.E. Wachs, *J. Phys. Chem.*, 1989, **93**, 6796.

characterised by ^1H m.a.s. n.m.r. spectroscopy.[2436] ^{13}C T_1 has been determined for $[Mo(CO)_6]$ physisorbed on alumina.[2437] The correlation times for γ-alumina supported $Mo(CO)_3$ have been determined by ^{13}C n.m.r. spectroscopy.[2438] ^{95}Mo n.m.r. spectroscopy has been used to study the interaction of heptamolybdate solutions with alumina and silica.[2439] Changes in the vicinity of impurity Fe^{3+} in various zeolites have been studied by ^1H n.m.r. relaxation of adsorbed molecules.[2440] Anchored ruthenium and rhodium clusters have been characterised by ^{13}C n.m.r. spectroscopy.[2441] ^{13}C m.a.s. n.m.r. spectroscopy has been used to characterise $[H_3Os_4(CO)_{12}]^-$ and $[H_2Os_4(CO)_{12}]^{2-}$ on MgO.[2442] $Rh(CO)_2$ species on SiO_2 and Al_2O_3 have been identified by ^{13}C n.m.r. spectroscopy.[2443] $Rh(CO)_2$ species formed by adsorption of ^{13}CO onto a number of Rh-Y zeolites have been examined by ^{13}C n.m.r. spectroscopy.[2444] The motions of adsorbed $Rh(CO)_2$ species in Y zeolites have been studied by ^{13}C n.m.r. spectroscopy.[2445] A stable μ-peroxo complex of Rh(II) intercalated in the lamellar spaces of montmorillonite has been identified by ^{27}Al, ^{29}Si, and ^{31}P n.m.r. spectroscopy.[2446] The influence of pH on rhodium and platinum catalysts supported on γ-alumina has been studied by ^{195}Pt n.m.r. spectroscopy.[2447] The electronic properties of small supported platinum particles have been investigated by ^{195}Pt n.m.r. spectroscopy.[2448] Surface species on platinum colloids have been observed using ^{195}Pt n.m.r. spectroscopy.[2449] The sorption mechanism for $[Pt(SnCl_3)_5]^{3-}$ on γ-Al_2O_3 has been examined using ^{119}Sn and ^{195}Pt n.m.r. spectroscopy.[2450] The ^{113}Cd n.m.r. spectra have been observed for Cd^{2+} adsorbed on montmorillonites.[2451]

The translational and rotational dynamics of methane in ZSM-5 have been investigated by ^1H n.m.r. spectroscopy.[2452] ^1H n.m.r. spectroscopy has been used to

2436 B.M. Reddy, K. Narsimha, P.K. Rao, and V.M. Mastikhin, *J. Catal.*, 1989, **118**, 22 (*Chem. Abstr.*, 1989, **111**, 141 456).
2437 W.M. Shirley and J.S. Frye, *Spectrochim. Acta, Part A*, 1989, **45A**, 763 (*Chem. Abstr.*, 1989, **111**, 186 151).
2438 G.W. Wagner and B.E. Hanson, *J. Am. Chem. Soc.*, 1989, **111**, 5098.
2439 P. Sarrazin, B. Mouchel, and S. Kasztelan, *J. Phys. Chem.*, 1989, **93**, 904.
2440 H. Winkler, B. Birner, and V. Bosacek, *Zeolites*, 1989, **9**, 293 (*Chem. Abstr.*, 1989, **111**, 121 478).
2441 U. Kiiski, A. Tapani, and O. Krause, *J. Mol. Catal.*, 1989, **50**, 143 (*Chem. Abstr.*, 1989, **110**, 180 097).
2442 H.H. Lamb, L.C. Hasselbring, C. Dybowski, and B.C. Gates, *J. Mol. Catal.*, 1989, **56**, 36.
2443 A.M. Thayer and T.M. Duncan, *J. Phys. Chem.*, 1989, **93**, 6763.
2444 P.F. Molitor, R.K. Shoemaker, and T.M. Apple, *J. Phys. Chem.*, 1989, **93**, 2891.
2445 P. Molitor and T. Apple, *J. Phys. Chem.*, 1989, **93**, 7055.
2446 M.M.T. Khan, M.R.H. Siddiqui, and S.A. Samad, *Inorg. Chem.*, 1989, **28**, 4427.
2447 J.H.A. Martens and R. Prins, *Appl. Catal.*, 1989, **46**, 31 (*Chem. Abstr.*, 1989, **110**, 122 315).
2448 J.P. Bucher and J.J. Van der Klink, *Phys. Rev. B: Condens. Matter*, 1988, **38**, 11 038 (*Chem. Abstr.*, 1989, **110**, 87 210); J.P. Bucher, J. Buttet, J.J. Van der Klink, and M. Graetzel, *Surf. Sci.*, 1989, **214**, 347 (*Chem. Abstr.*, 1989, **111**, 64 700).
2449 R.D. Newmark, M. Fleischmann, and B.S. Pons, *J. Electroanal. Chem. Interfacial Electrochem.*, 1988, **255**, 325 (*Chem. Abstr.*, 1989, **110**, 87 202).
2450 T.V. Kireeva, V.P. Doronin, L.Ya. Alt, and V.K. Duplyakin, *React. Kinet. Catal. Lett.*, 1989, **38**, 7 (*Chem. Abstr.*, 1989, **110**, 122 305).
2451 S. Bank, J.F. Bank, and P.D. Ellis, *J. Phys. Chem.*, 1989, **93**, 4847.
2452 H. Jobic, M. Bee, and G.J. Kearley, *Zeolites*, 1989, **9**, 312 (*Chem. Abstr.*, 1989, **111**, 181 361); V.M. Mastikhin, I.L. Mudrakovskii, A.G. Pel'menshchikov, and K.I. Zamaraev, *Khim. Fiz.*, 1988, **7**, 1096 (*Chem. Abstr.*, 1989, **110**, 153 643).

investigate the diffusion of ethane in silicalite-1.[2453] ^{2}H T_1 measurements have been used to investigate a n-butane film on graphite.[2454] The self-diffusion of molecules on the surface of pyrogenic silica has been studied by n.m.r. spectroscopy.[2455] The molecular motion of benzene, n-hexane, and cyclohexane in K zeolite L has been studied by ^{2}H n.m.r. spectroscopy.[2456] The physical properties of cyclohexane confined within porous silica have been investigated by n.m.r. spectroscopy.[2457] The diffusion of linear paraffins and cyclohexane in NaX and 5A zeolite crystals has been studied by n.m.r. spectroscopy.[2458] A model has been proposed to explain the ^{1}H and ^{13}C n.m.r. spectra observed for C_5-C_{16} *n*-alkanes adsorbed in CaA, CaX, or NaX zeolites.[2459] The diffusion of water and *n*-decane in porous alumina has been studied using pulsed field gradient n.m.r. spectroscopy.[2460] Solid state ^{2}H n.m.r. spectroscopy has been used to show that cyclooctane-d_n and C_6D_6 rapidly rotate in $[\{(Ph_3P)_2IrH_2\}_3PW_{12}O_{40}]$.[2461]

^{13}C n.m.r. spectroscopy has been used to study coke formation through the reaction of C_2H_4 over mordenite.[2462] Y-faujasite catalysts coked with C_2H_4 have been studied by ^{13}C n.m.r. spectroscopy.[2463] The dynamics of C_2D_4 sorbed in the zeolite NaY have been studied by ^{2}H n.m.r. spectroscopy.[2464] The interaction of palladium ions with C_2H_4 over Pd-Y zeolite catalysts has been investigated by ^{23}Na n.m.r. spectroscopy.[2465] ^{2}H n.m.r. spectroscopy has been used to study the dynamics of C_2H_4 in Ag exchanged X type zeolites.[2466] The oligomerisation reactions of propene on zeolite HY have been studied by variable temperature ^{13}C n.m.r. spectroscopy.[2467] ^{2}H n.m.r. spectroscopy has been used to study the rotation and diffusion kinetics of propene and benzene in NaX and AgNaX zeolites.[2468] The mechanism of the aromatization of lower olefin oligomers chemisorbed on ZSM-5 has been studied by ^{13}C n.m.r. spectroscopy.[2469] An n.m.r.

2453 N. Van-den-Begin, L.V.C. Rees, J. Caro, M. Buelow, M. Hunger, and J. Kaerger, *J. Chem. Soc., Faraday Trans. 1*, 1989, **85**, 1501.
2454 B. Boddenberg and G. Neue, *Mol. Phys.*, 1989, **67**, 385.
2455 V.V. Brei and A.A. Chuiko, *Teor. Eksp. Khim.*, 1989, **25**, 99 (*Chem. Abstr.*, 1989, **111**, 141 018).
2456 B.G. Silbernagel, A.R. Garcia, J.M. Newsam, and R. Hulme, *J. Phys. Chem.*, 1989, **93**, 6506.
2457 J.C. Dore, M. Dunn, T. Hasebe, and J.H. Strange, *Colloids Surf.*, 1989, **36**, 199 (*Chem. Abstr.*, 1989, **111**, 161 269).
2458 M. Eic and D.M. Ruthven, *Zeolites*, 1988, **8**, 472 (*Chem. Abstr.*, 1989, **110**, 13 993).
2459 L.G. Davydov and A.G. Lundin, *Zh. Strukt. Khim.*, 1989, **30**, 76 (*Chem. Abstr.*, 1989, **111**, 202 521).
2460 K.J. Packer and F.O. Zelaya, *Colloids Surf.*, 1989, **36**, 221 (*Chem. Abstr.*, 1989, **111**, 160 788).
2461 A.R. Siedle, *New J. Chem.*, 1989, **13**, 719; A.R. Siedle, R.A. Newmark, M.R.V. Sahyun, P.A. Lyon, S.L. Hunt, and R.P. Skarjune, *J. Am. Chem. Soc.*, 1989, **111**, 8346.
2462 J.P. Lang, A. Gutsze, J. Allgeier, and H.G. Karge, *Appl. Catal.*, 1988, **45**, 345 (*Chem. Abstr.*, 1989, **110**, 195 047).
2463 G.W. Zajac, T.H. Fleisch, B.L. Meyers, G.J. Ray, and J.T. Miller, *J. Catal.*, 1989, **115**, 254 (*Chem. Abstr.*, 1989, **110**, 160 933).
2464 R. Burmeister, B. Boddenberg, and M. Verfuerden, *Zeolites*, 1989, **9**, 318 (*Chem. Abstr.*, 1989, **111**, 181 362).
2465 X. Bai, G. Wang, and L. Zheng, *Proc. - Int. Congr. Catal., 9th*, 1988, **1**, 340 (*Chem. Abstr.*, 1989, **111**, 153 145).
2466 B. Boddenberg and R. Burmeister, *Zeolites*, 1988, **8**, 480 (*Chem. Abstr.*, 1989, **110**, 13 994).
2467 J.F. Haw, B.R. Richardson, I.S. Oshiro, N.D. Lazo, and J.A. Speed, *J. Am. Chem. Soc.*, 1989, **111**, 2052.
2468 B. Boddenberg and R. Burmeister, *Zeolites*, 1988, **8**, 488 (*Chem. Abstr.*, 1989, **110**, 29 373).
2469 A.S. Medin, V.Yu. Borovkov, and V.B. Kazanskii, *Kinet. Katal.*, 1989, **30**, 177 (*Chem. Abstr.*, 1989, **111**, 77 444).

relaxation study of benzene and cyclohexane adsorbed on charcoals has been reported.[2470] Fast adsorption-desorption kinetics of hydrocarbons in silicalite-1 have been measured by the n.m.r. pulsed field gradient technique.[2471] The permeation and diffusion characteristics of benzene and cyclohexane in inorganic porous membranes have ben investigated.[2472] ^{2}H n.m.r. spectroscopy has been used to investigate benzene, *p*-xylene, and ethene sorbed in zeolites NaX, AgNaX, and NaY. The dynamics of anisotropic reorientation and site exchange motion was assessed.[2473] ^{13}C n.m.r. spectroscopy has been used to study ethyne adsorbed on platinum on alumina.[2474]

^{1}H n.m.r. spectroscopy has been used to study benzene adsorbed on η-Al_2O_3.[2475] The diffusion of C_6H_6 in ZSM-5 has been studied by pulsed field gradient n.m.r. spectroscopy.[2476] ^{1}H and ^{2}H n.m.r. results of 1,4-$(D_3C)_2C_6H_4$, $C_6H_5CD_3$, and C_6D_6 sorbed on ZSM-5 have been reported.[2477] ^{2}H and ^{13}C n.m.r. spectroscopy has been used to investigate the ordering and dynamics of benzene derivatives in a lamellar lyotropic mesophase.[2478] ^{1}H and ^{2}H n.m.r. spectra of benzene adsorbed on η-Al_2O_3 and Pt/η-Al_2O_3 catalysts have been reported.[2479] The ZSM-5/*p*-xylene system has been investigated by ^{29}Si n.m.r. spectroscopy.[2480] High pressure ^{2}H n.m.r. spectroscopy has been used to study pyridine-d_5 intercalated $CdPS_3$.[2481] The adsorption of pyridine and 2,6-dimethylpyridine on silica, alumina, and zeolites NaY and HY has been studied by ^{13}C n.m.r. spectroscopy.[2482] The kaolinite-pyridine intercalation compound has been characterised by ^{29}Si n.m.r. spectroscopy.[2483] ^{1}H and ^{13}C n.m.r. spectroscopy has been used to model acetonitrile hydrogenation on γ-Mo_2N.[2484]

The differential ^{1}H and ^{13}C n.m.r. line broadening of methanol adsorbed on silica gel has been observed.[2485] Modified silicas have been studied by solid state n.m.r. spectro-

2470 Y. Du and Y. Kong, *Wuli Huaxue Xuebao*, 1989, **5**, 274 (*Chem. Abstr.*, 1989, **111**, 181 610).
2471 N. Van-Den-Begin, L.V.C. Rees, J. Caro, and M. Buelow, *Zeolites*, 1989, **9**, 287 (*Chem. Abstr.*, 1989, **111**, 181 360)
2472 S. Shinoda, I. Ohkouchi, J. Ichinose, and T. Yamakawa, *Chem. Express*, 1989, **4**, 149 (*Chem. Abstr.*, 1989, **111**, 160 931).
2473 R. Burmeister, H. Schwarz, and B. Boddenberg, *Ber. Bunsenges. Phys. Chem.*, 1989, **93**, 1309.
2474 Y.H. Chin and P.D. Ellis, *J. Am. Chem. Soc.*, 1989, **111**, 7653.
2475 C.F. Tirendi, G.A. Mills, and C.R. Dybowski, *J. Phys. Chem.*, 1989, **93**, 3282.
2476 M. Morgan, T. Cosgrove, and R. Richardson, *Colloids Surf.*, 1989, **36**, 209 (*Chem. Abstr.*, 1989, **111**, 160 787).
2477 I. Kustanovich, H.M. Vieth, Z. Luz, and S. Vega, *J. Phys. Chem.*, 1989, **93**, 7427 (*Chem. Abstr.*, 1989, **111**, 181 369).
2478 J.P. Caniparoli, T. Bredel, C. Chachaty, and J. Maruani, *J. Phys. Chem.*, 1989, **93**, 797.
2479 B. Boddenberg and B. Beerwerth, *J. Phys. Chem.*, 1989, **93**, 1435; 1440.
2480 B.F. Mentzen and F. Bosselet, *C.R. Acad. Sci., Ser. 2*, 1989, **308**, 1533 (*Chem. Abstr.*, 1989, **111**, 141 266).
2481 P.L. McDaniel, G. Liu, and J. Jonas, *Physica A (Amsterdam)*, 1989, **156**, 203 (*Chem. Abstr.*, 1989, **111**, 45 605).
2482 T. Sheng, Z. Xu, H. Shao, and S. Hao, *Gaodeng Xuexiao Huaxue Xuebao*, 1989, **10**, 317 (*Chem. Abstr.*, 1989, **111**, 161 078).
2483 Y. Sugahara, S. Satokawa, K. Yoshioka, K. Kuroda, and C. Kato, *Clays Clay Miner.*, 1989, **37**, 143 (*Chem. Abstr.*, 1989, **110**, 176 819).
2484 G.W. Haddix, A.T. Bell, and J.A. Reimer, *J. Phys. Chem.*, 1989, **93**, 5859.
2485 C.J. Hartzell, P.C. Stein, T.J. Lynch, L.G. Werbelow, and W.L. Earl, *J. Am. Chem. Soc.*, 1989, **111**, 5114; F. Heatley, *J. Chem. Soc., Faraday Trans. 1*, 1989, **85**, 917.

scopy.[2486] The reactivity of Pr^iOH on H-ZSM-5 has been monitored by ^{13}C and ^{129}Xe n.m.r. spectroscopy.[2487] ^{13}C n.m.r. spectroscopy has been used to investigate the intermediates formed when $Me_3{}^{13}COH$ is adsorbed on H-ZSM-5,[2488] and CO diffusion and site exchange on rhodium particles.[2489] The adsorbed states of CO on dispersed metals have been studied by ^{13}C n.m.r. spectroscopy.[2490] Surface intermediates of CO chemisorbed on SnO_2 have been identified by ^{13}C n.m.r. spectroscopy.[2491]

The control of gas adsorbability of zeolites by chemical vapour deposition of silanes has been investigated by ^{129}Xe n.m.r. spectroscopy.[2492] The adsorption behaviour of a silane coupling agent on a colloidal silica surface has been studied by ^{29}Si n.m.r. spectroscopy.[2493] The binding of organic silanes to silica surfaces has been investigated by ^{13}C and ^{29}Si n.m.r. spectroscopy.[2494] ^{13}C n.m.r. spectra have been recorded for benzylsilyl derivatives of silica.[2495] ^{119}Sn n.m.r. spectroscopy has been used to study alloying silica supported tin with nickel and palladium.[2496]

^{31}P n.m.r. spectroscopy has been used to study PMe_3 as a probe of acid sites in normal and dealuminated zeolite Y.[2497] The oxidation of PMe_3 in zeolite Y has been investigated by ^{31}P n.m.r. spectroscopy.[2498] The adsorption of H_3PO_4 on TiO_2 and ZrO_2 has been investigated by ^{31}P n.m.r. spectroscopy.[2499] Anodic oxidation of $[SPO_3]^{3-}$ has been studied by ^{31}P n.m.r. spectroscopy.[2500] The structure of the surface of silica modified by PCl_3 has been investigated by ^{31}P n.m.r. spectroscopy.[2501]

Oxygen chemisorption on Al_2O_3 supported molybdena has been studied by ^1H n.m.r. spectroscopy.[2502] The self-diffusion of H_2S on ZnO has been studied by pulsed field

2486 M. Sidqi, H. Balard, E. Papirer, A. Tuel, H. Hommel, and A.P. Legrand, *Chromatographia*, 1989, **27**, 311 (*Chem. Abstr.*, 1989, **111**, 160 997).

2487 M.E. Davis, P. Hathaway, D. Morgan, T. Glass, and H. Dorn, *Stud. Surf. Sci. Catal.* 1987, (Pub. 1988), **38**, 263 (*Chem. Abstr.*, 1989, **111**, 96 375).

2488 M.T. Aronson, R.J. Gorte, W.E. Farneth, and D. White, *J. Am. Chem. Soc.*, 1989, **111**, 840.

2489 T.M. Duncan, A.M. Thayer, and T.W. Root, *Phys. Rev. Lett.*, 1989, **63**, 62 (*Chem. Abstr.*, 1989, **111**, 161 046).

2490 T.M. Duncan, K.W. Zilm, D.M. Hamilton, and T.W. Root, *J. Phys. Chem.*, 1989, **93**, 2583.

2491 O. Gonen, P.L. Kuhns, J.S. Waugh, and J.P. Fraissard, *J. Phys. Chem.*, 1989, **93**, 504.

2492 Y. Teraoka, K. Kunitake, S. Kagawa, and M. Iwamoto, *Nippon Kagaku Kaishi*, 1989, 424 (*Chem. Abstr.*, 1989, **110**, 178 260).

2493 N. Nishiyama, K. Horie, and T. Asakura, *J. Colloid Interface Sci.*, 1989, **129**, 113 (*Chem. Abstr.*, 1989, **110**, 199 786).

2494 J.M.J. Vankan, J.J. Ponjee, J.W. De Haan, and L.J.M. Van de Ven, *J. Colloid Interface Sci.*, 1988, **126**, 604 (*Chem. Abstr.*, 1989, **110**, 14 018).

2495 D. Slotfeldt-Ellingsen, H.A. Resing, K. Unger, and J. Frye, *Langmuir*, 1989, **5**, 1324.

2496 T.H. Chang, C.P. Cheng, and C.T. Yeh, *Mater. Chem. Phys.*, 1989, **22**, 503 (*Chem. Abstr.*, 1989, **111**, 220 059).

2497 J.H. Lunsford, P.N. Tutunjian, P.J. Chu, E.B. Yeh, and D.J. Zalewski, *J. Phys. Chem.*, 1989, **93**, 2590.

2498 D.J. Zalewski, P.J. Chu, P.N. Tutunjian, and J.H. Lunsford, *Langmuir*, 1989, **5**, 1026.

2499 S.B. Randarevich, V.V. Strelko, V.N. Belyakov, V.Yu. Korovin, and A.I. Bortun, *Teor. Eksp. Khim.*, 1988, **24**, 633 (*Chem. Abstr.*, 1989, **110**, 29 462).

2500 A. Ngoviwatchai and D.C. Johnson, *Electroanalysis (N.Y.)*, 1989, **1**, 125 (*Chem. Abstr.*, 1989, **110**, 181 551).

2501 V.M. Bogatyrev, V.V. Brei, and A.A. Chuiko, *Teor. Eksp. Khim.*, 1988, **24**, 629 (*Chem. Abstr.*, 1989, **110**, 29 391).

2502 B.M. Reddy and V.M. Mastikhin, *Proc.-Int. Congr. Catal., 9th.*, 1988, **1**, 82 (*Chem. Abstr.*, 1989, **111**, 45 894).

gradient n.m.r. spectroscopy.[2503] The reaction of H_2S with the surface of Al_2O_3 and SiO_2 has been studied by ^1H n.m.r. spectroscopy.[2504] ^{27}Al n.m.r. spectroscopy has been used to study the adsorption of Cl_2 on γ-Al_2O_3.[2505]

The nuclear magnetism of atomic layers of ^3He adsorbed on graphite has been studied by n.m.r. spectroscopy.[2506] ^{129}Xe n.m.r. spectroscopy has been used to study absorbed xenon on zeolites as a method of investigation.[2507] The effects of pore size distribution on ^{129}Xe n.m.r. chemical shifts of adsorbed xenon have been studied.[2508] The correlation between the results of ^{129}Xe n.m.r. shift and ^1H n.m.r. pulsed field gradient measurements on zeolites has been discussed.[2509] Low temperature ^{129}Xe n.m.r. studies of steam dealuminated Y zeolites have been reported.[2510] Nickel exchanged NaY zeolites have been studied by ^{129}Xe n.m.r. spectroscopy.[2511] Platinum dispersion in zeolites has been assessed using ^{129}Xe n.m.r. spectroscopy.[2512] ^{129}Xe n.m.r. spectroscopy has been used to study xenon adsorbed on ferrierite.[2513] The adsorptive behaviour of xenon in zeolitic pores of faujasite has been investigated by ^{129}Xe n.m.r. spectroscopy.[2514] ^{129}Xe n.m.r. spectroscopy has been used to investigate SAPO-37,[2515] $AlPO_4$-11,[2516] and $AlPO_4$-5.[2517] The porous structure of agglomerated silica microspheres has been characterised by ^{129}Xe n.m.r. spectroscopy.[2518]

6 Group IIIB Compounds

Reviews entitled '^{11}B n.m.r. spectroscopy',[2519] 'N.m.r. spectroscopy of boron compounds containing two-, three-, and four-coordinate boron',[2520] 'Chemistry of basic

2503 H. Spindler and J. Kaerger, *Z. Phys. Chem. (Leipzig)*, 1989, **270**, 225 (*Chem. Abstr.*, 1989, **111**, 13 038).
2504 V.M. Mastikhin, I.L. Mudrakovskii, A.V. Nosov, and A.V. Mashkina, *J. Chem. Soc., Faraday Trans. 1*, 1989, **85**, 2819.
2505 M.C. Rohner, V.K. Sharma, and W. Richarz, *Can. J. Chem. Eng.*, 1989, **67**, 513 (*Chem. Abstr.*, 1989, **111**, 121 489).
2506 H. Godrin, R.R. Ruel, and D.D. Osheroff, *J. Phys., Colloq.*, 1988, **(C8, Proc. Int. Congr. Magn., 1988, Pt. 3)**, C8-2045 (*Chem. Abstr.*, 1989, **111**, 107 854).
2507 J. Fraissard, T. Ito, and M.A. Springuel-Huet, *J. Chim. Phys. Phys.-Chim. Biol.*, 1988, **85**, 747.
2508 T.T.P. Cheung, *J. Phys. Chem.*, 1989, **93**, 7549.
2509 J. Fraissard and J. Kaerger, *Zeolites*, 1989, **9**, 351 (*Chem. Abstr.*, 1989, **111**, 121 362).
2510 T.T.P. Cheung and C.M. Fu, *J. Phys. Chem.*, 1989, **93**, 3740.
2511 A. Gedeon, J.L. Bonardet, T. Ito, and J. Fraissard, *J. Phys. Chem.*, 1989, **93**, 2563; A. Gedeon, J.L. Bonardet, and J. Fraissard, *J. Chim. Phys. Phys.-Chim. Biol.*, 1988, **85**, 871 (*Chem. Abstr.*, 1989, **110**, 179 971).
2512 L. Petrakis, M.A. Springuel-Huet, T. Ito, T.R. Hughes, I.Y. Chan, and J. Fraissard, *Proc. - Int. Congr. Catal., 9th*, 1988, **1**, 348 (*Chem. Abstr.*, 1989, **111**, 64 833).
2513 T. Ito, M.A. Springuel-Huet, and J. Fraissard, *Zeolites*, 1989, **9**, 68 (*Chem. Abstr.*, 1989, **110**, 102 356).
2514 T. Yoshida, J. Koizumi, M. Ashida, and Y. Akai, *Nippon Kagaku Kaishi*, 1989, 458 (*Chem. Abstr.*, 1989, **110**, 176 062).
2515 N. Dumont, T. Ito, and E.G. Derouane, *Appl. Catal.*, 1989, **54**, L1 (*Chem. Abstr.*, 1989, **111**, 202 609).
2516 M.A. Springuel-Huet and J. Fraissard, *Chem. Phys. Lett.*, 1989, **154**, 299 (*Chem. Abstr.*, 1989, **110**, 160 874).
2517 Q.J. Chen, M.A. Springuel-Huet, and J. Fraissard, *Chem. Phys. Lett.*, 1989, **159**, 117 (*Chem. Abstr.*, 1989, **111**, 161 065).
2518 W.C. Conner, E.L. Weist, T. Ito, and J. Fraissard, *J. Phys. Chem.*, 1989, **93**, 4138.
2519 A.R. Siedle, *Annu. Rep. N.M.R. Spectrosc.*, 1988, **20**, 205.
2520 B. Wrackmeyer, *Annu. Rep. N.M.R .Spectrosc.*, 1988, **20**, 61.

commercial aluminium hydrolysis complexes' which contains ^{27}Al n.m.r. spectra,[2521] 'Multinuclear studies of aluminium compounds',[2522] and 'Thallium n.m.r. spectroscopy' have appeared.[2523]

A ^{11}B n.m.r. data base for personal computers has been described. It contains 3300 entries.[2524]

Boron Hydrides and Carboranes.——The reactions of acyclic aminophosphines with BH_3.THF have been studied by ^{11}B and ^{13}C n.m.r. spectroscopy.[2525] The ^{11}B antipodal effect in borane cages such as EB_9H_9 and $EB_{11}H_{11}$ has been examined theoretically by CNDO/2 and extended Hückel calculations.[2526] N.m.r. data have also been reported for [$LiBH_3CH_2CN$], (^{7}Li, ^{11}B, ^{13}C),[2527] [$(H_3BPPh_2)_2C{=}C(PPh_2)H$], (^{13}C),[2528] [2,6-$(EtO)_2C_6H_3$-$PMe_2BH_3$], (^{11}B, ^{13}C, ^{29}Si),[2529] [$(Me_3Si)_3CB(\mu\text{-}H)_3Li$.THF], (^{7}Li, ^{11}B, ^{13}C, ^{29}Si),[2530] [Me_3P-$BH_{3-n}Br_n$], (^{11}B, ^{13}C),[2531] [RBH_2L], (^{11}B),[2532] [$(MeO)_2P(O)CH_2NEt_2BH_2CN$], (^{11}B),[2533] [BH-Bu^{t}R], (^{11}B),[2534] [$N_3B_3H_5(CR^1{=}CHR^2)$], (^{11}B, ^{13}C),[2535] [$B_3N_3H_5X$], (^{11}B),[2536] [$HC(CH_2CH_2)_3$-$NBH(CH_2Ph)CN$], (^{11}B),[2537] [$Me_3PBH(PMe_2CH_2)_2$], (^{11}B, ^{13}C),[2538] [HB(3-Bu^{t}pyrazolyl)$_3$Tl], (^{13}C),[2539] [$(\eta^5\text{-}C_5H_5)Mo(\eta^5{:}\eta^1\text{-}C_5H_4)B_4H_7$], (^{11}B, ^{13}C),[2540] [BrB_5H_7]$^-$, (^{11}B),[2541] [(η^5-NC_4Me_2-R)$Co_2(\mu,\eta^5\text{-}Et_2C_2B_3H_3)Co(\eta^5\text{-}C_5Me_5)$], (^{11}B),[2542] [$R^1NB_6H_8R^2$]$^-$, (^{11}B, ^{13}C),[2543] [[2,3-$(Me_3Si)_2$-2,3-$C_2B_4H_5$][$Na(THF)_2$]]$_2$, (^{11}B, ^{13}C),[2544] *closo*-[1-Sn(η^5-C_5H_5)Fe(η^5-$C_5H_4CH_2$-

2521 J.J. Fitzgerald, *Cosmet. Sci. Technol. Ser.*, 1988, **7(Antiperspirants Deodorants)**, 119 (*Chem. Abstr.*, 1989, **110**, 44 725).
2522 J.W. Akitt, *Prog. Nucl. Magn. Reson. Spectrosc.*, 1988, **21**, 1.
2523 J.F. Hinton, K.R. Metz, and R.W. Briggs, *Prog. Nucl. Magn. Reson. Spectrosc.*, 1988, **20**, 423.
2524 H. Nöth and E. Striedl, *Mitteilungsbl. - Ges. Dtsch. Chem., Fachgruppe Chem.-Int.*, 1989, **15**, 3 (*Chem. Abstr.*, 1989, **111**, 56 524).
2525 R.K. Kanjolia, D.K. Srivastava, C.L. Watkins, and L.K. Krannich, *Inorg. Chem.*, 1989, **28**, 3341.
2526 S. Heřmánek, D. Hnyk, and Z. Havlas, *J. Chem. Soc., Chem. Commun.*, 1989, 1859.
2527 J.L. Peters, M. Mittakanti, T.J. Lofthouse, and K.W. Morse, *Inorg. Chem.*, 1989, **28**, 492.
2528 H. Schmidbaur, C. Paschalidis, O. Steigelmann, D.L. Wilkinson, and G. Müller, *Chem. Ber.*, 1989, **122**, 1857.
2529 J.A. Albanese, D.E. Gingrich, C.D. Schaeffer, jun., S.M. Coley, J.C. Otter, M.S. Samples, and C.H. Yoder, *J. Organomet. Chem.*, 1989, **365**, 23.
2530 S.S. Al-Juaid, C. Eaborn, M.N.A. El-Kheli, P.B. Hitchcock, P.D. Lickiss, M.E. Molla, J.D. Smith, and J.A. Zora, *J. Chem. Soc., Dalton Trans.*, 1989, 447.
2531 H. Schmidbaur, T. Wimmer, A. Grohmann, O. Steigelmann, and G. Müller, *Chem. Ber.*, 1989, **122**, 1607.
2532 M. Srebnik, T.E. Cole, P.V. Ramachandran, and H.C. Brown, *J. Org. Chem.*, 1989, **54**, 6085.
2533 M.P. Kaushik, M.R.M.D. Charandabi, M.L. Ettel, T.J. Lofthouse, and K.W. Morse, *Inorg. Chem.*, 1989, **28**, 897.
2534 H.C. Brown and P.V. Ramachandran, *J. Org. Chem.*, 1989, **54**, 4504.
2535 A.T. Lynch and L.G. Sneddon, *J. Am. Chem. Soc.*, 1989, **111**, 6201.
2536 J.T. Nelson and W.J. Pietro, *Inorg. Chem.*, 1989, **28**, 544.
2537 W.J. Mills, L.J. Todd, and J.C. Huffman, *J. Chem. Soc., Chem. Commun.*, 1989, 900.
2538 T. Wimmer, O. Steigelmann, G. Müller, and H. Schmidbaur, *Chem. Ber.*, 1989, **122**, 2109.
2539 A.H. Cowley, R.L. Geerts, C.M. Nunn, and S. Trofimenko, *J. Organomet. Chem.*, 1989, **365**, 19.
2540 P.D. Grebenik, M.L.H. Green, M.A. Kelland, J.B. Leach, and P. Mountford, *J. Chem. Soc., Chem. Commun.*, 1989, 1397.
2541 L. Barton, P.K. Rush, T. Zhu, P. Nevels, and M.H. Owens, *Inorg. Chem.*, 1989, **28**, 381.
2542 K.J. Chase and R.N. Grimes, *Organometallics*, 1989, **8**, 2492.
2543 S. Küpper and P. Paetzold, *Chem. Ber.*, 1989, **122**, 479.
2544 N.S. Hosmane, U. Sirawardane, G. Zhang, H. Zhu, and J.A. Maguire, *J. Chem. Soc., Chem. Commun.*, 1989, 1128.

$NMe_2)$-2,3-$(Me_3Si)_2$-2,3-$C_2B_4H_4]$, (^{11}B, ^{13}C, ^{119}Sn),[2545] [μ-{(η^5-C_5H_5)Fe(CO)(PPh_3)(η^1:μ-*trans*-MeC=CMe)}-2,3-$Et_2C_2B_4H_5$], (^{11}B),[2546] [(η^6-arene)Ru($Et_2C_2B_4H_4$)], (^{11}B),[2547] [(η^5-C_5H_5)Co($Et_2C_2B_4H_4$)], (^{11}B),[2548] [1-Sn(2,2′:6′,2″-$C_{15}H_{11}N_3$)-2,3-$(Me_3Si)_2$-2,3-$C_2B_4H_4$], (^{13}C),[2549] *arachno*-[7-(η^5-C_5H_5)Co-6,8-$S_2B_6H_8$], (^{11}B),[2550] *closo*-[1-Pb-2,3-$(Me_3Si)_2$-2,3-$C_2B_4H_4$], (^{11}B, ^{13}C, ^{207}Pb),[2551] *hypho*-$[S_2B_6H_9]^-$, (^{11}B),[2552] *hypho*-$[MeCNS_2B_7H_8]^-$, (^{11}B, ^{13}C),[2553] [5-F-2,4-$C_2B_5H_6$], (^{11}B),[2554] [1-Cl-3,5-Et_2-2,4-$C_2B_5H_4$], (^{11}B),[2555] [4-SB_8H_{12}], (^{11}B),[2556] [4-(η^5-C_5Me_5)-4-(MeNC)-*arachno*-4-IrB_8H_{12}], (^{11}B),[2557] $[4,5\text{-}C_2B_6H_{11}]^-$, ($^{11}B$),[2558] *hypho*-$[C_2B_6H_{13}]^-$, ($^{11}B$, ^{13}C),[2559] [μ-6,9-(Me_2N)-10-($PhMe_2P$)-5-(η^6-C_6Me_6)-*arachno*-5-RuB_9H_{10}], (^{11}B),[2560] [(η^5-C_5Me_5)$RhB_9(SMe_2)H_{10}RhB_9H_7(SMe_2)_2$], ($^{11}B$),[2561] [($\eta^5$-$C_5Me_5$)$RhIrB_9H_{11}$], ($^{11}B$),[2562] [10-($Ph_3P$)-2,2,2-$(H_3P)_2$($Ph_2P$-2-$C_6H_4$)-*closo*-2-$IrB_9H_7$-1], ($^{11}B$),[2563] $[B_{10}H_{11}]^-$, (^{11}B),[2564] [1,2-$P_2B_{10}H_{10}$], (^{11}B),[2565] [$(MeCN)_6$Yb(μ-H)$B_{10}H_{12}$], (^{11}B),[2566] [Rh($XB_{10}H_{10}$){S_2C-H(PPh_3)}$]_2$, (^{11}B),[2567] [1,1-$(Ph_3P)_2$-1-H-1,2,4-$IrC_2B_8H_{10}$], (^{11}B),[2568] $[B_{10}H_{10}(AlCl_3)_2]^{2-}$, ($^{27}Al$),[2569] [$B_{11}H_{11}PMe$], ($^{11}B$),[2570] [$MoFe_2(\mu_3\text{-}CC_6H_4\text{-}4\text{-}CH_3)$(μ-σ:σ′:$\eta^5$-$C_2B_9H_7Me_2$)$(CO)_8]^-$,

2545 N.S. Hosmane, J.S. Fagner, H. Zhu, U. Siriwardane, J.A. Maguire, G. Zhang, and B.S. Pinkston, *Organometallics*, 1989, **8**, 1769.
2546 M.G.L. Mirabelli, P.J. Carroll, and L.G. Sneddon, *J. Am. Chem. Soc.*, 1989, **111**, 592.
2547 J.H. Davis, jun., E. Sinn, and R.N. Grimes, *J. Am. Chem. Soc.*, 1989, **111**, 4776.
2548 J.H. Davis, jun., E. Sinn, and R.N. Grimes, *J. Am. Chem. Soc.*, 1989, **111**, 4784.
2549 U. Siriwardane, J.A. Maguire, J.J. Banewicz, and N.S. Hosmane, *Organometallics*, 1989, **8**, 2792.
2550 S.O. Kang, P.J. Carroll, and L.G. Sneddon, *Inorg. Chem.*, 1989, **28**, 961.
2551 N.S. Hosmane, U. Siriwardane, H. Zhu, G. Zhang, and J.A. Maguire, *Organometallics*, 1989, **8**, 566.
2552 S.O. Kang and L.G. Sneddon, *J. Am. Chem. Soc.*, 1989, **111**, 3281.
2553 S.O. Kang, G.T. Furst, and L.G. Sneddon, *Inorg. Chem.*, 1989, **28**, 2339.
2554 S.S. Durso, R.A. Sheeks, E.V. O'Gorman, and R.A. Beaudet, *Inorg. Chem.*, 1989, **28**, 1615.
2555 W. Nam, Z.J. Abdou, H. Lee, T. Banuelos, and T. Onak, *Inorg. Chem.*, 1989, **28**, 669.
2556 K. Baše, M.G.H. Wallbridge, X.L.R. Fontaine, N.N. Greenwood, J.H. Jones, J.D. Kennedy, and B. Štíbr, *Polyhedron*, 1989, **8**, 2089.
2557 K. Nestor, X.L.R. Fontaine, N.N. Greenwood, J.D. Kennedy, and M. Thornton-Pett, *J. Chem. Soc., Dalton Trans.*, 1989, 1465.
2558 T. Jelínek, B. Štíbr, S. Heřmánek, and J. Plešek, *J. Chem. Soc., Chem. Commun.*, 1989, 804.
2559 T. Jelínek, J. Plešek, S. Heřmánek, and B. Štíbr, *Main Group Met. Chem.*, 1989, **10**, 397 (*Chem. Abstr.*, 1989, **111**, 97 321).
2560 E.J. Ditzel, X.L.R. Fontaine, N.N. Greenwood, J.D. Kennedy, Z. Sisan, and M. Thornton-Pett, *J. Chem. Soc., Chem. Commun.*, 1989, 1762.
2561 E.J. Ditzel, X.L.R. Fontaine, N.N. Greenwood, J.D. Kennedy, and M. Thornton-Pett, *J. Chem. Soc., Chem. Commun.*, 1989, 1262.
2562 K. Nestor, X.L.R. Fontaine, N.N. Greenwood, J.D. Kennedy, and M. Thornton-Pett, *J. Chem. Soc., Chem. Commun.*, 1989, 455.
2563 J. Bould, P. Brint, X.L.R. Fontaine, J.D. Kennedy, and M. Thornton-Pett, *J. Chem. Soc., Chem. Commun.*, 1989, 1763.
2564 V.N. Mustyatsa, N.A. Votinova, K.A. Solntsev, and N.T. Kuznetsov, *Dokl. Akad. Nauk SSSR*, 1988, **301**, 1396 (*Chem. Abstr.*, 1989, **110**, 68 600).
2565 J.L. Little, J.G. Kester, J.C. Huffman, and L.J. Todd, *Inorg. Chem.*, 1989, **28**, 1087.
2566 J.P. White, tert., H.-B. Deng, and S.G. Shore, *J. Am. Chem. Soc.*, 1989, **111**, 8946.
2567 Faridoon, T.R. Spalding, G. Ferguson, J.D. Kennedy, and X.L.R. Fontaine, *J. Chem. Soc., Chem. Commun.*, 1989, 906.
2568 K. Nestor, X.L.R. Fontaine, N.N. Greenwood, J.D. Kennedy, J. Plešek, B. Štíbr, and M. Thornton-Pett, *Inorg. Chem.*, 1989, **28**, 2219.
2569 B. Bonnetot, A. Aboukhassib, and H. Mongeot, *Inorg. Chim. Acta*, 1989, **156**, 183.
2570 T.D. Getman, H.-B. Deng, L.-Y. Hsu, and S.G. Shore, *Inorg. Chem.*, 1989, **28**, 3612.

(^{11}B, ^{13}C),[2571] [$WIr(\mu_3-CC_6H_4-4-CH_3)(CO)_2(PEt_3)_2(\eta^5-C_2B_9H_9Me_2)]^+$, ($^{11}B$, ^{13}C),[2572] [WPt{$\mu_3-CC_6H_3-2,6-(CH_3)_2\}(CO)_2(PEt_3)(\mu-\sigma:\eta^5-C_2B_9H_8Me_2)$], ($^{11}B$, ^{13}C),[2573] [$Re(CO)_5B_{11}H_{14}$], (^{11}B),[2574] [3-($\eta^6-C_6Me_6$)-*closo*-3,1,2-$RuC_2B_9H_{11}$], (^{11}B),[2575] [$(Ph_3P)CuC_2B_9H_{11}]^-$, (^{11}B),[2576] [$RC_2B_9H_{10}GaMe_2]^-$, (^{11}B),[2577] [Tl{*commo*-3,3'-Ga(3,1,2-$GaC_2B_9H_{11})_2$}], (^{11}B),[2578] [W(≡C-C_6H_4-4-Me)$(CO)_2(\eta^6-C_2B_{10}H_{10}Me_2)]^-$, ($^{11}B$, ^{13}C),[2579] and [$IrCl(CO)(PPh_3)_2Ag(B_{11}CH_{12})$], ($^{11}B$).[2580]

Other Compounds of Boron.——The ^{13}C chemical shifts of trihalomethyl cations and related halocarbenium ions correlate well with ^{11}B chemical shifts of their isoelectronic boron halide derivatives.[2581] N.m.r. data have also been reported for [$Ph_2\overline{BCPh=CPh}]^-$, ($^{11}B$, ^{13}C),[2582] [$EtPh_2PC≡CB(CH_2Ph)_3$], (^{11}B, ^{13}C),[2583] [$Me_3\overline{PbCR^1=CR^2BR^2{}_2C}≡CR^1$], ($^{11}B$, ^{13}C, ^{207}Pb),[2584] [$Me\overline{BCMe=CMeBMeC}HMe$], ($^{11}B$, ^{13}C),[2585] [$Et_2\overline{BC=C(SiMe_3)SnMe_2C(SiMe_3)=C}Et$], ($^{11}B$, ^{13}C, ^{29}Si, ^{119}Sn),[2586] [(menthyl)$_2BCH_2$-$CH=CH_2$], (^{11}B, ^{13}C),[2587] [9-$Me_3SiCH_2CH_2$-9-borabicyclo[3.3.1]nonane], (^{13}C, ^{29}Si),[2588] [$(EtCMe_2)_2B(CH_2)_{10}CO_2SiMe_3$], ($^{11}B$, ^{13}C, ^{29}Si),[2589] [9-$R^1CH=C(SiMe_3)$-9-borabicyclo[3.3.1]-nonane], (^{13}C, ^{29}Si),[2590] [9-(9,10-dihydrophenanthyl)BR_2], (^{11}B, ^{13}C),[2591] [$Bu^t\overline{B=CBBu^tC}$-$(SiMe_3)_2$], (^{11}B, ^{13}C),[2592] [$R^1R^2BCR^3=CR^4SiR^5{}_3$], ($^{11}B$, ^{13}C, ^{14}N, ^{15}N, ^{17}O, ^{29}Si),[2593] [BR_3],

2571 D.D. Devore, C. Emmerich, J.A.K. Howard, and F.G.A. Stone, *J. Chem. Soc., Dalton Trans.*, 1989, 797.
2572 J.C. Jeffery, M.A. Ruiz, P. Sherwood, and F.G.A. Stone, *J. Chem. Soc., Dalton Trans.*, 1989, 1845.
2573 D.D. Devore, J.A.K. Howard, J.C. Jeffery, M.U. Pilotti, and F.G.A. Stone, *J. Chem. Soc., Dalton Trans.*, 1989, 303.
2574 H. Mongeot, A. Tangi, and B. Bonnetot, *Inorg. Chim. Acta*, 1989, **161**, 143.
2575 M. Bown, J. Plešek, K. Baše, B. Štíbr, X.L.R. Fontaine, N.N. Greenwood, and J.D. Kennedy, *Magn. Reson. Chem.*, 1989, **27**, 947.
2576 Y. Do, H.C. Kang, C.B. Knobler, and M.F. Hawthorne, *Report*, 1987, **TR-128**; Order No. AD-A181979, 16 pp. Avail. NTIS. From *Gov. Rep. Announce. Index (U.S.)*, 1987, **87**, Abstr. No. 745 216 (*Chem. Abstr.*, 1989, **110**, 24 027).
2577 V.I. Bregadze, A.Ya. Ustyatinskii, V.A. Antonovich, and N.N. Godovikov, *Izv. Acad. Nauk SSSR, Ser. Khim.*, 1988, 670 (*Chem. Abstr.*, 1989, **110**, 114 890).
2578 B.A. Bandman, C.B. Knobler, and M.F. Hawthorne, *Inorg. Chem.*, 1989, **28**, 1204.
2579 S.J. Crennell, D.D. Devore, S.J.B. Henderson, J.A.K. Howard, and F.G.A. Stone, *J. Chem. Soc., Dalton Trans.*, 1989, 1363.
2580 D.J. Liston, Y.J. Lee,W.R. Scheidt, and C.A. Reed, *J. Am. Chem. Soc.*, 1989, **111**, 6643.
2581 G.A. Olah, L. Heiliger, and G.K.S. Prakash, *J. Am. Chem. Soc.*, 1989, **111**, 8020.
2582 M.A. Kropp and G.B. Schuster, *J. Am. Chem. Soc.*, 1989, **111**, 2316.
2583 H.J. Bestmann, H. Behl, and M. Bremer, *Angew. Chem., Int. Ed. Engl.*, 1989, **28**, 1219.
2584 B. Wrackmeyer, K. Horchler, and R. Boese, *Angew. Chem., Int. Ed. Engl.*, 1989, **28**, 1500.
2585 J.K. Uhm, H. Roemich, H. Wadepohl, and W. Siebert, *Z. Naturforsch., B*, 1988, **43**, 306 (*Chem. Abstr.*, 1989, **110**, 212 889).
2586 B. Wrackmeyer, *J. Organomet. Chem.*, 1989, **364**, 331.
2587 H.C. Brown, K.S. Bhat, and R.S. Randad, *J. Org. Chem.*, 1989, **54**, 1570.
2588 J.A. Soderquist, I. Rivera, and A. Negron, *J. Org. Chem.*, 1989, **54**, 4051.
2589 G.W. Kabalka and D.E. Bierer, *Organometallics*, 1989, **8**, 655.
2590 J.A. Soderquist, J.C. Colberg, and L.D. Valle, *J. Am. Chem. Soc.*, 1989, **111**, 4873.
2591 R. Köster, W. Schüßler, and M. Yalpani, *Chem. Ber.*, 1989, **122**, 677.
2592 R. Boese, P. Paetzold, A. Tapper, and R. Ziembinski, *Chem. Ber.*, 1989, **122**, 1057.
2593 R. Köster, G. Seidel, and B. Wrackmeyer, *Chem. Ber.*, 1989, **122**, 1825.

(^{11}B, ^{13}C),[2594] [$(C_8H_{14})BCX{=}CHR$], (^{11}B),[2595] [(2,3,5,6-Me_4C_6H)$BCBu^t{=}CBCl$(2,3,5,6-Me_4C_6H)], (^{11}B, ^{13}C),[2596] [$(Me_3Si)_2C\{B$(2,3,5,6-Me_4C_6H)$\}_2C{=}B(OEt_2)$(2,3,5,6-Me_4C_6H)], (^{11}B, ^{13}C),[2597] [$(Me_3Si)_2C\{B$(2,3,5,6-Me_4C_6H)$\}_2CMe_2$], (^{11}B, ^{13}C),[2598] [$Et_2BCEt{=}CMeM^1Me_2NH$-M^2], (M^1 = Si, Sn, Pb; M^2 = Na, K; ^{11}B, ^{13}C, ^{29}Si, ^{119}Sn, ^{207}Pb),[2599] [$Bu_2BCBu{=}CPhPPh_2$], (^{31}P),[2600] [Et_2BOBEt_2.N-base], (^{11}B, ^{13}C),[2601] (64), (E = O, S, Se; ^{11}B, ^{13}C),[2602] [R_2B(1,2,3-triazole)]$_3$, (^{11}B, ^{13}C),[2603] [$(C_8H_{14})_2B_2(\mu$-3-Me-pz)(μ-S)], (^{11}B, ^{13}C),[2604] [$(C_8H_{14})B(NC_5H_4Me$-4)Se]$_2$, (^{11}B, ^{13}C, ^{77}Se),[2605] (65), (^{11}B),[2606] [(dibenzocyclopentadienyl)$_3BNR_2$], (^{11}B, ^{13}C),[2607] [$(F_3C)_2BNPr^i_2$], (^{11}B, ^{13}C),[2608] (66), (^{11}B, ^{13}C),[2609] [$PR^1BR^2_2]_2$, (^{11}B),[2610] $HPR^1BR^2_2$, (^{11}B),[2611]

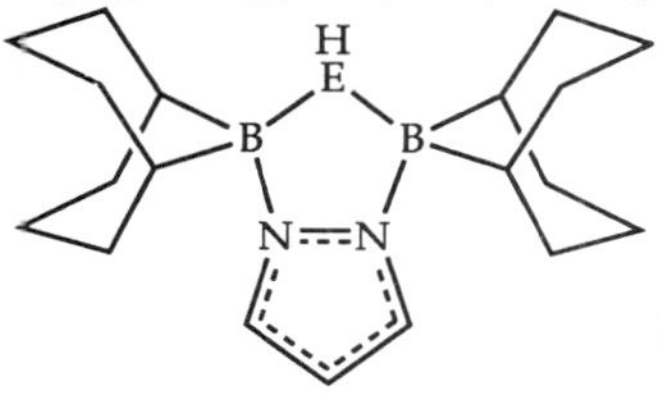

(64)

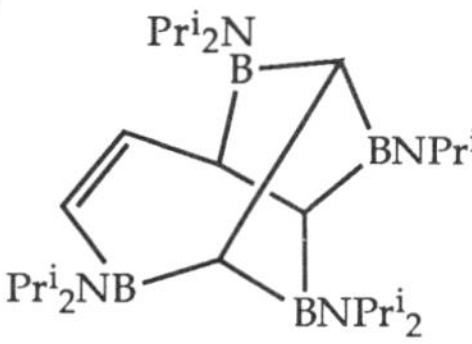

(65)

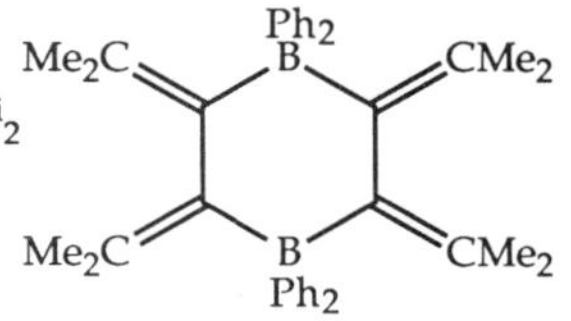

(66)

[$Bu^t_2BPBu^t_2$], (^{11}B),[2612] $R_2M(CH_2CH_2)_2BOMe$, (^{13}C),[2613] [$MeCNXeC_6F_5$][$(C_6F_5)_2BF_2$], (^{11}B, ^{13}C, ^{129}Xe),[2614] [(menthyl)BRCl], (^{11}B),[2615] [(menthyl)$_2$BX], (^{11}B, ^{13}C),[2616] [EtB(μ-NH)(μ-NMeCONMe)BEt], (^{11}B, ^{13}C),[2617] [$MeH_2NBMe(pz)_2$], (^{11}B, ^{13}C),[2618] *N*-boryliminophos-

2594 J.A. Soderquist and I. Rivera, *Tetrahedron Lett.*, 1989, **30**, 3919.
2595 H.C. Brown, C.D. Blue, D.J. Nelson, and N.G. Bhat, *J. Org. Chem.*, 1989, **54**, 6064.
2596 A. Höfner, B. Ziegler, W. Massa, and A. Berndt, *Angew. Chem., Int. Ed. Engl.*, 1989, **28**, 186.
2597 R. Hunold, M. Pilz, J. Allwohn, M. Stadler, W. Massa, P.v.R. Schleyer, and A. Berndt, *Angew. Chem., Int. Ed. Engl.*, 1989, **28**, 781.
2598 M. Pilz, M. Stadler, R. Hunold, J. Allwohn, W. Massa, and A. Berndt, *Angew. Chem., Int. Ed. Engl.*, 1989, **28**, 784.
2599 R. Köster, G. Seidel, B. Wrackmeyer, K. Horchler, and D. Schlosser, *Angew. Chem., Int. Ed. Engl.*, 1989, **28**, 918.
2600 S.G. Vul'fson, N.N. Sarvarova, A.S. Balueva, O.A. Erastov, and B.A. Arbuzov, *Izv. Akad. Nauk SSSR, Ser. Khim.*, 1988, 1445 (*Chem. Abstr.*, 1989, **110**, 95 395).
2601 M. Yalpani, J. Serwatowski, and R. Köster, *Chem. Ber.*, 1989, **122**, 3.
2602 M. Yalpani, R. Köster, and R. Boese, *Chem. Ber.*, 1989, **122**, 19.
2603 K. Niedenzu and K.R. Woodrum, *Inorg. Chem.*, 1989, **28**, 4022.
2604 M. Yalpani, R. Boese, and R. Köster, *Chem. Ber.*, 1989, **122**, 1231.
2605 R. Köster, G. Seidel, and M. Yalpani, *Chem. Ber.*, 1989, **122**, 1815.
2606 A. Meller, U. Seebold, W. Maringgele, M. Noltemeyer, and G.M. Sheldrick, *Chem. Ber.*, 1989, **111**, 8299.
2607 W. Maringgele, M. Noltemeyer, and A. Meller, *J. Organomet. Chem.*, 1989, **365**, 187.
2608 D.J. Brauer, H. Bürger, F. Dörrenbach, G. Pawelke, and W. Weuter, *J. Organomet. Chem.*, 1989, **378**, 125.
2609 A. Maercker, W. Brieden, T. Schmidt, and H.D. Lutz, *Angew. Chem., Int. Ed. Engl.*, 1989, **28**, 477.
2610 D.C. Pestana and P.P. Power, *J. Am. Chem. Soc.*, 1989, **111**, 6887.
2611 R.A. Bartlett, H.V.R. Dias, X. Feng, and P.P. Power, *J. Am. Chem. Soc.*, 1989, **111**, 1306.
2612 H.H. Karsch, G. Hanika, B. Huber, J. Riede, and G. Müller, *J. Organomet. Chem.*, 1989, **361**, C25.
2613 J.A. Soderquist and A. Negron, *J. Org. Chem.*, 1989, **54**, 2462.
2614 H.J. Frohn, S. Jakobs, and G. Henkel, *Angew. Chem., Int. Ed. Engl.*, 1989, **28**, 1506.
2615 H.C. Brown, M. Srebnik, and P.V. Ramachandran, *J. Org. Chem.*, 1989, **54**, 1577.
2616 M. Srebnik, N.N. Joshi, and H.C. Brown, *Isr. J. Chem.*, 1989, **29**, 229.
2617 L. Komorowski and K. Niedenzu, *Inorg. Chem.*, 1989, **28**, 804.
2618 C. Habben, L. Komorowski, W. Maringgele, A. Meller, and K. Niedenzu, *Inorg. Chem.*, 1989, **28**, 2659.

phoranes, (^{11}B, ^{13}C, ^{31}P),[2619] [$R^1B(NR^2)_2BR^1$], (^{11}B, ^{13}C),[2620] [$R^1R^2{}_2B_3N_3R^3{}_3$], (^{11}B, ^{13}C),[2621] *gem*-[$N_3P_3(NMe_2)_4(NHB_3N_3Me_5)_2$], ($^{11}B$),[2622] [{$Pr^i{}_2N\,BCH{=}CHB(NPr^i{}_2)$}$_2P$]$^+$, ($^{13}C$),[2623] [1,2-$Bu^t{}_2$-3-{$(Me_3Si)_2C$}-1,2-diborirane]$^{2-}$, ($^{11}B$, ^{13}C),[2624] [SeCEt=CEtC(O)NR^{1}B-R^2], (^{11}B, (^{13}C),[2625] [$Pr^i{}_2N{=}BXCH(SiMe_3)_2$], ($^{11}B$, ^{13}C),[2626] [R^1BPR^2]$_3$, (^{11}B),[2627] [MeCH=CH-$CH_2BO_2C_6H_4$], (^{13}C),[2628] [$Cl_2CHBOCHRCHRO$], (^{13}C),[2629] [$ArB(OH)_2$], (^{13}C),[2630] [(Ph-BO)$_3$L], (^{11}B),[2631] [RC=CRSSBX], (^{11}B, ^{13}C, ^{14}N, ^{15}N),[2632] 2-Cl-1-alkenyldichloroboranes, (^{11}B, ^{13}C),[2633] [$C_4H_4B_6Cl_6$], (^{11}B, ^{13}C),[2634] [$C_2B_5X_7$], (^{11}B),[2635] [$CH_2{=}CHCH_2BBr_2$], and [$R^1{}_3SnCH_2CH{=}CHR^2$], (^{11}B, ^{119}Sn).[2636]

$^3J(^{13}C-^{11}B)$ has been observed in the ^{13}C n.m.r. spectrum of [$PhCH_2NMe_2BX_3$].[2637] N.m.r. data have also been reported for [{$CH_2(CH_2CMe_2)_2N$}B(OMe){NRE(OMe)$_2$}], (E = As, Sb; ^{11}B, ^{13}C),[2638] mixed esters of boric acid with amino acids, (^{11}B),[2639] [(2-PhC_6H_4E)-$BNPr^i{}_2$], (E = O, S; ^{11}B, ^{13}C),[2640] (67), (^{11}B),[2641] (68), (^{11}B, ^{14}N),[2642] [$PhCH_2C{=}NNP{=}NBF_3$], (^{13}C, ^{15}N),[2643] [$(H_2P)_3BPH_2B(PH_2)_3$]$^-$, (^{11}B),[2644] [{(2,4,6-$Me_3C_6H_2)_2P$}$_2BX$], (^{11}B),[2645] [Pr^iO-$BOCMe_2CMe_2O$], (^{13}C),[2646] [$C_6H_4O_2BOC_6H_4Me$-4], (^{13}C),[2647] [$(RO)_4B$]$^-$, (^{13}C),[2648] [B_2F_2-

2619 B.L. Li, P. Mukherjee, and R.H. Neilson, *Inorg. Chem.*, 1989, **28**, 605.
2620 C. Habben, A. Meller, and S. Pusch, *Z. Naturforsch., B*, 1988, **43**, 959 (*Chem. Abstr.*, 1989, **111**, 115 573).
2621 L.A. Jackson and C.W. Allen, *J. Chem. Soc., Dalton Trans.*, 1989, 2423.
2622 M.F. Welker, I. Manners, M. Parvez, and H.R. Allcock, *J. Chem. Soc., Chem. Commun.*, 1989, 871.
2623 M. Driess, P. Frankhauser, H. Pritzkow, and W. Siebert, *Angew. Chem., Int. Ed. Engl.*, 1989, **28**, 217.
2624 H. Meyer, G. Schmidt-Lukasch, G. Baum, W. Massa, and A. Berndt, *Z. Naturforsch., B*, 1988, **43**, 801 (*Chem. Abstr.*, 1989, **111**, 57 819).
2625 C.D. Habben, *J. Organomet. Chem.*, 1989, **361**, 289.
2626 A. Tapper, T. Schmitz, and P. Paetzold, *Chem. Ber.*, 1989, **122**, 595.
2627 H.V.R. Dias and P.P. Power, *J. Am. Chem. Soc.*, 1989, **111**, 144.
2628 M. Satoh, Y. Nomoto, N. Miyaura, and A. Suzuki, *Tetrahedron Lett.*, 1989, **30**, 3789.
2629 R.W. Hoffmann, K. Ditrich, G. Köster, and R. Stürmer, *Chem. Ber.*, 1989, **122**, 1783.
2630 Y. Tamamoto, T. Seko, F.G. Rong, and H. Nemoto, *Tetrahedon Lett.*, 1989, **30**, 7191.
2631 J.F. Mariategui and K. Niedenzu, *J. Organomet. Chem.*, 1989, **369**, 137.
2632 C.D. Habben, *Z. Anorg. Allg. Chem.*, 1989, **573**, 199.
2633 D. Zhao, P.N. Gates, and P.S. Jones, *Sci. Sin., Ser. B (Engl. Ed.)*, 1987, **30**, 1233 (*Chem. Abstr.*, 1989, **110**, 154 349).
2634 W. Haubold, W. Keller, and G. Sawitzki, *J. Organomet. Chem.*, 1989, **367**, 19.
2635 W. Haubold and W. Keller, *J. Organomet. Chem.*, 1989, **361**, C54.
2636 P. Harston, J.L. Wardell, D. Marton, G. Tagliavini, and P.J. Smith, *Inorg. Chim. Acta*, 1989, **162**, 245.
2637 J. Atchekzaï, B. Bonnetot, J.-C. Duplan, B. Fenet, B. Frange, and H. Mongeot, *Magn. Reson. Chem.*, 1989, **27**, 699.
2638 A. Brandl, P. Kölle, and H. Nöth, *Chem. Ber.*, 1989, **122**, 419.
2639 G. Heller and K. Seeger, *Z. Naturforsch., B*, 1988, **43**, 547 (*Chem. Abstr.*, 1989, **111**, 57 818).
2640 W. Maringgele and A. Meller, *Z. Anorg. Allg. Chem.*, 1989, **572**, 140.
2641 V.P. Singh, R.V. Singh, and J.P. Tandon, *Synth. React. Inorg. Metal-Org. Chem.*, 1989, **19**, 669.
2642 A. Kendrick, H. Nöth, B. Stalla, and W. Storch, *J. Chem. Soc., Dalton Trans.*, 1989, 1311.
2643 M. Haddad, M.-T. Boisdon, L. Lopez, C. Malavaud, G. Pelletier, J. Barrans, G. Pfister-Guillouzo, and J.-P. Legros, *J. Chem. Res. (S)*, 1989, 250.
2644 M. Baudler, C. Block, H. Budzikiewicz, and H. Münster, *Z. Anorg. Allg. Chem.*, 1989, **569**, 7.
2645 H.H. Karsh, G. Hanika, B. Huber, K. Meindl, S. König, C. Krüger, and G. Müller, *J. Chem. Soc., Chem. Commun.*, 1989, 373.
2646 M.W. Andersen, B. Hildebrandt, G. Köster, and R.W. Hoffmann, *Chem. Ber.*, 1989, **122**, 1777.
2647 A.F. Abramov, V.I. Grachek, G.R. Motol'ko, and S.F. Naumova, *Zh. Obshch. Khim.*, 1988, **58**, 1571 (*Chem. Abstr.*, 1989, **110**, 212 886).

$(O_2)_2(HCO_4)_2]^{2-}$, (^{11}B, ^{13}C, ^{19}F),[2649] curcumin-BF_3, (^{11}B, ^{13}C),[2650] and BF_3 adducts, (^{13}C).[2651]

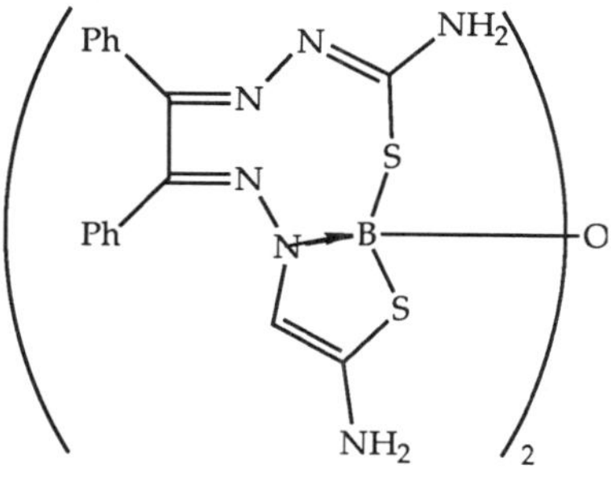
(67)

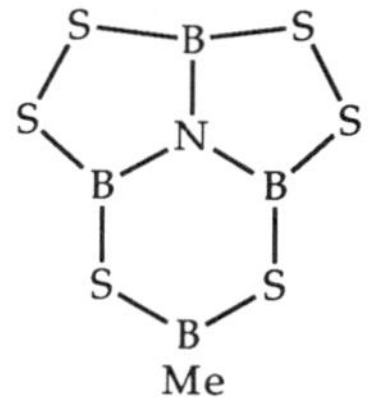
(68)

Complexes of Other Group IIIB Elements.——Through space coupling between ^{205}Tl and ^{19}F has been investigated in a number of fluorophenyl thallium compounds.[2652] $^1J(^{205}Tl\text{-}^{13}C)$ of 14636 and 13746 Hz for $[Tl(CN)]^{2+}$ and $[Tl(CN)_2]^-$ are the largest known values.[2653] N.m.r. data have also been reported for $[LiHAlBu^t_3]_2$, (^{13}C),[2654] $[Me_3AlPPh_3]$, (^{13}C),[2655] $[Me_2Sn(CH_2InMe_2)_2]$, (^{13}C, ^{119}Sn),[2656] $[Bu^n_{3-n}In(SR)_n]$, (^{13}C),[2657] [2-HO-3,5-$MeC_6H_2CH_2N\{CH_2CH_2N(CH_2C_6H_2O\text{-}2\text{-}Me_2\text{-}3,5)_2\}_2(CH_2)_2Ga]^+$, (^{13}C),[2658] $[Pr^i_{3-n}InCl_n]$, (^{13}C),[2659] $[Al\{CH(SiMe_3)_2\}_3]$, (^{13}C),[2660] $[Ga\{CH(SiMe_3)_2\}_3]$, (^{13}C),[2661] $[In\{CH(SiMe_3)_2\}_3]$, (^{13}C),[2662] $[\{(PhCH_2)_5C_5\}In]$, (^{13}C),[2663] $[(2,4,6\text{-}Me_3C_6H_2)_3Al.THF]$, (^{13}C),[2664] $[Me_2Al\text{-}(NSiMe_3)_2CMe]$, (^{13}C),[2665] $[Me_2InNR_2]_2$, (^{13}C),[2666] $[N(CH_2CH_2O)_3Al_2Me_3]_2$, (^{13}C),[2667]

2648 J.D. White, M.A. Avery, S.C. Choudhry, O.P. Dhingra, B.D. Gray, M.-c. Kang, S.-c. Kuo, and A.J. Whittle, *J. Am. Chem. Soc.*, 1989, **111**, 790.
2649 O.V. Brovkin and B.N. Chernyshov, *Zh. Neorg. Khim.*, 1989, **34**, 299 (*Chem. Abstr.*, 1989, **110**, 224 392).
2650 A.S. Rao, S. Divakar, and R. Seshadri, *Indian J. Chem., Sect. B*, 1988, **27B**, 926 (*Chem. Abstr.*, 1989, **111**, 153 487).
2651 S.E. Denmark, T.M. Willson, and N.G. Almstead, *J. Am. Chem. Soc.*, 1989, **111**, 9258.
2652 G.N. Pecksen and R.F.M. White, *Can. J. Chem.*, 1989, **67**, 1847.
2653 J. Blixt, B. Gyori, and J. Glaser, *J. Am. Chem. Soc.*, 1989, **111**, 7784.
2654 W. Uhl, *Z. Anorg. Allg. Chem.*, 1989, **570**, 37.
2655 D.A. Wierda and A.R. Barron, *Polyhedron*, 1989, **8**, 831.
2656 H. Schumann, R. Mohtachemi, and M. Schwichtenberg, *Z. Naturforsch., B*, 1988, **43**, 1510 (*Chem. Abstr.*, 1989, **111**, 78 215).
2657 R. Nomura, S. Inazawa, K. Kanaya, and H. Matsuda, *Polyhedron*, 1989, **8**, 763.
2658 D.A. Moore, P.E. Fanwick, and M.J. Welch, *Inorg. Chem.*, 1989, **28**, 1504.
2659 B. Neumüller, *Chem. Ber.*, 1989, **122**, 2283.
2660 W. Uhl, *Z. Anorg. Allg. Chem.*, 1989, **579**, 75.
2661 W. Uhl, M. Layh, and T. Hildenbrand, *J. Organomet. Chem.*, 1989, **364**, 289.
2662 W. Uhl, M. Layh, and W. Hiller, *J. Organomet. Chem.*, 1989, **368**, 139.
2663 H. Schumann, C. Janiak, F. Görlitz, J. Loebel, and A. Dietrich, *J. Organomet. Chem.*, 1989, **363**, 243.
2664 V.S.J. De Mel and J.P. Oliver, *Organometallics*, 1989, **8**, 827.
2665 R. Lechler, H.-D. Hausen, and J. Weidlein, *J. Organomet. Chem.*, 1989, **359**, 1.
2666 K.A. Aitchison, J.D.J. Backer-Dirks, D.C. Bradley, M.M. Faktor, D.M. Frigo, M.B. Hursthouse, B. Hussain, and R.L. Short, *J. Organomet. Chem.*, 1989, **366**, 11.
2667 M.D. Healy and A.R. Barron, *J. Am. Chem. Soc.*, 1989, **111**, 398.

$[(Me_3Si)_2AlPPh(SiMe_3)]_2$, ($^{13}C$),[2668] $[Pr^i{}_2GaBr.HEEt]$, (E = S, Se; ^{13}C),[2669] $[ClMe-InC_6H_3-2,6-(CH_2NEt_2)_2]$, ($^{13}C$),[2670] and poly[(acyloxy)aloxane], (^{27}Al).[2671]

^1H and ^2H T_1 has been determined for the protons in partially deuteriated $[Al(OH_2)_6]^{3+}$. Isotopic shifts and rotational correlation times were determined.[2672] ^{27}Al n.m.r. spectroscopy has been used to determine $^1H/^2H$ isotope shifts in $[Al(OH_2)_5(SO_4)]^+$ and $[Al(OH_2)_n\{OPMe(OMe)_2\}_{6-n}]^{3+}$.[2673] The structure of simple aluminosilicate anions formed by the reaction of silicate and aluminate anions has been studied by ^{27}Al and ^{29}Si n.m.r. spectroscopy.[2674] N.m.r. data have also been reported for $[Al(NEt_2)$(tetraphenylporphyrin)], (^{13}C),[2675] [AlCl(tetraphenylporphyrin)], (^{13}C),[2676] $[\{1,4,7-(O_2CCH_2)_3$-1,4,7-triaazacyclononane}Ga], (^{71}Ga),[2677] $[Br_2GaNHSiMe_3]_2$, (^{13}C, ^{29}Si),[2678] $[Cl_3Al-PMeCH(SiMe_3)C(SiMe_3)(SiMe_2Cl)]$, ($^{13}C$, ^{29}Si),[2679] $[Al(OPr^i)_{3-n}(OAr)_n]$, (^{27}Al),[2680] $[Al(O_2CCH_3)_3]$, (^{13}C, ^{27}Al),[2681] $[M(isomaltate)_3]$, M = Al, Ga, In; ^{27}Al),[2682] (69), (^{27}Al),[2683] $[Al_{13}O_4(OH)_{24}(OH_2)_{12}]^{7+}$, ($^{27}Al$),[2684] (70), ($^{13}C$, ^{27}Al, ^{29}Si),[2685] $[Ga(OH)X_2]$, (^{71}Ga),[2686] $[Tl_2Sn(OEt)_6]$, (^{119}Sn),[2687] In(III) complexes of 2-alkylaminocyclopentene-1-carbodithioic acid, (^{13}C, ^{14}N),[2688] $[In_2Se_{21}]^{4-}$, (^{77}Se),[2689] [AlCl], (^{27}Al),[2690] $[P(NHBu^t)(C_5Me_5)][AlCl_4]$, ($^{13}C$, ^{27}Al),[2691] $[C_6H_4S_2P][AlCl_4]$, (^{27}Al),[2692] and $[PhCCl_2][AlCl_4]$, (^{13}C, ^{27}Al).[2693]

2668 J.F. Janik, E.N. Duesler, W.F. McNamara, M. Westerhausen, and R.T. Paine, *Organometallics*, 1989, **8**, 506.
2669 G.G. Hoffmann and R. Fischer, *Inorg. Chem.*, 1989, **28**, 4165.
2670 H. Schumann, W. Wassermann, and A. Dietrich, *J. Organomet. Chem.*, 1989, **365**, 11.
2671 Y. Kimura, M. Furukawa, H. Yamane, and T. Kitao, *Macromolecules*, 1989, **22**, 79 (*Chem. Abstr.*, 1989, **110**, 62 329).
2672 J.W. Akitt, J.M. Elders, and O.W. Howarth, *J. Chem. Soc., Faraday Trans. 1*, 1989, **85**, 2035.
2673 J.W. Akitt and O.W. Howarth, *J. Chem. Soc., Faraday Trans. 1*, 1989, **85**, 121.
2674 A.V. McCormick, A.T. Bell, and C.J. Radke, *J. Phys. Chem.*, 1989, **93**, 1741.
2675 Y. Hirai, T. Aida, and S. Inoue, *J. Am. Chem. Soc.*, 1989, **111**, 3062.
2676 J.M. DeSimone, A.M. Hellstern, Y. Yoo, J.S. Riffle, and J.E. McGrath, *Polym. Prepr. (Am. Chem. Soc., Div. Polym. Chem.)*, 1989, **30**, 243 (*Chem. Abstr.*, 1989, **110**, 232 152).
2677 A.S. Craig, D. Parker, H. Adams, and N.A. Bailey, *J. Chem. Soc., Chem. Commun.*, 1989, 1793.
2678 W.R. Nutt, J.S. Blanton, F.O. Kroh, and J.D. Odom, *Inorg. Chem.*, 1989, **28**, 2224.
2679 E. Niecke, M. Leuer, and M. Nieger, *Chem. Ber.*, 1989, **122**, 453.
2680 T. Athar, R. Bohra, and R.C. Mehrotra, *Indian J. Chem., Sect. A*, 1989, **28A**, 492 (*Chem. Abstr.*, 1989, **111**, 224 075).
2681 J.W. Akitt, J.M. Elders, X.L.R. Fontaine, and A.K. Kundu, *J. Chem. Soc., Dalton Trans.*, 1989, 1897.
2682 T.G. Lutz, D.J. Clevette, S.J. Rettig, and C. Orvig, *Inorg. Chem.*, 1989, **28**, 715.
2683 W.O. Nelson, S.J. Rettig, and C. Orvig, *Inorg. Chem.*, 1989, **28**, 3153.
2684 S. Cheng and T.-C. Wang, *Inorg. Chem.*, 1989, **28**, 1283.
2685 F.J. Feher, T.A. Budzichowski, and K.J. Weller, *J. Am. Chem. Soc.*, 1989, **111**, 7288.
2686 M.G. Ivanov, S.D. Vashchenko, T.A. Denisova, and I.I. Kalinichenko, *Zh. Neorg. Khim.*, 1989, **34**, 329 (*Chem. Abstr.*, 1989, **110**, 184 731).
2687 M.J. Hampden-Smith, D.E. Smith, and E.N. Duesler, *Inorg. Chem.*, 1989, **28**, 3399.
2688 S.K. Singh, Y. Singh, A.K. Rai, and R.C. Mehrotra, *Polyhedron*, 1989, **8**, 633.
2689 M.G. Kanatzidis and S. Dhingra, *Inorg. Chem.*, 1989, **28**, 2024.
2690 M. Tacke and H. Schnoeckel, *Inorg. Chem.*, 1989, **28**, 2895.
2691 D. Gudat, M. Nieger, and E. Niecke, *J. Chem. Soc., Dalton Trans.*, 1989, 693.
2692 N. Burford and B.W. Royan, *Phosphorus Sulphur Silicon*, 1989, **41**, 37.
2693 U.S. Racherla, T. Daniel, P.R. Rajamohanan, and N.R. Ayyangar, *J. Am. Chem. Soc.*, 1989, **111**, 7659.

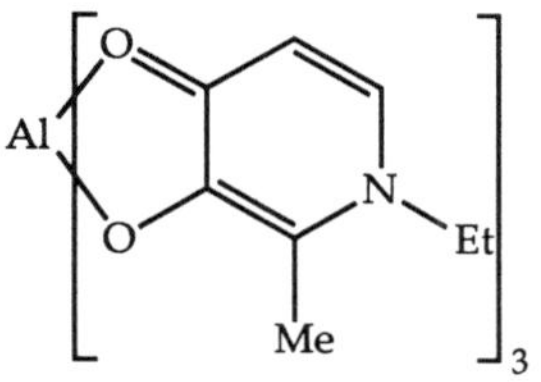

(69)

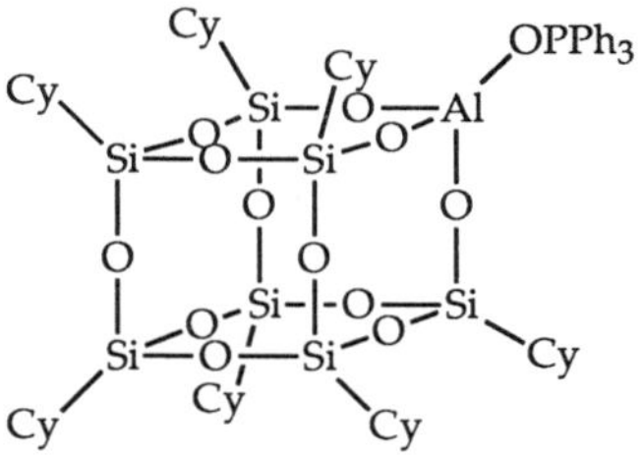

(70)

7 Group IVB Elements

Two reviews have appeared:- '(^{1}H, ^{13}C, ^{29}Si) N.m.r. spectroscopy of organosilicon compounds',[2694] and 'Use of ^{119m}Sn Mössbauer and ^{119}Sn n.m.r. spectroscopy in the study of organotin complexes'.[2695]

The nuclear quadrupole coupling constants of ^{2}H and ^{14}N have been calculated for HCN and HNC.[2696] A new approach to the analysis of the U-shaped dependence of the ^{29}Si chemical shifts on the sum of group electronegativities of ligands has been presented.[2697] Solvent effects on ^{29}Si chemical shifts of some chlorosilanes have been calculated and compared with experiment.[2698] *Ab initio* computation of ^{29}Si chemical shifts for a range of representative compounds has been reported.[2699] Deuterium isotope effects in ^{1}H, ^{13}C, and ^{29}Si n.m.r. spectra of $[Ph_nSiH_{4-n}]$ have been determined.[2700] The molecular structure of $[SiH_3CN]$ has been determined in a nematic liquid crystal by ^{1}H, ^{13}C, ^{15}N, and ^{29}Si n.m.r. spectroscopy.[2701] Variations in the ^{17}O nuclear quadrupole coupling constants and shielding constants have been evaluated for $[(H_3Si)_2O]$ and related compounds. ^{29}Si chemical shifts were also reported.[2702] N.m.r. data have also been reported for $[CH_n(SiH_3)_{4-n}]$, (^{13}C, ^{29}Si),[2703] $[Pr^iN(SiH_3)_2]$, (^{13}C, ^{29}Si),[2704] $[(F_3C)_3GeGeSiH_3]$, (^{13}C, ^{29}Si, ^{77}Se),[2705] [1-naphthyl-SiH_2-C_6H_4-2-CHMeNMe$_2$], (^{29}Si),[2706] (71), (R = H, Bu^t; ^{13}C, ^{29}Si),[2707] $[PhH_2SiSiHPhSiH_2Ph]$, (^{29}Si),[2708] $[H_2Si(OR)_3]^-$, (^{29}Si),[2709] [MeHSi{C(=CH$_2$)C-

2694 E.A. Williams, *Chem. Org. Silicon Compd.*, 1989, **1**, 511 (*Chem. Abstr.*, 1989, **111**, 232 913).
2695 A. Kaur and G.K. Sandhu, *J. Chem. Sci.*, 1986, **12**, 1 (*Chem. Abstr.*, 1989, **111**, 153 852).
2696 G. Prasad, V. Keshari, and P. Chandra, *J. Mol. Struct.*, 1989, **192**, 253.
2697 V.G. Malkin, O.V. Gritsenko, and G.M. Zhidomirov, *Chem. Phys. Lett.*, 1988, **152**, 44 (*Chem. Abstr.*, 1989, **110**, 106 805).
2698 T. Takayama, I. Ando, and T. Asakura, *Bull. Chem. Soc. Jpn.*, 1989, **62**, 1233.
2699 J.R. Van Wazer, C.S. Ewig, and R. Ditchfield, *J. Phys. Chem.*, 1989, **93**, 2222.
2700 E. Liepiņš, V. Gevorgyan, and E. Lukevics, *J. Magn. Reson.*, 1989, **85**, 170.
2701 P.D. Blair, A.J. Blake, R.W. Cockman, S. Cradock, E.A.V. Ebsworth, and D.W.H. Rankin, *J. Mol. Struct.*, 1989, **193**, 279.
2702 J.A. Tossell and P. Lazzeretti, *Phys. Chem. Miner.*, 1988, **15**, 564 (*Chem. Abstr.*, 1989, **110**, 49 977).
2703 R. Hager, O. Steigelmann, G. Müller, and H. Schmidbaur, *Chem. Ber.*, 1989, **122**, 2115.
2704 D.G. Anderson and D.W.H. Rankin, *J. Chem. Soc., Dalton Trans.*, 1989, 779.
2705 A. Haas, H.-J. Kutsch, and C. Krüger, *Chem. Ber.*, 1989, **122**, 271.
2706 J. Boyer, C. Breliere, F. Carre, R.J.P. Corriu, A. Kpoton, M. Poirier, G. Royo, and J.C. Young, *J. Chem. Soc., Dalton Trans.*, 1989, 43.
2707 L.S. Chang and J.Y. Corey, *Organometallics*, 1989, **8**, 1885.
2708 C. Aitken, J.-P. Barry, F. Gauvin, J.F. Harrod, A. Malek, and D. Rousseau, *Organometallics*, 1989, **8**,

$(=CH_2)\}_2SiHMe]$, (^{13}C),[2710] $[HSi(NRCH_2CH_2)_3N]$, (^{13}C, ^{29}Si),[2711] [1,4-$Me_3SiC_6H_4SiHMe$-OPr^i], (^{13}C),[2712] $[Ph_2HSiOSO_2CF_3]$, (^{29}Si),[2713] [2,4,6-$Bu^t{}_3C_6H_2SiH(OR)_2$], (^{13}C),[2714] $[HSi(OR)_4]^-$, (^{29}Si),[2715] [2,4,6-$Me_3C_6H_2SiHBu^tCl$], (^{13}C, ^{29}Si),[2716] and [1-Pr^i-2-HMeRSn-4-MeC_6H_9], (^{13}C, ^{119}Sn).[2717]

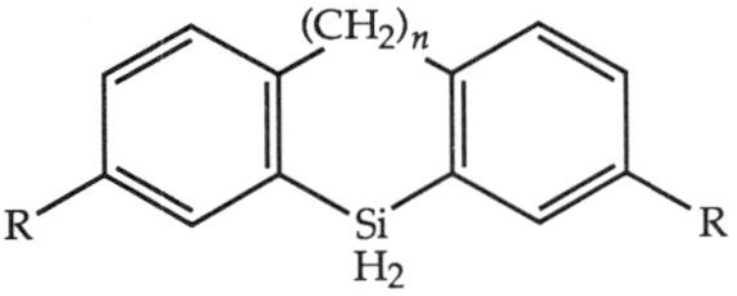

(71)

Natural abundance ^{29}Si imaging has been described using $[Me_4Si]$.[2718] ^{13}C and ^{29}Si n.m.r. spectra of substituted alkenyl- and alkynyl-$SiMe_3$ compounds have been reported and provide evidence of d_π-p_π bonding.[2719] Steric and electronic substituent effects on ^1H, ^{13}C, ^{29}Si, and ^{31}P n.m.r. spectra of $[(Me_3Si)_2C{=}PR]$ have been reported.[2720] ^{13}C and ^{29}Si n.m.r. spectra have been used to study intramolecular interactions in [RP=C-$(SiMe_3)_2$].[2721] ^{13}C n.m.r. spectra of $[(Me_3Si)_2C{=}E]$, E = O, S, have been discussed relative to MNDO calculations.[2722] The ^1H and ^{13}C n.m.r. spectra of $[(9\text{-triptycyl})_3Ge]^+$ are consistent with a static structure, with barriers to internal rotation in excess of 17 kcal mol^{-1}.[2723] N.m.r. data have also been reported for $[Me_3SiR]$, (^{13}C, ^{29}Si),[2724] (^{13}C),[2725]

1732.

2709 B. Becker, R.J.P. Corriu, C. Guérin, B. Henner, and Q. Wang, *J. Organomet. Chem.*, 1989, **368**, C25.

2710 A. Maercker, F. Brauers, W. Brieden, and B. Engelen, *J. Organomet. Chem.*, 1989, **377**, C45.

2711 D. Gudat and J.G. Verkade, *Organometallics*, 1989, **8**, 2772.

2712 H. Nagashima, K. Tatebe, and K. Itoh, *J. Chem. Soc., Perkin Trans.1*, 1989, 1707.

2713 W. Uhlig and A. Tzschach, *J. Organomet. Chem.*, 1989, **378**, C1.

2714 H. Weiss and H. Oehme, *Z. Anorg. Allg. Chem.*, 1989, **568**, 157.

2715 B. Becker, R. Corriu, C. Guerin, B. Henner, and Q. Wang, *J. Organomet. Chem.*, 1989, **359**, C33.

2716 G.R. Gillette, G.H. Noren, and R. West, *Organometallics*, 1989, **8**, 487.

2717 H. Schumann and B.C. Wassermann, *J. Organomet. Chem.*, 1989, **365**, C1.

2718 L.D. Hall and A.G. Webb, *J. Magn. Reson.*, 1989, **83**, 371.

2719 M.I. Al-Hassan, I.M. Al-Najjar, and I.M. Al-Oraify, *Magn. Reson. Chem.*, 1989, **27**, 1112.

2720 M.I. Povolotskii, A.S. Tarasevich, and I.E. Boldeskul, *Ukr. Khim. Zh. (Russ. Ed.)*, 1988, **54**, 1073 (*Chem. Abstr.*, 1989, **111**, 78 189).

2721 Yu.P. Egorov, M.I. Povolotskii, and V.D. Romanenko, *Teor. Eksp. Khim.*, 1988, **24**, 729 (*Chem. Abstr.*, 1989, **111**, 57 886).

2722 G. Barbarella and A. Bongini, *Tetrahedron*, 1989, **45**, 5137.

2723 J.M. Chance, J.H. Geiger, and K. Mislow, *J. Am. Chem. Soc.*, 1989, **111**, 2326.

2724 R.J. Mills, N.J. Taylor, and V. Snieckus, *J. Org. Chem.*, 1989, **54**, 4372; J. Wan and W.P. Weber, *Polym. Bull. (Berlin)*, 1988, **20**, 249 (*Chem. Abstr.*, 1989, **110**, 8781).

2725 A. Maercker and K.-D. Klein, *Angew. Chem., Int. Ed. Engl.*, 1989, **28**, 83; S.L. Buchwald and R.B. Nielsen, *J. Am. Chem. Soc.*, 1989, **111**, 2870; L.A. Paquette, G.D. Maynard, C.S. Ra, and M. Hoppe, *J. Org. Chem.*, 1989, **54**, 1408; J.P. Gilday, J.C. Gallucci, and L.A. Paquette, *J. Org. Chem.*, 1989, **54**, 1399; F.L. Koerwitz, G.B. Hammond, and D.F. Wiemer, *J. Org. Chem.*, 1989, **54**, 743; F.L. Koerwitz, G.B. Hammond, and D.F. Wiemer, *J. Org. Chem.*, 1989, **54**, 738; R. Walsh, S. Untiedt, M. Stohlmeier, and A. de Meijere, *Chem. Ber.*, 1989, **122**, 637; A. Castañeda, D.J. Kucera, and L.E. Overman, *J. Org. Chem.*, 1989, **54**, 5695; E.J. Ginsburg, C.B. Gorman, S.R. Marder, and R.H. Grubbs, *J. Am. Chem. Soc.*, 1989, **111**, 7621; J.B.M. Rewinkel, S. García-Granda, P.T. Beurskens, R. Raaijmakers, and B. Zwanenburg, *Recl. Trav. Chim. Pays-Bas*, 1989, **108**, 61; K. Miura, K. Oshima, and K. Utimoto, *Tetrahedron Lett.*, 1989, **30**, 4413; D.P. Curran and P.A. van Elburg, *Tetrahedron Lett.*, 1989, **30**, 2501; E. Schaumann, C. Friese, and G. Adiwidjaja, *Tetrahedron*, 1989, **45**, 3163; J.R. Hwu and P.S. Furth, *J. Am. Chem. Soc.*, 1989, **111**, 8834; 8842.

$[Me_3SiN{=}P(OCH_2CF_3)MeCH_2SiMe_3]$, (^{13}C),[2726] $[Me_3SiN{=}PMeX(CH_2SiMe_3)]$, (^{13}C),[2727] (72), (^{13}C),[2728] $[(Me_3Si)_2CHCH(SiMe_3)_2]Li_2$, $(^{7}Li$, ^{13}C, $^{29}Si)$,[2729] $[\{(Me_3Si)_2CH\}(H_2N$-$CPh{=}N)P{=}NCPh{=}NP\{CH(SiMe_3)_2\}]$, (^{13}C),[2730] $[(Me_3Si)_2CHGe]_6$, $(^{13}C$, $^{29}Si)$,[2731] $[(Me_3Si)$-$(Me_3Sn)CHCH_2P(O)Ph_2]$, $(^{29}Si$, $^{119}Sn)$,[2732] $[\{(Me_3Si)_3C\}_3SiX]$, $(^{13}C$, $^{29}Si)$,[2733] $[(Me_3Si)_3CX]$, $(^{13}C$, $^{29}Si)$,[2734] $[(Me_3Si)_2(Ph_2XSi)CY]$, $(^{13}C$, $^{29}Si)$,[2735] $[(Me_3Si)_3CTeX]$, (^{125}Te),[2736] $[ROCX^1{=}C(MMe_3)PX^2{}_2]$, (M = Si, Ge; $^{13}C)$,[2737] $[RCH{=}CHMMe_3]$, $(^{13}C$, ^{29}Si, ^{119}Sn, $^{207}Pb)$,[2738] $[(RO)_nMe_{3-n}SiCH{=}CHSiMe_{3-n}(OR)_n]$, (^{13}C),[2739] $[Ph_2PCH(SiMe_3)CH{=}CHSiMe_3]$, $(^{13}C$, $^{29}Si)$,[2740] $[Me_3SiCH{=}CHSiMe_2CH_2CH_2SiMe_2SiMe_3]$, (^{13}C),[2741] $[2\text{-}Me_3Si\text{-}1,3\text{-}C_4H_5]_n$, $(^{13}C$, $^{29}Si)$,[2742] $[(Me_3Si)_2C{=}CHSiMe_3]$, (^{29}Si),[2743] $[(Me_3Si)_3C_5H_2Li.L]$, $(^{6}Li$, ^{7}Li, ^{13}C, $^{29}Si)$,[2744] $[ArC_2SiMe_3]_n$, (^{13}C),[2745] $[Me_3SiCX{=}CHX]$, (^{13}C),[2746] $[CH_2{=}CMeSiMe_3]$, $(^{13}C$, $^{29}Si)$,[2747] $[(Pr^i{}_2N)_2PCH{=}NNBu^tP{=}C(SiMe_3)_2]$, (^{13}C),[2748] $[(Pr^i{}_2N)_2P{=}C(SiMe_3)_2]^+$, $(^{13}C$, $^{29}Si)$,[2749] $[Me_3SiC{\equiv}CR]$, (^{13}C),[2750] $[Me_3SiC{\equiv}CBu^n]$, (^{13}C),[2751] $[R_3MC{\equiv}CCHO]$, (M = Si, Ge; ^{13}C, $^{17}O)$,[2752] $[(Me_3SiC{\equiv}C)_2O]$, (^{13}C),[2753] $[2,6\text{-}(Me_3Si)_2C_6H_3SH]$, (^{13}C),[2754] $[Me_3SiMeSi{=}C(O$-

2726 G.M. Scheide and R.H. Neilson, *Phosphorus Sulphur Silicon*, 1989, **46**, 139.
2727 G.M. Scheide and R.H. Neilson, *Organometallics*, 1989, **8**, 1987.
2728 M. Ishikawa, H. Sakamoto, F. Kanetani, and A. Minato, *Organometallics*, 1989, **8**, 2767.
2729 A. Sekiguchi, T. Nakanishi, C. Kabuto, and H. Sakurai, *J. Am. Chem. Soc.*, 1989, **111**, 3748.
2730 J.-P. Majoral, C. Roques, M.-R. Mazières, J. Jaud, and M. Sanchez, *J. Chem. Soc., Chem. Commun.*, 1989, 1496.
2731 A. Sekiguchi, C. Kabuto, and H. Sakurai, *Angew. Chem., Int. Ed. Engl.*, 1989, **28**, 55.
2732 T.N. Mitchell and H.-J. Belt, *J. Organomet. Chem.*, 1989, **368**, 167.
2733 A.I. Al-Mansour, S.S. Al-Showiman, and I.M. Al-Najjar, *Spectrochim. Acta*, 1988, **44A**, 643 (*Chem. Abstr.*, 1989, **110**, 17 557).
2734 S.S. Al-Juaid, S.M. Dhaher, C. Eaborn, P.B. Hitchcock, C.A. McGeary, and J.D. Smith, *J. Organomet. Chem.*, 1989, **366**, 39.
2735 N. Wiberg, M. Link, and G. Fischer, *Chem. Ber.*, 1989, **122**, 409.
2736 K. Giselbrecht, B. Bildstein, and F. Sladky, *Chem. Ber.*, 1989, **122**, 1247.
2737 E.I. Lazhko, E.V. Luzikova, G.Yu. Mikhailov, M.A. Kazankova, and Yu.A. Ustynyuk, *Zh. Obshch. Khim.*, 1988, **58**, 1247.
2738 B. Wrackmeyer and K. Wagner, *Chem. Ber.*, 1989, **122**, 857.
2739 B. Marciniec, H. Maciejewski, J. Guliński, and L. Rzejak, *J. Organomet. Chem.*, 1989, **362**, 273.
2740 B.A. Boyd and R.H. Neilson, *Inorg. Chem.*, 1989, **28**, 3219.
2741 M. Ishikawa, T. Ono, Y. Saheki, A. Minato, and H. Okinoshima, *J. Organomet. Chem.*, 1989, **363**, C1.
2742 W. Jiang and W.P. Weber, *Polym. Bull. (Berlin)*, 1989, **21**, 335 (*Chem. Abstr.*, 1989, **111**, 40 046); W. Jiang, Y.X. Ding, and W.P. Weber, *Polym. Bull. (Berlin)*, 1989, **21**, 541 (*Chem. Abstr.*, 1989, **111**, 154 492).
2743 H. Schmidbaur and J. Ebenhoech, *Z. Naturforsch., B*, 1987, **42**, 1543 (*Chem. Abstr.*, 1989, **110**, 154 363).
2744 P. Jutzi, W. Leffers, S. Pohl, and W. Saak, *Chem. Ber.*, 1989, **122**, 1449.
2745 Y.S. Gal, S.K. Choi, and C.Y. Kim, *J. Polym. Sci., Polym. Chem. Ed.*, 1989, **27**, 31 (*Chem. Abstr.*, 1989, **110**, 193 483).
2746 K. Kamieńska-Trela, L. Kania, J. Sitkowski, and E. Bednarek, *J. Organomet. Chem.*, 1989, **364**, 29.
2747 D.G. Anderson, D.W.H. Rankin, H.E. Robertson, C.M.F. Frazão, and H. Schmidbaur, *Chem. Ber.*, 1989, **122**, 2213.
2748 A. Igau, A. Baceiredo, G. Bertrand, K. Kühnel-Lysek, and E. Niecke, *New J. Chem.*, 1989, **13**, 359.
2749 A. Igau, A. Baceiredo, H. Grützmacher, H. Pritzhow, and G. Bertrand, *J. Am. Chem. Soc.*, 1989, **111**, 6853.
2750 R. Brückner, *Chem. Ber.*, 1989, **122**, 193.
2751 H. tom Dieck, M. Mallien, and R. Diercks, *J. Mol. Catal.*, 1989, **51**, 53.
2752 A.S. Medvedeva, O.I. Margorskaya, I.D. Kalikhman, N.I. Golovanova, N.I. Shergina, and N.S. Vyazankin, *Izv. Akad. Nauk SSSR, Ser. Khim.*, 1988, 321 (*Chem. Abstr.*, 1989, **110**, 95 411).
2753 I.-H. Ooi and R.H. Smithers, *J. Org. Chem.*, 1989, **54**, 1479.
2754 E. Block, G. Ofori-Okai, and J. Zubieta, *J. Am. Chem. Soc.*, 1989, **111**, 2327.

SiMe$_3$)adamantyl], (^{13}C, ^{29}Si),[2755] [(Me$_3$Si)$_2$SiRCBu^{t}CBu^t=CBu^t], (^{13}C, ^{29}Si),[2756] (73), (^{13}C, ^{29}Si),[2757] [(Me$_3$Si)$_3$SiAlPh$_2$.THF], (^{13}C),[2758] [(Me$_3$Si)$_3$SiSi(OH)$_3$], (^{29}Si),[2759] [(Me$_3$Si)$_2$-SiSiMe$_2$Si(SiMe$_3$)$_2$SiMe$_2$SiMe$_2$], (^{13}C, ^{29}Si),[2760] [Ph$_n$Me$_{3-n}$SiSi{C$_6$H$_4$-2-C(CF$_3$)$_2$O}$_2$], (^{29}Si),[2761] [Me$_3$SiSiR^1=C(OLi)R^2], (^{29}Si),[2762] and [4-Me$_3$SiMe$_2$SiC$_6$H$_4$CHCH$_2$]$_n$, (^{13}C).[2763]

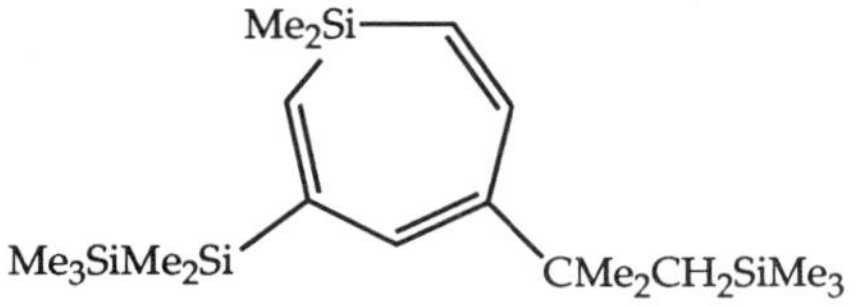

(72)

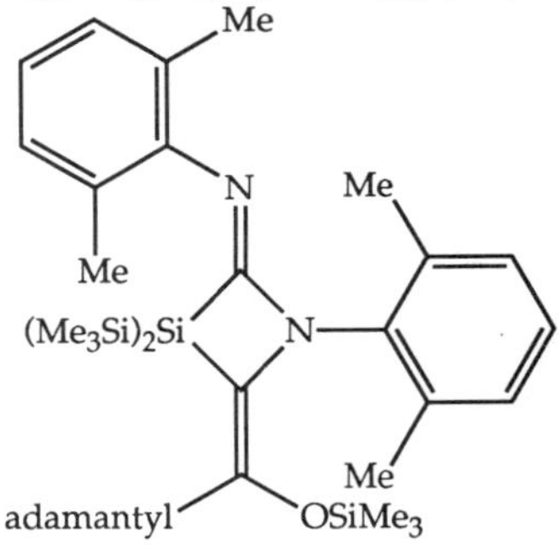

(73)

The conformation of dibenzochalcogenosilacins and -silanins has been determined from ^{1}H, ^{13}C, and ^{29}Si n.m.r. spectra.[2764] INADEQUATE has been applied to determine the ^{29}Si-^{29}Si framework in four cyclosilanes.[2765] The temperature dependence of the ^{1}H n.m.r. spectra of [PhMeSiSiMe$_2$SiMe$_2$SiMePh] has been determined.[2766] N.m.r. data have also been reported for [MeCHMeCHCHMeSiMe$_2$], (^{13}C),[2767] [CH$_2$CH=CPhSiMe$_2$], (^{13}C),[2768] [PhMe$_2$SiR], (^{13}C),[2769] (75), (^{13}C),[2770] [HOCH(CH$_2$)$_4$SiMe$_2$], (^{13}C, ^{29}Si),[2771] [Me$_2$SiR$_2$], (^{13}C),[2772] *cis*-[Bu^{t}SiMe$_2$CH=CHOSiMe$_3$], (^{13}C),[2773] [Bu^{t}SiMe$_2$R], (^{13}C),[2774]

2755 K.M. Baines, A.G. Brook, R.R. Ford, P.D. Lickiss, A.K. Saxena, W.J. Chatterton, J.F. Sawyer, and B.A. Behnam, *Organometallics*, 1989, **8**, 693.
2756 D.B. Puranik, M.P. Johnson, and M.J. Fink, *Organometallics*, 1989, **8**, 770.
2757 A.G. Brook, A.K. Saxena, and J.F. Sawyer, *Organometallics*, 1989, **8**, 850.
2758 M.L. Sierra, V.S.J. de Mel, and J.P. Oliver, *Organometallics*, 1989, **8**, 2312.
2759 S.S. Al-Juaid, N.H. Buttrus, R.I. Damja, Y. Derouiche, C. Eaborn, P.B. Hitchcock, and P.D. Lickiss, *J. Organomet. Chem.*, 1989, **371**, 287.
2760 J. Maxka, B.R. Adams, and R. West, *J. Am. Chem. Soc.*, 1989, **111**, 3447.
2761 M. Kira, K. Sato, C. Kabuto, and H. Sakurai, *J. Am. Chem. Soc.*, 1989, **111**, 3747.
2762 I.S. Biltueva, D.A. Bravo-Zhivotovskii, I.D. Kalikhman, V.Yu. Vitkovskii, S.G. Shevchenko, N.S. Vyazankin, and M.G. Voronkov, *J. Organomet. Chem.*, 1989, **368**, 163.
2763 K. Saigo, *J. Polym. Sci., Part A: Polym. Chem.*, 1989, **27**, 2203 (*Chem. Abstr.*, 1989, **111**, 115 808).
2764 S.N. Tandura, Yu.A. Strelenko, S.I. Androsenko, E.E. Masterov, N.V. Alekseev, V.I. Rakitskaya, O.G. Rodin, and V.F. Traven, *Zh. Obshch. Khim.*, 1988, **58**, 398 (*Chem. Abstr.*, 1989, **110**, 95 335).
2765 E. Hengge and F. Schrank, *J. Organomet. Chem.*, 1989, **362**, 11.
2766 B.D. Lavrukhin and A.A. Zhdanov, *Metalloorg. Khim.*, 1988, **1**, 453 (*Chem. Abstr.*, 1989, **110**, 213 061).
2767 R.T. Conlin, M. Namavari, J.S. Chickos, and R. Walsh, *Organometallics*, 1989, **8**, 168.
2768 R.T. Conlin, S. Zhang, M. Namavari, K.L. Bobbitt, and M.J. Fink, *Organometallics*, 1989, **8**, 571.
2769 I. Fleming, M. Rowley, P. Cuadrado, A.M. González-Nogal, and F.J. Pulido, *Tetrahedron*, 1989, **45**, 413.
2770 R.T. Conlin and M. Namavari, *J. Organomet. Chem.*, 1989, **376**, 259.
2771 R. Tacke, H. Hengelsberg, H. Zilch, and B. Stumpf, *J. Organomet. Chem.*, 1989, **379**, 211.
2772 G.D. Prestwich and C. Wawrzeńczyk, *Tetrahedron Lett.*, 1989, **30**, 403.
2773 L. Duhamel, J. Gralak, and B. Ngono, *J. Organomet. Chem.*, 1989, **363**, C4.

derivatives of (1,1-$Me_2SiCH_2CH_2CH$=$CHCH_3)_n$, (^{13}C, ^{29}Si),[2775] [($Me_4C_5H)SiMe_2(C_5Me_4$)-M]$^+$, (M = Ge, Sn, Pb; ^{11}B, ^{13}C, ^{29}Si, ^{119}Sn, ^{207}Pb),[2776] [2,5-($Me_2Si)_2$thiophene]$_n$, (^{13}C, ^{29}Si),[2777] [$Me_2Si(C{\equiv}CSiMe_2C{\equiv}C)_2SiMe_2$], ($^{13}C$, ^{29}Si),[2778] [$PhC{\equiv}CSiMe_2SiMe_2C{\equiv}CPh$], ($^{13}C$),[2779] [2-thiophenyl$SiMe_2SiMe_2$-2-thiophenyl], ($^{13}C$, ^{29}Si),[2780] [$Me_2Si(SiMe_2SiMe_2)_2$-SiMeF], (^{13}C, ^{29}Si),[2781] [MePhSiCH=CHSiMePh]$_n$, (^{13}C),[2782] [SiMeRSiMeRCH=CHC≡C]$_n$, (^{13}C),[2783] and [SiMePhSiMePhC≡C]$_n$, (^{13}C).[2784]

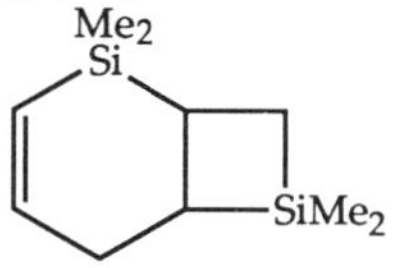

(75)

A method of analysing the ^1H n.m.r. spectra of strongly coupled 6-spin systems has been proposed and applied to [$Ph_2\overline{SiCH_2CH(OH)CH(OH)CH_2}$].[2785] The ^{29}Si n.m.r. signal of [$(C_5Me_5)_2Si$] is at δ -398. The ^{13}C n.m.r. spectrum was also reported.[2786] N.m.r. data have also been reported for [Et_3SiR], (^{13}C),[2787] [$CH_2C(SiEt_3)CHCH_2$]$_n$, (^{13}C),[2788] [$PhEt_2MLi$], (M = Si, Ge; ^{13}C),[2789] [(CH_2=$CH)_2Si(CH_2CH_2)_2PR$], (^{13}C),[2790] [Bu^{t}CH=CH-$SiPh_2CH(CH_2Bu^t)SiPh_2H$], ($^{13}C$),[2791] [(2,4,6-$Me_3C_6H_2)_2\overline{SiCH_2C}$=CHBu^t], ($^{13}C$),[2792] [(2,4,6-$Me_3C_6H_2$)Bu^{t}Si=SiBu^t(2,4,6-$Me_3C_6H_2$)], ($^{13}C$, ^{29}Si),[2793] [Ph_3SiR], (^{13}C),[2794] and [(THF)$_3$Li-

2774 J.S. Nowick and R.L. Danheiser, *J. Org. Chem.*, 1989, **54**, 2798; R.L. Danheiser, E.J. Stoner, H. Koyama, D.S. Yamashita, and C.A. Klade, *J. Am. Chem. Soc.*, 1989, **111**, 4407.
2775 Q. Zhou and W.P. Weber, *Macromolecules*, 1989, **22**, 1300 (*Chem. Abstr.*, 1989, **110**, 136 015); Q. Zhou and W.P. Weber, *Polym. Bull. (Berlin)*, 1989, **21**, 173 (*Chem. Abstr.*, 1989, **111**, 24 045); Q. Zhou and W.P. Weber, *Macromolecules*, 1989, **22**, 2987 (*Chem. Abstr.*, 1989, **111**, 40 023).
2776 F.X. Kohl, R. Dickbreder, P. Jutzi, G. Mueller, and B. Huber, *Chem. Ber.*, 1989, **122**, 871.
2777 S.S. Hu and W.P. Weber, *Polym. Bull. (Berlin)*, 1989, **21**, 133 (*Chem. Abstr.*, 1989, **111**, 78 721).
2778 E. Hengge and A. Baumegger, *J. Organomet. Chem.*, 1989, **369**, C39.
2779 K.A. Horn, R.B. Grossman, J.R.G. Thorne, and A.A. Whitenack, *J. Am. Chem. Soc.*, 1989, **111**, 4809.
2780 S.-S. Hu and W.P. Weber, *J. Organomet. Chem.*, 1989, **369**, 155.
2781 E. Hengge and M. Eibl, *J. Organomet. Chem.*, 1989, **371**, 137.
2782 J. Ohshita, D. Kanaya, M. Ishikawa, and T. Yamanaka, *J. Organomet. Chem.*, 1989, **369**, C18.
2783 J. Ohshita, K. Furumori, M. Ishikawa, and T. Yamanaka, *Organometallics*, 1989, **8**, 2084.
2784 M. Ishikawa, Y. Hasegawa, T. Hatano, A. Kunai, and T. Yamanaka, *Organometallics*, 1989, **8**, 2741.
2785 A.A. Panasenko, R.A. Sadykov, A.A. Fatykhov, and L.V. Spirikhin, *Izv. Akad. Nauk SSSR, Ser. Khim.*, 1988, 551 (*Chem. Abstr.*, 1989, **110**, 94 364).
2786 P. Jutzi, U. Holtmann, D. Kanne, C. Krueger, R. Blom, R. Gleiter, and I. Hyla-Kryspin, *Chem. Ber.*, 1989, **122**, 1629.
2787 R. Skoda-Földes, L. Kollár, and B. Heil, *J. Organomet. Chem.*, 1989, **366**, 275; P. Martinet, R. Sauvêtre, and J.-F. Normant, *J. Organomet. Chem.*, 1989, **367**, 1; F. Tellier, R. Sauvêtre, and J.-F. Normant, *J. Organomet. Chem.*, 1989, **364**, 17.
2788 W. Jiang and W.P. Weber, *Polym. Bull. (Berlin)*, 1989, **21**, 427 (*Chem. Abstr.*, 1989, **111**, 97 895).
2789 S.D. Pigarev, D.A. Bravo-Zhivotovskii, I.D. Kalikhman, N.S. Vyazankin, and M.G. Voronkov, *J. Organomet. Chem.*, 1989, **369**, 29.
2790 M.L.J. Hackney, R.C. Haltiwanger, P.F. Brandt, and A.D. Norman, *J. Organomet. Chem.*, 1989, **359**, C36.
2791 I.N. Jung, D.H. Pae, B.R. Yoo, M.E. Lee, and P.R. Jones, *Organometallics*, 1989, **8**, 2017.
2792 H. Saso, W. Ando, and K. Ueno, *Tetrahedron*, 1989, **45**, 1929.
2793 B.D. Shepherd, D.R. Powell, and R. West, *Organometallics*, 1989, **8**, 2664.
2794 J. Kron and U. Schubert, *J. Organomet. Chem.*, 1989, **373**, 203; G.A. Molander and K. Mautner, *J. Org. Chem.*, 1989, **54**, 4042.

$(SiPh_2)_2Li(THF)_3]$, (^7Li, ^{29}Si).[2795]

For $[Me_nGe(CH{=}CH_2)_{4-n}]$, a very good correlation has been found between the ^1H, ^{13}C, and ^{73}Ge chemical shifts and the corresponding ^1H, ^{13}C, and ^{29}Si chemical shifts for $[Me_nSi(CH{=}CH_2)_{4-n}]$.[2796] ^1H NOESY has been used to determine the structure of $[Me_2GeCH_2CHPhCHPhCH_2]$.[2797] N.m.r. data have also been reported for $[(2,6\text{-}Et_2C_6H_3)_2GeC(CO_2Me){=}C(CO_2Me)Ge(C_6H_3Et_2\text{-}2,6)CH_2]$, ($^{13}C$),[2798] $[(2,4,6\text{-}Me_3C_6H_2)_2\text{-}GeGe(2,4,6\text{-}Me_3C_6H_2)_2CH_2O]$, ($^{13}C$),[2799] (76), ($^{13}C$),[2800] (77), ($^{13}C$),[2801] and $[(Bu^t{}_2C_5H_3)M]\text{-}[BF_4]$, (M = Ge, Sn, Pb; ^{11}B, ^{13}C, ^{119}Sn, ^{207}Pb).[2802]

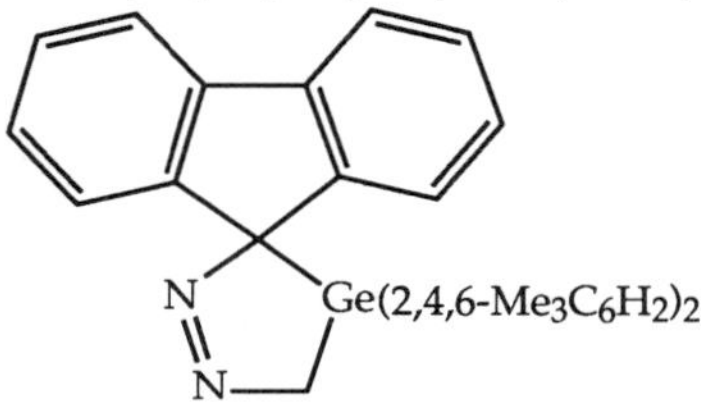

(76)

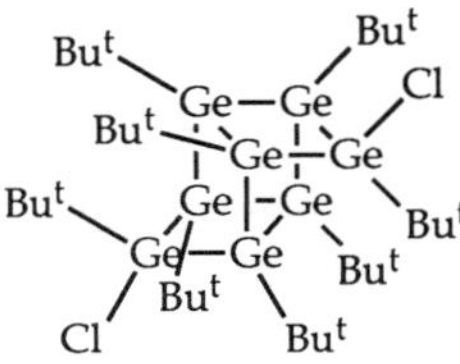

(77)

^{13}C and ^{119}Sn n.m.r. spectra of $[Ar_4Sn]$ and $[Ar_3SnX]$, Ar = 4-YC_6H_4, have been measured. The shifts correlate with σ_R or σ_R° for $[Ar_4Sn]$ and σ_p or σ_p° for $[Ar_3SnI]$. For $[Ar_3SnNCS]$, $^1J(^{119}Sn\text{-}^{14}N)$ has been observed.[2803] *Ab initio* calculations have been performed on $[Me_3MC{\equiv}CMe]$, M = Sn, Pb, and the ^{13}C, ^{119}Sn, and ^{207}Pb n.m.r. spectra were discussed.[2804] N.m.r. data have also been reported for (78), (^{119}Sn),[2805] $[Me_2Sn\text{-}(CH_2CH_2)_2P(E)R]$, (E = O, S, Se; ^{13}C, ^{119}Sn),[2806] $[1\text{-}Me_2N\text{-}C_6H_{10}\text{-}3\text{-}SnMe_3]$, ($^{13}C$, ^{119}Sn),[2807] $[Me_3SnR]$, (^{13}C),[2808] (^{13}C, ^{119}Sn),[2809] $[Me_3SnC(CO_2Me){=}CHCO_2Me]$, ($^{13}C$),[2810] $[Me_3SnC(O)\text{-}Ph]$, (^{13}C, ^{119}Sn),[2811] $[1\text{-}Bu^n{}_3Sn\text{-}2\text{-}H_{21}C_{10}\text{-}3\text{-}HOC_6H_9]$, ($^{13}C$),[2812] $[Bu^n{}_3SnR]$, (^{13}C),[2813] (^{13}C,

2795 G. Becker, H.M. Hartmann, E. Hengge, and F. Schrank, *Z. Anorg. Allg. Chem.*, 1989, **572**, 63.
2796 Y. Takeuchi, H. Inagaki, K. Tanaka, and S. Yoshimura, *Magn. Reson. Chem.*, 1989, **27**, 72.
2797 A.J. Shusterman, B.E. Landrum, and R.L. Miller, *Organometallics*, 1989, **8**, 1851.
2798 T. Tsumuraya and W. Ando, *Organometallics*, 1989, **8**, 2286.
2799 W. Ando and T. Tsumuraya, *J. Chem. Soc., Chem. Commun.*, 1989, 770.
2800 M.P. Egorov, S.P. Kolesnikov, O.M. Nefedov, and A. Krebs, *J. Organomet. Chem.*, 1989, **375**, C5.
2801 A. Sekiguchi, H. Naito, H. Nameki, K. Ebata, C. Kabuto, and H. Sakurai, *J. Organomet. Chem.*, 1989, **368**, C1.
2802 P. Jutzi and R. Dickbreder, *J. Organomet. Chem.*, 1989, **373**, 301.
2803 I. Wharf, *Inorg. Chim. Acta*, 1989, **159**, 41.
2804 M.V. Andreocci, M. Bossa, C. Cauletti, S. Stranges, B. Wrackmeyer, and K. Horchler, *Inorg. Chim. Acta*, 1989, **162**, 83.
2805 H.P. Abicht and H. Weichmann, *Z. Chem.*, 1988, **28**, 69 (*Chem. Abstr.*, 1989, **110**, 8330).
2806 H. Weichmann, B. Rensch, M. Dargatz, and J. Meunier-Piret, *Bull. Soc. Chim. Belg.*, 1989, **98**, 1.
2807 M. Ochiai, S. Iwaki, Y. Takaoka, and Y. Nagao, *Organometallics*, 1989, **8**, 1751.
2808 P. Knochel, M.C.P. Yeh, and C. Xiao, *Organometallics*, 1989, **8**, 2831.
2809 W. Adcock and H. Gangodawila, *J. Org. Chem.*, 1989, **54**, 6040.
2810 J.C. Cochran, L.E. Williams, B.S. Bronk, J.A. Calhoun, J. Fassberg, and K.G. Clark, *Organometallics*, 1989, **8**, 804.
2811 A. Capperucci, A. Degl'Innocenti, C. Faggi, G. Reginato, A. Ricci, P. Dembech, and G. Seconi, *J. Org. Chem.*, 1989, **54**, 2966.
2812 M. Ochiai, T. Ukita, S. Iwaki, Y. Nagao, and E. Fujita, *J. Org. Chem.*, 1989, **54**, 4832.
2813 J.M. Chong and E.K. Mar, *Tetrahedron*, 1989, **45**, 7709.

^{119}Sn),[2814] [$Bu^n{}_3SnAr$], (^{13}C),[2815] 1,5,9-tristannacyclododecanes, (^{119}Sn),[2816] [1-Me-2-R_3Sn-imidazole], (^{13}C, ^{119}Sn),[2817] [$R^1{}_3R^2Sn$], (^{13}C, ^{119}Sn),[2818] organopolystannanes, (^{119}Sn),[2819] [$R_2Sn]_3$, (^{13}C, ^{119}Sn),[2820] [$(C_6H_{11})_3SnR$], (^{13}C, ^{119}Sn),[2821] 2,2,4,4,5,5-(2,6-$Et_2C_6H_3)_6Sn_5$[1.1.1]-propellane, (^{13}C, ^{119}Sn),[2822] (79), (^{13}C),[2823] and [$(C_5Me_5)Pb(BF_4)$], (^{11}B, ^{13}C, ^{207}Pb).[2824]

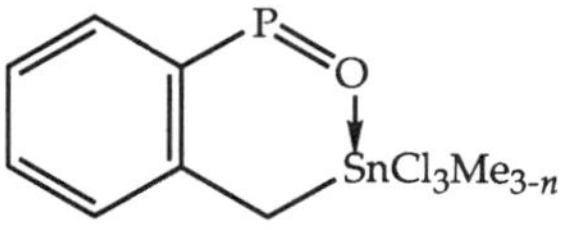

(78)

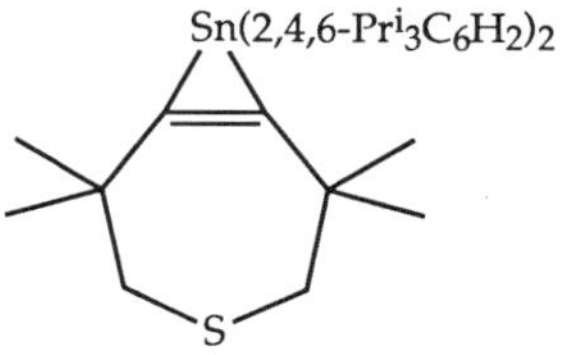

(79)

For compounds of the type [$Me_2MNBu^tSiMe_2SiMe_2NBu^t$], M = Sn, Pb, a linear relationship

$$\delta(^{207}Pb) = 3.30\ \delta(^{119}Sn) + 2336$$

has been found.[2825] INEPT INADEQUATE ^{29}Si n.m.r. spectroscopy has been applied to determine $^1J(^{29}Si\text{-}^{29}Si)$ in [$(2,4,6\text{-}Me_3C_6H_2)_2SiNPhOSi(C_6H_2Me_3\text{-}2,4,6)_2$].[2826] $^1J(^{31}P\text{-}^7Li)$ is observed in [$LiPPh(SiMe_3)(tmeda)]_2$ below 0 °C.[2827] N.m.r. data have also been reported for [$Me_3SiNPr^iC{=}PPNPr^i{}_2$], (^{13}C),[2828] [$Me_3SiNR_2$], (^{13}C),[2829] [$(Me_3SiNEt)_2SiMe_2$], (^{13}C, ^{29}Si),[2830] [$(Me_3Si)_2NCS_2R$], (^{13}C),[2831] [$\{(Me_3Si)_2N\}_2S$], (^{13}C, ^{15}N, ^{29}Si),[2832] [PhP(S)$NSiMe_3$], (^{13}C, ^{29}Si),[2833] [$MePhSiFNHSiBu^t{}_2OSiMe_3$], (^{13}C, ^{29}Si),[2834] [$R^1P\{N(SiMe_3)\}_2CHR^2$], (^{13}C, ^{29}Si),[2835] [$Me_3SiNC(O)CH_2CH_2CH_2$], (^{13}C, ^{29}Si),[2836] [Me_3SiNR_2], (^{13}C, ^{29}Si),[2837] [Me_3Si-

2814 S. Sharma and A.C. Oehlschlager, *J. Org. Chem.*, 1989, **54**, 5064.
2815 S.S. Al-Diab, *Inorg. Chim. Acta*, 1989, **160**, 93.
2816 K. Jurkschat, H.G. Kuivila, S. Liu, and J.A. Zubieta, *Organometallics*, 1989, **8**, 2755.
2817 K.C. Molloy, P.C. Waterfield, and M.F. Mahon, *J. Organomet. Chem.*, 1989, **365**, 61.
2818 J.C. Podestá, A.D. Ayala, A.B. Chopa, and N.N. Giagante, *J. Organomet. Chem.*, 1989, **364**, 39.
2819 S. Adams and M. Draeger, *Main Group Met. Chem.*, 1988, **11**, 151 (*Chem. Abstr.*, 1989, **111**, 57 786).
2820 L.R. Sita and R.D. Bickerstaff, *J. Am. Chem. Soc.*, 1989, **111**, 3769.
2821 D. Zhang, Q. Xie, and J. Zheng, *Huaxue Tongbao*, 1988, 30 (*Chem. Abstr.*, 1989, **111**, 78 218).
2822 L.R. Sita and R.D. Bickerstaff, *J. Am. Chem. Soc.*, 1989, **111**, 6454.
2823 L.R. Sita and R.D. Bickerstaff, *Phosphorus Sulphur Silicon*, 1989, **41**, 31.
2824 P. Jutzi, R. Dickbreder, and H. Nöth, *Chem. Ber.*, 1989, **122**, 865.
2825 B. Wrackmeyer, C. Stader, and K. Horchler, *J. Magn. Reson.*, 1989, **83**, 601.
2826 G.R. Gillette, J. Maxka, and R. West, *Angew. Chem., Int. Ed. Engl.*, 1989, **28**, 54.
2827 E. Hey, C.L. Raston, B.W. Skelton, and A.H. White, *J. Organomet. Chem.*, 1989, **362**, 1.
2828 E. Niecke, R. Streubel, M. Nieger, and D. Stalke, *Angew. Chem., Int. Ed. Engl.*, 1989, **28**, 1673.
2829 L.F. Tietz, M. Bratz, and M. Pretor, *Chem. Ber.*, 1989, **122**, 1955.
2830 D. Seyferth, J.M. Schwark, and R.M. Stewart, *Organometallics*, 1989, **8**, 1980.
2831 G. Schubert and G. Gattow, *Z. Anorg. Allg. Chem.*, 1989, **579**, 129.
2832 G. Schubert and G. Gattow, *Z. Anorg. Allg. Chem.*, 1989, **574**, 165.
2833 K. Dostál, J. Příhoda, S. Stoklásek, J. Loucký, J. Pinkas, G. Grossmann, and G. Ohms, *Z. Anorg. Allg. Chem.*, 1989, **576**, 54.
2834 D. Schmidt-Baese and U. Klingebiel, *J. Organomet. Chem.*, 1989, **364**, 313.
2835 C. Roques, M.-R. Mazières, J.-P. Majoral, M. Sanchez, and J. Jaud, *Inorg. Chem.*, 1989, **28**, 3931.
2836 I.D. Kalikhman, A.I. Albanov, O.B. Bannikova, L.I. Belousova, M.G. Voronkov, V.A. Pestunovich, A.G. Shipov, E.P. Kramarova, and Yu.I. Baukov, *J. Organomet. Chem.*, 1989, **361**, 147.
2837 C. Bessenbacher and W. Kaim, *J. Organomet. Chem.*, 1989, **362**, 37.

$N{=}PPh_2N(SiMe_3)_2]$, (^{29}Si),[2838] $[Sn\{N(SiMe_3)_2\}(\mu\text{-}OBu^t)]_2$, (^{13}C, ^{119}Sn),[2839] $[Pr^i{}_2N\text{-}P\{N(SiMe_3)\}_2P(=NSiMe_3)NPr^i{}_2]$, (^{13}C, ^{15}N, ^{29}Si),[2840] $[Me_3SiNArSnMe_3]$, (^{13}C, ^{15}N, ^{29}Si, ^{119}Sn),[2841] $[(R^1R^2N)(S)PSBRNSiMe_3]$, (^{11}B, ^{13}C, ^{29}Si),[2842] $[R_3SiN{=}Ge\{N(SiMe_3)_2\}_2]$, (^{13}C, ^{14}N, ^{15}N, ^{29}Si),[2843] $[R^1NCH{=}CR^2CHPhGeMe_2]$, (^{13}C),[2844] $[(Et_2N)MeSiCH{=}CMeCMe{=}CH]$, (^{13}C, ^{29}Si),[2845] $[1\text{-}Pr^i{}_3Si\text{-}3\text{-}HO_2C\text{-pyrrole}]$, (^{13}C),[2846] $[1,3\text{-}(Pr^i{}_3Si)_2\text{-pyrazole}]$, (^{13}C, ^{29}Si),[2847] $[(Bu^t{}_3SiNHSiBu^t{}_2CH)_2C{=}CH_2]$, (^{13}C),[2848] $[(Me_3Si)_2PPCPh{=}CBu^t]$, (^{13}C),[2849] trimethylsilyl-silicates, (^{29}Si),[2850] $[Me_3SiOR]$, (^{13}C),[2851] (^{13}C, ^{29}Si),[2852] (^{15}N, ^{29}Si),[2853] $[Me_{3-n}SiR\text{-}(OSiMe_3)_n]$, (^{13}C, ^{17}O, ^{29}Si),[2854] $[(CH_2{=}CHCH_2Me_2Si)_8Si_8O_{20}]$, (^{29}Si),[2855] $[HO(CH_2)_n\text{-}(SiMe_2O)_mSiMe_2(CH_2)_nOH]$, (^{13}C, ^{29}Si),[2856] $[Me_2Si(OMe)CH_2N\{C(O)Me\}N\{C(O)Me\}\text{-}CH_2Si(OMe)Me_2]$, (^{13}C, ^{29}Si),[2857] $[Bu^tMe_2SiOR]$, (^{13}C),[2858] $[R_fSiMe_2OSiMe_2OSiMe_2O\text{-}SiMe_2R_f]$, (^{29}Si),[2859] $[C_6H_3\text{-}1,3,5\text{-}(SiMe_2OSiMe_2)_3\text{-}1,3,5\text{-}C_6H_3]$, (^{13}C, ^{29}Si),[2860] $[Et_3SiOR]$, (^{13}C),[2861] $[Ph(C_6H_{11})Si(OH)CH_2N(CH_2CH_2)_2NMe_2]^+$, (^{13}C),[2862] $[Bu^tPh_2SiOR]$, (^{13}C),[2863] $[\{(C_5Me_5)_2Si\}_2(\mu\text{-}S)_2]$, (^{29}Si),[2864] $[MeESiMe_3]$, (E = Se, Te; ^{13}C, ^{77}Se, ^{125}Te),[2865] $[Me_3SiF_2]^-$,

2838 M. Witt, H.W. Roesky, M. Noltemeyer, and A. Schmidpeter, *New J. Chem.*, 1989, **13**, 403.
2839 M.J. McGeary, K. Folting, and K.G. Caulton, *Inorg. Chem.*, 1989, **28**, 4051.
2840 J. Böske, E. Niecke, M. Nieger, E. Ocando, J.-P. Majoral, and G. Bertrand, *Inorg. Chem.*, 1989, **28**, 499.
2841 B. Wrackmeyer and H. Zhou, *J. Organomet. Chem.*, 1989, **375**, 1.
2842 D. Fest, C.D. Habben, and A. Meller, *Chem. Ber.*, 1989, **122**, 861.
2843 J. Pfeiffer, W. Maringgele, M. Noltemeyer, and A. Meller, *Chem. Ber.*, 1989, **122**, 245.
2844 K. Bootz and W.P. Neumann, *Tetrahedron Lett.*, 1989, **30**, 6669.
2845 J.-P. Béteille, A. Laporterie, and J. Dubac, *Organometallics*, 1989, **8**, 1799.
2846 K.-P. Stefan, W. Schuhmann, H. Parlar, and F. Korte, *Chem. Ber.*, 1989, **122**, 169.
2847 F. Castan, A. Baceiredo, and G. Bertrand, *Angew. Chem., Int. Ed. Engl.*, 1989, **28**, 1250.
2848 M. Weidenbruch, B. Brand-Roth, S. Pohl, and W. Saak, *J. Organomet. Chem.*, 1989, **379**, 217.
2849 O. Wagner, M. Ehle, and M. Regitz, *Angew. Chem., Int. Ed. Engl.*, 1989, **28**, 225.
2850 E. Bertling and H.C. Marsmann, *Z. Anorg. Allg. Chem.*, 1989, **578**, 166.
2851 K. Sutowardoyo, M. Emziane, and D. Sinou, *Tetrahedron Lett.*, 1989, **30**, 4673; S. Hünig and C. Marschner, *Chem. Ber.*, 1989, **122**, 1329; M. Mitani, O. Tachizawa, H. Takeuchi, and K. Koyama, *J. Org. Chem.*, 1989, **54**, 5397; D.J. Pollart and H.W. Moore, *J. Org. Chem.*, 1989, **54**, 5444; T.C. Eisenschmid and R. Eisenberg, *Organometallics*, 1989, **8**, 1822.
2852 M. Bergdahl, E.-L. Lindstedt, and T. Olsson, *J. Organomet. Chem.*, 1989, **365**, C11; I.D. Kalikhman, O.B. Bannikova, L.I. Volkova, B.A. Gostevskii, T.I. Yushmanova, V.A. Lopyrev, and M.G. Voronkov, *Izv. Akad. Nauk SSSR, Ser. Khim.*, 1988, 460 (*Chem. Abstr.*, 1989, **110**, 75 608).
2853 K.A. Thorn, D.W. Folan, J.B. Arterburn, M.A. Mikita, and P. MacCarthy, *Sci. Total Environ,*. 1988, (Pub. 1989), **81**, 209 (*Chem. Abstr.*, 1989, **111**, 120 479).
2854 E. Liepins, I. Zicmane, I. Sleiksa, and E. Lukevics, *Latv. PSR Zinat. Akad. Vestis, Kim. Ser.*, 1988, 726 (*Chem. Abstr.*, 1989, **111**, 7483).
2855 D. Hoebbel, I. Pitsch, T. Reiher, W. Hiller, H. Jancke, and D. Müller, *Z. Anorg. Allg. Chem.*, 1989, **576**, 160.
2856 F. Braun, L. Willner, M. Hess, and R. Kosfeld, *J. Organomet. Chem.*, 1989, **366**, 53.
2857 I.D. Kalikhman, B.A. Gostevskii, O.B. Bannikova, M.G. Voronkov, and V.A. Pestunovich, *J. Organomet. Chem.*, 1989, **376**, 249.
2858 C. Siegel and E.R. Thornton, *J. Am. Chem. Soc.*, 1989, **111**, 5722; G.M. Wieber, L.S. Hegedus, B. Åkermark, and E.T. Michalson, *J. Org. Chem.*, 1989, **54**, 4649.
2859 B. Boutevin, Y. Pietrasanta, and B. Youssef, *J. Fluorine Chem.*, 1988, **39**, 61 (*Chem. Abstr.*, 1989, **110**, 192 903).
2860 A. Sekiguchi, T. Yatabe, C. Kabuto, and H. Sakurai, *Angew. Chem., Int. Ed. Engl.*, 1989, **28**, 757.
2861 A.B. Smith, tert., A.T. Lupo, jun., M. Ohba, and K. Chen, *J. Am. Chem. Soc.*, 1989, **111**, 6648.
2862 R. Tacke, H. Linoh, K. Rafeiner, G. Lambrecht, and E. Mutschler, *J. Organomet. Chem.*, 1989, **359**, 159.
2863 A.G.M. Barrett, T.E. Barta, and J.A. Flygare, *J. Org. Chem.*, 1989, **54**, 4246.
2864 P. Jutzi, A. Möhrke, A. Müller, and H. Bögge, *Angew. Chem., Int. Ed. Engl.*, 1989, **28**, 1518.
2865 B. Bildstein, K. Giselbrecht, and F. Sladky, *Chem. Ber.*, 1989, **122**, 2279.

(^{13}C),[2866] [$SiPh_3F$], (^{13}C),[2867] poly(silapropylene), (^{13}C, ^{29}Si),[2868] and (80), (^{13}C, ^{29}Si).[2869]

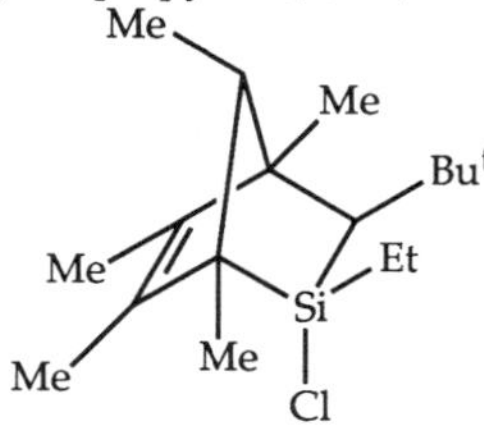

(80)

The INEPT pulse sequence has been used to obtain ^{15}N n.m.r. spectra from bis(amino)stannylenes and -plumbylenes. $^{1}J(^{117}Sn-^{15}N)$, $^{1}J(^{119}Sn-^{15}N)$, and $^{1}J(^{207}Pb-^{15}N)$ were observed.[2870] The dependence of $^{2}J(^{119}Sn-^{119}Sn)$ on the molecular structure of compounds containing the Sn—E—Sn moiety, E = O, S, Se, Te, has been investigated. A plot of $^{2}J(^{119}Sn-^{119}Sn)$ against the Sn—E—Sn bond angle gives a straight line.[2871] The acidities of a series of organotin halides towards $Ph_2P(O)CH_2CH_2P(O)Ph_2$ have been determined from ^{31}P n.m.r. spectroscopy. The ^{119}Sn n.m.r. spectra were also reported.[2872] Solvent effects on $^{2}J(^{119}Sn-^{117}Sn)$ in [$(R_3Sn)_2O$] have been studied.[2873] N.m.r. data have also been reported for [$H_2C=C=CHGeMe_2SCN$], (^{13}C),[2874] [$Bu^t{}_2C=CSGe(C_6H_2Me_3\text{-}2,4,6)_2$], (^{13}C),[2875] [$R_2CSGeMe_2$], (^{13}C),[2876] [$(CH_2=CHCH_2)ClGe(C_6H_2Me_3\text{-}2,4,6)_2$], (^{13}C),[2877] [$Bu^t{}_3SnP=P(C_6H_2Me_3\text{-}2,4,6)$], (^{119}Sn),[2878] [$PhCH_2Sb(SnMe_3)_2$], (^{13}C, ^{119}Sn),[2879] [$Me_3SnOC_6H_3Ph_2\text{-}2,6$], (^{13}C),[2880] [$(Me_3Sn)_2$(glutamate)], (^{13}C, ^{119}Sn),[2881] [$Bu^n{}_3SnER$], (E = O, S; ^{13}C, ^{119}Sn),[2882] [$Bu^n{}_3SnOR$], (^{13}C),[2883] [$(C_6H_{11})_3SnOR$], (^{119}Sn),[2884] (^{13}C, ^{119}Sn),[2885] [$R_3SnO(CH_2)_nC_5H_4N$], (^{13}C, ^{119}Sn),[2886] [$R_3Sn\{HCONHCH(CH_2Ph)CO_2\}$],

2866 T. Konishi, Y. Tamai, M. Fujii, Y. Einaga, and H. Yamakawa, *Polym. J. (Tokyo)*, 1989, **21**, 329 (*Chem. Abstr.*, 1989, **111**, 58 622).
2867 C. Clankenship and S.E. Cremer, *J. Organomet. Chem.*, 1989, **371**, 19.
2868 H.J. Wu, and L.V. Interrante, *Chem. Mater.*, 1989, **1**, 564 (*Chem. Abstr.*, 1989, **111**, 134 859).
2869 N. Auner, *J. Organomet. Chem.*, 1989, **377**, 175.
2870 C. Stader and B. Wrackmeyer, *Z. Naturforsch., B*, 1987, **42**, 1515 (*Chem. Abstr.*, 1989, **110**, 154 419).
2871 T.P. Lockhard, H. Puff, W. Schuh, H. Reuter, and T.N. Mitchell, *J. Organomet. Chem.*, 1989, **366**, 61.
2872 C.H. Yoder, S.M. Coley, S.P. Kneizys, and J.N. Spencer, *J. Organomet. Chem.*, 1989, **362**, 59.
2873 T.P. Lockhart, *Inorg. Chem.*, 1989, **28**, 4265.
2874 G. Billeb, H. Brauer, S. Maslov, and W.P. Neumann, *J. Organomet. Chem.*, 1989, **373**, 11.
2875 W. Ando and T. Tsumuraya, *Organometallics*, 1989, **8**, 1467.
2876 T. Tsumuraya, S. Sato, and W. Ando, *Organometallics*, 1989, **8**, 161.
2877 W. Ando, H. Itoh, and T. Tsumuraya, *Organometallics*, 1989, **8**, 2759.
2878 D. Hänssgen, H. Aldenhoven, and M. Nieger, *J. Organomet. Chem.*, 1989, **375**, C9.
2879 G. Becker, M. Meiser, O. Mundt, and J. Weidlein, *Z. Anorg. Allg. Chem.*, 1989, **569**, 62.
2880 G.D. Smith, P.E. Fanwick, and I.P. Rothwell, *J. Am. Chem. Soc.*, 1989, **111** , 750.
2881 F. Huber, B. Mundas-Glowacki, and H. Preut, *J. Organomet. Chem.*, 1989, **365**, 111.
2882 A. Lyčka, J. Holeček, and K. Handlíř, *Collect. Czech. Chem. Commun.*, 1989, **54**, 2386.
2883 P.P.L. Jacques and R.C. Poller, *J. Organomet. Chem.*, 1989, **365**, 47.
2884 S.J. Blunden, B.N. Patel, P.J. Smith, and B. Sugavanam, *Appl. Organomet. Chem.*, 1987, **1**, 241 (*Chem. Abstr.*, 1989, **110**, 8325).
2885 C. Cruzado, M. Bernabe, and M. Martin-Lomas, *J. Org. Chem.*, 1989, **54**, 465.
2886 R. Visalakshi, V.K. Jain, and G.S. Rao, *Indian J. Chem., Sect. A*, 1988, **27A**, 427 (*Chem. Abstr.*, 1989, **110**, 212 964).

(^{13}C, ^{119}Sn),[2887] [$Ph_3SnO_2C(CH_2)_2C(O)Ph$], (^{13}C),[2888] [R_3SnS-8-quinolinyl], (^{13}C, ^{15}N, ^{119}Sn),[2889] [Me_3SnF_2]$^-$, (^{13}C),[2890] [$(Bu^n{}_3SnF)_2F$]$^-$, (^{119}Sn),[2891] [$Me_2SnClC_6H_4NO_2$], (^{13}C, ^{119}Sn),[2892] [3,5-$Bu^t{}_2C_6H_3CMe_2CH_2Sn(C_6H_2Bu^t{}_3)_2X$], ($^{13}C$),[2893] 2,3,4,5-tetraaza-6-diphenylstannyl[3.4]bicyclonona-1,3-diene, (^{119}Sn),[2894] and [$R^1{}_2SnBrCHPhCHR^2CO_2Me$], ($^{13}C$, ^{119}Sn).[2895]

Structural studies on poly(dimethylsilylethylenediamine) using two dimensional ^{13}C-^{1}H and ^{29}Si-^{1}H n.m.r. spectra have been reported.[2896] ^{31}P n.m.r. spectra of [$(Me_2Si)_{6-n}P_4(SiEt_2)_n$] show a linear increase in ^{31}P chemical shift with an increasing number of ethyl groups.[2897] An incremental system for predicting the ^{17}O shift of variously substituted siloxanes has been published.[2898] The mobility and ^{29}Si relaxation in some siloxanes has been reported.[2899] ^{29}Si n.m.r. spectroscopy has been used to investigate the sequence distribution in polysiloxanes.[2900] ^{17}O, ^{29}Si, and ^{13}C n.m.r. spectroscopy has been used to study motion in [$Me_2SiO]_4$.[2901] A detailed structural analysis of polysiloxane antifoam agents has been performed using ^{13}C and ^{29}Si n.m.r. spectroscopy.[2902] N.m.r. data have also been reported for 2,2,3-trihydrocarbylbenzo-1-oxa-3-aza-2-silacyclohexen-4-ones, (^{29}Si),[2903] [$Me_2Si(NBu^t)_2(OBu^t)SnBr_2$], ($^{13}C$),[2904] [{$Me_2Si(\mu$-$CH_2)(\mu$-$NCMe_3)$}$_2Sn$], ($^{13}C$, ^{15}N, ^{29}Si, ^{119}Sn),[2905] poly(1,2-dimethylsilazane), (^{29}Si),[2906] (81), (^{13}C, ^{29}Si),[2907] (82), (^{13}C, ^{119}Sn),[2908] [$(THF)_2Li(FSiBu^t{}_2PPh)_2Li(THF)_2$], ($^{13}C$, ^{29}Si),[2909] [$R^1R^2Si(OCH_2CH_2)_2O$], (^{29}Si),[2910] poly(dimethylvinylsiloxane), (^{13}C),[2911] a

2887 G.K. Sandhu, G. Kaur, J. Holeček, and A. Lyčka, *J. Organomet. Chem.*, 1989, **365**, 215.
2888 S.W. Ng, V.G.K. Das, and A. Syed, *J. Organomet. Chem.*, 1989, **364**, 353.
2889 A. Lyčka, J. Jirman, J. Holeček, and I. Tkáč, *J. Organomet. Chem.*, 1989, **372**, 327.
2890 L. Heuer, L. Ernst, R. Schmutzler, and D. Schomburg, *Angew. Chem., Int. Ed. Engl.*, 1989, **28**, 1507.
2891 S.J. Blunden and R. Hill, *J. Organomet. Chem.*, 1989, **371**, 145.
2892 T.A.K. Al-Allaf, U. Kobs, and W.P. Neumann, *J. Organomet. Chem.*, 1989, **373**, 29.
2893 M. Weidenbruch, K. Schäfers, S. Pohl, W. Saak, K. Peters, and H.G. von Schnering, *Z. Anorg. Allg. Chem.*, 1989, **570**, 75.
2894 M.F. Mahon, K.C. Molloy, and P.C. Waterfield, *J. Organomet. Chem.*, 1989, **361**, C5.
2895 A.B. Chopa, L.C. Koll, J.C. Podestá, and T.N. Mitchell, *J. Organomet. Chem.*, 1989, **376**, 283.
2896 Y.H. Mariam and P. Abrahams, *Polym. Prepr. (Am. Chem. Soc., Div. Polym. Chem.)*, 1988, **29**, 364 (*Chem. Abstr.*, 1989, **110**, 8875).
2897 G. Fritz and J. Reuter, *Z. Anorg. Allg. Chem.*, 1989, **575**, 39.
2898 U. Scheim, K. Rühlmann, J.W. Kelly, and S.A. Evans, jun., *J. Organomet. Chem.*, 1989, **375**, 33.
2899 R. Kosfeld, C. Kreuzburg, and R. Krause, *Makromol. Chem.*, 1988, **189**, 2077 (*Chem. Abstr.*, 1989, **110**, 8944).
2900 G.W. Gray, W.D. Hawthorne, D. Lacey, M.S. White, and J.A. Semlyen, *Liq. Crystals*, 1989, **6**, 503.
2901 J. Kowalewski and E. Berggren, *Magn. Reson. Chem.*, 1989, **27**, 386.
2902 I.R. Herbert and A.D.H. Clague, *Macromolecules*, 1989, **22**, 3267 (*Chem. Abstr.*, 1989, **111**, 98 033).
2903 M. Wang, D. Zhang, G. Li, and Q. Xie, *Gaodeng Xuexiao Huaxue Xuebao*, 1988, **9**, 410 (*Chem. Abstr.*, 1989, **110**, 114 922).
2904 M. Veith, V. Huch, R. Lisowsky, and P. Hobein, *Z. Anorg. Allg. Chem.*, 1989, **569**, 43.
2905 C. Stader and B. Wrackmeyer, *Z. Naturforsch., B*, 1988, **43**, 707 (*Chem. Abstr.*, 1989, **111**, 7604).
2906 Y.H. Mariam, P. Abrahams, and K. Feng, *Polym. Mater. Sci. Eng.*, 1988, **59**, 105 (*Chem. Abstr.*, 1989, **110**, 58 427).
2907 P. Arya, J. Boyer, F. Carré, R. Corriu, G. Lanneau, J. Lapasset, M. Perrot, and C. Priou, *Angew. Chem., Int. Ed. Engl.*, 1989, **28**, 1016.
2908 A. Krebs, A. Jacobsen-Bauer, E. Haupt, M. Veith, and V. Huch, *Angew. Chem., Int. Ed. Engl.*, 1989, **28**, 603.
2909 D. Stalke, M. Meyer, M. Andrianarison, U. Klingebiel, and G.M. Sheldrick, *J. Organomet. Chem.*, 1989, **366**, C15.

copolymer of $[MePhPN]_n$ and $[Me_2SiO]_n$, (^{29}Si),[2912] $[Me_2Si(CH_2CH_2)_2O]$, (^{13}C),[2913] $[MeXSi(OSiMe_2)_2O]$, (^{29}Si),[2914] $[R^1R^2Si\{OSiMe(CH_2)O\}_2SiR^1R^2]$, ($^{13}C$, ^{29}Si),[2915] $[Bu^t{}_2Si(OH)(OLi)]$, (^{13}C, ^{29}Si),[2916] $[(MeO)_2Si(CMe{=}CMe)_2Si(OMe)_2]$, ($^{13}C$),[2917] and $[(C_5Me_5)_2SiSC({=}NPh)Si(C_5Me_5)_2S]$, ($^{29}Si$).[2918]

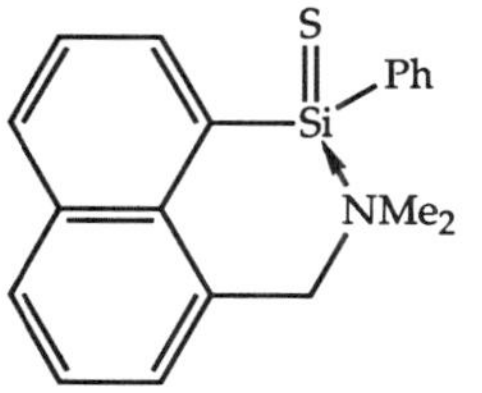

(81)

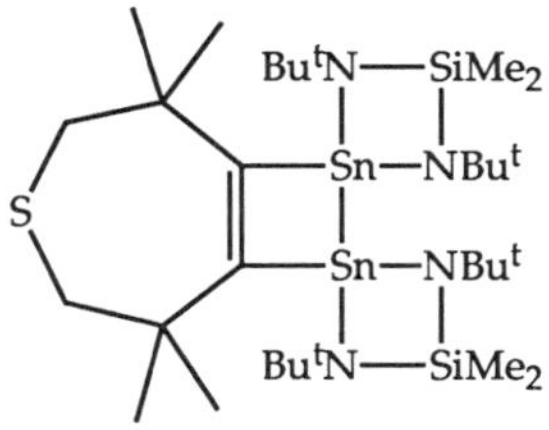

(82)

The ^{13}C and ^{119}Sn n.m.r. spectra of (83) have been reported. $^1J(^{119}Sn\text{-}^{13}C)$ of 1215 Hz is the largest such coupling so far reported.[2919] A correlation between $\delta(^{119}Sn)$ and the electronegativity of E has been found for $[Bu^t{}_2SnE]_2$, E = S, Se, and Te. The ^{13}C n.m.r. spectra were also recorded.[2920] N.m.r. data have also been reported for $[Me_2SnECX{=}NN{=}CHC_6H_4O\text{-}2]$, (E = O, S; X = Ph, SMe; ^{13}C),[2921] Me_2Sn Schiff base complexes, (^{13}C, ^{119}Sn),[2922] $[Bu_2SnN(C_6H_4Cl\text{-}2)CMeCHCPhO]$, ($^{13}C$),[2923] $[HN(CH_2CO_2)_2SnR_2]$, (^{13}C),[2924] $[(CH_2{=}CH)_2Sn(8\text{-hydroxyquinolinate})_2]$, ($^{13}C$, ^{15}N, ^{119}Sn),[2925] $[4\text{-}R^1{}_2NC_6H_4C(O)R^2SnR^3{}_2Cl_2]$, ($^{13}C$, ^{119}Sn),[2926] $[R_2Sn(PPh)_2CH_2]$, (^{119}Sn),[2927] $[Bu^t{}_2SnPBu^t]_2$, (^{119}Sn),[2928] $[R^1N(CH_2CO_2)_2SnR^2{}_2]$, ($^{13}C$),[2929] $[Bu_2SnXOSnYBu_2]$, (^{119}Sn),[2930]

2910 V.M. D'yakov, A.N. Kir'yanova, L.N. Kireeva, A.E. Chernyshev, V.N. Bochkarev, and S.I. Androsenko, *Zh. Obshch. Khim.*, 1988, **58**, 539 (*Chem. Abstr.*, 1989, **110**, 23 956).
2911 P.A. Batkina, I.G. Kolokol'tseva, and V.D. Lobkov, *Kauch. Rezina*, 1988, 11 (*Chem. Abstr.*, 1989, **111**, 39 898).
2912 P. Wisian-Neilson and M.S. Islam, *Macromolecules*, 1989, **22**, 2026 (*Chem. Abstr.*, 1989, **110**, 173 914).
2913 R.H. Cragg and R.D. Lane, *Main Group Met. Chem.*, 1987, **10**, 315 (*Chem. Abstr.*, 1989, **110**, 231 694).
2914 R. Gewald, K. Rühlmann, U. Scheim, and A. Porzel, *J. Organomet. Chem.*, 1989, **377**, 9.
2915 A.A. Zhdanov, T.V. Astapova, and B.D. Lavrukhin, *Izv. Akad. Nauk SSSR, Ser. Khim.*, 1988, 657 (*Chem. Abstr.*, 1989, **110**, 115 107).
2916 D. Schmidt-Bäse and U. Klingebiel, *Chem. Ber.*, 1989, **122**, 815.
2917 G. Maier, H.P. Reisenauer, K. Schöttler, and U. Wessolek-Kraus, *J. Organomet. Chem.*, 1989, **366**, 25.
2918 P. Jutzi and A. Möhrke, *Angew. Chem., Int. Ed. Engl.*, 1989, **28**, 762.
2919 M. Careri, A. Mangia, G. Predieri, and C. Vignali, *J. Organomet. Chem.*, 1989, **375**, 39.
2920 H. Puff, G. Bertram, B. Ebeling, M. Franken, R. Gattermayer, R. Hundt, W. Schuh, and R. Zimmer, *J. Organomet. Chem.*, 1989, **379**, 235.
2921 M.F. Iskander, L. Labib, M.M.Z. Nour El-Din, and M. Tawfik, *Polyhedron*, 1989, **8**, 2755.
2922 A.K. Varshney, S. Varshney, and J.P. Tandon, *J. Prakt. Chem.*, 1989, **331**, 511.
2923 P. Dixit and J.P. Tandon, *J. Prakt. Chem.*, 1989, **331**, 659.
2924 F.L. Lee, E.J. Gabe, L.E. Khoo, W.H. Leong, G. Eng., and F.E. Smith, *Inorg. Chim. Acta*, 1989, **166**, 257.
2925 J. Jirman, A. Lyčka, J. Holeček, K. Handlíř, M. Nadvornik, and S.T.M. Said, *Sb. Ved. Pr., Vys. Sk. Chemickotechnol. Pardubice*, 1987, **50a**, 57 (*Chem. Abstr.*, 1989, **110**, 24 019).
2926 A.S. Kothiwal, A. Singh, A.K. Rai, and R.C. Mehrotra, *Indian J. Chem., Sect. A*, 1988, **27A**, 507 (*Chem. Abstr.*, 1989, **110**, 231 757).
2927 R. Bohra, P.B. Hitchcock, M.F. Lappert, and W.-P. Leung, *J. Chem. Soc., Chem. Commun.*, 1989, 728.
2928 D. Hänssgen, H. Aldenhoven, and M. Nieger, *J. Organomet. Chem.*, 1989, **367**, 47.
2929 A. Meriem, M. Gielen, and R. Willem, *J. Organomet. Chem.*, 1989, **365**, 91.
2930 D.C. Gross, *Inorg. Chem.*, 1989, **28**, 2355.

$[\{Bu^n_2Sn(O_2CC_5H_4N)\}_2O]_2$, ($^{13}C$, ^{119}Sn),[2931] $[Ag(dppm)_2][SnPh_2(NO_3)_3]$, ($^{119}Sn$),[2932] $[Me_2SnClS_2N_2]^-$, (^{119}Sn),[2933] $[(MeO_2CCH_2CH_2)_2Sn(S_2CNMe_2)_2]$, ($^{13}C$),[2934] $[R_2SnX_2]$ complexes with secondary phosphines, (^{119}Sn),[2935] $[Bu^t_8Sn_4(SPh)_2]$, (^{119}Sn),[2936] polymer bound $[R_2SnCl_2]$, (^{119}Sn),[2937] and $[(C_5Me_5)_2SnBr_2]$, (^{13}C, ^{119}Sn).[2938]

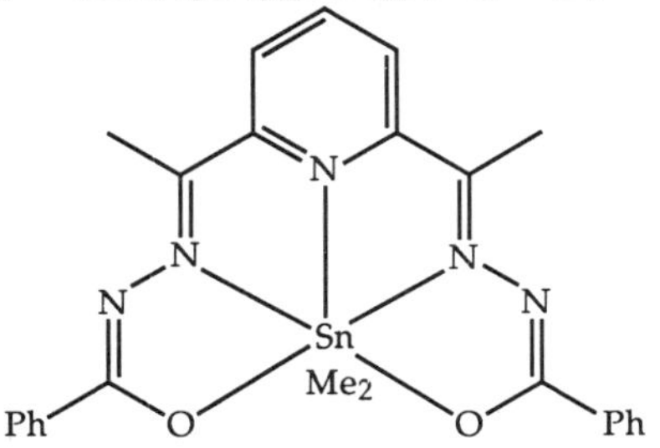

(83)

^1H n.m.r. spectroscopy has been used to determine the conformation of $[NCCH_2CH_2Si(OEt)_3]$.[2939] N.m.r. spectroscopy has been used to study the structure and orientation of $[C_6H_5SnCl_3]$ and $[C_6H_5SeCl]$ in a nematic phase.[2940] N.m.r. data have also been reported for $[MeSi(NRCH_2CH_2)_3N]$, (^{13}C, ^{29}Si),[2941] $[H_2NCH_2CH_2NH(CH_2)_n Si(OCH_2CH_2)_3N]$, ($^{13}C$),[2942] $[R^1Si(NR^2CH_2CH_2)_3N]$, (^{29}Si),[2943] $[PhCH_2SiO_{1.5}]_8$, (^{13}C, ^{29}Si),[2944] $[RSi(OEt)_3]$, (^{13}C),[2945] $[Bu^tSiP]_4$, (^{29}Si),[2946] $[PhSi(O_2C_6H_3NO_2)_2]^-$, ($^{29}Si$),[2947] $[ArSiO_{1.5}]_8$, (^{13}C, ^{29}Si),[2948] $[EtGe(SePh)_3]$, (^{77}Se),[2949] $[CH_2\{CH(GeCl_3)CH_2\}_2C{=}NPh_2][GeCl_3]$, ($^{13}C$),[2950] $[Ph_5C_5MCl_3]$, (M = Ge, Sn; ^{13}C),[2951] (84), (^{13}C, ^{119}Sn),[2952]

2931 C.S. Parulekar, V.K. Jain, T.K. Das, A.R. Gupta, B.F. Hoskins, and E.R.T. Tiekink, *J. Organomet. Chem.*, 1989, **372**, 193.

2932 D. Franzoni, G. Pelizzi, G. Predieri, P. Tarasconi, F. Citali, and C. Pelizzi, *J. Chem. Soc., Dalton Trans.*, 1989, 247.

2933 J. Fait and K.J. Schmidt, *Inorg. Chem.*, 1989, **28**, 3018.

2934 S.W. Ng, C. Wei, V.G.K. Das, G.B. Jameson, and R.J. Butcher, *J. Organomet. Chem.*, 1989, **365**, 75.

2935 D. Dakternieks and C.L. Rolls, *Inorg. Chim. Acta*, 1989, **161**, 105.

2936 H. Puff, B. Breuer, G. Gehrke-Brinkmann, P. Kind, H. Reuter, W. Schuh, W. Wald, and G. Weidenbrück, *J. Organomet. Chem.*, 1989, **363**, 265.

2937 D.E. Bergbreiter and S.A. Walker, *J. Org. Chem.*, 1989, **54**, 5138.

2938 P. Jutzi and K.-H. Schwartzen, *Chem. Ber.*, 1989, **122**, 287.

2939 L.W. Jenneskens, J.N. Louwen, E.M.M. van den Berg, A.A. van Beuzekom, and J. Lugtenburg, *Recl. Trav. Chim. Pays-Bas*, 1989, **108**, 230.

2940 T. Turki and E. Haloui, *J. Soc. Chim. Tunis.*, 1987, **2**, 9 (*Chem. Abstr.*, 1989, **110**, 114 970).

2941 D. Gudat, L.M. Daniels, and J.G. Verkade, *J. Am. Chem. Soc.*, 1989, **111**, 8520.

2942 M.G. Voronkov, V.P. Baryshok, N.F. Lazareva, V.V. Saraev, T.I. Vakulskaya, P. Hencsei, and I. Kovacs, *J. Organomet. Chem.*, 1989, **368**, 155.

2943 D. Gudat, C. Lensink, H. Schmidt, S.-K. Xi, and J.G. Verkade, *Phosphorus Sulphur Silicon*, 1989, **41**, 21.

2944 F.J. Feher and T.A. Budzichowski, *J. Organomet. Chem.*, 1989, **373**, 153.

2945 Z. Foltynowicz and B. Marciniec, *J. Organomet. Chem.*, 1989, **376**, 15.

2946 M. Baudler, G. Scholz, K.-F. Tebbe, and M. Fehér, *Angew. Chem., Int. Ed. Engl.*, 1989, **28**, 339.

2947 G. Cerveau, C. Chuit, R.J.P. Corriu, L. Gerbier, and C. Reyé, *Phosphorus Sulphur Silicon*, 1989, **42**, 115.

2948 F.J. Feher and T.A. Budzichowski, *J. Organomet. Chem.*, 1989, **379**, 33.

2949 H.J. Gysling and H.R. Luss, *Organometallics*, 1989, **8**, 363.

2950 S.P. Kolesnikov, S.L. Povarov, A.I. Lutsenko, V.A. Svyatkin, V.V. Samoshin, and N.S. Zefirov, *Zh. Org. Khim.*, 1988, **24**, 1660 (*Chem. Abstr.*, 1989, **111**, 78 208).

$[\overline{ClSnCMe_2CH_2CH_2N}Me_2]_2$, ($^{13}C$, ^{119}Sn),[2953] $[(MeO_2CCH_2CH_2)(Me_2NCS_2)SnS]_n$, ($^{13}C$, ^{119}Sn),[2954] $[MeO_2CCH_2CH_2Sn(S_2CNMe)_2Cl_2]$, ($^{13}C$, ^{119}Sn),[2955] $[Bu^nSn(OH)\{O_2P(OBu^t)_2\}_3O][O_2P(OBu^t)_2]$, ($^{119}Sn$),[2956] $[\{Bu^nSn(OH)O_2PCy_2\}_3O][\{Bu^nSnCl_2O_2PCy_2\}_2OH]$, ($^{119}Sn$),[2957] and $[Pr^iSnCl_3]$, (^{13}C).[2958]

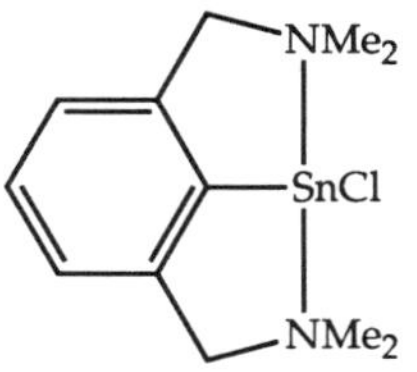

(84)

$^1J(^{119}Sn-^{31}P)$ has been shown to be negative for $[(R_3P)_2SnX_4]$ complexes. Hence the Fermi contact contribution calculations are valid, unlike for other types of Sn—P interaction. ^{13}C and ^{119}Sn n.m.r. spectra were reported.[2959] ^{29}Si multiple quantum filtered n.m.r. spectra of ^{29}Si isotopically enriched K_4SiO_4 solution show six silicate anions are present.[2960] $^{16}O/^{18}O$ isotope effects on ^{29}Si chemical shifts have been measured. The magnitude depends on the nature of the Si—O bonds.[2961] The interactions between phosphate and silicate ions in aqueous solution have been investigated by ^{31}P n.m.r. spectroscopy.[2962] Polysilicate intermediates have been characterised by ^{29}Si INADEQUATE n.m.r. spectroscopy.[2963] Species present during the hydrolysis of $[Si(OMe)_4]$ have been characterised by ^{29}Si n.m.r. spectroscopy.[2964] ^{17}O n.m.r. spectra have been obtained from aqueous silicate solutions.[2965] Silica gel dissolution in aqueous alkali metal hydroxides has been studied by ^{29}Si n.m.r. spectroscopy.[2966] Polymers derived from the hydrolysis of $[Si(OEt)_4]$ have been characterised by ^{29}Si n.m.r. spectroscopy.[2967] The angular momentum correlation time has been determined for

2951 C. Janiak, M. Schwichtenberg, and F.E. Hahn, *J. Organomet. Chem.*, 1989, **365**, 37.
2952 J.T.B.H. Jastrzebski, P.A. Van der Schaaf, J. Boersma, G. Van Koten, M.C. Zoutberg, and D. Heijdenrijk, *Organometallics*, 1989, **8**, 1373.
2953 K. Jurkschat, C. Klaus, M. Dargatz, A. Tzschach, J. Meunier-Piret, and B. Mahieu, *Z. Anorg. Allg. Chem.*, 1989, **577**, 122.
2954 O.-S. Jung, J.H. Jeong, and Y.S. Sohn, *Polyhedron*, 1989, **8**, 2557.
2955 O.-S. Jung, J.H. Jeong, and Y.S. Sohn, *Polyhedron*, 1989, **8**, 1413.
2956 C.G. Schmid, R.O. Day, and R.R. Holmes, *Phosphorus Sulphur Silicon*, 1989, **41**, 69.
2957 K.C. Kumara Swamy, R.O. Day, and R.R. Holmes, *Phosphorus, Sulphur, Silicon*, 1989, **41**, 291.
2958 H. Puff and H. Reuter, *J. Organomet. Chem.*, 1989, **368**, 173.
2959 W. McFarlane and N.H. Rees, *Polyhedron*, 1989, **8**, 2047.
2960 C.T.G. Knight, R.J. Kirkpatrick, and E. Oldfield, *J. Chem. Soc., Chem. Commun.*, 1989, 919.
2961 Ē. Kupče, E. Liepiņš, I. Zicmane, and E. Lukevics, *J. Chem. Soc., Chem. Commun.*, 1989, 818.
2962 V.V. Malyarenko, N.F. Chuiko, V. Yu. Tretinnik, and S.V. Pakhovchishin, *Ukr. Khim. Zh. (Russ. Ed.)*, 1988, **54**, 686 (*Chem. Abstr.*, 1989, **110**, 14 279).
2963 W.G. Klemperer, V.V. Mainz, S.D. Ramamurthi, and F.S. Rosenberg, *Mater. Res. Soc. Symp. Proc.*, 1988, **121**, 15 (*Chem. Abstr.*, 1989, **110**, 119 924).
2964 R.A. Assink and B.D. Kay, *J. Non-Cryst. Solids*, 1988, **107**, 35 (*Chem. Abstr.*, 1989, **111**, 27 642).
2965 C.T.G. Knight, A.R. Thompson, A.C. Kunwar, H.S. Gutowsky, E. Oldfield, and R.J. Kirkpatrick, *J. Chem. Soc., Dalton Trans.*, 1989, 275.
2966 P.W.J.G. Wijnen, T.P.M. Beelen, J.W. De Haan, C.P.J. Rummens, L.J.M. Van de Ven, and R.A. Van Santen, *J. Non-Cryst. Solids*, 1989, **109**, 85 (*Chem. Abstr.*, 1989, **111**, 62 743).
2967 H. Inoue and T. Okada, *Toso Kenkyu Hokoku*, 1989, **33**, 39 (*Chem. Abstr.*, 1989, **110**, 213 733); L.W.

SiF_4 in liquid argon.[2968] T_1 and T_2 of ^{73}Ge in $[GeBr_4]$ and $[GeI_4]$ have been measured. T_1 is dominated by quadrupolar relaxation while T_2 is dominated by scalar coupling and quadrupolar relaxation.[2969] N.m.r. data have also been reported for $[Ge(NCS)_4(bipy)]$, (^{14}N, ^{73}Ge),[2970] $[RN(CH_2CH_2O)_2Ge(OCH_2CH_2O)]$, ($^{73}Ge$),[2971] $[Sn(NEt_2)_n(OEt)_{4-n}]$, (^{119}Sn),[2972] $[SnCl_2(imidazole)_2]$, (^{13}C, ^{119}Sn),[2973] $[CH_2(3,5\text{-}Me_2C_3HN_2)_2SnX_4]$, ($^{13}C$, ^{119}Sn),[2974] $[(2,4,6\text{-}Bu^t{}_3C_6H_2PH)_2SiX_2]$, (X = Cl, Br; ^{29}Si),[2975] $[Si_8O_{20}]^{8-}$, (^{29}Si),[2976] derivatives of silicic acid, (^{29}Si),[2977] $[(Me_4C_2O_2)_2SiOMe]^-$, (^{13}C, ^{29}Si),[2978] polyiodosiloxane, (^{29}Si),[2979] [Sn(diethyl tartrate)$_2$]$_2$, (^{119}Sn),[2980] $[Sb\{Sn(OR)_3\}_3]$, $[Pb\{Sn(OR)_3\}_2]$, (^{13}C, ^{119}Sn),[2981] $[Sn\{Sb(OR)_4\}_2]$, (^{13}C, ^{119}Sn),[2982] $[Sn(O_2C_6H_2Bu^t{}_2)X_2]$, ($^{13}C$),[2983] $[Pb(O_2CR)_2]$, (^{207}Pb),[2984] $[Pb_4O(OBu^t)_6]$, (^{207}Pb),[2985] $[(C_6H_4S_2)_2Ge]$, (^{13}C),[2986] $[SnCl_n(OH_2)_{6-n}]^{4-n}$, ($^{119}Sn$),[2987] and $[SnCl_mBr_nI_{3-m-n}]^-$, (^{119}Sn).[2988]

8 Group VB Elements

A review has appeared:- 'The reactions of phosphazenes with difunctional and polyfunctional nucleophilic reagents'.[2989]

The ^{14}N chemical shift, and T_1, and T_2 values of N_2 have been measured, and the

Kelts and N.J. Armstrong, *Mater. Res. Soc. Symp. Proc.*, 1988, **121**, 519 (*Chem. Abstr.*, 1989, **110**, 140 261).

2968 D.N. Shchepkin, N.S. Golubev, and S.M. Melikova, *Vestn. Leningr. Univ., Ser. 4: Fiz., Khim.*, 1989, 89 (*Chem. Abstr.*, 1989, **111**, 240 277).

2969 T. Harazono, K. Tanaka, and Y. Takeuchi, *Inorg. Chem.*, 1989, **28**, 1813.

2970 Ē. Kupče, L. Ignatovitch, and E. Lukevics, *Latv. PSR Zinat. Akad. Vestis, Kim. Ser.*, 1988, 625 (*Chem. Abstr.*, 1989, **110**, 68 528); Ē. Kupče, L.M. Ignatovich, and E. Lukevics, *J. Organomet. Chem.*, 1989, **372**, 189.

2971 Ē. Kupče, E. Lukevics, O.D. Fild, N.A. Viktorov, and T.K. Gar, *J. Organomet. Chem.*, 1989, **372**, 187.

2972 M. Froneman, T.A. Modro, and S.M. Vather, *Inorg. Chim. Acta*, 1989, **164**, 17.

2973 S. Vasnin and R.A. Geanangel, *Inorg. Chim. Acta*, 1989, **160**, 167.

2974 G.G. Lobbia, F. Bonati, A. Cingolani, D. Leonesi, and A. Lorenzotti, *J. Organomet. Chem.*, 1989, **359**, 21.

2975 K. Hassler, F. Mitter, and B. Reiter, *J. Organomet. Chem.*, 1989, **376**, C1.

2976 I. Hasegawa, S. Sakka, Y. Sugahara, K. Kuroda, and C. Kato, *J. Chem. Soc., Chem. Commun.*, 1989, 208.

2977 E. Bertling and H.C. Marsmann, *Phys. Chem. Miner.*, 1988, **16**, 295 (*Chem. Abstr.*, 1989, **111**, 186 126).

2978 R.R. Holmes, R.O. Day, and J.S. Payne, *Phosphorus Sulphur Silicon*, 1989, **42**, 1.

2979 W.N. Rowlands and P.L. Timms, *J. Chem. Soc., Chem. Commun.*, 1989, 1432.

2980 J. Augé and G. Bourleaux, *J. Organomet. Chem.*, 1989, **377**, 205.

2981 T. Athar, R. Bohra, and R.C. Mehrotra, *Synth. React. Inorg. Metal-Org.*, 1989, **19**, 195.

2982 T. Athar, R. Bohra, and R.C. Mehrotra, *Main Group Met. Chem.*, 1987, **10**, 399 (*Chem. Abstr.*, 1989, **110**, 106 914).

2983 T.A. Annan, B.R. McGarvey, A. Ozarowski, D.G. Tuck, and R.K. Chadha, *J. Chem. Soc., Dalton Trans.*, 1989, 439.

2984 C.G. Bazuin, D. Guillon, A. Skoulios, A.M. Amorim da Costa, H.D. Burrows, C.F.G.C. Geraldes, J.J.C. Teixeira-Dias, E. Blackmore, and G.J.T. Tiddy, *Liq. Cryst.*, 1988, **3**, 1655 (*Chem. Abstr.*, 1989, **110**, 163 969).

2985 R. Papiernik, L.G. Hubert-Pfalzgraf, and M.-C. Massiani, *Inorg. Chim. Acta*, 1989, **165**, 1.

2986 J. Pfeiffer, M. Noltemeyer, and A. Meller, *Z. Anorg. Allg. Chem.*, 1989, **572**, 145.

2987 X. Mao, X. You, and A. Dai, *Inorg. Chim. Acta*, 1989, **156**, 177.

2988 J.M. Coddington and M.J. Taylor, *J. Chem. Soc., Dalton Trans.*, 1989, 2223.

2989 R.A. Shaw, *Phosphorus Sulphur Silicon*, 1989, **45**, 103.

reorientation studied.[2990] One-bond ^{2}H isotope effects on ^{14}N and ^{15}N shielding of $[NH_4]^+$ has been investigated for different anions.[2991] The feasibility of ultra-high resolution ^{31}P n.m.r. spectra has been examined.[2992] Based on CNDO MO calculations for $[CH_3PH_2]$ and $[(H_3C)_3P]$, an equation was derived relating $^{1}J(^{31}P\text{-}^{13}C)$ with P—C—H bond angles.[2993]

The ^{13}C n.m.r. spectra of $[R_3As]$ and $[R_4As]^+$ have been discussed using the Hammett-Taft dual substitution parameter analysis.[2994] The ^{13}C n.m.r. spectra of $[N_3P_3Cl_{6-n}Ph_n]$ and $[N_3P_3Ph_2\{X(CH_2)_nY\}Cl_2]$ show temperature dependent multiplets due to second order effects. Inductive and resonance effects were calculated.[2995] Extensive use has been made of COSY to analyse the ^{31}P n.m.r. spectra of $[P_{10}Bu^t_6]$[2996] and $[P_{12}Bu^t_8]$.[2997] N.m.r. data have also been reported for $[F_3CPHD]$, (^{13}C),[2998] $[As(CH_2Bu^t)_nBr_{3-n}]$, (^{13}C),[2999] (85), (^{13}C),[3000] $[(4\text{-}MeC_6H_4)_3SbC_6H_4C(CF_3)_2O]$, ($^{13}C$),[3001] $[Bi(C_6H_4F\text{-}4)_3Cl_2]$, ($^{13}C$),[3002]

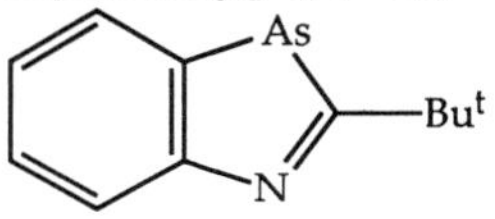

(85)

$[(C_6F_5)_3BiF_2]$, (^{13}C),[3003] $[Ph_4P_2N_3MeSPh]^+$, (^{13}C),[3004] $[1,5\text{-}Ph_4P_2N_4SMe]^+$, ($^{13}C$),[3005] $[P_3N_3Cl_4\{CH(CN)_2\}_2]$, ($^{13}C$),[3006] $[Ph_2P(NSeMeN)_2PPh_2]$, (^{77}Se),[3007] $[NaAs(CF_3)_2]$, (^{13}C),[3008] $[(RO)_2PS_2SbPh_2]$, (^{13}C),[3009] $[P_4(C_5Me_5)_2]$, (^{13}C),[3010] $[P_{14}Pr^i_4]$, (^{13}C),[3011] $[P_4C_4Bu^t_4]$, (^{13}C),[3012] $[P_2C_3Bu^t_3]^-$, (^{13}C),[3013] poly(methylphenylphosphazene), (^{13}C),[3014] [C(OPh)=NP(OPh)$_2$=N-

2990 D.D. McIntyre, A.W. Apblett, P. Lundberg, K.J. Schmidt, and H.J. Vogel, *J. Magn. Reson.*, 1989, **83**, 377.
2991 P.E. Hansen and A. Lycka, *Acta Chem. Scand.*, 1989, **43**, 222.
2992 S.R. Maple, J.E. Carson, and A. Allerhand, *J. Am. Chem. Soc.*, 1989, **111**, 7293.
2993 Yu.Yu. Samitov and R.K. Safiullin, *Zh. Obshch. Khim.*, 1988, **58**, 983 (*Chem. Abstr.*, 1989, **110**, 212 926).
2994 C.L. Watkins, L.K. Krannich, R.K. Kanjolia, and D.K. Srivastava, *Magn. Reson. Chem.*, 1989, **27**, 616.
2995 W.F. Deutsch, H.G. Parkes, and R.A. Shaw, *Magn. Reson. Chem.*, 1989, **27**, 207.
2996 M. Baudler, Ch. Wiaterek, and K. Kazmierczak, *Z. Anorg. Allg. Chem.*, 1989, **579**, 7.
2997 M. Baudler, M. Schnalke, Ch. Wiaterek, S. Opiela, and J. Hahn, *Z. Anorg. Allg. Chem.*, 1989, **578**, 7.
2998 J. Grobe, D. Le Van, and R. Demuth, *J. Fluorine Chem.*, 1988, **39**, 385 (*Chem. Abstr.*, 1989, **111**, 23 597).
2999 J.C. Pazik and C. George, *Organometallics*, 1989, **8**, 482.
3000 J. Heinicke, *J. Organomet. Chem.*, 1989, **364**, C17.
3001 Y. Yamamoto, H. Fujikawa, H. Fujishima, and K.-y. Akiba, *J. Am. Chem. Soc.*, 1989, **111**, 2276.
3002 A. Schmuck and K. Seppelt, *Chem. Ber.*, 1989, **122**, 803.
3003 W. Tyrra and D. Nasumann, *Can. J. Chem.*, 1989, **67**, 1949.
3004 T. Chivers, J. Fait, and S.W. Liblong, *Inorg. Chem.*, 1989, **28**, 2803.
3005 T. Chivers, G.Y. Denes, S.W. Liblong, and J.F. Richardson, *Inorg. Chem.*, 1989, **28**, 3683.
3006 H. Köhler, S. Ahmed, L. Jäger, and A. Kolbe, *Z. Anorg. Allg. Chem.*, 1989, **576**, 197.
3007 T. Chivers, D.D. Doxsee, and J.F. Fait, *J. Chem. Soc., Chem. Commun.*, 1989, 1703.
3008 R. Minkwitz and A. Liedtke, *Inorg. Chem.*, 1989, **28**, 1627.
3009 R. Karra, Y.P. Singh, and A.K. Rai, *Phosphorus Sulphur Silicon*, 1989, **45**, 145.
3010 P. Jutzi, R. Kroos, A. Müller, and M. Penk, *Angew. Chem., Int. Ed. Engl.*, 1989, **28**, 600.
3011 M. Baudler, H. Jachow, B. Lieser, K.-F. Tebbe, and M. Fehér, *Angew. Chem., Int. Ed. Engl.*, 1989, **28**, 1231.
3012 T. Wettling, J. Schneider, O. Wagner, C.G. Kreiter, and M. Regitz, *Angew. Chem., Int. Ed. Engl.*, 1989, **28**, 1013.
3013 A.H. Cowley and S.W. Hall, *Polyhedron*, 1989, **8**, 849.
3014 P. Wisian-Neilson and R.R. Ford, *Macromolecules*, 1989, **22**, 72 (*Chem. Abstr.*, 1989, **110**, 58 219); P. Wisian-Neilson, M.S. Islam, S. Ganapathiappan, D.L. Scott, K.S. Raghuveer, and R.R. Ford,

$P(OPh)_2{=}N]_n$, (^{13}C),[3015] $[Ph_3PC(CN)CNPClRP(NPr^i{}_2)CPPh_3]^+$, ($^{11}B$, ^{13}C),[3016] polyphosphazene electrolytes, (^{13}C),[3017] [PhPSePPhSePPh], (^{77}Se),[3018] arsenical sulphydryl adducts, (^{13}C),[3019] $[Pr^i{}_2NAsP(PPh_3)C(NPr^i{}_2)NPClNPr^i{}_2]^+$, ($^{11}B$, ^{13}C),[3020] $[PrSb]_5$, (^{13}C),[3021] and $[Bu^tPN{=}S{=}NPR^1R^2]$, (^{13}C).[3022]

The collapse of the P_B parts of the ^{31}P n.m.r. spectra of α-$[P_4S_3(^{14}NCS)_2]$, α-$[P_4S_3(^{14}NCS)I]$, and α-$[P_4S_3(^{14}NCS)Cl]$ has been attributed to ^{14}N relaxation.[3023] The structure of N_2O_5 in HNO_3 has been studied using ^1H, ^{14}N, and ^{17}O n.m.r. spectroscopy.[3024] N.m.r. data have also been reported for $[N_4P_4Cl_{8-2n}\{O(CH_2)_mO\}_n]$, ($^{13}C$),[3025] $[O{=}CNMePCl_3NCCl_2CHCl_2]$, ($^{13}C$),[3026] $[PhN\{P(O)ClNPh\}_2P(O)Cl]$, ($^{13}C$),[3027] 1,3,2-benzothiazarsolium tetrachloroaluminate, (^{13}C),[3028] $[Bi(NPh_2)_3]$, (^{13}C),[3029] esters of phosphorus acids, (^{13}C, ^{17}O),[3030] $[MSb(OR)_6]$, (^{13}C),[3031] $[Sb(O_2CCH_3)_2(S_2POCH_2CEt_2CH_2O)]$, ($^{13}C$),[3032] $[Bi(OBu^t)_3]$, (^{13}C),[3033] and $[PPAs_2Se_3]$, (^{77}Se).[3034]

9 Compounds of Groups VI and VII and Krypton and Xenon

A review has appeared:- 'Sulphur and selenium iodine compounds: from non-existence to significance'.[3035]

^{125}Te chemical shifts of tellurapyrans and related compounds have been discussed.[3036] ^{13}C and ^{77}Se T_1 data have been determined for CSe_2 at several magnetic fields to deter-

Macromolecules, 1989, **22**, 4382 (*Chem. Abstr.*, 1989, **111**, 215 051).
3015 I. Manners, H.R. Allcock, G. Renner, and O. Nuyken, *J. Am. Chem. Soc.*, 1989, **111**, 5478.
3016 H. Grützmacher and H. Pritzkow, *Chem. Ber.*, 1989, **122**, 1411.
3017 G. Nazri, D.M. MacArthur, and J.F. Ogara, *Chem. Mater.*, 1989, **1**, 370 (*Chem. Abstr.*, 1989, **110**, 216 231).
3018 P.T. Wood and J.D. Woollins, *Phosphorus Sulphur Silicon*, 1989, **41**, 51.
3019 R.J. O'Connor, E.L. McGown, K. Dill, and S.F. Hallowell, *Magn. Reson. Chem.*, 1989, **27**, 669.
3020 H. Grützmacher and H. Pritzkow, *Chem. Ber.*, 1989, **122**, 1417.
3021 M. Ates, H.J. Breunig, S. Güleç, W. Offermann, K. Häberle, and M. Dräger, *Chem. Ber.*, 1989, **122**, 473.
3022 M. Herberhold, S.M. Frank, and B. Wrackmeyer, *Z. Naturforsch., B*, 1988, **43**, 985 (*Chem. Abstr.*, 1989, **111**, 115 331).
3023 B.W. Tattershall, *J. Chem. Soc., Chem. Commun.*, 1989, 216.
3024 G.V. Lagodzinskaya, N.G. Yunda, E.P. Kirpichev, A.I. Kazakov, Yu.I. Rubstov, and V.O. Zavel'skii, *Khim. Fiz.*, 1989, **8**, 236 (*Chem. Abstr.*, 1989, **110**, 160 916).
3025 A.H. Alkubaisi and R.A. Shaw, *Phosphorus Sulphur Silicon*, 1989, **45**, 7.
3026 M.Yu. Dmitrichenko, V.G. Rozinov, and V.I. Donskikh, *Zh. Obshch. Khim.*, 1989, **59**, 715 (*Chem. Abstr.*, 1989, **111**, 194 899).
3027 O. Johnson, M. Murray, and G. Woodward, *J. Chem. Soc., Dalton Trans.*, 1989, 821.
3028 N. Burford, B.W. Royan, and P.S. White, *J. Am. Chem. Soc.*, 1989, **111**, 3746.
3029 W. Clegg, N.A. Compton, R.J. Errington, N.C. Norman, and N. Wishart, *Polyhedron*, 1989, **8**, 1579.
3030 V.V. Vasil'ev, V.E. Dmitriev, V. Mets, V.Ya. Komarov, V.V. Gavrilov, and B.I. Ionin, *Zh. Obshch. Khim.*, 1989, **59**, 821 (*Chem. Abstr.*, 1989, **111**, 194 907).
3031 T. Athar, R. Bohra, and R.C. Mehrotra, *Indian J. Chem., Sect. A*, 1989, **28A**, 302.
3032 S.K. Pandey, G. Srivastava, and R.C. Mehrotra, *Synth. React. Inorg. Metal.-Org. Chem.*, 1989, **19**, 795.
3033 W.J. Evans, J.H. Hain, jun., and J.W. Ziller, *J. Chem. Soc., Chem. Commun.*, 1989, 1628.
3034 B.W. Tattershall, R. Blachnik, and H.P. Baldus, *J. Chem. Soc., Dalton Trans.*, 1989, 977.
3035 T. Klapoetke and J. Passmore, *Acc. Chem. Res.*, 1989, **22**, 234.
3036 M.R. Detty, W.C. Lenhart, P.G. Gassman, and M.R. Callstrom, *Organometallics*, 1989, **8**, 861.

mine the relative chemical shift anisotropy and spin rotation coupling constant.[3037] ^{1}H, ^{13}C, ^{15}N, and ^{77}Se chemical shifts and coupling constants have been determined for *N*,*N*-diethylbenzeneseleneamide and related compounds. Solvent effects were examined.[3038] N.m.r. data have also been reported for [Me_4Te], (^{13}C, ^{125}Te),[3039] [Me_2TeCF_3]$^+$, (^{13}C, ^{125}Te),[3040] (86), (^{13}C),[3041] metalla-2,5-dichalcogenacyclopentenes, (^{77}Se),[3042] [PhSeR], (^{13}C),[3043] [$Bu^t_2C=CN=NCBu^t_2Se$], (^{13}C, ^{77}Se),[3044] (87), (^{13}C),[3045] [$MeTeCH_2CH=CH_2$], (^{13}C,

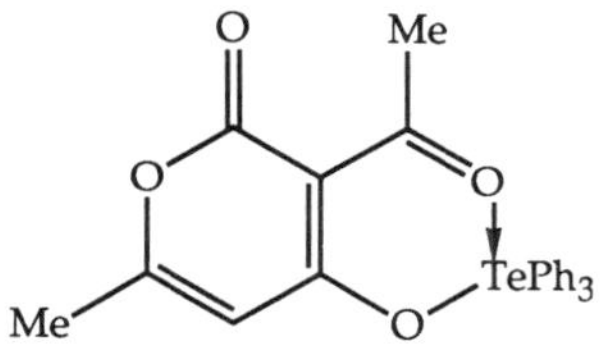

(86)

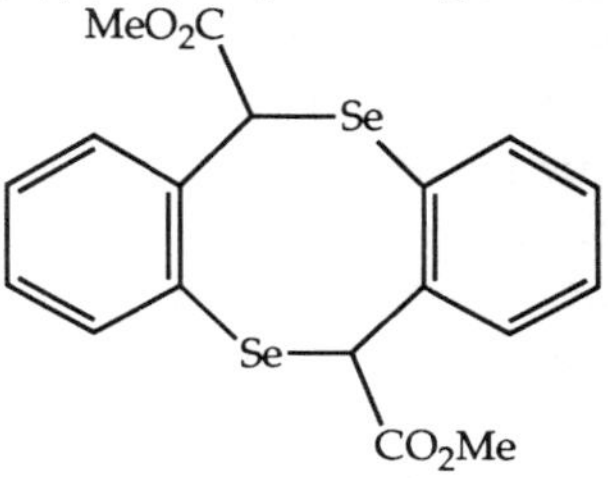

(87)

^{125}Te),[3046] [1,2-$(MeTe)_2C_6H_4$], (^{13}C, ^{125}Te),[3047] [MeTeC≡CTeMe], (^{13}C, ^{125}Te),[3048] [(2-RN=$CHC_6H_4)_2Te$], (^{125}Te),[3049] [$F_4S=CCF_2OSO_2$], (^{13}C),[3050] [$ArCH_2Se_nCH_2Ar$], (^{77}Se),[3051] (88), (^{13}C, ^{77}Se),[3052] (89), (^{13}C),[3053] [SeAr]$^-$, (^{77}Se),[3054] [$Se=C(NR^1R^2)_2$], (^{13}C, ^{77}Se),[3055] [$Se=CHC_6H_2Bu^t_3$-2,4,6], (^{13}C, ^{77}Se),[3056] coordination compounds with thioseleno and diseleno ligands, (^{125}Te),[3057] bis(di-isobutylselenocarbamoyl) triselenide, (^{13}C, ^{77}Se),[3058]

3037 A. Pulkkinen, J. Jokisaari, and J. Kowalewski, *J. Magn. Reson.*, 1989, **85**, 388.
3038 P. Ruostesuo, A.M. Hakkinen, R. Liias-Lepisto, and U. Salminen, *Spectrochim. Acta, Part A*, 1988, **44A**, 1105 (*Chem. Abstr.*, 1989, **110**, 127 305).
3039 R.W. Gedridge, jun., D.C. Harris, K.T. Higa, and R.A. Nissan, *Organometallics*, 1989, **8**, 2817.
3040 D. Naumann, J. Fischer, and B. Wilkes, *Z. Anorg. Allg. Chem.*, 1989, **576**, 225.
3041 K.M. Prasad, N.C. Rao, M.C. Ganorkar, and C.M. Chary, *J. Organomet. Chem.*, 1989, **376**, 53.
3042 T.M. Klapoetke, *Phosphorus Sulphur Silicon*, 1989, **41**, 105 (*Chem. Abstr.*, 1989, **111**, 115 267).
3043 M. Tiecco, L. Testaferri, M. Tingoli, and D. Bartoli, *Tetrahedron*, 1989, **45**, 6819; G. Pandey, J. Rao, and U.T. Bhalerao, *J. Chem. Soc., Chem. Commun.*, 1989, 416; C. Betancor, E.I. León, T. Prange, J.A. Salazar, and E. Suárez, *J. Chem. Soc., Chem. Commun.*, 1989, 450.
3044 R.H. Berg, N. Harrit, E. Larsen, and A. Holm, *Acta Chem. Scand.*, 1989, **43**, 885.
3045 S. Yamazaki, K. Kohgami, M. Okazaki, S. Yamabe, and T. Arai, *J. Org. Chem.*, 1989, **54**, 240.
3046 K.T. Higa and D.C. Harris, *Organometallics*, 1989, **8**, 1674.
3047 T. Kemmitt and W. Levason, *Organometallics*, 1989, **8**, 1303.
3048 R.W. Gedridge, jun., K.T. Higa, D.C. Harris, R.A. Nissan, and M.P. Nadler, *Organometallics*, 1989, **8**, 2812.
3049 I.D. Sadekov, A.A. Maksimenko, A.G. Maslakov, G.K. Makhrotra, G.S. Borodkin, O.E. Kompan, Yu.T. Struchkov, M.A. Fedotov, and V.I. Minkin, *Metalloorg. Khim.*, 1989, **2**, 298 (*Chem. Abstr.*, 1989, **111**, 232 217).
3050 R. Winter, D.H. Peyton, and G.L. Gard, *Inorg. Chem.*, 1989, **28**, 3766.
3051 F. Hansen, L. Henriksen, S. Larsen, and L. Teuber, *Acta Chem. Scand.*, 1989, **43**, 450.
3052 H. Fujihara, R. Akaishi, T. Erata, and N. Furukawa, *J. Chem. Soc., Chem. Commun.*, 1989, 1789.
3053 W. Mayer and H. Meier, *Chem. Ber.*, 1989, **122**, 509.
3054 J.M. Ball, P.M. Boorman, J.F. Fait, A.S. Hinman, and P.J. Lundmark, *Can. J. Chem.*, 1989, **67**, 751.
3055 A.M. Boccanfuso, D.W. Griffin, R.B. Dunlap, and J.D. Odom, *Bioorg. Chem.*, 1989, **17**, 231 (*Chem. Abstr.*, 1989, **111**, 152 955).
3056 R. Okazaki, N. Kumon, and N. Inamoto, *J. Am. Chem. Soc.*, 1989, **111**, 5949.
3057 W. Dietzsch, R.M. Olk, E. Hoyer, W. Meiler, and W. Robien, *Magn. Reson. Chem.*, 1989, **27**, 500.
3058 Y. Mazaki and K. Kobayashi, *Tetrahedron Lett.*, 1989, **30**, 2813.

[$RTeCl_3$], (^{13}C),[3059] oxatellurolylium halides, (^{125}Te),[3060] and [XTeAr], (^{13}C).[3061]

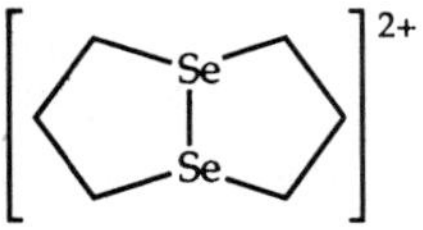

(88)

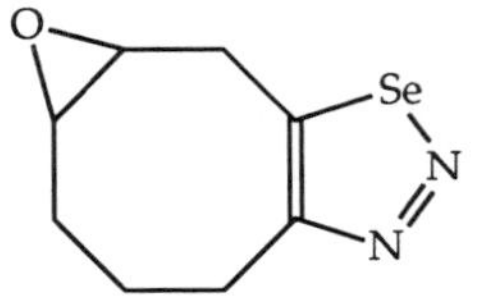

(89)

^{77}Se n.m.r. spectroscopy has been used to show that SeO_2, H_2SeO_3, and Na_2SeO_3 dissolve in 100% H_2SO_4 to yield $[SeO_2H]^+$, $[Se_2O_4H]^+$, $[Se_3O_6H]^+$, and possibly other higher polymeric species.[3062] ^{19}F and ^{125}Te n.m.r. spectroscopy has been used to study the interactions of $[TeF_5]^-$ with several solvents.[3063] N.m.r. data have also been reported for [$N_2S_2SeCl_2$], (^{14}N),[3064] [F_3CS=NSN], (^{13}C, ^{14}N),[3065] [(tosylN=S=N)$_2$S], (^{13}C),[3066] $[Se_2N_2S]^{2+}$, (^{77}Se),[3067] [$(Bu^t{}_2P)_2E$], (E = Se, Te; ^{13}C),[3068] [$F_5TeOOTeF_5$], (^{17}O),[3069] and [Se_nS_{8-n}], (^{77}Se).[3070]

Two dimensional multiple quantum ^{35}Cl n.m.r. spectra have been obtained for Cl^- in an oriented lyotropic liquid crystal.[3071] The ^{17}O n.m.r. spectrum of $[BrO_2]^-$ has been measured.[3072]

The n.m.r. frequency of polarised ^{83}Kr and ^{129}Xe has been measured in the presence of spin polarised alkali metal atoms.[3073] Solvent effects on the ^{129}Xe chemical shift of dissolved xenon gas have been interpreted.[3074] A ^{129}Xe n.m.r. spin-echo study of self-diffusion in xenon has been reported.[3075] N.m.r. data have also been reported for $[C_6F_5Xe]^+$, (^{129}Xe),[3076] and [$FXeOIOF_4$], (^{129}Xe).[3077]

3059 M.E.S. Ali, M.A. Malik, and B.C. Smith, *Inorg. Chim. Acta*, 1989, **162**, 157.
3060 M.R. Detty, W.C. Lenhart, P.G. Gassman, and M.R. Callstrom, *Organometallics*, 1989, **8**, 866.
3061 K.Y. Abid, N.I. Al-Salim, M. Greaves, W.R. McWhinnie, A.R. West, and T.A. Hamor, *J. Chem. Soc., Dalton Trans.*, 1989, 1697.
3062 R.J. Gillespie and R. Kapoor, *Indian J. Chem., Sect. A*, 1989, **28A**, 298 (*Chem. Abstr.*, 1989, **111**, 65 111).
3063 Yu.V. Kokunov, V.M. Afanas'ev, M.P. Gustyakova, and Yu.A. Buslaev, *Koord. Khim.*, 1989, **15**, 348 (*Chem. Abstr.*, 1989, **110**, 219 976).
3064 A. Apblett, T. Chivers, and J.F. Fait, *J. Chem. Soc., Chem. Commun.*, 1989, 1596.
3065 A. Haas and M. Rieland, *Z. Anorg. Allg. Chem.*, 1989, **571**, 82.
3066 R. Lux and G. Kresze, *Liebigs Ann. Chem.*, 1989, 605.
3067 A. Haas and J. Kasprowski, *Chimia*, 1988, **42**, 343 (*Chem. Abstr.*, 1989, **110**, 87 300).
3068 W.-W. du Mont, R. Hensel, W. McFarlane, I.J. Colquhoun, M.L. Ziegler, and O. Serhadli, *Chem. Ber.*, 1989, **122**, 37.
3069 J.C.P. Sanders and G.J. Schrobilgen, *J. Chem. Soc., Chem. Commun.*, 1989, 1576.
3070 R.S. Laitinen, P. Pekonen, Y. Hiltunen, and T.A. Pakkanen, *Acta Chem. Scand.*, 1989, **43**, 436; R. Steudel, D. Jensen, and M. Papavassiliou, *Phosphorus Sulphur Silicon*, 1989, **41**, 349.
3071 Y. Xu, T.M. Barbara, W.D. Rooney, and C.S. Springer, jun., *J. Magn. Reson.*, 1989, **83**, 279.
3072 W. Levason, J.S. Ogden, M.D. Spicer, M. Webster, and N.A. Young, *J. Am. Chem. Soc.*, 1989, **111**, 6210.
3073 S.R. Schaefer, G.D. Cates, T.R. Chien, D. Gonatas, W. Happer, and T.G. Walker, *Phys. Rec. A: Gen. Phys.*, 1989, **39**, 5613 (*Chem. Abstr.*, 1989, **111**, 165 943).
3074 M. Luhmer, A. Dejaegere, and J. Reisse, *Magn. Reson. Chem.*, 1989, **27**, 950.
3075 P.W.E. Peereboom, H. Luigjes, and K.O. Prins, *Physica A (Amsterdam)*, 1989, **156**, 260 (*Chem. Abstr.*, 1989, **111**, 121 296).
3076 H.J. Frohn and S. Jakobs, *J. Chem. Soc., Chem. Commun.*, 1989, 625; D. Naumann and W. Tyrra, *J. Chem. Soc., Chem. Commun.*, 1989, 47.
3077 R.G. Syvret and G.J. Schrobilgen, *Inorg. Chem.*, 1989, **28**, 1564.

10 Appendix

This appendix contains a list of papers in which the use of nuclei other than ^{1}H, ^{19}F, and ^{31}P, has been described. The nuclei are ordered by increasing atomic number and mass.

^{2}H 30, 41, 925, 1407, 1445, 1451, 1555, 1710, 1808, 1810, 1811, 1842, 1843, 1852, 1950, 1967, 1969, 1972, 1976, 1981, 2013, 2018, 2031, 2333, 2365, 2367, 2391, 2412, 2416, 2423, 2454, 2456, 2461, 2464, 2466, 2468, 2473, 2477-2479, 2481, 2672, 2673, 2696, and 2991.

^{3}H 2367.

^{3}He 2181 and 2506.

^{6}Li 14, 15, 18-20, 26, 33, 35, 1645, 2041, 2043, and 2744.

^{7}Li 13, 16, 19, 22, 25, 27, 29-32, 34-38, 62, 64, 67, 1137, 1425, 1427-1429, 1465, 1469, 1470, 1474-1476, 1487-1490, 1496-1499, 1502, 1581, 1582, 1807, 1827, 1882, 1946-1949, 1951, 1961, 1970, 1971, 1990, 2038, 2041-2044, 2075, 2105, 2108, 2116, 2120, 2122, 2161, 2311, 2527, 2530, 2729, 2744, 2795, and 2827.

^{8}Li 2214.

^{9}Be 1221 and 2096.

^{11}B 36, 174, 182, 134, 135, 331, 365, 369, 383, 384, 397, 552, 554, 558, 579, 697, 698, 724, 725, 754, 782, 805, 828, 883, 905-907, 913, 1002, 1075, 1112, 1144, 1226, 1242, 1388, 1389, 1392, 1603-1606, 1770-1774, 1844, 1990, 2022, 2052, 2102, 2211-2213, 2215, 2216, 2219, 2240-2242, 2244, 2288, 2519, 2524-2527, 2529-2538, 2540-2548, 2550-2568, 2570-2587, 2589, 2591-2599, 2601-2612, 2614-2622, 2624-2627, 2631-2642, 2644, 2645, 2649, 2650, 2776, 2802, 2824, 2842, 3016, and 3020.

^{12}B 2214.

^{13}C 5, 11, 12, 14, 15, 16, 18-24, 28, 32-34, 36-38, 45-55, 59-89, 92-128, 130-132, 134-162, 164-166, 168, 171-180, 182, 183, 185-189, 193, 194, 196, 198, 199, 204-265, 267-360, 363-372, 374-376, 380-399, 401-404, 406-415, 417-424, 426-438, 444-447, 449, 455-457, 464, 467-470, 472-477, 479-492, 494, 496-516, 518-529, 532-537, 547, 548, 551, 553, 555-557, 559-566, 572-598, 600-602, 604, 605, 607-649, 652-656, 659-705, 707-753, 755, 757-781, 784-795, 798-804, 807-818, 823, 826, 828-834, 838, 840-843, 845, 846, 848-858, 860-881, 884-905, 907-912, 914-932, 937, 940, 943-951, 953-970, 972-986, 989, 990, 994, 995, 997, 999-1034, 1037-1056, 1059-1066, 1068, 1070-1078, 1082-1091, 1093-1098, 1100, 1101, 1103, 1105, 1106, 1108, 1109, 1112-1122, 1125, 1131, 1132, 1136, 1137, 1138, 1141-1163, 1165-1172, 1174, 1176-1183, 1185-1192, 1194, 1198, 1200-1203, 1205, 1206, 1208, 1217-1221, 1223, 1226-1229, 1232-1235, 1238, 1239, 1243, 1246, 1247, 1249-1257, 1259, 1261-1263, 1265-1269, 1274, 1277, 1285, 1286, 1290, 1291, 1293, 1295, 1297, 1300-1305, 1307, 1310-1314, 1317, 1319-1325, 1327-1330, 1332, 1335, 1338, 1344, 1346, 1348-1350, 1352-1356, 1358, 1361, 1363, 1366, 1370, 1372, 1374, 1378, 1384, 1385, 1388, 1391, 1393-1396, 1398, 1399, 1402, 1403, 1405, 1408, 1410-1412, 1418, 1419, 1448, 1464-1468, 1470, 1471, 1477, 1479, 1483, 1484, 1508, 1523, 1528, 1532, 1537, 1543, 1544, 1548, 1558, 1567, 1569, 1576, 1580-1582, 1590, 1593, 1598, 1600, 1605, 1617, 1625, 1636, 1638, 1645, 1648-1650, 1653, 1655, 1657, 1661-1663, 1668, 1669, 1671, 1674, 1688, 1690, 1692, 1693, 1695-1697, 1702, 1705, 1706, 1708, 1709, 1711, 1713, 1714, 1718, 1721-1724, 1728, 1731, 1733, 1735, 1736, 1740, 1742, 1745, 1746, 1748, 1752-1754, 1756, 1759-1761, 1776, 1782, 1791, 1792, 1794, 1821, 1840, 1853, 1854, 1860, 1862, 1869-1871, 1873, 1874, 1879, 1886, 1887, 1891, 1909, 1977, 1978, 1994, 2002, 2022, 2026, 2053, 2071, 2103, 2119, 2134, 2138-2140, 2143, 2146, 2152-2155, 2157, 2158, 2160, 2174, 2176, 2177, 2182, 2183, 2185-2187, 2189, 2190, 2195, 2284, 2286, 2299, 2325, 2329, 2330, 2333, 2336, 2339-2341, 2344, 2345, 2347, 2350, 2351, 2374, 2398, 2404, 2406, 2408, 2416, 2437, 2438, 2441-2445, 2459, 2462, 2463, 2467, 2469, 2474, 2478, 2482, 2484, 2485, 2487, 2488, 2490, 2491, 2494, 2495, 2525, 2527-2531, 2535, 2538-2540, 2543-2545, 2549, 2551, 2553, 2559, 2571-2573, 2579, 2581-2594, 2596-2599,

^{54}Mn 2149.
^{55}Mn 461, 470, 2147, and 2148.
^{57}Fe 652, 2077, and 2078.
^{59}Co 41, 765, 859, 887, 905, 907, 934-936, 942, 952, 971, 988, 1563, 1737, and 2147.
^{63}Cu 1917, 2087, 2089, 2091, 2093-2096, 2098, 2148, 2190, 2196, and 2199.
^{65}Cu 2087, 2091, 2093, 2095, 2098, 2115, 2148, 2194, 2196, and 2199.
^{67}Zn 1552.
^{71}Ga 452, 453, 1457, 2237, 2326, 2677, and 2686.
^{73}Ge 1635, 2796, and 2969-2971.
^{77}Se 17, 133, 384, 459, 460, 777, 800, 1133, 1164, 1184, 1269, 1276, 1410, 1418, 1419, 1657, 1696, 1764, 1798, 2040, 2117, 2122, 2136, 2374, 2410, 2412, 2605, 2689, 2705, 2865, 2949, 2949, 3007, 3018, 3034, 3037, 3038, 3042, 3044, 3051, 3052, 3054-3056, 3058, 3062, 3067, and 3070.
^{79}Br 1638, 2021, and 2199.
^{81}Br 1602 and 2011.
^{83}Kr 3073.
^{87}Rb 42, 1474, 1503, 1511-1513, 1957, 2045, and 2141.
^{89}Y 55, 67, 81, 1436, 2095, 2096, and 2099.
^{93}Nb 184.
^{95}Mo 4, 330, 373, 376, 378, 1271, 1542, 1552, 1698, 2126, 2142, and 2439.
^{97}Mo 58.
^{99}Tc 462, 530, 531, 538, and 539.
^{103}Rh 836, 837, 839, 864, 883, 907, 911, 929, 974, 975, and 986.
^{107}Ag 1138.
^{109}Ag 1336 and 1985.
^{113}Cd 1173, 1184, 1185, 1193-1199, 1206, 1207, 1517, 1519, 1601, 1683, 2205, 2207, and 2451.
^{115}In 2236.
^{117}Sn 2377, 2870, and 2873.
^{119}Sn 242, 369, 465, 567, 605, 655, 765, 844, 1035, 1058, 1297, 1370, 1405, 1411, 1636, 1694, 1695, 1793-1795, 2083, 2190, 2373-2377, 2380-2382, 2450, 2496, 2545, 2586, 2599, 2636, 2656, 2687, 2695, 2717, 2732, 2738, 2776, 2802-2807, 2809, 2811, 2814, 2816-2822, 2825, 2839, 2841, 2870-2873, 2878, 2879, 2881, 2882, 2884-2887, 2889, 2891, 2892, 2894, 2895, 2905, 2908, 2919, 2920, 2922, 2925-2928, 2930-2933, 2935-2938, 2952-2957, 2959, 2972-2974, 2980-2982, 2987, and 2988.
^{125}Te 393, 454, 706, 777, 800, 1134, 1204, 1420, 1642, 1762, 1799, 1800, 2374, 2736, 2865, 3036, 3039, 3040, 3046-3049, 3057, 3060, and 3063.
^{127}I 1433, 1951, 2046, 2199, and 2393.
^{129}Xe 2487, 2492, 2507-2518, 2614, and 3073-3077.
^{133}Cs 43, 44, 1425, 1474, 1480, 1514, 1515, 1827, 1957, 1996, 2032, 2038, 2171, 2179, 2194, 2212, and 2331.
^{137}Ba 2110.
^{139}La 1437, 1526, 1527, 1917, 2080, and 2088.
^{141}Pr 2084.
^{153}Eu 2086.
^{165}Ho 2072 and 2074.
^{171}Yb 56 and 74.
^{183}W 259, 266, 398, 428, 442, 443 and 451.
^{187}Os 571.
^{195}Pt 229, 248, 251-253, 286, 348, 420, 996, 998, 1035, 1053, 1058, 1059, 1069, 1076, 1078, 1084, 1099-1105, 1107, 1109-1111, 1114, 1123, 1124, 1126-1131, 1134, 1218, 1349, 1362, 1364, 1369, 1370, 1381, 1449, 1575, 1576, 1677, 1756, 1760, 1762, 1764, 1944, and 2447-

2450.

^{199}Hg 569, 599, 980, 1035, 1168, 1169, and 1175.

^{203}Tl 2100 and 2238.

^{205}Tl 1262, 1425, 1522, 1626-1628, 2056, 2057, 2100, 2132, 2238, 2239, 2652, and 2653.

^{207}Pb 1637, 2239, 2376, 2388, 2551, 2584, 2599, 2738, 2776, 2802, 2804, 2824, 2825, 2870, 2984, and 2985.

^{235}U 57.

2
Nuclear Quadrupole Resonance Spectroscopy

BY K. B. DILLON

1 Introduction

This chapter reports on the pure nuclear quadrupole resonance (n.q.r.) spectra of quadrupolar ($I > \frac{1}{2}$) nuclei in inorganic or organometallic solids. Interest in n.q.r. investigations of higher-temperature superconductors continues unabated, with more than 60 papers on transition metals and lanthanides published this year. N.q.r. and n.m.r. studies of magnetism and superconductivity in high-T_c oxides have been extensively reviewed,[1] as have recent n.q.r. and n.m.r. results for high-T_c oxide superconductors,[2] and $^{63,65}Cu$ n.q.r. (and ^{17}O n.m.r.) data for $La_{1.85}Sr_{0.15}CuO_4$ and $YBa_2Cu_3O_7$.[3] Ligand influence in octahedral alkyltin(IV) halide complexes, as deduced mainly from halogen n.q.r. frequencies and also from X-ray data, has been reviewed.[4] More specialised reviews have appeared on n.q.r. double resonance methods (using ^{17}O and ^{14}N nuclei) for the study of ferroelectric phase transitions, exemplified by work on $PbHPO_4$ and diglycine nitrate,[5] the use of n.q.r. for studying incommensurate and commensurate ferroelectrics, illustrated by literature results for $C_8H_{17}NH_3Cl$ and $MeNH_3HgCl_3$,[6] and n.q.r. studies of incommensurate phases under high pressure, exemplified by published work on the pressure (p)-dependence of bromine resonance from Rb_2ZnBr_4 and arsenic resonance from proustite, Ag_3AsS_3.[7] A variational correlated ground state wavefunction theory has been developed for predicting and interpreting the p-dependence of the n.q.r. frequency and isotopic shift in deuterated and undeuterated H-bonded crystals.[8] The predictions of the theory were compared with experimental results from the literature for several H-bonded compounds, and isotope effects on the n.q.r frequency shift and pressure coefficients were discussed. Two patent applications involving n.q.r. have been reported, one on methods and apparatus for detecting narcotics, explosives and other compounds containing n.m.r.-active and n.q.r.-active nuclei and capable of exhibiting double resonance,[9] and one on a method of identifying n.q.r. spectra.[10]

The normal format is adopted in the more detailed sections, *i.e.* results for main group elements (Groups I, III, V, VI and VII), followed by those for transition metals and lanthanides.

2 Main-group Elements

2.1 Group I (Sodium-23 and Potassium-39). - ^{17}O resonance frequencies and lineshapes in natural abundance at 150 K have been measured by a double resonance technique from polycrystalline samples of $(MeCOO)_2KH$ and $(MeCOO)_2NaH$.[11] In the sodium compound, only one chemically inequivalent O site was found, showing that the O-H--O H-bonds are symmetric. Signals were not observed from the non-H-bonded O sites. In the potassium compound, two inequivalent O sites were detected, showing an asymmetric H-bond with different OH distances. The ^{23}Na and ^{39}K resonance frequencies at 150 K were also measured. Two signals were found for the sodium compound, in agreement with X-ray data showing two inequivalent sites. A single frequency only was observed for the potassium compound, showing that all the K sites were crystallographically equivalent within the experimental resolution.

2.2 Group III (Boron-11 and Gallium-69 and -71). - Two-dimensional spin echo double resonance (SEDOR) has been used to obtain the ^{11}B and ^{35}Cl n.q.r. frequencies at 77 K from the mixed boron halides BCl_2Br and $BClBr_2$, which only exist in equilibrium with each other and with the binary halides BCl_3 and BBr_3.[12] The continuous decrease in the ^{11}B and ^{35}Cl frequencies in the sequence $BCl_3 > BCl_2Br > BClBr_2$ was explained in terms of intramolecular variations of electron density, caused by the differing electronegativity of Cl and Br.

Gallium and chlorine n.q.r. (including isotope studies to confirm the assignments) of Ga_2Cl_6 and some $GaCl_3$ complexes with ketones $R_1R_2C{=}O$, including temperature (T)-dependence measurements from 77 to 303 K, have been recorded.[13] The ligands coordinate through oxygen, and the degree of charge transfer could be estimated from the frequency shifts. Most of the complexes gave three ^{35}Cl resonances, except for that with methyl 2-naphthyl ketone, which gave two resonances in a 2:1 intensity ratio. Two types of complex could be distinguished, depending on the splitting of the ^{35}Cl resonances and the T-dependence behaviour of the ^{69}Ga resonance, which showed an anomalous (positive) T-coefficient in some of the complexes. This behaviour was associated with the asymmetry of the electronic environment of the Ga atom, and depended on the symmetry of the donor molecule. N.q.r. (Ga and/or halogen) frequencies at 77 K have been measured for 1:1 complexes of $GaCl_3$ with MeCN, $ClCH_2CN$, EtCN, PhCN, p-MeC_6H_4CN, Pr_3N, dimethylcyclohexylamine and dimethylaniline, for 1:1 complexes of $AlBr_3$ with Pr_3N, MeCN, $ClCH_2CN$, EtCN, PhCN, p-MeC_6H_4CN and p-BrC_6H_4CN, and for the 2:1 complexes of $GaCl_3$ and $AlBr_3$ with terephthalonitrile.[14] T-dependence studies on $GaCl_3$ complexes with Pr_3N and PhCN to 300 K were also reported. The main contribution to the electric field gradient (e.f.g.) at the halogen atom was deduced to be intramolecular, not crystallographic. A correlation was established between the multiplicity of the n.q.r. spectra and the symmetry of the donor molecule. Electron distribution in the complexes was discussed on the basis of the results, and a semi-empirical approach was developed through which the orbital populations for the central metal atom after complexation could be calculated.

2.3 Group V (Nitrogen-14, Arsenic-75, Antimony-121 and -123, and Bismuth-209). - The T-dependence of the ^{14}N resonances from NH_4NO_3, where observable, has been determined from 223 to 383 K, covering phases II (high temperature), III, IV and V (below 257 K), using a double resonance technique.[15] No distinct ^{14}N n.q.r. frequency was observed from the NH_4^+ ions in phase II, which were deduced to be undistorted within the experimental resolution. A single frequency only was seen from the NO_3^- ion, indicating that the asymmetry parameter $\eta = 0$ and that the ion was of trigonal symmetry. In phase III the NO_3^- ion was still trigonal, but the NH_4^+ ion was strongly distorted. In phase IV the quadrupolar coupling constant for N in NO_3^- was smaller than in phases III or V, while the coupling constant for N in NH_4^+ was higher than in phase III, in agreement with the proposed crystal structure. In phase V, three frequencies were observed from NO_3^- and three from NH_4^+, indicating crystallographic equivalence and disagreeing with the suggested crystal structure. T-dependence studies from 353-473 K of the intensity, frequency and lineshape of ^{14}N resonance from potassium tetracyanoquino-dimethane (K-TCNQ) showed a first order transition at 403 K on heating, with hysteresis.[16] Above 403 K, the high temperature phase containing fluctuating charge-density waves became incommensurate, transformed into a multisolitonic phase with narrow solitons above 414.5 K, and near the decomposition temperature changed towards the single-soliton limit. On cooling from above 423 K, K-TCNQ transformed directly from the multisolitonic phase to the high temperature commensurate phase at 414.4 K, that coexisted with the low temperature phase from 401 K and only disappeared below 363 K. The previously undetected phase transition at 414.5 K was attributed to a change in the effective dimensionality in the ordering of the charge-density waves along the TCNQ stacks. The results indicated that the Peierls transition was probably a complex two-step process.

The possibilities of variation of the signal/noise ratio in pulsed n.q.r. for I=1 nuclei with saturation of the quadrupole system at an adjacent transition have been examined, and calculations have been carried out for the two limiting cases.[17] The results were compared with experimental data for ^{14}N in $NaNO_2$ at 77 K, and good agreement was obtained. A convenient tuning method for n.m.r./n.q.r. spectrometers has been described, that uses the piezoelectric resonance from quartz crystals.[18] The method was applied to the determination of the ^{14}N resonance frequency in $NaNO_2$. An n.q.r. spectrometer operating from 2-50 MHz. has been constructed by modification of a Robinson type spectrometer.[19] Its use was demonstrated by detecting the ^{14}N signal from $NaNO_2$ under pressure, and the ^{35}Cl signal from $CrCl_3$, at 292 K.

^{75}As n.q.r. measurements at 77 K on betaine arsenate, $Me_3NCH_2COO.H_3AsO_4$, which is ferroelectric below T_c 119 K, have shown two resonances, attributed to two crystallographically inequivalent sites for the As atoms.[20] An increase in T caused a rapid decrease in signal intensity, and these faded out above 100 K, probably because of stochastic modulation of the e.f.g. at each As site by reorienting Me groups in the betaine moieties. A single ^{75}As resonance was detected in the high temperature phase (above 160 K). The e.f.g. and apparent ν_Q at the As site were calculated by a point charge model for three plausible sets of proton configurations at room temperature, and were compared with the experimental data. Possible proton configurations above and below T_c were suggested from the results. T-dependence studies on ^{75}As n.q.r. in CsH_2AsO_4 below the phase transition temperature T_c (143 K) have shown the occurrence of a second, weaker line at higher frequency, as well as the strong resonance associated with the ferroelectric phase of the crystal.[21] Below 110 K only the lower frequency line was observed. The higher frequency line was attributed to antiferroelectric ordering, with two phases coexisting between 110 and 143 K, accompanied by an anomaly in the spin-lattice relaxation time T_1. ^{75}As n.q.r. at 77 K has been used to probe the bonding arrangements in the metal chalcogenide glass systems Cu-As-S and Cu-As-Se.[22] In $Cu_x(As_{0.4}Z_{0.6})_{1-x}$ (Z = S or Se), the measurements were consistent with the appearance of As-As bonds, the density of which increased with x. For $(Cu_{0.67}Z_{0.33})_y(As_{0.4}Z_{0.6})_{1-y}$, no evidence was found for the presence of As-As bonds from n.q.r. The results were consistent with a model proposed to explain the local structural order in these materials.

^{121}Sb and ^{123}Sb n.q.r. frequencies, mainly at 77 K but including a few values at 298 K, have been measured for the fluoroantimonates(III) $(CN_4H_7)_2SbF_5$, $(CN_4H_7)Sb_2F_7$, $RSbF_4$ (R = CN_4H_7, Bu_4N or Li), $RSbF_3Cl$ (R = CN_3H_6 or CN_4H_7), $(CN_3H_6)SbF_3(NO_3)$, $RSbF_3(SCN)$ (R = CN_3H_6, NH_4 or Cs), $(CN_4H_8)SbF_3(SO_4)$, $MSbF_2SO_4$ (M = Rb or Cs), $(CN_3H_6)SbF_2C_2O_4.H_2O$, $RSbF_2(C_6H_6O_7)$ (R = K, Rb or NH_4), and $SbF_3.C_3H_7ON$.[23] The asymmetry parameter η and quadrupolar coupling constant were evaluated in each case. R_2SbF_5 with R = aminoguanidinium contained two non-equivalent $SbF_5{}^{2-}$ polyhedra. Nuclear electric hexadecapole interaction terms have been derived from accurate n.q.r. frequency measurements near 273 K on $^{121,123}Sb$ nuclei in Sb_2S_3, and ^{93}Nb nuclei (two crystal sites) in NbF_5.[24] The theory of n.q.r. multiple pulse spin-locking as a method of studying molecular motions in solids has been developed.[25] Experimental measurements on ^{123}Sb nuclei in $SbCl_3$, and ^{35}Cl nuclei in $P(CCl_3)Cl_4$, showed that the theory was confirmed for librations and reorientations. An unexpected T-independence of the multiple pulse spin-locking relaxation time was observed, however, in the range of ultra-slow molecular reorientations, requiring further theoretical and experimental examination. ^{35}Cl n.q.r. spectra at 77K have been recorded for three tricyclic molecules of known structure containing $SbCl_3$ fragments, and several analogous compounds have been shown to have similar structures by a combination of ^{35}Cl and $^{121,123}Sb$ n.q.r. at 77 K.[26] The T-dependence of the ^{35}Cl frequencies from 77 K to room temperature was also determined for some of the compounds; all the resonances showed normal (negative) T-coefficients. The ^{35}Cl n.q.r. frequencies for compounds containing $SbCl_3$ fragments were found to be less sensitive to bond length than those for related complexes containing $SbCl_4$ groups. Other results for ^{121}Sb nuclei are described in the sub-section on Group VII.[27]

^{209}Bi and ^{63}Cu n.q.r. have been recorded for various samples of $Bi_2Sr_2Ca_{n-1}Cu_nO_{4+2n}$ (n = 2 or 3), obtained by varying the Sr/Ca ratio, or by partially substituting Bi by Pb.[28] All the samples showed a strong ^{63}Cu resonance around 22 MHz. as part of a complex spectrum with a large frequency spread (18-28 MHz.). ^{209}Bi showed a very broad resonance at 140 K between 42 and 85 MHz. The n.q.r. spectroscopic properties were correlated with the observed features of X-ray powder diffraction patterns, and with the onset of superconductivity.

<u>2.4 Group VI (Oxygen-17)</u>. - Results for this nucleus have been described in the sub-section on Group I.[11]

<u>2.5 Group VII (Chlorine-35 and -37, Bromine-79 and -81, and Iodine-127)</u>. - Evidence from ^{35}Cl n.q.r. frequency shifts and splittings at 77 K, among other techniques, has shown that CH_2Cl_2 and $CH_2Cl.CH_2Cl$ can act as coordinating solvents in the compounds $Ag_2(CH_2Cl_2)_4Pd(OTeF_5)_4$ (I), where CH_2Cl_2 acts as a bidentate ligand, $[AgOTeF_5(CH_2Cl.CH_2Cl)]_2$ (II), and $[AgOTeF_5(CH_2Cl_2)]_2$ (III).[29] The X-ray structure of I at 253 K was also determined. The T-dependence of both ^{37}Cl resonances from $HgCl_2$ has been followed from 77 to 310 K.[30] The results were interpreted on the basis of the theories of Bayer-Kushida, Brown and Brunetti. Rapid cooling and tempering of molten $PCl_4(NMe)_2CCCl_3$ (II), which has six-coordinate phosphorus, led to the formation of the isomer $PCl_3(NMe)_2C(Cl)CCl_3$ (I), with five-coordinate phosphorus.[31] This was confirmed by recording the ^{35}Cl n.q.r. spectra and relaxation times of both isomers at 77 K. The nitrogen atoms in the four-membered heterocycle were deduced to occupy axial and equatorial positions in the tbp in I. Isomer I was stable at 77 K, but heating to 293 K for 15 min. showed some formation of II. Repetition of this heating cycle led to an increase in the amount of II, and a further repeat produced complete conversion. The T-dependence of ^{35}Cl n.q.r. for $SnCl_4.2Me_2O$ from 77-350 K has shown the existence of three crystal modifications.[32] One form, obtained by

recrystallisation, gave a wide separation of the two resonances observed, and these faded out at 350 K. It was assigned a cis structure. Form 2, prepared by sublimation above 323 K, gave one line only where observable, and therefore had a trans configuration. Form 2 was unstable at room temperature, and changed to another modification (form 3) overnight, which also yielded one resonance only and therefore had a trans structure. Form 2 was converted immediately to form 3 by cooling to near 270 K. When form 2 was cooled to 77 K, absorption lines for form 1 appeared. Form 3 was also unstable, being transformed to form 1 in 1 d. at room temperature and immediately below 210 K. $SnCl_4.2Et_2O$ showed two close lines which disappeared near 280 K, and a trans structure was assigned; this was confirmed by Zeeman experiments on single crystal samples. The T-dependence of the ^{35}Cl n.q.r. from $SnCl_4.2THF$, which is known to have a trans geometry in the solid, was also shown. The Zeeman effect was observed on single crystals of cis-$SnCl_4.2Me_2O$ and trans-$SnCl_4.2Et_2O$ at room temperature and 235 K respectively, enabling information on the bond angles and bond character to be obtained, and the η values to be determined.

A new 2-D n.q.r. experiment has been described, in which the first time period (t_1) is the duration of the RF exciting pulse, and the second (t_2) is the normal free precession of a quadrupolar nucleus at zero field.[33] After double Fourier transformation, the result was a 2-D spectrum, in which the first frequency dimension was the nutation spectrum for the quadrupolar nucleus at zero field. The method could be applied to single crystal and polycrystalline samples; for the latter, η values could be obtained. Theoretical spectra and some experimental examples were presented, including $NaClO_3$ single crystal (the experiment failed with powdered $NaClO_3$), powdered $KClO_3$, $NaClO_2$, $NaClO_2.3H_2O$ and various organochloro-compounds. The reliability of the method and its experimental requirements were discussed. A multimode averager with a built-in hardware Fourier transform for pulsed n.q.r. or n.m.r. spectrometers has been described, and its n.q.r. use illustrated by recording the ^{35}Cl signal from $Pb(ClO_3)_2$ at room temperature.[34] The advantages of the system were discussed. A computer controlled pulse programmer for pulsed n.q.r. experiments, including relaxation time measurements, has been reported, and has been used to obtain ^{35}Cl data from $[Me_2NH_2]_2SnCl_6$.[35] The n.q.r. responses of ^{35}Cl nuclei in $KClO_3$ and $NaClO_3$ powders to a spin-locking sequence (SLS) and to a phase-alternated multiple-pulse sequence (PAPS) have been investigated experimentally.[36] SLS yielded spin-locking for arbitrary values of the pulse angle θ and the phase ϕ as long as $\tau \leq T_2$, and the resonance lines were found to be narrowed. The observation that spin-locking was insensitive to ϕ in powder specimens could not be explained on the basis of existing theories of multiple-pulse spin-locking.

Line narrowing was also observed with PAPS for arbitrary ϕ values and $\tau \leq T_2$. The multiple-echo decay time constants in this case were more sensitive to θ than for SLS.

Two independent T-dependence studies on T_1 values for ^{35}Cl nuclei in Rb_2ZnCl_4, which has known phase transitions at 302 K, 192 K and 75 K, have been reported.[37,38] The T-dependence of T_1 in the paraelectric, incommensurate and commensurate phases was examined for both amplitude and Goldstone modes.[37] The results agreed with theoretical predictions. It was deduced that the bulk of the incommensurate phase consisted of very broad solitons, which were very well approximated by the plane-wave modulation model. An increase in the phason gap close to the lock-in transition temperature T_c was observed, and was attributed to multisoliton lattice effects. The T-dependence of T_1 from 393 to 153 K has been measured.[38] In the incommensurate phase, the phason and amplitudon components were determined separately. The theory of Blinc was extended to include contributions to $1/T_1$ from terms both linear and quadratic in the modulation wave amplitude , for the special case that the linear term was smaller than the quadratic term. The phason gap in the incommensurate region was estimated from the data. The results were in good agreement with those from other methods. The T-dependence from 173 to 313 K of the n.q.r. frequency and T_1 for ^{35}Cl nuclei in $[Rb_{1-x}(NH_4)_x]ZnCl_4$ (x = 0.00, 0.01 or 0.04) has been determined.[39] The results clearly showed that close to the lock-in transition temperature T_c, T_1 became T-dependent, reflecting the T-dependence of the acoustic-like branch in the multisoliton lattice. For higher x, the steps in the plot of T_1 as a function of T showed that the wave vector q locked for certain temperature intervals. The phason gap of the optic-like branch was calculated. The results were in agreement with ^{87}Rb n.m.r. measurements.

The effects of pressure up to 300 MPa on the ^{35}Cl n.q.r. spectrum (6 lines in 2 groups of 3) from the ferroelectric phase of $Me_4NH(Cl_3CCOO)_2$ at 77 K have been studied.[40,41] The separation between the average ^{35}Cl frequencies of the two CCl_3COO^- groups was shown to be a sensitive monitor of the proton deviation from the centre of the H-bond.[40] The resonance line separation decreased with increasing p. The results could be interpreted by using the theory of Matsushitu and Matsubara, and reflected proton migration within the asymmetric H-bond.[41] The T-dependence of the ^{35}Cl n.q.r. frequency and relaxation time T_1 for chlorine nuclei in pyridinium tetrachloroaurate, $pyHAuCl_4$, has been found from 400 K until the signals faded out at *ca*. 230 K.[42] The dominant relaxation mechanism below 350 K was attributed to fluctuations of the e.f.g. caused by motion of the pyH^+ ions. Above 360 K, the relaxation time decreased rapidly with increasing T, suggesting the onset of motion of the $[AuCl_4]^-$ ions, with an activation energy estimated as

67 kJ mol^{-1}. The T-dependence of the ^{35}Cl n.q.r. frequency and the relaxation times T_1 and T_2 for chlorine nuclei in $RbAuCl_4$ and $CsAuCl_4$ has been similarly determined from 82 K to 340 and 320 K respectively, where the signals faded out.[43] Two resonances were observed for each compound. In both cases the spin-lattice relaxation below 270 K was interpreted as mainly determined by rotary lattice vibration around the pseudo-C_4 (C'_4) axis and each Cl-Au-Cl diagonal axis of the anion, whereas above 270 K it was shown to be determined by reorientation of the anions about the C'_4 axis. The activation energy of this process was estimated for both compounds, and the anion dynamics were discussed.

The structures of 2,6-Br_2-4-$ClC_6H_2OPCl_4$ (I) and 2,4,6-$Br_3C_6H_2OPCl_4$ (II) have been established by ^{35}Cl n.q.r. at 77 K, which showed that the organic group was equatorial, while the equatorial chlorines were equivalent but the axial chlorines were inequivalent.[44] T-dependence studies on I from 77-380 K[44,45] indicated that the signals from the equatorial chlorines and one axial chlorine faded out at ca. 270 K, whereas the signal from the second axial chlorine only faded out near 380 K. Anomalies in the T-dependence of T_1 were also seen at these temperatures. Two simultaneous independent exchange processes were thought to take place, one involving both equatorial chlorines and one axial chlorine, and a slower one involving the second (lower frequency) axial chlorine. The activation energies for both processes were evaluated, and the possible nature of the chlorine atom rearrangements was discussed. Similar studies on $PhPCl_4$ and 4-$ClC_6H_4PCl_4$ from 77 K to their melting points showed evidence of exchange motion (pseudorotation) in these compounds also, involving three or four chlorine atoms.[45,46] The activation energies for the motions were evaluated. A further investigation of the T-dependence from 77-330 K of ^{35}Cl n.q.r. from Cs_2HgCl_4 has been carried out, because of conflicting literature reports.[47] Eight lines were observed at 77 K, with no change in multiplicity to 164 K, although two of the lines had positive T-coefficients. A first order phase transition occurred near 164 K, with hysteresis of about 2 K, and two lines only were seen up to 170 K. No signals were detected between 170 and 174 K, when a possible first order phase transition occurred, leading to the appearance of 8 low intensity lines from an intermediate phase. At 180 K an incommensurate phase appeared, and no signals were seen from 180 to 208 K. Between 208 and 210 K the three resonances of the paraelectric phase became evident, after a second order phase transition. The signals faded out above 330 K.

The T-dependence between 64 and 220 K of the ^{35}Cl n.q.r. from $[Me_2NH_2]_2TeCl_6$ has been reinvestigated, using a pulsed spectrometer.[48] Several new resonances were observed, which were not detectable in the previous (SRO) study. One new line was found for the intermediate phase (between 98 and 171 K), giving 4 overall, and two new lines in the low temperature phase (below 98 K), giving 6 in all. The resonance frequencies could be precisely measured, even near the phase transition. The T-dependence of T_1 for both ^{35}Cl and ^{37}Cl nuclei was also observed. From 64-160 K, the results suggested that relaxation arose mainly from libration of the complex anion. Above 160 K, T_1 decreased very rapidly with increasing T, interpreted in terms of the onset of reorientation of the anion as a whole, which was responsible for the signal fade-out near 220 K. The nature of the phase transition and the anion dynamics were discussed in detail. The T-dependence of the ^{35}Cl and ^{37}Cl resonances from $(NH_4)_2PbCl_6$ between 77 and ca. 380 K has been determined, together with the T-dependence of the spin-lattice relaxation rate.[49] The results indicated a structural phase transition at 80 K, similar to that observed previously in $(NH_4)_2TeCl_6$ at 85 K. The transition was characterised by a discontinuity in the T-dependence plot and a cusp in the spin-lattice relaxation rate, but no splitting of the resonance was observed below T_c, indicating a transition that left the Cl sites equivalent, such as a small angle rotation of the MX_6^{2-} octahedron about the three-fold axis. Some ^1H n.m.r. measurements were also performed, to investigate the influence of motion of the NH_4^+ ion on the chlorine n.q.r. In the mixed crystal $(NH_4:K)_2PbCl_6$, containing 10 mol % K^+, the phase transition was shifted to higher temperature by about 5 K. For cubic alkali metal hexachlorometallates $A_2[MCl_6]$ (M = Sn, Te, Pb, Pt), a correlation has been found between $\nu^{35}Cl$ and $1/r^3$, where r is the distance between the ions in the lattice and the Cl nucleus.[50] A similar relationship between the average ^{35}Cl frequency and $1/r^3$ was observed in compounds with organic cations such as pyridinium, where distortion of the MCl_6^{2-} octahedra by H-bonding and decrease of the lattice symmetry affected the individual chlorine frequencies. The average of the M-Cl bond lengths or the n.q.r. frequencies over all 6 Cl atoms was not influenced by the nature of the cation, but only by its size or the size of the unit cell, respectively. The increase of $\nu^{35}Cl$ with increasing size of the cation was considered to originate from a decrease in the lattice energy.

Complex formation between $SnCl_4$ and various esters has been studied by means of ^{35}Cl n.q.r. at 77 K.[51] Octahedral 1:2 complexes were formed by $SnCl_4$ with MeCOOR (R = Pr^i, Ph or CH_2CH_2Cl), $ClCH_2COOR$ (R = Me, Et or Pr^i), $BrCH_2COOMe$ and $Cl_2CHCOOMe$, but a 1:1 tbp complex with $Cl_3CCOOEt$. $SnCl_4.2MeCOOPr^i$ was deduced to have a trans structure, but the spectra of the other octahedral complexes were described as not in keeping with the concepts regarding the spectra of cis and trans isomers. Possible structures were discussed. The compound $SnCl_4.2ClCH_2COOEt$, which contained two different octahedral forms, reverted on standing at low temperature to a trans-octahedral complex. No complexation occurred with CF_3COOEt. Similar ^{35}Cl

n.q.r. measurements at 77 K on complexes of $SnCl_4$ with 2-$MeC_6H_4OPOCl_2$,[52] 2,6-$Me_2C_6H_3OPOCl_2$,[52] 2,4,6-$Cl_3C_6H_2OPOCl_2$[52] and 2-ClCO-4-$ClC_6H_3OPOCl_2$[53] have shown that the 1:1 adducts have tbp structures, but 1:2 ratios of the reactants give octahedral species. The donor centre was the phosphoryl oxygen. Cis-coordination was deduced in the octahedral complexes, with the structure strongly distorted in some cases. ^{35}Cl data at 77 K for some alkoxyphenyltrichlorostannanes have been interpreted in terms of intramolecular coordination of O to Sn when the alkoxy substituent was in the 2-position, giving a very distorted tbp structure.[54] No such interaction was observed for 4-alkoxy-substituted compounds. A weaker interaction, causing less structural distortion from tetrahedral, was found for 2,6-$(MeO)_2C_6H_3SnCl_3$. From the ^{35}Cl n.q.r. spectra at 77 K of eight compounds containing the fragment $Cl_3GeCNC(O)X$, the Ge atom was deduced to be usually pentacoordinate, with an intramolecular O $\rightarrow$ Ge interaction.[55] A satisfactory correlation was found between the n.q.r. frequency and the Ge-Cl bond length for two of the compounds in this series, and others of the type Cl_3GeX. The charges on the chlorine atoms in the compounds were estimated from the experimental frequencies. Values of η at 77 K have been determined for ^{35}Cl nuclei in some octahedral complexes of $SnCl_4$ with organic ligands, in the expectation that the electron distribution for the equatorial chlorines would show a greater divergence from axial symmetry than that for the axial chlorines.[56] Most results were in agreement with this supposition, allowing the lines to be assigned, but the n.q.r. was not clear-cut for $SnCl_4$.2MeCN, which was known to have a cis structure from X-ray data. Structures and bonding in the complexes were discussed on the basis of the results. Similar determinations of η at 77 K for ^{35}Cl nuclei in 1:1 complexes of $SiCl_4$ with NMe_3 and $[Me_2N]_2CO$, $GeCl_4$ with NMe_3, $[Me_2N]_2CO$ and $[Me_2N]_3PO$, and $SnCl_4$ with $MeNO_2$, and in the parent tetrachlorides, have been reported.[57] The lines from the 5-coordinate complexes were assigned to axial or equatorial chlorines, and the ionic (i) and π contributions to the bonding were estimated from a Townes-Dailey approach.

The first attempt to apply the Jeener-Broekaert pulse sequence to an inhomogeneous pure quadrupole resonance as a means to study motional processes has been made, using single crystal and polycrystalline samples of K_2OsCl_6 at 77 and 295 K.[58] The experiments were performed for various separations τ_1 of the preparation pulse pair as a function of the delay time τ_2 of the observed pulse. The observed decays could be represented by double exponential functions. For short τ_1, the dominant response was that characteristic of dipolar order, and was reasonably well understood. For long τ_1, the response was that of a stimulated echo, and was characteristic of a system where no spin temperature could be defined. Although this decay was characterised, the underlying mechanism was not identified. Sample-dependent aspects of the results were attributed to different defect structures in the polycrystalline and single crystal samples. A new design for an n.q.r. probe to study solids subjected to high hydrostatic pressure at low T has been described.[59] The probe was used to measure T_1 for ^{35}Cl nuclei at 3 pressures (up to 345 MPa) in a temperature range just above the structural phase transition (42 K) in K_2OsCl_6. The data were influenced by the softening of a rotary lattice mode and the evolution of a dynamic central peak. The results indicated that p had a small effect on the correlated fluctuations associated with the soft mode, but no detectable effect on those associated with the dynamic disorder component of the central peak. Sample differences in relation to the contribution of the dynamic disorder component were also noted.

The T-dependence of the halogen n.q.r. frequencies, T_1 for ^{35}Cl, and the ^1H n.m.r. second moment, have been observed for [pyH][$SbCl_4$], [pyH][$SbBr_4$], and their N-deuterated analogues.[27] Some ^{121}Sb measurements were also reported. The β-phase of the chloro-complex showed a second order phase transition at 232 K, while the bromo-compound gave a similar transition at 253 K. The transition temperature was lowered by 4 and 5 K respectively on deuteration. A minimum in T_1 for ^{35}Cl was observed at ca. 125 K for the chloro-species, attributed to reorientation of the pyridinium cation. Not all the lines could be detected from the deuterated compounds because of low intensity. The chloro-complex had two crystal forms at room temperature; the low T (α) phase gave 4 lines, and the high T (β) phase gave two signals. The bromo-compound gave 4 lines at lower T and two lines at room temperature, which could be assigned from the frequencies to bridging or terminal halogens. One of the terminal halogen signals in the low temperature phase showed an anomalous T-dependence. In the chloro-complex the signals of the low T phase disappeared at 316 K, and a further increase in T gave the β phase, in a first order transition. The β (high T) phase was deduced to transform into the α phase via a metastable phase, whereas the high T phase of the bromo-compound transformed directly into the low T phase. At low T, one of the two terminal halogen atoms in SbX_4^- formed H-bonds with the NH group in the cation. T-dependence studies of the halogen resonance frequencies in $(NH_4)_2TeX_6$ (X = Cl, Br or I) have been carried out from 4.2-430 K for X = Cl, 77-291 K for X = Br, and 77-380 K for X = I.[60] Some relaxation time measurements on Cl nuclei were also reported. A second-order phase transition was located at 89 K for X = Cl, with the spectrum possibly consisting of a closely-spaced multiplet below this temperature. A first order phase transition was found at 194 K for X = Br, with 3 signals from 77-194 K and two lines in a 2:1 intensity ratio from 194 to 212 K, in addition to the phase transition reported previously, which was accurately located at 212 K. One resonance was seen above this temperature. The lowest frequency line in the low temperature phase had a positive T-coefficient. Except for the highest frequency resonances for ν_1 and ν_2, the lines from X = I showed an anomalous T-dependence, which could be interpreted by considering partial breaking of Te-I---H H-bonds due to the motion of the NH_4^+ ion. A phase transition at 449 K was located by thermal measurements. Halogen (^{35}Cl, ^{79}Br or ^{127}I) n.q.r. frequencies have been measured at various temperatures between 77 and 300 K for 2-chloropyridinium bromide and iodide, 3-chloropyridinium bromide and 3,5-dichloropyridinium iodide.[61] The T-dependence of T_1 for both Cl and I nuclei (80-350 K) was also found for 2-chloropyridinium iodide, and for ^{35}Cl nuclei in the other salts. 3,5-dichloropyridinium iodide yielded two ^{35}Cl resonances at all temperatures, showing the presence of two crystallographically inequivalent chlorines. In 2-chloropyridinium iodide, the relaxation mechanism was thought to be governed by librational modes below 200 K, but the plot deviated from linearity above this temperature, suggesting considerable anharmonicity in these modes. Similar ^{35}Cl T_1 results were obtained for the other compounds. T_1 for ^{127}I in 2-chloropyridinium iodide decreased with increasing T, and the results could be well interpreted by the ionic model of Van Kranendonk. Other ^{35}Cl results have been described in the sub-sections on Group III[12-14] and Group V.[19,25,26]

The single crystal Zeeman effect has been studied on the ^{81}Br resonances from $(NH_4)_4Sb^{III}Sb^{V}Br_{12}$ at room temperature.[62] From the n.q.r. results the $SbBr_6^{3-}$ ions possessed D_{2d} symmetry and the $SbBr_6^-$ ions possessed D_{4h} (nearly O_h) symmetry, differing from the X-ray conclusions which indicated D_{2d} for $SbBr_6^-$ and O_h for $SbBr_6^{3-}$ ions. The compound was known to undergo a structural phase transition at 214 K. The greater distortion in the $SbBr_6^{3-}$ ions was attributed to Br---H-N H-bonding, even in the high temperature phase. The synthesis and characterisation of a new molybdenum bromide, $MoBr_{2.5}$, have been described.[63] $^{79,81}Br$ n.q.r. measurements at 77 K and room temperature were interpreted as characteristic of bridging bromine atoms only. Three types of Br atom were seen at both temperatures, in a 1:3:1 intensity ratio. The n.q.r. linewidths, T_1 and T_2 of ^{79}Br nuclei in anilinium bromide have been measured at various temperatures between 100 and 445 K.[64] T_1 showed an unusual T-variation over a wide range, both well below and above the known phase transition temperature (297 K). The mechanism of this anomaly was explained in terms of the soft acoustic mode reported previously to exist in the compound. Two independent bromine n.q.r. investigations of $ThBr_4$ have been described.[65,66] The T-dependence of the bromine frequencies of α- and β-$ThBr_4$ has been determined from 63-300 K, together with the frequency for the α-form at 4.2 K.[65]. Some relaxation time measurements were also made. The line shape in the incommensurate phase, the phase transition in the β form

(95.5 K), nuclear relaxation in the normal phase both near to and far from the transition temperature, and strong phason relaxation in the I phase were discussed. The T-dependence of the ^{79}Br frequency has been found for β-$ThBr_4$ from 293 to 2.5 K, together with line shape and T_1 measurements.[66] In addition, the theory of the effects of Raman processes on amplitudon and phason spin-lattice relaxation in incommensurate systems has been extended, by obtaining general expressions for the spectral densities and phason gap that were valid at all temperatures in the I phase. By measuring T_1 selectively for different parts of the broadened resonance in β-$ThBr_4$, the phason and amplitudon contributions were obtained separately. Excellent agreement was found between the theory and the experimental data, demonstrating that spin-lattice relaxation in this compound was dominated by Raman processes, and the phason gap was determined.

The T-dependence of the ^{79}Br or ^{127}I n.q.r. frequencies in the range 77-420 K has been found for $[H_3N(CH_2)_2NH_3]^{2+}$ $2X^-$ (X = Br or I), $[H_3N(CH_2)_3NH_3]^{2+}$ $2X^-$ (X = Br or I), piperazinium dibromide monohydrate and piperazinium monoiodide.[67] The three iodides showed phase transitions, at 400 and 404 K for $[H_3N(CH_2)_2NH_3]^{2+}2I^-$, 366 K for $[H_3N(CH_2)_3NH_3]^{2+}2I^-$, and 196 K for the piperazinium compound. No phase transitions were observed for the bromides. The crystal structures of the piperazinium compounds were also determined, and H-bonding in these compounds was discussed, as well as the structures of the other salts. The crystal structures have been determined of glycyl-ℓ-alanine hydriodide monohydrate (I) and of the corresponding hydrobromide monohydrate (II).[68] The T-dependence of the ^{127}I n.q.r. frequency from 77 to 370 K of I and its deutero-analogue was also measured, and no phase transitions were detected. The frequency was strongly influenced by H-bonding, as proved by the frequency shift on deuteration. The bromine n.q.r. results reported previously for II were in agreement with the crystal structure. $^{63,65}Cu$ n.q.r. frequencies at 77 K have been measured for several complexes of copper(I) halides with alkyl pyridines, of known crystal structure.[69] A few halogen (^{79}Br or ^{127}I) frequencies were also reported. The results confirmed the trends observed previously for analogous complexes with PPh_3 ligands. These trends were briefly discussed in terms of the electronic environment of the Cu nucleus in the complexes. Other results for bromine nuclei have been given in the sub-section on Group III.[14]

^{127}I n.q.r. at various temperatures between 77 and 297 K has been obtained from a large single crystal of DIO_3, and compared with the results for HIO_3.[70] The observed differences in the n.q.r. parameters were explained in terms of very small changes in the electron population. Bonding in the compounds, including the effects of H-bonding, was discussed. In an extension of previous work, the ^{127}I n.q.r. of $NH_4IO_3.2HIO_3$ has been recorded at 77 K.[71]

Six lines were observed for each transition. The spectrum was interpreted and the lines were assigned on the basis of the known structure at 300 K, and a proposed model of the phase transition at 213 K (T_c). There was deduced to be a doubling of the unit cell volume at T_c, and the possible low temperature structure was discussed. Structure and bonding in two modifications of $CsSnI_3$ has been studied by ^{127}I n.q.r., X-ray powder diffraction and thermal measurements.[72] Three pairs of lines were detected for the low T modification at widely differing frequencies, as expected from the crystal structure, which showed terminal, bridging and triply bridging iodines. The high T modification showed two very broad lines between 88 and 95 MHz for ν_1 in a 2:1 intensity ratio, consistent with a tetragonal perovskite structure. The T-dependence of the ^{127}I frequency for the $\frac{1}{2} \leftrightarrow \frac{3}{2}$ transition in a single crystal of Cs_2HgI_4 has been followed from 77 to 525 K.[73] Three signals were seen at room temperature, in good agreement with the published values. These had a normal T-dependence up to 525 K, where the signals became undetectable, attributed to the start of reorientation involving the HgI_4^{2-} ions. There was a thermochromic transformation at 448 K, accompanied by a change in the T-dependence of the signal amplitude. No changes occurred in the spectrum on natural cooling to 77 K. If the specimen was heated to 593 K and then rapidly quenched in liquid nitrogen, the n.q.r. spectrum at 77 K consisted of 16 pairs of lines of approximately equal intensity, indicating 16 non-equivalent positions of the iodine atoms. These signals were all undetectable by 200 K, and three resonances corresponding to the normal phase appeared at 255 K. The transition at 200 K was reversible, but that at 255 K was irreversible, suggesting a possible incommensurate phase between 200 and 255 K. The addition of small quantities of water had no effect on the spectrum. ^{127}I n.q.r. frequencies at 77 and 296 K have been measured for NbI_5 and TaI_5.[74] The resonances could be assigned to bridging, axial or equatorial atoms, and the results were compared with crystal structure data. The compounds were not isostructural. A broad resonance at 77 K was also observed for ^{181}Ta from TaI_5, allowing the average values of e^2Qq/h and η for this nucleus to be evaluated.

3 Transition Metals and Lanthanides

3.1 Manganese-55. - ^{55}Mn n.q.r. has been obtained at 0.3 K ($\frac{3}{2} \leftrightarrow \frac{5}{2}$ transition) from $Y_{1-x}Sc_xMn_2$, together with the T-dependence of T_1 at low temperature.[75] ^{45}Sc n.m.r. measurements were also reported. Good agreement was found between the T-dependence of the relaxation of Sc nuclei with the self-consistent renormalisation theory for nearly antiferromagnetic metals, but was less good for Mn nuclei, possibly related to large spin fluctuation at the Mn sites.

3.2 Copper-63 and -65. - A marginal type n.q.r. spectrometer has been constructed, using field effect transistors, and has been used to obtain an excellent absorption derivative curve for the broad ^{63}Cu resonance from Cu_2O at 301.5 K.[76] As last year, many papers have appeared on $^{63,65}Cu$ n.q.r.[77-122] from high-T_c superconducting oxides and related materials, again resulting in a good deal of duplication. The main point of disagreement between different research groups now seems to be as to whether superconductivity and antiferromagnetic order can co-exist below T_c in certain materials. The results from particular research groups have been reported together, where possible.

N.q.r. data have been obtained for a homogeneous and highly ordered sample of $YBa_2Cu_3O_{6.7}$, that contained a substantial subset of Cu sites with no measurable metallic character and magnetically inactive, i.e. essentially $3d^{10}$ Cu(I).[77] These were identified as Cu(1) chain sites with twofold O coordination. The spin-lattice relaxation rates were reduced by more than 3 orders of magnitude compared with similar sites in $YBa_2Cu_3O_7$. The generation of these Cu(1) sites by the ordering of oxygen vacancies onto "empty" chains increased the carrier concentration, and might help to maintain the observed 60 K plateau in the variation of T_c with oxygen concentration. Spin dynamics at planar sites in $YBa_2Cu_3O_{6+x}$ have been studied by means of T_1 measurements for $^{63,65}Cu$ n.q.r., and ^{89}Y n.m.r.[78] A sharp decrease of the ^{63}Cu nuclear relaxation below 100 K in $YBa_2Cu_3O_{6.7}$ (T_c 60 K) suggested the possible onset of spin pairing in the individual planes well above the transition to 3-d bulk superconductivity. Comparison with the data for $YBa_2Cu_3O_7$ in the range 100-300 K showed that the planar spin dynamics were strongly affected by changes in the carrier concentration. The effects of fluorination on $Ba_2YCu_3O_7$ have been studied by various physical techniques, including the ^{63}Cu n.q.r. of $Ba_2YCu_3O_{6.2}F_{1.3}$ at 100 K.[79,80] The resonances were essentially identical in position and width with those of the unfluorinated material. It was deduced from various measurements that the fluorine was not incorporated into the superconducting phase, but into a secondary amorphous phase. A sample of $YBa_2Cu_3O_{6.64}$ ($T_c \sim 60$ K) has been prepared isotopically enriched in ^{63}Cu, and its n.q.r. has been obtained at 100 K.[81] Some relaxation times were also measured. The spectrum showed a narrow slowly relaxing line at 31.4 MHz., attributed to two-coordinate Cu(1) sites, a rapidly relaxing doublet at 22.0-22.6 MHz. due to four-coordinate Cu(1) sites, and a broad complex absorption with three rapidly relaxing lines at 27.5, 28.8 and 30.6 MHz., due to planar Cu(2) sites. The nuclear spin-lattice relaxation rates exhibited a strong reduction of $(T_1T)^{-1}$ between 100 K and T_c, while the behaviour resembled that of $YBa_2Cu_3O_7$ below T_c.

The $^{63,65}Cu$ n.q.r. frequencies at 1.3 K for $(La_{0.925}Sr_{0.075})_2CuO_4$ have been measured, together with the T-dependence of $1/T_1$ from 1.3 to 200 K.[82,83] The T-dependence of $1/T_1$ was similar to that of the Cu(II) (CuO_2 plane) sites in $YBa_2Cu_3O_{7-\delta}$, and a similar relaxation mechanism was suggested, dominated by antiferromagnetic spin fluctuations of the nearly localised Cu-3d holes. Cu n.q.r. has been observed from the material of nominal composition $Bi_{1.6}Pb_{0.4}Sr_2Ca_3Cu_4O_y$ (T_c 109 K) in the range 1.3-150 K.[83,84] Two ^{63}Cu and two ^{65}Cu resonances were detected, and the peaks were assigned to different CuO_2 layers in the structure. The T-dependence of $1/T_1$ was similar to that of the plane site in $YBa_2Cu_3O_7$, demonstrating that Cu d-spin correlation played an important role in the Cu relaxation process. The T-dependence of n.q.r. and n.m.r. for the chain and planar Cu sites has been investigated in $YBa_2Cu_3O_x$ ($6.1 \leq x \leq 6.4$).[85,86] A divergence of the spin-echo decay rate for n.q.r. at the chain sites was observed at ca. 20 K, irrespective of x, for $x \geq 6.2$. Below this temperature, only line broadening became significant, without any resonance shifts. This result, together with n.m.r. data, indicated that the secondary transition associated with the moments on the Cu(2) sites took place at ca. 20 K, well below the Néel temperature. The magnetic structure below the transition temperature was discussed. No such transition was observed for x = 6.1 down to 1.3 K. The magnetic phase diagram was derived from these and other measurements. The T-dependence of $1/T_1$ for $(La_{1-x}Sr_x)_2CuO_4$ (x = 0.075, 0.125 or 0.15), where the system is transformed from a high T_c superconductor to the normal metallic state, has been examined.[87] The n.q.r. frequency increased and the spectrum was broadened with increasing x. For the non-superconducting compound with x = 0.15, each component of $1/T_1$ was suppressed, compared with those for the superconducting material with x = 0.075, and obeyed the Korringa relationship from 1.4 to 90 K. Short range antiferromagnetic (AF) spin correlation was deduced to be present even in the superconducting materials, and an increase in the Sr content reduced the magnetic coherent length of the AF spin fluctuation, yielding a simultaneous decrease in T_c. The Cu(1) and Cu(2) sites have been assigned in $MBa_2Cu_3O_{7-\delta}$ (M = Sm, Nd or Gd) from T_1 measurements on the Cu n.q.r. lines.[88] In $YBa_2(Cu_{0.98}Zn_{0.02})_3O_{7-\delta}$, $1/T_1$ followed the Korringa law below 90 K, although it was saturated above 100 K.

Cu n.q.r. signals have been observed from $La_{2-x}Sr_xCuO_4$ at 1.5 K; T-dependence studies of $1/T_1$ and $1/T_2$ were also reported.[89] The spectra were split into two peaks for each Cu nucleus at low temperature and the separation decreased with increasing x, indicating that the hyperfine field due to the magnetic ordering of the Cu moments existed in the superconducting and normal metallic regions. The relaxation rate $1/T_1$ was suppressed, apparently by the superconducting energy gap, for $x \leq 0.20$, while the T-dependence obeyed the Korringa relationship in the normal region ($x \geq 0.25$). $1/T_2$ increased with increasing hyperfine field, indicating dominant indirect nuclear spin coupling via virtual excitation of the antiferromagnetic spin waves. The authors concluded that the coexistence of superconductivity and magnetic ordering of the Cu d-electrons with a tiny moment was demonstrated by the n.q.r. results. The magnetic order was suggested to be of the long-range type. The Cu n.q.r. signals at 1.5 K from $Bi_2Sr_2(Ca_{1-x}Y_x)Cu_2O_{8+\delta}$ were spread over a wide frequency range between 18 and 32 MHz., indicating a large distribution of e.f. gradients at the Cu nuclei in the CuO_2 planes, and reflecting the structural modulation with displacement of Cu and Bi atoms in the Bi system.[90,91] T_2 decreased rapidly with Y concentration, showing that antiferromagnetic spin fluctuations became dominant for large x. T_c decreased rapidly for $x > 0.3$, and superconductivity disappeared at $x > 0.5$. The Cu n.q.r. spectra were much broader than in the La or Y systems, taken as supporting evidence for the existence of the proposed modulation structure. Cu n.q.r. at 1.5 K has been observed from the 2223 phase of $Tl_2Ba_2Ca_2Cu_3O_{10}$ (T_c 121 K) and $Bi_{1.6}Pb_{0.4}Sr_2Ca_2Cu_3O_{10}$ (T_c 110K).[92] Both materials gave two sets of signals, corresponding to Cu(I) sites (lower frequency), and Cu(II) sites (higher frequency). The Cu(I) sites were assigned to Cu-O pyramids and the Cu(II) sites to Cu-O square planes. The results were compared with those for other high-T_c superconductors. Cu n.q.r. at 1.4 K in $La_{2-x}Sr_xCuO_4$ has been obtained for a wide range of x (up to 0.80).[93] The T-dependence of T_1 was also measured. $1/T_1$ was suppressed by the superconducting energy gap for $0.12 < x < 0.2$, while the T-dependence obeyed the Korringa relationship in the normal region ($x > 0.4$). The spectrum profile also changed markedly with increasing x beyond 0.3. The results indicated a strong correlation between high T_c superconductivity and antiferromagnetic fluctuations of the Cu spins. The T-dependence of the nuclear relaxation rates for Cu nuclei in $YBa_2(Cu_{1-x}Fe_x)_3O_7$ indicated the appearance of magnetic order without destroying the superconductivity.[94] T_c decreased smoothly with increasing x. $1/T_1$ showed a peak at 2.5 K for $x = 0.005$, and at 4 K for $x = 0.015$, for both Cu sites. Superconductivity was concluded to coexist in some sense with the antiferromagnetic order of the Cu moments induced by magnetic impurities.

Cu n.q.r. and n.m.r. have been observed for metallic but non-superconducting $La_4BaCu_5O_{13+x}$ ($0.2 < x < 0.4$).[95] Only one ^{63}Cu (and one ^{65}Cu) resonance was detected, and the frequency was T-independent between 1.3 and 300 K. There was no magnetic broadening, hence magnetic ordering did not exist. The electric quadrupole interaction was similar to that observed for $YBa_2Cu_3O_7$. The spin-lattice relaxation rate was much lower at 100 K than that for the normal state of $YBa_2Cu_3O_7$, indicating that the behaviour of the antiferromagnetic copper spin fluctuations of the non-superconducting and superconducting oxides were very different. The relaxation rate obeyed the Korringa relationship up to 180 K, where an additional relaxation mechanism set in. N.q.r. measurements on Cu nuclei have been carried out for

$La_{2-x}Sr_xCuO_4$ (x = 0.15, 0.20 or 0.30), together with n.m.r. studies on uniaxially aligned samples.[96] No evidence was found for magnetic ordering down to 1.3 K. The existence of two sets of $^{63}Cu/^{65}Cu$ resonances from nonequivalent Cu planar sites was deduced from the spectra and relaxation measurements at low temperature. These resonances were ascribed to the effect of the charge differentiation introduced by substitution of Sr for La. N.q.r. and n.m.r. measurements on the Cu(1) and Cu(2) sites of $YBa_2Cu_3O_y$ ($6.0 < y < 7.0$) have revealed that the electronic state of the Cu depends on its coordination number for oxygen.[97] The relaxation time T_1 for the Cu(2) site was characterised by antiferromagnetic spin fluctuation in the normal state, and did not show an abnormal behaviour due to S-wave character below T_c. To understand the Cu spin dynamics in the normal (metallic) state of high T_c oxides and related compounds, the n.q.r. frequencies and T-dependence of T_1 for ^{63}Cu nuclei in $YBa_2Cu_3O_y$, $La_{2-x}Sr_xCuO_4$ and various other metallic copper oxides have been measured.[98] A sizeable charge differentiation in the Cu-O plane sites was found, even in the metallic phases of high T_c materials, where a highly enhanced relaxation process was found to exist. The data showed the opening of a gap in the low-lying excitations, with an energy comparable to the superconducting gap. Cu n.q.r. has been used to characterise the local oxygen coordination of inequivalent Cu sites in $YBa_2Cu_3O_{6+x}$ ($0 \leq x \leq 0.91$).[99,100] Signals from the Cu(1) sites were observed for x = 0, in agreement with other work. Four distinct resonances were found at intermediate O content, ascribed to 2,3 and 4 oxygen-coordinated Cu sites in the Cu-O chains and 5 oxygen-coordinated planar sites. The relative intensities depended on the O content, giving a microscopic understanding of the Cu chemistry. An anomalous T-dependence of T_1 was observed for both Cu(1) and Cu(2) sites with x = 0.91, and was discussed in connection with antiferromagnetic spin fluctuations in the normal state.[100]

Cu n.q.r. frequencies at room temperature have been measured for 11 samples of $YBa_2Cu_3O_x$, with x ranging from 7.00 to 6.00.[101] Some T_1 measurements were also made at the resonance frequencies of selected lineshape maxima. The signals had either very short (< 0.3 ms) or much longer (> 60 ms) relaxation times. The Cu sites with short T_1 were identified as conducting structural elements, while those with long T_1 were nonconducting substructures. When x was close to 7.0, all the Cu atoms were in conducting environments at room temperature; when $x < 6.4$, no conducting substructures were present. For intermediate x ($6.4 < x < 6.8$), insulating and conducting structures coexisted. In the $7.0 > x > 6.9$ range (T_c 90 K), the shapes of the n.q.r. spectra were independent of x, but the overall intensity decreased with decreasing O content. In the range $6.8 > x > 6.4$ (T_c 55 K), narrow peaks from the non-conducting substructures were observed, together with distinctive short T_1 lineshapes, indicating that the superconducting state with this stoichiometry was essentially different from that at x = 7. The results were consistent with a superordered structure of O vacancies in the conducting substructure. Six Cu resonance frequencies were identified between 20 and 32 MHz., some of which could be assigned to specific local crystal structures. In several instances when the O content was lowered, the n.q.r. signal from metallic Cu was replaced by a nonmetallic signal (different T_1 behaviour) at the same frequency. The overall integrated n.q.r. intensity gradually decreased from a value corresponding to three Cu atoms per formula unit for x = 7 to one Cu for x = 6, thought to be caused by antiferromagnetic electron spin interactions. Cu n.q.r. was also given for $ErBa_2Cu_3O_7$ and $ErBa_2Cu_3O_6$; the data confirmed the line assignments for $YBa_2Cu_3O_7$. For x = 6, a single ^{63}Cu signal (and ^{65}Cu signal) from the Cu(1) sites was seen, in agreement with other work, despite a big frequency difference from the Cu(1) sites for x = 7. The room temperature n.q.r. of Cu nuclei (including T_1 and T_2 measurements) has been reported for the materials $(Y_{1-y}Ca_y)Ba_2Cu_3O_{6+\delta}$, with $0 < y < 0.5$ and $\delta << 1$.[102,103] Superconduction was observed for y > 0.1, with $T_c \sim 50$ K. For y values below 0.25, the spectra consisted of two components, one signal identical to that of Cu(1) in $YBa_2Cu_3O_6$, showing that the chain sites were not affected by the hole doping that accompanied the substitution of Ca for Y. The other signal, which could be assigned to Cu(2) sites, was identical for all superconducting samples, and was not observed when y < 0.1. The relaxation times indicated that Cu(1) atoms were in insulating environments, and Cu(2) sites were conducting. The significance of the observed behaviour was discussed.

The linewidths of the Cu resonances at room temperature have been measured as a function of δ for values from 0.018 to 0.071 in $YBa_2Cu_3O_{7-\delta}$.[104] The data were compared with a computer simulation based on a random distribution of vacancies over the chain oxygen sites. The simulated data were almost the same for Cu(1) and Cu(2), and described quite well the behaviour of the Cu(1) signals, but a discrepancy was apparent for the experimentally much broader Cu(2) signals. Reasons for this behaviour were discussed. The T-dependence of the $^{63,65}Cu$ n.q.r. frequencies, linewidths, T_1 and T_2 between 6 and 750 K has been reported for chain and planar Cu sites in $YBa_2Cu_4O_8$ (T_c 81 K).[105] Above T_c, the e.f.g. tensors differed slightly but subtly from those in $YBa_2Cu_3O_7$. At the Cu(2) site the derivative of the n.q.r. frequency with respect to T changed sign around 200 K. There was no evidence for magnetic ordering at the Cu sites, or for Bardeen-Cooper-Schriefer (BCS) type relaxation below T_c. Above T_c, the Cu(1) relaxation rates suggested the presence of quasiparticles and phonon interactions, while the Cu(2) rates could perhaps be explained by the model of Monien et al, postulating the coupling of Cu nuclei to itinerant interacting quasiparticles. There was no evidence for a structural phase transition at T_c, and the relatively small Cu linewidth seemed to be in accord with fully occupied oxygen sites. The effect of hydrostatic pressure up to 0.6 GPa has been studied on the Cu n.q.r. frequency from the Cu(1) sites in $YBa_2Cu_3O_6$, and on the resonance frequencies and T_1 values of both Cu sites in $YBa_2Cu_3O_7$.[106] In $YBa_2Cu_3O_6$ the frequency increased linearly with p. In $YBa_2Cu_3O_7$, T_1 was independent of p above T_c, but the Cu(1) frequency decreased and the Cu(2) frequency increased with p.

T_1 and T_2 of ^{63}Cu nuclei (31.5 MHz. resonance) at 4.2 and 120 K have been measured for polycrystalline samples of $YBa_2Cu_3O_{7-\delta}$, with δ 0.11-0.43.[107] A sharp change in relaxation rate was observed at δ 0.3. Models for the electronic structure of the Cu(1) and Cu(2) centres were proposed, on the basis of analysis of the experimental data and calculation of the e.f.g. tensors. A model has been proposed for nuclear spin relaxation of the planar Cu nuclei in $YBa_2Cu_3O_{7-\delta}$, that explained the experimental n.q.r. data for the T-dependence of $1/T_1$ and $1/T_2$ at $T > T_c$ and $T < T_c$.[108] The T-dependence of T_1 and T_2 for ^{63}Cu nuclei at planar sites in $YBa_2Cu_3O_{7-\delta}$ showed structural features near 35 K.[109] It was suggested from these and other results that either a second order phase transition or a restructuring of the dynamic state of intrinsic defects could occur at this temperature. Various physical methods including ^{63}Cu n.q.r. have been used to study the state of the Cu atoms in $YBa_2Cu_3O_{6+x}$ ($0.05 \leq x \leq 0.95$).[110] The behaviour of the n.q.r. frequency near T_c was discussed. Various methods, including measurement of the resonance frequency and the T-dependence of T_1 for ^{63}Cu nuclei in planar sites, have been used to examine samples of $YBa_2Cu_3O_{6.95}$ irradiated by fast neutrons at 77 K, together with unirradiated samples.[111] Irradiation lowered T_c from 93 K to 71 K. The results showed that even for small degrees of disorder there existed localised states in these systems. Localisation caused T_c to decrease, until the superconductivity was entirely suppressed in the region of strong localisation. A theoretical interpretation of the experimentally observed effects was given.

The Cu n.q.r. at 4.2 K from $Bi_2Sr_2CaCu_2O_x$ (T_c 70 K) consisted of a broad resonance extending from 18 to 30 MHz.[112] The large width was possibly due to poorly resolved resonances originating from Cu sites with close e.f. gradients, or to variation in chemical composition. T_1 and T_2 were measured at 22.5 MHz. N.q.r. measurements at 4.2 K have been reported for Cu nuclei in $YBa_2(Cu_{1-x}Ga_x)_3O_{7-\delta}$ for x = 0.00, 0.01, 0.02, 0.03 and 0.05.[113] The resonances broadened with increasing Ga^{3+} concentration, suggesting a distribution of e.f. gradients at the Cu sites. Both types of site were affected, although the Ga ions occupied only chain sites. T_1 was sharply reduced with increase of x, strongly suggesting that certain Cu atoms acquired a magnetic moment as a consequence of the introduction of non-magnetic Ga atoms. Various nuclear resonance studies on high-T_c superconductors have been described, including ^{63,65}Cu n.q.r. at 4.2 K for $YBa_2Cu_3O_{7-x}$, and some T_1 measurements.[114] The difference in the observed relaxation times for the two Cu sites was discussed. N.q.r. at 4.2 K for Cu nuclei has been observed from the Bi(Pb)-Sr-Ca-Cu-O system containing mainly (80-95%) either the 2212 or 2223 superconducting phases.[115] Both phases showed a broad distribution of resonance frequencies in the range 15-29 MHz., which could be fitted to a pair of relatively broad Cu lines. Possible line assignments were discussed, and the results were compared with data for similar samples. The Cu n.q.r. spectra of $YBa_2Cu_3O_y$ ($6.0 < y < 6.25$) have been claimed to show two different antiferromagnetic structures with a different stacking sequence.[116] The magnetic moments in both of these were located at the Cu(2) sites perpendicular to the c-axis. One structure revealed strong T-dependent magnetic fluctuations, but none was found in the second structure, which showed a constant internal field of about 0.2 T.

The ^{63}Cu n.q.r. frequencies for the planar Cu(2) sites at 1.3 K (20 K for Ho) have been measured for $RBa_2Cu_3O_{7-\delta}$ (R = Nd, Sm, Eu, Gd, Dy, Ho, Er or Yb).[117] The frequencies were found to decrease monotonously towards the heavy end of the rare earth elements. This was explained by the volume change, and the on-site charge at the Cu(2) site was estimated to be almost R-independent. The T-dependence of $1/T_1$ to > 100 K for this resonance was also obtained, and was explained as reflecting the fluctuation spectra of the R^{3+} paramagnetic moments, strongly modified by crystal field interaction. Four groups of R^{3+} ions were classified, and the behaviour of each type was discussed in some detail. N.q.r. and high field n.m.r. spectra have shown the absence of signals from one Cu site in $YBa_2Cu_3O_{6.2}$, taken as evidence for a localised magnetic moment on this site.[118] The other site had a strongly T-dependent spin-lattice relaxation rate, and the magnetic character of the relaxation was proved by comparing the rates for the ^{63}Cu and ^{65}Cu isotopes. The results could be explained by an antiferromagnetic transition, as proposed by several authors. An unambiguous site assignment was not possible. Cu n.q.r. at 170 K for a superconducting sample of $YBa_2Cu_3O_{7-\delta}$ (T_c 89.5 K), and for a second sample obtained by annealing the previous sample in an oxygen atmosphere, have been recorded, together with the ^{57}Fe Mössbauer spectra from compounds doped with ^{57}Fe.[119] The results were compared and the Cu sites were assigned. The role of oxygen vacancies in the hyperfine interactions was discussed.

The effect of magnetic impurities on ^{63}Cu n.q.r. from $YBa_2Cu_3O_7$ has been studied.[120] In this system Fe ions substituted for Cu at both Cu(1) and Cu(2) sites, but Co ions substituted preferentially at Cu(1) sites. The T-dependence of the relaxation behaviour of Cu indicated the appearance of magnetic ordering without destroying the superconductivity for a 0.5% Fe doped sample. The effects of oxygen deficiencies and magnetic impurities on nuclear relaxation anomalies in $La_{2-x}M_xCuO_4$ (M = Ba or Sr) were also investigated via the T-dependence of $1/T_1$ for ^{139}La nuclei.[120,121] The results revealed an unexpected enhancement of this rate at $T_c^* \sim 10$ K, suggesting the occurrence of magnetic instability in the low T region. In undoped La_2CuO_4, the oxygen deficiencies strongly affected the nuclear relaxation rate. Magnetic impurities such as Fe or Co suppressed the enhancement of $1/T_1$, but non-magnetic Zn did not. The results also indicated that the fluctuation time of the Cu moments changed drastically with hole-doping, but did not slow down with decreasing T in the superconducting state. ^{139}La and ^{63}Cu n.q.r. relaxation measurements in $La_{2-x}Sr_xCuO_4$ ($0 \leq x \leq 0.3$) from 1.6 to 450 K have indicated that over most of the x and T range the relaxation was controlled by the Cu^{2+}-Cu^{2+} in plane (2D) magnetic correlation length.[122] Long range magnetic correlation was strongly limited by mobile charge defects introduced by Sr doping. Relaxation and specific heat measurements suggested that at low T the antiferromagnetic order was restored at $T_N(x)$, as a consequence of the freezing of the mobile defects. Other results for Cu nuclei have been described in the sub-sections on Group V[28] and Group VII.[69]

3.3 Niobium-93. - The results for this nucleus have been given in the sub-section on Group V.[24]

3.4 Lanthanum-139. - The effects of oxygen deficiencies and magnetic impurities on the n.q.r. of ^{139}La in $La_{2-x}M_xCuO_4$ (M = Ba or Sr) have been investigated.[123] A decrease of oxygen deficiency led to a considerable increase in the relaxation rate below 70 K. $1/T_1$ of samples doped with Ba or Sr was enhanced, and diverged at a temperature T_c^* for concentrations near x = 0.02, at which the antiferromagnetic (AF) order disappeared. Magnetic impurities such as Fe suppressed the enhancement of $1/T_1$, but non-magnetic Zn did not. The results suggested the occurrence of magnetic instability in the low T region. A new phase diagram was proposed, involving the re-entrant behaviour of the 3D-AF order for small x, and some kind of spin-glass state between the AF and superconducting states. In the superconducting region, the fluctuations of the Cu moments at low T seemed to remain extremely fast, as at high T. The T-dependence of T_1 from 3 to 100 K has been determined for ^{139}La nuclei ($\frac{5}{2} \leftrightarrow \frac{7}{2}$ transition) in $(La_{1-x}Sr_x)_2CuO_4$ with x = 0.045, 0.075 or 0.10 (high-T_c superconductors) and x = 0.15 (non-superconductor).[124] For the non-superconductor, T_1 conformed to the Korringa relation. T_1 values for x = 0.075 and 0.10 in the normal state showed a stronger T-dependence and were shorter than for x = 0.15. Fluctuation of the Cu d-spins was considered to play an important role in determining T_1 for the superconducting materials, and the relaxation behaviour in the compound with x = 0.045 was discussed.

^{139}La n.q.r. has been observed at various temperatures from 1.4 to 139 K for $La_2NiO_{4+\delta}$.[125] The spectrum at 1.4 K was found to depend on the annealing procedures, and hence on δ. Well-characterised spectra were obtained for oxygenated ($\delta \geq 0.066$) and deoxygenated ($\delta \sim 0.022$) samples with different structures, showing the existence of internal magnetic fields parallel with and perpendicular to the basal Ni-O plane. These were significantly larger than in La_2CuO_4, and could not be explained simply by the dipole field caused by the Ni spin moments. Antiferromagnetic ordering at low T was deduced for both types of sample. N.q.r. indicated the existence of another different but fractional magnetic phase, which could not be identified.

The T-dependence to 100 K of $1/T_1$ for ^{139}La nuclei in $(La_{1-x}Sr_x)_2CuO_4$ (x = 0.10 or 0.075) has been measured.[126] $1/T_1$ decreased rapidly in the superconducting state, indicating superconducting energy gap formation, but there were no signs of $1/T_1$ enhancement just below T_c. In the normal state, $1/T_1$ showed a slightly stronger T-dependence than expected from the Korringa relationship, which was seen in the non-superconducting compound with x = 0.15. ^{139}La n.q.r. at 4.2 K has been recorded for $La_{2-x}CuO_{4-y}$ (x = 0.05 or 0.10), with $T_c \sim 30$ K, synthesised by high T and high p oxygen enrichment.[127] The results were compared with those for non-superconducting La_2CuO_4. A weak and broad additional resonance was found around 6.35 MHz. for the oxygen-enriched samples, corresponding to the frequency in stoichiometric La_2CuO_4 when the internal field at the La sites was absent. It could only be seen for samples with a large superconducting fraction (> 0.3%). There was no significant change to the four main resonances, in contrast with Sr-doped samples. The results indicated that superconductivity occurred in regions where the antiferromagnetic ordering was sufficiently suppressed by carrier doping. The main magnetic phase in the oxygen-enriched sample was deduced to be essentially stoichiometric La_2CuO_4. The intensity of the new line increased with increasing external field up to 100 Oe. It has been suggested from accurate control measurements of the ^{139}La n.q.r. frequencies at 77 K from $La_2CuO_{4-\delta}$ that the results accorded completely with the simple antiferromagnetic structure found by neutron diffraction, when the directions of the local magnetic field and the principal axis of the e.f.g. tensor were correctly taken into account.[128] Other results for ^{139}La nuclei have been described in the sub-section on $^{63,65}Cu$.[120-2]

3.5 Praseodymium-141. - Optically detected multipulse n.q.r. at 2 K has been observed from Pr^{3+} doped into a single crystal of LaF_3, by using a Raman heterodyne optical detection technique.[129] The pulse sequence used produced a lengthening of the coherence decay time from 17 to 51 μ sec. in the absence of an external field, and from 37 to 372 μ sec. in a static field of 80 G. Optical ion doublets from Pr^{3+} (0.5 atom %) doped into LaF_3 have been detected in optical hole-burning experiments.[130] N.q.r. spectroscopy ($\frac{3}{2} \leftrightarrow \frac{5}{2}$ transition) of the optical doublets was obtained for the first time, and was used to assign the doublets to specific ion pairs.

3.6 Tantalum-181. - Results for this nucleus have been given in the sub-section on Group VII.[74]

3.7 Rhenium-187. - The T-dependence of T_1 up to 4.2 K for the n.q.r. of ^{187}Re nuclei ($\frac{1}{2} \leftrightarrow \frac{3}{2}$ transition) in Re metal has been reported, together with the T-dependence to 150 K of T_1 for n.m.r. of the same isotope.[131] The dominant relaxation mechanism was found to be due to magnetic rather than quadrupolar hyperfine interactions. The data were analysed in terms of the electronic structure of Re metal, confirming that Re was a "weakly enhanced" transition metal with a nuclear relaxation rate dominated by the s-contact hyperfine interaction.

References

1 Y. Kitaoka, K. Ishida, K. Fujiwara, K. Asayama, H. Katayama-Yoshida, Y. Okabe and T. Takahashi, IBM J. Res. Dev., 1989, 33, 277.

2 K. Asayama, Y. Kitaoka and Y. Kohori, J. Phys., Colloq. 1988 (C8, Proc. Int. Congr. Magn., 1988, Pt. 3), C8-2125.

3 Y. Kitaoka, K. Ishida, K. Asayama, H. Katayama-Yoshida, Y. Okabe and T. Takahashi, in High T_c Supercond., Proc. Int. Symp. Electron. Struct. High T_c Supercond. 1988 (Pub. 1989), Pergamon, Oxford, 111.

4 Yu. A. Buslaev, E.A. Kravchenko, M. Yu. Burtsev and L.A. Aslanov, Coord. Chem. Revs., 1989, 93, 185.

5 J. Seliger and R. Blinc, Ferroelectrics, 1988, 78, 223.

6 F. Milia, Ferroelectrics, 1988, 78, 283.

7 I.P. Aleksandrova, Ferroelectrics, 1988, 78, 11.

8 M. Mačkowiak, J. Mol. Struct., 1989, 192, 189.

9 J.A.S. Smith, Brit. UK Pat. Appl. GB 2,200,462; Chem. Abstr., 1989, 111, 92226.

10 V.A. Shishkin, USSR SU 1,448,257; Chem. Abstr., 1989, 110, 164865.

11 J. Seliger, V. Žagar and R. Blinc, Chem. Phys. Lett., 1989, 164, 405.

12 O. Masching, A. Lötz and J. Voitländer, J. Mol. Struct., 1989, 192, 279.

13 L.A. Popkova, E.N. Gur'yanova, V.I. Muromtsev and A.P. Zhukov, J. Gen. Chem. USSR, 1988, 58, 2225.

14 L.A. Popkova, E.N. Gur'yanova, V.I. Muromtsev and A.P. Zhukov, J. Gen. Chem. USSR, 1989, 59, 25.

15 J. Seliger, V. Žagar and R. Blinc, Z. Phys. B, 1989, 77, 439.

16 A. Colligiani, J. Murgich, R. Angelone and R. Ambrosetti, Phys. Rev. B, 1989, 39, 6765.

17 V.S. Grechishkin, G.V. Mozzhukin, N. Ya. Sinyavskii and E.V. Yurepina, Izv. Vyssh. Uchebn. Zaved., Fiz. (Engl. transl.), 1988, 31, 647.

18 J.-G. Yoon, I. Yu and S.-I. Kwun, Sae Mulli, 1986, 26, 298.

19 S. Cho and S.H. Choh, J. Korean Phys. Soc., 1988, 21, 399.

20 N. Nakamura, Ferroelectrics, 1988, 78, 191.

21 A.P. Zhukov and W. Zapart, Ferroelectrics, 1988, 80, 91.

22 Z.M. Saleh, G.A. Williams and P.C. Taylor, Phys. Rev. B, 1989, 40, 10557.

23 L.A. Zemnukhova, R.L. Davidovich and G.A. Fedorishcheva, Koord. Khim., 1988, 14, 641; Chem. Abstr., 1989, 110, 127320.

24 T.-J. Wang and J.-C. Yang, J. Mol. Struct., 1989, 192, 387.

25 G.B. Furman, G.E. Kibrik, A. Yu. Poljakov and I.G. Shaposhnikov, J. Mol. Struct., 1989, 192, 207.

26 C. Gérard-Dion and E.A.C. Lucken, Inorg. Chim. Acta, 1989, 155, 207.

27 T. Okuda, Y. Aihara, N. Tanaka, K. Yamada and S. Ichiba, J. Chem. Soc. Dalton Trans., 1989, 631.
28 G. Guidi, C. Bucci, P. Carretta, R. de Renzi, R. Tedeschi, G. Calestani and C. Vignali, Physica C, 1989, 162-4, 189.
29 T.D. Newbound, M.R. Colsman, M.M. Miller, G.P. Wulfsberg, O.P. Anderson and S.H. Strauss, J. Am. Chem. Soc., 1989, 111, 3762.
30 A. Erginsav and N. Deliktas, Rev. Fac. Sci. Univ. Istanbul, Ser. C, 1987, 52, 35.
31 E.A. Romanenko, Teor. Eksp. Khim., (Engl. transl.), 1989, 25, 216.
32 M. Mishima and T. Okuda, Bull. Chem. Soc. Jpn., 1989, 62, 2263.
33 G.S. Harbison, A. Slokenbergs and T.M. Barbara, J. Chem. Phys., 1989, 90, 5292.
34 P. Kamasa, F.I. Toth, J. Kasprzak and J. Pietrzak, Nauchn. Appar., 1988, 3 (3), 31.
35 K. Horiuchi, Bull. Coll. Sci. Univ. Ryukyus, 1987, 45, 53.
36 N. Reddy and P.T. Narasimhan, J. Mol. Struct., 1989, 192, 345.
37 F. Milia and G. Papavassiliou, Phys. Rev. B, 1989, 39, 4467.
38 S. Chen and D.C. Ailion, Solid State Commun., 1989, 69, 1041.
39 F. Milia, G. Papavassiliou and E. Giannakopoulos, Phys. Rev. B, 1989, 39, 12349.
40 M. Zdanowska-Frączek and M. Maćkowiak, Ferroelectrics, 1988, 77, 145.
41 M. Zdanowska-Frączek, M. Maćkowiak and R. Jakubas, Ferroelectrics, 1988, 80, 87.
42 A. Ishikawa, Y. Ito, K. Horiuchi, T. Asaji and D. Nakamura, J. Mol. Struct., 1989, 192, 221.
43 A. Ishikawa, T. Asaji, D. Nakamura and R. Ikeda, Z. Naturforsch., 1989, 44a, 125.
44 E.S. Kozlov, I.A. Kyuntsel', V.A. Mokeeva and G.B. Soifer, Russ. J. Inorg. Chem., 1988, 33, 1634.
45 I.A. Kyuntsel', V.A. Mokeeva, G.B. Soifer and I.G. Shaposhnikov, J. Mol. Struct., 1989, 192, 215.
46 G.E. Kibrik, E.S. Kozlov, I.A. Kyuntsel', V.A. Mokeeva, A. Yu. Polyakov and G.B. Soifer, Russ. J. Phys. Chem., 1989, 63, 535.
47 A.A. Boguslavskii, D.L. Zagorskii, I.S. Zheludev, V.V. Kirilenko, R. Sh. Lotfullin and V.I. Pakhomov, Sov. Phys. Crystallogr., 1989, 34, 453.
48 A. Ishikawa, K. Horiuchi, R. Ikeda and D. Nakamura, J. Mol. Struct., 1989, 192, 237.
49 C. Dimitropoulos and J. Pelzl, Z. Naturforsch., 1989, 44a, 109.
50 D. Borchers and A. Weiss, Ber. Bunsenges. Phys. Chem., 1989, 93, 559.
51 V.P. Feshin, G.V. Dolgushin, I.M. Lazarev and M.G. Voronkov, J. Gen. Chem. USSR, 1988, 58, 2017.
52 V.P. Feshin, G.V. Dolgushin, I.M. Lazarev and M.G. Voronkov, J. Gen. Chem. USSR, 1989, 59, 656.
53 V.P. Feshin, G.V. Dolgushin, I.M. Lazarev and M.G. Voronkov, J. Gen. Chem. USSR, 1989, 59, 659.
54 V.P. Feshin, G.V. Dolgushin, I.M. Lazarev and M.G. Voronkov, Russ. J. Struct. Chem., 1989, 30, 144.
55 V.P. Feshin, P.A. Nikitin, T.K. Gar, O.A. Dombrova and N.A. Viktorov, Teor. Eksp. Khim. (Engl. transl.), 1989, 25, 355.
56 V.P. Feshin, G.V. Dolgushin, I.M. Lazarev, M.G. Voronkov, E.A. Kravchenko, V.G. Morgunov and M. Yu. Burtsev, Dokl. Akad. Nauk SSSR, (Engl. transl.), 1988, 301, 743.
57 Yu. A. Buslaev, E.A. Kravchenko, V.G. Morgunov, M. Yu. Burtsev, V.P. Feshin, G.V. Dolgushin, I.M. Lazarev and M.G. Voronkov, Dokl. Akad. Nauk SSSR, (Engl. transl.), 1988, 301, 754.
58 M.A. Singh and R.L. Armstrong, J. Magn. Reson., 1989, 84 448.
59 M. Krupski and R.L. Armstrong, Can. J. Phys., 1989, 67, 566.
60 Y. Furukawa and D. Nakamura, Ber. Bunsenges. Phys. Chem., 1989, 93, 13.
61 T. Suto, T. Asaji and D. Nakamura, J. Mol. Struct., 1989, 192, 229.
62 H. Terao and T. Okuda, Chem. Lett., 1989, 649.

63 S.A. Ryabov, O.G. Ellert, V.D. Butskii, V.S. Morgunov, M. Yu. Burtsev and V.S. Pervov, Russ. J. Inorg. Chem., 1988, 33, 1855.
64 K. Horiuchi and D. Nakamura, Ber. Bunsenges. Phys. Chem., 1989, 93, 909.
65 R.J.C. Brown, A. Christides, M. Gourdji and L. Guibé, J. Mol. Struct., 1989, 192, 355.
66 S. Chen and D.C. Ailion, Phys. Rev. B, 1989, 40, 2332.
67 J. Hartmann, S.-Q. Dou and A. Weiss, Z. Naturforsch, 1989, 44a, 41.
68 A. Kehrer, S.-Q. Dou and A. Weiss, Z. Naturforsch, 1989, 44a, 659.
69 A. Habiyakare and E.A.C. Lucken, J. Mol. Struct., 1989, 213, 231.
70 Y. Nishiyama and K. Watanabe, J. Phys. Soc. Jpn., 1989, 58, 2213.
71 D.F. Baisa, A.I. Barabash, E.A. Shadchin, A.I. Shanchuk and V.A. Shishkin, Sov. Phys. Crystallogr., 1989, 34, 621.
72 K. Yamada, T. Tsuritani, T. Okuda and S. Ichiba, Chem. Lett., 1989, 1325.
73 A.A. Boguslavskii, D.L. Zagorskii, R. Sh. Lotfullin, V.I. Pakhomov, V.V. Kirilenko and G.K. Semin, Russ. J. Inorg. Chem., 1989, 34, 1023.
74 G.K. Semin, S.I. Kuznetsov and T.L. Khotsyanova, Russ. J. Inorg. Chem., 1989, 34, 37.
75 H. Nakamura, Y. Kitaoka, K. Yoshimura, Y. Kohori, K. Asayama, M. Shiga and Y. Nakamura, J. Phys., Colloq. 1988, (C8, Proc. Int. Congr. Magn., 1988, Pt. 1), C8-257.
76 H. Niki, M. Hashimoto and K. Mano, Kaiho-Kagaku PC Kenkyukai, 1989, 11, 3.
77 W.W. Warren Jr., R.E. Walstedt, G.F. Brennert, R.J. Cava, B. Batlogg and L.W. Rupp, Phys. Rev. B, 1989, 39, 831.
78 W.W. Warren Jr., R.E. Walstedt, G.F. Brennert, R.J. Cava, R. Tycko, R.F. Bell and G. Dabbagh, Phys. Rev. Lett., 1989, 62, 1193.
79 S.M. Zahurak, D.W. Murphy, S. Nakahara, W.W. Warren Jr., D.M. Krol, A.M. Thayer, D.C. Douglass, G.F. Brennert, S.H. Glarum, E.M. Gyorgy, S.M. Fine and M. Greenblatt, Solid State Ionics, 1989, 32-3, 1104.
80 S.M. Zahurak, S.M. Fine, D.W. Murphy, M. Greenblatt, S. Nakahara, W.W. Warren Jr., D.M. Krol, A.M. Thayer, D.C. Douglass and G.F. Brennert, Report 1988 AD-A 197176.
81 W.W. Warren Jr., R.E. Walstedt, G.F. Brennert, R.F. Bell, G.P. Espinosa and R.J. Cava, Physica C, 1989, 162-4, 179.
82 K. Ishida, Y. Kitaoka and K. Asayama, J. Phys. Soc. Jpn., 1989, 58, 36.
83 Y. Kitaoka, K. Ishida, F. Fujiwara, T. Kondo, K. Asayama, M. Horvatic, Y. Berthier, P. Butaud, P. Segransan, C. Berthier, H. Katayama-Yoshida, Y. Okabe and T. Takahashi, Springer Ser. Solid-State Sci., (Strong Correlation and Superconductivity), 1989, 89, 262.
84 K. Fujiwara, Y. Kitaoka, K. Asayama, H. Sasakura, S. Minamigawa, K. Nakahigashi, S. Nakanishi, M. Kogachi, N. Fukuoka and A. Yanase, J. Phys. Soc. Jpn., 1989, 58, 380.
85 M. Matsumura, H. Yamagata, Y. Yamada, K. Ishida, Y. Kitaoka, K. Asayama, H. Takagi, H. Iwabuchi and S.-I. Uchida, J. Phys. Soc. Jpn., 1989, 58, 805.
86 Y. Kitaoka, K. Ishida, K. Asayama, H. Takagi, H. Iwabuchi and S. Uchida, J. Phys., Colloq. 1988 (C8, Proc. Int. Congr. Magn., 1988, Pt. 3), C8-2157.
87 K. Ishida, T. Kondo, Y. Kitaoka and K. Asayama, J. Phys. Soc. Jpn., 1989, 58, 2638.
88 Y. Kohori, H. Shibai, Y. Oda, Y. Kitaoka, T. Kohara and K. Asayama, J. Phys. Colloq. 1988 (C8, Proc. Int. Congr. Magn., 1988, Pt. 3), C8-2237.
89 K. Kumagai and Y. Nakamura, Physica C, 1989, 157, 307.
90 T. Oashi, K. Kumagai, Y. Nakajima, T. Tomita and T. Fujita, Physica C, 1989, 157, 315.
91 K. Kumagai, T. Oashi, H. Nakajima, T. Tomita and T. Fujita, Physica C, 1989, 162-4, 183.
92 T. Oashi, K. Kumagai, H. Nakajima, M. Kikuchi and Y. Syono, Physica C, 1989, 161, 367.
93 Y. Nakamura and K. Kumagai, Physica C, 1989, 162-4, 187.
94 Y. Nakamichi, K. Kumagai, H. Nakajima and T. Fujita, Physica C, 1989, 162-4, 185.

95 I. Imai, H. Yasuoka, T. Shimizu, Y. Ueda and K. Kosuge, J. Phys. Soc. Jpn., 1989, 58, 1528.
96 K. Yoshimura, T. Imai, T. Shimizu, Y. Ueda, K. Kosuge and H. Yasuoka, J. Phys. Soc. Jpn., 1989, 58, 3057.
97 H. Yasuoka, Oyo Butsuri, 1989, 58, 739; Chem. Abstr., 1989, 111, 246525.
98 H. Yasuoka, T. Imai and T. Shimizu, Springer Ser. Solid-State Sci., (Strong Correlation and Superconductivity), 1989, 89, 254.
99 H. Yasuoka, S. Sasaki, T. Imai, T. Shimizu, Y. Ueda and K. Kosuge, Phase Transitions, 1989, 15, 183.
100 H. Yasuoka, T. Shimizu, T. Imai, S. Sasaki, Y. Ueda and K. Kosuge, Hyperfine Interact., 1989, 49, 167.
101 A.J. Vega, W.E. Farneth, E.M. McCarron and R.K. Bordia, Phys. Rev. B, 1989, 39, 2322.
102 A.J. Vega, M.K. Crawford, E.M. McCarron and W.E. Farneth, Phys. Rev. B, 1989, 40, 8878.
103 A.J. Vega, M.K. Crawford, E.M. McCarron and W.E. Farneth, Physica C, 1989, 162-4, 165.
104 H. Schiefer, M. Mali, J. Roos, H. Zimmermann, D. Brinkmann, S. Rusiecki and E. Kaldis, Physica C, 1989, 162-4, 171.
105 H. Zimmermann, M. Mali, D. Brinkmann, J. Karpinski, E. Kaldis and S. Rusiecki, Physica C, 1989, 159, 681.
106 K. Müller, M. Mali, J. Roos and D. Brinkmann, Physica C, 1989, 162-4, 173.
107 M.V. Eremin, I. É. Graboĭ, A.V. Egorov, V.V. Naletov, M.S. Tagirov and M.A. Teplov, JETP Lett., 1989, 49, 511.
108 A. Yu. Zavidonov, M.V. Eremin, A.V. Egorov, V.V. Naletov, M.S. Tagirov, M.A. Teplov and N.M. Chebotaev, JETP Lett., 1989, 50, 201.
109 A.V. Bondar', S.M. Ryabchenko, Yu. V. Fedotov and A.A. Motuz, JETP Lett., 1989, 50, 146.
110 B.A. Aleksashin, A.M. Bogdanovich, S.V. Verkhovskii, Yu. I. Zhdanov, V.L. Kozhevnikov, K.N. Mikhalev, K.A. Okulova, A.A. Romanyukha, V.V. Serikov, A.M. Sorkin, V.A. Tsurin, S.M. Cheshnitskii, Yu. A. Shvachko, V.I. Anisimov, M.I. Katsnel'son, A.I. Likhtenshtein, A.V. Trefilov and S.A. Turzhevskii, JETP Lett, 1988, 48, 289.
111 B.A. Aleksashin, V.P. Voronin, S.V. Verkhovskii, B.N. Goshchitskii, S.A. Davydov, Yu. I. Zhdanov, A.E. Kar'kin, V.L. Kozhevnikov, A.V. Mirmel'shtein, K.N. Mikhalev, M.V. Sadovskii, V.V. Serikov and S.M. Cheshnitskii, Sov. Phys. JETP, 1989, 68, 382.
112 A.K. Rajarajan, V.R. Palkar, N.C. Mishra, M.S. Multani, R. Vijayaraghavan and L.C. Gupta, Solid State Commun, 1989, 71, 835.
113 A.K. Rajarajan, L.C. Gupta, R. Vijayaraghavan and N.C. Mishra, Physica C, 1989, 162-4, 163.
114 R. Vijayaraghavan, A.K. Rajarajan and L.C. Gupta, Stud. High Temp. Supercond., 1989, 2, 79.
115 R. Michalak, M. Rosenberg and H. Lütgemeier, Physica C, 1989, 162-4, 267.
116 H. Lütgemeier and B. Rupp, J. Phys., Colloq. 1988 (C8, Proc. Int. Congr. Magn., 1988, Pt. 3), C8-2147.
117 M. Itoh, K. Karashima, M. Kyogoku and I. Aoki, Physica C, 1989, 160, 177.
118 L. Mihály, Phys. Scr., 1989, T25, 80.
119 J. Danon, Quim. Nova, 1988, 11, 92.
120 K. Kumagai, Y. Nakamichi, Y. Nakamura, I. Watanabe, H. Nakajima and T. Fujita, Hyperfine Interact., 1989, 50, 583.
121 K. Kumagai, I. Watanabe, Y. Nakamura and H. Nakajima, J. Phys., Colloq. 1988 (C8, Proc. Int. Congr. Magn., 1988, Pt. 3), C8-2145.
122 F. Borsa, M. Corti, T. Rega, A. Rigamonti and J. Ziolo, Helv. Phys. Acta, 1989, 62, 796.
123 I. Watanabe, K. Kumagai, Y. Nakamura and H. Nakajima, J. Magn.Magn. Mater., 1988, 76-7, 599.
124 T. Kobayashi, S. Wada, Y. Kitaoka, K. Asayama and K. Amaya, J. Phys. Soc. Jpn., 1989, 58, 779.
125 S. Wada, T. Kobayashi, M. Kaburagi, K. Shibutani and R. Ogawa, J. Phys. Soc. Jpn., 1989, 58, 2658.

126 T. Kobayashi, S. Wada, Y. Kitaoka and K. Asayama, J. Phys. Soc. Jpn., 1989, 58, 2662.
127 T. Kobayashi, S. Wada, K. Shibutani and R. Ogawa, J. Phys. Soc. Jpn., 1989, 58, 3497.
128 V.D. Doroshev and M.M. Savosta, JETP Lett., 1989, 50, 363.
129 L.E. Erickson, Phys. Rev. B, 1989, 39, 6342.
130 M. Lukac and E.L. Hahn, Opt. Commun., 1989, 70, 195.
131 C. Dimitropoulos, J.P. Bucher, F. Borsa and M. Corti, Phys. Rev. B, 1989, 39, 7232.

3
Rotational Spectroscopy

BY J. H. CARPENTER

1 Introduction

I have kept in the main to the organisation of the material used by my predecessor. I have used abbreviations for some commonly used terms, and have defined them when first used. Several books and reviews of relevance to rotational spectroscopy have been published in the last year or so. Among these are the proceedings of conferences: 'The Structure of Small Molecules and Ions',[1] 'Structure and Dynamics of Weakly Bound Molecular Complexes',[2] 'The Physics and Chemistry of Interstellar Molecular Clouds: mm and Sub-mm Observations in Astrophysics'[3] and 'Structure and Dynamics of the Interstellar Medium'.[4] There are also reviews on the spectroscopy of the H_2^+ ion;[5] the IR predissociation spectrum of the H_3^+ ion;[6] IR spectroscopy[7,8] of weakly bound complexes; theoretical methods for rovibrational states of floppy molecules;[9] spectroscopic properties of heavy *p*-block diatomics;[10] and carbon molecules, ions and clusters.[11]

2 Weakly Bound Complexes

A spectroscopic study[12] of Ne.HF showed a weak fundamental and strong combination band when studied by IR difference frequency laser (DFL) spectroscopy. The anomalous intensities were ascribed to the large amplitude internal rotor motion of the HF in the isotropic potential. An extended study[13] of Ar.HF using a slit nozzle gave spectroscopic constants for nine of the eleven vibrational states below the Ar + HF(v=1,J=0) dissociation limit. A novel IR pump, UV probe laser induced fluorescence (LIF) method was used[14] to study the vibrational overtone of ArOH. There was a slight increase in the R_{cm} (cm=centre of mass) distance in v=2 compared with the ground state.

A pulsed jet Fourier-Transform Microwave (FTMW) study[15] of the floppy Ar.HCN complex showed that the ^{14}N quadrupole coupling tensor element χ_a varies linearly with $J(J+1)$, implying a decrease in the average Ar--HCN bending amplitude on increasing J. Higher-J states were observed in a MW and IR investigation[16] of Ar.HCN, and the large amount of centrifugal distortion seen also confirmed the increasing alignment of the HCN axis with the cm axis. The ν_1 (CH stretch) band showed a large but different centrifugal distortion, while the $\nu_1+\nu_5$ band was perturbed. In another study[17] of Ar.HCN and $(HCN)_2$, a slit jet was used with a continuous wave (cw) direct absorption spectrometer. Although not as sensitive as FT methods, it is widely tuneable, easily automated and extends to higher frequencies. Centrifugal terms to J^{14} were required to fit the observed lines!

The T-shaped C_{2v} structure of $Ar.C_2H_2$ was confirmed by rotational spectroscopy[18] despite the very low dipole moment of the complex. The centrifugal distortion constants allowed estimates of the low frequency stretching and bending vibrations between C_2H_2 and Ar of 36.6 cm^{-1} and 8.9 cm^{-1} respectively.

Complexes of both Ar and Kr with OCS were investigated[19] in the CO stretching region using a tuneable diode laser (TDL) with a pulsed jet; little change in the structure was observed on going from the ground state to the excited vibrational state. A similar study[20] of Ne, Ar and Kr complexed to BF_3, near the ν_3 band of the monomer, showed the C_{3v} symmetry of the complex and gave vibrational shifts and rotational constants. Pulsed jet FTMW enabled the dipole moment, rotational and ^{83}Kr quadrupole coupling constants to be obtained[21] for $Ar.PF_3$ and $Kr.PF_3$. These are asymmetric tops with the noble gas atom probably lying on a PF_2 face. No evidence for internal rotation was observed. In $Ar.CH_3OH$, on the other hand, two tunnelling states were observed in the MW spectrum.[22] A rotational analysis of the A tunnelling state indicated a T-shaped heavy atom skeleton similar to $Ar.CH_3Cl$, with no evidence of hydrogen bonding between OH and argon.

Pulsed jet FTMW studies of complexes of Ar and Kr with pyridine[23] and Ar with pyrrole[24] were reported. In both cases the noble gas atom lies above the plane of the molecule. The dipole moment of $Ar.C_4H_4N$ was less than in C_4H_4N itself; the negative end of the dipole was rotated towards the argon.

Two-photon laser pump-probe spectroscopy was used[25] to determine the structure and dissociation dynamics of the $Ne_2.Cl_2$ trimer. Rotational constants from the $\tilde{B}$-$\tilde{X}$ transition showed a distorted tetrahedron, with r(ClCl) the same as in the corresponding states of Cl_2 and r(NeNe) = 323 pm, R(cm Ne_2-cm Cl_2) = 312 pm in both states. The tetramer Ne_3Cl_2 was also seen; the three neon atoms encircle the Cl-Cl bond. The $Ar_2.DCl$ and $Ar_3.DCl$ complexes were observed using jet FTMW spectroscopy.[26]

The first spectroscopic observation[27] of the $CO.CO_2$ complex used IR TDL spectroscopy to detect ν_3 of CO_2, then FTMW to obtain rotational constants. The complex is T-shaped with the CO along the C_2 axis of CO_2 and the two carbon atoms contiguous; it is therefore isostructural as well as isoelectronic with $N_2.CO_2$ and $HCN.CO_2$ (but see below) and is the weakest complex of the three. Similar information on the isostructural $HCN.CO_2$ was obtained[28] from IR TDL and FTMW spectra. A linear isomer of $HCN.CO_2$ has also been reported;[29] this was observed in helium or He/Ne carriers but not in argon. A substitution structure for the linear species was determined, and torsional amplitudes for the CO_2 and HCN units. A well depth of 590 cm^{-1} was obtained from centrifugal distortion constants, implying that the linear form is the more stable, in disagreement with the intensity information. The $HCCH.CO_2$ complex showed[30] yet another geometry in which the molecule is planar with C_{2v} symmetry and the two molecular axes parallel. An IR molecular beam spectrum with optothermal detection gave much more information on this species than previous studies; the dipole moment was 0.167 D. The same method was used[31] to investigate the IR spectra of $CO_2.HF$ and OCS.HF; FTMW spectra were also taken. The spectra were analysed in terms of a quasilinear molecule whose A rotational constant (or ν_3 bending mode) value of 2.7 cm^{-1} indicates that the molecule is more bent than previous studies suggest.

Several studies of the linear ONN.HF and bent NNO.HF isomers were carried out. An IR study[32] of the HF stretching vibration using a slit jet determined the ONN.HF sequence in the linear complex. Isotopic substitution gave R_e(cm) = 445.6 pm, and the dissociation energy of 886 cm^{-1} was between the weaker $N_2.HF$ and stronger HCN.HF complexes. In the bent isomer the broad linewidths made rotational analysis difficult; the authors conclude that the linear molecule has a relatively rigid hydrogen bond whereas the bent isomer has a stronger hydrogen bond but with a softer bending potential. The FTMW studies of the bent NNO.HF complex[33] and both bent and linear $N_2O.DF$ isomers[34] were consistent; the R(F-cm) distance of 350.70 pm in the bent isomer of DF was 1.2 pm shorter than in HF.

The IR spectra of the OC.HF complex[35] and the OC.HCl species[36] were obtained with a computer-controlled TDL spectrometer and a cw planar (slit) jet. The rotational analyses gave B and D_J in both states; the vibrational shift from CO itself was 24.4 cm^{-1} in OC.HF and 12.2 cm^{-1} in OC.HCl, both to higher frequency. The ^{14}N quadrupole coupling constants found[37] from the FTMW spectrum of N_2.HBr indicated average torsional amplitudes of 20.05° for N_2 and 25.45° for HBr. The *r*(N--Br) distance of 394.74(2) pm in N_2.HBr was shortened by 1.6 pm on deuteration. In N_2.H_2O evidence for the complex tunnelling motion was seen[38] in the FTMW spectrum. The average structure has N-N--H-O nearly linear.

The FTIR spectra of $(HF)_2$ and its deuterated isomers were analysed[39] in terms of the tunnelling splitting for hydrogen bond rearrangements which were found to be mode specific. Analysis of the *K*=1-0 sub-band of the HF stretch in HF.DF yielded[40] rovibrational constants. An optothermal molecular beam electric resonance (MBER) MW and IR study[41] of HF.HCl revealed, in addition to the HF--HCl dimer, a metastable HCl--HF species. This is L-shaped with the H of HCl at right angles to the Cl--H-F axis. A detailed analysis[42] of the FTIR spectrum of $(HCl)_2$ between 40 and 200 cm^{-1} showed eight sub-bands including two from $H^{35}Cl.H^{37}Cl$. The tunnelling splitting of 463979.2 MHz in $(HCl)_2$ was measured directly[43] by far-IR laser sideband spectroscopy in the K_a=1 states. The tunnelling splitting was also measured[44] in the IR spectrum of $H^{35}Cl.H^{37}Cl$ from bands which are forbidden in the symmetric dimers. The splitting for $(DCl)_2$ was just over a third of the $(HCl)_2$ value.

The analysis[45] of the perpendicular rotation-tunnelling *K*=1-2 spectrum of $(H_2O)_2$ in the far-IR favoured its interpretation as one band with tunnelling terms rather than two vibrational bands. The RF and MW MBER spectrum[46] showed Coriolis perturbations in the *K*=0 and 1 levels of the E_2^+ states. Further work[47,48] on the microwave spectrum of $(H_2O)_2$ was reported. A study[49] of $(D_2O)_2$ showed the tunnelling splitting varied markedly with (*B*-*C*) in different symmetry states. All four OH stretching vibrations in $(H_2O)_2$ have been examined[50] and characterised as essentially independent H-donor or H-acceptor motions.

The structure of $H_2O.SO_2$ was deduced from its MW spectrum.[51] It is nonplanar with an angle of 44° between the H_2O and SO_2 molecules with the O of H_2O closest to the S and the hydrogens above the oxygens of SO_2. Splitting due to internal rotation of the H_2O was seen. Some work on $(SO_2)_2$ was also reported.

Two reports on ternary complexes give information about third-body interactions. The jet FTMW spectra[52] of $(CO_2)_2.H_2O$ show that it has a C_2 axis coincident with that of the H_2O. The CO_2 molecules are parallel to each other with 57° between their axes and the C--C line as in the $(CO_2)_2$ dimer. The H_2O sits above the $(CO_2)_2$ plane with the hydrogens away from the CO_2 molecules and its plane almost parallel to them. The CH stretch at 3308 cm^{-1} of the HCN.$(HF)_2$ trimer was observed[53] using optothermal detection and an IR colour centre laser (CCL). Rotational constants agree with a calculated structure in which the N atom is hydrogen bonded to the inner hydrogen of the $(HF)_2$ dimer.

Finally in this section, rotational ^PQ and ^RQ structure was seen[54] in one of the bands in the IR vibrational predissociation spectrum of the $NH_4^+(NH_3)_4$ ion cluster, formed in a corona discharge in a mixture containing NH_3 followed by a supersonic expansion. The line spacing implied free internal rotation of a NH_3 molecule attached to the central NH_4^+ ion.

3 Triatomic molecules and ions

The $2\nu_3$ overtone of H_3^+ has been seen[55] in the emission from a hollow-cathode discharge. The results were added to those from a previously observed Jupiter spectrum and assigned with the aid of *ab initio* calculations.

The rotational data obtained[56] in the MW spectrum of HBO were added to others to produce a cubic force field and equilibrium structure: r_e(BH) = 116.667 pm, r_e(BO) = 120.068 pm.

The FTIR spectrum[57] of the ν_2 bending vibration in HCN gave rotational constants and band intensities for the fundamentals of HCN, $H^{13}CN$ and $HC^{15}N$ as well as five hot bands. Another FTIR study in the near-IR analysed[58] 28 overtone bands between 5400 and 15100 cm^{-1}. The Stark effect suggested that the molecule is still quasilinear rather than bent in these overtones. A re-analysis[59] of the 190 nm band system of HCN and DCN was interpreted in terms of transitions to a single state, $\tilde{A}^1A''$, not two as previously suggested, and an equilibrium structure in this state was determined.

Laser excitation spectroscopy was employed to investigate the $ClCN^+$ ion[60] and $BrCN^+$ ion.[61] Substitution structures were obtained for the $\tilde{X}^2\Pi$ states. The C-Hal bondlengths are shorter and CN bondlengths longer than in the corresponding molecules.

Measurements[62] on HCS^+ in the 3080-3150 cm^{-1} region enabled vibrations involving all three modes to be analysed. From these B_e = 0.71520 cm^{-1} was obtained. All bands except ν_1 showed Fermi and/or Coriolis perturbations.

The C_3 molecule is quasilinear with a low bending frequency. Hot bands of ν_3 in C_3 were observed[63] using IR TDL kinetic spectroscopy. The bending vibration was estimated as 66 cm^{-1} in (010) and 45 cm^{-1} in (011). A stimulated emission pumping (SEP) scheme was used by two groups to study high vibrational levels of C_3. One group analysed[64] the $(0\nu_20)$ vibrational states to give rotational constants and ω_2 = 54.6 cm^{-1}; the other group investigated[65] the $(0\nu_21)$ and $(6\nu_21)$ states and found ω_2=51.3 cm^{-1} for $(0\nu_21)$ and 155.3 cm^{-1} for $(6\nu_21)$. Strong anharmonicity was observed and large l-doubling constants indicated strong rotation-vibration coupling. A new band system was seen[66] between 266 and 302 nm in a resonance enhanced multiphoton ionisation (REMPI) study of jet-cooled C_3. Rotational analysis indicated $\Sigma_u{}^+-\Sigma_g{}^+$ structure originating from the ground vibronic level. The upper electronic state was assigned as $^1\Pi_g$ or $^1\Delta_g$ coupled by Herzberg-Teller coupling to a $^1\Sigma_u{}^+$ state. The same electronic system was also observed in a LIF study;[67] information on the potential energy surface in the $\tilde{X}^1\Sigma_g{}^+$ state was obtained.

Rotational constants of three vibrational states of $\tilde{X}^2\Pi$ and four of $\tilde{A}^2\Pi$ were obtained[64,68] from the LIF spectrum of NCS. The SEP scheme showed that the (020) - (001) multiplet was perturbed by Fermi resonance and split by Renner-Teller and spin-orbit coupling.

Two-step absorption $\tilde{C}^2\Sigma_g{}^+ - \tilde{A}^2\Pi_u - \tilde{X}^2\Pi_g$ in beams of $CO_2{}^+$ was used[69] to characterise the $\tilde{C}^2\Sigma_g{}^+$ state. A rotational analysis showed that the $\tilde{C}$ state was perturbed, probably by another $^2\Sigma_g{}^+$ state. The bondlength in $\tilde{C}$ was shorter than in $\tilde{X}$. A corresponding study[70] of the predissociative $\tilde{C}^2\Sigma_g{}^+$ state of $CS_2{}^+$ showed that r_0 in $\tilde{C}$ was again shorter than in the ground state. Rotational constants were obtained[71] from the (001)-(000) and (011)-(010) bands of $^{14}CO_2$ using FTIR and TDL spectrometers. Metal-insulator-metal diodes were used[72] to mix two CO_2 lasers to study the high-J rotational transitions in all vibrational states of OCS below 2000 cm^{-1} with an accuracy of 200 kHz.

The first spectral characterisation of $HOSi^+$ and $DOSi^+$ used a hollow-cathode discharge. The ν_1(OH) vibration at 3662.3 cm^{-1} was observed with a DFL spectrometer,[73] while a TDL was used[74] for ν_3(OSi) at 1127.0 cm^{-1}. From the rotational constants the r_0 structure was obtained: r(HO) = 94.0 pm, r(OSi) = 153.7 pm. The first high-resolution IR spectra of SiH_2 in its ground $\tilde{X}^1A_1$ state were obtained[75] by TDL kinetic spectroscopy; rotational constants were obtained for both the ground vibrational and ν_2 state. In addition to a r_0 structure and harmonic force field, the r_e

structure was estimated: r_e(SiH) = 151.40 pm, θ_e(HSiH) = 92.08°. The measurements[76] of the 1_{01}-0_{00} rotational transition in SiC_2 were added to astronomical data to give rotational constants. The structure confirmed previous work with the very small CSiC angle of 40.3°. The dipole moment is $\mu = \mu_a$ = 2.393D. The analysis of the *b*-type transitions in the MW spectrum of three isotopic species of $SiCl_2$ yielded[77] r_0(SiCl) = 207.0 pm and θ_0(ClSiCl) = 101.25°. The chlorine quadrupole coupling tensor showed high asymmetry caused predominantly by π-electron back donation from Cl to Si. A LIF study[78] of the $\tilde{A}^1B_1(060)$-$\tilde{X}^1A_1(000)$ band of $SiCl_2$ gave the structure in both $\tilde{X}$ and $\tilde{A}$ states; there is a large angle opening and a small decrease in bondlength on excitation.

A new tuneable far-IR spectrometer with a cw planar jet was used[79] to produce high concentrations of NH_2, and one spin-rotation component of the 1_{11}-0_{00} transition was resolved. Another study[80] of jet-cooled NH_2 using a CCL investigated the kinetics of the system. An IR DFL allowed the first high-resolution study[81] of NH_2^+ in ν_3 of the $\tilde{X}^3B_1$ state. The ion is quasilinear; its spectra were fitted as a doubly degenerate state of a linear rotor and as an asymmetric rotor.

The ν_1 spectrum of the HNO^- ion was investigated[82] using high-resolution autodetachment spectroscopy. The rotational constants obtained gave some information on the structure.

Rotational constants for several states of N_2O have been obtained with high accuracy using heterodyne techniques on vibrational[83,84] transitions. Linewidth[85] and linestrength[85,86] studies have been reported. An anharmonic force field for N_2O up to sextic terms and an equilibrium structure have been calculated[87] from rovibrational data. A cw krypton laser was used[88] to dissociate N_2O^+ ions *via* predissociation of the $\tilde{A}^2\Sigma^+$ state; the photofragments were detected. The bands were analysed in terms of a large amplitude bending vibration. The Zeeman effect due to the earth's magnetic field was observed[89] in the millimetre-wave (mmw) spectrum of NO_2. The optical-optical double resonance (OODR) spectroscopy of jet-cooled NO_2 enabled[90] separate determination of the splitting in the upper and lower states; Lamb dip spectra were also taken. The ν_1 fundamental and several overtones and combinations of ONCl were the subject of FTIR studies.[91,92]

The vibrational spectrum of H_2O continues to be extensively investigated to obtain line positions,[93] intensities[94] and pressure broadened linewidths.[95-97] A near-IR InGaAsP laser was used[98] between 6380 and 6600 cm^{-1} to study overtones of HDO and D_2O. A full analysis including Fermi and Coriolis interactions was possible for HDO but not for D_2O. Rotational parameters up to J^{10} were obtained[99] from a FTIR study of ν_2 of $D_2{}^{18}O$. Predissociation of the $\tilde{C}^1B_1$ state of H_2O was observed[100] with the (3+1)REMPI scheme $\tilde{X}^1A_1(000)$-$\tilde{C}^1B_1(000)$-$H_2O^+(\tilde{X})$ in a supersonic jet. Linewidths and intensities depend on the rotational state in $\tilde{C}$. The ν_2 and $2\nu_2-\nu_2$ bands of the ground state ($\tilde{X}^2B_1$) of H_2O^+ were measured[101] using a TDL; the states involved were unperturbed.

A r_e structure for HOF was obtained from FTIR studies of the $2\nu_2$ state[102] and the interacting ν_1 and $2\nu_2+\nu_3$ states.[103] Pure rotational spectra between 11 and 786 GHz of HOBr and DOBr were analysed[104] to give rotational and quadrupole coupling constants. A hybrid r_0/r_s structure gave r(OBr) = 183.4 pm, r(OH) = 96.1 pm and θ(HOBr) = 102.3°. The principal axis of the bromine quadrupole coupling tensor is at 1.15° to the bond.

The ν_2 band of $^{16}O^{18}O^{16}O$ and $^{16}O^{16}O^{18}O$ was measured[105] by FTIR spectroscopy while the very weak $2\nu_2$ and $\nu_1+\nu_2-\nu_3$ bands of $^{16}O_3$ were observed[106] at 0.01 cm^{-1} resolution in solar occultation spectra. In order to obtain hyperfine structure free of perturbations, the mmw spectrum of the $\tilde{X}^2A''$ state of OOF was investigated.[107] Rotational structure, but not individual lines, was seen when the $\tilde{A}^2A_2$-$\tilde{X}^2B_1$ absorption of cw jet cooled ClO_2 was measured[108] at 0.1 cm^{-1} resolution using FTUV spectroscopy.

Rotational analysis[109] of the ν_2 band of HDS seen by FTIR required terms up to J^{10}.

4 Tetra-atomic ions and molecules

Isotopically enriched $^{10}BF_3$ spectra[110] of various vibrational bands gave rotational constants and r_e(BF) = 130.70 pm.

Stark and FT MW spectra of $HCN^{17}O$ enabled both ^{14}N and ^{17}O quadrupole coupling constants to be determined.[111] IR and MW spectra have characterised *iso*cyanogen CNCN, an isomer of cyanogen NCCN and the already known CNNC. In the first report[112] it was formed by flash pyrolysis of norbornadienone. The product was stable in the glass cells used to study the ν_1, ν_2 and ν_5 vibrations with FTIR, but less stable in the metal waveguide cell used for MW work on rotational transitions in the ground, ν_4 and ν_5 states. The linear structure was $r(C_1N_2)$ = 117.5 pm, $r(N_2C_3)$ = 131.4 pm and $r(C_3N_4)$ = 116.0 pm. In the second paper,[113] CNCN was produced from N-cyano-2,3-diphenylcyclopropeneimine; higher-J rotational and additional vibrational transitions were studied.

In the sub-mmw laser Stark spectrum of $^{14}ND_3$, dipole-forbidden *a*-*a* and *s*-*s* rotation-inversion transitions were observed.[114] The effective dipole moment decreased slightly as *K* increased. The FTIR spectra of ν_1, ν_3 and $2\nu_4$ of NH_3 have been partially analysed.[115] Perturbation-allowed $\Delta k = +2$ transitions gave the *l*-type (2,-1) term. Another FTIR study[116] found rovibrational constants for three bands of NH_3 in the 4300 cm^{-1} region. A mainly theoretical paper[117] on NH_3 included the rotational analysis of some microwave-optical double resonances (MWODR) of a large number of NH stretching overtones. No *J* odd, *K*=0 lines were seen[118] in the IR DFL spectrum of ν_3 of NH_3^+, thus confirming the planar D_{3h} structure. The bondlength, r_e(NH) = 101.4 pm, was close to that in the NH_3 molecule.

The FTMW spectra of $H^{14}N_3$, $H^{15}N^{14}N_2$ and $H^{14}N_2{}^{15}N$ were analysed[119] to obtain both rotational and ^{14}N quadrupole coupling constants. The A_0 values confirm the structure previously obtained from *B* and *C*, and the quadrupole values agree with other azides. In the FTIR spectrum[120] of ν_4 of HN_3, perturbations by ν_3, $\nu_5+\nu_6$ and other states were seen. The dissociative lifetime and level mixing in high overtones (n_{NH} = 5 or 6) of HN_3 were investigated[121] by observing the LIF of the NH radical produced by photodissociation of the highly vibrationally excited HN_3 molecules.

Two papers[122,123] look at the A_1-A_2 splittings in ν_2/ν_4 of PH_3. Because of the small value of $\nu_2-\nu_4$, the *x*-*y* Coriolis coupling causes enhanced splittings and anomalous intensities.

The pure inversion transitions of H_3O^+ were measured[124] around 55 cm^{-1} with far-IR laser sideband spectroscopy. In the DFL spectrum[125] of H_3S^+ the ν_1 and ν_3 states, which interact strongly by Coriolis coupling, were analysed simultaneously. Perturbation-allowed $\Delta(k-l)=3$ lines enabled C_0 as well as B_0 to be determined, giving r_0(SH) = 135.85 pm and θ(HSH) = 94.113°, very similar to those in PH_3.

The staggering of levels observed[126] in the far-IR spectrum of H_2O_2 gave V_{cis} = 2563 cm^{-1}. The *trans* barrier is 387.1 cm^{-1} and the potential minimum is at a dihedral angle of 111.9° from the *cis* configuration. The v = 4 local mode overtones of the OH stretching vibrations in H_2O_2 were studied[127] in both a static sample and in a supersonic jet. The static spectrum was complex, indicating much torsional motion and/or other interactions. The much simpler jet spectra indicated an antisymmetric (*B*) upper state whose rotational constants were determined. A torsional potential has also been obtained[128] for H_2S_2; this gave V_{cis} = 2800 cm^{-1} and

V_{trans} = 1990 cm^{-1}. Further torsional and rotational spectra of HSSH were also reported.[129] The tunnelling splitting of 7.8904 MHz found[130] in the sub-mmw spectrum of v_t=1 of $HS^{34}SH$ is close to the ground state value.

In a flow reactor, chlorine atoms reacted with Cl_2O to form ClO radicals which then dimerised. The *b*-type sub-mmw spectrum and intensity pattern observed[131] indicated that this dimer was ClOOCl, with r_0(OCl) = 170.44 pm, r_0(OO) = 142.59 pm, θ_0(ClOO) = 110.07° and dihedral angle = 81.03°. The torsional frequency of 127 cm^{-1} was estimated from the intensities.

The FTIR spectrum[132] of SO_3 showed that ν_2 and ν_4 interacted by Coriolis coupling. The rovibrational parameters agreed well with those deduced from a force field, as did r_e(SO) = 141.73 pm.

5 Penta-atomic ions and molecules

Calcium monomethyl, $CaCH_3$, was formed by the reaction of excited calcium atoms with $SnMe_4$ or $HgMe_2$, and investigated[133] by LIF on the $\tilde{A}^2E$-$\tilde{X}^2A_1$ transition. Both A_0 and B_0 could be determined in each state giving a r_0 structure with only one constraint. The quadrupole hyperfine structure of ^{11}B in BF_2OH and BF_2OD was studied[134] using FTMW on a static sample. The complete tensor could be determined.

The linear C_5 molecule has previously been observed only in matrix isolation. The ν_3 vibration (highest asymmetric stretch) at 2169 cm^{-1} has now been reported in the gas phase from three separate groups: in the circumstellar absorption spectrum[135] of a carbon star and in subsequent laboratory spectra on molecules in a hollow-cathode discharge[136] and in a supersonic jet after laser vaporisation from graphite.[137] The rovibrational constants from all three studies are consistent; they imply an average *r*(CC) of 128.3 pm and indicate that, unlike C_3, this is not a floppy molecule. The *l*-doubling constants from measurements[138] on hot bands of C_5 indicate that the lowest bending vibration is at about 118 cm^{-1}. The new molecule C_4Si was detected[139] in an astronomical source and in a glow discharge. The results indicate a linear molecule with B_0 value close to that predicted from an *ab initio* structure: $r(SiC_1)$=168.9, $r(C_1C_2)$ = 127.5, $r(C_2C_3)$ = 130.3 and $r(C_3C_4)$ = 127.8 pm.

Several transitions were measured[140] in the ν_2/ν_4 diad of SiH_4 with better than 1 MHz accuracy using a CO_2 waveguide laser. The vibrationally induced dipole in this diad in $^{28}SiH_4$ gave rise to a linear Stark effect.[141] IRIRDR on the N=2 and N=3 (where N is the number of quanta of Si-H stretches excited) using a near-IR dye laser at 0.2 cm^{-1} resolution gave only partial resolution of the rotational structure.[142] The N=3 and N=5 overtones were investigated[143] at 0.01 cm^{-1} resolution using FTIR and the pseudo-symmetric top structure characteristic of local modes was seen. The N=7 and N=8 overtones of $SiHD_3$ were unperturbed;[144] rotational structure was not resolved in the intracavity laser absorption spectra but the changes in rotational constants between ground and upper state could be determined. A similar picture emerged[145] for the N=6 overtone, while for SiH_4 itself this overtone exhibited well-resolved clusters showing vibrational perturbations. Both IR and IRMWDR were employed[146] for the ν_1/ν_3 diad of SnH_4. A cubic force field was obtained using these and other data.

In the FTIR spectra[147] of $^{74}GeH_3F$ the ν_2/ν_5 and ν_3/ν_6 bands were analysed; Coriolis, *l*-type and *r*-coupling were observed, and A_0 and D_K could be determined from perturbation-allowed lines. A TDL was used[148] to investigate the ν_2/ν_5 band of $^{116}SnH_3{}^{79}Br$; rovibrational constants were obtained. The laser Stark spectrum of the ν_4 vibration of SiD_3F showed[149] some remarkable perturbations because of the field dependent $(\Delta l, \Delta k) = (2,-1)$ interactions caused by the vibrationally induced dipole. Transitions with $\Delta(k-l) = 3$ enabled C_0 to be determined, as well as other rovibrational constants, μ_0 and μ_4. The ν_3 band of SiF_4 has been studied[150] at sub-Doppler resolution using a laser sideband spectrometer and IRRFDR.

The MW spectrum of CF_2=NBr was observed[151] in a flow system. Analysis of the spectrum yielded rotational and quadrupole coupling constants. The structure is similar to that of CF_2=NCl but the results depended on the constraints used. The MW spectrum of CH_2=PF showed[152] that it is planar. With transfer of data from CH_2=PCl, bondlengths of r(CP) = 164.4 pm, r(PF) = 159.8 pm and angle θ(CPF) = 104.16^{o} indicated a shortening of r(C=P) and an increase in θ(CPX) compared with CH_2=PH.

A number of papers report details of the IR spectrum of HNO_3, mainly for use in atmospheric studies. They include the ν_3/ν_4 bands studied by FTIR spectroscopy[153] and a TDL spectrometer.[154] FTIR studies are also reported on the ν_6, ν_7 and ν_8 fundamentals[155] and the $\nu_8+\nu_9$ combination band.[156] A high-resolution rotational Zeeman study of HNO_3 and DNO_3 enabled the complete determination of the magnetic and quadrupole tensors.[157] The ^2H hyperfine structure in $D^{15}NO_3$ has been analysed.[158]

The mmw spectrum of HSSSH was measured[159] in states in which one or two quanta of the lowest bending vibration were excited. These were estimated at 240 cm^{-1} (which compares with a matrix isolation value of 211 cm^{-1}) and 480 cm^{-1}.

6 Molecules containing six or more atoms

The ^{11}B quadrupole coupling tensor was determined[160] in a FTMW study of $C_6H_5{}^{11}BF_2$. The MW spectrum of five isotopic species of 5-fluoro-2,4-dicarbaheptaborane(7), $C_2B_5H_6F$, was measured[161]. The structure is a pentagonal bipyramid with apical BH groups and an equatorial BF group. The short r(B-F) distance of 133 pm and slightly oversize framework imply double bond character in the B-F bond with back-donation by fluorine relieving the electron deficiency of the framework.

The cyclic compounds 3H-3-azacyclotriboroxane BH-O-BH-NH-BH-O (cyclic) and 3-oxacylotriboroxane BH-NH-BH-NH-BH-O (cyclic) were identified in the reaction products of B_2H_6 and NO in a high temperature flow tube by their MW spectra.[162] Both molecules were identified from the form of their spectra and by isotopic substitution; structures for the heavy atom skeletons were deduced.

The degenerate fundamental ν_{12} of CH_3SiH_3 showed[163] interesting torsional effects: there was a Coriolis interaction with $3\nu_6$ (ν_6 = torsion) and the splittings in the l=+1 and l=-1 levels were almost identical. Rotational spectra[164] of six isotopic species of CH_3SiH_2I were added to previous data to give a r_s structure. The methyl group was collinear with the C-Si bond, and the ^{127}I quadrupole coupling tensor showed a tilt of only 0.68° of its z-axis towards the H atoms. The mmw spectrum of $CH_3Si^{35}Cl_3$ in its ground and two lowest E vibrational states was analysed.[165] The l-resonance gave values for the zeta constants while a local resonance between the (kl-1)=-3 and one l-doublet in one of the E vibrations enabled C_v to be determined. A re-investigation[166] of the MW spectra of twelve isotopic species of $(CH_3)_2SiH_2$ and eleven of $(CH_3)_2S$ gave more information on their structures and internal rotation. In $(CH_3)_2SiH_2$ the internal rotor axes coincide with the Si-C bond whereas those for $(CH_3)_2S$ show a 2.72° tilt of the methyl groups towards the lone pairs on the sulphur atom. Both SiF_3CN and SiF_3NC were detected by MW spectroscopy;[167] this is the first detection of SiF_3NC. The r_z structures for the C_{3v} molecules were obtained, and μ = 1.620 D for SiF_3CN and 1.13 D for SiF_3NC.

Three isotopic species of H_3GeN_3 were studied[168] in the MW region. The Ge-N bondlength is 4-5 pm longer than in H_3GeNCS and H_3GeNCO, and the GeH_3 group tilts 4.0° away from the azide group. The dipole moment and torsional splitting were also measured. The MW spectra of no fewer than seventeen isotopic species of CH_3GeH_2F were analysed[169] to give rotational, centrifugal distortion and internal rotation parameters.

The ^{14}N quadrupole hyperfine structure in $NH_2.NH_2$ was studied[170] using FTMW jet spectroscopy. The tensor elements were obtained and it was noted that erroneous conclusions would be obtained if the ordinary semirigid rotor approach were used on this molecule in which significant tunnelling occurs.

The structure of PH_2F_3 was determined by a combined electron diffraction (ED) and MW investigation.[171] The trigonal bipyramidal structure had $\theta(F_ePF_a)$ = 91.9°, $\theta(HPF_a)$ = 89.1° and θ(HPH) = 125.8°. A similar treatment of PHF_4 in the same paper used previously-obtained MW data and ED data reported here. The Q-branch of ν_3 of PF_5 was resolved[172] by using saturation (Lamb dip) spectroscopy. The band origin and rotational constants were determined; no evidence was found for the triplet structure expected from Berry pseudorotation or for Stark splitting in fields up to 3000 V/cm.

The analysis of the MW spectra of $(CH_3)_3PS$ gave[173] a r_0 structure with some assumptions. A similar study was reported174 for $(CH_3)_2NPF_2$; in this a planar NC_2 framework bisects the PF_2 angle. The out of plane NC_2 wag was estimated at 35 cm^{-1} and the torsion at 115 cm^{-1} from intensity measurements; the torsional barrier was V_2 = 4356 cm^{-1}. In $(CH_3)_2CHPF_2$ both *trans* and *gauche* isomers were observed;[175] differences between their structures were explained in terms of nonbonded repulsions.

The first rotationally resolved IR spectrum of $Ni(CO)_4$ was reported;[176] a TDL was used with a supersonic jet to observe ν_5. P-, Q- and R-branches were seen with the expected intensity variation due to nuclear spin statistics. The band origin at 2061.305 cm^{-1} is very different from that deduced from the room temperature spectrum. The separation at low-J is close to the value of $2B$ calculated from electron diffraction data; this implies that ζ_5 is close to zero.

Rotational PQR structure was also observed[177] in the $\nu_1+\nu_3$ and $\nu_2+\nu_3$ combination bands of NpF_6. The Coriolis constant ζ was estimated as -0.25 for $\nu_1+\nu_3$. Rotational transitions up to J = 40 were assigned in the Q-branch.

Table: Diatomic Molecules and Ions

Mol.	Ref	Technique	State(s)	Information derived
a) 1-electron species				
H_2^+	5			Review
	178	Microwave	$2p\sigma_u$-$1s\sigma_g$	Rovib. energy
HD^+	179	RFIRDR	X, v=17,21	1st DR for mol.ion beam
D_2^+	180	IRMWDR	$2p\sigma_u$-$1s\sigma_g$	Rovib. consts, r_0'=770 pm
b) 2-electron species				
H_2	181	Raman	X, J=0-3	Rot. linewidths in aq.soln
	182	FTIR	Rydberg	Rot. assignment v'=0-3
HeH^+	183	IR DFL	X, v=1-0	Rot. constants
NaH	184	FTIR	X, v''=0,1,2	Dunham coeffs, trans. mom.
Li_2	185	FTIR + LIF	$2^1\Sigma_u^+$-$2^1\Sigma_g^+$	Rot. constants
	186	FTIR + LIF	$1^3\Sigma_g^+$-$a^3\Sigma_u^+$	Eq. rot. constants
NaLi	187	FTIR + LIF	A-X, B-A	Rot. consts, r_e=337.08pm
Na_2	188	FTIR + LIF	A-X, B-A	Dunham coeffs, RKR curve
	189	UV exc. sp.	C-X	Dunham coeffs, Λ-doubling
	190	UV exc. sp.	A-X, b-X	Hyperfine (hf) structure
	191	OODR	$1^3\Delta_g$	Rot. consts, r_e=345.7 pm
	192	OODR	$2^3\Pi_g$	Rot. consts, r_e=444.7 pm
	193	OODR	$4^3\Sigma_g^+$	hf structure
NaK	194	Vis. abs. sp.	c-X, b-X	Rot. consts for c, r_e=410pm
	195	Exc. sp.	d-X	Rot. constants
	196	LIF + Stark	B-X	Lifetimes, $\mu(r_e)$=2.4 D
K_2	197	OODR	Rydberg	Rot. constants
Cs_2	198	Pol. sp., OODR	B-X	Dunham coeffs, $r_e(B)$=485pm
c) 3-electron species				
CaH	199	IR TDL	X, v''=0-4	Dunham & mass-invar.consts
BaH	200	FTIR + LIF	A', A, B-X	Rot. consts, spin-orbit
d) 4-electron species				
He_2	201	Emission	G, N-A	Rot. constants
	202	Absorption	a-b	Rot. structure PQR
GaH	203	TDL IR	X, v''=0-4	Dunham coeffs, r_e=166.212pm
InH	204	TDL IR	X, v''=0-4	Dunham coeffs, r_e=183.776pm
TlH	205	TDL IR	X, v''=0-2	Dunham coeffs
e) 5-electron species				
CH	206	Solar FTIR	X, v''=0-3	Rot. consts
	207	intermod. fluor	$A^2\Delta$	Stark, μ=0.77 D
CD	208	Far-IR LMR	X, v=0	Rot. + spin consts, hfs
SiH	209	REMPI	F	State identification.
SnH	210	IR LMR	X, v=0	Rot. & hf consts
f) 6-electron species				
NH	211	FTIR emission	$X, \Delta v$=1	Intensities, trans. moment
BiH	212	TDL IR	X, v''=0-2	Dunham coeffs, r_e=180.867pm
SH^+	213	TDL IR	$X^3\Sigma^-, v''$=1-3	Rot consts, r_e=136.36 pm
	214	photofrag. sp.	$c^1\Pi$-$b^1\Sigma^+$	Rot. constants

Table (continued): **Diatomic Molecules and Ions**

g) 7-electron species				
LiO	215	mmw sp.	$X, v=0,1$	Rot. & hf consts
NaO	216	mmw sp.	$X, v=0-3$	Rot. & hf consts
OH	217	Chemilumin.	$X, v=0$	Rot. distrib. from $H+O_2$
	218	CCL IR	$X, v=0$	Intensities, $\mu(r)$
TeH	219	FTIR emission	$X_2-X_1, v=0-2$	Rot. consts, r_e=165.592 pm
HF^+	220	CCL IR	$X, v''=0$	Rot, hf + Λ-doubl. consts
	221	laser abs	$X, v=0,1$	^{19}F hf consts
HCl^+	222	far-IR LMR	$X, v=0$	Rot. & hf consts
HBr^+	223	far-IR LMR	$X, v=0$	Rot. & hf consts
h) 8-electron species				
LiBr	224	TDL IR	$X, \Delta v=2$	Dunham coeffs
NaCl	225	TDL IR	$X, v''=0-4$	Dunham coeffs
MgS,CaS	226	mmw sp.	$X, v''=0-3$	Rot. constants
CaO	227	TDL IR	$X, v''=0-3$	Dunham coeffs,state distr.
	228	Fluorescence	$c-a(0,0)$	Rot. assignment & consts
	229	LIF + OODR	$C'-A'$	Rot. consts
BaO	230	LIF	$A-X, A'-X, v'=10-17$	Rot. consts
BN	231	Abs. sp.	$?-1^3\Sigma^-$	New band, rot. consts
C_2	232	(1+1)REMPI	$1^1\Delta_u-B^1\Delta_g$	New band, band origin
SiC	233	mmw sp.	X	Rot. consts, lab. & astro.
	234	FTIR emission	$A-X(0,0)$	Rot. consts, $r_e(A)$=181.36pm
HF,DF	235	(1+1)REMPI	$B-X, C-X$	Rot. constants
DF	236	Emission	$B-X, v'=0-5, v''=16-26$	Rot. consts, D_e
ArH^+	237	far-IR Zeeman	$X^1\Sigma^+$	Rot. g-factors, μ=3.0 D
MnH	238	TDL IR	$X, v''=0-2$	Dunham coeffs, hf consts
i) 9-electron species				
MgCl	239	mmw sp.	$X^2\Sigma^+$	Dunham coeffs
SrF	240	LIF	$A-X$, $B-X$	hf consts, dipole moments
SrI	241	Pol. & exc. sp.	$A-X(0,0)$	Rot. consts, A B perturb., re=295.5pm(A),294.6pm(B)
	242	MW-Opt. pol. sp.	$B-X(0,0)$	hf consts
AlO	243	MW sp.	$X, v=0$	Rot. & hf consts
	244	Emission	$B-X$	Rot. & spin consts
CN	245	Emission	$B-X$	Rovib. distr. from Ar+BrCN
	246	FTUV emission	$B-X$	Rot. constants, Tvib, Trot
	247	OODR	$B-A, B-X$	Transition moments
CP	248	mmw sp.	$X, v=0$	Rot. & hf consts
CO^+	249	OODR	$A-X, B-X$	Collisional energy transf.
$^{13}C^{18}O^+$	250	Emission	$A-X$	Rot consts, r_e', r_e''=111.6pm
ScO	251	LIF	$A-X$	$A'-A$ perturb. $r_e(A')$=172pm
j) 10-electron species				
MgAr	252	LIF	$C-X$	Rot.consts, r_e', r_e''=111.6pm
BF	253	mmw sp.	$X, v=0-2$	Rot. & hf consts, r_e
BBr	254	Emission	$a-X$	Rot. constants
	255	Emission	$A-X$	Rot. constants

Table (continued): **Diatomic Molecules and Ions**

AlCl	256	Emission	A-X	Rot. consts, predissoc.
	257	Emission	a-X, $b^3\Delta$-a	Rot. constants
InF	258	FTIR emission	X, v''=0-5	Dunham coeffs
InCl	259	mmw sp.	X, v=0-6	Rot.constants, r_e=240.119pm
InBr	260	mmw sp.	X	Dunham coeffs, r_e=254.318pm
TlCl	261	FTIR	X	Rot. constants
TlI	262	LIF	A-X	Dunham coeffs
CO	263	IR heterodyne	X	Dunham coeffs
	264	2-phot. exc.+LIF	$^1\Delta$--$X^1\Sigma^+$	Rot. constants
	265	Optogalv. sp.	$W^1\Pi$-$B^1\Sigma^+$(0,0)	Rot. consts, r_0'=125.7pm
	266	FTUV emission	B-e, d, a'	Perturbations
	267	Emission	C-A	Rot. constants
	268	2-phot. exc.+LIF	B--X, C--X	Polarisation ratios
SiF^+	269	TDL IR	X, v''=0,1	Rot.constants, r_e=152.652pm
GeS	270	FTIR emission	X, v''=0-2	Dunham coeffs
PbO	271	TDL IR	X, v''=0-3	Dunham coeffs
N_2	272	Emission	B-A	Rot. consts, predissoc.
	273	VUV emission	C'-X	
	274	FTUV emission	C-B	Rot. consts, perturb.
$^{15}N_2$	275	FTIR/VIS emiss.	B-A, B'-B	Rot. constants
VN	276	Emission	A-X(0,0)	Rot. consts, r_0''=156.6pm
	277	LIF	d-X(0,0)	Rot. consts, r_0'=157.4pm
NbN	278	Fluor. sp.	B-X(0,0)	Rot. & hf consts
CoH	279	CO-Faraday-LMR	X, v''=0	Rot & magn. consts
	280	LIF	var-X	Rot consts, s.o. splitt.
k) 11-electron species				
SiF	281	TDL IR	X, v''=0	Rot consts, Λ-doubling, r_e
NO	282	2-photon sp.	A--X	Fine & hf consts for A
	283	OODR, MPI+LIF	Ryd.-D-X	Rot consts, mixing
$AsCl^+$	284	Emission	A-X, v'=0,1	Rot.consts, r_e''=203.1pm, r_e'
TiCl	285	Emission	$^2\Phi$-$^2\Delta$	Rot. constants
NiH	286	OODR-Zeeman	B-X	Rot. assignments
NiD	287	CO-Faraday-LMR	X, v''=0	Rot. constants
PtD	288	Emission	$^2\Delta$-X, $^2\Phi$-X	Rot. constants
l) 12-electron species				
O_2	289	Ioniz. sp.	$X(O_2^+)$-$X(O_2)$	Rot. structure, I.P.
	290	(2+1)REMPI	Ryd.-A	Rot. constants, predissoc.
$^{16}O^{18}O$	291	Absorption	B-X	Rot. & spin consts
SeTe	292	near-IR emiss.	b-X	Rot. consts, r_0''=236.11 pm
CrO	293	Fluor. + Stark	B-X	μ'=4 D, μ''=3.9 D
AgH	294	TDL IR	X, v''=0-2	Dunham coeffs, r_e=161.780pm
m) 13-electron species				
SCl	295	TDL IR	X, v''=0,1	Rot. constants
Cl_2^+	296	Emission	A-X	Rot. consts, Λ-doubling
n) 14-electron species				
FI	297	TDL IR	X, v''=0,1	Dunham coeffs
	298	Emission	A-X	Rot. consts, RKR curve

Table (continued): **Diatomic Molecules and Ions**

Cl_2	299	OODR	1_g-A-X	Rot. constants
Br_2	300	OODR	$1_u, 0_u^{-}-A-X$	Dunham coeffs, RKR curve
BrI	301	TDL near-IR	$A-X$	Rot. constants
I_2	302	2-step exc. sp.	$1_g, 0_g^{+}-B-X$	hf consts
	303	3-photon abs.	$F, F', D---X$	Dunham coeffs, assign.(D)
	304	LIF	$D-X$	Rot. consts (high v')
	305	FM sat. sp.	$B-X$	hfs, $eQq(v)$
	306	Abs. sp.	$B-X$	hfs, intensities
FeO	307	Fluor. + Stark	$\Omega'=2-X$	$\mu'=2.6$ D, $\mu''=4.7$ D
	308	OODR	$\Omega'=4-X$	Perturbations
SmO	309	Emission		Rot. consts
o) Species with more than 14 electrons				
CuCl	310	LIF	$D-X, E-X$	Rot. & hf constants
ErO	311	Emission	$?, \Omega+1-\Omega$	Rot. constants
NiCu	312	(2+1)REMPI	$A..F-X$	Rot. consts, $r_e(F)=235$ pm
Cu_2	313	LIF	$A^1\Sigma_u^{+}-X$	Rot. consts,electr.assign.

References

1. 'The Structure of Small Molecules and Ions', ed. R. Naaman and Z. Vager, Plenum Press, New York, 1988. ISBN 0-306-43016-9.
2. 'Structure and Dymnamics of Weakly Bound Molecular Complexes', ed. A. Weber, Reidel, Dordrecht, 1987. ISBN 90-277-2584-5.
3. 'The Physics and Chemistry of Interstellar Molecular Clouds: mm and Sub-mm Observations in Astrophysics', ed. G. Winnewisser and J.T. Armstrong, Springer-Verlag, Berlin, 1989. ISBN 3-540-51297-7.
4. 'Structure and Dynamics of the Interstellar Medium', ed G. Tenorio-Tagle, M. Moles and J. Melnick, Springer-Verlag,Berlin,1989. ISBN 3-540-51956-4.
5. A. Carrington, I.R. McNab and C.A. Montgomerie, *J. Phys. B*, 1989, **22**, 3551.
6. A. Carrington and I.R. McNab, *Acc. Chem. Res.*, 1989, **22**, 218.
7. R.E. Miller, *Science*, 1989, **240**, 447.
8. R.J. Saykally, *Acc. Chem. Res.*, 1989, **22**, 295.
9. Z. Bačić and J.C. Light, *Ann. Rev. Phys. Chem.*, 1989, **40**, 469.
10. K. Balasubramanian, *Chem. Rev.*, 1989, **89**, 1801.
11. W. Weltner Jr and R.J. Van Zee, *Chem. Rev.*, 1989, **89**, 1713.
12. D.J. Nesbitt, C.M. Lovejoy, T.G. Lindemann, S.V. ONeil and D.C. Clary, *J. Chem. Phys.*, 1989, **91**, 722.
13. C.M. Lovejoy and D.J. Nesbitt, *J. Chem. Phys.*, 1989, **91**, 2790.
14. K.M. Beck, M.T. Berry, M.R. Brustein and M.I. Lester, *Chem. Phys. Letter*, 1989, **162**, 203.
15. T.D. Klots, C.E. Dykstra and H.S. Gutowsky, *J. Chem. Phys.*, 1989, **90**, 30.
16. G.T. Fraser and A.S. Pine, *J. Chem. Phys.*, 1989, **91**, 3319.
17. R.E. Bumgarner and G.A. Blake, *Chem. Phys. Lett.*, 1989, **161**, 308.
18. Y. Ohshima, M. Iida and Y. Endo, *Chem. Phys. Lett.*, 1989, **161**, 202.
19. G.D. Hayman, J. Hodge, B.J. Howard, J.S. Muenter and T.R. Dyke, *J. Mol. Spectrosc.*, 1989, **133**, 423.
20. Y. Matsumoto, Y. Ohshima, M. Takami and K. Kuchitsu, *J. Chem. Phys.*, 1989, **90**, 7017.
21. A. Taleb-Bendiab, M.S. LaBarge, L.L. Lohr, R.C. Taylor, K.W. Hillig II, R.L. Kuczkowski and R.K. Bohn, *J. Chem. Phys.*, 1989, **90**, 6949.
22. R.D. Suenram, F.J. Lovas, G.T. Fraser, J.Z. Gillies, C.W. Gillies and M. Onda, *J. Mol. Spectrosc.*, 1989, **137**, 127.
23. T.D. Klots. T. Emilsson, R.S. Ruoff and H.S. Gutowsky, *J. Phys. Chem.*, 1989, **93**, 1255.
24. R.K. Bohn, K.W. Hillig II and R.L. Kuczkowski, *J. Phys. Chem.*, 1989, **93**, 3456.

25. S.R. Hair, J.I. Cline, C.R. Bieler and K.C. Janda, *J. Chem. Phys.*, 1989, **90**, 2935.
26. T.D. Klots and H.S. Gutowsky, *J. Chem. Phys.*, 1989, **91**, 63.
27. A.C. Legon and A.P. Suckley, *J. Chem. Phys.*, 1989, **91**, 4440.
28. A.C. Legon and A.P. Suckley, *Chem. Phys. Lett.*, 1989, **157**, 5.
29. T.D. Klots, R.S. Ruoff and H.S. Gutowsky, *J. Chem. Phys.*, 1989, **90**, 4216.
30. Z.S. Huang and R.E. Miller, *Chem. Phys.*, 1989, **132**, 185.
31. G.T. Fraser, A.S. Pine, R.D. Suenram, D.C. Dayton and R.E. Miller, *J. Chem. Phys.*, 1989, **90** , 1330.
32. C.M. Lovejoy and D.J. Nesbitt, *J. Chem. Phys.*, 1989, **90**, 4671.
33. S.G. Kukolich and D.J. Pauley, *Chem. Phys.*, 1989, **131**, 403.
34. S.G. Kukolich and D.J. Pauley, *J. Chem. Phys.*, 1989, **90**, 3458.
35. Z. Wang and J.W. Bevan, *J. Chem. Phys.*, 1989, **91**, 3335.
36. Z. Wang, M. Eliades and J.W. Bevan, *Chem. Phys. Lett.*, 1989, **161**, 6.
37. N.W. Howard and A.C. Legon, *J. Chem. Phys.*, 1989, **90**, 672.
38. H.O. Leung, M.D. Marshall, R.D. Suenram and F.J. Lovas, *J. Chem. Phys.*, 1989, **90**, 700.
39. K. von Puttkamer and M. Quack, *Chem. Phys.*, 1989, **139**, 31.
40. G.T. Fraser and A.S. Pine, *J. Chem. Phys.*, 1989, **91**, 633.
41. G.T. Fraser and A.S. Pine, *J. Chem. Phys.*, 1989, **91**, 637.
42. N. Moazzen-Ahmadi, A.R.W. McKellar and J.W.C. Johns, *J. Mol. Spectrosc.*, 1989, **138**, 282.
43. G.A. Blake and R.E. Bumgarner, *J. Chem. Phys.*, 1989, **91**, 7300.
44. M.D. Schuder, C.M. Lovejoy, D.D. Nelson Jr and D.J. Nesbitt, *J. Chem. Phys.*, 1989, **91**, 4418.
45. K.L. Busarow, R.C. Cohen, G.A. Blake, K.B. Laughlin, Y.T. Lee and R.J. Saykally, *J. Chem. Phys.*, 1989, **90**, 3937.
46. T.A. Hu and T.R. Dyke, *J. Chem. Phys.*, 1989, **91**, 7348.
47. G.T. Fraser, R.D. Suenram, L.H. Coudert and R.S. Frye, *J. Mol. Spectrosc.*, 1989, **137**, 244.
48. G.T. Fraser, R.D. Suenram and L.H. Coudert, *J. Chem. Phys.*, 1989, **90**, 6077.
49. R.D. Suenram, G.T. Fraser and F.J. Lovas, *J. Mol. Spectrosc.*, 1989, **138**, 440.
50. Z.S. Huang and R.E. Miller, *J. Chem. Phys.*, 1989, **91**, 6613.
51. K. Matsumura, F.J. Lovas and R.D. Suenram, *J. Chem. Phys.*, 1989, **91**, 5887.
52. K.I. Peterson, R.D. Suenram and F.J. Lovas, *J. Chem. Phys.*, 1989, **90**, 5964.
53. D.C. Dayton and R.E. Miller, *Chem. Phys. Lett.*, 1989, **156**, 578.
54. J.M. Price, M.W. Crofton and Y.T. Lee, *J. Chem. Phys.*, 1989, **91**, 2749.
55. W.A. Majewski, P.A. Feldman, J.K.G. Watson, S. Miller and J. Tennyson, *Astrophys. J.*, 1989, **347**, L51.
56. Y. Kawashima, Y. Endo and E. Hirota, *J. Mol. Spectrosc.*, 1989, **133**, 116.
57. G. Duxbury and Y. Gang, *J. Mol. Spectrosc.*, 1989, **138**, 541.
58. A.M. Smith, S.L. Coy, W. Klemperer and K.K. Lehmann, *J. Mol. Spectrosc.*, 1989, **134**, 134.
59. A. Meenaskshi, K.K. Innes and G.A. Bickel, *Mol. Phys.*, 1989, **68**, 1179.
60. M. Rösslein and J.P. Maier, *J. Phys. Chem.*, 1989, **93**, 7342.
61. M. Rösslein, M.A. Hanratty and J.P. Maier, *Mol. Phys.*, 1989, **68**, 823.
62. N.H. Rosenbaum, J.C. Owrutsky and R.J. Saykally, *J. Mol. Spectrosc.*, 1989, **133**, 365.
63. K. Kawaguchi, K. Matsumura, H. Kanamori and E. Hirota, *J. Chem. Phys.*, 1989, **91**, 1953.
64. F.J. Northrup and T.J. Sears, *Chem. Phys. Lett.*, 1989, **159**, 421.
65. E.A. Rohlfling and J.E.M. Goldsmith, *J. Chem. Phys.*, 1989, **90**, 6804.
66. G.W. Lemire, Z. Fu, Y.M. Hamrick, S. Taylor and M.D. Morse, *J. Phys. Chem.*, 1989, **93**, 2313.
67. E.A. Rohlfling, *J. Chem. Phys.*, 1989, **91**, 4531.
68. F.J. Northrup and T.J. Sears, *J. Chem. Phys.*, 1989, **91**, 762.
69. T. Wyttenbach, D.D. Evard and J.P. Maier, *J. Chem. Phys.*, 1989, **90**, 4645.
70. D.D. Evard, T. Wyttenbach and J.P. Maier, *J. Phys. Chem.*, 1989, **93**, 3522.
71. S. Dobos, G. Winnewisser, F. Kling and J. Mink, *Z. Naturforsch. A*, 1989, **44**, 633.
72. M.D. Vanek, D.A. Jennings, J.S. Wells and A.G. Maki, *J. Mol. Spectrosc.*, 1989, **138**, 79.
73. H.E. Warner, A. Fox, T. Amano and D.K. Bohme, *J. Chem. Phys.*, 1989, **91**, 5310.

74. N. Moazzen-Ahmadi, A.R.W. McKellar, H.E. Warner and T. Amano, *J. Chem. Phys.*, 1989, **91** , 5313.
75. C. Yamada, H. Kanamori, E. Hirota, N. Nishiwaki, N. Itabashi, K. Kato and T. Goto, *J. Chem. Phys.*, 1989, **91**, 4582.
76. R.D. Suenram, F.J. Lovas and K. Matsumura, *Astrophys. J.*, 1989, **342**, L103.
77. M. Tanimoto, H. Takeo, C. Matsumura, M.Fujitake and E. Hirota, *J. Chem. Phys.*, 1989, **91**, 2102.
78. G. Meijer, J. Heinze, W.L. Meerts, J.J. ter Meulen and J.T. Hougen, *J. Mol. Spectrosc.*, 1989, **138**, 251.
79. R.C. Cohen, K.L. Busarow, C.A. Schuttenmaer, Y.T. Lee and R.J. Saykally, *Chem. Phys. Lett.*, 1989, **164**, 321.
80. R.F. Curl, K.K. Murray, M. Petri, M.L. Richnow and F.K. Tittel, *Chem. Phys. Lett.*, 1989, **161**, 98.
81. M. Okumura, B.D. Rehfuss, B.M. Dinelli, M.G. Bawendi and T.Oka, *J. Chem. Phys.*, 1989, **90** , 5918.
82. H.C. Miller, J.L. Hardwickand J.W. Farley, *J. Mol. Spectrosc.*, 1989, **134**, 329.
83. M.D. Vanek, M. Schneider, J.S. Wells and A.G. Maki, *J. Mol. Spectrosc.*, 1989, **134**, 154.
84. A.G. Maki, J.S. Wells and M.D. Vanek, *J. Mol. Spectrosc.*, 1989, **138**, 84.
85. P. Varanasi and S. Chudamani, *J. Quant. Spectrosc. Rad. Transf.*, 1989, **41**, 351.
86. L-W. Tang, S.Nadler and S.J. Daunt, *J. Quant. Spectrosc. Rad. Transf.*, 1989, **41**, 97.
87. J-L. Teffo and A. Chédin, *J. Mol. Spectrosc.*, 1989, **135**, 389.
88. M. Larzillière and Ch. Jungen, *Mol. Phys.*, 1989, **67**, 807.
89. M. Semmond-Monnanteuil, J-M. Colmont, A. Perrin, J-M. Flaud and C. Camy-Peyret, *J. Mol. Spectrosc.*, 1989, **134**, 176.
90. F. Bylicki, G. Persch, E. Mehdizadeh and W. Demtröder, *Chem. Phys.*, 1989, **135**, 255.
91. M. Cheikh and C. Alamichel, *Spectrochim. Acta A*, 1989, **45**, 153.
92. J. Vergès, J. Chevillard, C. Alamichel and J-R. Etcheberry, *Chem. Phys. Lett.*, 1989, **159**, 315.
93. J-P. Chevillard, J-Y. Mandin, J-M. Flaud and C. Camy-Peyret, *Canad. J. Phys.*, 1989, **67**, 1065.
94. B.E. Grossman and E.V. Browell, *J. Mol. Spectrosc.*, 1989, **136**, 264.
95. J-Y. Mandin, J-P. Chevillard, J-M. Flaud and C. Camy-Peyret, *J. Mol. Spectrosc.*, 1989, **138**, 430.
96. B.E. Grossman and E.V. Browell, *J. Mol. Spectrosc.*, 1989, **138**, 562.
97. J-Y. Mandin, J-P. Chevillard, C. Camy-Peyret and J-M. Flaud, *J. Mol. Spectrosc.*, 1989, **138**, 272.
98. T. Ohshima and H. Sasada, *J. Mol. Spectrosc.*, 1989, **136**, 250.
99. G. Di Lonardo and L. Fusina, *J. Mol. Spectrosc.*, 1989, **135**, 250.
100. H-H. Kuge and K. Kleinermanns, *J. Chem. Phys.*, 1989, **90**, 46.
101. P.R. Brown, P.B. Davies and R.J. Stickland, *J. Chem. Phys.*, 1989, **91**, 3384.
102. H. Bürger, G. Pawelke, S. Sommer, A. Rahner, E.H. Appelman and I.M. Mills, *J. Mol. Spectrosc.*, 1989, **136**, 197.
103. H. Bürger, G. Pawelke, A. Rahner, E.H. Appelman and L. Halonen, *J. Mol. Spectrosc.*, 1989, **138**, 346.
104. Y. Koga, H. Takeo, S. Kondo, M. Sugie, C. Matsumura, G.A. McRae and E.A. Cohen, *J. Mol. Spectrosc.*, 1989, **138**, 467.
105. J-M. Flaud, C. Camy-Peyret, A. N'Gom, V.M. Devi, C.P. Rinsland and M.A.H. Smith, *J. Mol. Spectrosc.*, 1989, **133**, 217.
106. J-M. Flaud, C. Camy-Peyret, C.P. Rinsland, M.A.H. Smith and V.M. Devi, *J. Mol. Spectrosc.*, 1989, **134**, 106.
107. M. Bogey, P.B. Davies, C. Demuynck, J.L. Destombes and T.J. Sears, *Mol. Phys.*, 1989, **67**, 1033.
108. E.C. Richard, C.T. Wickham-Jones and V. Vaida, *J. Phys. Chem.*, 1989, **93**, 6346.
109. C. Camy-Peyret, J-M. Flaud, A. N'Gom and J.W.C. Johns, *Mol. Phys.*, 1989, **67**, 693.
110. E. Zeisberger and A. Ruoff, *J. Mol. Spectrosc.*, 1989, **136**, 295.

111. M.C.L. Gerry, S.J. Howard, N. Heineking and H. Dreizler, *Z. Naturforsch. A*, 1989, **44**, 1187.
112. F. Stroh and M. Winnewisser, *Chem. Phys. Lett.*, 1989, **155**, 21.
113. K.M.T. Yamada, M.W. Markus, G. Winnewisser, W. Joentgen, R. Kock, E. Vogel and H-J. Altenbach, *Chem. Phys. Lett.*, 1989, **160**, 113.
114. R.L. Bhattacharjee, L.H. Johnston, G.R. Sudhakaran and J.C. Sarker, *J. Mol. Spectrosc.*, 1989, **138**, 38.
115. G. Guelachvili, A.H. Abdullah, N. Tu, K.N. Rao, S. Urban and D. Papoušek, *J. Mol. Spectrosc.*, 1989, **133**, 345.
116. Š. Urban, N. Tu, K.N. Rao and G. Guelachvili, *J. Mol. Spectrosc.*, 1989, **133**, 312.
117. S.L. Coy and K.K. Lehmann, *Spectrochim. Acta A*, 1989, **45**, 47.
118. M.G. Bawendi, B.D. Rehfuss, B.M. Dinelli, M. Okumura and T. Oka, *J. Chem. Phys.*, 1989, **90**, 5910.
119. M.C.L. Gerry, M. Heineking, H. Mäder and H. Dreizler, *Z. Naturforsch. A*, 1989, **44**, 1079.
120. J. Bendtsen and F.M. Nicolaisen, *J. Mol. Spectrosc.*, 1989, **133**, 193.
121. B.R. Foy, M.P. Casassa, J.C. Stephenson and D.S. King, *J. Chem. Phys.*, 1989, **90**, 7037.
122. Y-T. Chen and T. Oka, *J. Mol. Spectrosc.*, 1989, **133**, 148.
123. D. Papoušek, H. Birk, U. Magg and H. Jones, *J. Mol. Spectrosc.*, 1989, **135**, 105.
124. P. Verhoeve, M. Versluis, J.J. ter Meulen, W.L. Meerts and A. Dymanus, *Chem. Phys. Lett.*, 1989, **161**, 195. 125. T. Nakanaga and T. Amano, *J. Mol. Spectrosc.*, 1989, **133**, 201.
126. J-M. Flaud, C. Camy-Peyret, J.W.C. Johns and B. Carli, *J. Chem. Phys.*, 1989, **91**, 1504.
127. C. Donketis and J.P. Reilly, *J. Chem. Phys.*, 1989, **91**, 5239.
128. E. Herbst and G. Winnewisser, *Chem. Phys. Lett.*, 1989, **155**, 572.
129. S. Urban, E. Herbst, P. Mittler, G. Winnewisser, K.M.T. Yamada and M. Winnewisser, *J. Mol. Spectrosc.*, 1989, **137**, 327.
130. P. Mittler, G. Winnewisser and K.M.T. Yamada, *Z. Naturforsch. A*, 1989, **44**, 718.
131. M. Birk, R.R. Friedl, E.A. Cohen, H.M. Pickett and S.P. Sander, *J. Chem. Phys.*, 1989, **91** , 6588.
132. J. Ortigoso, R. Escribano and A.G. Maki, *J. Mol. Spectrosc.*, 1989, **138**, 602.
133. C.R. Brazier and P.F. Bernath, *J. Chem. Phys.*, 1989, **91**, 4548.
134. K. Vormann and H. Dreizler, *Z. Naturforsch. A*, 1989, **44**, 1191.
135. P.F. Bernath, K.H. Hinkle and J.J. Keady, *Science*, 1989, **244**, 562.
136. N. Moazzen-Ahmadi, A.R.W. McKellar and T. Amano, *Chem. Phys. Lett.*, 1989, **157**, 1.
137. J.R. Heath, A.L. Cooksy, M.H.W. Gruebele, C.A. Schuttenmaer and R.J. Saykally, *Science*, 1989, **244**, 564.
138. N. Moazzen-Ahmadi, A.R.W. McKellar and T. Amano, *J. Chem. Phys.*, 1989, **91**, 2140.
139. M.Ohishi, N. Kaifu, K. Kawaguchi, A. Murakami, S. Saito, S. Yamamoto, S-I. Ishikawa, Y. Fujita, Y, Shiratori and W.M. Irvine, *Astrophys. J.*, 1989, **345**, L83.
140. J. Skrzypczak, G. Pierre, F. Herlemont and J. Lemaire, *J. Mol. Spectrosc.*, 1989, **138**, 311.
141. H. Prinz, W. Höhe, W.A. Kreiner, M. Loëte, J. Hilico, G. Pierre, G. Magerl and W. Schupita, *J. Mol. Spectrosc.*, 1989, **135**, 144.
142. M. Chevalier and A. De Martino, *J. Chem. Phys.*, 1989, **90**, 2077.
143. Q. Zhu, B. Zhang, Y. Ma and H. Qian, *Chem. Phys. Lett.*, 1989, **164**, 596.
144. A. Campargue, M. Chenevier and F. Stoeckel, *Chem. Phys.*, 1989, **137**, 249.
145. A. Campargue, M. Chenevier and F. Stoeckel, *Chem. Phys.*, 1989, **138**, 405.
146. L. Jörissen, Y. Ohshima, Y. Mutsumoto, M. Takami and K. Kuchitsu, *J. Chem. Phys.*, 1989, **90** , 2109.
147. H. Bürger, R. Eujen, M. Litz and S. Cradock, *J. Mol. Spectrosc.*, 1989, **138**, 332.
148. H. Bürger, M. Betzel-Laufs, S. Sommer, G. Blanquet and J. Walrand, *J. Mol. Spectrosc.*, 1989, **134**, 214.

149. K. Harada, I. Nagano, S. Kimura, K. Tanaka and T.Tanaka, *J. Mol. Spectrosc.*, 1989, **138**, 230.
150. L. Jörissen, H. Prinz, W.A. Kreiner, Ch. Wenger, G. Pierre, G. Magerl and W. Schupta, *Canad. J. Phys.*, 1989, **67**, 532.
151. P. Groner, H. Nanaie, J.R. Durig, D.D. DesMarkan and C.W. Banknight Jr, *J. Chem. Phys.*, 1989, **91**, 5934.
152. O.I. Osman, B.J. Whitaker, N.P.C. Simmons, D.R.M. Walton, J.F. Nixon and H.W. Kroto, *J. Mol. Spectrosc.*, 1989, **137**, 373.
153. A. Perrin, O. Lado-Bordowsky and A. Valentin, *Mol. Phys.*, 1989, **67**, 249.
154. R.D. May and C.R. Webster, *J. Mol. Spectrosc.*, 1989, **138**, 383.
155. A.G. Maki and W.B. Olson, *J. Mol. Spectrosc.*, 1989, **133**, 171.
156. A. Maki, *J. Mol. Spectrosc.*, 1989, **136**, 105.
157. L. Albinus, J. Spieckermann and D.H. Sutter, *J. Mol. Spectrosc.*, 1989, **133**, 128.
158. E. Fliege, H. Dreizler, S. Gerber and H. Huber, *J. Mol. Spectrosc.*, 1989, **137**, 24.
159. D. Mauer, G. Winnewisser and K.M.T. Yamada, *J. Mol. Spectrosc.*, 1989, **136**, 380.
160. K. Vormann and H. Dreizler, *Z. Naturforsch. A*, 1989, **44**, 84.
161. S.S. Durso, R.A. Sheeks, E.V. O'Gorman and R.A. Beaudet, *Inorg. Chem.*, 1989, **28**, 1615.
162. Y. Kawashima, H. Takeo and C. Matsumura, *Inorg. Chem.*, 1989, **28**, 666.
163. N. Moazzin-Ahmadi, I. Ozier and W.L. Meerts, *J. Mol. Spectrosc.*, 1989, **137**, 166.
164, M. Hayashi, M. Fujitake and K. Kishioka, *J. Mol. Spectrosc.*, 1989, **135**, 305.
165. J.G. Smith, *J. Mol. Spectrosc.*, 1989, **136**, 109.
166. M. Hayashi, N. Nakata and S. Miyazaki, *J. Mol. Spectrosc.*, 1989, **135**, 270.
167. A.P. Cox, A.D. Couch, I.M. Hedgecock and S.G. Clement, *Chem. Phys. Lett.*, 1989, **158**, 6.
168. P. Groner, G.M. Attia, A.B. Mohamad, J.F. Sullivan, Y.S. Li and J.R. Durig, *J. Chem. Phys.*, 1989, **91**, 1434.
169. M. Hayashi, S. Kaminaka, M. Fujitake and S. Miyazaki, *J. Mol. Spectrosc.*, 1989, **135**, 289.
170. L.H. Coudert, J.T. Hougen and R.D. Suenram, *J. Mol. Spectrosc.*, 1989, **135**, 314.
171. D. Christen, J. Kadel, A. Liedtke, R. Minkwitz and H. Oberhammer, *J. Phys. Chem.*, 1989, **93**, 6672.
172. H. Prinz and W.A. Kreiner, *J. Mol. Spectrosc.*, 1989, **137**, 204.
173. J.R. Durig and K.K. Chatterjee, *Inorg. Chem.*, 1989, **28**, 298.
174. J.R. Durig, R.J. Harlan and P. Groner, *J. Phys. Chem.*, 1989, **93**, 3041.
175. J.R. Durig, M-S. Cheng, Y.S. Li and P. Groner, *J. Phys. Chem.*, 1989, **93**, 3492.
176. P.B. Davies, N.A. Martin, M.D. Nunes, D.A. Pape and D.K. Russell, *Chem. Phys. Lett.*, 1989, **156**, 553.
177. K.C. Kim and R.N. Mulford, *Chem. Phys. Lett.*, 1989, **159**, 327.
178. A. Carrington, I.R. McNab and C.A. Montgomerie, *Chem. Phys. Lett.*, 1989, **160**, 237.
179. A. Carrington, I.R. McNab and C.A. Montgomerie, *Mol. Phys.*, 1989, **66**, 519.
180. A. Carrington, I.R. McNab and C.A. Montgomerie and R.A. Kennedy, *Mol. Phys.*, 1989, **67**, 711.
181. D.G. Taylor III and H.L. Strauss, *J. Chem. Phys.*, 1989, **90**, 768.
182. Ch. Jungen, I. Dabrowski, G. Herzberg and D.J.W. Kendall, *J. Chem. Phys.*, 1989, **91**, 3926.
183. M.W. Crofton, R.S. Altman, N.H. Haese and T. Oka, *J. Chem. Phys.*, 1989, **91**, 5882.
184. A.G. Maki and W.B. Olson, *J. Chem. Phys.*, 1989, **90**, 6887.
185. C. Linton, F. Martin, R. Bacis and J. Vergès, *J. Mol. Spectrosc.*, 1989, **137**, 235.
186. C. Linton, T.L. Murphy, F. Martin, R. Bacis and J. Vergès, *J. Chem. Phys.*, 1989, **91**, 6036.
187. C.E. Fellows, *J. Mol. Spectrosc.*, 1989, **136**, 369.
188. O. Babaky and K. Hussein, *Canad. J. Phys.*, 1989, **67**, 912.

189. G-Y. Yan, B.W. Sterling, T. Kalka and A.L. Schawlow, *J. Opt. Soc. Amer. B*, 1989, **6**, 1975; G-Y. Yan and A.L. Schawlow, *J. Opt. Soc. Amer. B*, 1989, **6**, 2309.
190. H. Katô, M. Otami and M. Baba, *J. Chem. Phys.*, 1989, **91**, 5124.
191. L. Li, Q. Zhu and R.W. Field, *J. Mol. Spectrosc.*, 1989, **134**, 50; L. Li, A.M. Lyyra and W.C. Stwalley, *J. Mol. Spectrosc.*, 1989, **134**, 113.
192. X. Xie, R.W. Field, L. Li, A.M. Lyyra, J.T. Bahns and W.C. Stwalley, *J. Mol. Spectrosc.*, 1989, **134**, 119.
193. L. Li, Q. Zhu and R.W. Field, *Mol. Phys.*, 1989, **66**, 685.
194. P. Kowalczyk, *J. Chem. Phys.*, 1989, **91**, 2779.
195. P. Kowalczyk, *J. Mol. Spectrosc.*, 1989, **136**, 1.
196. J. Derouard, H. Debontribe, T.D. Nguyen and N. Sadeghi, *J. Chem. Phys.*, 1989, **90**, 5936.
197. H. Wang, L. Li, A.M. Lyyra and W.C. Stwalley, *J. Mol. Spectrosc.*, 1989, **137**, 304.
198. U. Diemer, R. Duchowicz, M. Erbel, E. Mehdizadeh and W. Demtröder, *Chem. Phys. Lett.*, 1989, **164**, 419.
199. D. Petitprez, B. Lemoine, C. Demuynck, J-L. Destombes and B. Macke, *J. Chem. Phys.*, 1989, **91**, 4462.
200. A. Bernard, C. Effantin, J. d'Incan, C. Fabre, R. Stringat and R.F. Barrow, *Mol. Phys.*, 1989, **67**, 1.
201. D.S. Ginter and M.L. Ginter, *J. Mol. Spectrosc.*, 1989, **138**, 29.
202. R.L. Brooks, J.L. Hunt and D.W. Tokaryk, *J. Chem. Phys.*, 1989, **91**, 7408.
203. R-D. Urban, U. Magg and H. Jones, *Chem. Phys. Lett.*, 1989, **164**, 135.
204. A.H. Bahnmaier, R-D. Urban and H. Jones, *Chem. Phys. Lett.*, 1989, **155**, 269.
205. R-D. Urban, A.H. Bahnmaier, U. Magg and H. Jones, *Chem. Phys. Lett.*, 1989, **158**, 443.
206. F. Mélen, N. Grevesse, A.J. Sanval, C.B. Farmer, R.H. Norton, H. Bredohl and I. Dubois, *J. Mol. Spectrosc.*, 1989, **134**, 305.
207. T.C. Steimle, D.F. Nachman, D.A. Fletcher and J.M.Brown, *J. Mol. Spectrosc.*, 1989, **138**, 222.
208. J.M. Brown and K.M. Evenson, *J. Mol. Spectrosc.*, 1989, **136**, 68.
209. R.D. Johnson III and J.W. Hudgens, *J. Phys. Chem.*, 1989, **93**, 6268.
210. W. Zimmermann, Th. Nelis, R. Bachem, R. Pahnke and W. Urban, *Mol. Phys.*, 1989, **68**, 199.
211. C. Chackerian Jr, G. Guelachvili, A. López-Piñeiro and R.H. Tipping, *J. Chem. Phys.*, 1989, **90**, 641.
212. A.M.R.P. Bopegedera, C.R. Brazier and P.F. Bernath, *Chem. Phys. Letter*, 1989, **162**, 301.
213. S. Civiš, C.E. Blom and P. Jensen, *J. Mol. Spectrosc.*, 1989, **138**, 69.
214. A.P. Levick and P.J. Sarre, *J. Mol. Spectrosc.*, 1989, **133**, 227.
215. C.Yamada, M. Fujitake and E. Hirota, *J. Chem. Phys.*, 1989, **91**, 137.
216. C.Yamada, M. Fujitake and E. Hirota, *J. Chem. Phys.*, 1989, **90**, 3033.
217. G.E. Hall and T.J. Sears, *Chem. Phys. Lett.*, 1989, **158**, 184.
218. D.D. Nelson Jr, A. Schiffman and D.J. Nesbitt, *J. Chem. Phys.*, 1989, **90**, 5455.
219. E.H. Fink, K.D. Setzer, D.A. Ramsay and M. Vervloet, *J. Mol. Spectrosc.*, 1989, **138**, 19.
220. D.C. Hovde, E.R. Keim and R.J. Saykally, *Mol. Phys.*, 1989, **68**, 599.
221. J.V. Coe, J.C. Owrutsky, E.R. Keim, N.V. Agman, D.C. Hovde and R.J. Saykally, *J. Chem. Phys.*, 1989, **90**, 3893.
222. K.G. Lubic, D. Ray, D.C. Hovde, L. Veseth and R.J. Saykally, *J. Mol. Spectrosc.*, 1989, **134**, 1. 223. K.G. Lubic, D. Ray, D.C. Hovde, L. Veseth and R.J. Saykally, *J. Mol. Spectrosc.*, 1989, **134**, 21. 224. C.R. Brazier, N.H. Oliphant and P.F. Bernath, *J. Mol. Spectrosc.*, 1989, **134**, 421.
225. H. Uehara, K. Horiai, K. Nakagawa and T. Fujimoto, *J. Mol. Spectrosc.*, 1989, **134**, 98.
226. S. Takano, S. Yamamoto and S. Saito, *Chem. Phys. Lett.*, 1989, **159**, 563.
227. H.G. Hedderich and C.E. Blom, *J. Chem. Phys.*, 1989, **90**, 4660.
228. J.B. Norman, K.J. Cross, H.S. Schweda, M. Polak and R.W. Field, *Mol. Phys.*, 1989, **66**, 235.
229. D.P. Baldwin and R.W. Field, *J. Mol. Spectrosc.*, 1989, **133**, 90.
230. N. Furio and J.G. Pruett, *J. Mol. Spectrosc.*, 1989, **136**, 120.

231. R.D. Verma, *J. Phys. B*, 1989, **22**, 3689.
232. P.M. Goodwin and T.A. Cool, *J. Mol. Spectrosc.*, 1989, **133**, 230.
233. J. Cernicharo, C.A. Gottlieb, M. Guelin, P. Thaddeus and J.M. Vrtilek, *Astrophys. J.*, 1989, **341**, L25.
234. C.R. Brazier, L.C. O'Brien and P.F. Bernath, *J. Chem. Phys.*, 1989, **91**, 7384.
235. L.M. Tashiro, W. Ubacho and R.N. Zare, *J. Mol. Spectrosc.*, 1989, **138**, 89.
236. J.A. Coxon and P.G. Hajigeorgiou, *J. Mol. Spectrosc.*, 1989, **133**, 45.
237. K.B. Laughlin, G.A. Blake, R.C. Cohen and R.J. Saykally, *J. Chem. Phys.*, 1989, **90**, 1358.
238. R-D. Urban and H. Jones, *Chem. Phys. Lett.*, 1989, **163**, 34.
239. M. Bogey, C. Demuynck and J-L. Destombes, *Chem. Phys. Lett.*, 1989, **155**, 265.
240. J. Kändler, T. Martell and W.E. Ernst, *Chem. Phys. Lett.*, 1989, **155**, 470.
241. W.E. Ernst, J.O. Schröder and B. Zeller, *J. Mol. Spectrosc.*, 1989, **135**, 161.
242. W.E. Ernst, J.O. Schröder and B. Schaal, *Chem. Phys. Lett.*, 1989, **155**, 47.
243. T. Törring and R. Herrmann, *Mol. Phys.*, 1989, **68**, 1379.
244. M.D. Saksena, G.S. Ghodgaonkar and M. Singh, *J. Phys. B*, 1989, **22**, 1993.
245. K. Kanda, H. Ito, K. Someda, K. Suzuki, T. Kondow and K. Kuchitsu, *J. Phys. Chem.*, 1989, **93**, 6020.
246. E.C. Richard, D.J. Donaldson and V. Vaida, *Chem. Phys. Lett.*, 1989, **157**, 295.
247. N. Furio, A. Ali, P.J. Dagdigian and H-J. Werner, *J. Mol. Spectrosc.*, 1989, **134**, 199.
248. S. Saito, S. Yamamoto, K. Kawaguchi, M. Ohishi, H. Suzuki, S-I. Ishikawa and N. Kaifu, *Astrophys. J.*, 1989, **341**, 1114.
249. A.V. Dentamaro and D.H. Katayama, *J. Chem. Phys.*, 1989, **90**, 91.
250. C.V.V. Prasad and S.P. Reddy, *J. Chem. Phys.*, 1989, **90**, 3010; S.P. Reddy and C.V.V. Prasad, *J. Chem. Phys.*, 1989, **91**, 1972.
251. S.F. Rice, W.J. Childs and R.W. Field, *J. Mol. Spectrosc.*, 1989, **133**, 22.
252. R.R. Bennett, J.G. McCaffrey, I. Wallace, D.J. Funk, A. Kowalski and W.H. Breckenridge, *J. Chem. Phys.*, 1989, **90**, 2139.
253. G. Cazzoli, L. Cludi, C. Degli Esposti and L. Dore, *J. Mol. Spectrosc.*, 1989, **134**, 159.
254. C. Destoky, H. Bredohl and I. Dubois, *J. Mol. Spectrosc.*, 1989, **134**, 314.
255. C. Destoky, I. Dubois and H. Bredohl, *J. Mol. Spectrosc.*, 1989, **136**, 216.
256. E. Mahieu, I. Dubois and H. Bredohl, *J. Mol. Spectrosc.*, 1989, **134**, 317.
257. E. Mahieu, I. Dubois and H. Bredohl, *J. Mol. Spectrosc.*, 1989, **138**, 264.
258. H. Uehara, K. Horiai, T. Mitani and H. Suguro, *Chem. Phys. Letter*, 1989, **162**, 137.
259. J. Hoeft and K.P.R. Nair, *Chem. Phys. Lett.*, 1989, **155**, 273.
260. J. Hoeft and K.P.R. Nair, *Chem. Phys. Lett.*, 1989, **164**, 33.
261. A.G. Maki, *J. Mol. Spectrosc.*, 1989, **137**, 147.
262. J. Kieckhäfer, H. Knöckel and E. Tiemann, *Chem. Phys. Lett.*, 1989, **160**, 570.
263. M. Schneider, K.M. Evenson, M.D. Vanek, D.A. Jennings, J.S. Wells, A. Stahn and W. Urban, *J. Mol. Spectrosc.*, 1989, **135**, 197.
264. C. Kittrell and B.A. Garetz, *Spectrochim. Acta A*, 1989, **45**, 31.
265. S. Sekine, Y. Adachi and C. Hirose, *J. Chem. Phys.*, 1989, **90**, 5346.
266. J-I. Choe, D-K. Lee, A.C. Le Floch and S.G. Kukolich, *J. Mol. Spectrosc.*, 1989, **136**, 173.
267. R. Kępa, *J. Mol. Spectrosc.*, 1989, **135**, 119.
268. P.J.H. Tjossem and K.C. Smyth, *J. Chem. Phys.*, 1989, **91**, 2041.
269. Y. Akiyama, K. Tanaka and T. Tanaka, *Chem. Phys. Lett.*, 1989, **155**, 15.
270. H. Uehara, K. Horiai, K. Sueoka and K. Nakagawa, *Chem. Phys. Lett.*, 1989, **160**, 149.
271. K. Horiai and H. Uehara, *J. Mol. Spectrosc.*, 1989, **136**, 56.
272. P.K. Carroll and E.J. Croke, *J. Phys. B*, 1989, **22**, 2717.
273. J-Y. Roncin, F. Launay and K. Yoshimo, *J. Mol. Spectrosc.*, 1989, **134**, 390.
274. F. Roux, F. Michaud and M. Vervloet, *Canad. J. Phys.*, 1989, **67**, 143.
275. F. Roux and F. Michaud, *J. Mol. Spectrosc.*, 1989, **136**, 205.
276. S.L. Peter and T.M. Dunn, *J. Chem. Phys.*, 1989, **90**, 5333.

277. B. Simard, C. Masoni and P.A. Hackett, *J. Mol. Spectrosc.*, 1989, **136**, 44.
278. Y. Azuma, J.A. Barry, M.P.J. Lyne, A.J. Merer, J.O. Schröder and J-L. Féménias, *J. Chem. Phys.*, 1989, **91**, 1.
279. K. Lipus, Th. Nelis, E. Bachem and W. Urban, *Mol. Phys.*, 1989, **68**, 1171.
280. T.D. Varberg, E.J. Hill and R.W. Field, *J. Mol. Spectrosc.*, 1989, **138**, 630.
281. K. Tanaka, Y. Akiyama and T. Tanaka, *J. Mol. Spectrosc.*, 1989, **137**, 55.
282. R.J. Miller, W.L. Glab and B.A. Bushaw, *J. Chem. Phys.*, 1989, **91**, 3277.
283. A. Fujii, T. Ebata and M. Ito, *J. Chem. Phys.*, 1989, **90**, 6993.
284. J.A. Coxon, P.G. Hajigeorgiou and P.I. Presunka, *Spectrochim. Acta A*, 1989, **45**, 281.
285. J.G. Phillips and S.P. Davis, *Astrophys. J. Suppl. Ser.*, 1989, **71**, 163.
286. M. Li and R.W. Field, *J. Chem. Phys.*, 1989, **90**, 2967.
287. K. Lipus, U. Simon, E. Bachem, Th. Nelis and W. Urban, *Mol. Phys.*, 1989, **67**, 1431.
288. G. Gustafsson and R. Scullman, *Mol. Phys.*, 1989, **67**, 981.
289. R.G. Tonkyn, J.W. Winniczek and M.G. White, *Chem. Phys. Lett.*, 1989, **164**, 137.
290. H. Park, L. Li and W.A. Chupka, *Chem. Phys. Letter*, 1989, **162**, 317.
291. A.S-C. Cheung, K. Yoshimo, D.E. Freeman, R.S. Friedman, A. Dalgarno and W.H. Parkinson, *J. Mol. Spectrosc.*, 1989, **134**, 362.
292. E.H. Fink, K.D. Setzer, D.A. Ramsay, M. Vervloet and G.Z. Xu, *J. Mol. Spectrosc.*, 1989, **136**, 218.
293. T.C. Steimle, D.F. Nachman, J.E. Shirley, C.W. Bauschlicher Jr and S.R. Langhoff, *J. Chem. Phys.*, 1989, **91**, 2049.
294. H. Birk and H. Jones, *Chem. Phys. Lett.*, 1989, **161**, 27.
295. E. Tiemann, H. Kanamori and E. Hirota, *J. Mol. Spectrosc.*, 1989, **137**, 278.
296. J.C. Choi and J.L. Hardwick, *J. Mol. Spectrosc.*, 1989, **137**, 138.
297. U. Magg, H. Birk, K.P.R. Nair and H. Jones, *Z. Naturforsch. A*, 1989, **44**, 313.
298. B.D. McFeeters, E.A. Walters and P.J. Wolf, *J. Mol. Spectrosc.*, 1989, **135**, 1.
299. T. Ishiwata, A. Ishiguro, K. Obi and I. Tanaka, *Chem. Phys. Lett.*, 1989, **159**, 594.
300. T. Ishiwata, K. Obi and I. Tanaka, *J. Chem. Phys.*, 1989, **91**, 3306.
301. M. Suzuki and T. Fujiwara, *J. Mol. Spectrosc.*, 1989, **133**, 233.
302. H. Knöckel, H. Richter, M. Siese and E. Tiemann, *Mol. Phys.*, 1989, **68**, 917.
303. G.W. King and T.D. McLean, *J. Mol. Spectrosc.*, 1989, **135**, 207.
304. M. Bartels, R.J. Donovan, A.J. Holmes, P.R.R. Langridge-Smith, M.A. MacDonald and T. Ridley, *J. Chem. Phys.*, 1989, **91**. 7355.
305. A. Morinaga, K. Sugiyama, N. Ito and J. Helmcke, *J. Opt. Soc. Amer. B*, 1989, **6**, 1656.
306. M. Wakasugi, M. Koizuma, T. Horiguchi, H. Sakata and Y. Yoshizawa, *J. Opt. Soc. Amer. B*, 1989, **6**, 1660.
307. T.C. Steimle, D.F. Nachman, J.E. Shirley and A.J. Merer, *J. Chem. Phys.*, 1989, **90**, 5360.
308. Y. Azuma and A.J. Merer, *J. Mol. Spectrosc.*, 1989, **135**, 194.
309. G. Bujin and C. Linton, *J. Mol. Spectrosc.*, 1989, **137**, 114.
310. I. Burghardt, C.R. Zink, D.A. Fletcher, J.M. Brown and I.R. Beattie, *Mol. Phys.*, 1989, **67**, 1401.
311. L.A. Kaledin and E.A. Shenyavskaya, *J. Mol. Spectrosc.*, 1989, **133**, 469.
312. Z. Fu and M.D. Morse, *J. Chem. Phys.*, 1989, **90**, 3417.
313. J.G. McCaffrey, R.R. Bennett, M.D. Morse and W.H. Breckenridge, *J. Chem. Phys.*, 1989, **91**, 92.

4
Characteristic Vibrations of Compounds of Main-group Elements

BY G. DAVIDSON

1. Group I

The vibrational/rotational Raman spectra of H_2, HD and D_2 have been recorded in argon, krypton and xenon matrices at 9K.[1]

An infrared band near 390 cm^{-1} in B_2O_3-$x$$Li_2O$ glasses, where $0 < x < 1$, has been assigned as νLi-O.[2] ^{6}Li/^{7}Li isotopic shifts in the infrared spectrum of $LiAl(MoO_4)_2$ show an νLiO_6 mode at 320 cm^{-1}.[3] Li-O stretching modes were assigned for $LiGaO_2$ and $LiInO_2$.[4] Polarised Raman spectra of single crystals of $^7Li_2SeO_4$ and $^6Li_2SeO_4$ show that many modes are sensitive to lithium isotopic substitution in the range 290 - 440 cm^{-1}.[5]

Lithium intercalated in MoS_2 gives rise to a band at about 205 cm^{-1} due to the lithium vibrating with respect to the host lattice.[6] ω_e for $^7Li^{79}Br$ and $^7Li^{81}Br$ were found to be at 562.28564 cm^{-1} and 561.71538 cm^{-1}, respectively.[7]

The harmonic stretching wavenumber for Na_2 in the $2^3\Pi_g$ state is at 94.35(1) cm^{-1}.[8] The infrared spectrum of gaseous NaH shows the fundamental vibration at 1171.76082(559) cm^{-1}.[9] $[Na(15\text{-crown-}5)]^+$, as a counterion to $[M(Se)_4)_2]^{2-}$, gives rise to νNa-O bands at 385 and ca.480 cm^{-1}.[10] Low-wavenumber bands in the infrared spectra of Li^+ and Na^+ salts of tetramethyl-o-phthalate or ethylmaleate were assigned to motions of M^+ in e.g. $^-O...Na^+...O^-$ units (80 - 110 cm^{-1}).[11]

Diode laser spectra of NaCl give the following vibrational wavenumbers: $Na^{35}Cl$ 364.6871(32) cm^{-1}; $Na^{37}Cl$ 360.7578(15) cm^{-1}.[12] Similar experiments on RbH produced νRbH at 937.1046(29) cm^{-1} (^{85}Rb) or 936.9792(13) cm^{-1} (^{87}Rb).[13]

2. Group II

$Be(IO_3)_2.2HIO_3.6H_2O$ gives infrared bands consistent with the presence of a $Be(H_2O)_4^{2+}$ cation, e.g. $\nu_3(BeO_4)$ 860 cm^{-1} and $\delta(Be\text{-}OH_2)$ 960 cm^{-1}.[14] The $[Be_3(OH)_3(H_2O)_6]^{3-}$ anion has the following characteristic infrared bands: ν_{as}BeO 850 cm^{-1}; ν_sBeO 425 cm^{-1}; $\nu Be\text{-}OH_2$ 960 cm^{-1}.[15]

Far-infrared and resonance Raman spectra of several

chlorophylls gave evidence for a number of bands in which Mg plays a major role. These give characteristic shifts in the aggregate state, the five-coordinate monomeric state and the six-coordinate state.[16]

νCaN, νCaO(H_2O) and νCaO(ring) modes were assigned from the infrared spectra of $Ca(NCS)_2$.L, where L = dibenzo-18-crown-6 and its 2,3-(4-bromo)-substituted derivative.[17]

A tunable diode laser was used to obtain the infrared absorption spectrum of CaO. The fundamental vibrational wavenumber is at 732.0314(7) cm^{-1}.[18] νCaO modes in Ca_2LnEuO_6, where E = Nb or Ta, were assigned using $^{40}Ca/^{44}Ca$ isotopic shifts, for Ln = La - Lu. All were in the range 197 - 260 cm^{-1}.[19] Low-wavenumber infrared bands of CaX_2.18-crown-6, where X = NCS, ClO_4 or I, and related species showed that νCaO(of crown ether) were in the region 176 - 197 cm^{-1}, with $\nu_{as}CaX_2$ at 291 cm^{-1} (NCS) or 245 cm^{-1} (ClO_4).[20]

F.T. infrared spectra of CaX_2 vapours trapped in matrices showed bands due to monomers, dimers and (for X = F) a trimer. D_{2h} symmetry was established for Ca_2X_4 and D_{2d} for Ca_3F_6. The following assignments were suggested for the CaF_2 monomer (in an argon matrix): ν_1 487, ν_2 157, and ν_3 560 (^{40}Ca), 548 (^{44}Ca) cm^{-1}.[21]

The gas-phase infrared spectrum of ^{88}SrH in its ground electronic state ($^2\Sigma^+$) shows that the vibrational wavenumber is 1206.8912(15) cm^{-1}.[22]

3. Group III

3.1 Boron:- Raman spectroscopy was used to study the formation of B-H units in passivated crystalline silicon.[23] Highly vibrationally-excited states of the BH_2 radical were identified during plasma dissociation of B_2H_6.[24]

Infrared and Raman spectra of $Me_2NEt.BH_3$ and its deuteriated (BD_3) analogue were assigned. These showed the presence of trans and gauche forms in the fluid phases, but only the gauche conformer in the solid.[25] νBH modes in $[BH_3(CH_2CN)]^-$ were identified at 2278 and 2235 cm^{-1}.[26] The infrared and Raman spectra of diethyl ether solutions of $LiMH_4$, where M = B or Al, show the presence of $(Et_2O)_2LiMH_4$ solvates at low concentrations, together with bridged associates at higher concentrations.[27]

The ions $B_2H_7^-$ and $B_2D_7^-$ in solution in glycol ethers give infrared and Raman spectra that are consistent with C_2 (bent) symmetry. An approximate normal coordinate analysis was reported.[28]

Low-temperature (10K) Raman spectra of $Cs_2B_6H_6$ and ^{11}B and ^{10}B enriched derivatives show well-resolved bands due to B_6-cage modes (ν_1 (a_{1g}), ν_3 (e_g) and ν_9 (t_{2g})) for all isotopomers $^{11}B_n{}^{10}B_{6-n}H_6{}^{2-}$, for n = 0 - 6.[29] The vibrational spectra of $B_6H_6X^-$, where X = Me, Et, Pr or Bu;[30] or H, Cl, Br or I;[31] all show νBH modes about 100 cm^{-1} higher than in the analogous $B_6H_5X^{2-}$ ions.

Variable-temperature Raman and F.T. infrared spectra for solid o-carbaborane reveal the existence of several phases at ambient pressure. Variable-pressure Raman measurements also showed a phase transition at 10.1 kbar.[32]

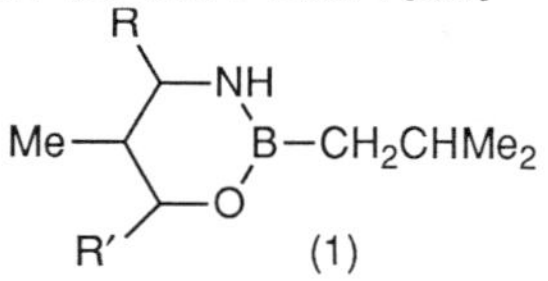

(1)

The infrared spectra of cis and trans stereoisomers of (1), where R = Me, R' = H or R = H, R' = Me, show νBN at 1488 - 1500 cm^{-1} and νBO 1305 - 1320 cm^{-1}.[33] The Raman spectrum of a tetrahydrofuran solution of $Na^+[B(PH_2)_4]^-$ shows $\nu_s BP_4$ at 536 cm^{-1}.[34]

The Raman spectrum of sodium metaborate at pH 11.5 shows the presence of $B(OH)_4^-$.[35] $[B_2(\mu\text{-}O_2)_2(OH)_2(O_2H)_2]^{2-}$ has $\nu_s B(O_2)B$ at 935 cm^{-1}. For $[B_2(\mu\text{-}O_2)_2(OH)_4]^{2-}$, $\nu_{as} B(O_2)B$ is at 711 cm^{-1} and $\nu_s B(OH)_2$ at 968 cm^{-1}.[36] Similar assignments were made for such peroxoborates, making use of ^{10}B, ^{11}B and D substitution.[37] Gas-phase infrared spectra of $Cl_nB(OMe)_{3-n}$, where n = 0, 1 or 2, with CH_3, CD_3 and CHD_2 substitution, gave evidence for the presence of 3 kinds of CH bond (two weaker, one stronger).[38] Ammonium mandelatoborate, $NH_4[(HO)_2B(mandelato)].H_2O$ has νBO bonds in the range 970 - 985 cm^{-1}.[39] Infrared and Raman assignments have been proposed for 2-, 2,4- and 2,5-substituted dioxaborinanes, especially the

ν_s and ν_{as} modes of B-O bonds.[40]

Infrared and Raman spectra were obtained and assigned for the following chain metaborates containing only condensed BO_3 groups: $LiBO_2$, CaB_2O_4 and SrB_2O_4 (including $^{10}B/^{11}B$, $^{6}Li/^{7}Li$ and $^{40}Ca/^{44}Ca$ data).[41] The infrared and Raman spectra of $M_2B_2O_5$, where M = Mg, Ca, Sr or Cd, could all be interpreted in terms of isolated $B_2O_5^{4-}$ anions ($^{10}B/^{11}B$ shifts were obtained for the Mg^{2+} salt).[42] Vibrational spectra of borate glasses containing varying amounts of Li_2O, LiCl or Cs_2O, were analysed in terms of both BO_3 and BO_4 groups. There was evidence for strong coupling of B-O modes involving different groups.[43]

The Raman and infrared spectra of glasses $xNa_2O.(1-x)B_2O_3$, where x = 0.35 - 0.75, show that for $x < 0.6$ there are large borate units built up of BO_4 components. For $x > 0.6$ the structure contains mainly pyro- and orthoborate groups.[44] Raman spectroscopy was used to study the structures of glasses $0.2Na_2O.(0.8-x)SiO_2.xB_2O_3$, showing the presence of bridging and non-bridging oxygens, and boron in 3- and 4-coordinate sites.[45] Low-wavenumber Raman spectra of a variety of alkali borate glasses contain a peak near 130 cm^{-1} due to librations of BO_3 and BO_4 units.[46] The Raman spectra of the systems B_2O_3-Na_2O-NaF or AlF_3 and B_2O_3-NaF-Al_2O_3 or AlF_3 show that four-coordinate $B(O,F)_4$ and $Al(O,F)_4$ units are present.[47]

The structural units present in the systems MgO-M_2O_3-B_2O_3, where M = Al [48] or Ga [49], were identified from their infrared and Raman spectra. The presence of two ν_{as}BO stretches in the infrared spectra of boralites suggests that two kinds of three-coordinate boron are present.[50] Vibrational spectra of PbO-Cr_2O_3-B_2O_3 glasses all showed that BO_3 and BO_4 units were present.[51] Infrared data for 35PbO-25Fe_2O_3-40B_2O_3 show that the BO_3 and BO_4 units are linked to produce boroxol rings, together with tetraborate and diborate units in the network.[52]

The following accurate vibrational wavenumbers were obtained for $^{10}BF_3$ from its high-resolution F.T. infrared spectrum: ν_2 719.285004(16), ν_3 1505.780448(14), ν_4 481.128612(16), $\nu_1+\nu_4$ 1363.658466(22) and $2\nu_2-\nu_2$ 721.95351(7) cm^{-1}.[53] Van der Waals complexes of BF_3 with Ne, Ar or Kr give characteristic shifts in the wavenumber of ν_3 of BF_3.[54-5] Infrared evidence has been found for the formation of contact ion-pairs and ion-triplets in $LiBF_4$ solutions in dipolar aprotic solvents.[56] The assignment of the vibrational spectrum of $NF_4^+BF_4^-$ has been re-assessed, taking into account disorder in the crystal structure.[57]

Vibrational constants have been calculated as follows for $^{11}B^{79}Br$: $X^1\Sigma^+$ ω_e 687.03(1.83) cm^{-1}, ω_ex_e 4.86(74) cm^{-1}; $A^1\Pi$ ω_e 628.11(2.51) cm^{-1}, ω_ex_e 9.46(1.23) cm^{-1}.[58] $PhC(NSiMe_3)_2EX_2$, where E = B, X = Br; E = Al, Ga or In, X = Cl, have the following νEX wavenumbers: νBBr 650 cm^{-1}; νECl 605 (Al), 385 (Ga), 280 (In) cm^{-1}.[59]

3.2 Aluminium:- Aluminium and methane react photochemically in solid methane to form CH_3AlH, with νAlH at 1764 and 1746 cm^{-1} (two lattice sites) and νAlC 610 cm^{-1}.[60] Al and H_2 in a krypton matrix react photochemically to give AlH_2, with ν_1 at 1766 cm^{-1}, ν_2 760 cm^{-1} and ν_3 1799 cm^{-1}. AlD_2 has analogous features at 1275, 560 and 1320 cm^{-1}.[61] Libration modes of the AlH_4^- group in $MAlH_4$, where M = Li, Na, K, Rb or Cs, were assigned from i.r. and Raman spectra.[62] The i.r. and Raman spectra of Li_3AlH_6, Na_3AlH_6 and Li_3AlD_6 show that AlH_6^{3-} is distorted from O_h geometry in the Na^+ but not in the Li^+ salt.[63] Vibrational data were also reported for $Li_3[AlH_nD_{6-n}]$, for n = 0, 1, 2, 4, 5 or 6, and assigned in detail. For n = 2 or 4 only the trans isomers were observed.[64] These data were used to perform a normal coordinate analysis on these isotopomers.[65]

(2), where E = Al or Ga, gave the following i.r. features: νAlAl 373, ν_sAlC_2 492, $\nu_{as}AlC_2$ 518 cm^{-1}; νGaGa 337, ν_sGaC_2 479, $\nu_{as}GaC_2$ 501 cm^{-1}.[66]

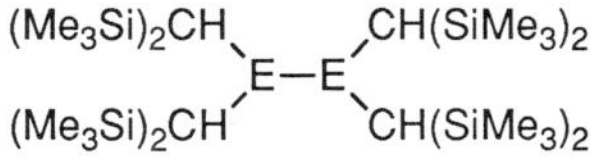

(2)

Aluminium and C_2H_4 in an argon matrix react to form AlC_2H_4, for which νAlC_2 were seen at 352 (a_1) and 226 (b_2) cm^{-1}.[67] $ClAl[CH(SiMe_3)_2]_2$ has i.r. bands due to νAlCl and νAlC_2 (doubtless extensively mixed) in the ranges 334 - 450 cm^{-1} and 520 - 550 cm^{-1} respectively.[68] $Me_2Al(NSiMe_3)_2CMe$ has similar features as follows: $\nu_{as}AlC_2$ 690 cm^{-1}, ν_sAlC_2 589 cm^{-1} and

ν_sAlN$_2$ 423 cm^{-1}.[69] The i.r. and Raman spectra of MeE[N(SiMe$_3$)$_2$]$_2$ show the presence of monomeric molecular units (E = Al, Ga or As).[70]

νAl-O bands due to Cl$_2$SO→AlCl$_3$ and [(Cl$_2$SO→)$_2$AlCl$_2$]$^+$AlCl$_4^-$ were observed in the AlCl$_3$-SOCl$_2$ system.[71] The cation [Al$_2$Li(OH)$_6$]$^+$ has a Raman spectrum consistent with D$_3$ site symmetry for the AlO$_6$ unit.[72]

The i.r. spectra of MgAl$_{2-x}$Cr$_x$O$_4$, FeAl$_{2-x}$Cr$_x$O$_4$ and related systems showed characteristic shifts in the symmetric mode seen at 770 cm^{-1} in MgAl$_2$O$_4$.[73] [AlO$_4$W$_{12}$O$_{36}$]$^{5-}$ has νAlO at 805 cm^{-1} and δAlO modes at 475, 490 and 540 cm^{-1}.[74] I.r. data (1200 - 400 cm^{-1}) on kaolinite, Al$_2$O$_3$.2SiO$_2$.2H$_2$O, were used to characterise its dehydration processes.[75] The i.r. spectra of A- and P$_1$-type zeolites containing Co^{2+} show that νAlO/νSiO modes are shifted compared with unsubstituted derivatives.[76]

The i.r. and Raman spectra of AlF$_3$-YF$_3$-MF$_2$, where M = Ca, Ba, Mg or Sr, glasses show the presence of aluminium as AlF$_6^{3-}$ and AlF$_4^-$ at low YF$_3$ concentrations, but only as AlF$_4^-$ at high YF$_3$ contents.[77] An analysis of the Raman spectra of fluoroaluminate glasses suggests that a band near 550 cm^{-1} is due to isolated AlF$_6$ octahedra, with features near 620 and 500 cm^{-1} due to non-bridging and bridging Al-F stretches of corner-linked AlF$_6$ octahedra.[78] Glasses in the system K$_2$NbOF$_3$-AlF$_3$ have Raman spectra consistent with octahedral aluminium coordination for AlF$_3$ contents of 21 - 29% but tetrahedral coordination at higher AlF$_3$ contents.[79]

The vibrational energies of the X$^1\Sigma^+$ and A$^1\Pi$ states of AlCl were determined from its electronic spectrum. They are 481.40(58) and 441.6(2.3) cm^{-1} respectively.[80] Cl$_2$ adsorbed on clean Al(111) surfaces gives a coverage-dependent νAlCl band at 745 - 760 cm^{-1}.[81] AlCl and diethyl ether in a pentane matrix form AlCl.xEt$_2$O, with νAlCl at 450 cm^{-1}. This subsequently disproportionates to AlCl$_3$.Et$_2$O, with ν_{as}AlCl at 545 cm^{-1}, ν_sAlCl 415 cm^{-1}.[82]

AlCl$_3$/dioxan complexes, AlCl$_3$.nC$_4$H$_8$O$_2$, where n = 1 or 2, have νAlCl characteristic of covalent AlCl$_3$ adducts (near 510 cm^{-1}).[83] The i.r. spectra of AlCl$_3$/M$_2$B$_{10}$H$_{10}$, where M = Et$_4$N, Bu$_4$N, K or Na, are consistent with the formation of [B$_{10}$H$_{10}$(AlCl$_3$)$_2$]$^{2-}$, with localisation of negative charge in the vicinity of each aluminium centre.[84] The F.T.i.r. spectra of

$AlCl_4^-$ in the presence of alkaline earth chlorides show the presence of C_{2v} and C_{3v} $AlCl_4^-$ anions.[85] A normal coordinate analysis has been carried out for $AlCl_4^-$, on the basis of literature data. Estimates were made of changes in Al-Cl bond lengths and ClAlCl angles in different crystalline salts.[86] Shifts in the wavenumbers of Al-Cl stretches and deformations on adding $AlCl_3$ to MCl-CuCl, where M = Na, K, Rb or Cs, melts are consistent with the formation of $CuAlCl_4$ species.[87] The Raman spectra of haloaluminate melts containing oxides show bands due to oxohalides at 270 - 280 cm^{-1} (Cl) or near 185 cm^{-1} (Br). These bands clearly involve motion of the X as well as O atoms.[88]

3.3 Gallium:- The vibrational wavenumbers of GaH were measured from diode laser spectroscopy: ^{69}Ga 1603.9566(20) cm^{-1}; ^{71}Ga 1603.6323(19) cm^{-1}.[89] The first definite report has been made of the preparation of digallane, Ga_2H_6, including detailed assignments of several vibrational modes (from i.r. data only), e.g. νGaH_t 1998 cm^{-1} (b_{1u}), 1981/1970 cm^{-1} (b_{3u}); νGaH_{br} 1278/1267 cm^{-1} (b_{2u}), 1205/1195 cm^{-1} (b_{3u}).[90] An ab initio m.o. calculation has also been made of the vibrational wavenumbers of Ga_2H_6.[91]

Oxygen-doped GaAs gives an i.r. band at 715 cm^{-1}, which shifts on both O and Ga isotopic substitution, i.e. it is a νGaO localised mode.[92] The Raman spectra of M_2O-Ga_2O_3 glasses, where M = Na, K or Cs, show that the gallium is tetrahedrally coordinated.[93]

The compounds $^iPr_2GaBr.E(H)Et$, where E = S or Se, have νGa-E bands at 393 cm^{-1} (S) or 259 cm^{-1} (Se).[94] I.r. and Raman spectra were used to detect phase transitions in $TlGaS_2$ (105, 120K) and $TlInS_2$ (near 190K).[95]

The reaction products of Al, ^{69}Ga or ^{71}Ga atoms and Cl_2 were studied by matrix (Ar) i.r. spectroscopy. There was evidence for MCl, MCl_2 and MCl_3. The νMCl bands were assigned for the first time, and they were consistent with non-linear (C_{2v}) molecules, e.g. for $^{69}GaCl_2$:ν_1 373.0, 368.6, 364.8 cm^{-1} ($^{35}Cl_2$, $^{35}Cl^{37}Cl$, $^{37}Cl_2$ respectively) and ν_3 415.1, 412.0 and 408.6 cm^{-1} (as before). These figures are consistent with a ClGaCl bond angle of approximately 117°.[96]

The Raman spectra of the ambient temperature liquids Ga_2Cl_6 +

1-methyl-3-imidazolium chloride mixtures show the presence of $GaCl_4^-$ and $Ga_2Cl_7^-$, i.e. the behaviour is very similar to that of the related $AlCl_3$ melt system.[97]

3.4 Indium:- The i.r. spectra of ^{113}InH and ^{115}InH gave ω_e values of 1475.5443(40) cm^{-1}, 1475.4343(24) cm^{-1} respectively.[98] $In(CH_2Ph)_3$ has a νInC band in the i.r. at 430 cm^{-1}.[99] νInC in $[Et_2InPtBu_2]_2$ is at 463 cm^{-1}.[100] Assignments of νInC were also made for $Me_2Sn[CH_2InMe_2]_2$ (438, 595 cm^{-1}) and $[Me_2Sn\{CH_2In[OCH_2CH_2NMe_2]_2\}_2$ (472 cm^{-1}).[101]

νInN, νInO and νInX modes were assigned for some indium(III) catecholato complexes with N-donor ligands, e.g. $(Cl_4C_6O_2)InCl.tmen$: νInN 480, 454 cm^{-1}; νInO 400 cm^{-1}; νInCl 270 cm^{-1}.[102] I.r. bands due to νInN (550 cm^{-1}) and νInS (350 cm^{-1}) were consistent with N,S-chelation in (3), where R = CH_3 or CH_2NEt_2.[103]

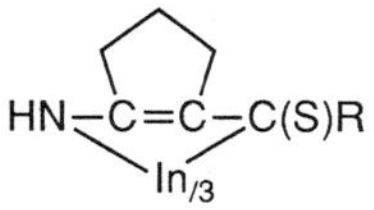

(3)

The Raman and i.r. spectra of $CdBaIn_4O_8$ show no coincidences. Thus the crystal structure is centrosymmetric.[104] The Raman spectra of MIn_2S_4, where M = Mn, Fe, Co or Ni, show resonance enhancement of some bands and more bands than expected, i.e. the symmetry is lowered.[105] The Raman and i.r. spectra of layered single crystals of $TlInS_2$ are consistent with the space group C_{2h}^6 in the temperature range 6 - 300K.[106]

Raman spectroscopy was used to study equilibria in the system

$$2In + InCl_3 \rightleftharpoons 3InCl$$

in aqueous solutions containing $ZnCl_2$. The presence of $ZnCl_2$ shifts the equilibrium towards InCl due to the decrease in water activity.[107] Phase transitions in Cs_2KInCl_6 were followed by Raman spectroscopy.[108]

The Raman spectrum of "In_5Br_7" shows that it contains an $In_2^{II}Br_6^{2-}$ unit, with νInIn at 139 cm^{-1}. The compound was therefore formulated as $In_3^{I}[In_2^{II}Br_6]Br$.[109]

3.5 Thallium:- ω_e values were determined from TlH from high- resolution i.r. spectroscopy, i.e. 1391.2681(22) cm^{-1} (^{205}Tl), 1391.3012(25) cm^{-1} (^{203}Tl).[110] $TlMe_2(L)$, where HL = cyclopentanone thiosemicarbazone, has ν_{as}, $\nu_s TlC_2$ at 520, 470 cm^{-1} respectively and νTlN and νTlS (of ligand) at 440, 270 cm^{-1} respectively.[111]

The high-resolution i.r. spectrum of TlCl at 450°C gave the following vibrational wavenumbers: $^{205}Tl^{35}Cl$ 284.71102(23) cm^{-1}; $^{203}Tl^{35}Cl$ 284.91534(23) cm^{-1}; $^{205}Tl^{37}Cl$ 278.06376(23) cm^{-1}; $^{203}Tl^{37}Cl$ 278.27297(24) cm^{-1}.[112]

Skeletal stretching wavenumbers have been assigned for the following Tl(III) species: $TlClBrI_2L_2$, where L = pyO, 2-, 3-, 4-$CH_3C_5H_4NO$, 4-Cl-, 4-CN-, 4-NO_2-, 4-$MeOC_5H_4NO$; [113] $TlBrCl_2L_2$, where L = pyO and a range of substituted pyridine-N-oxides;[114] $TlXI_2L_2$, where X = Cl, Br or I; L = R-substituted quinoline-N-oxide, R = H, 2-Me, 4-Me, 4-Cl or 4-NO_2;[115] and TlX_3.nL, where X = Cl, Br or I; L = N-methylpyrazole, N-methylimidazole, n = 1 - 3.[116]

4. Group IV

4.1 Carbon:- An analysis of a large number of vibrational/rotational lines of CH in the solar spectrum has given accurate molecular constants for the $X^2\Pi$ ground state of CH.[117] The high-resolution i.r. spectrum of HCS^+ gives the following values: ν_1 3160.6542(20) cm^{-1}; ν_2 775.4762(10) cm^{-1}.[118]

Changes in the Raman band width of ν_1 for CH_3MCl_3, where M = C or Si, were related to solute-solvent dispersion forces. For M = Ge or Sn, the changes were related to solvent dipole interactions.[119] The F.T.i.r. spectrum of $CH^{35}ClF_2$ in a supersonic jet showed that ν_4 is at 809.2726 cm^{-1}. Estimates were also made for $\nu_4(CH^{37}ClF_2)$ 804.5092 cm^{-1}; $2\nu_6(CH^{35}ClF_2)$ 829.062 cm^{-1}; $2\nu_6((CH^{37}ClF_2)$ 820.900 cm^{-1}.[120]

The first detailed measurement has been made of the i.r. spectral intensities in the ν_3 band of $^{12}CH_3{}^{37}Cl$.[121] ν_3 for CH_3Br was assigned as follows; (^{79}Br) 611.11857(4) cm^{-1}; (^{81}Br) 609.93709(4) cm^{-1}.[122] A high-resolution analysis of the $\nu_3+\nu_6$ band of $CH_3{}^{79}Br$ shows the band origin to be at 1561.8982(26) cm^{-1}.[123]

The following band centres were measured for CH_3I: $2\nu_2$ 2486.4568(2) cm^{-1}; $\nu_2+\nu_5$ 2682.4656(22) cm^{-1}; $\nu_2+\nu_3+\nu_6$ 2654.5356(196) cm^{-1} and $3\nu_6$ 2643.848 cm^{-1}.[124] The band centre for ν_3 of CH_3I is at 533.216849(25) cm^{-1}.[125] The ν_2 band of CD_3I is centred at 949.360 cm^{-1}.[126]

The ν_2/ν_4 dyad has been studied for $^{12}CH_4$ and $^{13}CH_4$. The following band origins were determined: ν_2 1533.332586(20) cm^{-1} (^{12}C); 1533.492791(15) cm^{-1} (^{13}C); ν_4 1310.7616216(97) cm^{-1} (^{12}C); 1302.780775(12) cm^{-1} (^{13}C).[127]

The Raman spectra of diamond-like amorphous carbon films show bands at 1400 and 1530 cm^{-1} which are indicative of the proportion of sp^3-hybridised carbon atoms in the films.[128] Ab initio calculations have been reported for the vibrational wavenumbers of the astrophysically important molecules C_3O and HC_3O^+.[129] Several studies have been made of the ν_3 mode of the linear C_5 molecule, near 2164 cm^{-1}.[130-2]

A comprehensive study has been made of the i.r. and Raman spectra of the CN^- fundamental and overtones in alkali metal cyanides.[133] Table 1 summarises the vibrational assignments for $F_2C{=}NH$, which is only stable below 260K.

Table 1 Vibrational assignments for $F_2C{=}NH$ (/cm^{-1}; data for $F_2C{=}ND$ in brackets)

νNH	3402	(2519)
νCN	1785	(1780)
$\nu_{as}CF_2$	1307	(1280)
δCNH	1031	(836)
$\nu_s CF_2$	947	(954)
δCF_2	618	(609)
ρCF_2	542	(492)
τNH	832	(614)
γCF_2	695	(705)

Accurate measurements have been made on 26 overtone and combination bands of HCN in the range 5400 - 15100 cm^{-1}.[135] The van der Waals bending vibration (ν_5) of Ar...HCN is estimated to be at 7.8 cm^{-1}.[136] The i.r. spectra of the isomers HNCO, HOCN, HCNO and HONC were reported (from argon-matrix and gas-phase experiments) and compared with ab initio calculations.[137]

The vibrational spetrum of isocyanogen, CNCN has been reported: ν_1 2302.001 cm^{-1}; ν_2 2059.731 cm^{-1} and ν_5 194.753 cm^{-1}.[138] νCS and νCN modes have been assigned for NCS^- as Bu_4N^+, NH_4^+ or Li^+ salts in H_2O-DMF mixtures. The former gives three features, due to hydrated free NCS^-, ion-pairs M-NCS (M = Li or NH_4) and free NCS^- surrounded by DMF molecules. νCN shows features corresponding to the first two only.[139] Raman (liquid) and i.r. (gas, matrix) spectra for BrNCO produced the assignments which are listed in Table 2.[140]

Table 2 Vibrational assignments for BrNCO (/cm^{-1})

ν_1	2196.0
ν_2	1290.9
ν_3	691.1
ν_6	572.2
ν_4	506.0 (^{79}Br), 504.6 (^{81}Br)
ν_5	137.4

The vibrational/rotational spectrum of the first excited state of CO ($a^3\Pi$) has been reported.[141] The high-resolution i.r. spectrum of the CO dimer shows that the band origin of ν_2 is at 2167.69904(11) cm^{-1}.[142] Similar data for $^{14}C^{16}O_2$ gave a value of 2214.74458(30) cm^{-1} for the band origin of ν_3.[143] The vibrational/rotational spectrum of the $N_2...CO_2$ van der Waals complex in the region of ν_3 for CO_2 is consistent with the T-shaped structure (4).[144] The ν_4 rovibrational band of CO_2...Ar (the CO_2 antisymmmetric stretch), near 2350 cm^{-1} was studied by diode-laser spectroscopy. These results were also consistent with a T-shaped geometry.[145] CO_2 on silver films dosed with small amounts of potassium forms $K^+CO_2^-$, with SERS bands at 755 and 1220 cm^{-1}, due to the δ and ν_s modes of CO_2^-.[146] Evidence has been found for CO_2^+ and CO_2^- in solid neon. ν_3 of CO_2^+ is at 1421.7 cm^{-1}, and ν_3 of CO_2^- at 1658.3 cm^{-1}.[147]

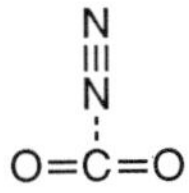

(4)

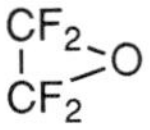

(5)

The ν_2 band of OCS was found to be centred at 520.42207(17) cm^{-1}, from heterodyne frequency measurements.[148] A similar experiment gave the following band centres: $^{16}O^{12}C^{32}S$ 00^02-00^00 1710.97625(9) cm^{-1}, 01^12-01^10 1697.60663(10) cm^{-1}; 00^03-00^01 1697.02429(13) cm^{-1}; $^{16}O^{12}C^{34}S$ 00^02-00^00 1688.75687(13) cm^{-1}; $^{16}O^{13}C^{32}S$ 00^02-00^00 1701.79097(15)cm^{-1}.[149] The i.r. spectra of van der Waals complexes show very small shifts to lower wavenumber for the free OCS band at 2062.20127 cm^{-1}.[150]

The i.r spectrum of gaseous carbonyl fluoride, $F_2{}^{12}C^{18}O$, gave the following fundamental wavenumbers: (a_1) 1895.2 (average of Fermi doublet), 947.3, 576.5 cm^{-1}; (b_1) 1242.3, 604.9 cm^{-1}; (b_2) 770.0 cm^{-1}.[151] The Raman and i.r. spectra of (5), and its ^{18}O analogue, gave a complete vibrational assignment, including revisions to some earlier assignments.[152]

An F.T.i.r. study of the photo-oxidation of CF_3I in low-temperature matrices confirmed the participation of $CF_3OO\bullet$ and $CF_3O\bullet$ radicals.[153] Published vibrational data were used to carry out a normal coordinate analysis for the oxalyl halides, $(COX)_2$, where X = F, Cl or Br. Many assignments were confirmed by new matrix-isolation i.r. data for oxalyl chloride, $(COCl)_2$.[154] Matrix-isolation F.T.i.r. data for photo-excited trans-oxalyl chloride gave evidence for the presence of a new cis-conformer. This in turn was involved in a cis $\rightleftharpoons$ gauche isomerisation.[155] The i.r. and Raman spectra of methyl thiofluoroformate, $FC(O)SCH_3$, show the presence of the trans (F,CH_3) conformer, of C_s symmetry, in both vapour and liquid phases. Wavenumbers were reported for all except torsional modes.[156]

Very precise measurements have been reported for ν_3 for $^{12}CS_2$ and $^{13}CS_2$; the values were 1535.35623(10) and 1485.33123(9) cm^{-1} respectively.[157] Changes in the intensities and frequency separation of the components of the Fermi doublet ($\nu_1/2\nu_2$) in liquid CS_2 with pressure have been measured by Raman spectroscopy.[158] A high-resolution (F.T.i.r.) analysis of ν_7 and $\nu_4-\nu_2$ of the SCCCS molecule confirms its linearity.[159]

The i.r. absorption intensity has been measured for the $^PR_1(20)$ line at 1240.50 cm^{-1} of CF_2.[160] ν_1 (νCH), ν_3 and ν_8

(ωCF_2) have been measured very accurately for $CHClF_2$ in a supersonic free jet.[161] Raman spectra of CF_3X, where X = Cl, Br or I, in liquefied xenon solutions, gave full vibrational assignments for all three, including a direct observation of ν_6 of CF_3I, at 267 cm^{-}.[162] Several overtones and combination bands of CF_3Cl have been subjected to very accurate analysis.[163-5] The i.r. spectrum of solid CF_3Cl suggests that the crystal structure is orthorhombic, with a C_{2v} factor group, two molecules per unit cell and a site symmetry of C_s for the molecules.[166] Analysis of the i.r. spectrum of gaseous CF_2Cl_2 showed that the bands at 671 and 1102 cm^{-1} are A_1, while those at 922 and 1160 cm^{-1} are of B type.[167]

Both CF_3Br and CF_3I (including isotopomers) have been the subject of a number of high-resolution i.r. studies.[168-72] The BrCCl bending mode of $BrClCH_2$ was shown by gas-phase Raman spectroscopy to be at 224.3 ± 0.2 cm^{-1}.[173] The Raman spectrum of solid CCl_4 at 90K has been analysed in detail, including a large number of isotopic wavenumbers.[174]

4.2 Silicon:- νSiH bands were detected at the surface of β-SiC. It was suggested that several SiH_x species were present.[175] Amorphous SiN_x:H films (from an R.F. discharge of SiH_4+N_2+H_2 mixtures at 300°C) give νSiH at about 2100 cm^{-1} and νSiN features at 750, 840 and 960 cm^{-1} in the i.r. spectrum.[176] I.r. bands due to νSiH, νSnH, νSiO and (possibly) νSnO were seen for amorphous hydrogenated $Si_{1-x}Sn_x$:H films.[177] An i.r. band at 898 cm^{-1} in Si-implanted hydrogenated GaAs was believed to be due to an Si-H fragment.[178]

Several band assignments have been proposed for $Me_{3-n}(Me_2SiNH)_nSiH$ and $(RMe_2SiNH)_3SiH$, where R = ^{n}Bu, Et or Ph, especially νSiH, νNH, δNH and ν_{as}SiNSi. The last was near 925 cm^{-1}.[179] The silyl rocking mode was assigned at 591.3800(57) cm^{-1} in CH_3SiH_3. Perturbations to this mode were ascribed to the influence of ν_6, the torsional mode, estimated to lie at 3.0167(48) cm^{-1}.[180] A vibrational/rotational analysis has been published for several fundamentals of vinylsilane, CH_2=$CHSiH_3$.[181] The Raman and i.r. spectra of SiH_3GeH_3 gave a quite detailed vibrational assignment, in terms of C_{3v} symmetry. The data were used for a normal coordinate analysis.[182] I.r. data were listed without assignment for $NR(SiH_3)_2$, where R = ^{i}Pr or ^{t}Bu.[183] I.r. spectra of SiH_4,

GeH_4, H_2S and H_2Se in liquid krypton solutions were analysed in the overtone and combination regions.[184]

νSiC modes were assigned for compounds with a $SiMe_3$ group attached to a cyclopropane ring, an allyl or a vinyl group.[185] The new anion $[Si(NCSe)_6]^{2-}$ has νSiN at 474 cm^{-1} in the i.r. spectrum.[186] The i.r. and Raman spectra of the silatranes, (6), where R = H, Me, CH_2Cl, MeCHCl, Ph or 4-ClC_6H_4, confirm the existence of the N→Si coordinate bond.[187]

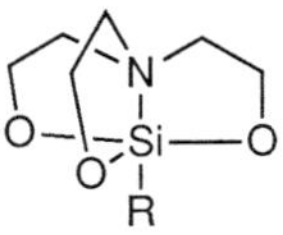

(6)

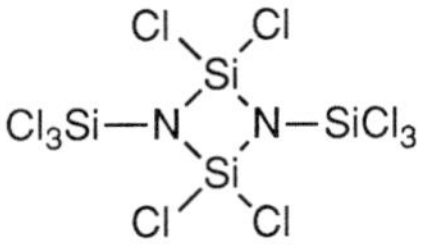

(7)

The i.r. and Raman spectra of $Cl_3SiN(SiCl_3)_2NSiCl_3$ confirm the structure (7).[188]

The intensities of i.r. bands due to SiO (1226 cm^{-1}) and $(SiO)_2$ (767.5 cm^{-1}) were measured in the Si+SiO_2 system.[189] The i.r. spectra of matrix-isolated $(SiO)_2$, together with ^{29}Si, ^{18}O isotopic variants, gave assignments consistent with a D_{2h} geometry. The b_{2u} and b_{3u} modes are at 803.2 and 768.2 cm^{-1} for $(^{28}Si^{16}O)_2$. There was, however, also some evidence for Si_2OO, of C_{2v} symmetry, with b_1 and a_1 modes at 789.5 and 750.4 cm^{-1} respectively.[190] The i.r. spectra of $R_2Si=O$ were observed in Ar matrices at 12K, where R = MeO or Ph. Bands due to $(MeO)_2Si=O$ are at 1471, 1247, 1199, 1174 and 888 cm^{-1}.[191]

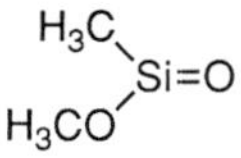

(8)

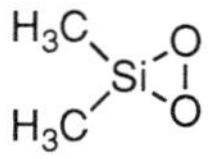

(9)

The species (8) and (9) have been obtained in inert gas matrices. (8) has νSiO at 1090 cm^{-1} and νSi=O near 1240 cm^{-1}, while (9) has νSiO bands at 1006 and 1021 cm^{-1}.[192] The i.r. spectra from Si/non-aqueous electrolyte surfaces gave evidence for clusters of Si-O-Si units in the early stages of silicon oxidation.[193] Assignments have been proposed for νSiO stretches, δSiOSi and δOSiO bending modes for SiO_2 amorphous films.[194] The i.r. spectra of quartz and quartz-glass were used to probe inter- and intramolecular bonding.[195] An i.r. method has been established for determining amounts of quartz and cristobalite in high-temperature modifications of silica.[196]

Several i.r. bands due to edge-shared SiO_4 tetrahedra were seen in dehydroxylated silica samples.[197] New calculations of the wavenumbers for compressed SiO_2 glass reproduce the observed behaviour of the Raman bands at 495 and 606 cm^{-1}.[198] The Raman spectra of sodium silicate glasses were interpreted in terms of the structural units SiO_2, $Na_2O.4SiO_2$, $Na_2O.2SiO_2$, $0.42Na_2O.0.58SiO_2$, $Na_2O.SiO_2$, $0.55Na_2O.0.45SiO_2$ etc.[199] $^{28}Si/^{30}Si$ isotopic substitution was used to identify νSiO modes in synthetic beryls (1211, 910 and 815 cm^{-1}), leaving νBeO_4 modes 730 - 787 cm^{-1}.[200]

The Raman spectra of garnets in the system $Mg_3Al_2Si_3O_{12}$-$Mg_4Si_4O_{12}$ show the presence of ordered $(Mg,Si)O_6$ units.[201] $BaCuSi_2O_6$ contains cyclic Si_4O_{12} units, with a characteristic in-plane ring breathing mode at 506 cm^{-1}.[202] The Raman spectra of BaO-M_2O_3-SiO_2, where M = Y or La, gave estimated quantities of $\equiv SiO$, $=SiO_2$ and SiO_4 units in the structures.[203]

Molecular SiS_2 in an argon matrix has $\nu_{as}SiS$ at 918 cm^{-1} (for $^{28}Si^{32}S_2$; 907 cm^{-1} for $^{29}Si^{32}S_2$; 897 cm^{-1} for $^{30}Si^{32}S_2$; 914 cm^{-1} for $^{28}Si^{32}S^{34}S$).[204]

The fundamental vibrational band origin for the SiF radical was shown by diode-laser spectroscopy to be at 847.7205 ± 0.0005 cm^{-1}.[205] The i.r. diode laser spectrum of SiF^+ was analysed to give ω_e 1050.3757 ± 0.0013 cm^{-1}, and $\omega_e x_e$ 4.9462 ± 0.0004 cm^{-1}.[206]

The ν_3 fundamental of $^{28}SiF_4$ was shown to be centred at 1031.5444126(21) cm^{-1}.[207] Ab initio calculations have been made for ω_e values for SiF_4^+, PO^+ and NS^+.[208] νSiF mode assignments were proposed for $SiF_4.2NH_3$ and $SiF_4.2ND_3$, i.e. at 725, 715 cm^{-1} respectively. These, together with δSiF assignments, were consistent with trans octahedral (D_{4h}) symmetry.[209]

Ab initio calculations of the vibrational wavenumbers for EX_2, where E = Si or Ge; X = Cl or Br, were in good agreement with experiment.[210]

4.3 Germanium:- $CF_3N{=}GeH_2$ has been prepared, with νGeH at 2329 cm^{-1}, νGe=N 1030 cm^{-1}, and CF_3 modes in the expected regions.[211] A high-resolution i.r. study om monoisotopic $H_3{}^{70}Ge^{35}Cl$ gives the following wavenumbers: ν_1 2120.078(14) cm^{-1}, and ν_2 2129.4721(3) cm^{-1}.[212]

There is i.r. evidence that photolysis of $Me_2Ge(N_3)_2$ in matrices forms monomeric Me_2Ge:, with νGeC at 541 and 527 cm^{-1}, as well as the known dimer $Me_2Ge{=}GeMe_2$.[213] Skeletal mode asignments in $Me_2Ge[S_2CNEt_2]Cl$ are consistent with the germanium atom being at the centre of a distorted trigonal bipyramid, with the two C atoms and one S occupying equatorial positions. The axial Ge-Cl bond is weaker than a normal 'tetrahedral' Ge-Cl bond.[214]

The i.r. spectra of $[GeCl_2L_2]Cl_2$, where L = $H_2N(CH_2)_nNH_2$, n = 7,8,9, or 10, show that there are octahedral $GeCl_2N_4$ units present, with trans chlorine atoms.[215]

νGeO has been assigned for $[GeO_2(OH)_2]^{2-}$ and $[GeO_3(OH)]^{3-}$.[216] The assignments for ν_{as} and ν_s Ge-E-Ge modes for polymeric $[(CF_3)_2GeE]_n$ are summarised in Table 3, where E = O, S or Se.[217]

Table 3 Assignment of Ge-E-Ge stretching modes for $[(CF_3)_2GeE]_n$ (/cm^{-1})

E =	O	S	Se
ν_{as}	924	445	352
ν_s	844	412	321

The i.r. and Raman spectra of compounds containing the group $Cl_3GeCNC(O)$ show that all have five-coordinate germanium, with C=O→Ge coordination.[218] I.r. and Raman data were reported and a normal coordinate analysis performed for the scheelite-type germanates $MGeO_4$, where M = Zr, Hf or Th.[219] The Raman spectra of low-alkali sodium germanate species were largely interpreted in terms of $8GeO_2.Na_2O$, $4GeO_2.Na_2O$ and $2GeO_2.Na_2O$.[220] I.r. data of $M_2(GeTe)O_6$, where M = K, Rb or Cs, are consistent with a pyrochlore structure. Bands were assigned as ν(Ge/Te-O) as follows: 620 cm^{-1} (K), 615 cm^{-1} (Rb) or 580 cm^{-1} (Cs).[221]

$Me_2Ge=S$ has been detected in low-temperature matrices. The νGe=S stretch was seen as a very strong i.r. band at 605 cm^{-1}, with ν_sGeC_2 at 516 cm^{-1}.[222] The Raman spectra of $Ge_{1-x}Sn_xSe$, where $0.00 \leqslant x \leqslant 0.70$, contain characteristic features due to stretches of the units $Ge(Se_{1/2})_4$ and $Sn(Se_{1/2})_4$.[223] Polymorphic changes in $RbGeI_3$ on heating were followed by Raman spectroscopy.[224]

4 4 Tin and Lead:- Ab initio m.o. calculations for SnH_4 gave a calculated vibrational spectrum in fair agreement with experiment.[225] A high-resolution i.r. study on ν_1 and ν_3 of SnH_4 gave values of the band centres for the isotopomers containing ^{116}Sn, ^{117}Sn, ^{118}Sn, ^{119}Sn and ^{120}Sn.[226] The i.r. spectrum of SnH_3Br was studied in the ν_2/ν_5 region, giving the following band centres: ν_2 686.65524(10) cm^{-1}, and ν_5 702.54292(14) cm^{-1}.[227] Vibrational wavenumbers have been calculated using ab initio m.o. techniques for Sn_2H_6.[228]

$Me_2Sn(S_2PMe_2)_2$ has ν_{as} and ν_sSnC_2 at 555 and 442 cm^{-1} respectively in the i.r. spectrum, showing that the Me-Sn-Me unit is non-linear. $Me_3SnS_2PMe_2$ gives two νSnC_3 bands in the i.r., and so the $SnMe_3$ fragment must be non-planar.[229] The i.r. spectra of R_2Sn-L-cysteinate complexes, where R = Me, Et or Bu, suggest that there is a trigonal-bipyramidal tin environment, with two C (R) atoms and the S atom equatorial, and an O (carboxylate) and an N (NH_2) atom apical.[230]

$SnEt_2(dapt)$, where H_2dapt = 2,6-diacetylpyridinebis-(2-thenoylhydrazine), has νSnC at 530, 500 cm^{-1}, and νSnO at 480, 430 cm^{-1}.[231] The i.r. spectra of Et_2SnCl_2L, where L = $Ph_2P(O)CH_2CH_2P(O)Ph_2$ or cis-$Ph_2P(O)CH=CHP(O)Ph_2$, show that the tin atom is approximately octahedrally coordinated, with the two Cl, two O and two C atoms in the cis,cis,trans arrangement.[232] $SnCl_2L_2$ and $SnCl_4L_2$, where L = pyrido[1,2-a]benzimidazole, both have νSnN in the range 280 - 315 cm^{-1}, and νSnCl 320 - 340 cm^{-1}.[233]

$[Tl_2Sn(OEt)_6]_n$ has νSnO as a single broad i.r. band at 528 cm^{-1}, shifting to 541 cm^{-1} in solution. This suggests that there are only bridging alkoxides in solution.[234] νSnX modes were assigned for $Sn(L)X_2$, where L = tetra-p- or tetra-m-tolylporphyrin, X = OH, $OClO_3$, F, Cl or Br.[235] $(L-L)SnCl_nMe_{4-n}$, where L-L = bis(3,5-dimethylpyrazol-1-yl)methane, n = 2 - 4, have νSnCl near 330 cm^{-1} in every case.[236] ν_1, ν_2 and ν_3 were assigned for SnI_2 and PbI_2 in xenon or argon matrices, and also in the gas phase (temperature range 400 - 1000°C).[237]

$(Me_3Pb)_2L$, where L = acetylenedicarboxylate or terephthalate, have i.r. bands near 500 and 470 cm^{-1}, which are believed to be due to ν_{as} and ν_s PbC_3 respectively.[238] Vibrational wavenumbers were obtained for PbO from its diode-laser i.r. spectra. The following band centres were deduced: (^{208}Pb) 720.95540(34) cm^{-1}; (^{207}Pb) 721.07986(34) cm^{-1}; and (^{206}Pb) 721.20560(34) cm^{-1}.[239] Skeletal PbS_4 vibrations were assigned for $Pb(S_2CNEt_2)_2$, although considerable mixing with ligand modes was suggested.[240]

5. Group V

5.1 Nitrogen:-The Raman spectrum of solid N_2 in the temperature range 4 - 40K shows an $\alpha \rightarrow \beta$ phase transition at 35.6K.[241] νN_2 (2328 cm^{-1}) for N_2 physisorbed on Ag(111) at 15K is very close to that for the α- phase of bulk solid nitrogen.[242]

The absolute i.r. intensity for ν_1 of HN_2^+ was measured by direct laser absorption spectroscopy.[243] The origin of the ν_4 band of HN_3 was shown to be at 1147.4 cm^{-1}.[244] The ν_3 fundamental of NH_2^+ is at 3359.9317(20) cm^{-1}.[245]

Table 4 Vibrational assignments for H_2NX, where X = F or Cl ($/cm^{-1}$)

	X =	F	Cl
a'	$\nu_s NH_2$	3234 (2250)	3297
	δNH_2	1564 (1145)	1534
	ωNH_2	1233 (970)	1056
	νNX	891 (882)	679/674
a"	$\nu_{as} NH_2$	3346 (2500)	3374
	δNHX	1241 (965)	1063

(Figures in brackets are for D_2NF)

Complete vibrational assignments have been proposed for H_2NF, D_2NF [246] and H_2NCl.[247] These are summarised in Table 4.

The ν_3 band of NH_3^+ has been observed at 3388.6453(63) cm^{-1}.[248] Gas phase Raman spectra were obtained for NH_nD_{3-n}, where $0 \leqslant n \leqslant 3$. The new data, together with previous i.r. results, enabled a normal coordinate analysis to be carried out, including anharmonicity corrections.[249] High resolution i.r. analyses have been made for the ν_1, ν_3, $2\nu_4$ [250] and $\nu_2+2\nu_4$, $\nu_1+\nu_2$ and $\nu_2+\nu_3$ [251] regions for $^{14}NH_3$.

An argon-matrix i.r. spectrum of the $NH_3.HCN$ complex shows that it has C_{3v} symmetry. The perturbed vibrations of each sub-molecule were identified.[252] The i.r. spectra of supersonic molecular beams of $(NH_3)_n$ clusters gave bands at 976.9 and 1003.4 cm^{-1} for n = 2, 1016.3 cm^{-1} for n = 3. The latter was consistent with a cyclic geometry.[253] A high-resolution study of the ν_4 band of NH_4^+, using i.r. diode laser velocity modulation spectroscopy, showed that the band centre was at 1447.2158(50) cm^{-1}.[254] The experimental i.r. spectra of $N_2H_{4-n}D_n$, where n = 0 - 4, were compared with calculated values from ab initio m.o. studies. This led to more definite assignments than had hitherto been possible.[255]

The $HNCN^-$ ion, as the Na^+ salt, gave the following vibrational assignments: νNH 3295 cm^{-1}; $\nu_{as}NCN$ 2160 cm^{-1}; δCNH 1258 cm^{-1}; ν_sNCN 1155 cm^{-1}; and δNCN 620 and 585 cm^{-1}. NCN^{2-}, again as the Na^+ salt, has $\nu_{as}NCN$ at 1970 cm^{-1}, and δNCN at 660 cm^{-1}.[256] The i.r. and Raman spectra of $(CF_3)_2N^-$ can be assigned in terms of overall C_{2v} symmetry. Comparison with $(CF_3)_2O$ shows that the C-N bonds in the former are weaker than the C-O bonds in the latter.[257] The ν_1 (νNH) band of the nitroxide ion, HNO^-, is centred at 2750.782650(5300) cm^{-1}.[258]

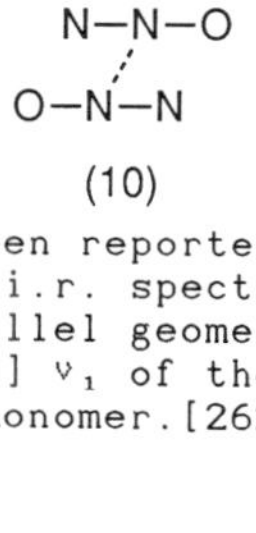

(10)

Detailed studies have been reported for ν_2 [259] and ν_3 [260] of N_2O. High-resolution i.r. spectra of $(N_2O)_2$ show that the dimer has a slipped parallel geometry with a centre of symmetry, e.g. (10).[261] ν_1 of the $(N_2O)_2$ dimer is 5.9126(5) cm^{-1} lower than in the monomer.[262-3]

Several high-resolution i.r. studies of gaseous HNO_3 have given precise values for the band centres of ν_3, ν_4,[264] ν_6, ν_7, ν_8,[265] ν_9, [266] and $\nu_8+\nu_9$.[267]

Resonance Raman spectra of $NOCl/Cl_2$ mixtures suggest that NOCl photodissociation is not as fast as previously thought.[268] High-resolution i.r. studies have been made for all the fundamentals, and many overtones and combination bands of NOCl.[269-72] The F.T.i.r. spectrum of gaseous nitryl bromide, as a transient species in the reactions N_2O_5 + NaBr and N_2O_5 + NOBr, contains bands at 787, 1292 and 1660 cm^{-1}.[273]

F.T.i.r. spectra of solid NO_2 prepared under a variety of conditions gave bands due to N_2O_4 (D_{2h}), to O=N-O-NO_2, to $NO^+NO_3^-$, and to NO_2 monomer.[274] Ab initio m.o. calculations for the isomeric anions NSO^-, SNO^-, SON^- and cyclic-NSO^- gave only moderate agreement with experimental data for the known NSO^-.[275] The i.r. spectrum of thiazyl bromide, NSBr, shows ν_1 (νNS) at 1311 cm^{-1}.[276] The analogous band for NSI is at 1295 cm^{-1}.[277]

5.2 Phosphorus:- The i.r. and Raman spectra of F_3C-P(H)D gave the following assignments: νPH 2337 cm^{-1}, νPD 1700 cm^{-1}, νPC 416 cm^{-1}, δHPD 944 cm^{-1}, δCPH 828 cm^{-1} and δCPD 634 cm^{-1}.[278] The Raman spectrum of $(CF_3S)PH_2F^+$, as the AsF_6^- salt, includes these bands: $\nu_{as}PH_2$ 2541 cm^{-1}, ν_sPH_2 2495 cm^{-1} and νPS 545 cm^{-1}.[279] Complete vibrational assignments (in terms of C_2 symmetry) were proposed from the i.r. and Raman spectra of $CH_2(PH_2)_2$ and $CH_2(PD_2)_2$.[280] Matrix i.r. spectra of PH_3F_2, PHF_2 and PH_2F were identified from PH_3/PF_3 reactions. The vibrational assignments for PH_2F are summarised in Table 5.[281]

Table 5 A vibrational assignment for PH_2F (/cm^{-1})

PH_2F	PHDF	PD_2F	Assignment
2310	2306	1680	ν_5(a") νPH(D)
2304	1676	1673	ν_1(a') νPH(D)
1090			ν_2(a') δPH_2
934		701	ν_3(a') $\delta PH_2(D_2)$
795		798	ν_4(a') νPF

Table 6 A vibrational assignment for PH_3F_2 (/cm^{-1})

I.r.(4K,solid)	Raman(195K,HF soln.)	Assignment
2538	2540, dp.	ν_5 (e') $\nu_{as}PH_3$
-	2486, pol.	ν_1 (a_1') ν_sPH_3
1287	-	ν_3 (a_2") γPH_3
-	1168, dp.	ν_8 (e") ρPF_2
984	975, dp.	ν_6 (e') δPH_3
680	-	ν_4 (a_2") $\nu_{as}PF_2$
-	636, pol.	ν_2 (a_1') ν_sPF_2
377	390, dp.	ν_7 (e') δPF_2

A complete vibrational assignment was proposed for PH_3F_2, based on low-temperature i.r. and Raman data (Table 6).[282]

The i.r. spectrum of phenylmethylidynephosphine, PhC≡P contains ν(C≡P) at 1569 cm^{-1}.[283] The analogous mode in P≡C-N($SiMe_3$)$CHMe_2$ is at 1588 cm^{-1}.[284] The i.r. and Raman spectra of gaseous, liquid and solid P(CH=CH_2)$_3$ suggest that there are at least two conformers, and possibly as many as 4, in the fluid phases.[285]

Published vibrational data were used for normal coordinate analyses of MeP$(CN)_2$ and MeAs$(CN)_2$. Some revised vibrational assignments were suggested.[286] Detailed assignments of the i.r. and Raman spectra of $(CH_3)_2CHPF_2$, $CH_3(CD_3)CHPF_2$ and $(CD_3)_2CDPF_2$ show that there are gauche and trans conformers in fluid phases, but only the gauche form in the solid.[287] The i.r. and Raman spectra of $(CH_3)_{3-n}(CD_3)_nPS$, for n = 0 - 3, and $(CH_3)_3P^{34}S$, were assigned under C_{3v} symmetry for n = 0 or 3, C_s for n = 1 or 2.[288] Quite detailed assignments were made for the P-Si and Si-C stretching modes for $(SiMe_3)_{3-n}P(SiMe_xPh_{3-x})_n$, for n, x = 0 - 3.[289]

The i.r. and Raman spectra of Me_2NPF_2, $(CH_3)(CD_3)NPF_2$ and $(CD_3)_2NPF_2$ gave complete vibrational assignments in each case. These were consistent with microwave results showing that the N, P and C atoms are coplanar.[290] Ab initio m.o. calculations on P_5^- suggested some reassignments of the experimental vibrational wavenumbers.[291]

I.r. spectra of matrix-isolated HOP=O, from PCl_3 + H_2O, were reported.[292] The Raman spectrum of MeOP(O)Cl_2 shows trans and gauche forms in the liquid, but only the trans form in the solid.[293]

A careful analysis of ν_sPO of $H_2PO_4^-$ in aqueous $NH_4H_2PO_4$ shows the presence of monomeric, dimeric and polymeric units.[294] Raman intensity measurements have been reported for aqueous solutions of K_3PO_4 and $K_4P_2O_7$.[295] $\nu_{as}PO_4$ in $PW_{11}O_{39}^{2-}$ shows a noticeable dependence on the nature of the cation.[296] A vibrational assignment was suggested for the $H_2P_2O_7^{2-}$ ion, from i.r. and Raman spectra of $K_2H_2P_2O_7$ and its hemihydrate.[297]

The i.r. spectra of TeO_2-P_2O_5 glasses shows that νP=O is absent, due to coordination of the P=O unit to TeO_4 polyhedra.[298] Several reports have been made of the vibrational wavenumbers of $P_3O_{10}^{3-}$ ion.[299-301] The i.r. and Raman spectra of the $P_4O_{12}^{4-}$ ion shows that it has S_4 site symmetry in $Mg_2P_4O_{12}$ [302] and in $MNdP_4O_{12}$, where M = Li or K.[303] I.r. spectra were obtained for P_4O_n, where n = 6 - 10, νP=O amd P_4O_6 cage modes were assigned in each case.[304]

νPS and νSS modes were used to probe short range order in P_xS_{1-x} glasses.[305] I.r. and Raman spectra of CH_2=CHP(S)Cl_2 suggest that the most stable conformer has P=S and C=C bonds nearly eclipsed.[306] The Raman spectrum of S=PBr_2F was reported and assigned.[307] (11) has bands due to the S-P-S fragment between 665 and 525 cm^{-1}.[308] The i.r. and Raman spectra of $[(RO)_2P(S)S]_3P$, where R = Me or Et, showed that there was no $P^{III}\leftarrow$S interaction involving the thione-type sulphur atom.[309]

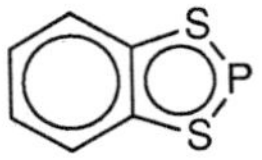

(11)

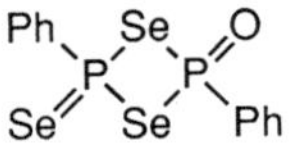

(12)

Thin films of amorphous $FePS_3$ gave i.r. bands due to the PS_3 unit at about 570 cm^{-1}, with νP-P near 415 cm^{-1}.[310] The Raman spectrum of P_4S_{10}, measured over a range of temperature and pressure, gave evidence for weak intermolecular coupling.[311] νP=Se in (12) is at 508 cm^{-1}.[312]

A high-resolution i.r. study of ν_3 of PF_5 showed no splitting due to Berry pseudorotation.[313] PCl_3/O_3 and PBr_3/O_3 mixtures in solid argon photolyse to OPX_3 molecules, as shown by i.r.

spectroscopy.[314]

5.3 Arsenic:- Arsine and ozone react in an argon matrix to give a complex $AsH_3.O_3$, for which several i.r. bands could be assigned. Small amounts of cis- and trans-H_2AsOH were also formed. Subsequent reactions produced H_3AsO and $HOAsO_2$, all of which had characteristic i.r. bands.[315] I.r. spectra of $CH_2(AsCl_2)_2$ as a pure liquid, or in solution, show the presence of two conformers, of C_2 and C_s symmetry, in polar media. In non-polar media, however, only the less-polar C_2 form is present.[316] The i.r. and Raman spectra of (13), where R = Ph, Cl, Br, p-Br- or p-O_2N-C_6H_4, were used to determine their conformations.[317]

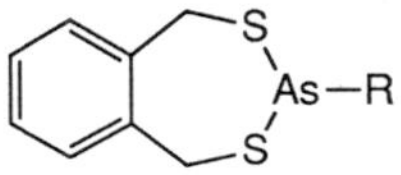

(13)

The i.r. and Raman spectra of As_7^{3-}, as a Cs^+ salt, gave the following vibrational wavenumbers associated with the cage: (a_1 modes) 264, 250, 231 and 167 cm^{-1}; (e modes) 257, 239, 175, 150 and 107 cm^{-1}.[318] I.r spectra of the photolysis products of As_4 and O_3 in argon matrices gave evidence for AsO (as an $(As_3)(AsO)$ complex) (ν 921.7 cm^{-1}) and bridge-bonded As_4O_n, where n = 1 - 5.[319]

The i.r. and Raman spectra of $Ca_2As_2O_7$ show that the As-O-As bridge deviates slightly from linearity.[320] I.r. and Raman data for $PbCuAs_2O_7$ were recorded and assigned.[321] $[As_6^{III}V_{15}^{V}O_{42}(H_2O)]^{6-}$ has νAsO at 635 cm^{-1}.[322]

Table 7 Vibrational assignments for $[M_2SBr_6]^{2-}$, where M = As or Sb (/cm^{-1})

M =	As	Sb
ν_{as}MSM	393	353
ν_sMSM	359	328
νMBr_2(t)	212,197	183
νMBrM(br)	143,127	131,116

Resonance Raman spectra of As_2S_3 glasses contain νAsAs at 230 cm^{-1}, and νSS at 490 cm^{-1}.[323] The i.r. spectra of

$[M_2SBr_6]^{2-}$, where M = As or Sb, are consistent with the presence of two octahedra bridging via one S and two Br atoms. The proposed assignments are given in Table 7.[324]

$[As_3S_3I_4]^-$ has several νAsS modes in the range 300 - 400 cm^{-1}, with νAsI near 200 cm^{-1}.[325] $AsBr_4^+$, as the AsF_6^- salt, gives the following vibrational wavenumbers: ν_2 88 cm^{-1}, ν_4 115 cm^{-1}, ν_1 244 cm^{-1} and ν_3 349 cm^{-1}.[326] $As_6Br_8^{2-}$ gives 7 i.r. and 7 Raman bands in the range 100 - 300 cm^{-1}. There are no coincidences, suggesting the point group D_{3d}, but no specific assignments were given.[327]

5.4 Antimony and Bismuth:- There is i.r. evidence for the formation of stibylene ozonide, $HSbO_3$, and hydroxystibine, H_2SbOH, as reaction products from SbH_3 + O_3 in argon matrices.[328] SbX_2(DpClBH), where X = Cl, Br or I; DpClBHH = p-chlorobenzoylhydrazone of di-(2-pyridyl)ketone, have νSbN in the range 540 - 560 cm^{-1}, νSbO 405 - 420 cm^{-1}, and νSbX at 335 cm^{-1} (Cl), 300 cm^{-1} (Br) or 295 cm^{-1} (I).[329] $Sb_2(SnMe_3)_4$ has a strong Raman band due to νSbSb and ν_sSnSb_2 at 161 cm^{-1}. $\nu_{as}SnSb_2$ is at 180 cm^{-1}.[330]

νSbO modes were assigned from the i.r. and Raman spectra of $A_2^{II}Sb_2O_7$ and $KLnSb_2O_7$, where A = Ca or Sr; Ln = Ho, Er, Y or Lu.[331] The i.r. and Raman spectra of (14), where R = Me, Et or Me_3C, show them to be polymeric in the solid, but monomer/dimer mixtures in solution or the pure liquids.[332]

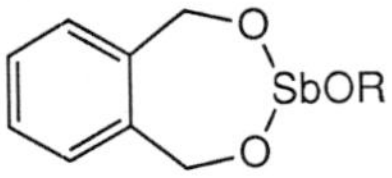

(14)

Skeletal modes were assigned for the pyramidal molecules $SBCl_{3-n}(OAr)_n$, where Ar = p-$C_6H_4CMe_3$, 2,6-$Me_2C_6H_3$, 2,4,6- or 2,3,6-$Me_3C_6H_2$, n = 1 - 3.[333] The i.r. spectrum of $K_2SeO_4SbF_2$ can be interpreted in terms of an SbF_3O_4E environment, where E represents a lone pair of electrons.[334] The Raman spectra of antimony dissolved in alkaline sulphide solutions are consistent with the presence of SbS_4^{2-} or $Sb_4S_7^{2-}$ (probably the former) in 0.1M Sb solutions. More dilute solutions contain SbS_2^- or SbS_3^{3-}.[335]

Raman bands at 65, 460 and 634 - 655 cm^{-1} in $Bi_2(Sr_{1-x}$-

$Ca_x)_2CuO_{6+\delta}$ and $Bi_2(Sr_{1-x}Ca_x)_3Cu_2O_{8+\delta}$ were assigned to modes involving BiO motion.[336] νBiO is at 480 cm^{-1} in $HBi(SO_4)_2.3H_2O$.[337]

6. Group VI

6.1 Oxygen:- The vibrational wavenumber of $^{18}OH^+$ was found to be at 2947.102(4) cm^{-1}.[338] I.r. laser absorption spectra of H_2O^+ show that ω_2 is at 1431.198 cm^{-1}.[339] ν_2 of $D_2{}^{18}O$ is centred at 1170.1572050(1156) cm^{-1}.[340] νOD bands of D_2O in H_2O (5 - 90°C) at concentrations of 10 - 100 mol% were consistent with a continuous model for the water structure.[341] I.r. spectra of H_2O in krypton matrices gave evidence for rotating and non-rotating monomers, dimers and trimers.[342]

The isotopic ($^{16}O/^{18}O$) shift for the 192 cm^{-1} mode of H_2O (going to 177 cm^{-1} for $H_2{}^{18}O$) confirms its assignment as a symmetric hydrogen-bonded stretching mode.[343] Several overtone and combination bands of H_2O vapour were identified in the range 8000 - 9500 cm^{-1}.[344] A high-pressure Raman study of the ice $I_h \rightarrow$ IX transition shows that it occurs at 150K, even though the existing phase diagram suggests that the transition should be $I_h \rightarrow$ II.[345] Polarised Raman spectra have been observed for films of ice I_c and I_a at 90K. Evidence was found for macroscopic extension of the resonant coupling for both crystalline and amorphous ice.[346] An i.r. band has been seen for the first time for $H_5O_2{}^+$.[347]

Band centres have been measured very precisely for ν_2 [348] and $2\nu_2$ [349] for O_3. A theoretical analysis of the ν_1 resonance Raman band progressions for $O_3{}^-$ and $S_3{}^-$ suggests that it is possible to determine excited state molecular geometries from the intensity distributions of overtone progressions at differing exciting wavelengths.[350] Potassium atoms and O_2 in an argon matrix form $K^+O_4{}^-$. New i.r. data gave more assignments for the tetroxide anion species. For $^{16}O_4{}^-$ the following wavenumbers have now been extablished: νO-O (a_g) 1291.5 cm^{-1}; νO-O (b_u) 993 cm^{-1}; νO_2-O_2 (a_g) 305 cm^{-1} and bend (b_u) 131 cm^{-1}. The data are consistent with the geometry (15).[351] The vibrational spectrum of the cyclic O_{12} molecule has been predicted by S.C.F.-M.O. calculations.[352] I.r absorption intensity has been measured for O_2F in a very long pathlength F.T.i.r. cell.[353]

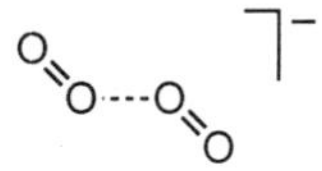

(15)

6.2 Sulphur and selenium rings and chains:- The torsional wavenumber in HSSH has been measured at 417.476765(7) cm^{-1}.[354] The Raman and i.r. spectra of CF_3SSCl are consistent with C_1 symmetry. 14 of the 15 fundamentals were assigned, including νSS at 531 cm^{-1} and νSCl 485 cm^{-1}.[355] The Raman spectrum of the new cation $CF_3SSCl_2^+$ shows that it possesses C_s symmetry, with νSS 535 cm^{-1}, $\nu_{as}SCl$ 477 cm^{-1} and ν_sSCl 455 cm^{-1}.[356] The νSS band in bicyclic 1,5-$Ph_4P_2N_3(NMe)S_2^+$ is at 280 cm^{-1}, compared to 268 cm^{-1} in the neutral 1,5-$Ph_4P_2N_4S_2$.[357]

The millimetre-wave spectrum of cis-H_2S_3 gave an estimated value of 240 ± 30 cm^{-1} for the SSS bending vibration.[358] Photolysis of S/NH_3 solutions was followed by Raman spectroscopy, giving evidence for S_3N^- and S_4N^- species.[359] The i.r. and Raman spectra of $[S_4N_3]Cl$ and $[S_3N_2]Cl$, including ^{15}N isotopic data, gave almost complete vibrational assignments in each case. For $S_4N_3^+$ νSS is at 429 cm^{-1}, with ν_sSN at 1001 cm^{-1}. For $S_3N_2^+$, νSS is at 403 cm^{-1}, νSN at 1014, 933 and 728 cm^{-1}.[360] S_{15} gave Raman bands in the range 409 - 480 cm^{-1} due to νSS, with deformation modes below 270 cm^{-1}.[361]

The Raman spectra of Se_2X_2, where X = Cl or Br, shows the existence of the equilibria:

$$2Se_2X_2 \rightleftharpoons SeX_2 + Se_3X_2$$
$$3Se_2X_2 \rightleftharpoons SeX_2 + Se_4X_2$$

Mixtures of Se_2Br_2 and Se_2Cl_2 contain Se_2BrCl.[362] The i.r. spectrum of solid Se_4N_4 shows νSeN at 800/788 cm^{-1} (e), 570 cm^{-1} (e), and 534 cm^{-1} (b_2); $\delta SeSeN$ at 425 cm^{-1} (e), 310 cm^{-1} (b_2) and 270/260 (e) cm^{-1}; and $\nu SeSe$ 147 cm^{-1}.[363]

6.3 Other sulphur and selenium comounds:- The following ω_e values have been obtained: SH^+ 2547.1711(51) cm^{-1}; SD^+ 1829.096 cm^{-1}.[364] The vibrational modes of EH^-, where E = S, Se or Te, centres in alkali halide matrices are split by external electric fields.[365] High-resolution F.T.i.r. measurements on

ν_2 of $HD^{32}S$ show that the band centre is at 1032.7152 cm^{-1}.[366] Similar results for $D_2{}^{32}S$ gave ν_1 1896.4319 cm^{-1}; ν_2 855.4042 cm^{-1} and ν_3 1910.1843 cm^{-1}.[367] The ν_1 and ν_3 fundamentals of SH_3^+ were shown to be at 2521.04816(40) cm^{-1} and 2525.74710(32) cm^{-1} respectively.[368]

The i.r. spectrum of SF_5CN has νS-CN at 630 cm^{-1}.[369] The i.r. and Raman spectra of the new compounds $CF_3SClF^+MF_6^-$, where M = As or Sb, were assigned in terms of C_1 symmetry, with νSC near 440 cm^{-1}, νSF 850 cm^{-1} and νSCl 546 cm^{-1}.[370] The Raman spectrum of the newly-prepared cation $(CF_3)_2SF^+$ contains νSF at 855 cm^{-1}, with νSC_2 modes at 465 cm^{-1} (a') and 459 cm^{-1} (a").[371]

Analysis of some very weak i.r. absorption bands of SO_2 gave the energies of the (021) state of $^{32}SO_2$ (2388.91551 cm^{-1}) and the (200) state of $^{34}SO_2$ (2281.4703 cm^{-1}).[372] The force field of SO_2 was calculated using ab initio m.o. methods.[373] SO_2 forms complexes with 1,2,3,4-thiatriazol-5-thiolate at several coordination sites, giving a complicated pattern of νSO_2 bands.[374] A low temperature Raman spectrum of sulphuryl chloride fluoride gave a number of revised vibrational assignments.[375]

The concentrations of each ion in the HSO_4^--SO_4^{2-}-$S_2O_8^{2-}$ system were followed by quantitative Raman spectroscopy, using characteristic bands of each anion.[376] Similar experiments were used to estimate the concentrations of SO_4^{2-}, HSO_4^-, $S_2O_6^{2-}$ and $S_4O_6^{2-}$ in solutions obtained by the aerial oxidation of SO_2.[377] A room-temperature polarised Raman spectrum of a single crystal of $Na_2(SeO_4)_{0.15}(SO_4)_{0.85}$ was used in a factor-group analysis of the internal modes of SO_4^{2-} and SeO_4^{2-}.[378]

The Raman spectra of $CH_3(CF_3)SX^+$, where X = F, Cl, Br or I included the following νSX wavenumbers: 834 cm^{-1} (F), 548 cm^{-1} (Cl), 458 cm^{-1} (Br) and 395 cm^{-1} (I).[379] I.r. data were reported for the $(Me_2N)_3S^+$ salt of OSF_5^-, and a normal coordinate analysis performed.[380] Numerous overtone and combination bands of SF_6 were assigned from a high-resolution i.r. study.[381]

The Raman spectra of $CF_3C(O)SX$, where X = Cl, Br or I, were assigned in terms of C_s symmetry. νSX wavenumbers were as follows: 522 cm^{-1} (Cl), 446 cm^{-1} (Br) and 397 cm^{-1} (I).[382]

Quite detailed assignments were reported from the i.r. and Raman spectra of $C_6F_5SX_2^+AsF_6^-$, where X = Cl or Br.[383] SCl_3^+, as the SbF_6^- salt, has νSCl modes at 545 and 518 cm^{-1}, with δSCl_2 at 283 and 217 cm^{-1}.[384]

A factor-group analysis has been made of both internal (SeO_4^{2-}) and external modes from a polarised Raman spectrum of a single crystal of Na_2SeO_4.[384] The Raman spectra of Se in SeO_2/HBr(aq., 8.72M) show the formation of Se_2Br_2 and $SeBr_4^{2-}$. In 4.8M aq. HBr, Se_2Br_2 and $SeOBr_3^-$ are seen.[386]

6.4 Tellurium:- νTeH in $PPh_4^+TeH^-$ is at 1975 cm^{-1} in the i.r.[387] Shifts in νTeO modes in glasses containing TeO_2 and lanthanide oxides were measured and related to the glass structures.[388] νTeO_6 modes were assigned from the i.r. and Raman spectra of $Te(OH)_6.(NH_4)_2SO_4$.[389] The vibrational constants for the $b0^+$ and X_10^+ states of $^{130}Te^{80}Se$ include ω_e at 289.8(6) cm^{-1} and 311.6(4) cm^{-1} respectively.[390] The Raman spectra of $TeBr_3Cl_3^-$ and $TeBr_4Cl_2^-$ were assignable in terms of C_{3v} and D_{4h} symmetries respectively.[391]

7. Group VII

There nave been many reports of i.r. spectra of weak complexes of HF: D_2...HF;[392] OC...HF and CO...HF;[393-4] HCN...HF;[395] OCO...HF, SCO...HF;[396] ONN...HF (linear) and NNO...HF (bent);[397-8] $(CD_3)_2O$...HF; [399] T-shaped (HC≡CH)...HF, (16);[400] amide...HF complexes;[401] and 18-crown-6...HX, where X = F or Cl.[402]

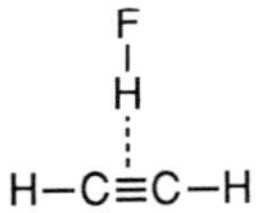

(16)

The Raman spectrum of $NMe_4^+HF_2^-$, containing symmetrical HF_2^-, shows ν_1 (Σ_g^+) at 596 cm^{-1}.[403] A high-resolution F.T.i.r. study of $H^{16}OF$ showed the band centres of $2\nu_2$ and ν_2 to be at 2686.9246(1) cm^{-1} and 1353.40469(4) cm^{-1} respectively.[404] The i.r. spectra of HOF in low-temperature matrices show ν_1 and ν_2 perturbed by adjacent base molecules, with appearance of librational modes of the D...H-OF unit. U.v. photolysis produced O...$(HF)_2$ and O_2...$(HF)_2$.[405] Anharmonic force fields were calculated by ab initio methods for HOF and F_2O.[406] The matrix-isolation technique was used to study reactions of F_2 with sulphur and phosphorus bases. Thus, in the F_2/H_2S system, the initial product appears to be H_2SF_2.[407]

CARS for supersonic expansions of $(HCl)_2$ gave values for ν_2^+ and ν_2^- H-Cl stretching modes of 2853 and 2842 cm^{-1} respectively.[408] The far i.r. spectrum of the $(HCl)_2$ dimer shows that the origin of the out-of- plane bending mode (ν_6) of $(H^{35}Cl)_2$ is at 160.78 cm^{-1}.[409] The i.r. spectrum of $(HCl)_3$ clusters in solid neon showed bands due to all possible H/D isotopic variants.[410] There is far i.r. evidence for the formation of van der Waals complexes Ar...HCl and HCl...Ar.[411] νH-Cl stretches have been identified for the 1:1 hydrogen-bonded complexes of H_2, O_2 and N_2 with HCl.[412]

I.r. line intensities were measured within the v = 0-1 fundamental of ClO.[413] I.r. band intensities were also reported for $^{35}Cl^{16}O$ and $^{37}Cl^{16}O$.[414] Absolute Raman scattering intensities were measured for ClO^-, ClO_2^-, ClO_3^- and ClO_4^-. The polarisabilities can be related to the numbers of valence electrons in non-bonding and antibonding orbitals.[415] Dimeric $(ClO)_2$ is generated in a gaseous discharge flow system. Bands at 752.6, 649.8 and 647.6 cm^{-1} were assigned to ClOOCl, and previous assignments to $(ClO)_2$ seem to be erroneous.[416] The Raman depolarisation ratio has been measured for ν_1 of ClO_4^- in $LiClO_4/H_2O$/ethylene carbonate mixtures. Depolarisation occurs because of fluctuating local fields due to the solvent.[417]

Matrix-i.r. spectra have been obtained for 1:1 molecular complexes of ClF and Cl_2 with 18-crown-6. The halogen stretch decreases by about 100 cm^{-1} compared to the free molecule value.[418] Bands due to ClF_2^- in $CsClF_2$ were assigned thus: $\nu_{as}(\Sigma_u^+)$ 635 cm^{-1}; $\delta(\Pi_u)$ 445 cm^{-1}; $\nu_s(\Sigma_g^+)$ 475 cm^{-1}.[419] The Raman and i.r. spectra of the NMe_4^+ salts of ClF_4^-, BrF_4^- and BrF_6^- were in good agreement with previous data for these anions.[420] The Raman spectrum of PBr_4Cl_3 shows that it is $PBr_4^+Cl_3^-$. A band at 267 cm^{-1} is due to Cl_3^-.[421]

The Raman spectrum of $NaBrO_2.3H_2O$, as solid or in aqueous solution, gave the following bands due to BrO_2^-: 728, 680 and 320 cm^{-1}.[422] $(Thiourea)_2X_2$, where X = Br or I, give very strong Raman bands at 255 and 215 cm^{-1} due to ν_sS-X-S for X =

Br, I respectively.[423] Diode laser spectra of BrF and BrCl gave the following values for the vibrational wavenumbers: (BrF) 669.9002(5) cm^{-1} (^{79}Br), 668.2937(6) cm^{-1} (^{81}Br); (BrCl) 444.2622(8) cm^{-1} ($^{79}Br^{35}Cl$), 442.5736(10) cm^{-1} ($^{81}Br^{35}Cl$), 435.8670(10) cm^{-1}, ($^{79}Br^{37}Cl$), 434.1468(11) cm^{-1} ($^{81}Br^{37}Cl$).[424]

I.r. and Raman spectra were reported for IF_4O^- salts, for which νIO lay in the range 864 - 896 cm^{-1}, and $\nu_s IF_4$ (in- and out-of-phase) 460 - 570 cm^{-1}.[425] Quite a large number of assignments were proposed from the Raman spectra of cis,cis-$Xe(OIOF_4)_2$, cis- and trans-$FXeOIOF_4$, e.g. νI=O near 890 cm^{-1}, νXeF near 527 cm^{-1}.[426] The i.r. spectrum of gas-phase IF has been seen for the first time, with ω_e at 610.2771(17) cm^{-1}.[427] The Raman spectrum of $[Co(acetamide)_4]I_2.4I_2$ shows that the I_5^- ion is present, with characteristic νII wavenumbers of 252, 190, 151 and 140 cm^{-1}. The two highest are due to terminal I-I stretches, the other two to internal I-I stretches.[428]

References

1 M.E.Alikhani, B.Silvi, J.P.Perchard and V.Chandrasekharan, J. Chem. Phys., 1989, 90, 5221.
2 C.Julien, M.Massot, A.Balkanski, A.Krol and W.Nazarewicz, Mater. Sci. Eng., B, 1989, B3, 307.
3 M.E.Poloznikova, O.I.Kondratov, N.G.Chaban, V.V.Safonov and V.V.Fomichev, Russ. J. Inorg. Chem., 1989, 34, 364.
4 H.Sobotta, G.Kühn, H.Neumann and V.Riede, Cryst. Res. Technol., 1989, 24, 459.
5 R.Frech and S.B.Kim, Spectrochim. Acta, Part A, 1988, 44A, 1387.
6 T.Sekine, C.Julien, M.Jouanne, M.Weber and M.Balkanski, Mater. Res. Soc., Symp. Proc., 1989, 135, 461.
7 C.R.Brazier, N.H.Oliphant and P.F.Bernath, J. Mol. Spectrosc., 1989. 134, 421.
8 X.Xie, R.W.Field, L.Li, M.Lyyra, J.T.Bahns and W.C.Stwalley, J. Mol. Spectrosc., 1989, 134, 119.
9 A.G.Maki and W.B.Olson, J. Chem. Phys., 1989, 90, 6887.
10 J.Adel, F.Weller and K.Dehnicke, Z. Naturforsch., B, 1988, 43b, 1094.
11 B.Brzezinski, G.Zundel and R.Krämer, J. Phys. Chem., 1988, 92, 7012.
12 H.Uehara, K.Horiai, K.Nakagawa and T.Fujimoto, J. Mol. Spectrosc., 1989, 134, 98.
13 U.Magg, H.Birk and H.Jones, Chem. Phys. Lett., 1988, 151, 503.
14 M.P.Maneva and M.P.Georgiev, Polyhedron, 1989, 8, 357.
15 M.P.Maneva and M.P.Georgiev, Russ. J. Inorg. Chem., 1989, 34, 803.
16 M.Fujiwara, H.Hayashi and M.Tasumi, Croat. Chim. Acta, 1988, 61, 435.
17 O.E.Raevskii, S.V.Trepalin, V.E.Zubareva and T.N.Kudrya, Koord. Khim., 1989, 15, 181.
18 H.G.Hedderich and C.E.Blom, J. Chem. Phys., 1989, 90, 4660.
19 V.B.Vyazovov, O.I.Kondratov and V.V.Fomichev, Russ. J. Inorg. Chem., 1988, 33, 1130.
20 O.E.Raevskii, S.V.Trepalin and V.E.Zubareva, Koord. Khim., 1988, 14, 1188.
21 F.Ramondo, L.Bencivenni, S.N.Cesaro and K.Hilpert, J. Mol. Struct., 1989, 192, 83.
22 U.Magg, H.Birk and H.Jones, Chem. Phys. Lett., 1988, 151, 263.
23 C.P.Herrero and M.Stutzmann, Phys. Rev., B, 1988, 38, 12668.
24 D.C.Miller, J.J.O'Brien and G.H.Atkinson, J. Appl. Phys., 1989, 65, 2645.
25 J.R.Durig, N.E.Lindsay, T.J.Hizer and J.D.Odom, J. Mol. Struct., 1988, 189, 257
26 J.L.Peters, M.Mittakanti, T.J.Lofthouse and K.W.Morse, Inorg. Chem., 1989, 28, 492.
27 S.V.Sigalova, I.P.Romm, E.N.Gur'yanova, R.R.Shifrina, E.S.Shcherbakova, L.N.Margolin, L.P.Ivanov and A.I.Gorbunov, Zh. Obshch. Khim., 1989, 59, 348.
28 Y.Matsui and R.C.Taylor, Spectrochim. Acta, Part A, 1989, 45A, 299.
29 W.Preetz and J.Thesing, Z. Naturforsch., B, 1989, 44b, 121.
30 A.Heinrich and W.Preetz, Z. Naturforsch., B, 1988, 43b, 1327.
31 W.Preetz, A.Heinrich and J.Thesing, Z. Naturforsch., B, 1988, 43b, 1319.
32 R.M.Paroli, N.T.Kawai, G.Lord, I.S.Butler and D.F.R.Gilson, Inorg. Chem., 1989, 28, 1819.

33 A.R.Kalyuskii, V.N.Kuznetsov, S.P.Krasnoshchekaya and A.I.Gren, Dokl. Akad. Nauk Ukr.SSR, Ser. B, 1989, 41.
34 M.Baudler and C.Block, Z. Anorg. Allg. Chem., 1988, 567, 7.
35 S.Padmaja, V.Ramakrishnan, J.Rajaram and J.C.Kuriacose, J. Chem. Soc., Farad. Trans., 1, 1989, 85, 2249.
36 J.Flanagan, W.P.Griffith, R.D.Powell and A.P.West, J. Chem. Soc., Dalton Trans., 1989, 1651.
37 J.Flanagan, W.P.Griffith, R.D.Powell and A.P.West, Spectrochim. Acta, Part A, 1989, 45A, 951.
38 D.C.McKean and A.M.Coats, Spectrochim. Acta, Part A, 1989, 45A, 409.
39 D.G.Berniyazova, E.M.Shvarts and V.G.Kalacheva, Russ. J. Inorg. Chem., 1989, 34, 335.
40 V.V.Kuznetsov, L.I.Alekseenko, A.I.Staikov and A.I.Gren, Ukr. Khim. Zh., (Russ. Ed.), 1988, 54, 1315.
41 A.Rulmont and M.Almou, Spectrochim. Acta, Part A, 1989, 45A, 603.
42 Yu.N.Il'in, V.V.Kravchenko, K.I.Petrov and Yu.M.Golovin, Russ. J. Inorg. Chem., 1988, 33, 1224.
43 W.Soppe and H.W.den Hartog, J. Non-Cryst. Solids, 1989, 108, 260.
44 E.I.Kamitsos and M.A.Karakassides, Phys. Chem. Glasses, 1989, 30, 19.
45 T.Assih, L.Beys, P.Hilaire and J.Phalippou, J. Raman Spectrosc., 1989, 20, 215.
46 W.Soppe, W.Ebens and H.W.den Hartog, J. Non-Cryst. Solids, 1988, 105, 251.
47 J.Fukunaga, R.Ota and N.Yoshida, Mater. Sci. Forum, 1988, 32-3, 441.
48 J.Fukunaga, R.Ota, M.Shiroyama and N.Yoshida, Nippon Kagaku Kaishi, 1988, 1971 (Chem. Abs., 1989, 110, 43435).
49 J.Fukunaga, R.Ota, M.Shiroyama and N.Yoshida, Nippon Seramikkusu Kyokai Gakujutsu Ronbunshi, 1989, 97, 22 (Chem. Abs., 1989, 111, 11365).
50 J.Datka and Z.Piwowarska, J. Chem. Soc., Farad. Trans., 1, 1989, 85, 47.
51 S.Ram and K.Ram, J. Mater. Sci., 1988, 23, 4521.
52 S.Ram and K.Ram, Infrared Phys., 1989, 29, 895.
53 E.Zeisberger and A.Ruoff, J. Mol. Spectrosc., 1989, 136, 295.
54 Y.Matsumoto, Y.Ohshima, M.Takami and K.Kuchitsu, J. Chem. Phys., 1989, 90, 7017.
55 Y.Matsumoto, Y.Ohshima, M.Takami and K.Kuchitsu, Reza Kogaku Kenkyu, 1988, 10, 1 (Chem. Abs., 1989, 110, 201880).
56 I.S.Perelygin and M.A.Klimchuk, Zh. Prikl. Spektrosk., 1989, 50, 280.
57 M.Arif and S.F.A.Kettle, Inorg. Chem., 1989, 28, 3765.
58 C.Destoky, I.Dubois and H.Bredohl, J. Mol. Spectrosc., 1989, 136, 216.
59 C.Ergezinger, F.Weller and K.Dehnicke, Z. Naturforsch., B, 1988, 43b, 1621.
60 J.M.Parnis and G.A.Ozin, J. Phys. Chem., 1989, 93, 1204.
61 J.M.Parnis and G.A.Ozin, J. Phys. Chem., 1989, 93, 1215.
62 V.E.Gorbunov, K.S.Gavrichev and G.A.Sharpataya, Russ. J. Inorg. Chem., 1988, 33, 1537.
63 J.C.Bureau, Z.Amri, P.Claudy and J.M.Letoffe, Mater. Res. Bull., 1989, 24, 23.
64 J.C.Bureau, Z.Amri, P.Claudy and J.M.Letoffe, Mater. Res. Bull., 1989, 24, 133.
65 J.C.Bureau, Z.Amri, P.Claudy and J.M.Letoffe, Mater. Res. Bull., 1989, 24, 267.

66 W.Uhl, M.Layh and T.Hildenbrand, J. Organometal. Chem., 1989, 364, 289.

67 L.Manceron and L.Andrews, J. Phys. Chem., 1989, 93, 2964.

68 W.Uhl, Z. Naturforsch., B, 1988, 43b, 1113.

69 R.Lechler, H.-D.Hausen and J.Weidlein, J. Organometal. Chem., 1989, 359, 1.

70 H.Krause and J.Weidlein, Z. Anorg. Allg. Chem., 1988, 563, 116.

71 P.A.Mosier-Boss, R.D.Bos, C.J.Gabrial, S.Szpak, J.J.Smith and R.J.Nowak, J. Chem. Soc., Farad. Trans.,1, 1989, 85, 11.

72 P.K.Dutta and M.Puri, J. Phys. Chem., 1989, 93, 376.

73 J.M.Malezieux and B.Piriou, Bull. Mineral., 1988, 111, 649.

74 H.-J.Lunk, M.Salmen, H.Weiner, C.Baeker, W.Wilde and D.Müller, Z. Anorg. Allg. Chem., 1989, 568, 171.

75 J.F.Lambert, W.S.Millman and J.J.Fripiat, J. Am. Chem. Soc., 1989, 111, 3517.

76 Yu.I.Tarasevich, S.A.Gogolyuk, G.V.Lantukh and V.S.Rak, Teor. Eksp. Khim., 1988, 24, 753.

77 H.Chen, L.Zhang and F.Gan, Guangxue Xuebao, 1988, 8, 791 (Chem. Abs., 1989, 110, 124256).

78 H.Inoue, T.Nanba, H.Hagihara, T.Kanezawa and I.Yasui, Mater. Sci. Forum, 1988, 32-3, 403.

79 C.Hua, Y.Qihua and C.Wangqiu, J. Non-Cryst. Solids, 1989, 107, 219.

80 E.Mahieu, I.Dubois and H.Bredohl, J. Mol. Spectrosc., 1989, 134, 317.

81 V.M.Bermudez and A.S.Glas, Langmuir, 1989, 5, 316.

82 M.Tacke and H.Schnöckel, Inorg. Chem., 1989, 28, 2895.

83 V.N.Lubkova, L.A.Shelyudkova, K.A.Kholdanidi and I.I.Yakovlev, Russ. J. Inorg. Chem., 1989, 34, 591.

84 B.Bonnetot, A.Aboukhassib and H.Mongeot, Inorg. Chim. Acta, 1989, 156, 183.

85 M.A.Einarsrud and E.Rytter, Mikrochim. Acta, 1987, 2, 381.

86 V.A.Narayanan, J. Raman Spectrosc., 1989, 20, 77.

87 E.Christensen, R.W.Berg and J.H.von Barner, Polyhedron, 1989, 8, 325.

88 B.P.Gilbert, R.W.Berg and N.J.Bjerrum, Appl. Spectrosc., 1989, 43, 336.

89 R.D.Urban, U.Magg and H.Jones, Chem. Phys. Lett., 1989, 154, 135.

90 A.J.Downs, M.J.Goode and C.R.Pulham, J. Am. Chem. Soc., 1989, 111, 1936.

91 C.Liang, R.D.Davy and H.F.Schaefer, Chem. Phys. Lett., 1989, 159, 393.

92 J.Schneider, B.Dischler, H.Seelewind, P.M.Mooney, J.Lagowski, M.Matsui, D.R.Beard and R.C.Newman, Appl. Phys. Lett., 1989, 54, 1442.

93 F.Miyaji and S.Sakka, Nippon Seramikkusu Gakujutsu Ronbunshi, 1989, 97, 267.

94 G.G.Hoffmann and R.Fischer, Inorg. Chem., 1989, 28, 4165.

95 Yu.L.Durnev, B.S.Kol'buzhev, V.I.Togashiev and Yu.I.Yuzyuk, Izv. Akad. Nauk SSSR, Ser. Fiz., 1989, 53, 1300.

96 E.D.Samsonova, S.B.Osin and V.F.Shevel'kov, Russ. J. Inorg. Chem., 1988, 33, 1598.

97 S.P.Wicelinski, R.J.Gale, S.D.Williams and G.Mamantov, Spectrochim. Acta, Part A, 1989, 45A, 759.

98 A.H.Bahnmaier, R.D.Urban and H.Jones, Chem. Phys. Lett., 1989, 155, 269.

99 A.R.Barron J. Chem. Soc., Dalton Trans., 1989, 1625.

100 N.W.Alcock, I.A.Degnan, M.G.H.Wallbridge, H.R.Powell, M.McPartlin and

G.M.Sheldrick, J. Organometal. Chem., 1989, 361, C33.
101 H.Schumann, R.Mohtachemi and M.Schwichtenberg, Z. Naturforsch., B, 1988, 43b. 1510.
102 T.A.Annan and D.G.Tuck, Can. J. Chem., 1988, 66, 2935.
103 S.K.Singh, Y.Singh, A.K.Rai and R.C.Mehrotra, Polyhedron, 1989, 8, 633.
104 A.K.Vazhnov. N.V.Porotnikov and K.I.Petrov, Russ. J. Inorg. Chem., 1989, 34, 428.
105 H.D.Lutz, W.Becker, B.Müller and M.Jung, J. Raman Spectrosc., 1989, 20, 99.
106 K.R.Allakhverdiev, S.S.Babaev, N.M.Tagiev and M.M.Shirinov, Phys. Stat. Solidi, 1989, 152, 317.
107 V.F.Kozin and I.A.Sheka, Ukr. Khim. Zh., (Russ. Ed.), 1988, 54, 904.
108 T.Guedira, J.P.Wignacourt. M.Drache, A.Lorriaux-Rubbens and F.Wallart, Phase Transitions, 1988, 13, 81.
109 T.Staffel and G.Meyer, Z. Anorg. Allg. Chem., 1988, 563, 27.
110 R.D.Urban, A.H.Bahnmaier, U.Magg and H.Jones, Chem. Phys. Lett., 1989, 158, 443.
111 A.Macias, M.C.Rodriguez-Argüelles, M.I.Suárez, A.Sánchez, J.S.Casas and U.Englert, J. Chem. Soc., Dalton Trans., 1989, 1787.
112 A.G.Maki, J. Mol. Spectrosc., 1989, 137, 147.
113 M.R.Bermejo, A.Fernandez, M.Gayoso, A.Castiñeiras, W.Hiller and J.Strähle, Polyhedron, 1988, 7, 2561.
114 M.R.Bermejo, M.I.Fernandez, J.Tajes and A.G.Deibe, An. Quim., Ser. B, 1988, 84, 298.
115 A.Syngollitou-Kouracou and A.Christofides, Gazz. Chim. Ital., 1989, 119, 289.
116 M.R.Bermejo, M.Gayoso, M.I.Fernandez, A.Hermida and E.Gomez, An. Quim., Ser. B, 1988, 84, 303.
117 F.Mélen, N.Grevesse, A.J.Sauval, C.B.Farmer, R.H.Norton, H.Bredohl and I.Dubois, J. Mol. Spectrosc., 1989, 134, 305.
118 N.H.Rosenbaum, J.C.Owrutsky and R.J.Saykally, J. Mol. Spectrosc., 1989, 133, 365.
119 W.P.Hu, A.A.Rodriguez and M.Schwartz, Spectrochim. Acta, Part A, 1989, 45A, 929.
120 A.J.Ross, A.Amrein, D.Luckhaus and M.Quack, Mol. Phys., 1989, 66, 1273.
121 G.Blanquet, J.Walrand and D.N.Mai, J. Mol. Spectrosc., 1989, 133, 471.
122 R.Lakanen, M.Koivusaari, V.-M.Horneman and J.Kauppinen, J. Mol. Spectrosc., 1989, 135, 188.
123 M.Ouahman, N.Ben Sari-Zizi and C.Alamichel, Spectrochim. Acta, Part A, 1989, 45A, 175.
124 F.Lattanzi, C.DiLauro and G.Guelachvili, Mol. Phys., 1989, 66, 509.
125 S.Alanko, V.-M.Horneman and J.Kauppinen, J. Mol. Spectrosc., 1989, 135, 76.
126 H.G.Cho and R.H.Schwendeman, J. Mol. Spectrosc.,1989, 133, 383.
127 J.P.Champion, J.C.Hilico, C.Wenger and L.R.Brown, J. Mol. Spectrosc., 1989, 133, 256.
128 M.Yoshikawa, G.Katagiri, H.Ishida, A.Ishitani and T.Akamatsu, J. Appl. Phys., 1988, 64, 6464.
129 P.Botschwina, J. Chem. Phys., 1989, 90, 4301
130 N.Moazzen-Ahmadi, A.R.W.McKellar and T.Amano, J. Chem. Phys., 1989, 91, 2140.

131 J.R.Heath, A.L.Cooksy, M.H.W.Gruebele, C.A.Schuttenmaier and R.J.Saykally, Science, 1989, 244, 564.
132 M.Vala, T.A.Chandrasekhar, J.Szczepanski, R.van Zee and W.Weitner, J. Chem. Phys., 1989, 90, 595.
133 D.Durand, L.C.Scavarda do Carmo and F.Luty, Phys. Rev.,B, 1989, 39, 6096.
134 H.Bürger and G.Pawelke, J. Chem. Soc., Chem. Comm., 1988, 105.
135 A.M.Smith, S.L.Coy, W.Klemperer and K.C.Lehman, J. Mol. Spectrosc., 1989, 134, 134.
136 G.T.Fraser and A.S.Pine, J. Chem. Phys., 1989, 91, 3319.
137 J.H.Teles, G.Maier, B.A.Hess, L.J.Schaad, M.Winnewisser and B.P.Winnewisser, Chem. Ber., 1989, 122, 753.
138 F.Stroh and M.Winnewisser, Chem. Phys. Lett., 1989, 155, 21.
139 S.Koda, J.Goto, T.Chikusa and H.Nomura, J. Phys. Chem., 1989, 93, 4959.
140 M.Gerke, G.Schatte and H.Willner, J. Mol. Spectrosc., 1989, 135, 359.
141 M.Havenith, W.Bohle, J.Werner and W.Urban, Mol. Phys., 1988, 64 1073.
142 Z.Wang and J.W.Bevan, J. Chem. Phys., 1989, 91, 3335.
143 S.Dobos, G.Winnewisser, F.Kling and J.Mink, Z. Naturforsch., B, 1989, 44a, 633.
144 M.A.Walsh, T.R.Dyke and B.J.Howard, J. Mol. Struct., 1988, 189, 111.
145 S.W.Sharpe, R.Sheeks, C.Wittig and R.A.Beaudet, Chem. Phys. Lett., 1988, 151, 267.
146 K.J.Maynard and M.Moskovits, J. Chem. Phys., 1989, 90, 6668.
147 M.E.Jacox and W.E.Thompson, J. Chem. Phys., 1989, 91, 1410.
148 M.Schneider, A.G.Maki, M.D.Vanek and J.S.Wells, J. Mol. Spectrosc., 1989, 134, 349.
149 J.S.Wells, M.D.Vanek and A.G.Maki, J. Mol. Spectrosc., 1989, 135, 84.
150 G.D.Hayman, J.Hodge, B.J.Howard, J.S.Muenter and T.R.Dyke, J. Mol. Spectrosc., 1989, 133, 423.
151 N.C.Craig, Spectrochim. Acta, Part A, 1988, 44A, 1225.
152 N.C.Craig, Spectrochim. Acta, Part A, 1989, 45A, 661.
153 K.C.Clemitshaw and J.R.Sodeau, J. Phys. Chem., 1989, 93, 3552.
154 G.Davidovics, M.Monnier, W.Schroeder, P.Verlaque, J.Pourcin and H.Bodot, J. Mol. Struct., 1989, 197, 213.
155 W.Schroeder, M.Monnier, G.Davidovics, A.Allouche, P.Verlaque, J.Pourcin and H.Bodot, J. Mol. Struct., 1989, 197, 227.
156 C.O.della Vedova, J. Raman Spectrosc., 1989, 20, 483.
157 J.S.Wells, M.Schneider and A.G.Maki, J. Mol. Spectrosc., 1988, 132, 422.
158 J.F.Bertran and B.La Serna, J. Raman Spectrosc., 1989, 20, 419.
159 F.Holland, M.Winnewisser and J.W.C.Johns, Mikrochim. Acta, 1987, 2, 417.
160 J.Wormhoudt, K.E.McCurdy and J.B.Burkholder, Chem. Phys. Lett., 1989, 158, 480.
161 A.Amrein, D.Luckhaus, F.Merkt and M.Quack, Chem. Phys. Lett., 1988, 152, 275.
162 G.Davidson and C.L.Davies, Spectrochim. Acta, Part A, 1989, 45A, 371.
163 L.Nencini and M.Snels, J. Mol. Spectrosc., 1989, 135, 131.
164 S.Giorganni, A.Baldacci, A.Gambi, R.Visinoni and S.Ghersetti, J. Mol. Spectrosc., 1988, 131, 185.
165 L.Nencini, M.Snels, H.Hollenstein and M.Quack, Mol. Phys., 1989, 67, 989.

166 D.O.Frazier, S.Bhatia and B.H.Loo, Spectrosc. Lett., 1988, 21, 819.
167 H.C.Jung, Bull. Korean Chem. Soc., 1988, 9, 275.
168 A.Amrein, H.Hollenstein, M.Quack and U.Schmitt, Infrared Phys., 1989, 29, 561.
169 A.Baldacci, S.Giorganni, R.Visinoni and S.Ghersetti, J. Mol. Spectrosc., 1989, 133, 36.
170 A.Baldacci, S.Giorganni, P.Stoppa and S.Ghersetti, Spectrochim. Acta, Part A, 1989, 45A, 329.
171 H.Bürger, A.Rahner, A.Amrein, H.Hollenstein and M.Quack, Chem. Phys. Lett., 1989, 156, 557.
172 P.B.Davies, N.A.Martin and M.D.Nunes, Spectrochim. Acta, Part A, 1989, 45A, 293.
173 Y.Niide, H.Tanaka and I.Ohkoshi, J. Mol. Spectrosc., 1989, 137, 104.
174 N.A.Chumaevskii, Russ. J. Inorg. Chem., 1989, 34, 798.
175 P.Quintard, G.Ramis, M.Cauchetier, G.Busca and V.Lorenzelli, J. Mol. Struct., 1988, 174, 369.
176 S.Hasegawa, H.Anbutsu and Y.Kurata, Philos. Mag.,B, 1989, 59, 365.
177 R.S.Katiyar, O.Resto, R.Perez, M.Gomez and Z.Weisz, Thin Solid Films, 1987, 164, 243.
178 Yu.A.Aleshchenko, L.K.Vodop'yanov, E.M.Omel'yanovskii, A.V.Pakhomov and A.Yu.Polyakov, Solid State Commun., 1989, 70, 123.
179 E.Popowski, H.Randow and H.Kelling, Z. Anorg. Allg. Chem., 1988, 565, 171.
180 N.Moazzen-Ahmadi, I.Ozier and W.L.Meerts, J. Mol. Spectrosc., 1989, 137, 166.
181 J.K.McDonald, V.F.Kalasinsky, J.A.S.Smith, T.J.Geyer and J.R.Durig, J. Mol. Struct., 1988, 174, 291.
182 S.Mohan, A.R.Prabakaran and F.Payami, J. Raman Spectrosc., 1989, 20, 119.
183 D.G.Anderson and D.W.H.Rankin, J. Chem. Soc., Dalton Trans., 1989, 779.
184 G.G.Devyatykh, P.G.Sennikov, K.G.Tokhadze and V.E.Shkrunin, Vysokochist. Veshchestva, 1988, 21 (Chem. Abs., 1989, 110, 104145).
185 A.Marchand, M.Grignon-Dubois and P.Gerval, Spectrochim. Acta, Part A, 1989, 45A, 769.
186 W.Heininger, K.Polborn and G.Nagorsen, Z. Naturforsch., B, 1988, 43b, 857.
187 Z.Wen, L.Jia, K.Lu and G.Wu, Guangpuxue Xu Guangpu Fenxi, 1988, 8, 15 (Chem. Abs., 1989, 110, 231704).
188 U.Wannagat, R.Flindt, D.J.Brauer, H.Bürger and F.Dorrenbach, Z. Anorg. Allg. Chem., 1989, 572m 33.
189 L.V.Serebrennikov, Vestn. Mosk. Univ., Ser. 2: Khim., 1989, 30, 53.
190 H.Schnöckel, T.Mehner, H.S.Plitt and S.Schunck, J. Am. Chem. Soc., 1989, 111, 4578.
191 V.N.Khabashevskii, Z.A.Kerzina and O.M.Nefedov, Izv. Akad. Nauk SSSR, Ser. Khim., 1988, 217.
192 A.Pattyk, W.Sander, J.Gauss and D.Cremer, Angew. Chem., Int. Ed. Engl., 1989, 28, 898.
193 F.Ozanam and J.N.Chazalviel, J.Electroanal. Chem. Interfacial Electrochem., 1989, 269, 251.
194 A.Borghesi, L.Colombo, S.Rojas and W.S.Wu, Nuovo Cimento Soc. Ital. Fis., D, 1988, 10D, 1487.
195 T.A.Sidorov, Izv. Akad. Nauk SSSR, Neorg. Mater., 1989, 25, 492.

196 O.P.Mchedlov-Petrosyan, L.I.Degtereva, I.A.Spirin and D.A.Uginchus, Silikattechnik, 1989, 40, 148.
197 B.C.Bunker, D.M.Haaland, K.J.Ward, T.A.Michalske, J.S.Binkley, C.F.Melius and C.A.Barre, Surf. Sci., 1989, 210, 406.
198 R.A.Murray and W.Y.Ching, Phys. Rev.,B, 1989, 39, 1320.
199 I.A.Mukhitdinov and O.V.Yanush, Fiz. Khim. Stekla, 1989, 15, 34.
200 M.V.Vladimirov, S.V.Gevk'yan and G.T.Ostapenko, Zh. Prikl. Spektrosk., 1989, 50, 144.
201 P.McMillan, M.Akaogi, E.Ohtani, Q.Williams, R.Nieman and R.Sato, Phys. Chem. Solids, 1989, 16, 428.
202 L.W.Finger, R.M.Hazen and R.J.Hemley, Am. Mineral., 1989, 74, 952.
203 H.Kozuka, Y.Li and S.Sakka, Boli Yu Tongci, 1988, 16, 7 (Chem. Abs., 1989, 110, 143826).
204 H.Schnöckel and R.Köppe, J. Am. Chem. Soc., 1989, 111, 4583.
205 K.Tanaka, Y.Akiyama and T.Tanaka, J. Mol. Spectrosc., 1989, 137, 55.
206 Y.Akiyama, K.Tanaka and T.Tanaka, Chem. Phys. Lett., 1989, 155, 15.
207 L.Jörissen, H.Prinz, W.A.Kreiner, O.Wenger, G.Pierre, G.Magerl and W.Schupita, Can. J. Phys., 1989, 67, 532.
208 K.A.Peterson and R.C.Woods, J. Chem. Phys., 1988, 89, 4929.
209 A.V.Sakharov, V.O.Gel'mbol'dt, V.F.Sukhoverkov and A.A.Ennan, Russ. J. Inorg. Chem., 1989, 34, 302.
210 J.M.Coffin, T.P.Hamilton, P.Pulay and I.Hargittai, Inorg. Chem., 1989, 28, 4092.
211 H.G.Ang and F.K.Lee, J. Chem. Soc., Chem. Comm., 1989,310.
212 F.Lattanzi, C.DiLauro, L.Henry, A.Valentin and H.Burger, Mol. Phys., 1989, 66, 1027.
213 J.Barrau, D.L.Bean, K.M.Walsh, R.West and J.Michl, Organometallics, 1989, 8, 2606.
214 R.K.Chadha, J.E.Drake and A.B.Sarkar, Inorg. Chim. Acta, 1988, 143, 31.
215 A.G.Martin and C.M.Rodriguez, Synth. React. Inorg. Met.-Org. Chem., 1989, 19, 33.
216 B.N.Ivanov-Emin, B.E.Zaitsev, A.V.Kostrikin, I.V.Lin'ko and A.I.Ezhov, Russ. J. Inorg. Chem., 1988, 33, 1606.
217 A.Haas, H.-J.Kutsch and C.Krüger, Chem. Ber., 1989, 122, 271.
218 N.A.Chumaevskii, D.A.Ivashchenko, T.K.Gar, O.A.Dombrova and N.A.Minaeva, Koord. Khim., 1988, 14, 1071.
219 M.T.Vandenborre, D.Michel and A.Ennaciri, Spectrochim. Acta, Part A, 1989, 45A, 721.
220 P.M.Valov, A.N.Zhilkin, I.A.Mukhitdinov, D.A.Zukhov and O.V.Yanush, Zh. Prikl. Spektrosk., 1989, 50, 120.
221 M.Amarilla, M.L.Velga, C.Pico, M.Gaitán and A.Jerez, Inorg. Chem., 1989, 28, 1701.
222 J.Barrau, V.Balaji and J.Michl, Organometallics, 1989, 8, 2034.
223 J.M.Mikrut and L.E.McNeil, J. Non-Cryst. Solids, 1989, 109, 237.
224 G.Thiele, H.W.Rotter and K.D.Schmidt, Z. Anorg. Allg. Chem., 1989, 571, 60.
225 J.F.Sanz, A.Marquez and C.Pouchan, Chem. Phys., 1989, 130, 451.
226 L.Jörissen, Y.Ohshima, Y.Matsumoto, M.Takami and K.Kuchitsu, J. Chem. Phys., 1989, 90, 2109.
227 H.Bürger, M.Betzel-Laves, S.Sommer, G.Blanquet and J.Walrand, J. Mol. Spectrosc., 1989, 134, 214.
228 J.F.Sanz and A.Marquez, J. Phys. Chem., 1989, 93, 7328.

229 C.Silvestru and I.Haiduc, J. Organometal. Chem., 1989, 365, 83.
230 A.Silvestri, D.Duca and F.Huber, Appl. Organomet. Chem., 1988, 2, 417.
231 C.Carini, G.Pelizzi, P.Tarasconi, C.Pelizzi, K.C.Molloy and P.C.Waterfield, J. Chem. Soc., Dalton Trans., 1989, 289.
232 G.Pelizzi, P.Tarasconi, C.Pelizzi, K.C.Molloy and P.C.Waterfield, Main Group Met. Chem., 1987, 10, 353.
233 T.G.Vasilenko, G.E.Ermolina, A.K.Molodkin, S.L.Kuznetsov, N.S.Prostakov, V.G.Pleshkov and B.Singh, Russ. J. Inorg. Chem., 1989, 34, 830
234 M.J.Hampden-Smith, D.E.Smith and E.N.Duesler, Inorg. Chem., 1989, 28, 3399.
235 K.M.Kadish, Q.Y.Y.Ku, G.B.Bhaskar, J.M.Barbe and R.Guilard, J. Chem. Soc., Dalton Trans., 1989, 1531.
236 G.G.Lobbia, F.Bonati, A.Cingolani, D.L.Fonesi and A.Lorenzotti, J. Organometal. Chem., 1989, 359, 21.
237 S.A.Zaitsev, S.B.Osin and V.F.Shevel'kov, Vestn. Mosk. Univ., Ser.2: Khim., 1988, 29, 564.
238 A.Glowacki, F.Huber and H.Preut, Rec. Trav. Chim. Pays-Bas, 1988, 107, 278.
239 K.Horiai and H.Uehara, J. Mol. Spectrosc., 1989, 136, 56.
240 N.Trandafilova, M..Tanasov, G.Nikolov, A.Yu.Tsivadze, V.A.Trofimov and I.K.Kireeva, Koord. Khim., 1988, 14, 1060.
241 G.Pangilinan, G.Guelachvili, R.Sooryakumar, K.N.Rao and R.H.Tipping, Phys. Rev., B, 1989, 39, 2522.
242 D.K.Veirs, V.K.F.Chia and G.M.Rosenblatt, Langmuir, 1989, 5, 633.
243 J.C.Owrutsky, E.R.Keim, J.V.Coe and R.J.Saykally, J. Phys. Chem., 1989, 93, 5960.
244 J.Bendtsen and F.M.Nicolaisen, J. Mol. Spectrosc., 1989, 133, 193.
245 M.Okumura, B.D.Rehfuss, B.M.Dinelli, M.G.Bawendi and T.Oka, J.Chem. Phys., 1989, 90, 5918.
246 R.Minkwitz and R.Nass, Z. Naturforsch., B, 1988, 43b, 1478.
247 R.Minkwitz, V.Wölfel, R.Nass, H.Härtner and W.Sawodny, Z. Anorg. Allg. Chem., 1989, 570, 127.
248 M.G.Bawendi, B.D.Rehfuss, B.M.Dinelli, H.Okumura and T.Oka, J. Chem. Phys., 1989, 90, 5910.
249 A.Loutellier and J.P.Perchard, J. Mol. Struct., 1989, 198, 51.
250 G.Guelachvili, A.H.Abdullah, N.Tu, K.N.Rao, Š.Urban and D.Papoušek, J. Mol. Spectrosc., 1989, 133, 345.
251 Š.Urban, N.Tu, K.N.Rao and G.Guelachvili, J. Mol. Spectrosc., 1989, 133, 312.
252 R.B.Bohn and L.Andrews, J. Phys. Chem., 1989, 93, 3974.
253 F.Huisken and T.Pertsch, Chem. Phys., 1988, 126, 213.
254 M.Polak, M.Gruebele, B.W.DeKock and R.J.Saykally, Mol. Phys., 1989, 66, 1193.
255 T.Tipton, D.A.Stone, K.Kubulat and W.B.Person, J. Phys. Chem., 1989, 93, 2917.
256 A.Harper and P.Hubberstey, J. Chem. Res., Synop., 1989, 194.
257 R.Minkwitz and A.Liedtke, Inorg. Chem., 1989, 28, 1627.
258 H.C.Miller, J.L.Hardwick and J.W.Farley, J. Mol. Spectrosc., 1989, 134, 329.
259 M.D.Vanek, M.Schneider, J.S.Wells and A.G.Maki, J. Mol. Spectrosc., 1989, 134, 154.
260 N.L.S.Yamazaki, C.Manzanares I., L.C.Baylor, G.C.Schatz and E.Weitz,

J. Phys. Chem., 1989, 93, 2204.
261 Z.S.Huang and R.E.Miller, J. Chem. Phys., 1988, 89, 5408.
262 Y.Ohshima, Y.Matsumoto, M.Takami and K.Kuchitsu, Chem. Phys. Lett., 1988, 152, 294.
263 Y.Ohshima, Y.Matsumoto, M.Takami and K.Kuchitsu, Reza Kagaku Kenkyu, 1988, 10, 8 (Chem. Abs., 1989, 110, 201881).
264 A.Perrin, O.Lado, B.Ordowsky and A.Valentin, Mol. Phys., 1989, 67, 249.
265 A.G.Maki and W.B.Olson, J. Mol. Spectrosc., 1989, 133, 171.
266 A.Goldman, J.B.Burkholder, C.J.Howard, R.Escribano and A.G.Maki, J. Mol. Spectrosc., 1988, 131, 195.
267 A.G.Maki, J. Mol. Spectrosc., 1989, 136, 105.
268 A.J.Bell, P.R.Pardon and J.G.Frey, Mol. Phys., 1989, 67, 465.
269 J.K.McDonald, V.F.Kalasinsky, T.J.Geyer and J.R.Durig, J. Mol. Spectrosc., 1988, 132, 104.
270 M.Cheikh and C.Alamichel, Spectrochim. Acta, Part A, 1989, 45A, 153.
271 M.Cheikh and A.Picard-Bersellini, Spectrochim. Acta, Part A, 1988, 44A, 1443.
272 J.Verges, J.Chevillard, C.Alamichel and J.R.Etcheverry, Chem. Phys. Lett., 1989, 159, 315.
273 B.J.Finlayson-Pitts, F.E.Livingston and H.N.Berko, J. Phys. Chem., 1989, 93, 4397.
274 A.Givan and A.Löwenschuss, J. Chem. Phys., 1989, 90, 6135.
275 S.P.So, Inorg. Chem., 1989, 28, 2888.
276 A.W.Allaf, G.Y.Matti, R.J.Suffolk and J.D.Watts, J. Electron Spectrosc. Relat. Phenom., 1989, 48, 411.
277 A.W.Allaf, G.Y.Matti, R.J.Suffolk and J.D.Watts, Chem. Phys. Lett., 1989, 155, 32.
278 J.Grobe, D.le Van and R.Demuth, J.Fluorine Chem., 1988, 39, 385.
279 R.Minkwitz and A.Liedtke, Z. Naturforsch., B, 1988, 43b, 1263.
280 P.C.Knüppel, G.Pawelke, H.Sommer, O.Stelzer, H.Oberhammer, M.F.Lappert, R.J.Suffolk and J.D.Watts, J. Organometal. Chem., 1988, 355, 55.
281 L.Andrews and R.Withnall, Inorg. Chem., 1989, 28, 494.
282 R.Minkwitz and A.Liedtke, Inorg. Chem., 1989, 28, 4238.
283 K.Ohno and H.Matsuura, Chem. Phys. Lett., 1989, 155, 443.
284 R.Appel and M.Poppe, Angew. Chem., Int. Ed. Engl., 1989, 28, 53.
285 J.R.Durig, W.Zhao and T.S.Little, J. Raman spectrosc., 1989, 20, 111.
286 H.G.M.Edwards and V.Fawcett, Spectrochim. Acta, Part A, 1989, 45A, 147.
287 J.R.Durig, M.S.Cheng, Y.S.Li, P.Groner and A.E.Stanley, J. Phys. Chem., 1989, 93, 3492.
288 J.R.Durig and K.K.Chatterjee, Inorg. Chem., 1989, 28, 298.
289 K.Hassler, Montash., 1988, 119, 863.
290 J.R.Durig, R.J.Harlan and P.Groner, J. Phys. Chem., 1989, 93, 3041.
291 T.P.Hamilton and H.F.Schaffer, Angew. Chem., Int. Ed. Engl., 1989, 28, 484.
292 H.Schnöckel and S.Schunck, Phosphorus Sulfur, 1988, 39, 89.
293 K.Taga, N.Hirabayashi and H.Okabayashi, J. Mol. Struct., 1988, 178, 89.
294 G.Liu and G.Wu, Spectrochim. Acta, Part A, 1988, 44A, 1007.
295 H.H.Eysel and K.T.Lim, J. Raman Spectrosc., 1988, 19, 535.
296 G.M.Maksimov, G.N.Kustova, K.L.Matveev and T.P.Lazarenko, Koord.

Khim., 1989, 15, 788
297 R.Ya.Mel'nikova, T.I.Barannikova, I.E.Malashonok, V.V.Pechkovskii, T.A.Tikhonovich and N.M.Ksenofontova, Russ. J. Inorg. Chem., 1988, 33, 1268.
298 N.Mochida, T.Sekiya, A.Ohtsuka and M.Tonokawa, Nippon Seramikkusu Kyokai Gakujutsu Ronbunshi, 1988, 96, 973 (Chem. Abs., 1988, 109, 235604).
299 I.E.Malashonok, R.Ya.Mel'nikova, V.V.Pechkovskii and N.V.Bulavkina, Russ. J. Inorg. Chem., 1988, 33, 1758.
300 I.E.Maloshonok, R.Ya.Mel'nikova, V.V.Pechkovskii, N.V.Bulavkina and T.A.Tikhonovich, Russ. J. Inorg. Chem., 1989, 34, 312.
301 L.S.Nashkevich, V.A.Lyutsko and M.V.Nikanovich, Koord. Khim., 1988, 14, 1085.
302 K.Viswanathan, V.U.Nayar and G.Aruldhas, Curr. Sci., 1989, 58, 188.
303 G.Sekar, G.Aruldhas and V.Ramakrishnan, Indian J. Pure Appl. Phys., 1988, 26, 570.
384 Z.Mielke and L.Andrews, J. Phys. Chem., 1989, 93, 2971.
305 L.Koudelka, M.Pisarčik, M.S.Gutenev and L.N.Blinov, J. Mater. Sci. Lett., 1989, 8, 933.
306 A.V.Chernova, G.M.Doroshkina, S.A.Katsyuba, R.R.Shagidullin, N.A.Khalkova and V.K.Khairullin, Izv. Akad. Nauk SSSR, Ser. Khim., 1988, 568.
307 S.Mohan, F.Payami and R.Madivanane, Curr. Sci., 1989, 58, 442.
308 N.Burford, B.W.Royan, A.Linden and T.S.Cameron, Inorg. Chem., 1989, 28, 144.
309 S.A.Katsyuba, E.A.Filippova, V.A.Al'fonsov, R.R.Shagidullin, V.N.Shegeda and D.A.Pudonik, Zh. Obshch. Khim., 1988, 58, 1744.
310 C.Gledel. J.P.Audiere, R.Clement and R.Cortes, J. Mater. Sci. Lett., 1988, 7, 1054.
311 J.R.Pollo and G.B.Burns, J. Phys. Chem. Solids, 1989, 50, 399.
312 P.T.Wood and J.D.Woollins, J. Chem. Soc., Chem. Comm., 1988, 1190.
313 H.Prinz and W.A.Kreiner, J. Mol. Spectrosc., 1989, 137, 204.
314 B.W.Moores and L.Andrews, J. Phys. Chem., 1989, 93, 1902.
315 L.Andrews, R.Wthnall and B.W.Moores, J. Phys. Chem., 1989, 93, 1279.
316 A.B.Remizov, F.G.Khalitov and F.G.Garieva, Zh. Strukt. Khim., 1988, 29, 54.
317 R.R.Shagidullin, I.Kh.Shakirov, P.N.Sobolev and N.A.Chadaeva, Izv. Akad. Nauk SSSR, Ser. Khim., 1988, 182.
318 M.Somer, W.Honle and H.G.von Schnering, Z. Naturforsch., B, 1989, 44b, 296.
319 L.Andrews and Z.Mielke, Inorg. Chem., 1989, 28, 4001.
320 A.C.Gonzalez-Baro, E.J.Baran and A.G.Nord, An. Asoc. Quim. Argent., 1988, 76, 121
321 E.J.Baran and S.B.Etcheverry, An. Asoc. Quim. Argent., 1988, 76, 397.
322 A.Müller and J.Döring, Angew. Chem., Int. Ed. Engl., 1988, 27, 1721.
323 H.Kawazoe, H.Yanagita, Y.Watanabe and M.Yamane, Phys. Rev., B, 1988, 38, 5661.
324 H.Sinning and U.Müller, Z. Anorg. Allg. Chem., 1988, 564, 37.
325 H.Sinning and U.Müller, Z. Anorg. Allg. Chem., 1989, 568, 49.
326 T.Klapotke, J.Passmore and E.G.Awere, J. Chem. Soc., Chem. Comm., 1988, 1426.
327 U.Müller and H.Sinning, Angew. Chem., Int. Ed. Engl., 1989, 28, 185.
328 L.Andrews, B.W.Moores and K.K.Fonda, Inorg. Chem., 1989, 28, 290.

329 I.A.Tossidis, C.A.Bolos, G.A.Katsoulos, M.P.Sigalas and C.A.Tsipis, Z. Anorg. Allg. Chem., 1988, 567, 161.
330 G.Becker, M.Meiser, O.Mundt and J.Weidlein, Z. Anorg. Allg. Chem., 1989, 569, 62.
331 A.M.Sych, M.I.Kabanova and S.G.Andreeva, Russ. J. Inorg. Chem., 1988, 33, 1585.
332 B.A.Arbuzov, R.R.Shagidullin, V.S.Vinogradova, Yu.M.Mareev, I.Kh.Shakirov and N.R.Fedotova, Izv. Akad. Nauk SSSR, Ser. Khim., 1987, 423.
333 T.Athar, R.Bohra and R.C.Mehrotra, J. Ind. Chem. Soc., 1989, 66, 189.
334 D.Mascherpa-Corral, B.Ducourant and S.Alberola, J. Solid State Chem., 1988, 76, 276.
335 S.A.Wood, Geochim. Cosmochim. Acta, 1989, 53, 237.
336 I.V.Aleksandrov, A.B.Bykov, A.F.Goncharov, V.N.Denisov, B.N.Mavrin, O.K.Mel'nikov, V.B.Podobedov and N.I.Romanova, Pis'ma Zh. Eksp. Teor. Fiz., 1989, 49, 150.
337 E.K.Chernova, A.B.Yaroslavtsev, Z.N.Prozorovskaya and V.F.Chuvaev, Russ. J. Inorg. Chem., 1989, 34, 38.
338 M.Gruebele, E.Keim, A.Stein and R.J.Saykally, J. Mol. Spectrosc., 1988, 131, 343.
339 P.R.Brown, P.B.Davies and R.J.Stickland, J. Chem. Phys., 1989, 91, 3384.
340 G.di Leonardo and L.Fusina, J. Mol. Spectrosc., 1989, 135, 250
341 V.Zhelyaskov, G.Georgiev, Zh.Nickolov and M.Miteva, J. Raman Spectrosc., 1989, 20, 67.
342 A.Engdahl and B.Nelander, J. Mol. Struct., 1989, 193, 101.
343 M.H.Brooker, G.Hancock, B.C.Rice and J.Shapter, J. Raman Spectrosc., 1989, 20, 683.
344 J.Y.Mandin, J.P.Chevillard, J.M.Flaud and C.Camy-Peyret, Can. J. Phys., 1988, 66, 997.
345 A.K.Garg, Phys. Stat. Solidi, A, 1988, 110, 467.
346 J.P.Devlin, J. Chem. Phys., 1989, 90, 1322.
347 V.D.Malorov and N.B.Librovich, Izv. Akad. Nauk SSSR, Ser. Khim., 1989, 967.
348 J.M.Flaud, C.Camy-Peyret, A.N'Gom, V.M.Devi, C.P.Rinsland and M.A.H.Smith, J. Mol. Spectrosc., 1989, 133, 217.
349 J.M.Flaud, C.Camy-Peyret, C.P.Rinsland, M.A.H.Smith and V.M.Devi, J. Mol. Spectrosc., 1989, 134, 106.
350 T.J.Dines, Spectrochim. Acta, Part A, 1988, 44A, 1087.
351 L.Manceron, A.M.Le Quere and J.P.Perchard, J. Phys. Chem., 1989, 93, 2960.
352 K.S.Kim, H.S.Kim, S.Kim, J.H.Jang and H.F.Schaeffer, J. Am. Chem. Soc., 1989, 111, 7746.
353 G.M.Campbell, J. Mol. Struct., 1988, 189, 301.
354 S.Urban, E.Herbst, P.Mittler, G.Winnewisser, K.M.T.Yamada and M.Winnewisser, J. Mol. Spectrosc., 1989, 137, 327.
355 C.O.della Vedova, J. Raman Spectrosc., 1989, 20, 279.
356 R.Minkwitz and J.Sawatzki, Z. Anorg. Allg. Chem., 1988, 564, 135.
357 T.Chivers, G.Y.Dénès, S.W.Liblong and J.F.Richardson, Inorg. Chem., 1989, 28, 3683.
358 D.Mauer, G.Winnewisser and K.M.T.Yamada, J. Mol. Spectrosc., 1989, 136, 380.
359 P.Dubois, J.P.Lelieur and G.Lepoutre, Inorg. Chem., 1989, 28, 2489.

360 I.Nevitt, H.S.Rzepa and J.D.Woollins, Spectrochim. Acta, Part A, 1989, 45A, 367.
361 R.Strauss and R.Steudel, Z. Naturforsch., B, 1988, 43b, 1151.
362 M.Lamoureux and J.Milne, Can. J. Chem., 1989, 67, 1936.
363 J.Adel, C.Ergezinger, R.Figge and K.Dehnicke, Z. Naturforsch., B, 1988, 43b, 639.
364 S.Civiš, C.E.Blom and P.Jenssen, J. Mol. Spectrosc., 1989, 138, 69.
365 J.Otto, Phys. Stat. Solidi, B, 1989, 151, 363.
366 C.Camy-Peyret, J.M.Flaud, A.N'Gom and J.W.C.Johns, Mol. Phys., 1989, 67, 693.
367 C.Camy-Peyret, J.M.Flaud, A.N'Gom and J.W.C.Johns, Mol. Phys., 1988, 65, 649.
368 T.Nakanaga and T.Amano, J. Mol. Spectrosc., 1989, 133, 201.
369 O.Lösking and H.Willner, Angew. Chem., Int. Ed. Engl., 1989, 28, 1255.
370 R.Minkwitz, G.Nowicki and H.Preut, Z. Anorg. Allg. Chem., 1989, 573, 185.
371 R.Minkwitz and A.Werner, J. Fluorine Chem., 1988, 39, 141
372 G.Guelachvili, O.V.Naumenko and O.N.Ulenikov, J. Mol. Spectrosc., 1988, 131, 400
373 G.B.Bacskay, A.P.L.Rendell and N.S.Hush, J. Chem. Phys., 1988, 89, 5721.
374 P.S.Santos and J.G.N.Santos, J. Raman Spectrosc., 1988, 19, 463.
375 N.C.Craig and K.Futamura, Spectrochim. Acta, Part A, 1989, 45A, 507.
376 A.Tasaka and Y.Takenaka, Denki Kagaku Oyobi Kogyo Butsuli Kagaku, 1988, 56, 745(Chem. Abs., 199, 110, 107241).
377 S.Milićev and A.Stegaršek, Spectrochim. Acta, Part A, 1989, 45A, 225.
378 Y.S.Park and R.Frech, Spectrochim. Acta, Part A, 1989, 45A, 795.
379 R.Minkwitz and A.Werner, Z. Naturforsch., B, 1988, 43b, 403.
380 W.Heilemann, R.Mews, S.Pohl and W.Saak, Chem. Ber., 1989, 122, 427.
381 C.Chapados and G.Birnbaum, J. Mol. Spectrosc., 1988, 132, 323.
382 R.Minkwitz and J.Sawatzki, Z. Anorg. Allg. Chem., 1988, 566, 151.
383 R.Minkwitz, H.Preut and J.Sawatzki, Z. Anorg. Allg. Chem., 1989, 569, 158.
384 R.Minkwitz and V.Gerhard, Z. Naturforsch., B, 1989, 44b, 364.
385 Y.S.Park and R.Frech, Spectrochim. Acta, Part A, 1989, 45A, 213.
386 J.Milne, Can. J. Chem., 1989, 67, 1056.
387 J.C.Huffman and R.C.Haushalter, Polyhedron, 1989, 8, 531.
388 R.A.El-Mallawany, Infrared Phys., 1989, 29, 781.
389 X.Matthew and V.U.Nayar, Spectrochim. Acta, Part A, 1988, 44A, 1061.
390 E.H.Fink, K.D.Setzer, D.A.Ramsay, M.Vervloet and G.Z.Xu, J. Mol. Spectrosc., 1989, 136, 218.
391 W.Abriel and H.Ehrhardt, Z. Naturforsch., B, 1988, 43b, 557.
392 C.M.Lovejoy, D.D.Nelson and D.J.Nesbitt, J. Chem. Phys., 1988, 89, 7180.
393 G.Schatte, H.Willner, D.Hoge, F.Knözinger and O.Schrems, J. Phys. Chem., 1989, 93, 6025.
394 K.McMillan, D.Bender, M.Eliades, D.Danzeiser, B.A.Wofford and J.W.Bevan, Chem. Phys. Lett., 1988, 152, 87.
395 B.A.Wofford, R.S.Ram, A.Quinonez, J.W.Bevan, W.B.Olson and W.J.Lafferty, Chem. Phys. Lett., 1988, 152, 299.
396 G.T.Fraser, A.S.Pine, R.D.Suenram, D.C.Dayton and R.E.Miller, J. Chem. Phys., 1989, 90, 1330.
397 C.M.Lovejoy and D.J.Nesbitt, J. Chem. Phys., 1989, 90, 4671.

398 L.Andrews, R.B.Bohn, R.T.Arlinghaus and R.D.Hunt, Chem. Phys. Lett., 1989, 158, 564
399 K.G.Tokhadze, M.A.Bernikov, A.Yu.Voronin and N.N.Tylets, Opt. Spektrosk., 1988, 65, 540.
400 Z.S.Huang and R.E.Miller, J. Chem. Phys., 1989, 90, 1478.
401 R.B.Bohn and L.Andrews, J. Phys. Chem., 1989, 93, 5684.
402 B.S.Ault, J. Phys. Chem., 1989, 93, 279.
403 W.W.Wilson, K.O.Christe, J.A.Feng and R.Bau, Can. J. Chem., 1989, 67, 1898.
404 H.Bürger, G.Pawelke, S.Sommer, A.Rahner, E.H.Appelman and I.M.Mills, J. Mol. Spectrosc., 1989, 136, 197.
405 E.H.Appelman, A.J.Downs and C.J.Gardner, J. Phys. Chem., 1989, 93, 598.
406 W.Thiel, G.Scuseria, H.F.Schaeffer and W.D.Allen, J. Chem. Phys., 1988, 89, 4965.
407 N.P.Machara and B.S.Ault, J. Mol. Struct., 1988, 172, 129.
408 A.Furlan, S.Wulfert and S.Leutwyler, Chem. Phys. Lett., 1988, 153, 291.
409 N.Moazzen-Ahmadi, A.R.W.McKellar and J.W.C.Johns, J. Mol. Spectrosc., 1989, 138, 282.
410 L.Andrews and R.B.Bohn, J. Chem. Phys., 1989, 90, 5205.
411 J.M.Hutson, J. Chem. Phys., 1988, 89, 4550.
412 R.B.Bohn, R.D.Hunt and L.Andrews, J. Phys. Chem., 1989, 93, 3979.
413 J.B.Burkholder, P.D.Hammer, C.J.Howard and A.Goldman, J. Geophys. Res., D, 1989, 94, 2225.
414 V.I.Lang, S.P.Sander and R.R.Friedl, J. Mol. Spectrosc., 1988, 132, 89.
415 H.H.Eysel, Spectrochim. Acta, Part A, 1988, 44A, 991.
416 B.M.Cheng and Y.P.Lee, J. Chem. Phys., 1989, 90, 5930.
417 S.Hyodo, Chem. Phys. Lett., 1989, 161, 245.
418 H.Bai and B.S.Ault, J. Mol. Struct., 1989, 196, 47.
419 Yu.M.Kiselev, A.I.Popov and S.A.Goryachenkov, Russ. J. Inorg. Chem., 1988, 33, 1218.
420 W.W.Wilson and K.O.Christe, Inorg. Chem., 1989, 28, 4172.
421 K.B.Dillon, M.P.Nisbet and R.N.Reeve, Polyhedron, 1988, 7, 1725.
422 W.Levason, J.S.Ogden, M.D.Spicer, M.Webster and N.A.Young, J. Am. Chem. Soc., 1989, 111, 6210.
423 D.L.A.de Feria, N.S.Gonçalves and P.S.Santos, Spectrochim. Acta, Part A, 1989, 45A, 643.
424 K.Nakagawa, K.Horiai, T.Konno and H.Uehara, J. Mol. Spectrosc., 1988, 131, 233.
425 K.O.Christe, W.W.Wilson and R.D.Wilson, Inorg. Chem., 1989, 28, 904.
426 R.G.Syvret and G.J.Schrobilgen, Inorg. Chem., 1989, 28, 1564.
427 U.Magg, H.Birk, K.P.R.Nair and H.Jones, Z. Naturforsch., A, 1989, 44a, 313.
428 L.Yu.Alikberova, M.G.Zaitseva, V.V.Kravchenko, N.S.Rukk, E.V.Savinkina and B.D.Stern, Russ. J. Inorg. Chem., 1989, 34, 852.

5

Vibrational Spectra of Transition-element Compounds

BY G. DAVIDSON

1. Scandium, Yttrium and the Lanthanides.

The Raman spectra of aqueous $ScCl_3$ and $Sc(ClO_4)_3$ solutions showed that the aquo-complexes are not octahedral, possibly seven coordinate.[1] Some vibrational assignments for ScF_6^{3-} are summarised in Table 1.[2]

Table 1 Some vibrational assignments for ScF_6^{3-} (/cm^{-1})

a_{1g}	ν_1	(stretch)	495
e_g	ν_2	(stretch)	375
t_{1u}	ν_3	(stretch)	458
t_{1u}	ν_4	(bend)	257
t_{2g}	ν_5	(bend)	235

I.r. and Raman spectra gave assignments for many vibrational modes for $KLnSb_2O_7$, where Ln = Y, Ho, Er, Yb or Lu.[3] Detailed vibrational assignments were proposed for $KLn(WO_4)_2$, where Ln = Y, La - Gd, Lu, from their i.r. and Raman spectra.[4] Glasses in the systems AlF_3-YF_3-MF_2, where M = Mg, Ca, Ba or Sr, give i.r. and Raman spectra which show the presence of eight-coordinate yttrium species.[5]

νLn-N(of phen) modes were seen in the range 300 - 320 cm^{-1} for $Ln(gly)_3(phen)_2(H_2O)$, where Ln = La, Ce, Pr, Nd, Sm, Eu, Gd, Tb, Dy, Ho, Er, Tm, Yb or Lu; glyH = glycine.[6] Low-wavenumber Raman spectra of $M_2(C_2O_4)_3.nH_2O$, where M = all lanthanides except Pm, show that for M = La to Er, n = 10, the compounds are all isostructural. Another isostructural group comprises M = Er to Lu, n = 6.[7] Lanthanide-oxygen stretches were also reported for these complexes.[8]

The νLn-O stretches of $[Ln(H_2O)_9](C_2H_5SO_4)_3$, where Ln = La - Lu, lie in the range 316 - 342 cm^{-1}, and decrease steadily with increasing ionic radius.[9,10] Stretching vibrations (LnO(crown), $LnO(OH_2)$ and LnBr) were identified from the i.r. spectra of $LnBr_3$(18-crown-6)$.nH_2O$, for Ln = La, Nd, Sm, Gd, Tb, Er or Y.[11] νLn-O modes were assigned for the dioxyapatites $Sr_8Ln_2(PO_4)_6O_2$, where Ln = La, Pr, Nd, Sm or Eu.[12] The νLnO modes in $LnCoO_3$, where L = La, Pr, Nd, Sm, Eu, Gd, Tb, Dy, Ho or Er, show that the covalency of the bond increases with increasing atomic number of the lanthanide.[13]

A band at 372 cm^{-1} in the Raman spectrum of $Ce(acac)_2L$, where L = tetra-p-chlorophenylporphin, was assigned to νCeO.[14] The i.r. and Raman spectra of $[M(AP)_6]I_3$, where M = Ce, Pr, Nd, Sm, Eu, Gd, Tb, Dy, Ho, Er, Tm, Yb, Lu or Y; AP = antipyrine, show that all have centrosymmetric crystal structures.[15] νNdO modes were assigned as follows in $Nd(ClO_4)_3.nPOCl_3$: 267 cm^{-1} for n = 5, 263 cm^{-1} for n = 3.[16]

The i.r. and Raman spectra of $Ln_2Ti_2O_7$, where Ln = Sm or Eu, show them to be isostructural with the known high-temperature ferroelectrics with Ln = La, Pr or Nd.[17] Vibronic spectra of EuL_6^{3+} complexes, where L = antipyrine, gave evidence for νEuO modes of the EuO_6 octahedron at 174 - 180 cm^{-1}.[18] I.r. and Raman spectra gave νEuO assignments for mixed chelates of Eu(III) with dibenzoylmethane and piperidine.[19] The i.r. spectrum of $EuNbO_4$ suggests that νNbO modes are in the range 500 - 850 cm^{-1}, with νEuO_8 200 - 400 cm^{-1}.[20] Some i.r. assignments were proposed for $LnVNb_2O_9$, where Ln = Gd, Tb, Dy or Ho.[21]

2. Titanium, Zirconium and Hafnium.

$K_2[TiO(O_2)Cl_2]$ has νTi=O at 970 cm^{-1} in solution. The solid is believed to be polymeric via -Ti-O-Ti-O- interactions.[22] The μ-oxo complex [Cl(acen)TiOTi(acen)Cl], where acen = N,N'-ethylenebis(acetylacetone- iminate), has ν_{as}TiOTi at 750 cm^{-1}.[23] νTiO bands were identified in $LnTiO_3$-$Ln_2Ti_2O_7$ systems, where Ln = La or Gd.[24] The i.r. and Raman spectra of $La_{1-x}Y_xTiO_3$, where $0.1 \leqslant x \leqslant 0.9$, are consistent with either considerable distortion of the TiO_6 coordination octahedron or the presence of both TiO_6 and TiO_4 units.[25] $[Ti_3O(S_2)_3Cl_6]^{2-}$ has νTi_3O at 850 - 910 cm^{-1}, with νTiCl/νTiS at 375, 330 and 270 cm^{-1}. $[TiSCl_4]^{2-}$ has νTi=S 529 cm^{-1} (A_1), $\nu_{as}TiCl_4$ 329 cm^{-1} (E) and $\nu_s TiCl_4$ 300 cm^{-1} (A_1).[26]

New data on the matrix i.r. spectra of TiF_4 and TiF_3 suggest that previous assignments to "TiF_2" were erroneous, removing evidence for a bent AB_2 molecule.[27] $Cp_2Ti(AsF_6)_2$ has νTi-F-As at 530 cm^{-1}.[28] The Raman spectra of aqueous solutions of $TiCl_4$ were interpreted in terms of the hydrolysis products of the $TiCl_4$, e.g. $[Ti(OH)_nCl_{6-n}]^{2-}$, where n = 0, 1, 2 or 3, $[Ti(OH)_n(H_2O)_{6-n}]^{(4-n)+}$ etc.[29] $Ti_2Cl_6(POCl_3)_2(PO_2Cl_2)_2$ has νTiCl(t) at 375 and 405 cm^{-1}.[30] A theoretical treatment of the resonance Raman ν_1 band progression of TiI_4 suggests that one can obtain information on excited state geometry from intensity distributions in the overtone progression.[31]

$CpCp^*Zr(H)OC(=CH_2)Si(SiMe_3)_3$ has νZrH at 1580 cm^{-1}.[32] νM-O bands were identified from the i.r. and Raman spectra of $MGeO_4$, where M = Zr, Hf or Th,[33] and Zr_3GeO_8.[34] νZr-L modes were assigned for ZrL_4, where L = acac or other β-diketonates, in the ranges 450 - 500 cm^{-1} and 678 - 702 cm^{-1} (both mixed with deformations of the chelate ring).[35] $\nu M(O_2)$ modes were seen in the range 590 - 650 cm^{-1} in $M(O_2)(H_2edta)$, where M = Zr or Th; $M(O)(O_2)(H_2edta)$, where M = W or U. The last two have νM=O at 910 cm^{-1} (U) or 955 cm^{-1} (W). All have νM-O and νM-N of the H_2edta ligand near 380 - 430, 265 - 320 cm^{-1} respectively.[36] Evidence was found for ZrF_6, ZrF_7 and ZrF_8 units, linked into chains, from the i.r. and Raman spectra of ZrF_4-BaF_2-LaF_3-ZnF_2 glasses.[37] The Raman spectra of ZrX_4-AlX_3, where X = Cl or Br, systems in the gas phase show that the dominant species are $ZrAlCl_7$, $ZrAl_2Cl_{10}$ and $ZrAlBr_7$.[38]

3. Vanadium, Niobium and Tantalum.

Earlier reference has been made to vibrational studies on $EuNbO_4$ [20] and $LnVNb_2O_9$, where Ln = Gd, Tb, Dy, Ho.[21]

F.T.i.r. of matrix-isolated species trapped from $CsVO_3$ vapour show the presence of both monomer and dimer, $(CsVO_3)_2$, units.[39] Polymeric $VOSeO_3$ has ν_sVO_3 at 950 cm^{-1}, $\nu_{as}VO_3$ at 865 cm^{-1}, νV_2O_2 at 625 cm^{-1}, together with other V-O-V bridging modes 560 - 505 cm^{-1}.[40] The i.r. and Raman spectra of $VO(O_2)_2F^{2-}$ contain νV=O at 950 cm^{-1}. νV-O and νV-F are strongly mixed, and give bands 480 - 650 cm^{-1}.[41]

$VOF(SO_4)^-$ has νV=O 975 cm^{-1}, νV-O 515 - 540 cm^{-1} and νV-F 480 - 486 cm^{-1}.[42] Skeletal mode assignments were also made for Cp^*VOX_2, where X = F, Cl or Br;[43] $[V_3O_3F_{12}]^{3-}$, $[V_2O_2F_9]^{3-}$ and $[V_2O_2F_8(H_2O)]^{2-}$.[44] $[V_2O(bipy)_4Cl_2]^{2+}$ shows pronounced resonance Raman enhancement of V_2O modes.[45] Other assignments of νV=O modes were made for $[Cl_3V(=O)$-O-$V(=O)Cl_3]^{2-}$,[46] $Cp^*VO(p\text{-}OC_6H_4CH_3)_2$,[47] vanadyl complexes with bipyridyl and a variety of substituted benzoic acid derivatives,[48] $[VO(O_2)(citrato)]^-$ and $[VO(O_2)(citrato)]_2^{2-}$,[49] [VO(salicylhydroximate)(MeOH)]$_3$,[50] $K_4(VO)_3(SO_4)_5$,[51] and $[VO(SO_4)_2(H_2O)_3]^{2-}$.[52]

$V(cat)(acac)_2$, cat = catecholato, has νV-O(of acac) at 462 cm^{-1}, compared to 493 cm^{-1} for $V(acac)_3^+$.[53] Characteristic i.r. and Raman bands were identified for $VW_5O_{19}^{3-}$.[54] The i.r. and Raman spectra of $(enH_2)^{2+}(VO_3)_2^{2-}$ contain ν_sVO_2 934, 891 cm^{-1} (Raman), 938, 910 cm^{-1} (i.r.); $\nu_{as}VO_2$ 834 cm^{-1} (Raman), 820 cm^{-1} (i.r.); $\nu_{as}VOV$ 639 cm^{-1} (Raman), 662 cm^{-1} (i.r.); and ν_sVOV 520 cm^{-1} (Raman), 521 cm^{-1} (i.r.).[55]

Raman spectra were used to identify specific oxovanadate species present when VO_4^{3-} is incorporated in layered lithium aluminate materials.[56] νVO modes were assigned for $LnVO_4$, where Ln = Er, Ho or Eu.[57] V-O stretching mode assignments were made for a large number of vanadate systems: $Li_2O-B_2O_3-V_2O_5$;[58] $V_2O_5-SnO_2$;[59] $Ba_2MV_3O_{11}$, where M = In, Sc, Y, La or 4f elements;[60] $V_5O_{13}^-$;[61] $(NH_4)_2V_6O_{16}$;[62] $PV_{14}O_{42}^{9-}$;[63] $[As_6^{III}V_{15}^{IV}O_{42}(H_2O)]^{6-}$;[64] and $[V_{19}O_{41}(OH)_9]^{8-}$.[65]

$Cp_2M(H)(\eta^2-N_2CPh)$ complexes have νMH at 1780 cm^{-1} (M = Nb) or 1858 cm^{-1} (Ta).[66] Skeletal mode assignments in the dichlorocyclopropenone complexes $NbCl_5(OC_3Cl_2)$ and $[NbOCl_3(OC_3Cl_2)]_4$ include $\nu Nb-OC_3Cl_2$ at 242, 284 cm^{-1} respectively. The latter complex also has νNbONb at 860 cm^{-1}.[67] $Cp_2Nb(O_2)Me$ has ν_s and ν_{as} for the Nb-(O_2) unit at 525, 545 cm^{-1} respectively.[68] $(NH_4)NbOBr_4$ contains polymeric anions, with $\nu_{as}NbONb$ at 793 cm^{-1}, $\nu_{as}NbBr_2$ 293 cm^{-1}, δNbONb 228 cm^{-1} and ν_sNbBr_4 210 cm^{-1}. The NbONb units appear to be linear.[69] Assignments were proposed for νE-O modes in $M_2^{II}LnEO_6$, where M^{II} = Ca, Sr or Ba, E = Nb or Ta.[70] Amorphous Nb_2O_5 shows no Raman band near 850 cm^{-1} corresponding to $\nu_{as}NbONb$, i.e. the NbONb units are linear.[71]

I.r. and Raman spectra were reported and partly assigned for glasses $xNb_2O_5.(1-x)(BaO.2B_2O_3)$, $0.5B_2O_3.(0.5-x)BaO.xNb_2O_5$;[72] and for samples in the system $SrO-La_2O_3-Nb_2O_5$.[73]

The i.r. and Raman spectra of $[Nb_2(S_2)_2(NCS)_8]^{4-}$ and $[Nb_2(SSe)_2(NCS)_8]^{4-}$ show that the dinuclear forms shown are correct.[74] The i.r. and Raman spectra for $[M(SCH=CHS)_3]^-$, where M = Nb or Ta, were compared to those for $[M(SCH_2CH_2S)_3]^-$, showing that the degree of twist of the MS_6 polyhedron varies in a complicated fashion, depending on crystal packing forces and electronic factors.[75]

Several vibrational assignments were made for $[Nb_6F_{12}]F_3$, based on i.r. and Raman data. These are listed in table 2.[76]

Table 2 Vibrational assignments for $[Nb_6F_{12}]F_3$ (/cm^{-1})

t_{1u}	νNbF	543, 515, 454
	νNbNb	287
	δFNbF	200, 180
a_{1g}	νNbNb	337
e_g	νNbNb	267
t_{2g}	νNbNb	215

The i.r. spectra of $Cp_2TaH_3.ZnCl_2$.THF and related complexes show one νTaH band near 1810 cm^{-1} and two in the range 1580 - 1730 cm^{-1}. These results are consistent with the structure (1).[77] Bands associated with Ta-H bonds were seen at 1739 and 1118 cm^{-1}, shifting to 1237 and 795 cm^{-1} on deuteriation, for (2), where R = ^{i}Pr.[78]

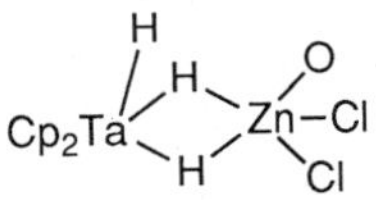

(1)

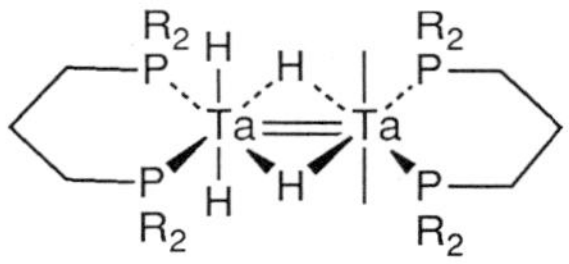

(2)

$[Cp^*TaN(Cl)]_4$ has νTaN at 960 cm^{-1} in the i.r. spectrum.[79] The Raman spectrum of single-crystal $LiTaO_3$ gave assignments for several normal modes in both ferro- and paraelectric phases. These were consistent with C_{3v} and D_{3d} site symmetries respectively.[80]

4. Chromium, Molybdenum and Tungsten.

Previous reference was made to vibrational data for $VW_5O_{19}^{3-}$.[54]

$Cr(CO)_5(H_2)$ is formed by the irradiation of $Cr(CO)_5(NH_3)$ in an Ar/H_2 matrix; it has $\nu_{as}Cr-(H_2)$ at 1380 cm^{-1}, and $\nu_sCr-(H_2)$ at 878 and 868 cm^{-1}. The symmetric mode is seen to shift as expected for the HD and D_2 analogues.[81] νM-N_4 modes were assigned from the low-wavenumber i.r. spectra of phthalocyanines of Cr, Fe, Co, Zn and Cd. The wavenumbers and intensities were related to deviations from D_{4h} symmetry of the MN_4 core unit.[82]

The Raman spectrum of CrO_2F_2 in SbF_3-HF shows the formation of a cationic species, probably CrO_2F^+, at high acidities.[83] Bands due to νCrO, νCrOFe and νFeO were assigned from the Raman spectrum of $(Fe,Cr)_3O_4$.[84] The time-dependent Raman spectra (νCrO) of $(NH_4)_2CrO_4$ were used to monitor the kinetics of decomposition at temperatures of 343 - 363K.[85] νCrO bands were in the range 860 - 890 cm^{-1} for $Na_2CrP_2O_7F$ and $Na_2(CrO)_2P_2O_7$.[86] The linear, oxo-bridged dimers $[L(tmpa)CrOCr(tmpa)L]^{2+}$, where tmpa = tris-(2-pyridyl-methyl)amine; L = NCS^-, NCO^-, N_3^-, CN^- or Cl^-, have ν_{as}CrOCr near 850 cm^{-1}.[87] Far-i.r. and Raman spectra were reported for $CaCrSe_3$, but no specific assignments were made.[88]

The i.r. and Raman spectra of CrF_2, CrF_3, CoF_2 and NiF_2 in argon or neon matrices showed that the difluorides had D_{h} symmetry, and CrF_3 D_{3h}.[89] Skeletal modes were observed and assigned, and normal coordinate analyses carried out for $[Cr(NH_3)_5X]^{2+}$, where X = F, Cl, Br or I. Most of the modes are not subject to extensive mixing. The νCrX assignments are : 540 cm^{-1} (F); 306 cm^{-1} (Cl); 206 cm^{-1} (Br) and 195 cm^{-1} (I).[90] The i.r. spectra of $(NH_4)_2[M^{III}F_5(H_2O)]$, where M = Cr or Fe, show that isolated octahedral anions are present.[91]

Values for some νCrCl modes were estimated from the vibronic spectrum of $(AsPh_4)(CrOCl_4)$.[92]

Table 3 Some vibrational assignments for $M_2Cl_4(PMe_3)_4$ (/cm^{-1})

M_2	νMCl	δMCl or νMP	νMM
Mo_2	329, 285	230, 181	354
MoW	315, 298	181	322
W_2	312, 291	184	-

νMoMo in $Mo_2(OAc)_2(Aza)_2$, where Aza = 7-azaindolyl, is at 377 cm^{-1}, compared to 406 cm^{-1} in $Mo_2(OAc)_4$, showing the weakening of the MoMo bond in the former.[93] Skeletal mode assignments for $M_2Cl_4(PMe_3)_4$, where M_2 = Mo_2, MoW or W_2, are given in Table

3.[94] I.r. bands were seen for the new polymeric molybdenum bromide, $MoBr_{2.5}$, at 275, 248 and 208 cm^{-1}. The last is probably νMoMo, the others νMoBr.[95]
νMoH modes lay in the range 1825 - 1852 cm^{-1} in $[Mo(Cp)_2H(R)]^+$, where R = NH_3, en, PMe_2Ph or PEt_2Ph.[96]
$Al_4[Mo(CN)_8]_3.24H_2O$ has νMo-CN at 505 and 580 cm^{-1}, with δMo-C-N 420, 460 cm^{-1}.[97]

νMoN, and other skeletal modes have been reported for $[MoNCl_3(L)]_2$, where L = 2,3,3-trichloroacrylonitrile;[98] $[Na(15\text{-crown-}5)]_2[MoNF_4]_2$;[99] $[Cl_3PNPCl_3]^+[MoNCl_4]^-$ (νMoN at the low value of 995 cm^{-1} due to interaction with the cation);[100] Mo_2NBr_5 (unsymmetrical bridge unit $Mo{\equiv}N^+{-}Mo^-$) and polymeric W_2NBr_7;[101] $\{MoNCl_3[N(SiMe_3)_2]\}^-$;[102] $[MoF_5(NCl)]^-$ (consistent with local C_{4v} symmetry for $MoNF_5$);[103] $[(\mu\text{-NSN})(MoF_5)_2]^{2-}$;[104] $[MoCl_4(NCl)]_2$ (consistent with structure (3));[105] and $[Cl_4Mo{=}N{=}SeCl]_2$.[106]

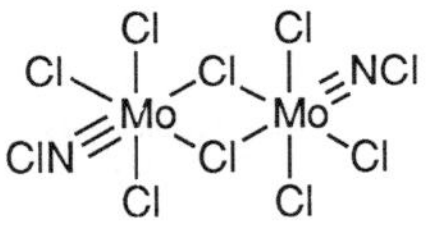

(3)

νM-N and νM-Cl modes were assigned for cis-$MCl_4(CH_3CN)_2$, for M = Mo or W, fac- and mer-$MoCl_3(CH_3CN)_3$. Thus, for $MoCl_4(CH_3CN)_2$, νMoN bands are at 428 and 400 cm^{-1}, νMoCl at 335 and 315 cm^{-1}.[107]

$Na_2[(\mu\text{-}O)_2Mo_2O_2L].mH_2O$, where H_4L = EDTA and related complexones, have νMo=O 900 - 965 cm^{-1} (two bands in each complex), with ν_sMoO_2Mo 450 - 500 cm^{-1} and $\nu_{as}MoO_2Mo$ 740 - 765 cm^{-1}.[108] $CpM(=O)(\eta^2\text{-}O_2)Cl$ gave the following M-O stretching wavenumbers: νM=O 934 cm^{-1} (M = Mo), 960, 949 cm^{-1} (W); νM-O 590, 571 cm^{-1} (Mo), 605, 571 cm^{-1} (W).[109] A number of vibrational assignments for $K_2[M_2O_3(O_2)_4(H_2O)_2].2H_2O$, where M = Mo or W, are given in Table 4.[110]

Table 4 Vibrational assignments for $K_2[M_2O_3(O_2)_4(H_2O)_2].2H_2O$ ($/cm^{-1}$)

M	νM=O	$\nu_{as}M_2O$	$\nu_sM\text{-}(O_2)$	$\nu_{as}M\text{-}(O_2)$	ν_sM_2O	$\nu M\text{-}OH_2$
Mo	967	715	575	531	454	314
W	960	765	615	560	450	317

Other assignments of νM=O modes, where M = Mo or W were reported for the following: $[O{=}M(Te_4)_2]^{2-}$;[111] $[MoO(SR)_4]^{n-}$, where n = 1, R = Ph or substituted phenyl groups; n = 2, R = C_6F_5 (showing higher wavenumbers for Mo(V) than for Mo(IV) complexes);[112] $[Mo_2OS_8]^{2-}$, $[Mo_2OS_9]^{2-}$;[113] MoOL, where H_2L = salicylaldehyde thiosemicarbazone or salicylaldehyde 4-phenylthiosemicarbazone (also νMoS near 375 cm^{-1});[114] $MoO_2(ClO_4)_3^-$, $MoO_2(ClO_4)_2$;[115] MoO_2(acac)(HL), where H_2L =

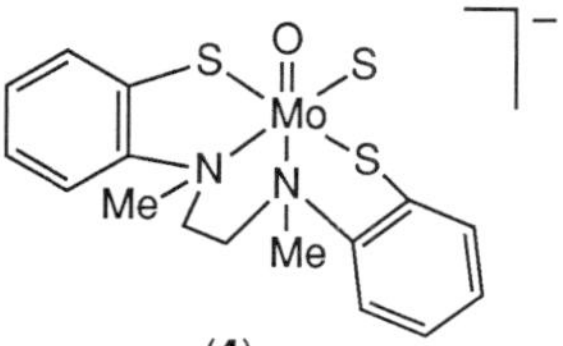

(4)

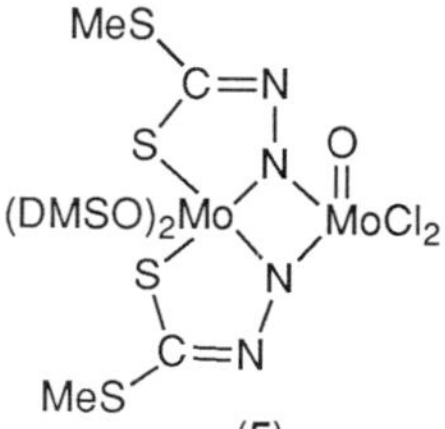

(5)

7,16-dihydro-6,8,15,17-tetramethyldibenzo[b,i][1,4,8,11]-tetra-azacyclotetradecane;[116] (4);[117] (5);[118] $[MoO(dmit)_2]^-$, where dmit = 4,5,-dimercapto-1,3-dithiol-2-dithionate;[119] $[MoO_2(5\text{-}SO_3ssp)]^-$, where ssp = 2-(salicylideneamino)-benzenethiolato(2-);[120] cis-$[MO_2L_2]^{2-}$, where H_2L = pyrogallol;[121] and $MoO_2(SCZ)_2$, where HSCZ = semicarbazones of furfuraldehyde, pyridine-2-aldehyde and related systems.[122]

Complexes of the naphthalene-2,3-diolato dianion (= ND), i.e. cis- $M_2O_5(ND)_2^{2-}$, where M = Mo or W, gave the skeletal vibrational assignments shown in Table 5.[123]

Table 5 Some vibrational assignments for cis-$M_2O_5(ND)_2^{2-}$. (/cm^{-1})

M	ν_sM_2O	ν_sMO_2	$\nu_{as}MO_2$	νM-O
Mo	496	943	886	624
W	499	902	867	630

The i.r. and Raman spectra of $LiAl(MoO_4)_2$, with $^{92}Mo/^{100}Mo$ isotopic replacement, show that ν_{as} and ν_s of the MoOMo bridge are at 947 and 566 cm^{-1} respectively.[124] I.r. spectra were used to characterise the double molybdates formed in the system UO_2MoO_4-K_2MoO_4.[125] R_2SnMoO_4, where R = Me, Et, Bu or octyl, have νMoO in the range 812 - 871 cm^{-1}, compared to 895 cm^{-1} in free MoO_4^{2-}.[126] The t_{1u} modes of octahedral MoO_6 were seen at 600 cm^{-1} (ν_3) and 365 cm^{-1} (ν_4) in the i.r. spectrum of $(BaLa)(MgMo)O_6$.[127]

Salts containing $[L_2Mo_2O_4]^{2+}$, where L = 1-thia-4,7-diazacyclononane, have νMoO bands as expected for a syn-Mo_2O_4 unit.[128] The i.r. and Raman spectra of polycrystalline $Na_2Mo_2O_7$ were assigned using $^{92}Mo/^{100}Mo$ isotopic shifts. It was possible to identify modes of MoO_4, MoO_6 and Mo-O-Mo groups.[129] The i.r. spectrum of the β-phase of $Bi_2Mo_2O_9$ shows that it is a distinct phase.[130] νMo=O (910 - 960 cm^{-1}) and νMo-S (290 - 360 cm^{-1}) modes were assigned for $Mo_2O_2S_9^{2-}$, $Mo_4O_4S_{18}^{2-}$ and related species.[131] The i.r. and Raman spectra of $Mo_2O_2F_9^{3-}$, $Mo_3O_6F_9^{3-}$ and $M_2O_4F_6(H_2O)^{2-}$, where M = Mo or W, contain the expected νM=O bands, together with νMF_t 560 - 600 cm^{-1}, νMF_{br} 420 - 460 cm^{-1}.[132]

The vibrational spectrum of $K_2Mo_3O_{10}$ was assigned with the help of $^{92}Mo/^{100}Mo$ isotopic shifts.[133] Characteristic bands of $Mo_2O_5^{2+}$ units were seen in the i.r. spectrum of $[Mo_4O_{10}(O_4C_6H_2)_2]^{2-}$, where $O_4C_6H_2^{2-}$ = 2,5-dihydroxybenzoquinonate.[134-5] $[Mo_6O_{18}(NNMePh)]^{2-}$ has $\nu Mo{=}O_t$ at 943 cm^{-1} and $\nu Mo{-}O_{br}$ at 780 cm^{-1}.[136] νMo=O and νMoOMo modes were also assigned for $[(HP)_5Mo_6O_{33}]^{10-}$;[137] $[Mo_8O_{16}(OMe)_8(C_2O_4)]^{2-}$;[138] and $S_2Mo_{18}O_{62}^{4-}$ [139]

I.r. and Raman spectra for α-MoS_3 and Li_xMoS_3, where $0 < x < 4.5$, show that the introduction of Li leads to a strengthening of the Mo-S bonds.[140] In situ Raman spectra of MO_3/γ-Al_2O_3, where M = Mo or W, under sulphiding conditions showed that MS_3-like phases were produced first, together with oxysulphides, while complete sulphiding gave MS_2 crystallites.[141] $[Mo_2Cu_5S_8(S_2CNMe_2)_3]^{2-}$ has $\nu Mo{=}S_t$ at 513 cm^{-1}. and $\nu Mo{-}S_{br}$ at 455, 444 and 422 cm^{-1}.[142]

νMoF bands in the i.r. spectrum of $[MoF_2(CO)_3]_n$ at 660, 620, 600 and 560 cm^{-1} were assigned to bridging fluorine atoms.[143] $[MoF_4Cl_2(N_3S_2)]^-$ has νMoF as strong i.r. bands at 570 and 503 cm^{-1}.[144] νMCl bands were assigned for $NH_4[MCl_4py]$, i.e. 315, 295 cm^{-1} (Mo), 310, 270 cm^{-1}.[145] Terminal and bridging stretching modes were in $[Mo_2(S_2)Cl_8]^{2-}$ and $[Mo_6Cl_{14}]^{2-}$.[146] $[Mo_2(O_2CPh)_4Br_2]^{2-}$ has νMoBr at the very low value of 85 cm^{-1}, corresponding to a very weak bond.[147]

$HW_2(\mu\text{-}CR)(\mu\text{-}C_4Me_4)(O^iPr)_4$, where R = Ph, $SiMe_3$ or ^tBu, all have $\nu W\text{-}H_t$ at 1875 cm^{-1}.[148] $W_2HCl_5py_4$ and $W_2HCl_5(Etpy)_4$ have $\nu_{as}WHW$ at 1510 and 1545 cm^{-1} respectively. The former complex has νWCl bands at 320, 309, 299, 275 and 267 cm^{-1}.[149] The dimeric complex $[WCl_4(HC{\equiv}CH)]_2$ gave the following skeletal mode assignments: ν_sWC_2 835 cm^{-1}, $\nu_{as}WC_2$ 580 cm^{-1}, νWCl_t 360, 330 cm^{-1} and νWCl_2W 277, 230 cm^{-1}.[150]

$[WNCl_3(NCPh)]_4$ has an i.r. band due to the WNW unit at 1082 cm^{-1}, and terminal chlorine atoms only (νWCl 368, 348 cm^{-1}).[151] νWN and/or νWNW modes were reported for $[W_2(^{15}N)Cl_9]_2$, $[W(^{15}N)Cl_4]^-$, $[W(^{15}N)Cl_3]_n$ and $[W_2(^{15}N)Cl_{10}]^-$.[152] (6) has νWNC at 1238 cm^{-1}, νWCl 386, 343 cm^{-1}, and $\nu MWCl$(trans to N) 285 cm^{-1}.[153]

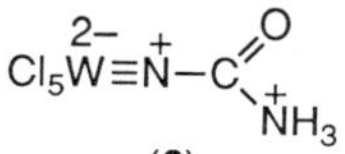

(6)

νWN modes were also assigned in $Cl_5W{=}N{=}CMe_2$;[154] $[Na(15\text{-}crown\text{-}5)][WF_5(NCl)]$;[155] $[WCl_4(NCl)]_2$;[156] and $\{[WCl_5(NCl)]_2Cl\}^{3-}$.[157]

νWO, together with other skeletal modes were assigned for $WO(O_2)_2(L\text{-}L)$, where L-L = phen or bipy;[158] $WOF_4.n(py)$, where n = 1 or 2;[159] and $[WOCl_4(OPPh_3)]^-$.[160] The crystallinity of amorphous WO_3 films was monitored by Raman bands due to the W-O stretches.[161] A vibrational analysis has been reported for $WO_3.H_2O$.[162] I.r bands were identified due to terminal (W=O) and bridging (WOW) units in $Na_4W_2O_3L_4$, where L = 2,3-dimercaptopropane sulphonate, together with νWS at 218 and 302 cm^{-1}.[163]

I.r. and Raman spectra have been obtained for $W_3OS_8^{2-}$ and $W_3S_9^{2-}$ and some assignments were given, e.g. $\nu W{=}O$ 975 cm^{-1}, $\nu W{=}S$ 527 cm^{-1}. WS_4^{2-} units give bands in the range 490 - 510 cm^{-1}.[164] The i.r. and Raman spectra of a wide range of $PW_{11}O_{39}^{2-}$ salts contained $\nu W{=}O$ bands whose positions were dependent on the strength of the interaction with the cation.[165] Characteristic stretching wavenumbers were established for $[AlO_4W_{12}O_{36}]^{5-}$;[166] $Na_{20}P_6W_{18}O_{79}.nH_2O$ and $Na_{12}P_4W_{14}O_{58}.nH_2O$.[167]

Raman spectra gave evidence for WS_3 formation in $WO_3/\gamma\text{-}Al_2O_3$

and NiO-WO_3/γ-Al_2O_3 catalysts after sulphurization.[168] The new complex $(Me_2NCS_2)_2WS(S_2)$ had νW=S at 498 cm^{-1} and νWS_2 442, 370 cm^{-1}.[169] $[(ClCu)_5Cl_2WS_4]^{4-}$ has i.r. bands due to bridged WS_4 groups at 450 and 420 cm^{-1}.[170] Skeletal mode assignments were made for $[(S_2)WO(\mu-S)_2WO-(\mu-S)_2W(S_2)]^{2-}$ and related species.[171]

Only a terminal νMCl was seen for $(Cl_3PO)WCl_4(PhC\equiv CPh)$.[172] Matrix- isolated tungsten chlorides and bromides have been prepared and some of their vibrational assignments are listed in Table 6.[173]

Table 6 Tungsten-halogen stretching assignments for matrix-isolated chlorides and bromides ($/cm^{-1}$)

WCl_6	(t_{1u})	367
WBr_6	(t_{1u})	305
WCl_5	(a_2'')	410
	(e')	365
WBr_5	(a_2'')	280
	(e')	245

5. Manganese, Technetium and Rhenium.

Skeletal mode assignments were made from the Raman spectra of polycrystalline $Mn(N_2H_4)_2Cl_2$, $Zn(N_2H_4)_2X_2$, where X = Cl or Br. The νMnN stretches were at 343 cm^{-1} (e_u) and 237 cm^{-1} (b_{1g}).[174] The i.r. and resonance Raman spectra of hydroxophthalocyaninato manganese(III) contain bands due to νMnN(α) at 298, 271 and 193 cm^{-1}.[175]

The F.T. Raman spectrum of crystalline $KMnO_4$ shows ν_1 at 841.5 cm^{-1}, ν_2 351 cm^{-1}, ν_3 903, 912 cm^{-1} and ν_4 396 cm^{-1}. The spectra were much better than any previously reported.[176] $Mn(OCH=CH_2)_4$ has νMn-O as a broad i.r. band 460 - 550 cm^{-1}.[177] $[LMnO]_2^{3+}$, where L = 1,4,8,11-tetra-azacyclotetradecane or 1,4,7,10-tetra-azacyclodecane, have νMn-O-Mn bridging modes near 685 cm^{-1}.[178] There is i.r. evidence for reversible coordination of O_2 (as a side-on peroxo group) to a solid $Mn(PEt_3)Br_2$ film. The adduct has ν_s and $\nu_{as}Mn-(O_2)$ at 498 and 489 cm^{-1} respectively.[179]

νM-O modes were assigned for ML_2 complexes, where L =

c-methoxybenzoate, M = Mn, Co, Ni, Zn or Cd, and also FeL_3.[180] The i.r. and Raman spectra of Mg_2SiO_4-Mn_2SiO_4 samples show largely octahedral Mn^{2+} coordination, but with some four- and five-coordinate units in Mn-rich samples.[181] The resonance Raman spectra of manganese X-protoporphyrin IX dimethyl ester complexes, where X = OAc, F, Cl, Br or I, show that changing the X ligands gives shifts only in modes involving the manganese atom.[182]

The Raman spectra of single crystals of MCr_2S_4, where M = Mn, Fe, Co, Zn or Cd, show resonance enhancements and frequency shifts of bands due to the tetrahedrally-coordinated metal ions.[183]

The Raman spectrum of $Na_3[(TCTA)Tc(\mu\text{-}O)_2Tc(TCTA)]$, where TCTA = 1,4,7-triazacyclononane-N,N',N"-triacetate, shows νTcTc at 354 cm^{-1} and νTcO_2Tc at 734 cm^{-1}.[184] The ground state of TcH is calculated to be of $^7\Sigma^+$ symmetry, with ω_e at 1633 cm^{-1}. Vibrational wavenumbers were also calculated for a large number of electronically excited states.[185] νTc≡N has been assigned in $[TcNX_3(CEP)]^-$, where CEP = $P(CH_2CH_2CN)_3$, X = Cl (1052 cm^{-1}) or Br (1050 cm^{-1}).[186]

The La^{3+} salt of $[Tc(OH)Cl_5]^{2-}$ has νTc-OH at 435, 460 cm^{-1}, νTc-Cl 310, 320 cm^{-1}.[187] $[Cp^*Tc_2O_3]_n$ has νTcO bands at 909 cm^{-1} (symmetric) and 880 cm^{-1} (antisymmetric).[188] The νTcO band at 962 cm^{-1} for $[Tc_2O_2(H_2ema)_4]^{2-}$, where H_4ema = N,N'-ethylenebis(2-mercaptoacetamide), coresponds to a Tc=O unit in a five-coordinate square-pyramidal geometry.[189]

νReH wavenumbers were assigned for $ReHX(NO)(PPh_3)_3$, where X = MeOD, OMe, F, Cl, Br, N_3, NCO etc.;[190] $TpReH_4(PPh_3)$, where Tp = tris(pyrazolyl)borate;[191] $Re_3(\mu\text{-}H)_4(CO)_{10}^-$ and related complexes.[192] νReC modes were assigned for $Cp^*Re(CO)_2X_2$, where X = Cl, Br or I (cis) or Br, I (trans). νReCp* was near 290 cm^{-1} and νReCl at 315 cm^{-1}.[193]

Both (mesityl)$Re(=O)_3$ [194] and $CpRe(=O)_3$ [195] have νRe=O at lower wavenumbers than for $MeRe(=O)_3$. A high-pressure Raman study of $MReO_4$, where M = K, Rb or Cs, up to 25 GPa gave evidence for several phase transitions in each case.[196] The i.r. spectra of $[Ln(ReO_4)_2(HMPA)_n]ReO_4$ show that for Ln = La - Nd, n= 2, the ReO_4 is bidentate, but that for Ln = Sm - Lu and Y, n = 4, the ReO_4 is unidentate.[197]

Other asignments of νReO and related modes were given for $Re(O)(OH)(MeC{\equiv}CMe)_2$;[198] $ReOMe_2Cl(PMe_2R)_2$, where R = Me or Ph, $ReOMe_2Cl(dmpe)$;[199] (7), where L = 1,4,7-triazacyclononane;[200] $Cp^*(O{=})Re(\mu\text{-}O)_2Re({=}O)Cp^*$;[201] $ReO_2(CH_2CMe_3)_2(SR)$, where R = Me or Ph;[202] $Cp^*Re(O)Me_2$ and $Cp^*Re({\equiv}CCMe_3)Cl_2$.[203]

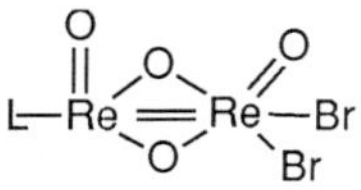

(7)

Quite detailed assignments have been made from the i.r. and Raman spectra of $Re_2X_{10}^{2-}$, where X = Cl or Br. The assignments for X = Cl are given in Table 7.[204]

Table 7 Vibrational assignments for $Re_2Cl_{10}^{2-}$ (/cm^{-1})

		i.r.	Raman
$\nu_sReCl_2(t)$	a_g	-	367
$\nu_{as}ReCl_2(t)$	b_{3u}	360	-
νCl_tReCl_t	b_{2g}	-	339
$\nu_{as}Cl_tReCl_t$	b_{1u}	336	-
$\nu ReCl_2(t)$	b_{2u}	332	-
$\nu_sCl_tReCl_t$	a_g	-	321
$\nu_{as}Cl_tReCl_t$	b_{3u}	318	-
$\nu_sRe_2Cl_2(b)$	a_g	-	278
$\nu_{as}Re_2Cl_2(b)$	b_{3u}	263	-
$\delta ReCl_2(t)$	a_g	-	178
$\delta ReCl_2(t)$	b_{3u}	170	-
δCl_tReCl_t	b_{1u}	153	-
$\pi ReCl_2(t)$	b_{2g}	-	135

The first rhenocene(VII) dichloride species has been prepared, $[Cp_2ReCl_2]^{3+}(SbF_6)_3^{3-}$, for which νReCl bands are at 340 and 296 cm^{-1}, compared to 314 and 295 cm^{-1} for the Re(V) analogue. The shifts are consistent with the higher oxidation state.[205] $[ReCl_2(CN^iPr)_2(PMePh_2)_2]^+$ has a single νReCl band at 295 cm^{-1} and a single νCN band at 2182 cm^{-1}, consistent with an all-trans geometry.[206] $ReBr_5(SCN)^{2-}$, $ReBr_5(NCS)^{2-}$, cis/trans-$ReBr_4(NCS)(SCN)^{2-}$ and related complexes all have

νReBr 187 - 226 cm^{-1}.[207]

6. Iron, Ruthenium and Osmium.

Earlier reference has been made to vibrational studies on iron(II) phthalocyanines;[82] $(Fe,Cr)_3O_4$;[84] $(NH_4)_2[Fe^{III}F_5(H_2O)]$;[91] FeL_3, where L = c-methoxybenzoate;[180] $FeCr_2S_4$;[191] trans-$[OsO_2L_2]^{2-}$ and trans-$[OsO_2py_2L]$, where H_2L = pyrogallol.[121]

The i.r. and Raman spectra of $Fe_2(CN)_{10}{}^{4-}$ (νFeFe and νFeC) are consistent with two $Fe(CN)_5$ units joined by an Fe-Fe bond in either D_{4h} or D_{4d} symmetry.[208] The reaction of Fe atoms and CH_2N_2 in argon matrices forms $FeCH_2$, for which $\nu Fe{=}CH_2$ is at 623.9 cm^{-1} (with expected shifts on isotopic substitution). U.v. photolysis forms HFeCH, with $\nu Fe{\equiv}C$ at 674.2 cm^{-1} and νFeH at 1681.6 cm^{-1}.[209] The μ-carbido iron phthalocyanine dimer $[(PcFe)_2C]$ has νFeCFe at 990 cm^{-1}. This decreases to about 940 cm^{-1} on the formation of adducts $[\{(L)PcFe\}_2C]$, where L = acetone, pyridine, piperidine etc.[210]

νFe-CO and νFe-His modes were assigned from resonance Raman spectra of asymmetrically Co-substituted Fe-Co hybrid haemoglobins and carbonmonoxy forms.[211] Studies on νFe-CO modes were also reported for a variety of bio-inorganic species and model compounds in their CO-adduct forms.[212-5]

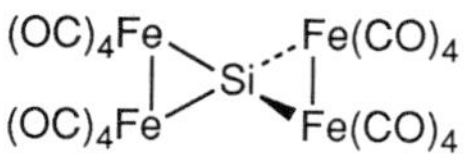

(8)

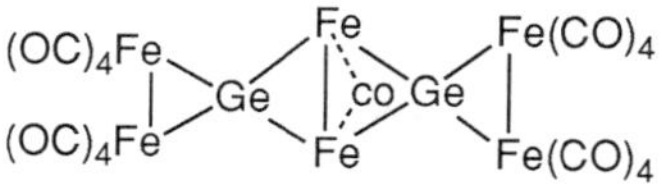

(9)

For (8), $\nu_s SiFe_4$ gives a Raman band at 196 cm^{-1}, while (9) has bands due to stretches of the heavy atom skeleton at 288, 232, 202 and 182 cm^{-1} (Raman), 278, 246 cm^{-1} (i.r.).[216]

The resonance Raman spectra of nitridoiron(V) porphyrin intermediates produced by laser photolysis of the corresponding azido complexes, e.g. NFe(OEP), contained $\nu Fe{\equiv}N$ bands near 870 cm^{-1}. The μ-nitrido species $[(OEP)Fe]_2N$ has ν_sFeNFe at 433 cm^{-1}.[217] $Fe(OEP)(OTeF_5)$ gives νFeN at 281 and 257 cm^{-1}, with νFeO at 415 cm^{-1}.[218] νFeN for haemoglobin derivatives with the histidine ligand replaced by tyrosine have been assigned.[219]

$Fe(O_2)$ species with C_s or C_{2v} (cyclic or bent linear) symmetries are formed by co-condensation of Fe and O_2 in argon or krypton matrices.[220] I.r. spectra of $CaO-P_2O_5-Fe_2O_3$ glasses showed a change in coordination from FeO_6 to FeO_4 units on increasing the concentration of iron.[221] $\nu M-O_2$ modes in $M(TPP)(piperidine)(O_2)$ were assigned from resonance Raman data, i.e. M = Fe(II) (575 cm^{-1}) or Co(II) (509 cm^{-1}).[222] Similar results for $Fe_2O(OAc)_2\{[OP(OEt)_2]_3CoCp_2\}$ shows that ν_sFeOFe is at 510 cm^{-1},[223] while for (10) it is at 518 cm^{-1}.[224]

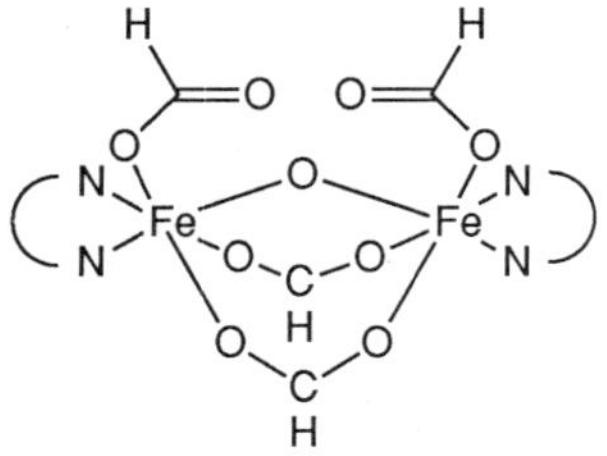

(10)

Resonance Raman enhancement was found for ν_sFeOFe (380 - 540 cm^{-1}) in a range of oxo-bridged dinuclear iron centres in proteins and model compounds.[225] Using a low-temperature bulk electrolysis cell for in situ resonance Raman spectroelectrochemistry it was possible to observe $\nu Fe=O$ of electrogenerated ferryl tetramesitylporphyrin at 841 cm^{-1}.[226] Other assignments of $\nu Fe=O$ in related complexes were reported: ferrylporphyrin cation radical;[227] and horseradish peroxidase compound I.[228]

Time-resolved resonance Raman spectra of O_2/cytochrome oxidase shows that $\nu Fe-O$ is at 589 cm^{-1} for an early intermediate in the reaction, i.e. consistent with the presence of an $a_3^{2+}-O_2$ complex.[229] The transient intermediate of cytochrome c oxidase generated by laser photolysis in the presence of O_2 gives a resonance Raman band at 477 cm^{-1}, due to $\nu FeOH$.[230] Resonance Raman data for the riboncleoside diphosphate reductase from E. coli includes a band at 493 cm^{-1} due to ν_sFeOFe of the B2 subunit.[231]

Skeletal mode assignments have been made for a number of iron-sulphur cluster ions, e.g. $Fe_4S_4X_4^{2-}$, where X = Cl or Br.[232] The i.r. and resonance Raman spectra of $[Fe_2S_2(MoS_4)_2]^{4-}$ show that there is strong coupling between the Fe_2S_2 and MoS_4 fragments - consistent with very delocalised bonding throughout the complex.[233] Resonance Raman data were described for several iron-sulphur proteins and model compounds.[234-7]

The dehydration of $Cs_2[FeCl_5(H_2O)]$ to an equimolar mixture of $Cs_3[FeCl_6]$ and $Cs_3[Fe_2Cl_9]$ was followed by Raman spectroscopy.[238] The Raman spectra of acidified aqueous $FeCl_3$ solutions were measured between 25 and 300°C. For Fe^{3+} concentrations of 0.75 - 1.0 mol kg^{-1}, with Cl^-/Fe^{3+} ratios of 3 - 9.5, the dominant species at 25°C is trans-$[Fe(H_2O)_4Cl_2]^+$.[239] Some new Raman data enable a normal coordinate analysis to be performed on Fe_2Cl_6 and $AlFeCl_6$.[240] The cyclic hydrazide complex $\{Fe_2Cl_4[(NSiMe_3)(CPh)_2N_2\}$ has νFeCl at 345 and 385 cm^{-1}, corresponding to an Fe(III) complex with terminal chlorine ligands.[241]

The resonance Raman and i.r. spectra of $\{[Ru(OEP)]_2\}^{n+}$, where OEP = octaethylporphyrin dianion, show νRuRu at the following values: 285 cm^{-1} (n = 0), 301 cm^{-1} (n = 1) and 310 cm^{-1} (n = 2). These are consistent with removal of electron density from the π^* M-M antibonding orbitals on oxidation.[242]

Trans-$Ru(H)Cl(dcpe)_2$, dcpe = 1,2-bis(dicyclohexyl-phosphino)ethane, has νRuH at 1985 cm^{-1}.[243] The resonance Raman spectrum of $(NC)_5Ru\text{-}CN\text{-}Ru(NH_3)_5^-$ shows νRuC(term.), νRuC(br.) and δRuCN contributing to bands at 565 and 539 cm^{-1}. $\nu Ru\text{-}NH_3$ is at 492 cm^{-1}, and $\nu Ru\text{-}NH_3$(axial) at 470 cm^{-1}.[244]

The influence of the X ligand in $M(NO)(NH_3)_2X_3$, where M = Ru, X = Cl or Br; M = Os, X = Cl, on the νM-N(NO) and $\nu M\text{-}N(NH_3)$ wavenumbers was measured. In X-Ox-NO, the strength of the Os-NO bond decreases in the order X = OH > Cl > Br > I.[245] Some skeletal mode assignments were made for trans-$[Ru(NO)(NH_3)_4X]^{2+}$, where X = OH^-, Cl^- or ReO_4^-, and $[Ru(NO)(NH_3)_4(H_2O)]^{3+}$, e.g. νRu-NO 615 - 630 cm^{-1}, νRu-X 595 cm^{-1} (OH), 317 cm^{-1} (Cl) or 460 cm^{-1} (H_2O).[246]

Time-dependent scattering theory was used to analyse the resonance and pre-resonance Raman spectra of $Ru(NH_3)_4(bipy)^{2+}$. This shows significant changes in both metal-ligand and intraligand bond lengths on going to the charge-transfer excited state.[247] Resonance Raman spectra of $Ru(CO)_3(^iPr_2CH\text{-}DAB)$, where DAB = 1,4-diaza-1,3-butadiene, shows enhancement of ν_sRu-C (372 cm^{-1}), ν_2Ru-N (242 cm^{-1}) together with some Ru-CO and Ru-DAB deformation modes.[248] νRu-N modes were assigned for RuL_3, where L = substituted derivatives of 2,2'-dipyridine. When the complexes were bound to a TiO_2 surface there was evidence for νRuO also (ca. 400 cm^{-1}), showing the presence of Ti-O-Ru bridges.[249]

F.T.i.r. spectra of matrix-isolated ruthenium oxides identified RuO, RuO_2, RuO_3 and RuO_4. RuO_2 is bent, with a bond angle of 149 ± 2°. RuO_3 is trigonal planar. Assignments were confirmed by $^{16}O/^{18}O$ isotopic shifts. These were the first spectra reported for isolated RuO_2 and RuO_3.[250] RuO_4 isotopomer mixtures were prepared from natural-abundance ruthenium isotopes and mixtures of ^{16}O and ^{18}O.[251]

νRu-OH was assigned to a band at 597 cm^{-1} in $[Ru(OH)(NH_3)_4NO](ReCl_4)_2$.[252] Skeletal modes for the new complexes trans-$[RuO_2py_2]^{2+}$, trans-$MO_2py_2X_2$, where M = Ru or Os, X = Cl or Br, and other related species.[253] The Raman spectrum of $Ru_2OCl_{10}^{4-}$ in aqueous H_2SO_4 solutions shows that ν_sRu-O-Ru (originally at 254 cm^{-1}) increases on aquation of the complex, e.g. $Ru_2OCl_6(H_2O)_4$ 305 cm^{-1}.[254] νRu=O modes were assigned for the following: $[Ru_3O(OAc)_6\{(pyrazine)Ru(NH_3)_5\}_3]^{6+}$;[255] $[Ru^{VI}(L)(O)_2](ClO_4)_2$, where L = N,N,N',N'-tetramethyl-3,6-dimethyl-diazaoctane-1,8-diamine;[256] $Ru^{VI}(OEP)(O)_2$ and $Ru^{IV}(OEP)(O)(EtOH)$.[257] νRuS was assigned for $Ru^{II}L_6^{2+}$, $Ru^{III}L_5Br^{2+}$ and $RuBr_3L_2$, where L = thiourea. For the last, the data were consistent with the formulation $[BrL_2Ru(\mu-Br)_3RuL_2Br]Br$.[258]

$Os_2(CO)_8(\mu_2-\eta^2-C_2H_4)$ has νOsOs at 146 cm^{-1}, with νOs-CO at 560, 496 cm^{-1}, and δOs_2C_2 266 cm^{-1}.[259] νOsH is at 2095 cm^{-1} in $OsHCl(CO)(P^iPr_3)_2(O_2)$.[260] Skeletal ($\nu OsC_2$, νOsO_2 and νOsL_2) modes were assigned for $[OsO_2(CN)_2L_2]^{2-}$, where L = OMe, NCO, NCS and SeCN.[261]

νOsN and νOsX modes were assigned for $[Os(NO)X_5]^{n-}$, where n = 1, X = F; n = 2, X = F, Cl or Br.[262-3] νOsN modes are at 370 and 360 cm^{-1}, with νOsO 324 cm^{-1} for $[Os(NH_3)_5(OsO_2CF_3)]^{2+}$.[264] A series of isomeric complexes has been prepared, i.e. $[Os(NCSe)_n(SeCN)_{6-n}]^{3-}$, for n = 0 - 3; νOs-Se is in the range 183 - 216 cm^{-1}, with νOs-N 224 - 246 cm^{-1}.[265]

The i.r. and Raman spectra of crystalline $K_2[OsO_2Cl_4]$ gave an assignment of all except one of the normal modes. The data were used to carry out a normal coordinate analysis.[266] $OsO_2L_4^{2-}$, where L = NCO^-, NCS^- or $SeCN^-$, have ν_sOsO_2 886 - 903 cm^{-1} and $\nu_{as}OsO_2$ 845 - 852 cm^{-1}. All data are consistent with trans geometry.[267] $[Os(thio)_6]^{2+}$, $[OsO_2(thio)_4]^{2+}$ and $[Os(thio)_5Cl]^{2+}$, where thio = thiourea, have νOsS bands between 263 and 270 cm^{-1}.[268] Far i.r. spectra of $OsCl_5I^{2-}$, cis-$OsCl_4I_2^{2-}$, fac-$OsCl_3I_3^{2-}$, $OsCl_5Br^{2-}$ and cis-$OsCl_4Br_2^{2-}$, all as Cs^+ salts, gave assignments of vibrational modes, consistent with the expected point groups. Normal coordinate analyses suggested that the OsX force constants are generally transferable from one compound to another, except for X-Os-Y axes, where distinct trans effects are found.[269]

7. Cobalt, Rhodium and Iridium.

Previous reference has been made to vibrational studies on cobalt(II) phthalocyanines;[82] CoF_2;[89] CoL_2 (L = o-methoxybenzoate);[180] $CoCr_2S_4$;[183] and Co(TPP)(piperidine)(O_2).[222]

$Co(C_2H_4)(PMe_3)_3$ has νCoC_2 at 440 cm^{-1}.[270] Skeletal modes have been assigned as follows for the isomeric complexes $Co(NH_3)_5NO_2^{2+}$ (νCo-NO_2 376 cm^{-1}; νCo-NH_3 510 cm^{-1}) and $Co(NH_3)_5ONO^{2+}$ (νCo-ONO 260 cm^{-1}; νCo-NH_3 488 cm^{-1}).[271] νM-N modes were assigned for $M(bipy)_3^{2+}$, where M = Co(II) or Ni(II), with polyhedral borane or carbaborane anions,[272] and for 1-(2-pyridylazo)-2-naphthol complexes of Co(II) and Co(III).[273]

The νM-O (of the ligand or H_2O) modes were asigned for $[M(m\text{-}NBH)_4(H_2O)_2]^{2+}$, where M = Co(II) or Ni(II), and m-NBH = m-nitrobenzoylhydrazine (coordinated via carboxyl oxygen).[274] νM-O bands were also identified for $M[P_2N_3SO_2(NH_2)_4]_2$, where M = Co (450 cm^{-1}) or Ni (445 cm^{-1}).[275] Co(II)-M-Y zeolites, where M = Na, K or Cs, in the presence of en and O_2, give resonance Raman spectra characteristic of superoxo-, μ-peroxo and μ-peroxo-μ-hydroxo complexes of cobalt(II).[276] The resonance Raman spectrum of anthracene pillared cofacial dicobalt(II) diporphyrin superoxo adduct has a band at 628 cm^{-1} due to ν_sCo-O, in the absence of a base. In the presence of a large base there was evidence for bridging and non-bridging adducts.[277]

The resonance Raman spectra of polymeric tris(cysteinato)-

cobalt(III) and of $\{Na_3Co^{III}(SCH_2COO\text{-}S)_3\}_n$ show characteristic differences in νCo-S modes, which relate to differences in bridging modes of the ligands.[278] νM-S bands assigned (300 - 400 cm^{-1}) in Co(II), Ni(II), Cu(II), Zn(II), Cd(II) and Hg(II) complexes of benzothiazole-2-thiol.[279]

I.r. and Raman spectra of $Rh_2(OAc)_4$, $Rh_2(^{18}O_2CCH_3)_4$, $Rh_2(O_2CCD_3)_4$ and $Rh_2(OAc)_4(H_2O)_2$ show that for the anhydrous complex νRhRh is near 352 cm^{-1}, with νRhO 319 - 389 cm^{-1} (Raman), 341 - 398 cm^{-1} (i.r.).[280] Resonance Raman spectra have been recorded and analysed for $Rh_2(O_2CR)_4(PPh_3)_2$, where R = H, CH_3, C_2H_5 and C_3H_7;[281] $Rh_2(O_2CCH_3)_4L_2$, where L = $AsPh_3$ or $SbPh_3$;[282] $Rh_2(OAc)_4(MPh_3)_2$, where M = P, As or Sb;[283] and $Rh_2(OAc)_4[S(CH_2Ph)_2]_2$.[284] The far-i.r. spectra of $Rh_2(OOCR)_4(N\text{-}N)_2$, where R = H, Me etc., N-N = Phen or bipy, gave assignments of the $Rh_2O_4N_4Cl_2$ skeletal modes, extensively mixed.[285] νRhRh in $Rh_2(CH_3CONH)_4L_2$ are at 273.5 cm^{-1} (L = PPh_3), 283.5 cm^{-1} ($AsPh_3$) and 294 cm^{-1} ($SbPh_3$). The analogues with CF_3CONH^- ligands gave very similar νRhRh values.[286]

νRhH was assigned, near 2000 cm^{-1}, in (11) and related complexes.[287] $[(triphos)Rh(\mu\text{-}H)_3Rh(triphos)]^{3+}$, where triphos = $H_3CC(CH_2PPh_2)_3$, has a band due to the $Rh(H)_3Rh$ unit at 1650 cm^{-1}.[288]

$CsRh(XO_4)_2.12H_2O$, where X = S or Se, give modes due to the RhO_6 unit of $Rh(H_2O)_6{}^{3+}$ as follows: ν_1 548 cm^{-1}; ν_2 533 cm^{-1} and ν_5 315 cm^{-1}.[289] Polymeric $RhO_{0.75}(ClO_4)_{2.5}$ has $\nu Rh{=}O_t$ at 1010 cm^{-1}, ν_{as}RhORh near 825 cm^{-1}, and νRh-O(ClO_4) 315 cm^{-1}.[290]

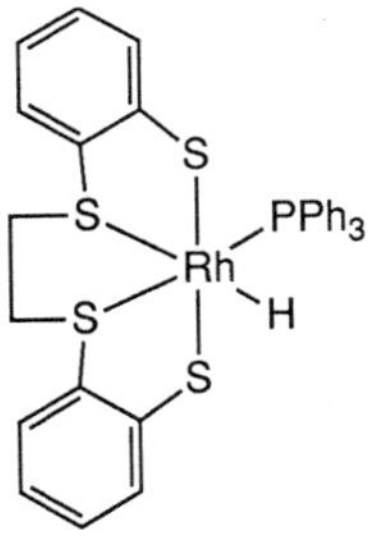

(11)

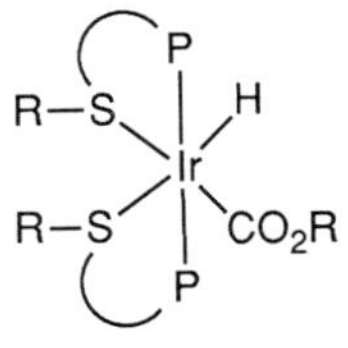

(12)

νIrH modes were assigned for (12), where R-S^P = $RSCH_2CH_2PPh_2$, R = Me or Et, R' = Me or Et.[291] νIrH bands are at 2220 and 2112 cm^{-1} in $Ir_2I_2(H)_2(CO)_2(dpm)_2$, where dpm = bis(diphenylphosphino)methane.[292]

$[Ir(NH_3)_5Cl][PtCl_4]$ has i.r. bands as follows: νIr-N 493, 477 cm^{-1}; νIr-Cl 306 cm^{-1}, and νPt-C 324 cm^{-1}.[293] Polynuclear

$IrCl_3$ has $\nu IrCl$ bands at 320 and 298 cm^{-1}, probably due to terminal and bridging Ir-Cl bands respectively.[294]

8. Nickel, Palladium and Platinum.

Earlier reference was made to vibrational studies on NiF_2;[89] NiL_2, where L = o-methoxybenzoate;[180] $Ni(bipy)_3^{2+}$ salts of polyhedral borane anions;[272] $[Ni(m\text{-}NBH)_4(H_2O)_2]^{2+}$, where m-NBH = m-nitrobenzoylhydrazine; [274], $Ni(P_2N_3SO_2(NH_2)_4]_2$;[275] and Ni(II) complexes of benzothiazole-2-thiol.[279]

νNiH was assigned for a large number of trans-$HNi(X)(PCy_3)_2$ complexes, where X = Ph, Me, CN, I, Br, OAc etc.[295] $\nu M\text{-}N$ modes were assigned for Ni(II) and Pd(II) complexes of N-(2-aminoethyl)- and N-(3-aminopropyl)-oxamide;[296] and for Ni and Cu complexes of N-(2-hydroxyethyl)- and N-(3-hydroxypropyl)oxamide.[297] Skeletal mode assignments were proposed for ML_2, where M = Ni or Cu, HL = imidazole or 2-methylimidazole;[298] and for $M(PAN)_2$, where M = Ni or Zn; HPAN = 1-(2-pyridylazo)-2-naphthol.[299]

$Ni(HMICdt)_2$, where HMICdt = hexamethyleneimine-carbodithioate has νNiS at 394 cm^{-1}. [300] In the fluorine-bridged complex $[(bipy)_2FNi\text{-}F\text{-}NiF(bipy)_2]^+$, the Ni-F(t) stretch is at 410 cm^{-1}, with $\nu NiF(br)$ at 380 and 360 cm^{-1}.[301] I.r. and Raman spectra of $NiCl_2$ in neon and argon matrices show that it is linear. [302]

Resonance Raman (including time-resolved) spectra of $M_2(dppm)_3$, where M = Pd or Pt; dppm = bis(diphenylphosphino)methane, show that νMM is at 120 cm^{-1} (Pd) or 102 cm^{-1} (Pt) in the ground state, but at 152 cm^{-1} (Pd) or 120 cm^{-1} (Pt) in the excited state. These figures are consistent with a shortening of about 10 pm in the M-M bond on electronic excitation.[303] The resonance Raman spectra of $M_2(dta)_3$, where M = Pd or Pt, dba = dibenzylideneacetone, show νMM at 76 cm^{-1} (Pd) or 72 cm^{-1} (Pt), i.e. the M-M interactions are very weak.[304]

Revised skeletal mode assignments are summarised in Table 8 for $[Pd(C_2H_4)Cl_3]^-$ and its C_2D_4 analogue.[305]

Lattice vibrations were observed and analysed for PdO and PtS, both of space group $P4_2/mmc$ - D_{4h}^9.[306] νM-O modes were assigned for $M(L-L)(ONO_2)_2$, where M = Pd or Pt, L-L = phen or bipy; all are in the range 340 - 363 cm^{-1}.[307] The far i.r. spectra of $Pd(HMTA)_2X_2$, where HMTA = hexamethylenetetramine, X = Cl, SCN or OAc, gave νPd-OAc at 296 cm^{-1}, νPd-SCN at 300 cm^{-1}, and νPd-Cl at 342 cm^{-1}.[308] $Pd(S_2N_2H)(L-L')$, where HL-L' = azobenzene, 2-phenylpyridine etc., give bands with contributions from νPdS between 320 and 360 cm^{-1}.[309] νPdS and νPdCl were assigned for $PdCl_2$ complexes with S-containing ligands, including piperidinomethyl dodecyl sulphide etc.[310]

Table 8 Skeletal vibrational assignments for $[Pd(C_2H_4)Cl_3]^-$ and its C_2D_4 analogue. ($/cm^{-1}$)

C_2H_4	C_2D_4	Assignment	
-	1380	a_1	νCC
428	387	b_1	$\nu_{as}PdC_2$
-	345	a_1	ν_sPdC_2
338	333	b_2	$\nu_{as}PdCl_2$
318	316	b_1	$\nu_{as}PdCl_2$
292	288	a_1	νPdCl(trans)
207	194	b_1	PdC_2 rock
178	178	b_2	$PdCl_2$ i.p. bend
155	155	b_1	$PdCl_2$ o.o.p. bend
149	147	a_1	$PdCl_2$ i.p. bend

νM-Cl assignments were made from far-i.r. and Raman spectra of MCl_2L_2, where M = Pd or Pt, L = PPh_3, PPh_2Cy, $PPhCy_2$ and PCy_3.[311] The i.r. and Raman spectra of $PdBr_6^{2-}$ as the $H_3NCH_2CH_2NH_3^{2+}$ salt are very difficult to assign, because of factor-group splitting and resonance effects.[312]

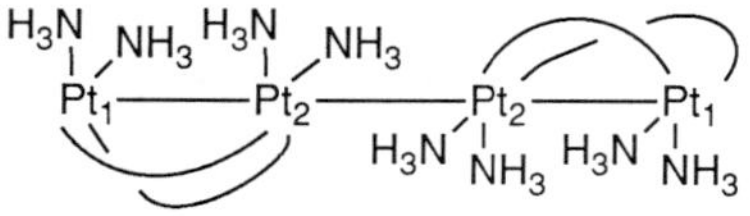

(13)

The resonance Raman spectrum of (13), with bridging α-pyridonate ligands, has νPt_2Pt_2 at 69 cm^{-1} (interdimer stretch) and νPt_1Pt_2 at 149 cm^{-1} (intradimer stretch).[313] νPtH modes were assigned in trans-$[PtH(CX_2)(PR_3)_2]^+$, where

X = OMe; R_3 = $BzPh_2$, Bz_3 and Cy_3; and trans-[$PtH(CO)(PR_3)_2]^+$, where R = Ph, Bz or Cy.[314] Three separate reports have been made of νPt-CO for CO adsorbed on Pt(111) surfaces.[315-7] νPt-C is near 600 cm^{-1} in [$Pt(C_6Cl_5)_2$-{(μ-Cl)Ag(PPh_2Me)}] and related complexes.[318] In [$Pt_2Ag(\mu\text{-}Cl)_2(C_6Cl_5)_4]_2^{2-}$, νPtC is at 625 cm^{-1}, and $\nu PtCl_{br}$ 252 cm^{-1}.[319]

I.r. and Raman data on νPtN and νPtS wavenumbers, and associated force constants, for $[Pt(SO_3)_n(NH_3)_{4-n}]^{(2n-2)-}$ show that the sulphite ligand has increased trans influence on increasing n from 0 to 4.[320] νPtX modes were assigned in $Pt(TAA)X_2$, where H_2TAA = (14), and X = N_3

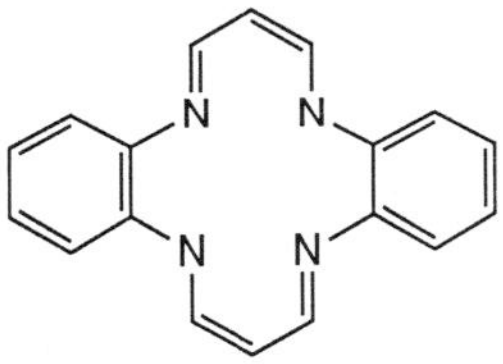

(14)

(νPtN 388 cm^{-1}), SCN (νPtS 311 cm^{-1}), Cl (νPtCl 321 cm^{-1}), Br (νPtBr 232 cm^{-1}) or I (νPtI 194 cm^{-1}).[321]

I.r spectra of Pt(II) complexes of dithiocarbamates derived from glycine, DL-alanine, DL-valine and DL-leucine all give νPtS in the range 320 - 340 cm^{-1}.[322] νPtS in $Pt(S_2C_2O_2)_2^{2-}$ is at 317 cm^{-1} (square planar PtS_4 unit).[323]

A very detailed study has been made of the series of mixed complexes $[PtF_nCl_{6-n}]^{2-}$, where n = 1 - 5, including data on ^{35}Cl and ^{37}Cl enriched species. Full assignments were made of all stretching and bending modes. For trans-^{35}Cl-Pt-^{37}Cl units the mutual exclusion rule broke down.[324-5] The Raman spectrum of $[PtCl_3(cytosine)]^-$ contains νPtCl bands at 314, 326 and 331 cm^{-1}.[326] I.r. and Raman spectra of a new platinum-urea complex were consistent with the presence of (15), (16) and the zwitterionic form (17).[327]

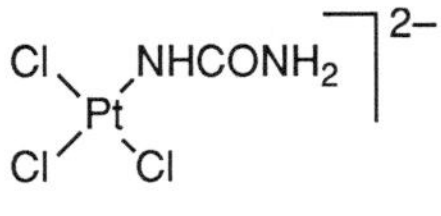

(15)

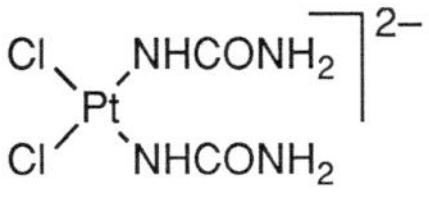

(16)

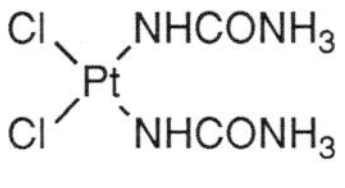

(17)

Cis-PtA_2Cl_2, where A = pyridine or γ-picoline, are transformed into the trans isomers on heating, as shown by changes in the i.r. spectra (νPtCl region).[328] νPtCl bands were seen at 317 and 294 cm^{-1} in the i.r. spectrum of $Cl_2Pt(PPh_2CH_2CH_2)_2NP(=O)NH(CH_2)_3$-O.[329] The resonance Raman spectra of $[Pt(pn)_2][Pt(pn)_2X_2](ClO_4)_4$, where X = Cl or Br, pn = 1,2-diaminopropane, and $[Pd(pn)_2][Pd(pn)_2Br_2](ClO_4)_4$ show that ν_1 (the symmetric MX stretch) has several components, whose relative intensities change with variation in excitation wavelength.[330]

Thermal isomerisations trans → cis for Pt(DMSO)-(amine)Br_2, where amine = NH_3, $MeNH_2$, $EtNH_2$, py, or piperidine, were followed by changes in νPtBr.[331] [Pt(4-Me-dien)I]-$[Pt(4\text{-}Me\text{-}dien)I_3]I_2$, where 4-Me-dien = 3-methyl-3-aza-pentane-1,5-diamine, has a resonance Raman progression in ν_s(I-Pt^{IV}-I) at 116.5 cm^{-1}. Note that this is a lower value than in the analogue with an unsubstituted dien ligand.[332]

9. Copper, Silver and Gold.

Earlier reference was made to vibrational studies on $[Mo_2Cu_5S_8(S_2CNMe_2)_3]^{7-}$;[142] $[(ClCu)_5Cl_2WS_4]^{4-}$;[170] Cu(II) complex of benzothiazole-2-thiol;[279] CuL_2^{2-}, where H_2L = N-(2-hydroxyethyl)- or N-(3-hydroxypropyl)oxamide;[297] CuL_2, where HL = imidazole or 2-methylimidazole;[298] $[Pt(C_6Cl_5)_2\{(\mu\text{-}Cl)Ag(PPh_2Me)\}_2$;[318] and $[Pt_2Ag(\mu\text{-}Cl)_2(C_6Cl_5)_4]_2^{2-}$.[319]

Cu and NH_3 react in argon matrices to form CuH (νCuH 1876 cm^{-1}), $HCuNH_2$ (νCuH 1851 cm^{-1}) and $CuNH_2$ (νCuN 748 cm^{-1}).[333] The i.r. spectra of hydrogen atoms on Cu(111) show the presence of H at two-fold bridging sites.[334] $Cu(C_2H_4)_n$ species have ν_sCuC at 320 cm^{-1} (n = 2) or 280 cm^{-1} (n = 3), compared to 420, 380 cm^{-1} for the nickel analogues.[335]

$Cu_2[PhC(NSiMe_3)_2]_2$ has νCuN as a single band at 271 cm^{-1}.[336] The i.r. spectrum of chloro[N,N-bis(2-hydroxyethyl)glycinato]copper(II) contains bands due to νCuN, νCuCl and νCuO(carboxylate) at 340, 317 and 275 cm^{-1}.[337] The i.r. and Raman spectra gave skeletal mode assignments for $Cu(PAN)_2$ and $[Cu(PAN)Cl]_2$, where HPAN = 1-(2-pyridylazo)-2-naphthol, i.e. νCuN(azo) 211 cm^{-1} and νCuO

327 cm^{-1} for the former (in which the pyridine N atom is not coordinated); νCuN(azo) 185 cm^{-1}, νCuN(py) 274 cm^{-1}, νCuO 337 cm^{-1}, νCuCl(br) 298, 236 cm^{-1} for the latter.[338] Skeletal vibrational assignments were proposed for CuXL, where L = 1-phenyl-4,6-dimethylpyrimidine-2-thione, X = Cl, Br, I, ClO_4 or BF_4.[339]

$Cu(O)(H_2O)_3$ and $Cu_2(O)Cl_2(H_2O)_2$ have νCu=O 1110, 1083 cm^{-1}; 970, 925 cm^{-1} respectively; νCu-OH_2 470 - 423 cm^{-1}; 440 - 395 cm^{-1} respectively. The data for the latter complex are consistent with the form (18).[340] The Raman spectrum of $BaCuSi_2O_6$ was consistent with square planar CuO_4 coordination.[341]

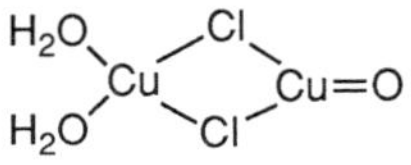

(18)

The Raman spectrum of polymeric $[Cu(mimtH)CN]_n$, where mimtH = 1-methylimidazoline-2-thione, contains a band at 190 cm^{-1} assigned as the bridging stretch, ν_sCuSCu.[342]

νCuF is at 480 cm^{-1} in $[CuF(SO_4)]^-$, as Na^+, K^+ and NH_4^+ salts. The band is very broad, and the fluorines are probably bridging.[343] XeF_2 and alkali metal fluorocuprates(II) form $Cu^{III}F_6^{3-}$, for which ν_3 (t_{1u}) lies in the range 470 - 490 cm^{-1}, ν_4 (t_{1u}) 225 - 245 cm^{-1}.[344]

SERS was observed for pyridine and Cl_2 adsorbed at a Cu(111) surface. This indicated extensive formation of CuCl units at the surface.[345] The i.r. and Raman spectra of $Cu(Me_4TMS)X$, where X = Cl, Br or I; Me_4TMS = tetramethylthiuram monosulphide, show that they have dimeric, nitrogen-bridged structures.[346] νCuX modes were assigned for $(Ph_3P)_2CuX$, where X = Cl (297 cm^{-1}), Br (218 cm^{-1}) or I (138 cm^{-1});[347] and LCuX, where L = tris(2,4,6-trimethoxyphenyl)phosphine, X = Cl (355 cm^{-1}) or Br (262 cm^{-1}).[348] The Raman spectra of equimolar melts MCl-CuCl, where M = Na, K, Rb or Cs, at 500°C contain one band, due to the linear $CuCl_2^-$ anion.[349] $[Cu(dpa)X_2]^-$, where dpa = di-(2-pyridyl)amine, has i.r. bands due to b_1 νCuX_2 at 214 cm^{-1} (Cl) or 160 cm^{-1} (Br).[350]

νCuCl is at 320 cm^{-1} in $[Cu(ept)Cl_2]_2$, where Hept =

$S{=}C(NH_2)NH{-}N{=}C(CH_3)C({=}O)OEt$.[351] Anion wavenumbers were assigned for $CuCl_4^{2-}$ as the 2-aminobenzothiazolium salt, and a normal coordinate analysis performed. There was some evidence for hydrogen-bonding between the cation and chlorine atoms of the anion.[352] The vibrational spectrum of $CuCl_6^{4-}$ has been reported for the first time. It was assignable in terms of D_{4h} symmetry. The a_{1g} stretch is at 262 cm^{-1}, with e_u at 278/273 cm^{-1}.[353]

The i.r. spectra of ethanolic solutions of Pr_4NBr + CuBr show that the main, and possibly only, bromocuprate species is the centrosymmetric, monomeric $CuBr_2^-$, with ν_3 at 323 cm^{-1}.[354] Trigonal planar $[(Ph_3P)CuBr_2]^-$ has $\nu_{as}CuBr_2$ at 195 cm^{-1}, and ν_sCuBr_2 at 150 cm^{-1}.[355] Only one νCuI band is observed in the far-i.r. of $(Ph_3P)_2CuI_2$ (118 cm^{-1}) and $[(Ph_3P)CuI_3Cu(PPh_3)]^-$ (127 cm^{-1}), even though two are expected each time.[356]

There has continued to be a very large number of reports on vibrational data for copper-containing superconducting and related systems. The vast majority concern Y/Ba/Cu/O species,[357-79], but others relate to Tl/Ca/Ba/Cu/O,[380-2] Bi/Sr/Ca/Cu/O,[383-5] lanthanide cuprates,[386-9] $Sr_2YCu_{3-x}Al_xO_{7-}$,[390] $Y_{1-x}Pr_xBa_2Cu_3O_{7-}$,[391] $YBa_{2-x}Sr_xCu_3O_{7-y}$,[392] $MBa_2Cu_3O_{7-}$, where M = Y, Sm, Gd or Eu,[393] $SmBa_2Cu_3O_7$,[394] $Nd_{1+x}Ba_{2-x}Cu_3O_y$, where x = 0 - 0.4, y ≈ 7,[395] $Pb_2Sr_2Y_{0.75}Ca_{0.25}Cu_3O_{8+}$,[396] and $Bi_2Sr_2CuO_6$.[397]

Aniline adsorbed on Ag or Au electrodes gave νM-N bands at 306, 226 cm^{-1} respectively.[398] The Raman and i.r. spectra of $[Me_3P.AgI]_4$ show ν_sAgP 122 cm^{-1}, $\nu_{as}AgP$ 175 cm^{-1}. The Ag_4I_4 modes are consistent with T_d symmetry for the cluster.[399] SERS for NCS^- adsorbed at a silver electrode showed a band assigned as νAgS in the range 190 - 225 cm^{-1} (dependent on the potential at the electrode).[400]

$O_2^+AgF_4^-$ gave bands as expected for D_{4h} AgF_4^- anions (470, 610 cm^{-1} for ν_2 (a_{2u}), ν_6 (e_u) respectively.[401] $XeF_5^+MF_4^-$, where M = Ag or Au, gave ν_sMF_4 at 554 (Ag), 582 (Au) cm^{-1}.[402]

$Au(2{-}C_6H_4CH_2NMe_2)(Ph)X$, where X = Cl, Br, I or CN, have νAuC bands near 490 cm^{-1} (phenyl) and near 510 cm^{-1} ($2{-}C_6H_4CH_2NMe_2$). When X = Cl, νAuCl is at 300 cm^{-1}.[403] νAuS for $Ph_3PAuSC_6H_4NO_2$ suggests that the Au-S bond is more polar than in corresponding types of complexes containing Hg or Pb.[404]

The Raman bands at 4K observed for AuXY(dibenzyl sulphide), where X, Y = Cl or Br, are assigned as shown in Table 9.[405-7]

Table 9 Vibrational assignments for AuXY(dibenzyl sulphide) ($/cm^{-1}$)

	$AuCl_2$(dbs)	AuClBr(dbs)	$AuBr_2$(dbs)
ν_sBrAuBr	-	188, 229	189, 232
ν_sClAuCl	319	-	-
ν_sSAuCl	339	332	-

10. Zinc, Cadmium and Mercury.

Zinc and cadmium phthalocyanines;[82] $Zn(N_2H_4)_2X_2$, where X = Cl or Br;[174] ML_2, where M =Zn or Cd, L = o-methoxybenzoate;[180] MCr_2S_4, where M = Zn or Cd;[183] Zn, Cd and Hg complexes of benzothiazole-2-thiol;[279] and $Zn(PAN)_2$, where HPAN = 1-(2-pyridylazo)-2-naphthol [299] have already been mentioned.

Detailed skeletal assignments have been proposed for $Zn(NH_3)_4^{2+}$ (with ^{15}N substitution) and $Cd(NH_3)_4^{2+}$ (with $^{110}Cd/^{116}Cd$ and H/D substitution).[408] Similar results, also based on extensive isotopic substitution, were given for $[Zn(en)_3]SO_4$;[409] and $[M(RNCSCSNR')]_n$, where M = Zn, Cd or Hg, R, R' = H or Me.[410] The low-wavenumber i.r. and Raman spectra of MX_2(4,4'-DTDP), where M = Zn, Cd or Hg, 4,4'-DTDP = 4,4'-dithiopyridine, have νMN bands in the range 224 - 268 cm^{-1}, and terminal νMX bands (301, 287 cm^{-1} for νZnCl, 318, 284 cm^{-1} for νHgCl).[411] Skeletal modes of the cations $[ML_2]^{2+}$, where M = Zn or Cd, L = propane-1,2-diamine, produce complex patterns which cannot readily be assigned to specific modes.[412] Assignments of νZnN bands for $[Zn(dmpd)_2]^{2+}$ as Cl^-, Br^- or I^- salts, where dmpd = 2,2-dimethylpropane-1,3-diamine, were consistent with D_{4h} factor group symmetry.[413]

νZnO and νZnCl modes were assigned from the i.r. and Raman spectra of $ZnCl_2(L).H_2O$, where L = 2,6-diethyl-2,4,6,8-tetra-azobicyclo[3.3.0]- octane-3,7-dione.[414] Skeletal mode assignments (νZnS, νZnX) for the dithio-oxamide complexes Zn(RR'CSCSNRR')X_2, where X = Cl, Br or I, R = R' = Me, R' = H, R = Me, were made using ^{64}Zn/^{68}Zn, NH/ND and CH_3/CD_3 isotopic substitution.[415] Similar modes were also assigned for M(bpdto)X_2, where M = Zn, X = Cl, Br or I; M = Cd, X = Cl or Br; and M_2(bpdto)X_4, where M = Cd, X = I; M = Hg, X = Cl, Br or I, bpdto = 1,8-bis(2-pyridyl)-3,6-dithiaoctane.[416] Zn(II) metallothionein has a Raman band at 138 cm^{-1} assigned as a δSZnS or δZnSC mode.[417] νMSe bands were assigned for $[M(Se)_4]^{2-}$, where M = Zn (195 cm^{-1}), Cd (177 cm^{-1}) or Hg (120 cm^{-1}).[418]

The Raman spectra of zinc(II) chloride hydrate melts $ZnCl_2.nH_2O$, where n = 1.8, 2.5, 3.0, 4.0 or 6.2, are consistent with the presence of tetrahedral species $[ZnCl_x(OH_2)_{4-x}]^{(2-x)+}$, with some evidence for these linking into aggregates via bridging Cl atoms.[419] Raman spectra of $ZnBr_2$ in DMSO solution show that $ZnBr_n^{(n-2)-}$, where n = 1 - 4, are all present. The formation constants for each species were estimated.[420] Raman data were also reported for glasses in the systems $ZnBr_2$-KBr, $ZnBr_2$-KBr-$CaBr_2$;[421] and $ZnBr_2$-KBr-$PbBr_2$.[422]

A possible νCdH mode is seen at 1285 cm^{-1} in hydrogen-implanted CdS.[423] The Raman spectrum of a single crystal of $CdSiF_6.6D_2O$ shows νCdO (of $Cd(OD_2)_6^{2+}$) at 345 cm^{-1}, with δOCdO at 171 cm^{-1}.[424] Assignments from i.r. and Raman spectra of $Cd(DH)_4X_2$, where X = Cl or Br; DH = dithio-oxamide, using ^{112}Cd/^{114}Cd and H/D substitution, are consistent with the presence of halide and dithio-oxamide bridges, and D_{4h} symmetry at the Cd.[425] The Raman spectrum of molten $CdCl_2$ shows that the Cd atoms are tetrahedrally coordinated.[426]

Normal coordinate analyses for $ClCH_2HgCl$ and $BrCH_2HgBr$, using published data, led to some re-assignments of the observed wavenumbers.[427] A correlation exists between the pattern of νHgC and νHgN bands in the Raman spectrum of the CH_3Hg complex of L-alanine, and the backbone conformation of the complex.[428] $(MeHg)_2$(tryptophan) had νHgC at 560 cm^{-1}, νHgN at 484 cm^{-1} in the Raman spectrum.[429] Complete force-field studies of $C(HgX)_4$, where X = Cl, Br, I, Me or CN, have been reported.[430]

Skeletal stretching modes were assigned from the i.r. and Raman spectra of (PTA)HgX_2, where PTA = 1,3,5-triazo-7-phospha-adamantane, X = Cl, Br, I, SCN or CN. Thus, for (PTA)$HgBr_2$, νHgP is at 146 cm^{-1}, νHgBr at 213 cm^{-1} (i.r.),

227 cm^{-1} (Raman).[431] ν_{as} and ν_s(SHgS) were reported for $HgCl_2[\mu-S(CH_2)_3NH(CH_3)_2]$, in which the Hg is four- coordinated by 2Cl and 2 bridging S atoms, at 308, 270 cm^{-1} respectively, with νHgCl at 233 cm^{-1}.[432] Normal coordinate analyses were reported for HgX_3Y^{2-}, where X ≠ Y = Cl, Br or I.[433]

11. Actinides.

Earlier reference was made to vibrational studies on $ThGeO_4$;[33] $Th(O_2)(H_2edta)$, $U(O)(O_2)(H_2edta)$;[36] and the $UO_2MoO_4-K_2MoO_4$ system.[125]

$Cp_2Th(BH_4)_2$ and $(indenyl)_2M(BH_4)_2$, where M = Th or U, have νM-ring 244 - 260 cm^{-1}. In $Cp_3Th(BH_4)$ and $(indenyl)_3M(BH_4)$ they are in the region 225 - 355 cm^{-1}.[434] νThF is at 355 cm^{-1} in $[Th(O_2)_2F_2(OH)_2]^{2-}$.[435]

An empirical equation has been set up to determine the U-O bond lengths in UO_2^{2+} compounds from the ν_1 wavenumbers of the UO_2 group.[436] A variety of UO_2^{2+}-containing species were identified from the Raman spectra of UO_2^{2+}-exchanged zeolite Y [437] and zeolite X.[438] Micro-Raman data were used to characterise UO_2F_2, $\gamma-UO_3$, $\alpha-UF_5$, $\beta-UF_5$ and UF_6.[439]

Raman spectra of $UO_2(NO_3)_2$ in H_2O/acetone mixtures gave evidence for $UO_2(H_2O)_5^{2+}$ in pure H_2O, $UO_2(NO_3)_2(H_2O)_2$ in pure acetone.[440] $^{235}U/^{238}U$ isotopic shifts were measured for $\nu_{as}UO_2$ of $UO_2(HCOO)_2.H_2O$ at room temperature and at low temperatures. They were found to be 0.63, 0.67 and 0.71 cm^{-1} at 290, 77 and 10K respectively.[441]

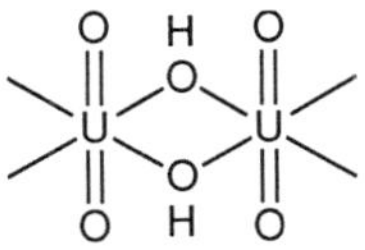

(19)

I.r spectra of dioxouranium complexes such as $(UO_2)_2(L)-(OH)_2(H_2O)(OAc)$, where HL = PhC(=NOH)$CH_3$, show the presence of the units (19).[442] Shifts in ν_3 of UO_2^{2+} in CsCl-CsF melts containing MCl_2, where M = Mg, Ca or Ba, show that replacing Cl by F in $UO_2Cl_nF_{5-n}^{3-}$ causes a decrease in ν_3. Shifts were also seen as the counterion M^{2+} changes (Mg > Ca > Ba).[443] $[UO_2(HABH)_3](NO_3)_2$, where HABH = acetone benzoylhydrazone, and related complexes all give UO_2 modes for a linear UO_2 unit and a total of 8-coordination.[444] νU-OH and νU-OD modes were assigned for the pyH^+ salts of $[UO_2(SO_4)OH]^-$ and its deuteriated analogue (νU-OH 805, 812 cm^{-1}; νU-OD 787, 770 cm^{-1}).[445] The i.r. and Raman spectra of $UO_2Cl_4(H_2O)^{2-}$, as the hexamethyltetraminium salt, shows that the H_2O is coordinated to U as shown, to give the uranium a coordination number of 7.[446]

Anharmonicity constants of the ν_3 mode of UF_6 were estimated from high-resolution i.r. studies of flow-cooled UF_6 after near-resonantly exciting it with a pulsed laser at 627.7 cm^{-1}.[447] A very long path length i.r. cell was used to study $3\nu_3$ of $^{238}UF_6$, suggesting a band centre of 1877.54(12) cm^{-1}.[448] The i.r. spectra of UCl_4 and $ThCl_4$ in neon matrices are consistent with regular T_d geometry. Detailed calculations of i.r. spectra of $Th^{35/37}Cl_4$ isotopomers, however, show that in krypton matrices $ThCl_4$ is not tetrahedral.[449]

ν_sNpO_2 and ν_sUO_2 were observed from the Raman spectra of hydrogen uranyl phosphate intercalated with NpO_2^{2+} ions.[450] $\nu_{as}NpO_2^{2+}$ modes were assigned for a number of β-diketonate adducts, NpO_2(A-A).L, where A-A = hexafluoroacetylacetonate, pivaloyltrifluoroacetonate etc; L = trimethylphosphine oxide and related complexes. They all lie in the range 912 - 950 cm^{-1}.[451] The i.r. spectra of NpF_6 at medium to high resolution gave data on $\nu_1+\nu_3$, $\nu_2+\nu_3$ and $3\nu_3$ bands.[452] A detailed analysis of ν_3 of PuF_6 showed that the band centre is at 619.7610(14) cm^{-1}.[453]

References

1 H.Kanno, T.Yamaguchi and H.Ohtaki, Kidorui, 1989, 14, 148 (Chem. Abs., 1989, 111, 122911).
2 S.Turrell, S.Hafsi, P.Conflant, P.Barbier, M.Drache and J.C.Champarnaud-Mesjard, J. Mol. Struct., 1988, 174, 449.
3 A.M.Sych, M.I.Kabanova and S.G.Andreeva, Russ. J. Inorg. Chem., 1988, 33, 1585.
4 M.V.Nikanovich, V.A.Lastochkina, N.M.Ksenofontova, R.A.Puko and T.I.Razvina, Zh. Prikl. Spektrosk., 1989, 50, 244.
5 H.Chen, L.Zhang and F.Gan, Guangxue Xuebao, 1988, 8, 791 (Chem. Abs., 1989, 110, 124256).
6 G.Bai, G.Chen, Z.Wang, L.Yuan, J.Kang and J.Gao, Wuji Huaxue, 1988, 4, 32 (Chem. Abs., 1989, 110, 87398).
7 D.E.Morris and D.E.Hobart, J. Raman Spectrosc., 1988, 19, 231.
8 X.Wang, J.Wu, N.Shi, T.Jin and G.Xu, Sci. China, Ser. B, 19189, 32, 293.
9 S.Yamauchi, H.Kanno and Y.Akama, Chem. Phys. Lett., 1988, 151, 315.
10 S.Yamauchi and H.Kanno, Kidorui, 1989, 14, 150 (Chem. Abs., 1989, 111, 122912).
11 Y.Liang, W.Zhao, S.Zhang and J.Ni, Wuli Huaxue Xuebao, 1988, 4, 466 (Chem. Abs., 1989, 110, 87378).
12 J.L.Lacout and M.Mikou, Ann. Chim., 1989, 14, 9.
13 Y.Chen, F.Ma and H.Lou, Huangzhou Daxue Xuebao, Ziran Kexueban, 1989, 16, 66 (Chem. Abs., 1989, 111, 86433).
14 J.Wang, C.Zhang and Z.Qin, Gaodeng Xuexiao Huaxue Xuebao, 1988, 9, 951 (Chem. Abs., 1989, 110, 127513).
15 V.V.Kravchenko, M.G.Zaitseva, N.Rukk and B.D.Stepin, Koord. Khim., 1988, 14, 622.
16 L.A.Aslanov, T.V.Galuzina, O.N.Gilyarov, A.L.Kamyshnyi, B.N.Kulkovskii, T.L.Novoderezhkina, V.A.Trofimov and A.Yu.Tsivadze, Russ. J. Inorg. Chem., 1988, 33, 1107.
17 N.V.Porotnikov and G.V.Bazuev, Russ. J. Inorg. Chem., 1989, 34, 142.
18 M.F.Berry, A.F.Kirby and F.S.Richardson, Mol. Phys., 1989, 66, 723.
19 M.Szyncel and S.Zachara, Acta Phys. Pol., A, 1989, A75, 773.
20 G.V.Bazuev and N.V.Porotnikov, Russ. J. Inorg. Chem., 1989, 34, 268
21 M.G.Zuev and A.A.Fotiev, Russ. J. Inorg. Chem., 1988, 33, 1382.
22 M.K.Chaudhuri and B.Das, Curr. Sci., 1989, 58, 339.
23 M.Mazzanti, J.M.Rosset, C.Floriani, A.Chiesi-Villa and C.Guastini, J. Chem. Soc., Dalton Trans., 1989, 953.
24 Ya.Yu.Nevyadomskaya, A.K.Molodkin and Yu.E.Bogatov, Russ. J. Inorg. Chem., 1988, 33, 1515.
25 G.V.Bazuev and N.V.Porotnikov, Russ. J. Inorg. Chem., 1989, 34, 140.
25 V.Krug, G.Koellner and U.Müller, Z. Naturforsch., B, 1988, 43b, 1501.
27 I.R.Beattie, P.J.Jones and N.A.Young, Angew. Chem., Int. Ed. Engl., 1989, 28, 313.
28 T.Klapötke and U.Thewalt, J. Organometal. Chem., 1988, 356, 173.
29 L.M.Sharygin, S.M.Vovk and V.F.Gonchar, Russ. J. Inorg. Chem., 1988, 33, 970.
30 P.P.K.Clarke, G.R.Willey and M.G.B.Drew, J. Chem. Soc., Dalton Trans., 1989, 57.
31 T.J.Dines, Spectrochim. Acta, Part A, 1988, 44A, 1087.
32 F.H.Elsner, T.D.Tilley, A.L.Rheingold and S.J.Geib, J. Organometal.

Chem., 1988, 358, 169.
33 M.T.Vandenborre, D.Michel and A,Ennaciri, Spectrochim. Acta, Part A, 1989, 45A, 712.
34 D.Michel, M.T.Vandenborre and A.Ennaciri, Adv. Ceram., 1988, 24A, 555.
35 N.B.Morozova, I.K.Igumenov, V.N.Mitkin, L.V.Kradenov, O.G.Potapova, V.B.Bazarev and Ya.Kh.Grinberg, Russ. J. Inorg. Chem., 1989, 34, 672.
36 M.T.H.Tarafder and A.A.M.A.Islam, Polyhedron, 1989, 8, 109.
37 L.Song and J.Cheng, Huadong Huagong Xueyuan Xuebao, 1988, 14, 358 (Chem. Abs., 1989, 110, 47630)
38 S.Boghosian and G.N.Papatheodorou, J. Phys. Chem., 1989, 93, 415.
39 G.Balducci, G.Bardi, L.Bencivenni, G.Gigli, M.Guido, S.N.Cesaro and F.Ramondo, High Temp. High Press., 1988, 20, 141.
40 A.L.Rocha and E.J.Baran, Z. Anorg. Allg. Chem., 1988, 564, 141.
41 E.M.Nour, A.B.El-Sayeed and S.A.Sadeek, Bull. Chem. Soc. Japan, 1989, 62, 317.
42 V.N.Krasil'nikov and M.P.Glazyrin, Russ. J. Inorg. Chem., 1989, 34, 191.
43 M.Herberhold, W.Kremnitz, M.Kuhnlein, M.L.Zeigler and K.Brunn, Z. Naturforsch., B, 1987, 42b, 1520.
44 M.Hilbers, M.Leimkühler and R.Mattes, Z. Naturforsch., B, 1989, 44b, 383.
45 H.E.Toma, P.S.Santos and F.T.P.Lellis, J. Coord. Chem., 1988, 18, 307.
46 F.Preuss, E.Fuchslocher, E.Leber and W.Towae, Z. Naturforsch., B, 1989, 44b, 271.
47 W.A.Herrmann, E.Herdtweck and G.Weichselbaumer, J. Organometal. Chem., 1989, 362, 321.
48 M.M.Islam, J. Bangladesh Acad. Sci., 1988, 12, 129.
49 C.Djordjevic, M.Lee and E.Sinn, Inorg. Chem., 1989, 28, 719.
50 V.L.Pecoraro, Inorg. Chim. Acta, 1989, 155, 171.
51 R.Fehrmann, S.Boghosian, G.N.Papatheodorou, K.Nielsen, R.W.Berg and N.J.Bjerrum, Inorg. Chem., 1989, 28, 1847.
52 M.N.Bhattacharjee, M.K.Chaudhuri and N.S.Islam, Inorg. Chem., 1989, 28, 2420.
53 C.J.Hawkins and T.A.Kabanos, Inorg. Chem., 1989, 28, 1084.
54 R.F.Klevtsova, E.N.Yurchenko, L.A.Glinskaya, L.V.Derkach and A.Marquez-Rios, Zh. Strukt. Khim., 1988, 29, 84.
55 C.M.Palopoli, S.B.Etcheverry and E.J.Baran, J. Raman Spectrosc., 1989, 20, 61.
56 J.Twu and P.K.Dutta, J. Phys. Chem., 1989, 93, 7863.
57 A.P.Nakhodnova, E.E.Belousova, Yu.I.Shuba and L.V.Zaslavskaya, Russ. J. Inorg. Chem., 1988, 33, 1535.
58 L.Zhi and H.Xihuai, Wuji Cailiao Xuebao, 1988, 3, 107 (Chem. Abs., 1989, 110, 62276).
59 X.Luo, N.Chen, Y.Chen and G.Luo. Cuihua Xuebao, 1988, 9, 351 (Chem. Abs., 1989, 110, 211905).
60 L.Kristallov, M.Ya. Khodos, L.A.Perelyaeva and N.D.Zhuravlev, Russ. J. Inorg. Chem., 1988, 33, 1134.
61 O.V.Koryakova, L.V.Zolotukhina, L.A.Perelyaeva and A.A.Ivakin, Russ. J. Inorg. Chem., 1989, 34, 675.
62 K.-J.Range, R.Zintl and A.M.Heyns, Z. Naturforsch., B, 1988, 43b, 309.
63 E.J.Baran and C.I.Cabello, J. Mol. Struct., 1988, 174, 401.
64 A.Müller and J.Döring, Angew. Chem., Int. Ed. Engl., 1988, 27, 1721.
65 A.Müller, M.Penk, E.Krickemeyer, H.Bögge and H.J.Walberg,

Angew. Chem., Int. Ed. Engl., 1988, 27, 1719.
66 J.F.Leboeuf, J.C.Leblanc and C.Moïse, J. Organometal. Chem., 1989, 364, C22.
67 U.Patt-Siebel, H.Thullen, U.Müller and K.Dehnicke, Z. Naturforsch., B, 1988, 43b, 795.
68 A.Antiñolo, J.M.de Ilarduya, A.Otero, P.Royo, A.M.M.Lanfredi and A.Tiripicchio, J. Chem. Soc., Dalton Trans., 1989, 2685.
69 M.Hörner, W.Hiller, J.Strähle, G.M.Sheldrick, Z. Naturforsch., B, 1988, 43b, 981.
70 V.B.Vyazovov, O.I.Kondratov and V.V.Fomichev, Russ. J. Inorg. Chem., 1988, 33, 1130.
71 T.Ikeya and M.Senna, J. Non-Cryst. Solids, 1988, 105, 243.
72 E.V.Kolobkova, Fiz. Khim. Stekla, 1989, 15, 533.
73 O.E.Averkova, V.V.Fomichev and V.P.Sirotinkin, Russ. J. Inorg. Chem., 1988, 33, 1391.
74 A.V.Mishchenko, V.E.Fedorov, B.A.Kolesov and M.A.Fedotov, Koord. Khim., 1989, 15, 200.
75 K.Tatsumi, I.Matsubara, Y.Sekiguchi, A.Nakamura and C.Mealli, Inorg. Chem., 1989, 28, 773.
76 G.Kliche and H.G.von Schnering, Z. Naturforsch., B, 1989, 44b, 74.
77 T.M.Arkhireeva, B.M.Bulychev, T.A.Sokolova, G.L.Soloveichik, V.K.Belsky and G.N.Boiko, Inorg. Chim. Acta, 1988, 141, 221.
78 M.D.Fryzuk and D.H.McConville, Inorg. Chem., 1989, 28, 1613.
79 H.Plenio, H.W.Roesky, M.Noltemeyer and G.M.Sheldrick, Angew. Chem., Int. Ed. Engl., 1988, 27, 1330.
80 C.Raptis, Phys. Rev., B, 1988, 38, 10007.
81 R.L.Sweany and A.Moroz, J. Am. Chem. Soc., 1989, 111, 3577.
82 G.Terzian, B.Moubaraki, M.Mossoyan-Deneux and D.Benlian, Spectrochim. Acta, Part A, 1989, 45A, 675.
83 J.Besida, T.A.O'Donnell and P.G.Eller, Can. J. Chem., 1989, 67, 2047.
84 S.Yang, W.Xu, G.Hu and X.Fu, Cuihua Xuebao, 1988, 9, 344 (Chem. Abs., 1989, 110, 138334).
85 D.De Waal, A.M.Heyns and K.J.Range, J. Solid State Chem., 1989, 80, 170.
86 P.G.Nagornyi, N.S.Slobodnyak, Z.I.Kornienko and E.S.Lugovskaya, Russ. J. Inorg. Chem., 1988, 33, 1618.
87 B.G.Gafford, C.O'Rear, J.H.Zhang, C.J.O'Connor and R.A.Hollwerda, Inorg. Chem., 1989, 28, 1720.
88 H.D.Lutz, B.Engelen, M.Fischer and M.Jung, Z. Anorg. Allg. Chem., 1988, 566, 55.
89 V.N.Bukhmarina, A.Yu.Gerasimov, Yu.B.Predtchenskii and V.G.Shklyarik, Opt. Spektrosk., 1988, 65, 876.
90 H.H.Schmidtke and M.Rosner, Inorg. Chem., 1989, 28, 2570.
91 U.Bentrup and K.Kolditz, J. Therm. Anal., 1987, 33, 827.
92 D.Collison, J. Chem. Soc., Dalton Trans., 1989, 1.
93 F.Allaire and A.L.Beauchamp, Inorg. Chim. Acta, 1989, 156, 241.
94 R.T.Carlin and R.E.McCarley, Inorg. Chem., 1989, 28, 280.
95 S.A.Ryabov, O.G.Ellert, V.D.Butskii, V.S.Morgunov, M.Yu.Burtsev and V.S.Pervov, Russ. J. Inorg. Chem., 1988, 33, 1855.
96 C.G.De Azevedo, A.R.Dias, A.M.Martins and C.C.Romao, J. Organometal. Chem., 1989, 368, 57.
97 D.I.Zubritskaya, D.I.Semenishin, O.Ya.Borovaya and I.S.Komarinskaya, Russ. J. Inorg. Chem., 1989, 34, 134.

98 A.Khabou, K.Dehnicke, K.Findeisen and D.Fenske, Z. Naturforsch., B, 1988, 43b, 1616.
99 K.Völp, K.Dehnicke and D.Fenske, Z. Anorg. Allg. Chem., 1989, 572, 26.
100 U.Patt-Siebel, A.Khabou, U.Müller and K.Dehnicke, Z. Anorg. Allg. Chem., 1989, 569, 91.
101 T.Godemeyer and K.Dehnicke, Z. Anorg. Allg. Chem., 1988, 566, 77.
102 D.Fenske, A.Frankenau and K.Dehnicke, Z. Anorg. Allg. Chem., 1989, 574, 14.
103 D.Fenske, K.Völp and K.Dehnicke, Z. Naturforsch., B, 1988, 43b, 1125.
104 A.El-Kholi, K.Völp, U.Müller and K.Dehnicke, Z. Anorg. Allg. Chem., 1989, 572, 18.
105 A.Frankenau and K.Dehnicke, Z. Naturforsch., B, 1989, 44b, 493.
106 J.Adel and K.Dehnicke, Chimia, 1988, 42, 413.
107 M.Scheer, A.Kolbe, E.Herrmann, V.P.Fedin, V.E.Fedorov, V.N.Ikorskii and M.A.Fedotov,Z. Anorg. Allg. Chem., 1988, 567, 111.
108 S.O.Oh and S.Y.Choi, Taehan Hwahakhoe Chi, 1989, 33, 90 (Chem. Abs., 1989, 111, 32626).
109 J.W.Faller and Y.Ma, J. Organometal. Chem., 1989, 368, 45.
110 N.J.Campbell, A.C.Dengel, C.J.Edwards and W.P.Griffith, J. Chem. Soc., Dalton Trans., 1989, 1203.
111 W.A.Flomer and J.W.Kols, Inorg. Chem., 1989, 28, 2513.
112 S.R.Ellis, D.Collison and C.D.Garner, J. Chem. Soc., Dalton Trans., 1989, 413.
113 D.Coucouvanis and S.-M.Koo, Inorg. Chem., 1989, 28,2.
114 M.N.Bhattarcharjee, M.K.Chaudhuri and R.N.D.Purkayastha, Inorg. Chem., 1989, 28, 3747.
115 V.P.Babaeva and V.Ya.Rosolovskii, Russ. J. Inorg. Chem., 1989, 34, 193.
116 S.Lee, C.Floriani, A.Chiesi-Villa and C.Guastini, J. Chem. Soc., Dalton Trans., 1989, 145.
117 R.Singh, J.T.Spence, G.N.George and S.P.Cramer, Inorg. Chem., 1989, 28, 8.
118 V.Schrenk and R.Mattes, Z. Anorg. Allg. Chem., 1988, 566, 71.
119 T.Nojo, G.Matsubayashi and T.Tanaka, Inorg. Chim. Acta, 1989, 159, 49.
120 J.A.Craig, E.W.Harlan, B.S.Snyder, M.A.Whitener and R.H.Holm, Inorg. Chem., 1989, 28, 2082.
121 A.M.El-Hendawy and W.P.Griffith, Inorg. Chim. Acta, 1989, 160, 67.
122 N.Kanoongo, R.Singh and J.P.Tandon, Transition Met. Chem., 1988, 13, 343.
123 A.M.El-Hendawy, W.P.Griffith, C.A.O'Mahoney and D.J.Williams, Polyhedron, 1989, 8, 519.
124 M.E.Poloznikova, O.I.Kondratov, N.G.Chaban, V.V.Safonov and V.V.Fomichev, Russ. J. Inorg. Chem., 1989, 34, 364.
125 C.Dion, Bull. Soc. Chim. France, 1989, 42.
126 P.J.Smith and B.N.Patel, Rec. Trav. Chim. Pays-Bas, 1988, 107, 167.
127 J.H.Choy and S.T.Hong, Bull. Korean Chem. Soc., 1989, 10, 8.
128 P.Hoffmann and R.Mattes, Inorg. Chem., 1989, 28, 2092.
129 M.E.Poloznikova, O.I.Kondratov and V.V.Fomichev, Russ. J. Inorg. Chem., 1988, 33, 1447.
130 M.M.El-Jamal, M.Forissier and A.Auroux, J. Chem. Soc., Farad. Trans., 1, 1988, 84, 3169.
131 A.I.Hadjikyriacou and D.Coucouvanis, Inorg. Chem., 1989, 28, 2169.

132 M.Leimkühler, N.Bucholz and R.Mattes, Z. Naturforsch., B, 1989, 44b, 389.
133 M.E.Poloznikova, O.I.Kondratov and V.V.Fomichev, Russ. J. Inorg. Chem., 1988, 33, 1140.
134 S.Liu, S.N.Shaikh and J.Zubieta, J. Chem. Soc., Chem. Comm., 1988, 1017.
135 S.Liu, S.N.Shaikh and J.Zubieta, Inorg. Chem., 1989, 28, 723.
136 H.Kang and J.Zubieta, J. Chem. Soc., Chem. Comm., 1988, 1192.
137 S.Himeno, T.Hori and A.Saito, Chem. Lett., 1989, 633.
138 Q.Chen, S.Liu and J.Zubieta. Angew. Chem. Int. Ed. Engl., 1988, 27, 1724.
139 S.Himeno, T.Hori and A.Saito, Bull. Chem. Soc. Japan, 1989, 62, 2184.
140 C.Sourisseau, O.Gorochov and D.M.Schleich, Mater. Sci. Eng., B, 1989, B3, 113.
141 E.Payen, S.Kasztelan and J.Grimplot, J. Mol. Struct., 1988, 174, 71.
142 X.Lei, Z.Huang, Q.Liu, M.Huong and H.Liu, Inorg. Chem., 1989, 28, 4302.
143 S.N.Misra and K.Venkatasubramanian, Bull. Chem. Soc. Japan, 1988, 61, 4482.
144 K.Völp, G.Baum, W.Massa and K.Dehnicke, Z. Naturforsch., B, 1988, 43b, 1235.
145 J.V.Brenčič and B.Čeh, Inorg. Chim. Acta, 1989, 160, 143.
146 V.P.Fedin, O.A.Ger'asko, V.E.Fedorov, Yu.L.Slovokhotov and V.T.Struchkov, Izv. Sib. Otd. Akad. Nauk SSSR, Ser. Khim. Nauk, 1988, 64.
147 H.Goesmann, D.Fenske, H.W.Swidersky and K.Dehnicke, Z. Anorg. Allg. Chem., 1988, 566, 83.
148 M.H.Chisholm, B.W.Eichhorn and J.C.Huffman, Organometallics, 1989, 8, 67.
149 R.T.Carlin and R.E.McCarley, Inorg. Chem., 1989, 28, 2604.
150 M.Kersting, A.El-Kholi, U.Müller and K.Dehnicke, Chem. Ber., 1989, 122, 279.
151 C.Ergezinger, A.El-Kholi, U.Müller and K.Dehnicke, Z. Anorg. Allg. Chem., 1989, 568, 55.
152 T.Godemeyer, K.Dehnicke and E.Fluck, Z. Anorg. Allg. Chem., 1988, 565, 41.
153 T.Ernst, A.El-Kholi, U.Müller and K.Dehnicke, Z. Anorg. Allg. Chem., 1988, 566, 7.
154 P.Klinzing, A.El-Kholi, U.Müller, K.Dehnicke and K.Findeisen, Z. Anorg. Allg. Chem., 1989, 569, 83.
155 A.Görge, K.Dehnicke and D.Fenske, Z. Naturforsch., B, 1989, 44b, 117.
156 A.Görge, K.Dehnicke and D.Fenske, Z. Naturforsch., B, 1988, 43b, 677.
157 A.Görge, U.Patt-Siebel, U.Müller, and K.Dehnicke, Z. Naturforsch., B, 1988, 43b, 1633.
158 R.Bhattacharyya, S.Biswas, J.Armstrong and E.M.Holt, Inorg. Chem., 1989, 28, 4297.
159 L.Arnaudet, R.Bougon, B.Ban, P.Charpin, J.Isabey, M.Lance, M.Nierlich and J.Vigner, Inorg. Chem., 1989, 28, 257.
160 M.Kersting, C.Friebel, K.Dehnicke, M.Krestel and R.Allyman, Z. Anorg. Allg. Chem., 1988, 563, 70.
161 Y.Shigesato, A.Murayama, T.Kamimoru and K.Matsuhiro, Appl. Surf. Sci., 1987, 33-4, 804.
162 D.Philip, G.Aruldhas and V.Ramakrishnan, Curr. Sci., 1989, 58, 560.

163 A.K.Ospanova, Kh.K.Ospanov, G.Kh.Shabikova and T.G.Khachturova, Russ. J. Inorg. Chem., 1989, 34, 41.
164 V.P.Fedin, Yu.V.Mironov and V.E.Fedorov, Russ. J. Inorg. Chem., 1988, 33, 1450
165 G.M.Maksimov, G.N.Kustova, K.L.Matveev and T.P.Lazarenko, Koord. Khim., 1989, 15, 788.
166 H.-J.Lunk, M.Salmen, H.Weiner, C.Baeker, W.Wilde and D.Müller, Z. Anorg. Allg. Chem., 1989, 568, 171.
167 J.Fuchs and R.Palm, Z. Naturforsch., B, 1988, 43b, 1529.
168 E.Payen, S.Kasztelan, J.Grimblot and J.P.Bonnelle, Catal. Today, 1988, 4, 57.
169 M.L.Ziegler, K.Blechschmitt, H.Bock, E.Guggolz and R.P.Korswagen, Z. Naturforsch., B, 1988, 43b, 590.
170 E.Sécheresse, J.-M.Manoli, C.Potvin and S.Marzak, J. Chem. Soc., Dalton Trans., 1988, 3055.
171 H.Zhu, R.Cong, X.Xin, A.Dai, Y.Duan and L.Ma, Wuji Huaxue, 1988, 4, 1 (Chem. Abs., 1989, 111, 145697).
172 I.Pauls, K.Dehnicke and D.Fenske, Chem. Ber., 1989, 122, 481.
173 A.K.Brisdon, E.G.Hope, W.Levason and J.S.Ogden, J. Chem. Soc., Dalton Trans., 1989, 313.
174 W..Personett, J.Szczepanski, K.Kubulat and M.Vala, J. Raman Spectrosc., 1989, 20, 147.
175 O.V.Quinazani, A.H.Jubert and P.J.Aymonino, J. Raman Spectrosc., 1989, 20, 141.
176 P.J.Hendra, P.Le Barazer and A.Crookell, J. Raman Spectrosc., 1989, 20, 35.
177 K.Jacob and K.-H.Thiele, Z. Anorg. Allg. Chem., 1988, 564, 81.
178 K.J.Brewer, A.Liegeois, J.W.Otvos, M.Calvin and L.O.Spreer, J. Chem. Soc., Chem. Comm., 1988, 1219.
179 H.D.Burkett and S.D.Worley, J. Am. Chem. Soc., 1989, 111, 5992.
180 Yu.Ya.Kharitonov and E.G.Khoshabova, Russ. J. Inorg. Chem., 1988, 33, 1660.
181 Q.Williams, P.McMillan and T.F.Cooney, Phys. Chem. Miner., 1989, 16, 352.
182 L.He and T.Jiang, Huaxue Wuli Xuebao, 1988, 1, 118 (Chem. Abs., 1989, 110, 239383).
183 H.D.Lutz, W.Becker, B.Müller and M.Jung, J. Raman Spectrosc., 1989, 20, 99.
184 K.E.Linder, J.C.Dewan and A.Davison, Inorg. Chem., 1989, 28, 3820.
185 J.Z.Wang and K.Balasubramanian, J. Mol. Spectrosc., 1989, 138, 204.
186 U.Abram, P.Mäding, P.Kirmse and K.Köhler, Z. Chem., 1989, 29, 183.
187 L.L.Zaitseva, A.A.Kruplov, A.V.Romanov, V.E.Samsonov and E.G.Teterin, Russ. J. Inorg. Chem., 1988, 33, 1452.
188 B.Kanellakopulos, B.Nuber, K.Raptis and M.L.Ziegler, Angew. Chem., Int. Ed. Engl., 1989, 28, 1055.
189 N.J.Bryson, D.Brenner, J.Lister-James, A.G.Jones, J.C.Dewan and A.Davison, Inorg. Chem., 1989, 28, 3825.
190 J.Y.Chen, K.R.Grundy and K.N.Robertson, Can. J. Chem., 1989, 67, 1187.
191 D.G.Hamilton, X.-L.Luo and R.H.Crabtree, Inorg. Chem., 1989, 28, 3198.
192 C.E.Anson, U.A.Jayasooriya, G.D'Alfonso and P.L.Stanghellini, J. Organometal. Chem., 1989, 372, 85.
193 G.Díaz, A.H.Klahn and C.Manzur, Polyhedron, 1988, 7, 2743.
194 W.A.Herrmann, M.Ladwig, P.Kiprof and J.Riede, J. Organometal. Chem.,

1989, 371, C13.
195 W.A.Herrmann, J.G.Kuchler, G.Weichselbaumer, E.Herdtweck and P.Kiprof, J. Organometal. Chem., 1989, 372, 351.
196 A.Jayaraman, G.A.Kouroukus, L.G.van Uitert, W.H.Grodkiewicz and R.G.Maines, Physica, A, 1989, 156, 325.
197 G.Vicentini, L.B.Zinner and S.M.F.Barreto, An. Acad. Bras. Cienc., 1987, 59, 341.
198 T.K.G.Erikson and J.M.Mayer, Angew. Chem., Int. Ed. Engl., 1988, 27, 1527.
199 D.M.Hoffman and D.A.Wierda, Polyhedron, 1989, 8, 959.
200 C.Pomp and K.Wieghardt, Polyhedron, 1988, 7, 2537.
201 W.A.Herrmann, M.Flöel, J.Kulpe, J.K.Felixberger and E.Herdtweck, J. Organometal. Chem., 1988, 355, 297.
202 S.Cai, D.M.Hoffman and D.A.Wierda, Inorg. Chem., 1989, 28, 3784.
203 J.K.Felixberger, P.Kiprof, E.Herdtweck, W.A.Herrmann, R.Jakobi and P.Gütlich, Angew. Chem. Int. Ed. ngl., 1989, 28, 334.
204 W.Kelm and W.Preetz, Z. Anorg. Allg. Chem., 1989, 568, 117.
205 P.Gowik, T.Klapötke and I.Tornieporth-Oetting, Chem. Ber., 1989, 122, 2273.
206 S.Warner and S.J.Lippard, Inorg. Chem., 1989, 28, 3008.
207 W.Kelm and W.Preetz, Z. Anorg. Allg. Chem., 1989, 568, 106.
208 B.Edwards, C.D.Garner and G.Stedman, Physica, B, 1989, 158, 223.
209 S.C.Chang, R.H.Hauge, Z.H.Kafafi, J.L.Margrave and W.E.Billups, J. Am. Chem. Soc., 1988, 110, 7975.
210 C.Ercolani, M.Gardini, V.L.Goedken, G.Pennesi, G.Rossi, U.Russo and P.Zanonato, Inorg. Chem., 1989, 28, 3097.
211 S.Kaminaka, T.Ogura, K.Kitagishi, T.Yonetani and T.Kitagawa, J. Am. Chem. Soc., 1989, 111, 3787.
212 A.Desbois, M.Momenteau and M.Lutz, Inorg. Chem., 1989, 28, 825.
213 D.Morikis, P.M.Champion, B.A.Springer and S.G.Sligar, Biochem., 1989, 28, 4791.
214 J.Ramsden and T.G.Spiro, Biochem., 1989, 28, 3125.
215 S.Dasgupta, D.L.Rousseau, H.Anni and T.Yonetani, J. Biol. Chem., 1989, 264, 654.
216 S.G.Anema, K.M.Mackay and B.K.Nicholson, Inorg. Chem., 1989, 28, 3158.
217 W.D.Agner and K.Nakamoto, J. Am. Chem. Soc., 1989, 111, 1590.
218 P.J.Kellett, M.J.Pawlik, L.F.Taylor, R.G.Thompson, M.A.Levstik, O.P.Anderson and S.H.Strauss, Inorg. Chem., 1989, 28, 440.
219 N.Nagai, Y.Yoneyama and T.Kitagawa, Biochem., 1989, 28, 2418.
220 L.V.Serebrennikov, Vestn. Mosk. Univ., Ser. 2: Khim., 1988, 29, 451.
221 A.M.Sanad, I.Kashif, M.A.Khaled, S.A.Aly and H.Farouk, Phys. Chem. Glasses, 1989, 30, 27.
222 K.Nakamoto, I.R.Paeng, T.Kuroi, T.Isobe and H.Oshio, J. Mol. Struct., 1988, 189, 293.
223 X.Feng, S.G.Bott and S.J.Lippard, J. Am. Chem. Soc., 1989, 111, 8046.
224 W.B.Tolman, A.Bino and S.J.Lippard, J. Am. Chem. Soc., 1989, 111, 8522.
225 J.Sanders-Loehr, W.D.Wheeler, A.K.Shiemke, B.A.Averill and T.M.Loehr, J. Am. Chem. Soc., 1989, 111, 8084.
226 R.S.Czernuszewicz and K.A.Macor, J. Raman Spectrosc., 1988, 19, 553.
227 J.R.Kincaid, A.J.Schneider and K.J.Paeng, J. Am. Chem. Soc., 1989, 111, 735.
228 K.J.Paeng and J.R.Kincaid, J. Am. Chem. Soc., 1988, 110, 7913.

229 C.Varotsis, W.H.Woodruff and G.T.Babcock, J. Am. Chem. Soc., 1989, 111, 6439.
230 S.Han, Y.C.Ching and D.L.Rousseau, J. Biol. Chem., 1989, 264, 6604.
231 G.Backes, M.Sahlin, B.M.Sjoeberg, T.M.Loehr and J.Sanders-Loehr, Biochem., 1989, 28, 1923.
232 A.Müller, N.H.Schladerbeck, F.Krickemeyer, H.Bögge, K.Schmitz, E.Bill and A.X.Trautwein, Z. Anorg. Allg. Chem., 1989, 570, 7.
233 A.Müller, A.Hildebrand, M.Penk, H.Bögge, E.Bill and A.X.Trautwein, Inorg. Chim. Acta, 1988, 148, 11.
234 S.Han, R.S.Czernuszewicz and T.G.Spiro, J. Am. Chem. Soc., 1989, 111, 3496.
235 S.Han, R.S.Czernuszewicz, T.Kimura, M.W.W.Adams and T.G.Spiro, J. Am. Chem. Soc., 1989, 111, 3505.
236 Y.A.Onate, S.J.Vollmer, R.L.Switzer and M.K.Johnson, J. Biol. Chem., 1989, 264, 18386.
237 J.F.Madden, S.Han, L.M.Siegel and T.G.Spiro, Biochem., 1989, 28, 5471.
238 M.Feist, K.Witke and D.Hass, Z. Anorg. Allg. Chem., 1989, 570, 152.
239 K.Murata, D.E.Irish and G.E.Toogood, Can. J. Chem., 1989, 67, 517.
240 L.Nalbandian, G.N.Papatheodorou, B.N.Cyvin, J.Brunvoll and S.J.Cyvin, Spectrosc. Lett., 1989, 22, 1.
241 W.Hiller, E.Hartmann and K.Dehnicke, Z. Naturforsch., B, 1989, 44b, 495.
242 C.D.Tait, J.M.Garner, J.P.Collma, A.P.Sattelberger and W.H.Woodruff, J. Am. Chem. Soc., 1989, 111, 7806.
243 A.Mezzetti, A.del Zotto, P.Rigo and N.B.Pahor, J. Chem. Soc., Dalton Trans., 1989, 1045.
244 S.K.Doorn and J.T.Hupp, J. Am. Chem. Soc., 1989, 111, 1142.
245 A.A.Svetlov, N.M.Sinitsyn and V.V.Kravchenko, Russ. J. Inorg. Chem., 1989, 34, 535.
246 E.Yu.Bobkova, N.L.Rogalevich, N.B.Borkovskii, G.G.Novitskii, V.N.Kokunova and N.M.Sintsyn, Russ. J. Inorg. Chem., 1989, 34, 524.
247 S.K.Doorn and J.T.Hupp, J. Am. Chem. Soc., 1989, 111, 4704.
248 H.K.van Dijk, J.J.Kok, D.J.Stufkens and A.Oskam, J. Organometal. Chem., 1989, 362, 163.
249 S.Umapathy, G.Lee-Son and R.E.Hester, J. Mol. Struct., 1989, 194, 107.
250 J.G.Kay, D.W.Green, K.Duca and G.L.Zimmerman, J. Mol. Spectrosc., 1989, 138, 49.
251 D.W.Green, J.G.Kay, G.L.Zimmerman and B.Balko, J. Mol. Spectrosc., 1989, 138, 62.
252 N.M.Sinitsyn, V.N.Kokunova, G.G.Novitskii and E.Yu.Bobkova, Russ. J. Inorg. Chem., 1988, 33, 1172.
253 A.M.El-Hendawy, W.P.Griffith, F.I.Taha and M.N.Moussa, J. Chem. Soc., Dalton Trans., 1989, 901.
254 N.G.Maksimov, O.K.Shpigunova, T.T.Babankova, N.M.Sinitsyn, E.D.Korniets, V.E.Volkov and I.D.Banilov, Russ. J. Inorg. Chem., 1989, 34, 234.
255 H.E.Toma, P.S.Santos and C.Cipriano, Spectrosc. Lett., 1988, 21, 909.
256 C.M.Che, W.T.Tang, M.H.W.Lam and T.C.W.Mak, J. Chem. Soc., Dalton Trans., 1988, 2885.
257 W.H.Leung and C.M.Che, J. Am. Chem. Soc., 1989, 111, 8812/
258 O.V.Rudnitskaya, V.N.Pichkov and G.G.Novitskii, Russ. J. Inorg. Chem., 1988, 33, 1332.
259 C.E.Anson, B.F.G.Johnson, J.Lewis, D.B.Powell, N.Sheppard,

A.K.Bhattacharyya, B.R.Bender, R.M.Bullock, R.T.Hembre and J.R.Norton, J. Chem. Soc., Chem. Comm., 1989, 703.
260 M.A.Esteruelas, E.Sola, L.A.Oro, U.Meyer and H.Werner, Angew. Chem., Int. Ed. Engl., 1988, 27, 1563.
261 C.Sartori and W.Preetz, Z. Anorg. Allg. Chem., 1989, 572, 151.
262 A.S.Salomov, Yu.N.Mikhailov, A.S.Kanishcheva, A.A.Svetlov, N.M.Sinitsyn, M.A.Porai-Koshits and N.A.Parpiev, Russ. J. Inorg. Chem., 1988, 33, 1496
263 E.Yu.Bobkova, N.L.Rogalevich, G.G.Novitskii, A.A.Svetlov, N.M.Sinitsyn and K.V.Kradenov, Russ. J. Inorg. Chem., 1989, 34, 539.
264 P.A.Lay, R.H.Magnuson and H.Taube, Inorg. Chem., 1989, 28, 3001.
265 K.Bütje and W.Preetz, Z. Naturforsch., B, 1988, 43b, 574.
266 S.A.Slivko, M.A.Sarukhanov, B.E.Zaitsev and V.A.Kagarlitskii, Russ. J. Inorg. Chem., 1989, 34, 82.
267 C.Sartori and W.Preetz, Z. Anorg. Allg. Chem., 1988, 565, 22.
268 M.G.Gribanovskaya, O.V.Rudnitskaya, V.N.Pichkov, I.V.Lipnitskii and G.G.Novitskii, Koord. Khim., 1989, 15, 991.
269 H.H.Schmidtke, C.Grzonska and T.Schönherr, Spectrochim. Acta, Part A, 1989, 45A, 129.
270 H.-F.Klein, G.Lull, B.Rodenhäuser, G.Cordier and H.Paulus, Z. Naturforsch., B, 1988, 43b, 1256.
271 A.M.Heyns and D.de Waal, Spectrochim. Acta, Part A, 1989, 45A, 905.
272 A.Kayumov, K.A.Solntsev, L.V.Goeva, N.T.Kuznetsov and O.G.Ellert, Russ. J. Inorg. Chem., 1988, 33, 1103.
273 M.S.Sastry, R.Ghose and A.K.Ghose, Transition Met. Chem., 1988, 13, 447.
274 A.G.Kayumov, K.A.Solntsev, L.V.Goeva, N.T.Kuznetsov and O.G.Ellert, Russ. J. Inorg. Chem., 1988, 33, 1007.
275 I.A.Rozanov, L.Ya.Medvedeva and L.V.Goeva, Russ. J. Inorg. Chem., 1988, 33, 1151.
276 P.K.Dutta and R.E.Zaykoski, J. Phys. Chem., 1989, 93, 2603.
277 L.M.Proniewicz, J.Odo, J.Góral, C.K.Chang and K.Nakamoto, J. Am. Chem. Soc., 1989, 111, 2105.
278 K.Kanamori, S.Ino, Y.Mizuno, I.Kanesaka, K.Kawai and J.Hidaka, Bull. Chem. Soc. Japan, 1989, 62, 2375.
279 N.V.Mel'nikova, Russ. J. Inorg. Chem., 1988, 33, 1785.
280 R.J.H.Clark and A.J.Hempleman, Croat. Chim. Acta, 1988, 61, 313.
281 R.J.H.Clark and A.J.Hempleman, Inorg. Chem., 1989, 28, 746.
282 R.J.H.Clark and A.J.Hempleman, Inorg. Chem., 1989, 28, 92.
283 K.S.Shin, R.J.H.Clark and J.I.Zink, J. Am. Chem. Soc., 1989, 111, 4244.
284 R.J.H.Clark and A.J.Hempleman, J. Mol. Struct., 1989, 197, 105.
285 F.Pruchnik, J.Hanuza, K.Hermanowicz, K.Wajda-Hermanowicz, H.Pasternak and M.Zuber, Spectrochim. Acta, Part A, 1989, 45A, 835.
286 S.P.Best, P.Chandley, R.J.H.Clark, S.McCarthy, M.B.Hursthouse and P.A.Bates, J. Chem. Soc., Dalton Trans., 1989, 581.
287 D.Sellmann, A.Fetz, M.Moll and F.Knoch, Polyhedron, 1989, 8, 613.
288 C.Bianchini, F.Laschi, D.Masi, C.Mealli, A.Meli, F.M.Ottaviani, D.M.Prospero, M.Sabat and P.Zanello, Inorg. Chem., 1989, 28, 2552.
289 S.P.Best, J.K.Beattie, R.S.Armstrong and G.P.Braithwiate, J. Chem. Soc., Dalton Trans., 1989, 1771.
290 J.L.Pascal and M.Fourati, Compt. Rend. Acad. Sci., Ser.2, 1989, 308, 281.

291 A.del Zotto, A.Mezzetti, G.Dolcetti, P.Rigo and N.B.Pahor, J. Chem. Soc., Dalton Trans., 1989, 607.
292 B.A.Vaartstra and M.Cowie, Inorg. Chem., 1989, 28, 3138.
293 B.Michelot, A.Ouali, M.J.Blaise, M.Guérin and C.Kappenstein, New J. Chem., 1988, 12, 293.
294 A.B.Venediktov and R.G.Kultyshev, Russ. J. Inorg. Chem., 1989, 34, 509.
295 M.Y.Darensbourg, M.Ludwig and C.G.Riordan, Inorg. Chem., 1989, 28, 1630.
296 S.P.Perlepes, F.Quaeyhaegens and H.P.Desseyn, Bull. Soc. Chim. Belg., 1989, 98, 153
297 F.Quaeyhaegens, S.P.Perlepes and H.O.Desseyn, Spectrochim. Acta, Part A, 1989, 45A, 809,
298 Z.A.Siddiqi, S.N.Qidwai, M.Jaria and M.Shakir, Transition Met. Chem., 1988, 13, 317.
299 P.M.Drożdżewski, Spectrochim. Acta, Part A, 1989, 45A, 455.
300 A.K.Singh, B.K.Puri and R.K.Rawlley, Ind. J. Chem., Sect. A, 1989, 28A, 59.
301 J.Emsley, M.Arif, P.A.Bates and M.B.Hursthouse, J. Chem. Soc., Dalton Trans., 1989, 1273.
302 V.N.Bukhmarina and Yu.B.Predtchenskii, Opt. Spektrosk., 1989, 66, 1026.
303 P.D.Harvey, R.F.Dallinger, W.H.Woodruff and H.B.Gray, Inorg. Chem., 1989, 28, 3057.
304 P.D.Harvey, F.Adar and H.B.Gray, J. Am. Chem. Soc., 1989, 111, 1312.
305 J.Mink, I.Pápai, M.Gál and P.L.Goggin, Pure Appl. Chem., 1989, 61, 973.
306 G.Kliche, Z. Naturforsch., B, 1989, 44a, 169.
307 S.Wimmer, P.Castain, F.L.Wimmer and N.P.Johnson, J. Chem. Soc., Dalton Trans., 1989, 403.
308 V.I.Efanov, I.A.Khartonik, T.M.Buslaeva, G.G.Novitskii and N.M.Sinitsyn, Russ. J. Inorg. Chem., 1989, 34, 596.
309 C.A.O'Mahoney, I.P.Parkin, D.J.Williams and J.D.Woollins, J. Chem. Soc., Dalton Trans., 1989, 1179.
310 R.A.Khisamutdinov, N.G.Afzaletdinova, Yu.I.Murinov, G.I.Nikiforova and A.M.Shakirova, Russ. J. Inorg. Chem., 1988, 33, 1648.
311 C.Xun, S.Guonyan and L.Tongxin, J. Mol. Struct., 1988, 174, 407.
312 R.J.H.Clark, V.B.Croud, H.M.Dawes and M.B.Hursthouse, Polyhedron, 1988, 7, 2611.
313 H.K.Mahtani and P.Stein, J. Am. Chem. Soc., 1989, 111, 1505.
314 R.A.Michelin, P.Ros, G.Guadalupi, G.Bombieri, F.Benetollo and G.Chapuis, Inorg. Chem., 1989, 28, 840.
315 D.Hoge, M.Tueshaus, E.Schweizer and A.M.Bradshaw, Chem. Phys. lett., 1988, 151, 230.
316 I.J.Malik and M.Trenaray, Surf. Sci., 1989, 214, L237.
317 R.Ryberg, Phys. Rev., B, 1989, 40, 5849.
318 R.Usón, J.Forniés, M.Tomás, I.Ara and J.M.Casas, Inorg. Chem., 1989, 28, 2388.
319 R.Usón, J.Forniés, M.Tomás and J.M.Casas, Angew. Chem., Int. Ed. Engl., 1989, 28, 748.
320 D.K.Breitinger and M.Vogt, J. Mol. Struct., 1988, 174, 423.
321 A.N.Boiko and V.I.Muzychenko, Russ. J. Inorg. Chem., 1989, 34, 393.
322 J.J.Criado, A.Carrasco, B.Macias, J.M.Salas, M.Medarde and M.Castillo,

Inorg. Chim. Acta, 1989, 160, 37.
323 C.Bellitto, M.Bonamico, Y.Fares, P.Imperatori and S.Patrizio, J. Chem. Soc., Dalton Trans., 1989, 719.
324 W.Preetz and P.Erlhöfer, Z. Naturforsch., B, 1989, 44b, 412.
325 P.Erlhöfer and W.Preetz, Z. Naturforsch., B, 1989, 44b, 619.
326 S.Jaworski, H.Schöllhorn, P.Eisenmann, U.Thewalt and B.Lippert, Inorg. Chim. Acta, 1988, 153, 31.
327 J.Anastassopoulou, P.K.Ganguli and T.Thephanides, Inorg. Chim. Acta, 1989, 159, 237.
328 V.A.Sokolenko, V.S.Bondarenko, E.D.Korniets, N.P.Kovtonyuk and N.B.Kovrova, Russ. J. Inorg. Chem., 1989, 34, 869.
329 J.Ellermann and N.Will, Z. Naturforsch., B, 1989, 44b, 127.
330 R.J.H.Clark and D.J.Michael, J. Mol. Struct., 1988, 189, 173.
331 V.Ya.Kukushkin, A.I.Moiseev and B.V.Zhdanov, Russ. J. Inorg. Chem., 1989, 34, 530.
332 F.P.Fanizzi, G.Natile, M.Lanfranchi, A.Tiripicchio, R.J.H.Clark and D.J.Michael, J. Chem. Soc., Dalton Trans., 1989, 1689.
333 D.W.Ball, R.H.Hauge and J.L.Margrave, Inorg. Chem., 1989, 28, 1599.
334 E.M.McCash, S.F.Parker, J.Pritchard and M.A.Chesters, Surf. Sci., 1989, 215, 363.
335 T.Merle-Méjean, S.Bouchareb and M.Tranquille, J. Phys. Chem., 1989, 93, 1197.
336 S.Maier, W.Hiller, J.Strähle, C.Ergezinger and K.Dehnicke, Z. Naturforsch., B, 1988, 43b, 1628.
337 H.Yamaguchi, M.Nagase, Y.Yukawa, Y.Inomata and T.Takeuchi, Bull. Chem. Soc. Japan, 1988, 61, 2763.
338 P.M.Drożdżewski, Spectrochim. Acta, Part A, 1988, 44A, 1297.
339 R.Battistuzzi, Transition Met. Chem., 1988, 13, 401.
340 E.M.Nour and M.J.Al-Thani, J. Phys. Chem. Solids, 1989, 50, 183.
341 L.W.Finger, R.M.Hazen and R.J.Hemley, Am. Mineral., 1989, 74, 952.
342 E.S.Raper, J.R.Creighton, J.D.Wilson, W.Clegg and A.Milne, Inorg. Chim. Acta, 1989, 155, 77.
343 M.N.Bhattacharjee, M.K.Chaudhuri and M.Devi, Polyhedron, 1989, 8, 457.
344 A.I.Popov and Yu.M.Kiselev, Russ. J. Inorg. Chem., 1988, 33, 1307.
345 K.N.El'tsov, G.Ya.Zueva, A.N.Klimov, V.V.Martynov and A.M.Prokhorov, Chem. Phys. Lett., 1989, 158, 271.
346 J.G.Contreras, J.A.Gnecco and H.V.Carbacho, J. Coord. Chem., 1989, 19, 371
347 G.A.Bowmaker, P.C.Healy, J.D.Kildea and A.H.White,Spectrochim. Acta, Part A, 1988, 44A, 1219.
348 G.A.Bowmaker, J.D.Cotton, P.C.Healy, J.D.Kildea, S.B.Silong, B.W.Skelton and A.H.White, Inorg. Chem., 1989, 28, 1462.
349 E.Christensen, R.W.Berg and J.H.van Barner, Polyhedron, 1989, 8, 325.
350 G.A.Bowmaker, P.C.Healy, D.L.Kepert, J.D.Kildea, B.W.Skelton and A.H.White,J. Chem. Soc., Dalton Trans., 1989, 1639.
351 M.B.Ferrari, G.G.Fava, P.Tarasconi and C.Pelizzi, J. Chem. Soc., Dalton Trans., 1989, 361.
352 H.O.Desseyn, B.Slootmakers, S.P.Perlepes and A.C.Fabretti, Inorg. Chim. Acta, 1989, 156, 113.
353 R.G.McDonald and M.A.Hitchman, Inorg. Chem., 1989, 28, 3996.
354 S.Andersson, M.Haakansson and S.Jagner, J. Crystallogr. Spectrosc. Res., 1988, 19, 147.
355 G.A.Bowmaker, B.W.Skelton, A.H.White and P.C.Healy, J. Chem. Soc.,

Dalton Trans., 1989, 2825.
356 G.A.Bowmaker, A.Camus, P.C.Healy, B.W.Skelton and A.H.White, Inorg. Chem., 1989, 28, 3883.
357 J.Hanuza, J.Klamut, R.Horyn and B.Jezowska-Trzebiatowska, J. Mol. Struct., 1989, 193, 57.
358 G.E.Blumberg, E.M.Fefer, T.A.Fimberg, E.Joon, A.Laht, R.R.Stern and L.A.Rebane, Pis'ma Zh. Eksp. Teor. Fiz., 1989, 49, 385.
359 C.Taliani, R.Zamboni, G.Ruani and F.C.Mattacotta, Synth. Met.. 1988, 29, F585.
360 A.K.Sood, K.Sankaran, Y.Haraihara, S.Vijayalakshmi, V.S.Sastri, S.Kalavathi and J.Janaki, Pramana, 1988, 31, 389 (Chem. Abs., 1989, 110, 124132).
361 E.L.Benitez, J.I.Lin, S.J.Poon, W.E.Farneth, M.K.Crawford and E.M.McCarron, Phys. Rev., B, 1988, 38, 5025.
362 Y.H.Kim, C.M.Foster, A.J.Heeger, S.Cox and G.Stucky, Phys. Rev., B, 1988, 38, 6478.
363 M.K.Crawford, W.E.Farneth and E.M.McCarron, Phys. Rev., B, 1988, 38, 11382.
364 T.Ruf, C.Thomsen, R.Liu and M.Cardona, Phys. Rev., B, 1988, 38, 11985.
365 F.E.Bates, Phys. Rev.,B, 1989, 39, 322.
366 A.Mascarenhas, H.Katama-Yoshida, J.Pankuve and S.K.Deb, Phys. Rev., B, 1989, 39, 4699.
367 C.Thomsen, H.Mattausch, M.Bauer, W.Bauhofer, R.Liu, L.Genzel and M.Cardona, Solid State Commun., 1988, 67, 1069.
368 M.Hangyo, S.Nakashima, M.Nishiuchi and A.Mitsuishi, Solid State Commun., 1988, 67, 1171.
369 D.Chen, C.Mou, Y.Qian, Z.Chen, T.He, J.Wang, H.Xin and F.Liu, Solid State Commun., 1988, 68, 263.
370 G.E.Blumberg, E.M.Fefer, T.Fimberg, E.Jooh, A.Laht, R.R.Stern and L.A.Rebane, Solid State Commun, 1989, 70, 647.
371 R.Zamboni, G.Ruani, A.J.Pal and C.Tallani, Solid State Commun., 1989, 70, 813.
372 Y.Morioka, A.Tokina, M.Kikuchi, V.Syono and K.Nagase, Solid Stae Commun., 1989, 70, 1127.
373 J.Chrzanowski, J.C.Irwin, R.R.Parsons and P.J.Mulhern, Can. J. Phys., 1988, 66, 695.
374 G.Ruani, C.Taliani, R.Zamboni, D.Cittone and F.C.Mattacotta, J.Opt. Soc. Am., B, 1989, 6, 409.
375 L.V.Gasparov, G.A.Emel'chenko, V.D.Kulakovskii, O.V.Misochko, V.B.Timofeev and E.I.Rashba, J. Opt. Soc. Am., B, 1989, 6, 440.
376 D.M.Krol, M.Stavola, L.F.Schneemeyer, J.V.Waszczak and S.A.Sunshine, J. Opt. Soc. Am., B, 1989, 6, 448
377 C.R.Huang, Y.S.Chang and M.C.Lee, Jpn. J. Appl. Phys., Part 1, 1989, 28, 598.
378 H.Monien and A.Zawadokski, Phys. Rev. Lett., 1989, 63, 911.
379 R.Feile, P.Leiderer, J.Kowalewski, W.Assmus, J.Schubert and U.Poppe, Z. Phys., B, 1988, 73, 155.
380 M.Krantz, H.J.Rosen, R.M.Macfarlane and V.Y.Lee, Phys. Rev., B, 1988, 38, 11962.
381 M.Krantz, H.J.Rosen, R.M.Macfarlane, W.Y.Lee, V.Y.Lee and R.Savoy, Solid Stat Commun, 1989, 69, 209.
382 I.Z.Kostadinov, V.G.Hadjiev, J.Tihov, M.Mateev, M.Mikhov, O.Petrov, V.Popov, E.Dinolova, Ts.Zheleva, G.Tyuliev and V.Kojouharov, Physica,

C, 1988, 156, 427.
383 V.N.Denisov, B.N.Mavrin, V.B.Podobedov, I.V.Aleksandrov, A.B.Bykov, A.F.Goncharov, O.K.Mel'nikov and N.I.Romanova, Solid State Commun., 1989, 70, 885.
384 G.Burns, G.V.Chandrashakhar, F.H.Dacol and P.Strobel, Phys. Rev., B, 1989, 39, 775.
385 A.K.Pradhan, Physica, C, 1989, 159, 277.
386 W.H.Weber, C.R.Peters, B.M.Wanklyn, C.Chen and B.E.Watts, Solid State Commun., 1988, 68, 61.
387 G.Burns, G.V.Chandrashekhar, F.H.Dacol and M.W.Shafer, Solid State Commun., 1988, 68, 67.
388 I.Ohana, M.S.Dresselhaus, Y.C.Liu, P.J.Picone, D.R.Grabbe, H.P.Janssen and G.Dresselhaus, Phys. Rev., B, 1989, 39, 2293.
389 V.A.Maroni, T.O.Brun, M.Grimsditch and C.K.Loong, Phys. Rev., B, 1989, 39, 4127.
390 M.Stavola, D.M.Krol, L.F.Schneemeyer, S.A.Sunshine, J.Waszczak and S.G.Kosinski, Phys. Rev., B, 1989, 39, 287.
391 H.B.Radoosky, K.F.McCarty, J.L.Peng and R.N.Shelton, Phys. Rev., B, 1989, 39, 12383.
392 Y.Matsuda, M.Yoshida and S.Hinotani, Jpn. J. Appl. Phys., Part 2, 1989, 28, L1128.
393 H.Ye, Z.Yu, W.Lu, J.Chen, X.Shen, N.Cao, J.Zheng, X.Chen and B.Miao, Wuli Xuebao, 1989, 38, 761 (Chem. Abs., 1989, 111, 104798).
394 J.E.Eldridge and F.E.Bates, Solid State Commun., 1989, 70, 153.
395 R.Nishitani, N.Yoshida, Y.Sasaki and Y.Nishina, Jpn. J. Appl. Phys., Part 2, 1989, 28, L569.
396 R.Liu, M.Cardona, B.Gegenheimer, E.T.Heyen and C.Thomsen, Phys. Rev., B, 1989, 40, 2684.
397 G.Burns, R.Strobel, G.V.Chandrashekhar, F.H.Dacol, F.Holtzberg and M.W.Shafer, Phys. Rev., B, 1989, 39, 2245.
398 Z.Tian, L.Lei and X.Jing, Wuli Huaxue Xuebao, 1988, 4, 458 (Chem. Abs., 1989, 110, 8772)
399 H.G.M.Edwards and D.W.Farwell, J. Mol. Struct., 1989, 197, 203.
400 Y.Huang and G.Wu, Spectrochim. Acta, Part A, 1989, 45A, 123.
401 Yu.M.Kiselev, A.I.Popov, K.V.Bukharin, A.A.Timakov and M.V.Korobov, Russ. J. Inorg. Chem., 1988, 33, 1852.
402 K.Lutar, A.Jesih, I.Leban, B.Zemva and N.Bartlett, Inorg. Chem., 1989, 28, 3467.
403 J.Vicente, M.T.Chicote, M.D.Bermúdez, M.J.Sanchez-Santano and P.G.Jones, J. Organometal. Chem., 1988, 354, 381.
404 L.M.Epshtein, E.S.Shubina, L.N.Saitkulova, E.I.Smyslova, K.I.Grandberg and D.N.Kravtsov, Metalloorg. Khim., 1988, 1, 860.
405 H.Tanino, K.Takahashi, M.Tajima, M.Kato and T.Yao, Phys. Rev., B, 1988, 38, 8327.
406 H.Tanino, M.Hanfland, M.Holtz, K.Syassen and K.Takahashi, Semicond. Sci. Technol. 1989, 4, 273.
407 H.Tanino, M.Holtz, M.Hanfland, K.Syassen and K.Takahashi, Phys. Rev., B, 1989, 39, 9992.
408 C.Tellez, Spectrosc. Lett., 1988, 21, 871.
409 A.M.A.Bennett, G.A.Foulds and D.A.Thornton, Spectrochim. Acta, Part A, 1989, 45A, 219.
410 P.Geboes, B.Slootmakers and H.O.Desseyn, Spectroscopy, 1988, &, 227.
411 J.G.Contreras and H.Cortés, Spectrochim. Acta, Part A, 1989, 45A,

1025.
412 G.Ciani, M.Moret, A.Sironi, S.Bruni, F.Cariati, A.Pozzi, T.Manfredini, L.Menabue and G.C.Pellacani, Inorg. Chim. Acta, 1989, 158, 9.
413 N.Masciocchi, M.Moret, A.Sironi, S.Bruni, F.Cariati, A.Pozzi, T.Manfredini, L.Menabue and A.Benedetti, Inorg. Chim. Acta, 1989, 159, 173.
414 V.B.Rybakov, L.A.Aslanov, M.G.Tsintsadze and A.Yu.Tsivadze, Zh. Strukt. Khim., 1989, 30, 175
415 P.Geboes and H.O.Desseyn, Spectrochim. Acta, Part A, 1988, 44A, 963.
416 A.Castiñeiras, G.Diaz, F.Florencio, S.Garcia-Blanco and S.Martinez-Carrera, Z. Anorg. Allg. Chem., 1988, 567, 101.
417 T.E.Elgre and D.E.Wilcox, Biochem. Biophys. Res. Commun., 1989, 163, 1093.
418 J.Adel, F.Weller and K.Dehnicke, Z. Naturforsch., B, 1988, 43b, 1094.
419 T.Yamaguchi, S.Hayashi and H.Ohtaki, J. Phys. Chem., 1989, 93, 2620.
420 J.van Heumen, T.Ozeki and D.E.Irish, Can. J. Chem., 1989, 67, 2030.
421 K.Kadono, H.Nakamichi and H.Tanaka, Mater. Sci. Forum, 1988, 32-3, 433.
422 K.Kadono, H.Nakamichi and H.Tanaka, Nippon, Serikkusu Kyokai Gakujutsu Ronbunshi, 1989, 97, 262 (Chem. Abs., 1989, 111, 27446).
423 J.Tatarkiewicz, A.Breitschwerdt and A.Witowski, Phys. Rev., B, 1989, 39, 3889.
424 G.Thakur and A.L.Verma, Spectrochim. Acta, Part A, 1989, 45A, 615.
425 D.Miernil, Monatsh. Chem., 1989, 120, 191.
426 Y.Takagi, N.Itoh and T.Nakamura, J. Chem. Soc., Farad. Trans., 1, 1989, 85, 493.
427 M.D.Armitage, H.G.M.Edwards and V.Fawcett, J. Organometal. Chem., 1989, 371, 1.
428 M.C.Corbeil and A.L.Beauchamp, J. Crystallogr. Spectrosc. Res., 1989, 19, 123
429 M.C.Corbeil and A.L.Beauchamp, Can. J. Chem., 1988, 66, 2458.
430 J.Mink, I.Pápai, G.Kerestury, D.K.Breitinger and P.L.Goggin, Teubner-Texte Phys., 1988, 20, 140.
431 E.C.Alyea, K.J.Fisher and S.Johnson, Can. J. Chem., 1989, 67, 1319.
432 I.Casals, P.González-Duarte, J.Sola, C.Miravitlles and E.Molina, Polyhedron, 1988, 7, 2509.
433 A.N.Pandey, A.Bigotto and S.Mohan, Spectrosc. Lett., 1988, 21, 859.
434 S.Brettonville and J.Goffart, J. Organometal. Chem., 1988, 356, 297.
435 C.R.Bhattacharjee, M.K.Chaudhuri and R.N.D.Purkayastha, Inorg. Chim. Acta, 1989, 160, 147.
436 J.R.Bartlett and R.P.Cooney, J. Mol. Struct., 1989, 193, 295.
437 J.R.Bartlett and R.P.Cooney, Spectrochim. Acta, Part A, 1989, 45A, 541.
438 J.R.Bartlett and R.P.Cooney, Spectrochim. Acta, Part A, 1989, 45A, 549.
439 D.P.Armstrong, R.J.Jarabek and W.H.Fletcher, Appl. Spectrosc., 1989, 43, 461.
440 M.E.D.G.Azenha, H.D.Burrows, S.J.Formosino, M.L.P.Leitão and M.de G.M.Miguel, J. Chem. Soc., Dalton Trans., 1988, 2893.
441 K.Yao, Y.Gao, X.Gao and F.Yang, He Huaxue Yu Fangshe Huaxue, 1988, 10, 238 (Chem. Abs., 1989, 110, 239388).
442 A.Yacouta-Nour, A.K.T.Maki and M.M.Mostafa, Spectrochim. Acta, Part A, 1988, 44A, 1291.

443 A.A.Khokhryakov and A.M.Khokhlova, Rasplavy, 1989, 106 (Chem. Abs., 1989, 111, 84921).
444 R.A.Lal, H.Askari and S.S.Bhattacharjee, Synth. React. Inorg. Met.-Org. Chem., 1988, 18, 717.
445 L.V.Kobets, N.G.Bend' and D.S.Umreiko, Russ. J. Inorg. Chem., 1988, 33, 1020.
446 L.V.Kobets, N.G.Bend' and D.S.Umreiko, Russ. J. Inorg. Chem., 1988, 33, 1350.
447 R.Harzer, G.Schweizer and K.Selter, J. Mol. Spectrosc., 1988, 132, 310.
448 B.J.Krohn, R.S.McDowell, C.W.Patterson, N.G.Nereson, M.J.Reisfeld and K.C.Kim, J. Mol. Spectrosc., 1988, 132, 285.
449 I.R.Beattie, P.J.Jones, K.R.Millington and A.D.Wiison, J. Chem. Soc., Dalton Trans., 1988, 2759.
450 P.K.Dorhout, R.J.Kissane, K.D.Abney, L.R.Avens, P.G.Eller and A.B.Ellis, Inorg. Chem., 1989, 28, 2926
451 N.R.Grebenshchikov, G.V.Sidorenko, D.N.Suglobov, L.L.Shcherbakova and V.M.Adamov, Inorg. Chim. Acta, 1988, 145, 315.
452 K.C.Kim and R.N.Mulford, Chem. Phys. Lett. 1989, 159, 327.
453 K.C.Kim, B.J.Krohn, R.A.Briesmeister and S.J.David, J. Mol. Spectrosc., 1988, 132, 207.

6
Vibrational Spectra of Some Co-ordinated Ligands

BY G. DAVIDSON

1. Carbon and Silicon Donors.

$Cp_2Ti[C{\equiv}CSiMe_3]_2$ has νCC as a weak i.r. band at 2009 cm^{-1}.[1] In the complex (1), νC=C is a very strong i.r. band at 1565 cm^{-1}.[2] The Raman spectrum of $Cp_2Ti(AsF_6)_2$ contains a νCH band of the Cp ligands at 3098 cm^{-1}, while the i.r. spectrum shows νAsF bands consistent with C_{4v} symmetry for the AsF_6^- ligands.[3] The ring substituents in (2) have νC=O at 1720 cm^{-1} and νC-O at 1158 cm^{-1}.[4]

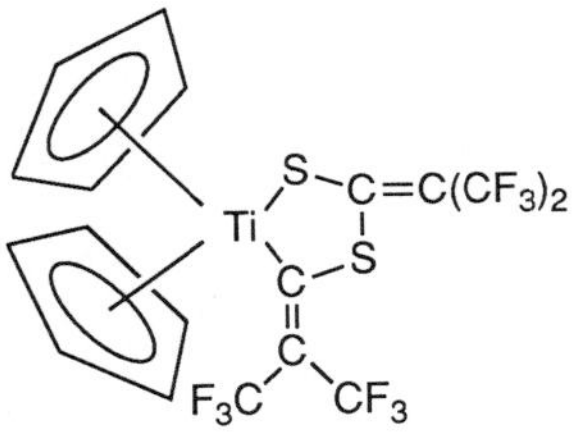

(1)

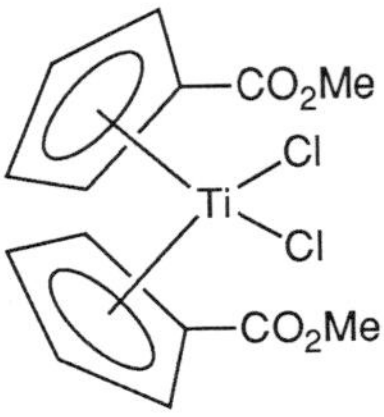

(2)

$Cp_2Zr(Ph)[-C(CHPPh_3)=O]$ and related species all have νC=O near 1560 cm^{-1}.[5] The η^2-iminoacyl group in $Cp_2Zr(Ph)$-$[\eta^2$-$N(CH_2Ph)=C(CHPPh_3)]$ gives a νC=N band at 1479 cm^{-1}.[6]

(3) and related systems have νC=N of the carbene ligand in the range 1650 - 1670 cm^{-1}.[7] The Raman spectrum of $Br(OC)_4CrCCHD_2$ over a range of temperatures showed the existence of two conformers, with non-equivalence of the CH bonds of the methyl group.[8] I.r. and Raman spectra of matrix-isolated Cr/C_6H_6 adducts provided evidence for additional species besides $Cr(C_6H_6)_2$, probably involving Cr_2 and Cr_3 groupings.[9-10]

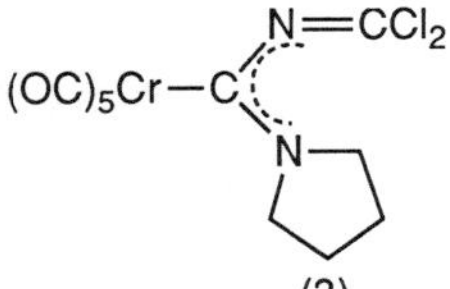

(3)

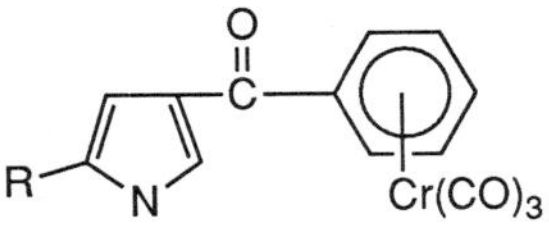

(4)

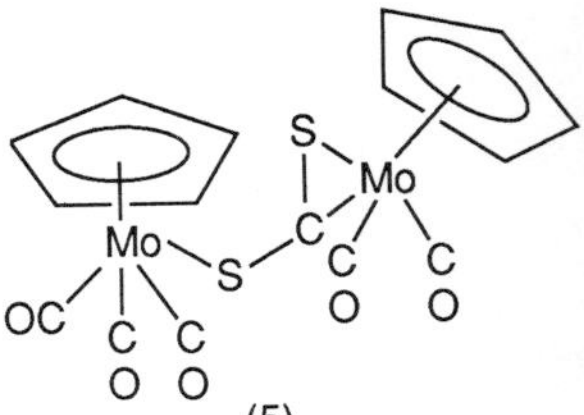

(5)

νC=O(ketone) in (4), where R = H, ^{i}Pr, Cl or I, lies between 1630 (^{i}Pr) and 1639 (Cl) cm^{-1}.[11] The νCS modes for (5) are at 1095 and 680 cm^{-1}; the latter is characteristic of the three-membered ring unit Mo-C-S.[12]

The amidinato ligand in (6) gives typical i.r. bands at 1592, 1492 and 1391 cm^{-1}.[13] The transient radical $[(\eta-C_7H_9)-(dppe)Mo-C\equiv C^nBu]^+$ has νC≡C at 2055 cm^{-1} in solution.[14] νCC for the complexed alkyne in (7) is at 1665 cm^{-1}, with νPO of the $OPCl_3$ at 1250 cm^{-1}, confirming O-coordination.[15] νC=C and νC=O are near 1460 and 1565 cm^{-1} respectively in (8), where R = Me or Et.[16]

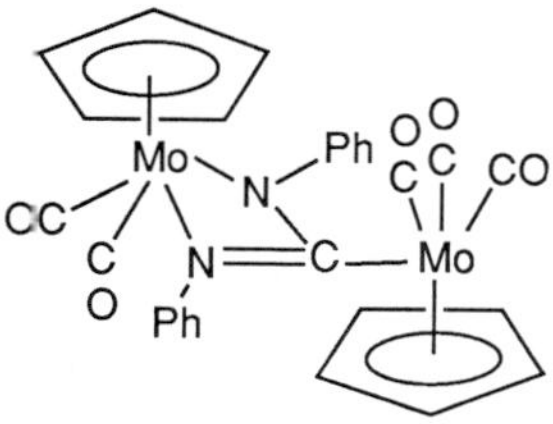

(6)

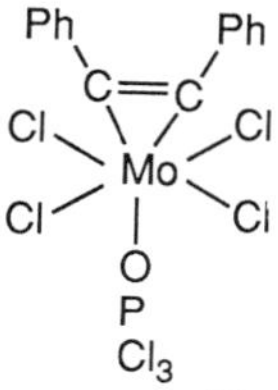

(7)

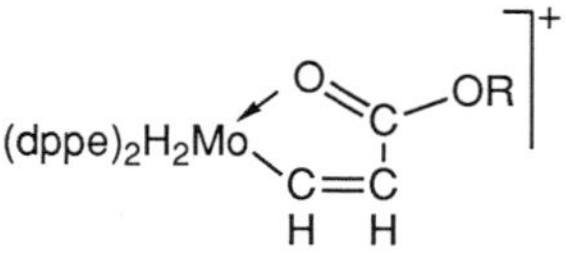

(8)

Some vibrational assignments were made for modes due to all the different ligands in (norbornadiene)$(OC)_3MoC(OEt)R$, where R = Ph, o-Me-, m-Me-, p-Me- or p-$MeOC_6H_4$. The data were consistent with a trans configuration.[17]

The resonance Raman spectra of $(OC)_5W=C(OCH_3)CH_3$ at 354.7, 396 and 416 nm, using a pulsed laser-excited Raman spectrometer, contained bands due to a laser-generated intermediate, possibly $(OC)_4W[S]=C(OCH_3)CH_3$, formed following CO photodissociation.[18] νCC of the methylthiovinyl ligand in (9) is at 1575 cm^{-1}.[19] νSO in (10) is at 1632 cm^{-1}.[20]

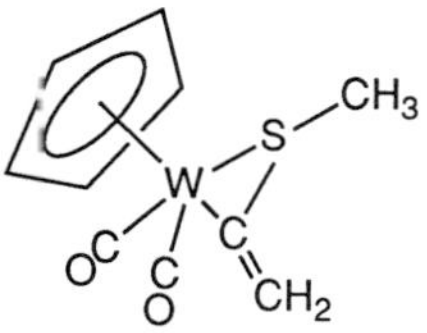

(9)

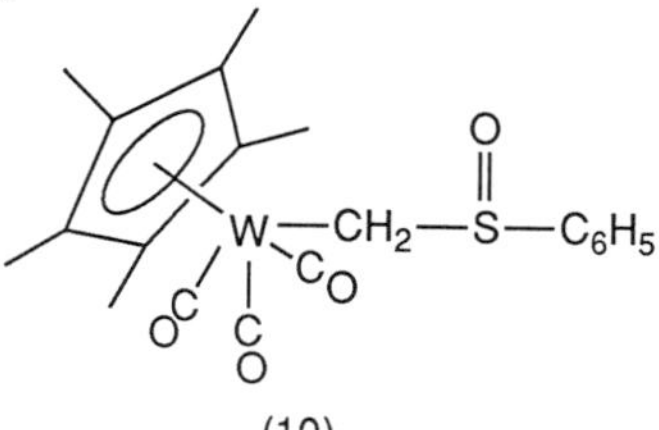

(10)

Table 1 Ligand mode assignments for $[WCl_4(XC\equiv CX)]_2$ (/cm^{-1})

	X =	H	D
A_1	ν_sCX	3097	2399
	νC=C	1628	-
	δ_sCX(i.p.)	956	740
B_1	γCX(o.o.p.)	675	587
B_2	ν_{as}CX	3075	2253
	δ_{as}CX(i.p.)	754	632

The dimeric complexes $[WCl_4(HC{\equiv}CH)]_2$ and $[WCl_4(DC{\equiv}CD)]_2$ have been reported. The assignments proposed from their i.r. spectra to modes of the complexed alkyne are listed in Table 1.[21] νCC modes have been assigned for a large number of tungsten alkyne complexes: $WI_2(CO)L_2(\eta^2\text{-}RC_2R)$, where L = various phosphines, R = Me or Ph;[22] $WI_2(CO)L(\eta^2\text{-}RC_2R)_2$, where L = $OC(NH_2)_2$, $SC(NH_2)Me$, $SC(NH_2)_2$ or $SC(NMe_2)_2$, R = Me or Ph;[23] $W(CO)[SC(NH_2)_2](dppm)(\eta^2\text{-}MeC_2Me)^+$ and related complexes containing substituted thiourea ligands;[24] and $[WI(CO)(NCMe)\text{-}\{Ph_2P(CH_2)_nPPh_2\}(\eta^2\text{-}RC_2R)]^+$, where n = 1 - 6, R = Me or Ph.[25] All show that the CC bond has been weakened very greatly, by both σ-donation and π-back-donation.

Detailed i.r. data have been reported for $MeM(CO)_5$ and $MeM(CO)_4(N_2)$ in matrices, as CH_3, CHD_2, CD_3 and $^{13}CH_3$ isotopomers, where M = Mn or Re. The replacement of one CO by N_2 leads to significant weakening of the C-H bonds, and strengthening of the $M\text{-}CH_3$ bond.[26] Raman and F.T.i.r. data for Cp_2M, where M = Mn, Fe or Ru, were used to carry out normal coordinate analyses for these complexes.[27] (11), where R is a very bulky group (CMe_3, cyclohexyl, 2,3,5-tri-t-butylphenyl) have a νCO (acyl) band near 1500 cm^{-1}, compared to about 1600 cm^{-1} in $RCORe(CO)_5$.[28]

A comprehensive study has been made of the CH_2 modes of $FeCH_2$ (from Fe + CH_2N_2 in argon matrices), involving ^{13}C and D substitution. Thus, for $Fe^{12}CH_2$ ν_sCH_2 is at 2941.6 cm^{-1}, $\nu_{as}CH_2$ at 3011.5 cm^{-1}. There was also evidence for N_2FeCH_2, with νNN at 1812 cm^{-1}, which may indicate 'side-on' coordination, as the value is much lower than normal for $M\text{-}N{\equiv}N$ coordination.[29] $Fe(CO)_4(CF_3)_2$ has νCF_3 at 1060 cm^{-1}, and δCF_3 at 710 cm^{-1}, in addition to the expected $Fe(CO)_4$ modes.[30]

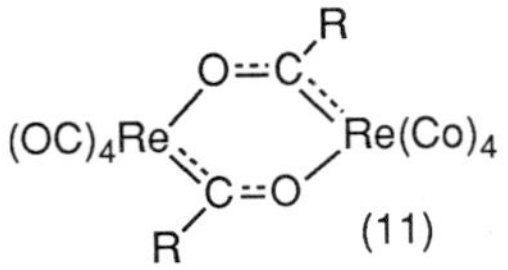

(11)

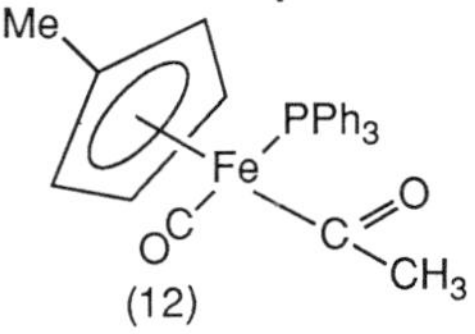

(12)

(12) has νC=O of $COCH_3$ at 1595 cm^{-1}, with νC=C of Ph at 1570 and 1585 cm^{-1}, νP-C of PPh_3 at 1090 cm^{-1} and νC≡O at 1910 cm^{-1}.[31] $(Indenyl)(OC)Fe(\mu\text{-}CO)_2Fe(indenyl)(COCH_3)^-$ gave the following νCO bands: 1906 cm^{-1} (terminal), 1702 cm^{-1} (bridging) and 1568 cm^{-1} (acyl).[32] The complex (13) has νC=O

at 1608 cm^{-1}, with νC≡O 1908 cm^{-1}. The Cp* analogue has values of 1585 and 1892 cm^{-1} respectively, showing the increased electron donor ability of Cp* compared to Cp itself.[33]

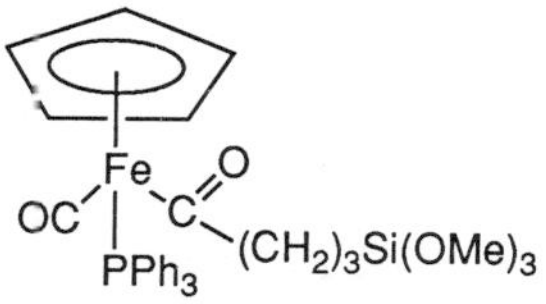

(13)

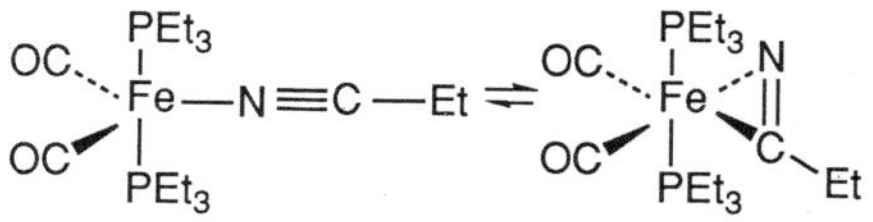

(14)

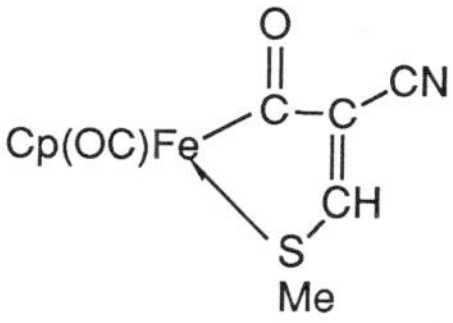

(15)

The presence of two sets of νCO and two of νCN shows the existence of an equilibrium mixture for (14). The η^2 form has νC=N at 1625 cm^{-1}, the η^1 form νC≡N at 2112 cm^{-1}.[34] (15) has νC=O at 1590 cm^{-1}, and νC=C at 1575 cm^{-1}.[35] Variable-temperature i.r. spectra of (η^4-norbornadiene)$Fe(CO)_3$ and related complexes showed the coalescence of νCO bands. This was interpreted in terms of CO site exchange.[36] The Raman and F.T.i.r. spectra of $(MeCp)_2Fe$ were used to carry out a normal coordinate analysis for this molecule (and also for the Ru and Os analogues).[37]

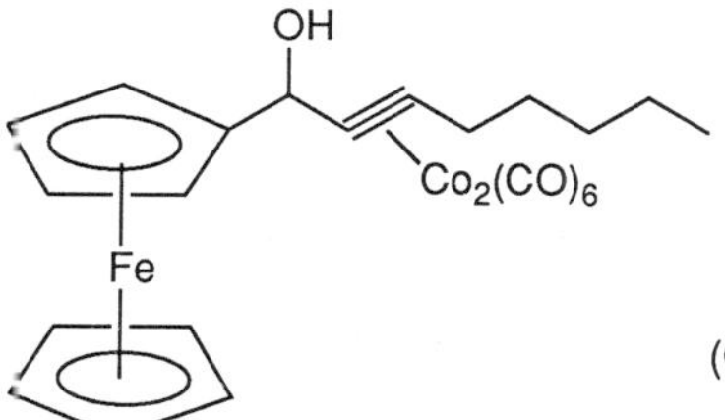

(16)

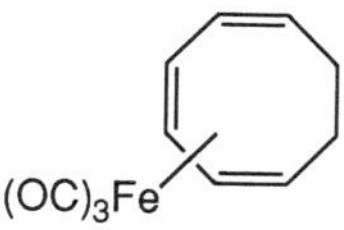

(17)

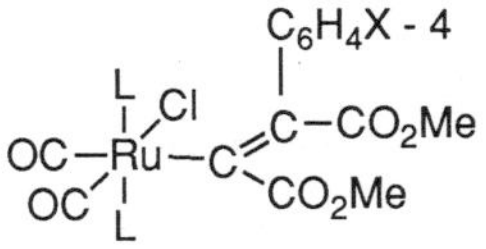

(18)

(16) has νOH at 3582 cm^{-1}.[38] I.r. spectra were reported for the bis(arene)iron(II) salts, $(Ar_2Fe)^{2+}(PF_6)_2^{2-}$, where Ar = mesitylene, 1,2,4,5-, 1,2,3,5- and 1,2,3,4-tetramethylbenzene, pentamethylbenzene or ethylbenzene. The νC=C band of the ring is near 1550 cm^{-1}, i.e. lower than in the free arene, but not so low as in $ArFeCp^+$.[39] The νC=C band (1670 cm^{-1}) appears in the i.r. spectrum of (cyclo-octatetraene)$Fe(CO)_3$ on electrolytic reduction, showing the production of a localised C=C bond, as in e.g. (17).[40]

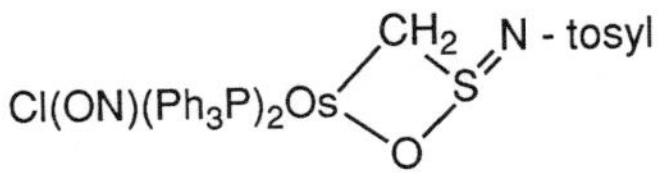

(19)

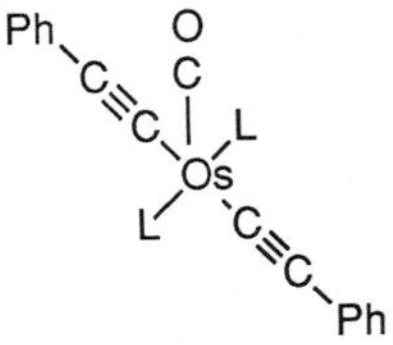

(20)

(18), where X = H, Cl, Me or OMe, have νC=O in the range 1695 - 1701 cm^{-1}.[41] The maleic anhydride complex $Ru(\eta^2-C_4H_2O_3)(CO)(CN-p-tolyl)(PPh_3)_2$ has νC=O for the π-bound maleic anhydride at 1789 and 1729 cm^{-1} (cf. 1850, 1780 cm^{-1} in the free ligand). νC≡O is much higher than in the precursor $Ru(CO)(CH-p-tolyl)(PPh_3)_2$, showing the good π-acceptor properties of the maleic anhydride.[42]

(19) has νNO at 1709 cm^{-1}, as expected for an Os(II) species.[43] The complex (20), where L = P^iPr_3, has only one νC≡C band in the i.r. spectrum (2060 cm^{-1}), showing the trans coordination of the two alkynyl ligands.[44] $[Os(NH_3)_5(\eta^2-PhC{\equiv}CPh)]^{n+}$ have νCC at 1910 cm^{-1} when n = 2, but 1815 cm^{-1}, when n = 3. The latter is characteristic of an electron-deficient diphenylacetylene complex.[45] A vibrational assignment for $Os_2(CO)_8(\mu_2-\eta^2-C_2H_4)$, with its deuteriated and ^{13}C-substituted analogues, is consistent with the 1,2-di-σ-bonding mode. Some of the assignments are listed in Table 2.[46]

Table 2 Some vibrational assignments for $Os_2(CO)_8(\mu_2-\eta^2-C_2H_4)$ ($/cm^{-1}$)

νCH_2	2936 (b_1), 2917 (a_2), 2910 (a_1), 2843 (b_1)
δCH_2	1445 (a_1), 1434 (b_1)
ωCH_2	1225 (b_1), 1144 (a_1)
τCH_2	1083 (a_2), 1012 (b_2)
νCC	981 (a_1)
ρCH_2	837 (a_2), 754 (b_2)

The i.r. spectrum of the coordinatively-unsaturated $CH_3CO-Co(CO)_3$ contains νC=O (acyl) at 1685.9 cm^{-1}, with a CH_3 deformation at 1344 cm^{-1}. νC≡O bands were at 2081.1, 2008 and 1977.3 cm^{-1}.[47]

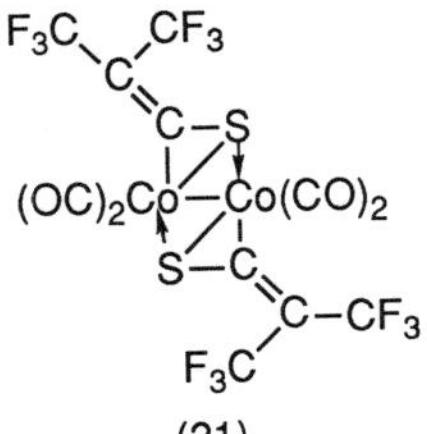

(21)

(21) has νC=C at 1630 cm^{-1}, showing C,S-coordination of the thioketene ligand.[48] $Co(C_2H_4)(PMe_3)_3$ has an i.r. band assigned as νC=C + δCH_2 at 1160 cm^{-1}, corresponding to strong bonding to the Co. ν_s and $\nu_{as}PC_3$ are at 655 and 702 cm^{-1} respectively.[49] $\{(PhC{\equiv}C)_2P[(\eta^2\text{-}C{\equiv}CPh)Co_2(CO)_5]\}_2$ has νCC of the η^2-alkyne-bridging ligand at 1556 cm^{-1} (compared to 2170 cm^{-1} in the free ligand).[50] The diolefin complexes $Co(L_2)(PMe_3)_2$, where L_2 = $[CH_2{=}CHSi(OMe)_2]_2O$, $[CH_2{=}CHSi(OEt)_2]_2O$, $[CH_2{=}CHCH_2]_2SiMe_2$ or $[(CH_2{=}CHCH_2)_2N]_2CH_2$, all have νC=C + δCH_2 at 1008 - 1060 cm^{-1}, and no νC=C band near 1600 cm^{-1}, showing coordination of both C=C bonds in each case.[51]

It was possible to identify an acylrhodium tetracarbonyl, $RC(O)Rh(CO)_4$, by in situ i.r. spectroscopy, as an intermediate in a rhodium-catalysed hydroformylation reaction at 25°C. For R = $-(C_2H_4)C(CH_3)_3$, νC=O (acyl) was at 1698 cm^{-1}, with νC≡O 2111, 2065, 2039 and 2020 cm^{-1}.[52] (22), where R = ^{t}Bu, has νC=O at 1675 and 1680 cm^{-1}.[53] (23), where M = Rh or Ir, have νCC at 1831 cm^{-1} (Rh), or 1830 cm^{-1} (Ir), showing strong π-coordination as indicated.[54]

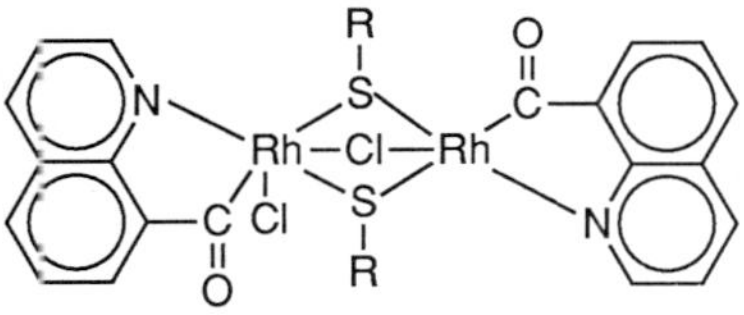

(22)

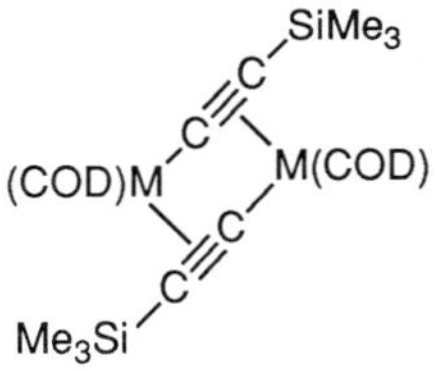

(23)

$\{[(\eta^5\text{-}C_5H_5)Rh(\mu\text{-}DMA)(\mu\text{-}Ph_2Ppy)Rh(CO)(\mu\text{-}Cl)]_2.CH_2Cl_2\}$, where DMA = dimethylacetylene dicarboxylate, Ph_2Ppy = 2-(diphenylphosphino)pyridine, was shown by X-ray diffraction to contain a μ-η^2-parallel cis-dimetalated (olefinic) bonding mode, and νCC was seen at 1615 cm^{-1}.[55] νCO and/or νCH i.r. bands were identified for $CpM(CO)_2$, $Cp^*M(CO)_2$, and $CpM(C_2H_4)_2$, where M = Rh or Ir, in alkali-metal Y zeolites. The results were consistent with internal confinement of the organometallics in the zeolite cages.[56]

νCO for the first reported Rh-Si bonded complex, $Rh[Si(CH_2CH_2PPh_2)_3](CO)$, is at 1940 cm^{-1}.[57]

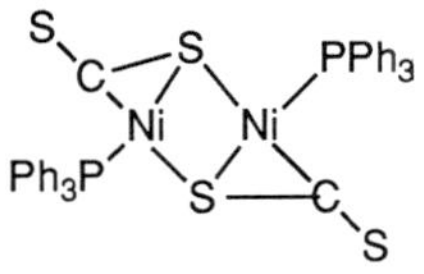

(24)

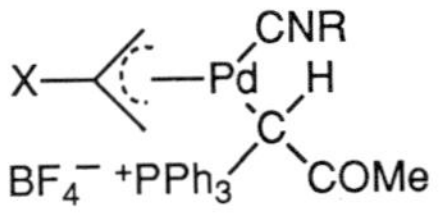

(25)

The η^2-CS_2 complex (24) has a characteristic νCS band at 1156 cm^{-1}.[58] The i.r. and Raman spectra for C_6H_6 on Ni/SiO_2 surfaces were consistent with a lowering of the symmetry of the molecule to between C_{6v} and C_{3v}.[59]

The complexes (25), where X = H, R = p-C_6H_4OMe or CMe_3; X = Me, R = Me, CMe_3, p-C_6H_4OMe, C_6H_{11} or p-$C_6H_4NO_2$, all have νC=O of the ylide ligand at 1630 - 1634 cm^{-1}, characteristic of Pd^{II}-(ylide) systems with keto-stabilised ylides.[60] The methoxycarbonyl ligands in $\{M(pnp)[C(=O)OMe]\}^+$, where M = Pd or Pt; pnp = 2,6-bis(diphenylphosphino-methyl)pyridine, have νC=O at 1665, 1600 cm^{-1} (Pd) or 1620 cm^{-1} (Pt), and νC-O-C at 1065 cm^{-1} (Pd) or 1028 cm^{-1} (Pt).[61] Some re-assignments were proposed from the i.r. and Raman spectra of $[Pd(C_2H_4)X_3]^-$, where X = Cl or Br, and their deuteriated derivatives.[62]

$\nu_s CF_3$ and $\nu_{deg} CF_3$ were assigned (near 1090 and two or three bands 940 - 1000 cm^{-1} respectively) in a wide range of Pt-CF_3 complexes, e.g. trans-$PtH(CF_3)(PR_3)_2$], where R_3 = Ph_3, $MePh_2$, Me_2Ph, Me_3 etc.[63] νCF modes in $Pt(CF_3)I(bipy)$ (1105, 1010 cm^{-1}) were shifted to higher wavenumbers than in $Pt(CF_3)(CH_3)(bipy)$ (1090, 960 cm^{-1}) - showing the sensitivity of the C-F bonds to variation of electron density at the platinum.[64] A normal coordinate analysis for $C_2H_4.PtCl_3$ - a model for Zeise's salt - suggested that there were only small kinematic couplings between C_2H_4 and skeletal modes.[65]

The kinetics of formation of ethylidyne on Pt(111) surfaces were followed by an increase in intensity of the $\delta_s CH_3$ mode at 1339 cm^{-1}.[66] $[(Ph_3P)(PhC{\equiv}C)Pt(AuPPh_3)_6]^+$ has νC≡C at 2044 cm^{-1}.[67] $[Pt_2(\mu\text{-}C_6F_5)_2(C_6F_5)_4]^{2-}$ and related species gave characteristic i.r. bands from the bridging and terminal C_6F_5 ligands.[68] $Pt(C_6F_5)I(CO)(PPh_3)$ has a νCF band at 960 cm^{-1}, compared to the value for $Pt_2(C_6F_5)_2(CO)_2(PPh_3)_2$. The increase was related to the higher oxidation state of platinum in the former.[69] Characteristic i.r. bands of η^1-C_6Cl_5 ligands were identified in $[Pt(C_6Cl_5)_2\{(\mu\text{-}Cl)Ag(PPh_2Me)\}_2]$ and related species;[70] and in $[Pt_2Ag(\mu\text{-}Cl)_2(C_6Cl_5)_4]_2^{2-}$.[71]

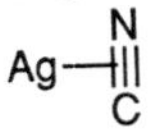

(26)

Co-condensation of Cu atoms and C_2H_4 or C_2H_4/Ar gave $Cu(C_2H_4)_n$, where n = 1, 2 or 3. νC=C modes were assigned as follows: 1518 (i.r.)/1530 (Raman) cm^{-1} (n = 3), 1510 (i.r.) cm^{-1} (2) or 1474 (i.r.) cm^{-1} (1). The species with n = 1 was believed to be dimeric.[72]

The coordinated dichloromethane in $Ag_2(CH_2Cl_2)_4Pd(OTeF_5)_4$ has $\nu_{as}CCl_2$ shifted by 18 cm^{-1} to lower wavenumber than for free CH_2Cl_2.[73] SERS for acrylonitrile adsorbed on a silver electrode gives two νCN bands, i.e. near 2100 cm^{-1} due to a π-type adsorbate, (26), and 2244 cm^{-1} due to a σ-type adsorbate, Ag-N≡C.[74] SERS of acetylene and other alkynes on Ag, Au or Au covered with a monolayer of under-potential deposited Ag show characteristic νC≡C modes in all cases.[75]

Overtones of νCH modes in $CdMe_2$ and $HgMe_2$ were observed using intracavity dye-laser photoacoustic spectroscopy.[76] A complete normal coordinate analysis has been carried out for $CH_3HgC{\equiv}CH$ and deuteriated derivatives.[77]

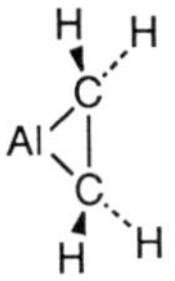

(27)

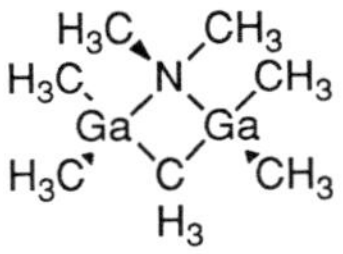

(28)

Al and CH_4 react photochemically in solid methane to form CH_3AlH, for which νCH bands are at 2938 and 2856 cm^{-1}, δCH_3 1162 cm^{-1}, and $\rho_r CH_3$ 730 cm^{-1}.[78] Several vibrational modes were assigned to AlC_2H_4, seen in the i.r. spectrum of an argon matrix containing Al and C_2H_4. These were all consistent with the form (27) (data included results for $Al^{13}C_2H_4$, AlC_2D_4 and $AlCH_2CD_2$.[79] δCH_3 bands in the i.r. spectrum of the product of thermal degradation of $Me_3Ga.NMe_3$ are consistent with its formulation as (28).[80]

$Me_2Ge{=}S$ was identified in low-temperature matrices, with ρCH_3 (in-phase) at 850 cm^{-1}, $\delta_{as}CH_3$ (in-phase) at 1390 cm^{-1}, $\delta_{as}CH_3$ (out-of-phase) at 1229 cm^{-1}, and other CH_3 bands at 761 and 753 cm^{-1}.[81] The i.r. and Raman spectra of $Me_3SnCCSnMe_3$, and its d_6-analogue, show that the molecule has D_{3d} symmetry.[82]

A complete assignment of $CH{=}CH_2$ ligand modes was given from the i.r. and Raman spectra of $P(CH{=}CH_2)_3$.[83] Small shifts in νC=C modes and in νCH (ring and methyl) were found on the formation of 2:1 adducts of $SbCl_3$ with methyl-substituted benzenes.[84]

2. Complexes containing CO, CS or CSe Ligands.

νCO for CO adsorbed on a Ni/TiO_2 is at 1969 cm^{-1} - possibly indicating bridged coordination.[85] $[CpHf(CO)_4]^-$ and $[Cp^*Hf(CO)_4]^-$ are the first recorded carbonyl anions of Hf(0), with νCO at 1920, 1780 cm^{-1} (Cp), 1915, 1774 cm^{-1} (Cp^*).[86]

I.r. spectroscopy (νCO) was used to monitor the photochemistry of $CpV(CO)_4$ in matrices at 12K. There was evidence for e.g. $(\eta^3-C_5H_5)V(CO)_4$, $(\eta^3-C_5H_5)V(CO)_5$ species (the latter in CO matrices), as well as $(\eta^5-C_5H_5)V(CO)_n$, where n = 1 - 3.[87] $(C_5Me_5)M(PMe_3)_2(CO)Cl_2$, where M = Nb or Ta, have νCO values consistent with a trans arrangement of the PMe_3 ligands.[88]

The νCO bands were used to monitor the formation of $\eta^5-C_5H_5-Cr(CO)_2L$, where L = unidentate tertiary phosphines, from the 17-electron species $\eta^5-C_5H_5Cr(CO)_3$.[89] $Cp_4Cr_4-(\mu_3-CO)_2Se_2$ has νCO at 1630 cm^{-1} from the triply-bridging carbonyls.[90]

The absolute intensities of νCO and νCS bands in $(\eta^6-RC_6H_5)-Cr(CO)_2(CS)$, where R = H, Me, Cl or CO_2Me, in CS_2 solution were shown to be dependent on the electronic nature of the substituent R in the arene ring. The intensities correlate better than wavenumbers with the Hammett σ° constants for R.[91] It was possible to follow isomerisation processes for stereochemically non-rigid $Cr(CO)_2(CX)[(MeO)_3P]_3$, where X = O, S or Se, by νCO i.r. bands.[92] Cotton-Kraihanzel force constants were calculated for a large number of (substituted arene)$Cr(CO)_3$ complexes. These were related to Hammett-Taft and Swain-Lipscomb parameters.[93] The low νCO wavenumbers (1823 and 1724 cm^{-1}) of $[As_7Cr(CO)_3]^{3-}$ show the powerful electron-donor properties of the As_7^{3-} ligand.[94]

Picosecond i.r. spectroscopy was used to identify the νCO bands of 'naked' $Cr(CO)_5$ from the photodissociation of $Cr(CO)_6$ in cyclohexane or THF. The C_{4v} form has a band near 1980 cm^{-1}, with a D_{3h} form characterised by one at 1920 cm^{-1}.[95-6] Irradiation of $Cr(CO)_5(NH_3)$ in Ar/H_2 matrices gave good yields of $Cr(CO)_5(H_2)$, which has νH_2 bands at 3087, 3027 and 2997 cm^{-1}.[97] Time-resolved i.r. was used to follow the kinetics of CO for C_2H_4 substitution in $(C_2H_4)Cr(CO)_5$.[98]

CO on Mo/SiO_2 catalysts forms Mo^{n+}-(CO), where $3 \leqslant n \leqslant 5$, for which νCO was at 2203, 2196 cm^{-1} (n = 5), 2190, 2181 cm^{-1} (4) or 2172 cm^{-1} (3), together with a number of more complicated species.[99]

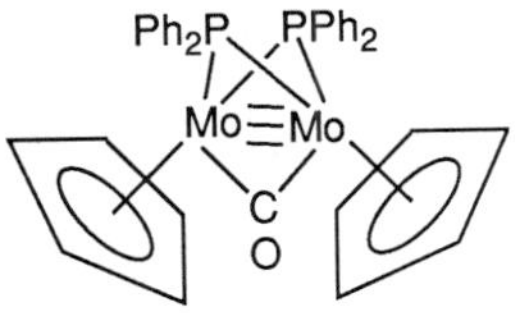

(29)

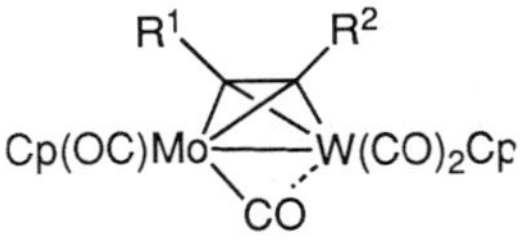

(30)

νCO from the bridging carbonyl in (29) is at the unusually low value of 1674 cm^{-1}.[100] The semi-bridging carbonyl group has νCO near 1825 cm^{-1} in all of the complexes (30), where $R^1 = R^2$ = Me or Ph; R^1 = Et or Ph, R^2 = Me.[101] νCO for $Mo(CO)_2(PBu_3)_2$(L-L), where L-L is an α-di-imine ligand, are all very low, showing the strong electron-donor properties of PBu_3.[102]

The νCO values of $MoX_2(CO)_3$ (probably polymeric), where X = F, Cl or Br, shift to lower wavenumbers with decreasing electronegativity of X.[103] In $[Fe_3S_4(SEt)_3Mo(CO)_3]^{3-}$ the νCO values of 1750 and 1864 cm^{-1} are consistent with the presence of Mo(0), and poor π-accepting properties for the $Fe_3S_4(SEt)_3{}^{3-}$ unit.[104] Five νCO bands are seen for (31), when R = H, as expected. When R = Me, however, only three i.r. νCO bands were seen. There was no obvious explanation for this observation.[105]

νCO bands were assigned and Cotton-Kraihanzel force constants calculated for cis-$Mo(CO)_4L_2$, where L = $Ph_2PC(O)Me$, $Ph_2P(2\text{-}C_4H_3S)$, $PhP(2\text{-}C_4H_3S)_2$, Ph_2PCN or PPh_3.[106] Thermal decomposition of the singly-bridged complexes $Mo_2(CO)_{10}$-[μ-$Ph_2P(CH_2)_nPPh_2$], where n = 4, 5 or 6, was followed by νCO bands in the i.r.[107]

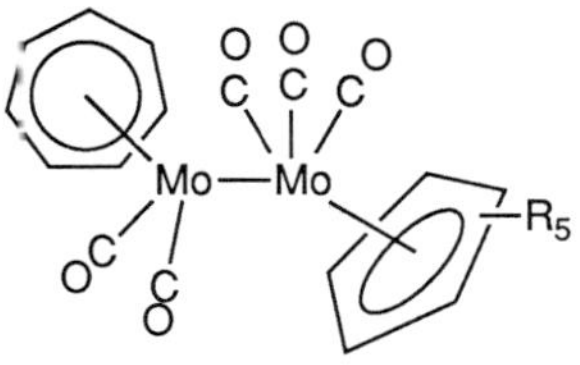

(31)

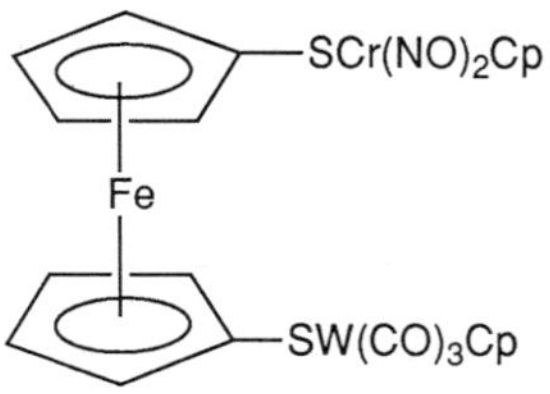

(32)

$[CpW(CO)_2(=CHPh)]^-$ has νCO at 1849 and 1713 cm^{-1}, as expected for such an anion.[108] νCO bands were reported for the new mer-tricarbonyl(η^2-olefin)(η^4-norbornadiene)-tungsten complexes, where olefin = (Z)-cyclo-octene or ethene.[109] νCO and νNO bands in ferrocene-1,1'-dithiolate complexes such as (32) show that the ferrocenyl dithiolate is a very strong donor ligand.[110]

Time-resolved i.r. spectra show that the laser photolysis of $W(CO)_6$ in the presence of L (= H_2, D_2 or N_2) forms $W(CO)_5L$, $W(CO)_4L$ and $W(CO)_4L_2$. The νCO bands for the species with L = H_2 or N_2 are very similar, showing that the W...L interactions are very similar in the two cases (probably a weak W→L donation).[111]

CO on a Rh-Mn/SiO_2 catalyst gave bands at 2055 cm^{-1} (terminal M-CO), 1815 cm^{-1} (M_2CO bridging) and 1713 cm^{-1} (M-OC form); these assignments were confirmed by ^{13}C substitution.[112] Photochemical reactions of $Ph_3SnM(CO)_3L$, where M = Mn or Re, L = α-di-imines, were followed by i.r. (νCO). The primary process is loss of CO.[113] νCO bands in (33) are at higher wavenumbers than for the precursor $[(C_5H_4Me)Mn(CO)_2]_2(\mu$-AsCl) - consistent with the positive charge on the former.[114] A correlation has been established between the pattern of νCO bands in the i.r. spectra and the substitution geometry in $M_2(CO)_{10-n}L_n$, where M = Mn or Re, L = CNR, PR_3 (R = alkyl or aryl), n = 1 - 3.[115]

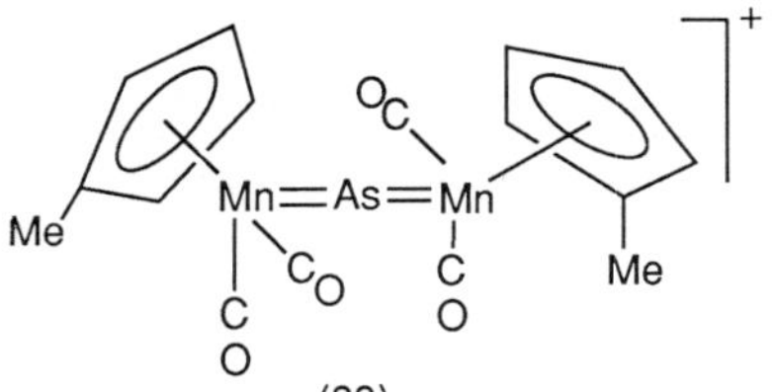

(33)

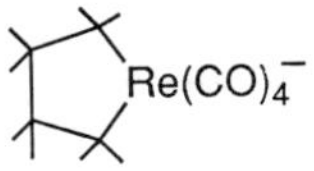

(34)

Photochemical cis/trans isomerisation of (η^5-C_5Me_5)Re-$(CO)_2X_2$, where X = Me, Cl, Br or I, was followed by i.r. (νCO).[116] The oxidation of the ferrocene rings in $FcRe(CO)_5$, fac-$(FcPPh_2)_2Re(CO)_3Cl$ and related species leads to an increase in νCO. The magnitude of the shift decreases with increased separation of the Fc and CO groups.[117] There is i.r. (νCO) evidence for the formation of cis-$Me_2Re(CO)_4^-$ and the unstable rhenacyclopentane (34) from $Re(CO)_4^{3-}$.[118]

Reactions of $Re(CO)_5^-$ with the dimers $Mn_2(CO)_{10}$, $Co_2(CO)_8$ and $Cp_2Mo_2(CO)_6$ were followed by i.r. (νCO) stopped-flow spectroscopy.[119] I.r. spectroscopy was used to identify intermediates in the matrix-isolation photochemistry of $Re_2(CO)_{10}$ in the presence of N_2. There was evidence for ax-$Re_2(CO)_9$, ax-$Re_2(CO)_9N_2$ etc.[120] The i.r. spectrum of $Re_2(CO)_{10}$ on an Al_2O_3 surface gave evidence for the slow formation of $(HO)_4Re_4(CO)_{12}$.[121] The νCO bands of $[Re_7C(CO)_{21}HgX]^{2-}$, where X = Cl, Br, CN, Ph, Me or CMe_3, are as expected for an Re_7M complex. The values of νCO increase with the electronegativity of X.[122]

F.T.i.r. data (νCO) for myoglobinCO show the existence of three major substates, each with a characteristic angle of tilt between the CO and the haem normal.[123] Time-resolved resonance Raman spectra were used to monitor the photo-dissociation of the CO adduct of haemoglobin.[124] The kinetics of recombination of myoglobin and CO were followed by fast time-resolved i.r., monitoring the νCO band of MbCO at 1945 cm^{-1}.[125] Low-temperature flash photolysis, with νCO i.r. studies, was used to probe the influence of the distal histidine (His-64(E7)) of sperm-whale myoglobin on the orientation of bound CO.[126] Changes in νCO, νCN and νNO for CO, CN^- or NO coordinated to E. coli sulphite reductase haemoprotein with iron oxidation state changes were observed using resonance Raman spectroscopy. Thus, νCO shifts from 1904 to 1920 cm^{-1} on going from the reduced to the oxidised form.[127]

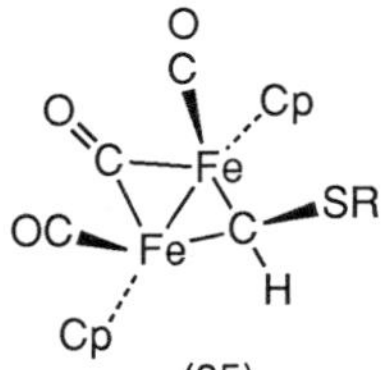

(35)

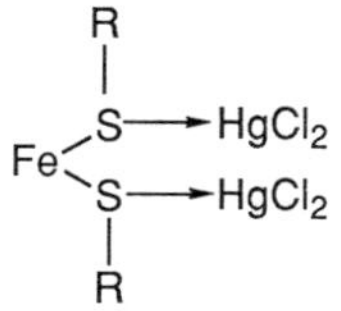

(36)

The i.r. (νCO) spectra of (35), where R = Bz, CH_3 or allyl, show that the predominant isomers have cis cyclopentadienyl groups as shown (with possibly a small amount of the trans form).[128] νCO for the $HgCl_2$ adducts of $Fe(CO)_2(dppe)(SR)_2$, where R = ${}^{s}Bu$, Ph, p-Me-, p-OMe or p-NHC(O)MeC_6H_4, dppe = $Ph_2PCH_2CH_2PPh_2$, are 30 - 50 cm^{-1} higher than in the original complexes. This is ascribed to bonding of the type (36).[129] $[CpFe(CO)_2(2\text{-methylthiophene})]^+BF_4^-$ had νCO of the $CpFe(CO)_2^+$ fragment at 2072 and 2031 cm^{-1}.[130] The i.r. νCO bands of $[In\{Fe(CO)_2(\eta\text{-}C_5H_4R)\}_3]$, where R = H or Me, show that they are not dissociated in solution, unlike the analogous Mo-containing species.[131]

Picosecond time-resolved i.r. was used to study the primary photoproducts of $[CpFe(CO)_2]_2$ in cyclohexane. There was evidence for a structure with Fe-CO bridge bonds broken, but $CpFe(CO)_2$ formation is not the primary process.[132] νCO of $[CoFe(CO)_2]_2$ on oxide surfaces gave evidence of hydrogen-bonding of bridging carbonyls with surface hydroxyl groups on SiO_2 or Al_2O_3.[133] $[Fe(CO)_3(PPh_3)]^{2-}$ has two νCO bands at the expected low wavenumbers (1784 and 1700 cm^{-1}). The presence of two bands is also consistent with local C_{3v} symmetry.[134]

$Fe(CO)_5$ adsorbed on dehydrated Na-Y zeolite gives i.r. (νCO) bands due to zeolite-$Fe(CO)_4$, trigonal-bipyramidal and square-pyramidal $Fe(CO)_5$ bonded to strong Lewis acid sites and $Fe(CO)_5$ in the supercages weakly bonded to Na^+.[135] Adsorption of $Fe(CO)_5$ on γ-Al_2O_3 gives molecules with approximately C_{3v} symmetry bonded via Fe-CO-Al^{3+} units. Bonds to Al^{3+} at tetrahedral sites give a 150 - 200 cm^{-1} decrease in νCO, while bonds to Al^{3+} at octahedral sites give a decrease of 100 - 150 cm^{-1}. νCO from non-bonded CO groups are raised with respect to the free molecule values.[136] Substitution reactions in the 33-electron dinuclear radical $Fe_2(CO)_7(\mu$-$PPh_2)$ was followed by monitoring i.r. νCO absorptions.[137]

CO on Al_2O_3-supported Ru gives νCO peaks in the i.r. due to $Ru^{n+}(CO)_2$, where n = 1 - 3, i.e. there is evidence for oxidative disruption of the Ru clusters.[138] Transient i.r. spectroscopy was used to identify the coordinatively unsaturated $Ru(CO)_x$, where x = 3 or 4. These are formed by excimer laser photolysis of gas-phase $Ru(CO)_5$.[139] Adsorption of CO on reduced titania-supported Ru gave i.r. evidence for the formation of some Ru(II) species, probably a $(TiO)_2Ru(CO)_3$ surface species.[140]

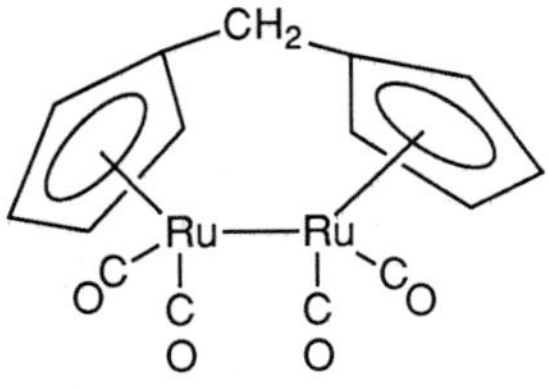

(37)

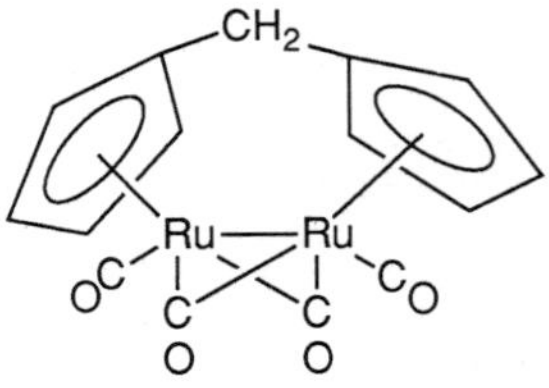

(38)

$[Ru_2(CO)_4(\eta^5{:}\eta^5$-$C_5H_4CH_2C_5H_4)]$ has terminal and (weak) bridging νCO bands in the i.r. This is consistent with the presence of two isomers, i.e. (37) and (38), with the former dominant.[141] ν(CO) (i.r.) was used to follow the thermal decomposition of $Ru_3(CO)_{12}$ on a carbon support. In the presence of H_2 there was evidence for the formation of $H_4Ru_4(CO)_{12}$.[142]

Decomposition routes for $Os_3(CO)_{12}$ were followed by dispersing the cluster on an oxygen-free carbon support, and measuring νCO by diffuse reflectance Fourier transform i.r. spectroscopy (DRIFTS).[143]

νCO (1916 cm^{-1}) for $[CoL(CO)]^+$, where L = 5,7,7,12,14,14-hexamethyl-1,4,8,11-tetra-azacyclotetradecane-4,11-diene, shows the presence of strong Co-CO back donation.[144] Co^{2+}-CO species were identified from the i.r. spectrum of Co adsorbed on partially-reduced Co_3O_4.[145] There was i.r. evidence for charge-transfer ion-pairs $X^+Co(CO)_4^-$, where X^+ = Cp_2Co^+, Na^+ or N-methylpyridinium, in solution, i.e. characteristic νCO bands of distorted $Co(CO)_4^-$, which are very similar to those for the corresponding crystalline salts.[146] νCO bands were used to obtain quantitative measures of the proportions of the components in the system $Co_4(CO)_{12}$- $Co_2(CO)_8$-$HCo(CO)_4$.[147]

νCO bands for CO adsorbed on Rh(III) surfaces [148] or on 0.5% Rh/10% Ce/Al_2O_3 [149] show that both terminal and bridged surface species are formed. Time-resolved i.r. (νCO) of the photolysis of $Cp^*Rh(CO)_2$ in liquid xenon or krypton gave evidence for the intermediate formation of $Cp^*Rh(CO)L$, where L = Xe or Kr. Addition of C_6H_{12} suggested that $Cp^*Rh(CO)(C_6H_{11})(H)$, is formed, perhaps via $Cp^*Rh(CO)$.[150] I.r. spectra of $RhL(CO)_2$, where HL = acetylacetone, 8-hydroxyquinoline or pivaloyltrifluoroacetone, containing ^{13}CO, show that the CO groups are non-equivalent.[151]

Intermediates in the photolysis of $Cp_2Rh_2(CO)_3$, $CpRh(CO)_2$ and $CpRh(C_2H_4)CO$ were followed by νCO (matrix i.r. or using time-resolved techniques).[152] F.T.i.r. in a rotating cryostat shows the Rh atoms and CO form $Rh(CO)_4$, $Rh_2(CO)_8$ and $Rh_4(CO)_{12}$. For $Rh(CO)_4$ $^{13}C/^{12}C$ data were consistent with D_{2d} geometry.[153] Adsorption of $Rh_6(CO)_{16}$ on partially hydroxylated MgO produces $Mg^{2+}[Rh_6(CO)_{15}]^{2-}$, which was characterised by in situ i.r. spectroscopy.[154] νCO bands for $PtRh_6(CO)_{16}^{2-}$ show considerable cation-dependence, because of ion-pairing.[155] $Rh_6(CO)_{16}$ trapped in NaY zeolite has νCO_t at 2096, 2046 and 2020 cm^{-1}, with ν(μ-CO) 1760 cm^{-1}.[156]

The i.r. spectrum of CO adsorbed on small Ir particles in NaY zeolite show the presence of a single, linear Ir-CO unit, with νCO at 2050 cm^{-1}.[157] νCO (i.r.) for $Na[HIr_4(CO)_{11}]$ show that in ether solution the salt is present largely as a contact ion-pair. In THF solution both free clusters and contact ion-pairs are present.[158]

I.r. (νCO) was used to follow the photochemical reactions of $Ni(CO)_2(^tBuDAB)$ and $Ni(CO)_2(2,6\text{-}^iPr_2C_6H_3DAB)$, where DAB = 1,4-diaza-1,3-butadiene. They have different structures, leading to different primary photoprocesses.[159] There was i.r. evidence (νCO only) for the formation of the dihydrogen complex $Ni(CO)_3(H_2)$ on u.v. irradiation of H_2/Ar matrices containing $Ni(CO)_4$.[160]

The i.r. spectrum of $Ni(CO)_4$, near ν_5 (νCO, t_2) has been measured in a pulsed molecular beam using diode laser spectroscopy. The band centre is at 2061.305(6) cm^{-1}.[161] Reaction of $Ni(CO)_4$ with an MgO oxide support produced high-nuclearity cluster anions, $Ni_5(CO)_{12}^{2-}$, $Ni_6(CO)_{12}^{2-}$ and $Ni_9(CO)_{18}^{2-}$, as shown by νCO bands in the i.r.[162] There is i.r. evidence for O-bonded adduct formation between $Ni(CO)_4$ and Bronsted acid centres at surfaces of zeolite Y or Al_2O_3. Thus, the νCO band is about 100 cm^{-1} lower than in free $Ni(CO)_4$.[163]

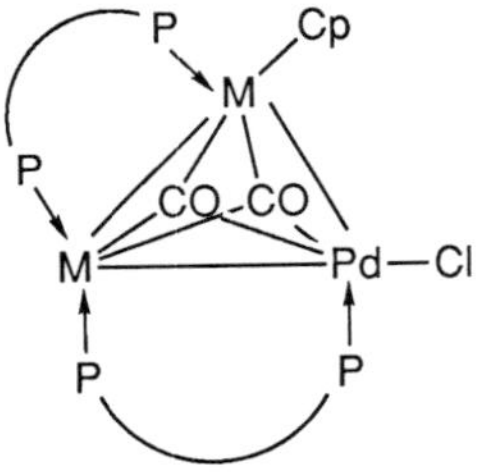

(39)

νCO bands were identified for palladium carbonyl clusters after adsorption of CO on Pd in sodium zeolite cages. Changes in the zeolite νOH bands indicated some interaction between the zeolite protons and the Pd carbonyl clusters.[164] (39), where M = Mo or W, M' = Pd; M = Mo, M' = Pt, P^P = $Ph_2PCH_2PPh_2$, all have νCO due to two triply semi-bridging carbonyls, at 1795 and 1755 cm^{-1}.[165] Several studies have been made of CO adsorbed on Pt surfaces.[166-8]

Adsorption of CO on copper-on-silica catalysts gave i.r. features due to νCO of Cu^{II}, Cu^I and Cu^0 species at the surface.[169] Time-resolved i.r. spectra were able to detect a Cu_B^+-CO bound transient intermediate in the photodissociation of a eukaryotic cytochrome oxidase-CO complex at ambient temperature.[170] Reaction of ^{63}Cu atoms with CO in inert hydrocarbon matrices at 77K was followed by i.r. spectroscopy. This showed the formation of D_{3h} $Cu(CO)_3$ as well as $Cu_2(CO)_6$.[171]

νCO for CO adorbed on silver electrodes shows only linear Ag-CO units. Similar data for gold electrodes showed linear Au-CO and also μ_2- and μ_3-CO ligands.[172] F.T.i.r. spectra of inert hydrocarbon matrices containing ^{197}Au and CO showed the presence of $Au(CO)_2$, AuCO, a gold microcrystallite-CO complex, and, after annealing, $Au_2(CO)_4$.[173]

I.r. spectra of Al and CO in an argon matrix gave evidence for the formation of $Al(CO)_2$, (OC)Al(OC) and, probably, Al(OC).[174] Indium atoms and CO in an argon matrix at 4K produce $In(CO)_2$, for which ν_sCO is at 2037 cm^{-1}, ν_{as}CO 1997 cm^{-1}.[175]

A theoretical calculation of the i.r. spectrum of $Si(CO)_2$, based on C_{2v} symmetry gave a good fit with experimental values, therefore suggesting a reassignment of the data from D_h to C_{2v} geometry for the latter.[176]

3. Boron Donors.

I.r. spectra of anionic BH_4^- complexes suggest that $[Sc(BH_4)_4]^-$ contains tridentate ligands, while $[Ti(BH_4)_4]^-$ contains both bi- and tridentate, and $[V(BH_4)_4]^-$ only bidentate ligands.[177] νBH modes of N_2H_4 adducts of $Ln(BH_4)_3$, where Ln = La, Gd or Lu, show that the hydrazine molecules are hydrogen-bonded to the terminal hydrogen atoms of the BH_4^- ligands.[178]

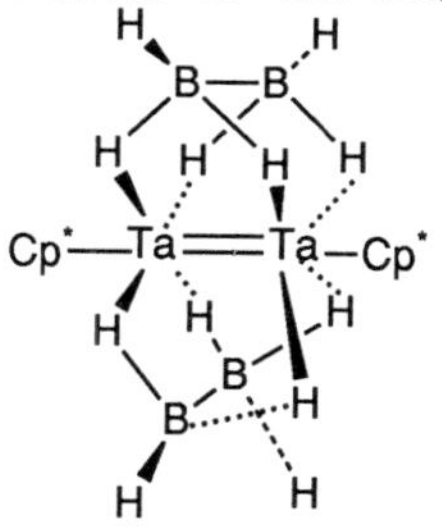

(40)

$Ti(\eta^2-BH_4)_2(dmpe)_2$ has νBH_t 2336 and 2299 cm^{-1}, νBH_{br} 2215 and 2170 cm^{-1}, i.e. typical of η^2-BH_4. $V(\eta^1-BH_4)_2(dmpe)_2$ has νBH bands at 2312 and 2095 cm^{-1}, with δBH_3 at 1057 cm^{-1}, characteristic of η^1-BH_4 (dmpe = $Me_2PCH_2CH_2PMe_2$).[179]

$V(BH_4)_3(PMe_3)_2$ has νBH at 2431, 2385 and 2087 cm^{-1} - indicative of bidentate BH_4^- ligands (confirmed by X-ray diffraction). Similar data were obtained for $[V(BH_4)H(PMe_3)_2]_2$ and $[V(BH_4)_2(PMe_3)_2]_2O$. [180]

(40) has νBH_t at 2417 cm^{-1} and νTaHB at 1784 cm^{-1} (which shift to 1800, 1328 cm^{-1}) when B_2D_6 is used. This is a rare example of coordinated B_2H_6 ligands.[181]

$Cp_2Th(BH_4)_2$ and $(indenyl)_2M(BH_4)_2$, where M = Th or U, all have νBH bands characteristic of bidentate BH_4 ligands.[182] $(\eta\text{-}C_6Me_6)U(BH_4)_3$ has νBH bands from tridentate BH_4 ligands at 2493, 2199 and 2122 cm^{-1}.[183]

4. Nitrogen Donors.

4.1 <u>Molecular Nitrogen, Azido- and Related Complexes</u>:- $Cp_2Zr(NSN)_2ZrCp_2$ and $CpCp''M(NSN)_2MCpCp''$, where M = Zr or Hf, containing the unit (41), all have νNSN bands in the range 1100 - 1195 cm^{-1}.[184]

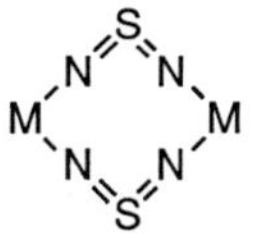

(41)

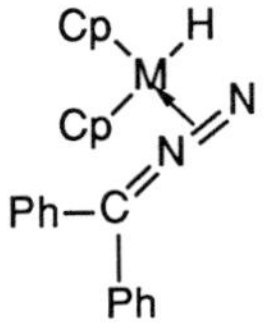

(42)

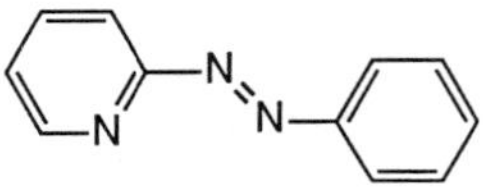

(43)

(42), where M = Nb or Ta, both have νNNC near 1430 cm^{-1}, compared to 2040 cm^{-1} in the free ligand, consistent with the π-bonding shown.[185]

νN=N in $M(CO)_4(2\text{-}PAP)$, where M = Cr, Mo or W, 2-PAP = (43), show a decrease of 121 - 136 cm^{-1} compared to the free ligand - consistent with extensive π-back-donation from M to the N=N group.[186] It is possible to separate cis and trans isomers of $Mo(N_2)_2(PMe_2Ph)_4$. The cis isomer has νN_2 (i.r.) at 2010 cm^{-1} (very strong) and 1935 cm^{-1} (very strong). For the trans form such modes are at 2010 cm^{-1} (medium) and 1923 cm^{-1} (very strong).[187] $[Mo_6O_{18}(NNMePh)]^{2-}$ has νN=N at 1589 cm^{-1}.[188] Such a mode is seen at 1560 cm^{-1} for trans-$[MoBr(N_2Ph)L]$, where L = $Me_8[16]aneS_4$.[189]

νNCl modes were reported for the following: $[MoF_5(NCl)]^-$;[190] $[MoCl_4(NCl)]_2$;[191] $[WF_5(NCl)]^-$;[192] and $[WCl_4(NCl)]_2$.[193]

νSN and νSF modes in $[Mn(CO)_5(NSF)]^+$ and $[CpFe(CO)_2(NSF)]^+$ gave weak i.r. bands 1366 - 1377 cm^{-1} and 640 - 650 cm^{-1} respectively.[194] νNN for the new dintrogen complexes (44), where R = Me, ^nBu, Cy, Ph or OMe, all show two bands due to the presence of more than one conformer in solution.[195]

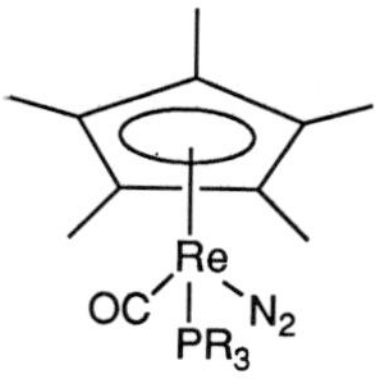

(44)

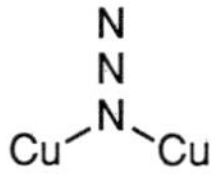

(45)

N_2 adsorbed on to a Ni(111) surface gives a characteristic νN_2 band near 2190 cm^{-1}, for a Ni-N_2 species.[196]

$[Ph_3PNPPh_3]_2[Cu(N_3)_4]$ has $\nu_{as}N_3$ in the i.r. at 2038 and 2045 cm^{-1}, consistent with C_i symmetry. $\nu_s N_3$ was estimated from combination bands to be near 1310 cm^{-1}.[197] The resonance Raman spectra of dinuclear copper(II)-azide complexes show ν_{as} of the Cu-NNN-Cu unit near 2025 cm^{-1}, while a unit (45) has $\nu_{as}N_3$ near 2065 cm^{-1}.[198]

I.r. data suggest that the adducts $Me_2Cd.N_2$, $Me_3Ga.N_2$ and $Me_2Cd.H_2$ all have unidentate structures, while $Et_3As.N_2$ and $Et_3As.H_2$ are bidentate.[199]

4 2. Amines and Related Ligands:- The i.r. spectra of (+)-, (-)- and (±)-$[M(en)_3]^{3+}$, where M = Cr or Co, contain characteristic band patterns for the $\Lambda(\delta\delta\delta)/\Delta(\lambda\lambda\lambda)$ conformation. Complexes with mixed ring conformations have more complex patterns, but these are not sufficient to identify specific conformations.[200]

Isomeric forms of $[Cl_2(NH^tBu)(PMe_2Ph)W\equiv]_2$ have νNH shifted significantly to lower wavenumber compared to $[Cl_2(NH^tBu)-(NH_2{}^tBu)W\equiv]_2$. [201] The isomers (46) and (47) gave characteristic νCN bands.[202]

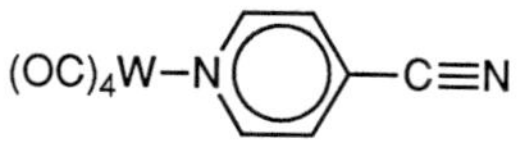

(46)

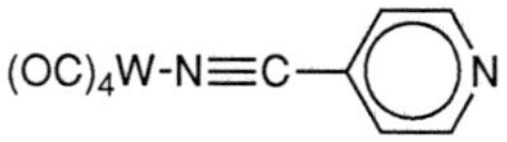

(47)

A large number of ligand modes were assigned for quinoline (=

Q) complexes, e.g. MQ_2Cl_2, where M = Mn, Fe, Co, Ni, Cu, Zn or Hg, and MQX_2, where M = Cd, Hg, X = Cl, Br, I etc.[203] The resonance Raman spectrum of $(OC)_4CoRe(CO)_3(bipy)$ contained resonantly-enhanced bands of the bipy ligand at 1607, 1563, 1492, 1323, 1276, 1175 and 1023 cm^{-1}.[204]

I.r. and Raman spectra gave numerous quite detailed ligand mode assignments for $M(2,6\text{-}L)Cl_2$, M = Fe or Cd, and $Cu(2,6\text{-}L)_2Cl_2$, where 2,6-L = 2,6-lutidine. All were based on C_{2v} symmetry for the coordinated ligand, although several bands were split in the square planar copper complex.[205]

There have been a large number of resonance Raman studies of Ru(II) complexes with bipyridyl and related ligands, largely to elucidate electronic changes on excitation.[206-13] The resonance Raman spectra of $[Ru_3O(OAc)\{(pyrazine)Ru(NH_3)_5\}_3]^{6+}$ showed enhancement of the totally symmetric pyrazine modes at 1605, 1230, 1085 and 700 cm^{-1}. These suggest that there is only weak coupling between the central and peripheral ruthenium atoms.[214]

SERS for pyridine adsorbed in rhodium-modified silver electrodes shows bands at 1046, 1020 and 646 cm^{-1} due to py at Rh sites (compared to 1036, 1010 and 625 cm^{-1} for py on silver alone).[215] Characteristic differences in ligand modes (especially the bipy ring deformation at about 400 cm^{-1}) were found between cis- and trans-$[Rh(bipy)_2Cl_2]^+$.[216]

The i.r. spectrum of Ni and NH_3 in argon or krypton matrices showed the presence of $NiNH_3$, $Ni(NH_3)_2$, $HNiNH_2$ and $HNiNH_2(NH_3)$.[217] $NiCl_2(NH_2NHCH_3)_2$ and its N-deuteriated analogue gave i.r. data which were used to carry out a normal coordinate analysis on the unit $NiNH_2NHCH_3Ni$.[218]

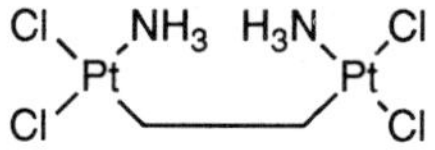

(48)

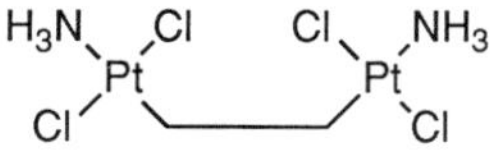

(49)

Characteristic differences were seen in the νNH band patterns for the isomers (48) and (49), with bridging 1,4-diamino-butane.[219]

I.r studies of the reactions of Cu and NH_3 in argon matrices gave the assignments listed in Table 3.[220]

Table 3 Vibrational assignments for $HCuNH_2$ (/cm^{-1})

	νNH	δNH_2	NH_2 wag
$HCuNH_2$	3329	1524	592
$HCu^{15}NH_2$	3325	1519	589
$DCuNH_2$	2444	1133	-

A very detailed study has been made of the i.r. spectra of $[Zn(en)_3]SO_4$, with ^{64}Zn, ^{68}Zn, N_2D_4 and C_2D_4 labelled analogues. It was possible to make a comprehensive assignment of the ligand modes of en.[221] I.r. and Raman spectra of $[M(1,3\text{-pn}))_3]^{2+}$, where M = Zn or Cd, 1,3-pn = 1,3-propanediamine, as halide salts, showed that both chair and boat conformation of the chelate rings were present.[222] Ligand mode vibrations in MX_2(4,4'-DTDP), where M = Zn or Cd, X = Cl, Br or I, 4,4'-DTDP = 4,4'-dithiopyridine, are consistent with bridging ligands, as in (50). νSS always occurs near 545 cm^{-1}.[223]

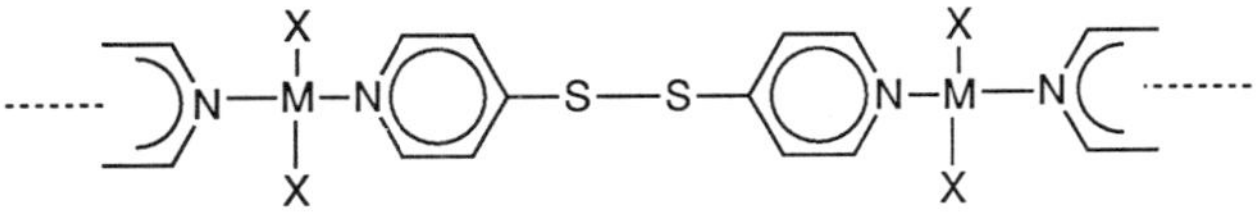

(50)

The following ND_3 modes were assigned in $SiF_3(ND_3)_2$: $\nu_{as}ND_3$ 2480 cm^{-1}; ν_sND_3 2370 cm^{-1}, $\delta_{as}ND_3$ 1163 cm^{-1}, δ_sND_3 1080 cm^{-1} and ρND_3 780, 665 cm^{-1}.[224]

4.3 Ligands containing >C=N< groups:- F.T.i.r. spectra for π-cation radicals produced by the oxidation of M(TPP), where M^{II} = Mg, Co, Cu or Zn, or M(TPP)Cl, where M^{III} = Mn or Fe, all show characteristic shifts in the macrocycle ligand modes. There was some evidence for the formation of a π-dication for Zn(TPP).[225]

Resonance Raman spectra of electrochemically-produced porphyrin radical cations derived from VO^{2+}, $FeCl^{2+}$, Ni^{2+} and Cu^{2+} compexes of OEP and TPP were used to give haem ligand assignments (with ^{15}N and D substitution). These show that the normal mode compositions are not significantly different from those for the neutral species, although there are appreciable shifts in wavenumbers.[226] Low-wavenumber deformation bands were identified from i.r. spectra of phthalocyanine complexes of Cr, Fe, Co and Zn.[227]

$Cl_5W=N=CMe_2$ has $\nu C=N$ of the aza-alkenylidene ligand at 1698 cm^{-1}. Other ligand modes were almost unshifted with respect to $Cl-N=CMe_2$.[228]

Resonance Raman effects for $Ph_3EM(CO)_3$(α-di-imine), where M = Mn, E = Ge, Sn or Pb; M = Re, E = Sn, α-di-imine = 1,4-diaza-1,3-butadiene, pyridine-2-carbaldehyde-imine or 2,2'-bipyridine, show that $\nu_s CN$ modes are very strongly affected. Hence, the lowest energy absorption band involves electronic transitions from M to the lowest π^* level of the α-di-imine.[229]

The i.r. and resonance Raman spectra of hydroxophthalocyaninatomanganese(III) gave assignments and enhancement profiles for a large number of ligand modes.[230]

Adducts of the meso-5,5,7,12,12,14-hexamethyl-1,4,8,11-N-tetra-azacyclotetra-deca-1,4,8,11-tetraene iron(II) complex with N-heterocyclic ligands, (51) (L = py, substituted pyridine etc.) give resonance Raman spectra which show that the electronic absorption band near 650 nm is a metal-di-imine charge transfer band. The band at 435 - 340 nm however, gave enhancement of ligand modes, so that this is a metal→N-heterocycle charge transfer band.[231]

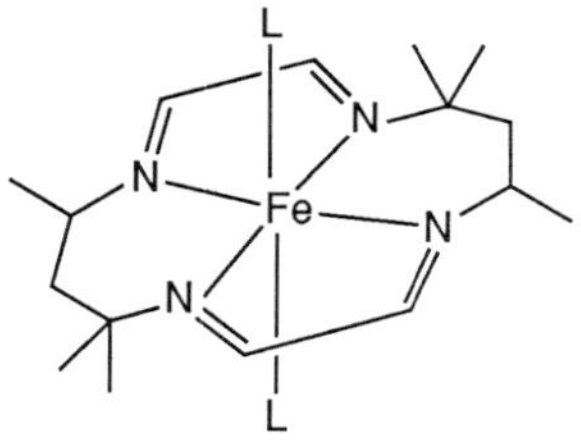

(51)

Transient u.v. resonance Raman spectroscopy was used to help elucidate the haemoglobin quaternary structure of photolysed carbonmonoxy haemoglobin.[232] Resonance Raman spectra of Fe(III), Co(III) and Cu(II) complexes of octaethylporphyrin and etioporphyrin I π-cation radicals all show that νC_aC_m, νC_aN and νC_aC_b decrease, while νC_bC_b increases on oxidation of the macrocycle by one electron.[233] Resonance Raman data, especially excitation profiles for ligand bands, for protoporphyrin IX in ferrocytochrome c suggest that vibronic coupling is weak for this molecule.[234]

The presence of H_2O or D_2O molecules near the cytochrome a of cytochrome c oxidase affects some resonance Raman bands in such a way as to suggest that they are adjacent to the formyl and vinyl groups of the cytochrome a.[235] Other resonance Raman results were reported and discussed for native bovine liver ferri-catalase;[236] metmyoglobin;[237] hydroxide form of ferri Aplysia myoglobin;[238] turnip cytochrome f;[239] horseradish peroxidase;[240] cyano- methaemoglobins;[241] human adult haemoglobin, sperm whale myoglobin etc.[242]

Time-resolved resonance Raman spectra of ruthenium(II) octaethylporphyrins gave evidence on the nature of photo-excited and photoproduct species. The behaviour of Ru(II)(CO)(OEP) in both coordinating and non-coordinating solvents is governed by both CO back-bonding and σ-donating character of the axial ligands.[243] OEP ligand modes in $\{[Ru(OEP)]_2\}^{n+}$, where n = 0, 1 or 2, are almost independent of n, and very similar to those found for monomeric Ru(OEP)(CO)L complexes. This shows that the oxidation is associated with metal electrons only.[244]

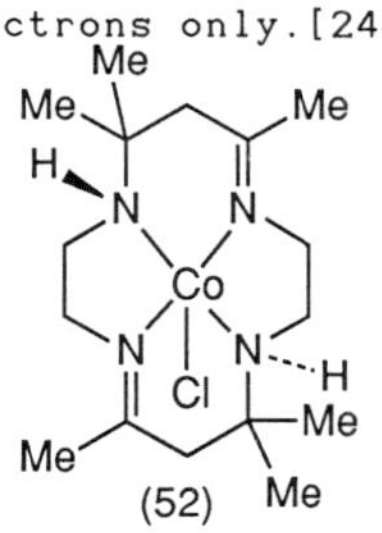

(52)

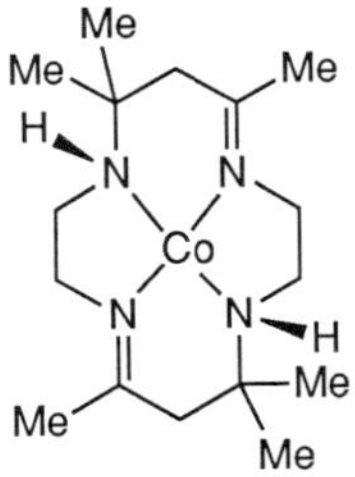

(53)

It is possible to distinguish between N-meso- and rac-isomers of $Co(L)Cl^+$ by characteristic i.r. bands near 1300 cm^{-1}. The isomers are (52) and (53).[245] The resonance Raman spectra of $Co(N\text{-}CH_3OEP)(OAc)$, where $N\text{-}CH_3OEP$ = N-methyl-octaethylporphyrin, and its isotopic analogues, show that $\nu C(\alpha)C(m)$ and $\nu C(\beta)C(\beta)$ are much lower than in Co(OEP). Several low-wavenumber out-of-plane deformations of the N-methylated pyrrole ring were also identified.[246] The resonance Raman spectra of CoPc (Pc = phthalocyaninato) at room temperature and at 80K gave excitation profiles which show that the low-wavenumber vibrations are influenced mainly by N-displacements, but the higher-wavenumber vibrations mainly by C(α) displacements.[247]

Core-size-sensitive resonance Raman bands for M(OEP) and M(PP), where M = Co, Ni, Cu or Zn; OEP = octaethylporphyrinato, PP = protoporphyrinato, show the expected relationships to complex structures.[248] The resonance Raman spectra of π-cation radicals of Co, Ni and Cu methyloctaethylchlorins gave evidence for marked core-size dependence for one band in the range 1500 - 1520 cm^{-1}, and another 1620 - 1650 cm^{-1}. The former is believed to involve C(b)C(b) and C(s)C(m) stretches, the latter C(s)C(m) only.[249]

The resonance Raman spectra of the water-soluble porphyrins M(TMpy-P4), where M = Co(III), Ni(II) or Cu(II); TMpy-P4 = tetrakis(4-N-methylpyridyl)porphyrinato cation, were used to probe their interactions with a number of nucleic acids.[250] Core-size marker bands were assigned in the protoporphyrin IX dimethyl esters of Co(II), Ni(II), Cu(II) or Zn(II); they show that the core sizes are Ni > Co; Cu > Zn.[251]

The resonance Raman spectra of $Ni(R\text{-}DAB)_2$, $Ni(CO)_2(R\text{-}DAB)$ and $Ni(CO)_2(R\text{-}pyca)$, where DAB = 1,4-diaza-1,3-butadiene; pyca = pyridine-2-carbaldehyde imine; R = ^{t}Bu, Cy etc., established characteristic shifts in spectra of tetrahedral and pseudo-planar complexes.[252] Characteristic resonance Raman bands were also seen for the p-phenylene- and m-phenylene-bridged dimers of NiII(TPP).[253] The resonance Raman spectra of the nickel tetrapyrrole cofactor of the S-methyl-coenzyme M methyl reductase enzyme from methanogenic bacteria showed the presence of both four- and six-coordinated forms.[254]

Nanosecond resonant CARS on Ni(OEP) gave evidence for photophysical transformations.[255] Resonance Raman spectra of Ni(OEP) in the triclinic and tetragonal forms have been reported. These crystalline forms contain respectively flat and ruffled porphyrins. In solution, the ligand bands show that the porphyrin has a ruffled structure, but the distinction from the planar form is not so great as in the 'ruffled' crystals.[256] Several other resonance Raman studies have been made on nickel-containing porphyrin species.[257-9]

The resonance Raman spectra of complexes of copper(II) with tetraphenylchlorin and tetraphenylbacteriochlorin (for several isotopic species) show that reduction of porphyrin to chlorin leads to a low effective symmetry, and hence a very complex vibrational pattern.[260]

Resonance Raman spectra of the anion radicals of Zn(II) complexes of TPP, TPP-β-d_8, TPP-(meso-$^{13}C)_4$ and TPP-($^{15}N)_4$, together with the neutral Cu(II) complexes of the same isotopic species, show that the reduction of the macrocycle gives Jahn-Teller distortion, so that the point group for the anion is D_{2h}.[261] Time-resolved resonance Raman data for the phenyl modes of zinc(II) tetraphenylporphyrin in the lowest triplet excited state show that they are unshifted compared to the ground (singlet) state. Thus there is no change in the geometry of the phenyl groups on excitation. The electronic changes are localised in the porphyrin ring.[262] Resonance Raman and SERS for $LnPc_2$, where Pc = phthalocyaninato, Ln = Lu or Yb, as thin solid films or films on silver are consistent with C_{4v} molecular symmetry.[264]

4.4 Cyanides, isocyanides and Related Complexes:- Solutions of $LiClO_4$ in CH_3CN gives a band at 2270 cm^{-1} due to CH_3CN coordinated to Li^+, probably as $Li(CH_3CN)_4^+$.[264]

$[Cp_2Ti(CH_3CN)_3]^{2+}$ has νCN of coordinated CH_3CN at 2293 cm^{-1}, compared with 2254 cm^{-1} for the free CH_3CN. This is consistent with a strong $CH_3CN \rightarrow Ti$ σ-donor interaction, but only a weak π-back donation.[265]

The i.r. spectra of $[PhC(NSiMe_3)_2MCl_3]_2$, where M = Ti or Zr, have νCN bands at about 1550 and 1360 cm^{-1}. These are consistent with benzamidino ligand coordination as (54).[266]

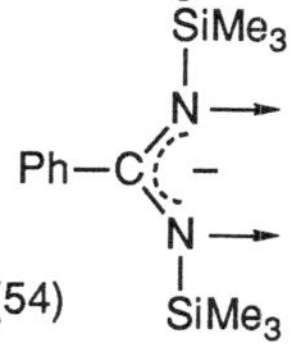

(54)

The i.r spectra of $Cr(CNR)_6^{n+}$, for R = Ph or substituted phenyls, n = 0 - 3, show that for Cr(III) complexes νCN decreases with increased electron donation due to σ(CNR) → Cr bonding. For decreased n, decreasing νCN values were found, due to increasing π-bonding effects.[267]

$Al_4[Mo(CN)_8]_3.24H_2O$ has an i.r. spectrum with νCN bands due both to non-bridging and bridging Mo-CN→Al cyano groups.[268] $[MoNCl_3(NC\text{-}C_2Cl_3)]_2$, where $NC\text{-}C_2Cl_3$ = 2,3,3-trichloroacrylonitrile, has νCN at 2275 cm^{-1}, compared to 2213/2240 cm^{-1} in the free ligand. This is due to Mo-NC bonding. Several other internal ligand modes were assigned.[269]

(55) has νCN bands consistent with the trans-geometry shown.[270] $[WNCl_3(NCPh)]_4$ has νCN of the benzonitrile ligand at 2280 cm^{-1}, showing W-NCPh coordination.[271] The formation of $CpW(CO)(CN^tBu)I_3$ from $CpW(CO)_2I(CNtBu)$ gave a characteristic shift in νCN of the CN^tBu ligand to higher wavenumbers due to the lower electron density at the metal.[272] νCN for the carbyne complexes $I_2(^tC_4H_9NC)_2L_2W{\equiv}CNEt_2^+$, where L_2 = bipy or.phen, is near 2169 cm^{-1}. The high value is due to the poor π-back-bonding due to the high oxidation state of the tungsten.[273] ReCl{CN(M)R}(dppe)$_2$, where M = $WCl_4(PPh_3)$, $ReOCl_3(PPh_3)$ or $CoCl_2(THF)$, R = ^tBu, have νCN in the range 1500 - 1600 cm^{-1}, consistent with their formulation as amino-carbyne-type complexes, i.e. (56).[274]

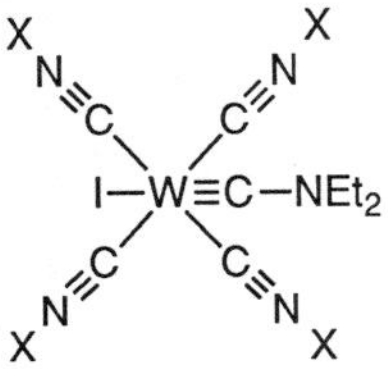

(55)

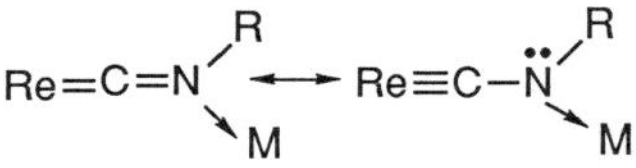

(56)

νCN of the coordinated trifluoromethylisocyanide ligand was assigned in $Cp^*Mn(CO)(CNCF_3)(\eta^2\text{-}C_8H_{14})$ (1768 cm^{-1}), $Cp^*Mn(CO)(CNCF_3)(THF)$ (1712 and 1692 cm^{-1}) and related species. Note the extremely low value in the presence of a purely donor ligand (THF).[275] $[ReCl_2(CN^iPr)_2(PMePh_2)_2]SbF_6$ has a single i.r. band due to νCN, at 2182 cm^{-1}, with a single νReCl band at 295 cm^{-1}, showing an all trans geometry.[276]

The number of νCN bands in anhydrous caesium nitroprusside shows the presence of two non-equivalent anions in each unit cell.[277] The resonance Raman spectrum of $Fe(CN)_5L^{3-}$, where L = 4-cyanopyridine, shows coordination via the pyridine N of the L ligand. The bridged complex $(NH_3)_5Ru\text{-}L\text{-}Fe(CN)_5^-$ shows a distinctly different resonance enhancement pattern from the mononuclear species.[278]

The formation of cyanide bridges in $\{[(bipy)(H_2O)PtNC]_2\text{-}M(CN)_4\}.nH_2O$, where M = Fe, Ru or Os, was shown by νCN in the i.r., above 2100 cm^{-1}, with the terminal νCN bands 2085 - 2090 cm^{-1}.[279] The νCN bands of (57), where M = Fe or Co, are much higher than in $Fe(CN)_6^{4-}$ - due to extensive electron donation to the bridging ligand, therefore cutting down back donation to CN^-.[280]

The post-resonance Raman spectrum of $(NC)_5Ru\text{-}CN\text{-}Ru(NH_3)_5^-$ contains νCN(bridge) at 2118 cm^{-1}, νCN (terminal) at 2077 cm^{-1}.[281]

The first transition metal isonitrilate has been reported: $K[Co(2,6\text{-}Me_2C_6H_3NC)_4]$, with νCN as broad bands at 1890 and 1315 cm^{-1}. The low values are due to extensive π-back-bonding from the electron-rich metal centre.[282]

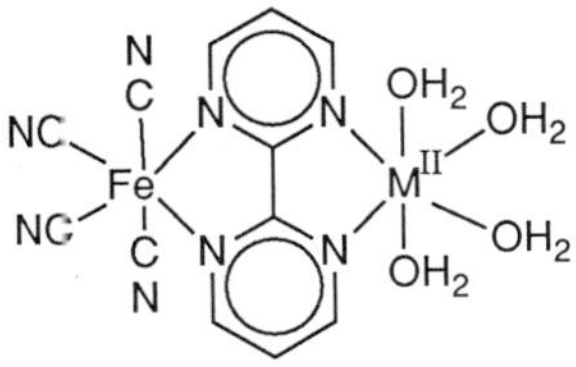

(57)

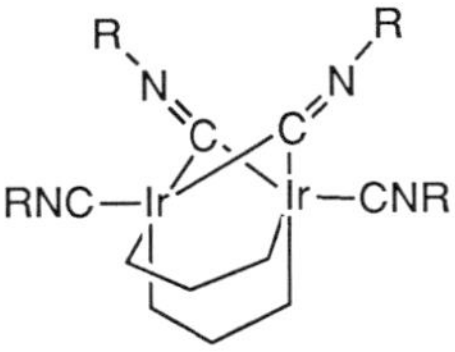

(58)

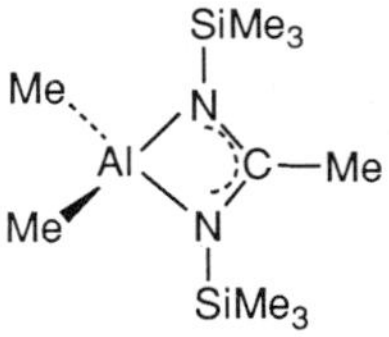

(59)

The novel unbridged dirhodium cation $Rh_2(CH_3CN)_{10}^{4+}$ has νCN at 2342, 2317 and 2300 cm^{-1}.[283] νCN bands for (58), where R = $2,6\text{-}Me_2C_6H_3$, and the bridging ligand is $Me_2PCH_2PMe_2$, are at 2038 and 1996 cm^{-1} (terminal) and 1564 and 1600 cm^{-1} (bridging). The very low values are consistent with a very electron-rich, zero-valent binuclear system.[284]

The series of compounds MX_2L_2, where M = Ni or Cu, X = $N(CN)_2$ or $C(CN)_3$, and L = pyrazole and substituted derivatives, all have ligand modes in the i.r. which show that they are polymeric, with bridging bidentate dicyanamide or tricyanmethide ligands.[285] In situ F.T.i.r. difference spectra of CN^- adsorbed on polished Pd or Pt electrodes show the formation of only linear $Pt\text{-}CN^-$(ads), while for Pd there was evidence for both linear and bridge-bound adsorbed CN^-.[286]

The bridging amidinato ligands in the copper(I) complex $Cu_2[PhC(NSiMe_3)_2]_2$ have νCN at 1470 and 1395 cm^{-1}, compared to the values of about 1530 and 1370 cm^{-1} for chelated ligands.[287] The i.r. and Raman spectra have been reported for a wide range of transition metals, including Cu, Ag and Au, with CN^- or SCN^- in liquid ammonia. There was evidence for many complex species, e.g. for Ag^I/CN^- at least 11 species, including linkage isomers, were identified.[288] $PhNC-Au-SiR_3$, where R_3 = Ph_2Tol, Ph_3 or Ph_2Me, all have νCN of the coordinated isocyanide at about 2173 cm^{-1}, compared with 2222 cm^{-1} in the precursor PhNC-Au-Cl. This is explained by the greater σ-donor power of SiR_3 than for Cl.[289]

$[Cd\{(CN)_2\}_2SO_2]_2^+$ contains bridging cyanogen ligands (νCN 2388 cm^{-1}), and an η^1-O-SO_2 ligand (bands at 1345 and 1165 cm^{-1}).[290] $[Cp_3U(NCMe)_2]^+$ shows only one νCN band in the i.r. spectrum, at 2275 cm^{-1}, whose position shows U-NCMe coordination.[291]

νCN modes in $PhC(NSiMe_3)_2EX_2$, where E = Al, Ga or In, X = Cl; E = B, X = Br, are consistent with N,N-chelation by the amidinato ligand.[292] $Me_2Al(NSiMe_3)_2CMe$, (59), has i.r. and Raman spectra consistent with C_{2v} symmetry, ν_sNCN 1450 cm^{-1}, and ν_{as}NCN 1468 cm^{-1}. These should be compared with the adducts $Me_3M.N(SiMe_3)=C=NSiMe_3$, where M = Ga or In, for which ν_{as}NCN is at 2190 cm^{-1}, ν_sNCN at about 1500 cm^{-1}.[293] ν_{as}NCN was assigned for $Ph_2P(E)-NCN-MR_3$, where E = O or S; M = Si, Sn or Pb, R = Ph; M = Sn, R = Me. All lie in the range 2130 - 2185 cm^{-1}.[294]

The i.r. spectrum of SF_5CN has νCN at 2235 cm^{-1};[295] while for SF_5NC the analogous band is at 2080 cm^{-1}.[296]

4.5 Nitrosyls:- The complexes $MCl_3(NO)(PMe_3)_3$ have νNO bands at 1620 cm^{-1} (Mo) or 1590 cm^{-1} (W), consistent with linear M-N-O and greater π-back-donation when M = W.[297] νNO bands at 1790 and 1670 cm^{-1} are also in agreement with linear Mo-N-O groups in $[LMo(NO)_2Cl]PF_6$, where L = 1-thia-4,7-diazacyclononane.[298] νNO in $[Mo(NO)L(A)_2$, where L = hydrotris(3,5-dimethylpyrazol-1-yl)borate, A = imidazole, pyrazole or pyridine, is close to 1620 cm^{-1}, compared to 1682 cm^{-1} in $Mo(NO)L(A)_2$, where A = pyrrol-1-yl. This shows the effect of changing from a 16-electron system (the latter) to 17-electron systems (the former).[299]

I.r. spectra gave νNO values for a large range of ReHX-(NO)$(PPh_3)_3$ complexes, where X = OMe, F, Cl, Br, N_3, NCO etc., and ReHX(CO)(NO)$(PPh_3)_2$, where X = F, Cl, Br, N_3, NCO, I or SCN.[300]

Fe$(\eta^5-C_5H_4S)_2$Mo(NO)[HB(3,5-$Me_2C_3N_2H)_3$] has νNO at 1634 cm^{-1}. This unusually low value is posibly explained by some Fe...ON-Mo interaction, but there was no confirmation of this from the Fe...O distance.[301]

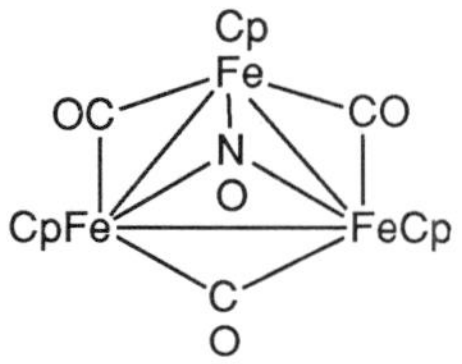

(60)

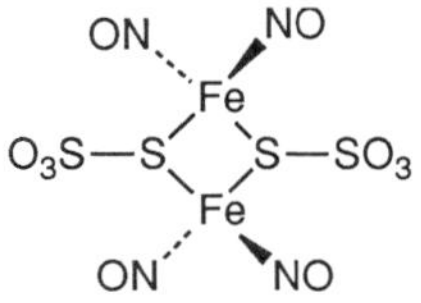

(61)

νNO from the μ_3-nitrosyl ligand in (60) is at 1290 cm^{-1} (Cp = C_5Me_5) or 1325 cm^{-1} (Cp = C_5H_4Me). The shift is consistent with the relative electron-donor properties of the two ligands.[302] (61) has νNO in the i.r. spectrum at 1787 and 1757 cm^{-1} (i.e. b_u and a_u modes). They shift to 1748 and 1718 cm^{-1} on ^{15}NO substitution.[303] The F.T.i.r. spectrum of NO adsorbed on the metalloporphyrins M(TPP), where M = Fe or Co, show νNO at 1698.6 cm^{-1} (Co) or 1883.8 cm^{-1} (Fe).[304]

νNO and other bands for $M(NO)(NH_3)_4(OH)^{2+}$ and M(NO)-$(NO_2)_4(OH)^{2-}$, where M = Ru or Os, show that they have trans octahedral structures.[305] Ligand modes were assigned for the cations $[Ru(NO)(NH_3)_4L]^{2+}$, where L = OH^-, Cl^- or ReO_4^-, and the dependence of the Ru-N-O force constants on the nature of the trans ligand calculated.[306] νNO in $[Ru(ReO_4)$-$(NH_3)_4(NO)](ReO_4)_2$ is at 1933 cm^{-1}, corresponding to weak σ- and π-trans-effects for ReO_4^-.[307]

The values of νNO for $Os(NO)F_5^{n-}$, where n = 1 or 2, are as follows: n = 1, 1924, 1885 cm^{-1}; n = 2, 1835, 1810 and 1790 cm^{-1}. The values are consistent with the changing oxidation state of the osmium.[308]

NO on Ni(111) at 85K gives νNO at 1460 cm^{-1} (low coverage) or 1508 and 1543 cm^{-1} (intermediate coverage). Heating NO-saturated Ni(111) surfaces to 320K gives a strong band at 1681 cm^{-1}, from linear Ni-N-O units.[309] NO adsorption on oxidised Ni or Ni/W catalysts supported on Al_2O_3 gives $\nu_s(NO)_2$

bands at 1865 cm^{-1} (Ni) or 1890 cm^{-1} (Ni/W).[310]

5. Phosphorus or Antimony Donors.

$(OC)_5Cr[^tBuP(NSN)_2PBu^t]Cr(CO)_5$ has ν_{as}, ν_s(NSN) at 1287 and 1154 cm^{-1} respectively.[311] I.r. spectra of $Cr(CO)_5L$, $W(CO)_5L$ and $Fe(CO)_4L$, where L = (62), show that all involve unidentate coordination of L to M, via the phosphorus atom only.[312] Many ligand modes were assigned for (63), where M = Cr, Mo or W, and related species with the ligands coordinated by two phosphorus and one nitrogen atoms.[313]

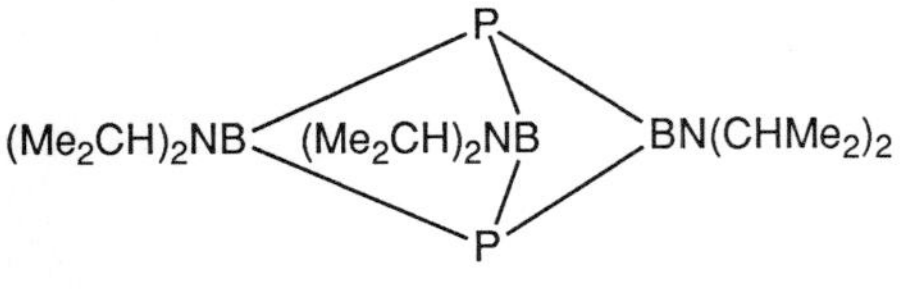

(62)

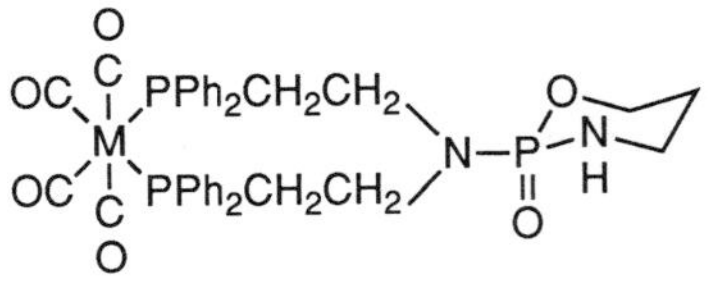

(63)

δCPC bends were assigned from far-i.r. spectra of $M_2Cl_4(PMe_3)_4$, Table 4.[314]

Table 4 δCPC deformation modes for $M_2Cl_4(PMe_3)_4$ (/cm^{-1})

M_2	=	Mo_2	351, 329
		MoW	343, 315
		W_2	338, 312

$W(^tBuPF_2)_6$ has four νPF bands in its i.r. spectrum, showing the breakdown of octahedral local symmetry for these species. [315] νNSN in (64), where M = Cr, Mo or W, is shifted by about 60 cm^{-1} (ν_{as}) or 10 cm^{-1} (ν_s) to lower wavenumber compared to the free ligand values.[316]

νNH of the coordinated bis(diphenylphosphino)amine (dppa) are seen in $Re_2(OAc)X_4(dppa)_2$, cis- and trans-$[Re_2(OAc)_2X_2(dppa)_2]X$, and $Re_2X_4(dppa)_2$, where X = Cl or Br. The values covered a wide range (3145 - 3446 cm^{-1}) due to differing amounts of hydrogen-bonding.[317]

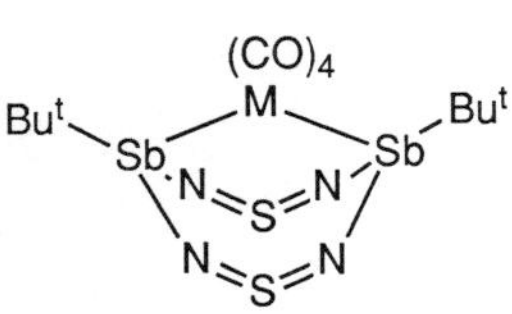

(64)

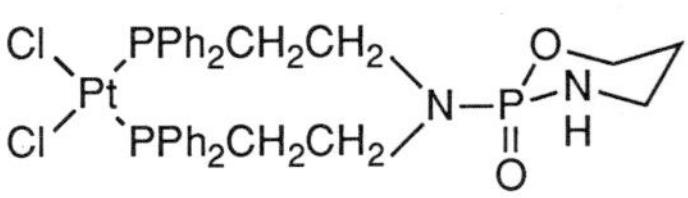

(65)

Characteristic bands of the phosphine ligands were seen for $Ru(SC_6F_5)_n(PPh_3)_2$, where n = 2 or 3, and $Ru(SC_6F_5)_3(PEt_2Ph)_2$.[318]

The product from heating $Pt_2Co_2(\mu_2\text{-}CO)_3(CO)_5(PPh_3)_2$ on a TiO_2 surface to 200°C still has νCH bands from coordinated PPh_3, even though all the CO ligands have been lost.[319] PF_3 adsorbed on a Pt(111) surface has ν_sPF_3 at 951 cm^{-1} and δ_sPF_3 at 562 cm^{-1}. The PF_3 has its C_3 axis oriented perpendicular to the surface.[320] The complex (65) has νNH at 3285 and 3210 cm^{-1}, νP=O 1231 cm^{-1} and $\nu P\text{-}CH_2$ 590 cm^{-1}.[321]

The Raman and i.r. spectra of $[Me_3PAgI]_4$ gave several assignments for the coordinated PMe_3 vibrational modes, e.g. ν_sPC_3 680 cm^{-1}, $\nu_{as}PC_3$ 742 cm^{-1}.[322]

The Raman spectrum of $Na^+[B(PH_2)_4]^-$ has νPH_2 modes at 2237 and 2255 cm^{-1}, with δPH_2 at 1016 and 1046 cm^{-1}.[323]

6. Oxygen Donors.

6.1 Molecular oxygen, Peroxo-, Aquo- and Related Complexes:- I.r bands due to OD groups on an MgO surface are at 2771, 2765 and 2756 cm^{-1} due to the presence of several different sites.[324] O_2 adsorbed on CeO_2 at room temperature gives i.r. bands which can be assigned to O_2^- and O_2^{2-} at the surface.[325-6]

The i.r. and Raman spectra of $[Ti(O_2)_2SO_4]^{2-}$ and $Ti(O_2)_2L$, where L = phen or bipy, show that the O_2^{2-} ligand is bidentate in all cases. The SO_4^{2-}, phen and bipy are all also bidentate.[327] The i.r. spectra of peroxo-complexes of Zr(IV), Th(IV), Mo(VI) and U(VI), containing bidentate organic ligands, show that νOO decreases with increased atomic number of the metal for any particular organic ligand.[328] νOO modes were seen in the range 800 - 860 cm^{-1} for $M(O_2)(H_2EDTA)$, where M = Zr or Th, and $M(O)(O_2)(H_2EDTA)$, where M = W or U.[329]

The i.r. spectra of Cp_2MO_2, where M = V, Cr, Mn, Fe, Co or Ni, from Cp_2M and O_2 in O_2-containing matrices, show 'end-on' coordination of O_2, i.e. M-O-O.[330] The i.r. and Raman spectra of $VO(O_2)_2F^{2-}$ contain ν_s and ν_{as} of O-O at 890, 875 cm^{-1}

respectively.[331] $[VO(O_2)(citrato)]^-$ has νOO at 933 cm^{-1}, showing that the O-O bond is quite strong. Several citrato ligand mode assignments were consistent with unidentate coordination of this.[332] νOO in $Cp_2Nb(O_2)Me$ is at 870 cm^{-1}.[333]

$\nu O\text{-}O$ bands were near 850 cm^{-1} in $K_2[M_2O_3(O_2)_4(H_2O)_2]$, where M = Mo or W,[334] and 864 (M = W) or 883 (Mo) cm^{-1} in $(\eta^5\text{-}C_5Me_5)M(=O)(\eta^2\text{-}O_2)Cl$.[335] νOH in $[M(CO)_3(\mu_3\text{-}OH)]_4{}^{4-}$ is at 3691 cm^{-1} (M = Mo) or 3676 cm^{-1} (W).[336] $WO(O_2)_2(L\text{-}L)$, where L-L = bipy or phen, have νOO near 840 cm^{-1}.[337]

Reversible coordination of O_2 to an $Mn(PEt_3)Br_2$ film (as a side-on peroxo group) was shown by i.r. spectroscopy for the first time. $\nu O\text{-}O$ is at 850.5 cm^{-1} fpr the $Mn(O_2)$ adduct.[338] Addition of excess KO_2 to $Mn^{III}TPP^+$ produces a species with i.r. bands at 920 and 970 cm^{-1}. These were assigned as νOO of a peroxomanganese porphyrin.[339]

νOH in $La_2[Tc(OH)Cl_5]_3$ is at 3100 cm^{-1}, with δOH 950 - 1140 cm^{-1}.[340] A similar mode is seen at 3332 cm^{-1} (with νOD at 2430 cm^{-1}) in $[Cp^*Re(OH)(\eta^2\text{-}PhC{\equiv}CPh)]^+$.[341] In (66), νOH is present as a sharp i.r. band at 3670 cm^{-1}, i.e. there is no hydrogen-bonding in this molecule.[342]

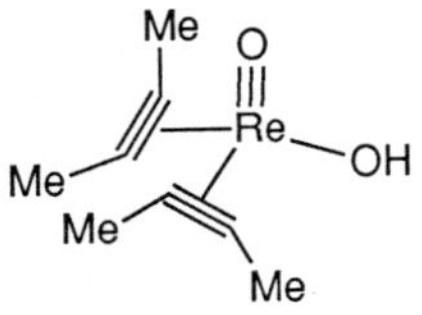

(66)

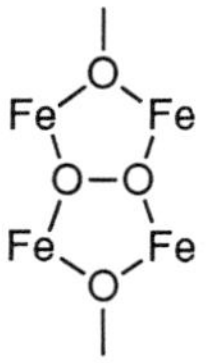

(67)

$O_2 + Fe_n$ aggregates, where n = 1, 2,..., have νOO of $Fe(O_2)$ at 956 cm^{-1}, and of $Fe(O_2)_3$ at 1098 cm^{-1}.[343] $Fe_6(O)_2(O_2)$-$(O_2CPh)_{12}(OH_2)_2$ has νOO for the μ_4-peroxide tetrairon unit, (67) at 853 cm^{-1}.[344] $[(Ph_3PO)_4FeOOFe(OPPh_3)_4]^{4+}$ has νOO of the bridging dioxygen at 882 cm^{-1} (Raman only). Its absence from the i.r. spectrum shows a centrosymmetric environment for the FeOOFe unit.[345]

The i.r. spectrum of Fe(salen), where salen = N,N'-ethylenebis-(salicylideniminato) dianion, cocondensed with O_2 or O_2/Ar mixtures, gives two $\nu O\text{-}O$ bands, at 1104 and 1001 cm^{-1}. The former has previously been assigned to a 'side-on' dioxygen adduct $Fe(salen)O_2$. The latter is thought to be due to the dimeric $[Fe(salen)]_2O_2$.[346] νO_2 bands were assigned from the i.r. spectra of $M(TPP)(piperidine)(O_2)$, where M = Fe(II) (1157

cm^{-1}) or Co(II) (1142 cm^{-1}).[347]

νOO due to end-on coordinated O_2 was seen for μ-oxo[meso-tetrakis(methoxyphenyl)porphyrinato]di-iron adsorbed on an electrode in the presence of O_2.[348] νOO (near 1100 cm^{-1}) features were assigned for several oxygenated, asymmetrically Co-substituted Fe-Co hybrid haemoglobins.[349] (68) has νOO at 862 cm^{-1}, showing very strong binding of the dioxygen to the osmium. This is probably due to strong σ-donor properties of the trans-H ligand.[350]

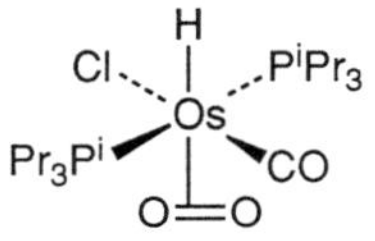

(68)

The O_2 adduct of a anthracene pillared cofacial dicobalt(II) diporphyrin complex in the presence of bases has νO_2 near 1138 cm^{-1}, typical of six-coordinate, non-bridging dioxygen adducts. In the absence of base, there was evidence for the O_2 ligand bridging the two porphyrin units, with νO_2 near 1080 cm^{-1}.[351] Resonance Raman spectra of O_2 adducts of imidazole complexes of a highly-protected cobalt porphyrin show that there is vibrational coupling of νOO with internal modes of the trans-coordinated imidazole ligand. These couplings are markedly dependent on hydrogen-bonding of the imidazole to other solution components.[352]

νO_2 is at about 825 cm^{-1} in (R)Ir(O_2)(CO)L_2, where R = Me, Ph or naphthyl, L = PPh_3, PPh_2Me, PMe_2Ph etc., i.e. there is 'side-on' coordination.[353] [NiR(PMe_3)(μ-OH)]$_2$, where R = CH_2SiMe_3, CH_2CMe_2Ph or CH_2Ph, all have νOH due to the bridging hydroxyl ligand near 3600 cm^{-1}.[354] νO-D of $Cd(OD_2)_6^{2+}$ was identified at 2565 cm^{-1} in a single crystal of $CdSiF_6.6D_2O$.[355]

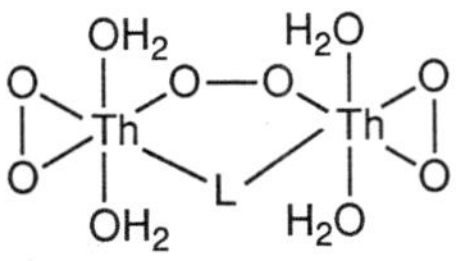

(69)

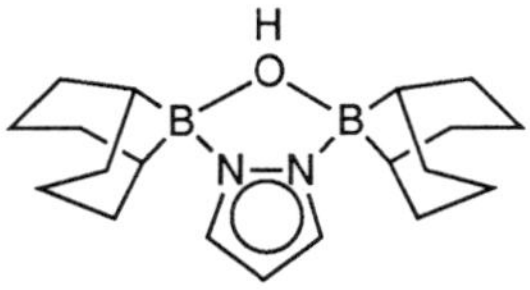

(70)

The presence of two νOO bands in the i.r. spectrum of $Th_2(O_2)_3(C_2O_4)(H_2O)_4$ is consistent with the geometry (69), where L = oxalato, i.e. there are both triangularly bidentate and bridging peroxo groups.[356]

(70) has νOH as a sharp i.r. band at 3580 cm^{-1}.[357] $[B_2(\mu-O_2)_2(OH)_2(O_2H)_2]^{2-}$ has νO-O of the hydrogenperoxo ligand at 860 cm^{-1}.[358] νOH in $Ga(OH)Br_2.H_2O$ is at 3780 cm^{-1}.[359]

6.2 Carboxylato and Related Complexes:- The Raman spectra of $Be^{2+}/CH_2(COOH)_2$ systems gave evidence for the formation of the following malonato complexes: $Be[CH_2(COO)_2]$, $Be[CH_2(COO)]_2^{2-}$ and $Be_3(OH)_3[CH_2(COO)_2]_3^{3-}$.[360]

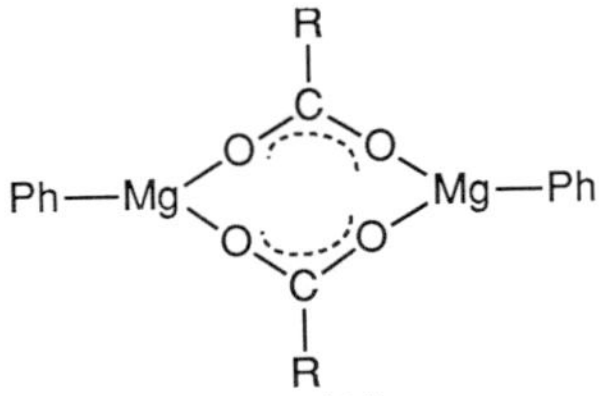

(71)

Phenylmagnesium carboxylates PhMgOOCR.THF, where R = Me, $ClCH_2$, Cl_2CH, Cl_3C, Et, ^iPr or CH_2Ph, all give carboxylate bands in the i.r. which are assignable to bridging ligands. The suggested structure is (71).[361]

The i.r. spectra of scandium(III) complexes of o-nitro-, o-chloro-, o-methyl, o-hydroxy- and o-aminobenzoates show that the carboxylato groups are unidentate.[362] There is i.r. evidence for the formation of bidentate carbonate by the adsorption of CO_2 on $LaCrO_2$.[363] The Raman spectra of $M_2(C_2O_4)_3.nH_2O$, where M = all lanthanides except Pm, n = 6 or 10, show that in every case the oxalato ligands are tetradentate, bridging two lanthanide atoms.[364] νCO_2 bands of $Ln(NCCH_2COO)_3(L-L)$, where Ln = La, Pr, Nd, Eu, Gd, Tb, Dy, Yb, Lu or Y; L-L = bipy or phen, are all consistent with unsymmetrically bidentate bonding of the carboxylate groups to Ln.[365] CO adsorption on CeO_2 produces carbonate species - there was i.r. evidence for both uni- and bidentate bonding.[366]

I.r. data for CO_2 adsorbed on monoclinic ZrO_2 reveal that bidentate carbonates are formed at 25°C, unidentate at 400°C.[367] $[Cp_2Zr(\mu-OCOC{\equiv}COCO)]_n$ gives carboxylate stretching modes which are consistent with the presence of one uni- and one bidentate carboxylate group, (72). The former has bands at 1650 and 1290 cm^{-1}, the latter at 1525 and 1450

cm^{-1}.[368]

I.r. bands show the presence of both uni- and bidentate carboxylates in $VO(O_2C^tBu)_3$, giving six-coordinate vanadium.[369] Bridging carboxylates were identified for $VOL_2.nH_2O$, however, where HL = benzoic acid and a range of substituted derivatives.[370]

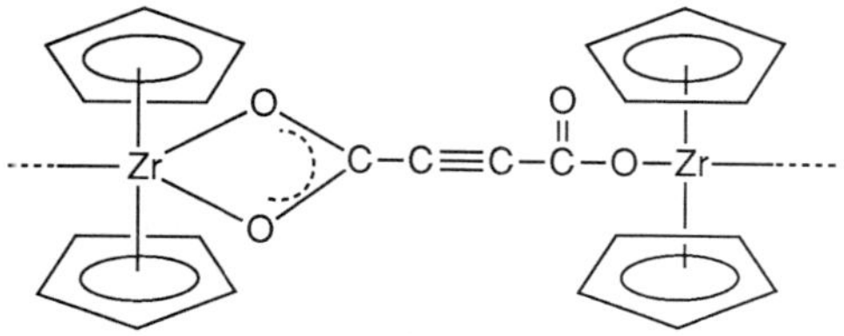

(72)

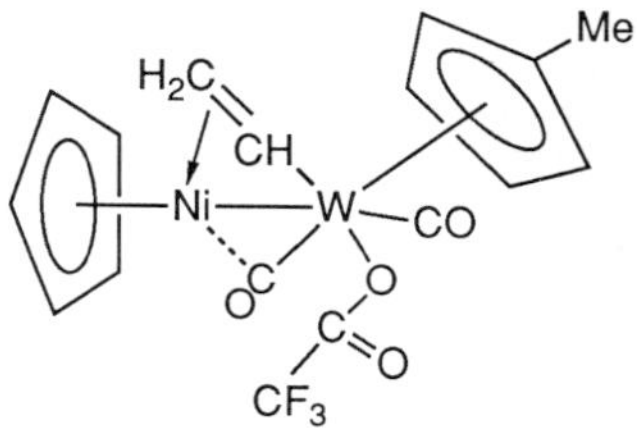

(73)

F T.i.r. evidence points to the occurrence of ionic, uni-, bidentate and bridging acetato groups in aqueous chromium(III) acetate solutions.[371] $[M(salen)(ox)]^-$, where M = Cr(III) or Fe(III), salen = N,N'-ethylenebis(salicylideneiminate, ox = oxalate, shows splittings of νCO_2 modes of oxalate due to bidentate chelation. Analogous modes in $Fe_2(salen)_2(ox)$, with bis(bidentate) bridging show no such splitting.[372]

Several new Mo(IV) complexes of acetylacetone have been prepared. Some of these show the presence of both uni- and bidentate ligands, e.g. $Mo(acac)_2(Hacac)(OH)_2(H_2O)$ has $\nu C{=}O$(free) 1700 - 1760 cm^{-1}.[373] νCO modes of (73) include one at 1697 cm^{-1} due to the trifluoroacetate ligand.[374]

$Fe_2Cl_5(OOCR).nRCOOH$, where R = Me or Et, n = 3; R = CH_2Cl, $CHCl_2$ or CCl_3, n = 2, all have νCO_2 bands due to bridging bidentate carboxylato groups.[375]

The i.r. spectrum of $Rh_2(acac)_4$ suggests that it is polymeric, with intermolecular coordination via oxygen atoms of neighbouring molecules.[376] I.r. data for $[(tn)_2Ni(\mu\text{-}C_2O_4)\text{-}Ni(tn)_2](ClO_4)_2$, where tn = 1,3-diaminopropane, show the presence of bis(bidentate) oxalato.[377]

I.r. spectra of $[Cu_2(\mu\text{-}IPHA)(bipy)_2]^{2+}$ and relatd systems, where IPHA = isophthalato, show that bidentate carboxylates are present. Analogues with four bipy ligands, however, have unidentate carboxylates.[378] $Zn_3(O_2CCH{=}CHMe)_6(quin)_2$, where quin = quinoline, has νCO_2 bands consistent with the presence of both uni- (1626 cm^{-1}) and bidentate (1568, 1409 cm^{-1}) crotonate ligands.[379]

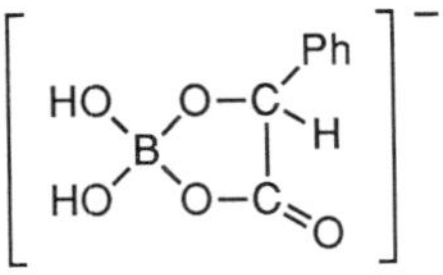

(74)

$\nu C{=}O$ of ammonium mandelatoborate is consistent with bidentate coordination of the mandelate group, (74).[380] The i.r. spectra of complex oxalates formed in Al^{3+}/oxalate solutions suggest C_{2v} or D_{2h} symmetries for the complexes.[381]

There is evidence for the formation of two types of carbonate (i.e. uni- and bidentate) when SnO_2 is exposed to dry air containing 20 - 50 ppm of CO.[382] $\nu_{as} - \nu_s$ (CO_2) for $(Ph_3Sn)_2ADC$, $(R_3Pb)ADC$, where R = Me or Ph, and $(Me_3Pb)_2TER$, where H_2ADC = acetylenedicarboxylic acid, H_2TER = terephthalic acid, show that bridging carboxylate coordination occurs in the solids, to give polymeric structures. In solution, dissociation occurs, to give species with unidentate carboxylate groups.[383]

I.r. spectra for a large number of triorganoantimony- and triorganobismuth dicarboxylates, $R_3M[O_2C(CH_2)_n\text{-}2\text{-}C_4H_3X]_2$, where M = Sb or Bi, n = 0 or 1, X = O, S, NH or NMe, all show the presence of unidentate carboxylates.[384]

6.3 Keto-, Alkoxy, Ether and Related Complexes:- The i.r. and Raman spectra of LiX(benzo-12-crown-4), where X = Cl, Br, NCS or ClO_4, and of $MNCS(benzo\text{-}12\text{-}crown\text{-}4)_2$, where M = Na or K, show greater deformation of the crown ether molecule by Li^+ than by Na^+ or K^+.[385] The Raman spectra of 12-crown-4, 15-crown-5, 18-crown-6 and their complexes with Li^+, Na^+, K^+ and Cu^{2+} have been reported. Analysis of νCH bands suggests that 15-crown-5 forms a 2:1 complex with K^+ and a 1:1 complex with Na^+. 12-crown-4 forms a 2:1 complex with Na^+.[386]

The δOCN mode of DMF was used to probe the solvation of Li^+, as NO_3^-, ClO_4^- salts, by the DMF.[387] $[K(18\text{-}crown\text{-}6)]_2[Hg_2I_6]$ and $[Hg(18\text{-}crown\text{-}6)][SCN]_2$ gave very similar i.r. bands due to the cationic complexes, despite the very great difference in the nature of the coordinating metal.[388]

$[MgCl(THF)_5]^+$, as the $FeCl_4^-$ salt, has bands due to the coordinated THF at 838, 927 and 1038 cm^{-1}.[389] The i.r. spectra of $CaX_2L.2H_2O$, where X = SCN or ClO_4; L = 2,3-(4'-acetyl)benzo-15-crown-5, show that L coordinates by ring O atoms only, and that the acetyl oxygen is not involved.[390]

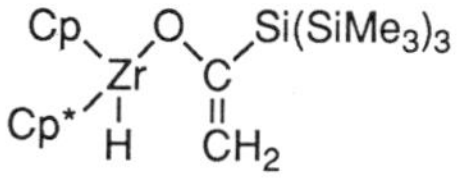

(75)

(75) has νC=C at 1540 cm^{-1}, νC-Si at 1350 cm^{-1}, and νC-O at 1180 cm^{-1}, shifting appropriately on ^{13}C substitution.[391] Dichlorocyclopropane as a ligand in tetrameric $[NbOCl_3(OC_3Cl_2)]_4$ has νC=O at 1835 cm^{-1} (1885 cm^{-1} in free OC_3Cl_2), νC=C at 1542 cm^{-1} (1615 cm^{-1}) and δCCC 731 cm^{-1} (693, 670 cm^{-1}).[392]

The i.r. spectra of $Cr(MDA)_3X_3$, where MDA = malondiamide, X = Cl or F, show that the malondiamide coordinates via the carbonyl oxygens (amide-I and amide-II bands both decrease with respect to the free ligand).[393] A large number of ligand modes were assigned from the i.r. and Raman spectra of naphthalene-2,3-diolato dianion (ND) complexes, cis-$M_2O_5(ND)_2^{2-}$, cis-$MO_2(ND)_2^{2-}$, where M = Mo or W; trans-$UO_2(ND)$ and trans-$OsO_2(ND)_2^{2-}$. They were consistent with bidentate bridging diolato ligands.[394]

Some ligand mode assignments for $Mn(OCH=CH_2)_4$ are summarised in Table 5.[395]

Table 5 Some vibrational assignments for $Mn(OCH=CH_2)_4$ (/cm^{-1})

$\nu_{as}CH_2$	2950
ν_sCH_2	2860
νC=C	1605
δCH_2	1460
νC-O	1260

The i.r. spectra of $[ML_4(H_2O)_4]I_2$, where M = Mn, Fe, Co or Ni, L = acetamide, all show that νC=O decreases by 23 - 28 cm^{-1} compared to the free ligand value, due to M→O=C coordination. [396] I.r. data for $MnCl_2L_2$, where L = diacetamide, are consistent with the coordination (76).[397] Shifts in $\nu C{=}O$ of gluconolactone in the presence of Mn^{2+} show strong coordination by C=O of the lactone ring to the metal.[398]

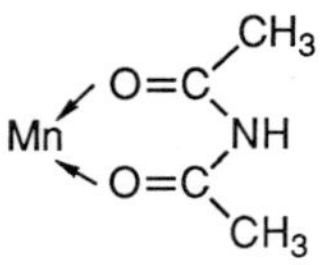

(76)

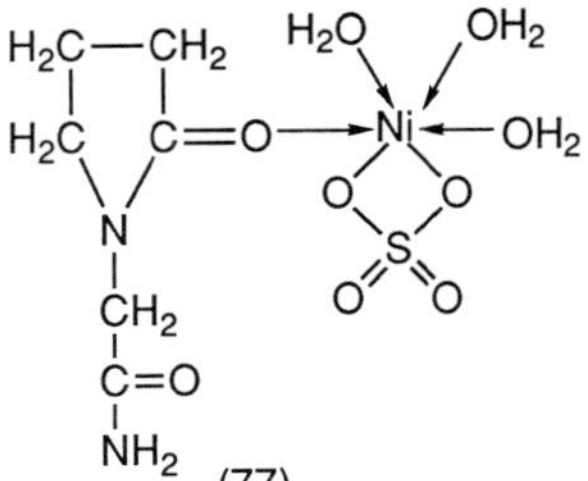

(77)

$Ni(PAM)SO_4.6H_2O$, where PAM = 2-oxo-1-pyrrolidinylacetamide, has an i.r. spectrum which indicates that the formula should be written as (77), with three water molecules of crystallisation.[399]

The i.r. spectra of $UO_2X_2.2EtOH$, where X = Cl^- or NO_3^-, show that the ethanol molecule is directly coordinated to the uranium.[400] $[UO_2L_n]^{2+}$, where L = mono- (n = 5), di- (n = 5) orr tetra- (n = 4) -methylurea, all have i.r. bands showing that the urea ligands are coordinated via carbonyl oxygen atoms.[401]

6.4 Ligands containing O-N or O-P Bonds:- $ML_n(NO_3)_3.mH_2O$, where M = Pr or Nd, L = R-substituted benzimidazole, R = H or 2-Me, n = 3, m = 3,4 or 5; R = 2-Et, 2-Pr, n = 2, m = 4 or 5, all give NO_3 bands due to both C_{2v} and D_{3h} nitrato units,[402] I.r. bands due to the NO_3^- ligands in $[M(NO_3)_5]^{2-}$, where M = Pr or Nd, show only symmetric bidentate nitrato ligands. There was also evidence for coordinated H_2O molecules.[403] The i.r. spectra of $Sm(NO_3)_5^{2-}$ salts show the presence of coordinated NO_3^-, probably bidentate.[404]

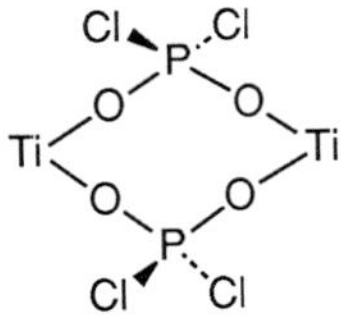

(78)

$Ti_2Cl_6(POCl_3)_2(PO_2Cl_2)_2$ has PO_2Cl_2 ligand bands consistent with bridging PO_2Cl_2 ligands, i.e. (78).[405] νPO modes shift to lower wavenumbers on going from $[Ph_2P(O)]_2CH_2$ (MDPO) to cis-$M(CO)_4$MDPO and fac-$M_2(CO)_6(MDPO)_3$, where M = Mo or W, showing M-O=P coordination.[406] $\nu P{=}O$ in $[WOCl_4(OPPh_3)]^-$ is at 1145 cm^{-1}, showing W←O=P coordination.[407]

The unusual 0,0'-chelating nitrite ligand in cis-$Re(O,O'-NO_2)(CO)_2(PPh_3)_2$ has ν_sNO_2 1241 cm^{-1}, $\nu_{as}NO_2$ 1180 cm^{-1} and δNO_2 887 cm^{-1} (all shift as expected on ^{15}N substitution).[408]

The i.r. spectra of Co(II) and Ni(II) complexes with ethyliminodimethylenephosphonic acid, $[EtN(CH_2PO_3H_2)_2 = H_4L]$ show that the L^{4-} ligand is coordinated in the betaine form.[409]

$M(L-L)(ONO_2)_2$, where M = Pd or Pt, L-L = bipy or phen, all give expected i.r. bands for unidentate nitrato ligands in a cis geometry.[410] $(NH_4)_4[Ni(P_3O_9)_2(H_2O)_4]$ has P_3O_9 ligand modes consistent with C_s local ligand symmetry.[411]

6.5 Ligands containing O-S or O-Te Bonds:- Raman spectra of DMSO vibrations for solutions containing DMSO and LiX, where X = Br or I, contain a band near 1006 cm^{-1}, due to $\nu S=O$ interacting with Li^+.[412]

νSO_4 modes in $VOF_4(SO_4)^-$ show bidentate coordination by the sulphato group.[413] In $[VO(SO_4)_2(H_2O)_3]^{2-}$, however, vibrational bands show that the sulphato groups are unidentate.[414]

$Fe(OEP)(OTeF_5)$ has νTeO at 852 cm^{-1} (812 cm^{-1} for the ^{18}O analogue), and $\nu_{as}TeF_4$ at 682 cm^{-1}. These are consistent with rather ionic $OTeF_5^-$ ligands.[415] $K_4[Ru_2(SO_4)_2(\mu-SO_4)_2(\mu-O)_2]$ has i.r. bands at 1160, 1030 and 920 cm^{-1} due to bidentate chelating SO_4 groups, and at 1244, 1124 and 950 cm^{-1} due to bridging SO_4 ligands.[416].

νSO_4 modes in $Cu(SO_4)F^-$, as Na^+, K^+ and NH_4^+ salts, clearly demonstrate the presence of bidentate SO_4 groups.[417] It has been suggested, from νSO_4 modes, that the sulphato ligands in $[UO_2(SO_4)OH]^-$ and $[UO_2(SO_4)OD]^-$ are tridentate bridging.[418] The i.r. spectra of $M_4M'(SO_4)_4.2H_2O$, where M = K or Rb; M' = U or Pu, suggest that the sulphato groups are bidentate, with some possibly chelating.[419]

The i.r. and Raman of the $AlCl_3$-$SOCl_2$ system showed the formation of $Cl_2SO \rightarrow AlCl_3$ and, at high $AlCl_3$ concentrations $[Cl_2Al(\leftarrow OSCl_2)_2]^+AlCl_4^-$.[420] The i.r. and Raman spectra of $SeO(SO_3CF_3)_2$ contain the following bands: $\nu_{as}SO_3$ 1385, 1240 and 1200 cm^{-1}; ν_sSO_3 + νSe=O 1020 cm^{-1}; νC-S 756 cm^{-1}; $\delta_{as}SO_3$ 620, 605 cm^{-1}; δ_sSO_3 570 cm^{-1}.[421]

6.6 Ligands containing O-Cl Bonds:- νP=O decreases by 30 cm^{-1} on going from $POCl_3$ to $Nd(ClO_4)_3.3POCl_3$. The perchlorate bands were split in such a way as to suggest bidentate coordination.[422] νClO_4 modes in $MoO_2(ClO_4)_2$ are consistent with bidentate chelating perchlorato groups only. For $MoO_2(ClO_4)_3^-$ there appear to be both uni- and bidentate ClO_4^- ligands, but unidentate only for $MoO_2(ClO_4)_4^{2-}$.[423]

Anhydrous $Mn(ClO_4)_2$ has been prepared; νClO_4 modes are consistent with tridentate perchlorato groups.[424] There is i.r. evidence for unidentate ClO_4^- ligands in M(hx)(hxH)-$ClO_4.H_2O$, where M = Fe or Ni, hxH = 6-oxopurine.[425]

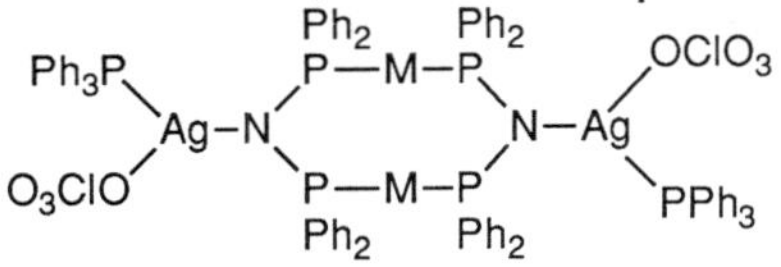

(79)

The i.r. spectrum for $RhO_{0.75}(ClO_4)_{2.5}$ shows that there are bridging bidentate ClO_4^- ligands.[426] (79), where M = Ag or Au, gives i.r. bands of unidentate perchlorato ligands.[427]

7. Sulphur and Selenium Donors.

νSS is at 557 cm^{-1} in $[Ti_3O(S_2)_3Cl_6]^{2-}$.[428] (80), where E^1 = E^2 = E^3 = E^4 = P, X = Cl, Br or CN; X = Cl, E^1 = E^2 = E^3 = P, E^4 = As, E^1 = E^3 = P, E^2 = E^4 = As, E^1 = E^2 = P, E^3 = E^4 = As, all have νSO bands indicative of SO_2 bridging i.e. near 1180 and near 1045 cm^{-1}.[429]

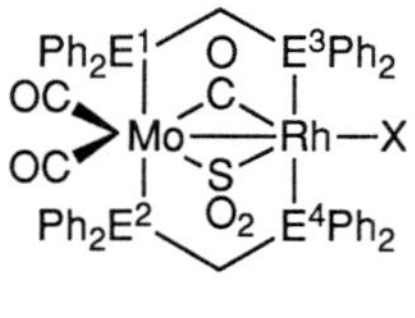

(80)

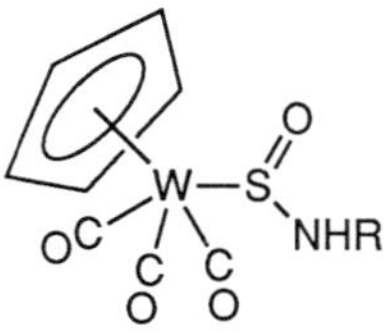

(81)

Shifts in ligand νC=C modes on going from $[MoO(dmit)_2]^{2-}$ to $[MoO(dmit)_2]^-$, where dmit = 4,5-dimercapto-1,3-dithiolato-2-thionate, suggest that oxidation is occurring at the ligand rather than at the metal centres.[430] The i.r. spectra of $Mo_2(R_2NCS_2)_4$, where R = Et, ${}^nPr$, Me_2CH or Me_2CHCH_2, are consistent with bidentate bridging $R_2NCS_2^-$ ligands.[431] νSS of side-bonded S_2 units are at 520 - 545 cm^{-1} for $Mo_2O_2S_9^{2-}$, $Mo_4O_4S_{18}^{2-}$ and related systems.[432]

$(Me_2NCS_2)_2WS(S_2)$ has νCN at 1530 cm^{-1}, νCS at 1150 cm^{-1}.[433] (81), where R = C_6H_4Me-2 and C_6H_4Me-4, have νS=O at 1007 cm^{-1}, 981 cm^{-1} respectively, showing that S=O character is maintained in the complexes.[434] (82) has a ketonic νC=O which is scarcely shifted from that of the free ligand.[435]

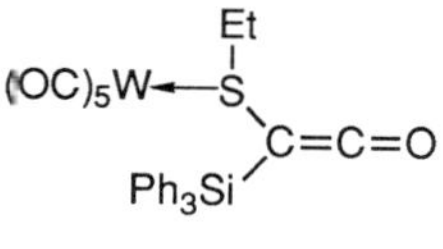

(82)

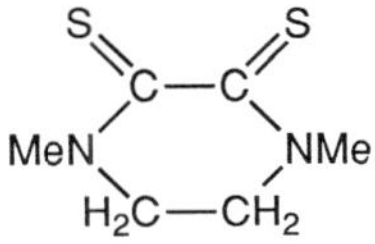

(83)

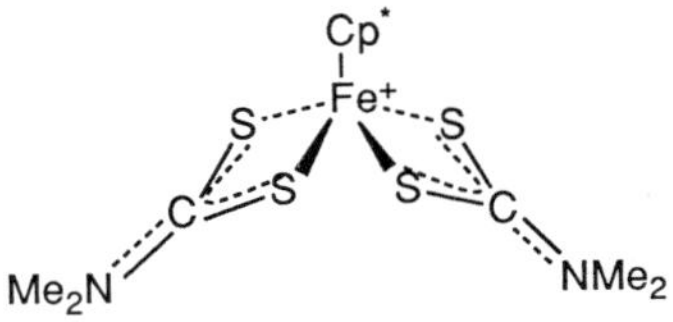

(84)

The resonance Raman spectrum of $Re(CO)_3$(cycl-DTO)Br, where DTO = (83), shows enhancement for νCC (1149 cm^{-1}) and νCO (CO - cis), the latter due to partial delocalisation of the MLCT band over the CO ligands.[436]

(84) has νCS at 1005 cm^{-1} and νCN at 1540 cm^{-1}.[437] Ph_3Fe(S-S), where S-S = dithiocarbamate, xanthate or dithiophosphate, all have i.r. bands consistent with η^1-S-coordination by these sulphur ligands.[438] The i.r. spectra of $Os(NO)Cl_2(S_2CNR_2)(PPh_3)$, where R = Me or Et, contain νCN bands of the dithiolate ligands at 1530, 1542 cm^{-1} respectively, i.e. higher than in the free ligand. This is due to increased C=N double bond character on coordination.[439] Characteristic thiourea (= thio) ligand modes in $[Os(thio)_6]^{2+}$, $[Os(thio)_5Cl]^{2+}$ and $[OsO_2(thio)_4]^{2+}$ all show S-coordination.[440]

The i.r. spectra of $[M(oximate)(ROCS_2)]^-$, where M = Co or Ni, or $[Cu(oximate)(ROCS_2)_2]^-$ show that the dithiocarbonate ligands are bidentate.[441] Vibrational data for $RhCl_3(SCl_2)_3$ and $[RhCl_3(SeCl_2)_2]_2$ show that the SCl_2 and $SeCl_2$ are coordinated via S, Se respectively. It is suggested that in $RhCl_3(TeCl_4)_3$ the $TeCl_4$ is coordinated through a chlorine atom.[442]

The i.r. spectra of metal(O) dimethyltetrathio-oxalate complexes, NiL_2 or MoL_3 are consistent with S,S-chelation, i.e. for the nickel complex, (85).[443] ν_s and $\nu_{as}CS_2$ bands were seen near 660 cm^{-1} and between 960 and 1029 cm^{-1} respectively for platinum(II) complexes of dithiocarbamates derived from glycine, DL-alanine, DL-valine and DL-leucine.[444] $[Trans\text{-}PtX_2\{(\mu\text{-}SC_6H_8)Pt(C_6F_5)_3\}_2]^{2-}$, where X = Cl, Br, I or C_6F_5, all give i.r. bands characteristic of bridging tetrahydrothiophene ligands.[445]

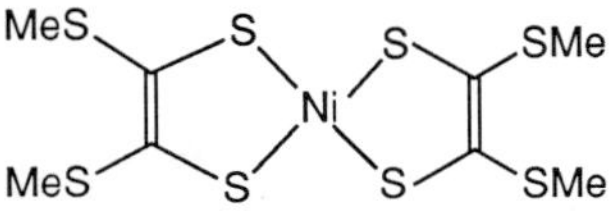

(85)

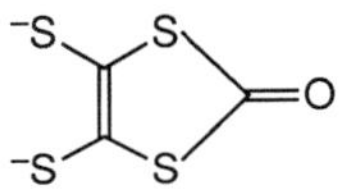

(86)

$Cu(S_2COBu)$ has an i.r. spectrum showing that it is polymeric, with bidentate bridging butylxanthate, while $Cu(S_2COBu)_2$ is monomeric, with the ligand bidentate via both sulphur atoms.[446] $\nu C{=}C$ bands for the metal-dithiolene complexes, $[M(dmid)_2]]^{2-}$, where M = Cu, Ni, Pd, Zn, Cd or Hg; dmid = (86), are all between 1425 and 1485 cm^{-1}.[447]

Zinc(II) forms distorted tetrahedral complexes with dithio-oxamates, which are S,S-bidentate. Detailed assignments were proposed for $Zn(CH_3NHCSCSNHCH_3)X_2$ and $Zn[(CH_3)_2NCSCSN(CH_3)_2]X_2$, where X = Cl, Br or I; these were assisted by NH/ND and CH_3/CD_3 isotopic substitution.[448] $[M(Se_4)_2]^{2-}$, where M = Zn, Cd or Hg, have $\nu SeSe$ bands in the range 240 - 340 cm^{-1}.[449] $[Cd(S_6)_2]^{2-}$ has i.r. bands characteristic of the metallahexasulphane ring at 450, 500 and 525 cm^{-1}.[450] The i.r. spectra of $HgX(R_2NCS_2)$, where X = Cl, Br or I, R_2 = Et_2 or piperidinyl, show bidentate dithiocarbamate ligands.[451]

The ligand modes in $Ph_2Ge[S_2CNEt_2]_2$ and related species are consistent with unidentate dithiocarbamate bonding to Ge.[452] The wavenumber differences between ν_{as} and ν_sPS_2 modes for $R_nSn(S_2PMe_2)_{4-n}$, where R = Me, ^nBu, CH_2Ph or Ph, n = 2; R = Me, Cy, CH_2Ph or Ph, n= 3, are all more than 100 cm^{-1}, showing unidentate coordination, (87).[453] The i.r. and Raman spectra of the $(thiourea)_2I_2$ charge-transfer complex show thiourea modes as expected for a bent $(thiourea)_2I^+$ cationic complex.[454]

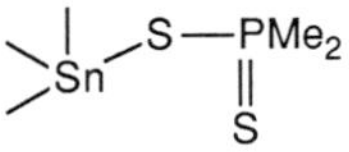

(87)

8. Potentially Ambident Ligands.

8.1 Cyanates, Thio- and Selenocyanates and their Iso-analogues:-$Cp_2Ti(NCO)_2$ has νNCO at 2200 cm^{-1}, while in $Cp_2Ti(NCO)(NCNSiMe_3)$ νNCO is at 2205 cm^{-1} and νNCN at 2110 cm^{-1}.[455]

The νNCS bands of $[Mo_2Cl_8(NCS)]^{3-}$ are consistent with an N-bonded bridged NCS^- ligand.[456] The bond isomers $ReCl_5(SCN)^{2-}$, $ReCl_5(NCS)^{2-}$, trans-$ReCl_4(SCN)_2{}^{2-}$, cis- and trans-$ReCl_4(SCN)(NCS)^{2-}$, cis- and trans-$ReCl_4(NCS)_2{}^{2-}$ and related species can be distinguished by the following ligand mode positions: νCN(S) > νCN(N); νCS(N) > νCS(S) and δNCS > δSCN.[457] Similar relationships were found in the bromo-analogues.[458]

$Ru(bipy)_2(NCS)_2$, as CH_3CN or $(CH_3)_2SO$ monosolvates, were found to have accidentally degenerate ν_s and ν_{as}CN modes.[459] $OsO_2(NCO)_4{}^{2-}$, $OsO_2(NCS)_4{}^{2-}$ and $OsO_2(SeCN)_4{}^{2-}$ have been reported, and ligand coordination modes identified from the i.r. and Raman band positions for νCO, νCS and νCSe respectively. All spectra were assigned under D_{4h} symmetry. [460] $[OsO_2(CN)_2L_2]^{2-}$, where L = NCO, NCS or SeCN, involve coordination via N, N and Se respectively, as shown by νCO, νCS or νCSe.[461] Isomers $[Os(NCSe)_n(SeCN)_{6-n}]^{3-}$ have been prepared and identified by i.r. spectroscopy (n = 1 - 3).[462]

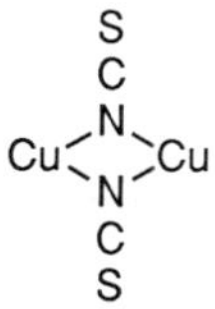

(88)

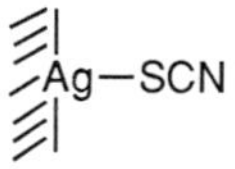

(89)

CO and NH_3 react at peroxidised Rh/SiO_2 surfaces to produce surface RhNCO and RhONC species (with i.r. bands at 2172, 2225 cm^{-1} respectively).[463]

$[Cu(paphy)(NCS)(SCN)]_2$, where paphy = pyridine-2-carbaldehyde 2'-pyridylhydrazone, gave νCN and νCS i.r. bands due to Cu-SCN, and a unit (88).[464] SERS for NCS^- adsorbed at a silver electrode revealed bands due to ν_{as} and ν_sNCS consistent with coordination (89).[465]

The Raman spectra of Cd(II)/NCS^- systems in DMF show that $Cd(NCS)^+$ and $Cd(NCS)_2$ are only N-bonded. For the tri- and tetrathiocyanato complexes, however, data are consistent with $Cd(NCS)_2(SCN)^-$ and $Cd(NCS)_3(SCN)^{2-}$ respectively. These results show that N-bonding is more favoured in DMF than in water, which was ascribed to hydrogen-bonding of the N atom of NCS^- in H_2O.[466] $(PTA)Hg(SCN)_2$, where PTA = 1,3,5-triaza-7-phospha-adamantane, has νNCS bands due to both terminal and bridging SCN ligands.[467] νCN (2059 cm^{-1}) in the adduct $[Ph_4P]_2(SCN)_2.Hg(CF_3)_2$ shows the presence of Hg-NCS bonds.[468]

νCN is at 2144 cm^{-1} in $[(SCN)B_6H_5]^-$, suggesting B-SCN coordination, as do the νCS and δNCS values of 702, 413 cm^{-1} respectively.[469] $[Si(NCSe)_6]^{2-}$ has been prepared for the first time, with νCN at 2091 cm^{-1}, νCSe 768 cm^{-1} and δNCSe at 405 cm^{-1}.[470]

8.2 Ligands containing N and O or P and O Atoms:- The Raman spectra of solutions of $LiAlCl_4$, $LiFeCl_4$, $NaFeCl_4$ or $KAlCl_4$ in $POCl_3$ can best be explained in terms of the formation of $[M(POCl_3)_4]^+$, where M = Li, Na or K.[471] The i.r. spectra of PAN (= 1-(2-pyridylazo)-2-naphthol) complexes with Ca(II) or Co(III) show coordination via O and pyridine N.[472]

The i.r. spectra of m-aminobenzoates of Y, La and the lanthanides, $Ln(m\text{-}NH_2C_6H_4COO)_3.nH_2O$, where n = 4, Ln = Ho or Tm; n = 5, Ln = Y, Sm, Dy, Er or Lu; n = 6, Ln = La - Nd, Eu, Gd, Tb or Yb, show that the ligands coordinate through the amino and carboxyl groups. The anhydrous complexes, however, coordinate only through bidentate carboxylates.[473] The i.r. spectra of $Ln_3(NO_3)_9.L.9H_2O$, where Ln = Y, Eu, Gd, Du, Er, Tm, Yb or Lu, L = (90), show that each crown ether coordinates to one Ln, with two nitrogen atoms coordinated to the third.[474] I.r. data for complexes of ε-aminocaproic acid, $LnCl_3$.-$3NH_2(CH_2)_5COOH.2H_2O$, where Ln = Pr or Dy, are consistent with coordination to Ln by the carboxyl oxygen alone.[475]

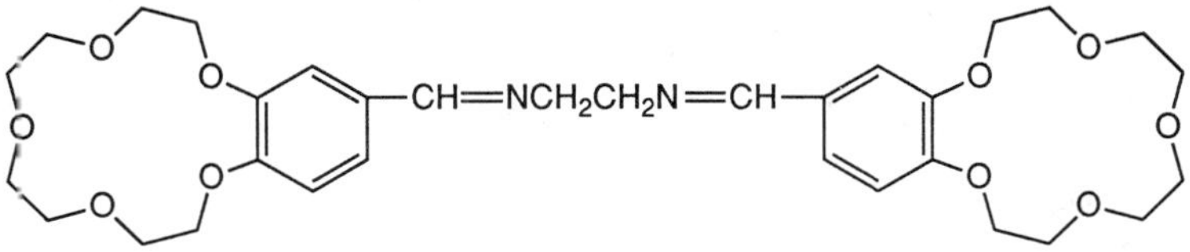

(90)

$MCpCp^*(NSO)_2$, where M = Zr or Hf, give characteristic i.r. bands of the M-NSO unit at 1250, 1080 and 520 cm^{-1}.[476] I.r. results for $M(HL)_2(H_2O)_2$, where M = Mn, Co, Ni, Cu or Zn, K_2L = 2-oximino-3-phenylpropionic acid, showed the HL^- is coordinated via the oxime N and the carbonyl O.[477] $Re(CO)_3$-$(C_{14}H_{21}NO)(C_{28}H_{39}NO_2)$, where the two ligands are (respectively) unidentate 3,5-di-tert-butyl-2-iminophenolate and bidentate 1-hydroxy-2,4,6,8-tetra-tert-butylphenoxazinyl, has an i.r. band at 3204 cm^{-1} due to νNH of the former ligand coordinated in the imino form.[478]

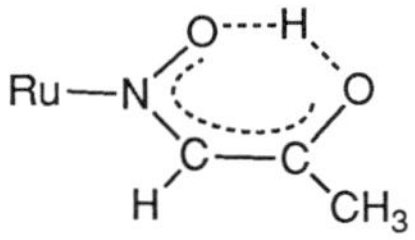

(91)

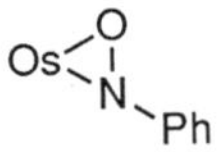

(92)

The i.r. spectrum of trans-$RuClpy_4[N(OH)CHC(O)CH_3]^+$, as the PF_6^- salt, is consistent with a major contribution from the tautomeric form (91).[479] νNO of O-bonded pyridine-N-oxide (pyO) in $OsO_4(pyO)$ and related complexes is near 1230 cm^{-1}.[480] $OsCl(NO)(ONPh)(PPh_3)_2$ has νNO (nitrosyl) at 1729 cm^{-1}, and νNO of (92) at 958 cm^{-1}.[481]

Characteristic ligand modes were assigned for the isomers $Co(NH_3)_5NO_2^{2+}$ and $Co(NH_3)_5ONO^{2+}$.[482] The i.r. spectra of $M(DMBH)(H_2O)_2$, where M = Co, Ni or Cu, H_2DMBH = PhC(O)NHN=C(Me)C(Me)=NOH, shows that coordination is via benzoyl N, azomethine N and oxime O.[483] $K[IrCl(gl)_2]$, where gl = glycinate, has $\nu_{as}CO_2$ at 1620 cm^{-1}, ν_sCO_2 at 1445 cm^{-1}, νNH at 3265, 3220 and 3128 cm^{-1}. These suggest the coordination of the glycinate through the amino group nitrogen and a carboxyl oxygen.[484] $[Ir(NO)(paa)(PPh_3)_2]^{2+}$, where paa = 2-pyridinecarbaldehyde azine, reacts with RCN (R = Me, Ph or Ph_2CH) to give cyclometallated complexes containing Ir-N-N=C-N(OH), with νNO at a characteristic value of about

$1300\ cm^{-1}$.[485]

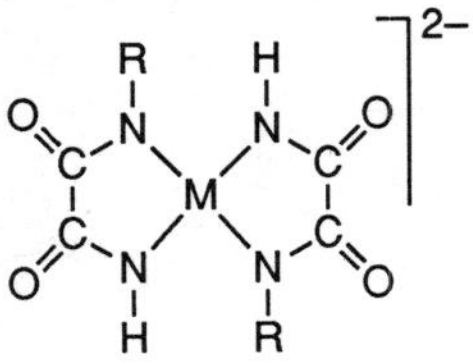

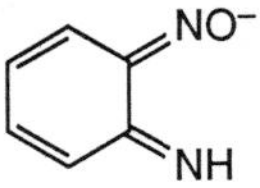

Detailed ligand mode assignments for (93), where R = CH_2CH_2OH or $CH_2CH_2CH_2OH$, M = Ni or Cu, were based on $^{58}Ni/^{62}Ni$, $^{63}Cu/^{65}Cu$, NH/ND and OH/OD isotopic substitution.[486] M(acac)-[$(Ph_2PO)_2H$], where M = Ni or Pd, have νPO at 1115 (Ni), 1100 (Pd) cm^{-1}, compared to 1200 cm^{-1} for free $Ph_2P(O)H$. The crystal structure shows that there is M-P coordination.[487] The i.r. spectrum of $Pt(PPh_3)(AH)(I)$, where AH = (94), is consistent with the presence of the ligand in the quinone-imine oxime form shown.[488]

I.r. data for $(ML)_2$, where M = Pd or Pt, H_2L = tetradentate Schiff bases derived from $PhC(=O)CH_2C(=O)Me$ and m-$C_6H_4(NH_2)_2$ or 2,6-diaminopyridine, show coordination via azomethine N atom and enolic O atoms.[489]

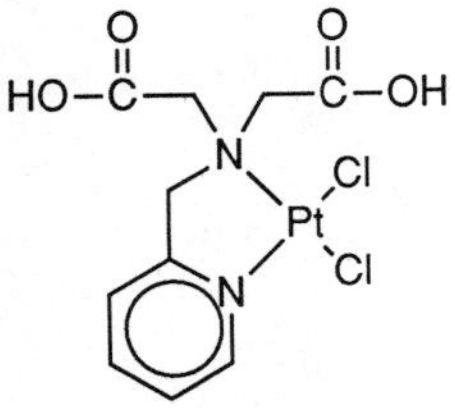

(95)

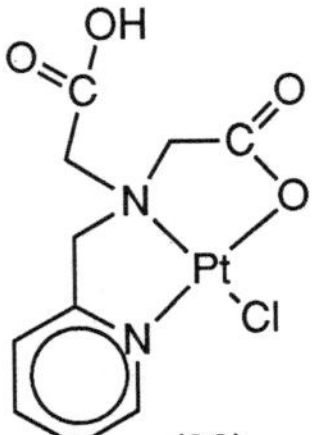

(96)

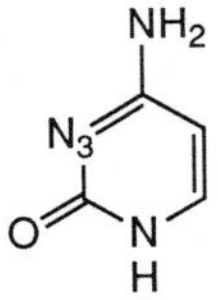

(97)

νC=O values for the pyridylmethylaminodiacetic acid (= $pidaH_2$) complex $Pt(pidaH_2)Cl_2$ shows that both carbonyl groups are free, (95). For Pt(pidaH)Cl, however, One νC=O band is at lower wavenumber, suggesting that the geometry is now (96).[490] The Raman spectrum of $[Cl_3Pt(L)]^-$, where L = cytosine, (97), shows that there is Pt-N_3 coordination.[491]

The Raman spectra of cis-$[(NH_3)_2Pt(1\text{-MeU})(1\text{-MeUH})]NO_3.2H_2O$ and cis-$[(NH_3)_2Pt(1\text{-MeUH})_2][PtCl_6].2H_2O$, where 1-MeUH = 1-methyluracil, confirm that the former contains two different, the latter two identical ligands.[492] I.r. and Raman spectra of the diammineplatinum(II) disodium 5'-uridine monophosphate blue complex show chelation of the amidate N(3)...O(2) fragment to Pt(II), and not N(4)...O(4) chelation. Splitting of $\nu_{as}PO_3$ of the PO_3^{2-} group of 5'-UMP suggests that this group plays an important role in the structure.[493]

The i.r. spectra of $CuL(H_2O)$ and $H_2[CuL_2]$, where H_2L = 7-hydroxy-5-methoxy-2-methylchromone aldehyde, show that L^{2-} acts as a tridentate, O-N-O ligand.[494] $CuLCl_2$, where L = (98) has an i.r. spectrum indicating bidentate L, via carbonyl O and heterocyclic N atoms.[495] Similarly, CuLBr, where HL = (99) has L^- coordinated through deprotonated amido N, ring N and nitro O atoms.[496]

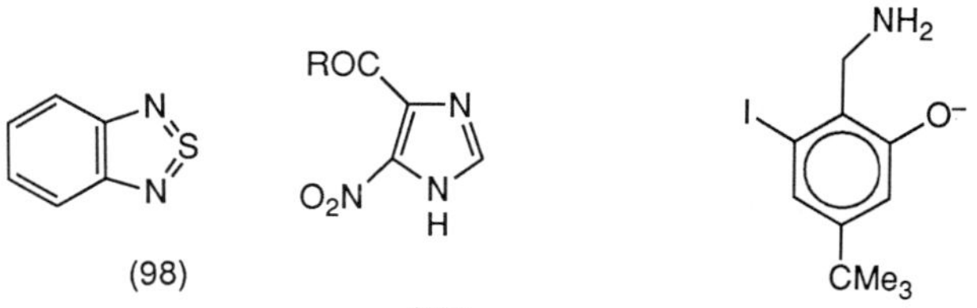

(98) (99) (100)

I.r. data for $Cu(HL)X.nH_2O$, where H_2L = benzoin thiosemicarbazone, X = NO_3, Cl, Br, SCN or ClO_4, n = 0 - 2, show HL^- coordinated via azine N, thiocarbonyl S and hydroxyl O.[497] Similar results for $Cu(MMSCl)_2.H_2O$, where MMSCl = S-methylmethionine chloride, show the presence of coordinated N and unidentate carboxylate.[498] I.r. results for $Cu(MK\text{-}H)_2$, where MK-H = (100), are consistent with chelation involving phenolate O and NH_2.[499]

$[Cu_2L(H_2L)(H_2O)_2](NO_3)_2$, where H_2L = N,N'-bis(2-pyridylmethyl)oxamide, has νCO bands due to H_2L (1530 cm^{-1}) and bridging oxamidato (L^{2-}) (1350 cm^{-1}).[500] Splittings of ν_s, $\nu_{as}CO_2$ in Cu(pydca)(tmpa), where pydca = pyridine-2,6-dicarboxylate; tmpa = N,N,N',N'-tetramethyl-1,3-diaminopropane, were ascribed to steric interaction between carboxylate groups and tertiary amine groups. The pydca is coordinated via N and two O atoms.[501] The presence of two non-equivalent ligands in (101), where [Fc] = ferrocenyl, was shown by the presence of two $\nu C{=}O$ bands in the i.r. (1615, 1650 cm^{-1}).[502]

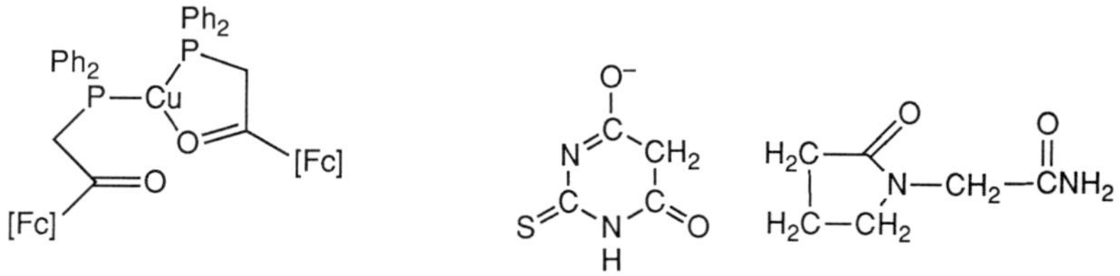

(101) (102) (103)

The i.r. spectrum of zinc thiobarbiturate, $Zn(TBA)_2.2H_2O$, shows that the ligand is present in the form (102), with tridentate coordination, via the thione S, the azomethine N and the O atom of C-O$^-$.[503] I.r spectra of Zn(II) complexes of pyracetam (PAM, (103)), e.g. $Zn(PAM)_3X_2$, where X = NCS or OAc, show coordination by both carbonyl oxygen atoms.[504]

$[UO_2(HABH)_3](NO_3)_2$ and related complexes have i.r. and Raman spectra showing that HABH (= acetone benzoylhydrazone) coordinates via carbonyl O and azomethine N.[505] The i.r. spectrum of $UO_2(NO_3)_2L_2$, where L = bis-(2-pyridyl-N-oxide)-disulphide suggest (i) that L coordinates via N-oxide oxygens, to form a 9-membered ring and (ii) that the nitrato groups are symmetrically bidentate.[506]

νNO bands shift 20 - 75 cm^{-1} to lower wavenumbers on forming the complexes $TlClBrIL_2$, where L = C_5H_5NO, 2-, 3- or 4-Me-C_5H_4NO or related species, showing Tl-ON coordination.[507]

Ligand vibrations in $SnCl_4L_2$, where L = p-HOC_6H_4CN or p-MeO-C_6H_4CN, and $SbCl_5$(p-HOC_6H_4CN) show that p-HOC_6H_4CN coordinates through the oxygen, whereas p-$MeOC_6H_4CN$ coordinates through N. All of the complexes were octahedral.[508] The i.r. and Raman spectra of $SnCl_4$(p-$RCO_2C_6H_4CONH_2$), where R = Me or Et, show coordination via the carbonyl oxygens of the amide and ester groups.[509]

The i.r. spectra of SbX_3 (X = Cl or Br) complexes with Schiff bases from the condensation of 2-aminothiazole or 2-aminobenzothiazole with salicylaldehyde or 2-hydroxy-1-naphthaldehyde, show that coordination occurs through the azomethine N atoms only.[510] (104) has i.r. bands consistent with the coordination mode shown, i.e. both azomethine N atoms and both carbonyl O atoms coordinated.[511]

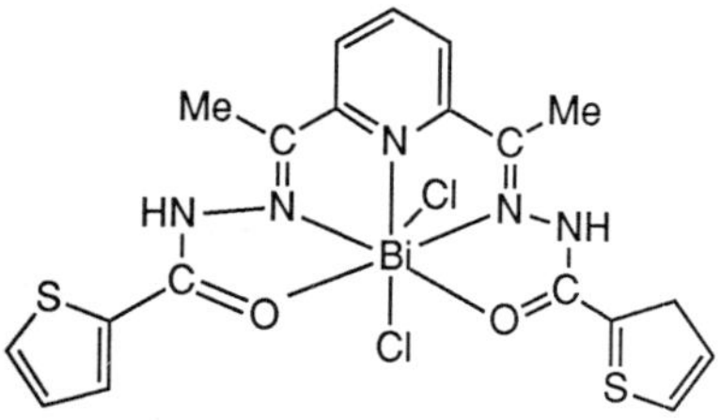

(104)

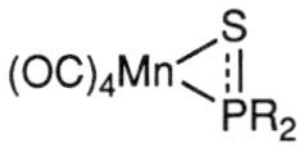

(105)

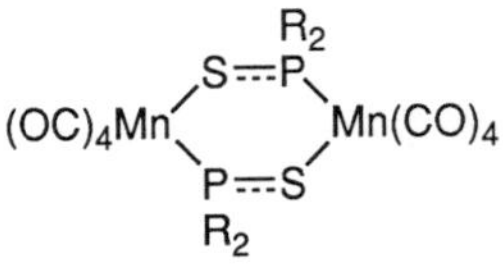

(106)

8.3 Ligands containing N and S or P and S Atoms:- The i.r. spectra of $TiCl_4L$, where L = $RCH{=}NCH_2CH_2NMe_2$, R = 2-thienyl, 3- or 5-methyl-2-thienyl, and related systems, show coordination via azomethine and amino N atoms only. The S atoms were not involved.[512]

I.r. data for $Mo_2O_3L_4.nH_2O$, where HL = thiosemicarbazide or the thiosemicarbazones of acetone, acetophenone, benzaldehyde or furfural, n = 2 - 4, show coordination of L via thioamide S and terminal hydrazido NH_2.[513] νPS bands in (105), where R = ^{i}Pr or Cy, and (106), where R = Me, Et, ^{n}Pr, ^{i}Pr or Cy, are all near 525 cm^{-1}, consistent with a PS bond order greater than one.[514] The presence of νNH (3312 cm^{-1}) and νC=O (1659 cm^{-1}) and the absence of νSH in $[Tc_2O_2(H_2ema)_4]^{2-}$, where H_4ema = N,N'-ethylenebis(2-mercaptoacetamide), show coordination by S atoms only, with uncoordinated amide groups.[515]

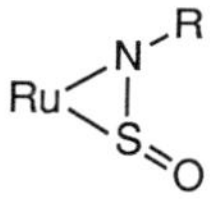

(107)

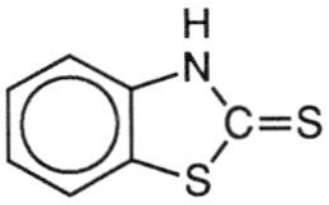

(108)

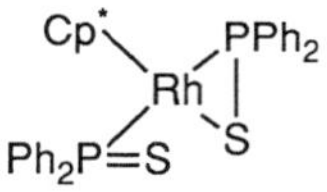

(109)

$I_3Fe[SC(NMe_2)_2]$ has ν_{as}CN of the coordinated tetramethylthiourea at 1558 cm^{-1}, compared to the free ligand value of 1508 cm^{-1}. There is, therefore, S-coordination.[516] $Ru(CO)(PPh_3)_2(L)(OSNR)$, where L = CO, $CN^{t}Bu$ or $CNC_6H_3Me_2$-2,6, R = 2-Me- or 4-Me-C_6H_4, give bands due to the RNSO ligand which are consistent with the η^2-coordination mode (107).[517] The i.r. spectra of $ML_2(OH).H_2O$, where M = Ru or Os, HL = p-tolylthiopicolinamide, show that L^- coordinates via the pyridine N and the S atom of the thioamide. $M(HL)_2(OH)_2Cl.H_2O$, however, involve unidentate HL, via the thioketone S atom only.[518]

I.r. spectra of $M(4\text{-PTNH})X_2$, where M = Co, Ni or Zn, X = Cl, Br or I, 4-PTNH = pyridine-4-carboxaldehyde thionicotinoylhydrazine, show that the 4-PTNH is a neutral bidentate (N,S) ligand.[519] The same conclusion was drawn for the DMPTA (= N,N-diethyl-5(3)-methylpyrazole-3(5)thiocarboxamide) ligand in $M(DMPTA)_2X_2$, where M = Co, Ni or Cu, X = Cl, Br, SCN, NO_3, ClO_4, BF_4 or ½SO_4.[520] The i.r. spectra of Co(II), Ni(II), Cu(II), Zn(II), Cd(II), Hg(II) and Pb(II) complexes of benzothiazole-2-thiol, (108), however, show coordination by S atoms only, and not N.[521] (109) has νPS, due to the P,S-chelate at 525 cm^{-1}, and due to the P-bonded ligand at 590 cm^{-1}.[522]

$Pd(S_2N_2H)(L\text{-}L')$, where HL-L' = azobenzene, 2-phenylpyridine etc., have νNH 3160 - 3300 cm^{-1}, and νNH 1105 - 1146 cm^{-1} (confirmed by ^{15}N and D substitution), as well as characteristic NS stretches in the range 680 - 1040 cm^{-1}.[523] The i.r. spectra of (110), where R' = H or Me; R" = Et, Ph, or ^nBu, show that νC=O is at higher wavenumbers than in the original zwitterionic ligand, due to increased localisation of electrons in the C=O unit.[524]

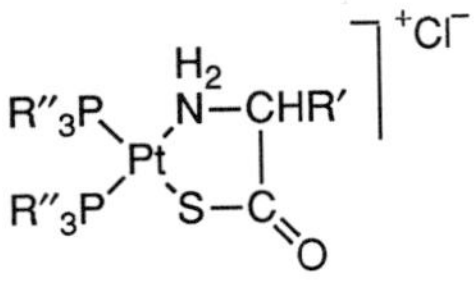

(110)

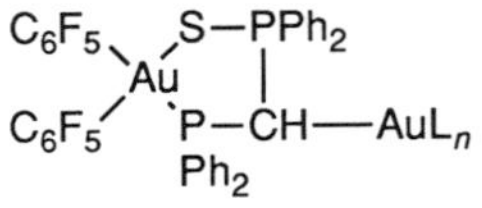

(111)

Shifts in the thioamide modes compared to the free ligand in polymeric $[Cu(mimtH)CN]_n$, where mimtH = 1-methylimidazoline-2-thione, show that there is interaction of the thiocarbonyl S atom with the Cu.[525] The i.r. spectra of $Cu_2L_2(\mu-X)_2$, $[CuL(PPh_3)_2]X$ and $[HgLX]_2$, where L = N'-(2-pyridyl)thiourea or N'-2-(4-phenylthiazolyl)thiourea; X = Cl or Br, show that the L ligands are bidentate via thionyl S and the ring N.[526] (111), where L_n = C_6F_5, Cl, $(C_6F_5)_3$ or $(C_6F_5)_2Cl$, all have νP-S in the range 555 - 572 cm^{-1}.[527]

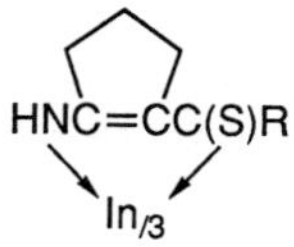

(112)

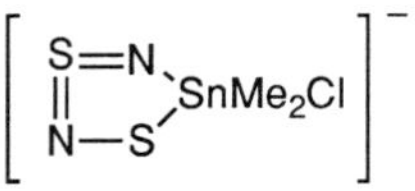

(113)

I.r. data for (112), where R = Me or CH_2NEt_2, confirm the proposed N,S-chelation.[528] $[SnMe_2Cl(S_2N_2)]^-$, containing the unit (113), has νS_2N_2 as a very strong Raman band near 880 cm^{-1} (somewhat cation dependent).[529]

8.4 Ligands containing S and O Atoms:- The Raman spectra of MX_2-$SOCl_2$-SO_2, where M = Ca, Sr or Ba, X = halide, are consistent with the formation of a complex between M^{2+} and SO_2.[530] The i.r. spectra of $[LnL_6](ClO_4)_3$, where L = dibenzyl sulphoxide, Ln = La, Pr, Nd, Sm, Gd, Tb, Dy, Ho or Yb, show that L is O-bonded to Ln.[531]

$[Cr(S_2CO)_3]^{3-}$ has νC=O at 1546, 1532 cm^{-1}, and νC-S 876 and 855 cm^{-1}. The splitting of the latter is consistent with unsymmetrical bidentate (S,S) bonding.[532] The i.r. spectra of $M(XBSOO)_2$, where M = Mn, Fe, Co, Ni, Cu, Zn, Cd or Hg; XBSOO = 4-X-benzenesulphinate, X = Cl, Br or OMe, show that the ligands are bonded through both O atoms, except in $Hg(MeO-BSOO)_2$, where S-coordination is postulated.[533]

$CpFe(CO)_2SCOR$, where R = 2-Me-, 2-CH_3COO-, 2-O_2N- or 2-F-C_6H_4 or 3,5-$(O_2N)_2C_6H_3$, have νC=O at 1598 - 1613 cm^{-1} and νC-S near 920 cm^{-1}.[534] Fac- and mer-isomers of $RuCl_3(DMSO)_3$ have been prepared. Their i.r. spectra show that the fac-isomer has all S-bonded DMSO, but that the mer-form has a mixture of S- and O-bonded ligands.[535] The O-coordinated triflato ligand in $[Os(NH_3)_5(OSO_2CF_3)](CF_3SO_3)_2$ has ν_{as}OSO at 1338 and 1320 cm^{-1}.[536]

$Ir_2(SO_2)_2(CO)_2Cl_2(\mu\text{-dpma})_2$, where dpma = bis((diphenylphosphino)methyl)phenylarsine, has νSO at 1184 and 1041 cm^{-1}, due to the η^1-S-bonded SO_2 ligands.[537]

SO_2 complexes with Ni(0) were obtained which gave νSO bonds characteristic of η^1-planar coordination (for $NiL_2(SO_2)_2$, $NiL_3(SO_2)$, where L = PPh_3, PPh_2Me or PPh_2Et), of η^2-S,O-(side-on) coordination (for $NiL_2(SO_2)$) or η^1-pyramidal coordination (for $NiL_3(SO_2)$, where L = $PPhMe_2$ or $PPhEt_2$).[538] A large range of nickel(II) sulphinato-complexes, $L_2Ni(Cl)SO_2R$, where R = Ph, p-tolyl, Me, NMe_2; L_2 = bipy, phen $(PBu_3)_2$ etc., give i.r. spectra consistent with O,O-chelation.[539]

νS=O for $M(DMSO)(N\text{-}N)Cl^+$, where M = Pd or Pt, N-N = bipy, phen or (for Pd only) en, show that there is S-bonding for the platinum complexes and the Pd/en species, but O-bonding for the Pd/bipy or phen complexes.[540] $[M(PR_3)_2(DMSO)_2]^{2+}$, where M = Pd or Pt, R_3 = $MePh_2$ or Ph_3, all have νSO due to O-bonded DMSO ligands.[541] $[Pd(R)Cl_2\{S(O)Me_2\}]$, where R = 2,4,6-$C_6H_2(NO_2)_3$, has νS=O at 1115 cm^{-1} due to S-bonded dimethylsulphoxide.[542]

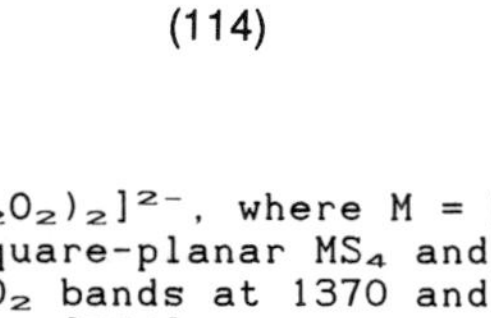

(114)

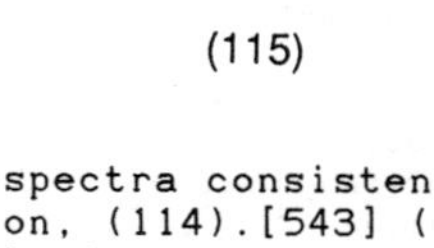

(115)

$[M(S_2C_2O_2)_2]^{2-}$, where M = Pd or Pt, have spectra consistent with square-planar MS_4 and S,S-coordination, (114).[543] (115) has νSO_2 bands at 1370 and 1360 cm^{-1} in the i.r. spectrum.[544]

References

1 G.L.Wood, C.B.Knobler and M.F.Hawthorne, Inorg. Chem., 1989, 28, 382.
2 J.Benecke and U.Behrens, J. Organometal. Chem., 1989, 363, C15.
3 T.Klapötke and U.Thewalt, J. Organometal. Chem., 1988, 356, 173.
4 M.D.Rausch, J.F.Lewison and W.P.Hart, J. Organometal. Chem., 1988, 358, 161.
5 G.Erker, U.Korek, R.Schlund and C.Krüger, J. Organometal. Chem., 1989, 364, 133.
6 G.Erker, U.Korek and J.L.Petersen, J. Organometal. Chem., 1988, 355, 121.
7 W.P.Fehlhammer and G.Beck, Chem. Ber., 1989, 122, 1907.
8 N.Q.Dao, G.P.Foulet-Fonseca, M.Jouen, E.O.Fischer, H.Fischer and J.Schmid, Compt. Rend. Acad. Sci., Ser. 2, 1988, 307, 245.
9 M.Dalibart, J.Derouault, H.Labarrère and M.Tranquille, J. Mol. Struct., 1988, 174, 17.
10 M.Dalibart and J.Derouault, Z.Phys., D, 1989, 121, 489.
11 N.J.Gogan, R.McDonald, H.J.Anderson and C.E.Loader, Can. J. Chem., 1989, 67, 433.
12 W.Deutsch, A.Gieren, C.Ruiz-Perez and K.E.Schwarzhans, J. Organometal. Chem., 1988, 355, 197.
13 H.Brunner, W.Meier, J.Wachter, I.Bernal and E.Raabe, J. Organometal. Chem., 1989, 362, 95.
14 R.L.Beddoes, C.Bitcon, A.Ricalton and M.W.Whiteley, J. Organometal. Chem., 1989, 367, C21.
15 I.Pauls, K.Dehnicke and D.Fenske, Chem. Ber., 1989, 122, 481.
16 N.Kashef and R.L.Richards, J. Organometal. Chem., 1989, 365, 309.
17 J.Yin, J.Chen, W.Xu and C.Nie, Huaxue Xuebao, 1988, 46, 1020 (Chem. Abs., 1989, 111, 194982).
18 S.E.J.Bell, K.C.Gordon and J.J.McGarvey, J. Raman Spectrosc., 1989, 20, 105.
19 F.R.Kreissl and N.Ullrich, Chem. Ber., 1989, 122, 1487.
20 H.-J.Kneuper, C.Zimmermann, K.Harms and G.Boche, Chem. Ber., 1989, 122, 1043.
21 M.Kersting, A.El-Kholi, U.Müller and K.Dehnicke, Chem. Ber., 1989, 122, 279.
22 E.M.Armstrong, P.K.Baker, M.E.Harman and M.B.Hursthouse, J. Chem. Soc., Dalton Trans., 1989, 295.
23 P.K.Baker and E.M.Armstrong, Polyhedron, 1989, 8, 351
24 P.K.Baker, K.R.Flower, P.A.Bates and M.B.Hursthouse, J. Organometal. Chem., 1989, 372, 263.
25 P.K.Baker, K.R.Flower, M.G.B.Drew and G.Forsyth, J. Chem. Soc., Dalton Trans., 1989, 1903.
26 S.Firth, A.Horton-Mastin, M.Poliakoff, J.J.Turner, D.C.McKean, G.P.McQuillan and J.Robertson, Organometallics, 1989, 8, 2876.
27 Y.Feng, J.Liu, W.Zhao, G.Zhang and Y.Liang, Huaxue Xuebao, 1988, 46, 131 (Chem. Abs., 1988, 109, 231253).
28 M.J.Schweiger, U.Nagel and W.Beck, J. Organometal. Chem., 1988, 355, 289.
29 S.C.Chang, R.H.Hauge, Z.H.Kafafi, J.L.Margrave and W.E.Billups, J. Am. Chem. Soc., 1988, 110, 7975.
30 W.Dukat and D.Naumann, J. Chem. Soc., Dalton Trans., 1989, 739.
31 C.Cabellero, J.A.Chávez, Ö.Göknur, I.Löchel, B.Nuber, H.Pfisterer and

M.L.Ziegler, J. Organometal. Chem., 1989, 371, 329.
32 T.C.Forschner, A.R.Cutler and R.W.Kullnig, J. Organometal. Chem., 1988, 356, C12.
33 M.Moran, C.Pascual, I.Cuadrado, J.R.Masaguer and J.Losada, J. Organometal. Chem., 1989, 363, 157.
34 R.Birk, H.Berke, H.-U.Hund, G.Huttner, L.Zsolnai, L.Dahlenburg, U.Behrens amd T.Sielisch, J. Organometal. Chem., 1989, 372, 397.
35 R.Kergoat, M.M.Kubicki, L.C.Gomes de Lima, H.Scordia, J.E.Guerchais and P.L'Haridon, J. Organometal. Chem., 1989, 367, 143.
36 F.-W.Grevels, J.Jacke and K.Seevogel, J. Mol. Struct., 1988, 174, 107.
37 J.Liu, Y.Feng, S.Jiang, W.Zhao and Y.Liang, Wuli Huaxue Xuebao, 1988, 4, 578 (Chem. Abs., 1989, 111, 57971).
38 L.L.Troitskaya, V.I.Sokolov, V.I.Bakhmutov, O.A.Reutov, M.Gruselle, C.Cordier and G.Jaouen, J. Organometal. Chem., 1989, 364, 195.
39 S.Abdul-Rahman, A.Houlton, R.M.G.Roberts and J.Silver, J. Organometal. Chem., 1989, 359, 331.
40 M.N.Golovin and M.J.Weaver, Inorg. Chim. Acta, 1988, 142, 177.
41 J.R.Crook, B.Chamberlain and R.J.Mawby, J. Chem. Soc., Dalton Trans., 1989, 465.
42 C.E.F.Rickards. W.R.Roper, L.J.Wright and L.Young, J. Organometal. Chem., 1989, 364, 391.
43 M.Herberhold and A.F.Hill, J. Organometal. Chem., 1988, 354, 227.
44 H.Werner, U.Meyer, M.A.Esteruelas, E.Sola and L.A.Oro, J. Organometal. Chem., 1989, 366, 187.
45 W.D.Harman, J.F.Wishart and H.Taube, Inorg. Chem., 1989, 28, 2411.
46 C.E.Anson, B.F.G.Johnson, J.Lewis, D.B.Powell, N.Sheppard, A.K.Bhattacharyya, B.R.Bender, R.M.Bullock, R.T.Hembre and J.R.Norton, J. Chem. Soc., Chem. Comm., 1989, 703.
47 R.L.Sweany, Organometallics, 1989, 8, 175.
48 K.Seitz, J.Benecke and U.Behrens, J. Organometal. Chem., 1989, 371, 247.
49 H.-F.Klein, G.Lull, B.Rodenhäuser, G.Cordier and H.Paulus, Z. Naturforsch., B, 1988, 43b, 1256.
50 H.Lang and L.Zsolnai, J. Organometal. Chem., 1989, 369, 131.
51 H.-F.Klein, M.Hellwig, U.Koch, G.Lull, M.Tadic, C.Krüger and P.Hofmann, Z. Naturforsch., B, 1988, 43b, 1427.
52 M.Garland and G.Bor, Inorg. Chem., 1989, 28, 410.
53 A.Shaver, H.L.Uhm, E.Singleton and D.C.Liles, Inorg. Chem., 1989, 28, 847.
54 J.Müller, M.Tschampel and J.Pickardt, J. Organometal. Chem., 1988, 355, 513.
55 E.Rotondo, S.Lo Schiavo, G.Bruno, C.G.Arena, R.Gobetto and F.faraone, Inorg. Chem., 1989, 28, 2944.
56 G.A.Ozin, D.M.Haddleton and C.J.Gill, J. Phys. Chem., 1989, 93, 6710.
57 D.E.Hendriksen, A.A.Oswald, G.B.Ansell, S.Leta and R.V.Kastrup, Organometallics, 1989, 8, 1153.
58 T.Hoffmann, M.Karabit, P.Zdunneck and E.Wenschuh, Z. Chem., 1988, 28, 415.
59 J.Mink and T.Szilyàgi, Croat. Chim. Acta, 1988, 61, 719.
60 G.Facchin, R.Bertani, L.Zanotto, M.Calligaris and G.Nardin, J. Organometal. Chem., 1989, 366, 409.
61 A.Sacco, G.Vasapollo, C.F.Nobile, A.Piergiovanni, M.A.Pellinghelli and M.Lanfranchi, J. Organometal. Chem., 1988, 356, 397.

62 J.Mink, I.Pápai, M.Gál and P.L.Goggin, Pure Appl. Chem., 1989, 61, 973.
63 R.A.Michelin and R.Ros, J. Chem. Soc., Dalton Trans., 1989, 1149.
64 T.G.Appleton, R.D.Berry, J.R.Hall and D.W.Neale, J. Organometal. Chem., 1989, 364, 249.
65 R.Řeřicha, B.N.Cyvin, J.Brunvoll and S.J.Cyvin, Coll. Czech. Chem. Commun., 1988, 53, 2377.
66 S.B.Mohsin, M.Trenary and H.J.Robota, Chem. Phys. Lett., 1989, 154, 511.
67 L.N.Ito, J.D.Sweet, A.N.Mueting, L.H.Pignolet, M.F.J.Schoondergang and J.J.Steggerda, Inorg. Chem., 1989, 28, 3696.
68 R.Usón, J.Forniés, M.Tomás, J.M.Casas and R.Navarro, J. Chem. Soc., Dalton Trans., 1989, 169.
69 R.Usón, J.Forniés, P.Espinet, C.Fortuño and M.Tomás, J. Chem. Soc., Dalton Trans., 1989, 1583.
70 R.Usón, J.Forniés, M.Tomás, I.Ara and J.M.Casas, Inorg. Chem., 1989, 28, 2388.
71 R.Usón, J.Forniés, M.Tomás and J.M.Casas, Angew. Chem., Int. Ed. Engl., 1989, 28, 748.
72 T.Merle-Méjean, S.Bouchareb and M.Tranquille, J. Phys. Chem., 1989, 93, 1197.
73 T.B.Newbound, M.R.Colsman, M.M.Miller, G.P.Wulfsberg, O.P.Anderson and S.H.Strauss, J. Am. Chem. Soc., 1989, 111, 3762.
74 H.Shindo, Chem. Phys. Lett., 1989, 159, 85.
75 H.Feichenfeld and M.J.Weaver, J. Phys. Chem., 1989, 93, 4276.
76 I.C.Manzanares, N.L.S.Yamazaki and E.Weitz, J. Phys. Chem., 1989, 93, 4733.
77 S.Mohan and F.Payami, Proc. Ind. Natl. Sci. Acad., Part A, 1988, 54, 984.
78 J.M.Parnis and G.A.Ozin, J. Phys. Chem., 1989, 93, 1204.
79 L.Manceron and L.Andrews, J. Phys. Chem., 1989, 93, 2964.
80 G.Jokuszies, R.Franzheld, W.Seifert and E.Butter, Z. Chem., 1988, 28, 342.
81 J.Barrau, V.Balaji and J.Michl, Organometallics., 1989, 8, 2034.
82 A.V.Belyakov, V.S.Nikitin, M.V.Polyakova, E.T.Bogoradovskii and V.S.Zavgorodnii, Zh. Fiz. Khim., 1989, 63, 652.
83 J.R.Durig, W.Zhao and T.S.Little, J. Raman Spectrosc., 1989, 20, 111.
84 H.Bettermann and H.H.Perkampus, Spectrochim. Acta, Part A, 1989, 45A, 735.
85 J.Gu, D.Liu and S.Zheng, Cuihua Xuebao, 1989, 10, 202 (Chem. Abs., 1989, 111, 219889).
86 S.R.Frerichs and J.E.Ellis, J. Organometal. Chem., 1989, 359, C41.
87 R.B.Hitam and A.J.Rest, Organometallics, 1989, 8, 1598.
88 H.G.Alt and H.E.Engelhardt, Z. Naturforsch., B, 1989, 44b, 367.
89 N.A.Cooley, P.T.F.MacConnachie and M.C.Baird, Polyhedron, 1988, 7, 1965.
90 L.Y.Goh and T.C.W.Mak, J. Organometal. Chem., 1989, 363, 77.
91 D.Albers, N.J.Coville, T.A.Ford, M.V.M.Hill and D.C.Lavendis, Spectrochim. Acta, Part A, 1989, 45A, 249.
92 A.A.Ismail, F.Sauriol and I.S.Butler, Inorg. Chem., 1989, 28, 1007.
93 J.Rawlings and C.A.L.Mahaffy, Spectrosc. Lett., 1988, 21, 597.
94 B.W.Eichhorn, R.C.Haushalter and J.C.Huffman, Angew. Chem., Int. Ed. Engl., 1989, 28, 1032.

95 L.Wang, X.Zhu and K.G.Spears, J. Am. Chem. Soc., 1988, 110, 8695.
96 L.Wang, X.Zhu and K.G.Spears, J. Phys. Chem., 1989, 93, 2.
97 R.L.Sweany and A.Moroz, J. Am. Chem. Soc., 1989, 111, 3577.
98 S.J.Duray, D.M.Becher and E.R.Grant, Laser Chem., 1988, 9, 63.
99 C.Louis, L.Marchese, S.Coluccia and A.Zecchina, J. Chem. Soc., Farad. Trans., 1, 1989, 85, 1655.
100 T.Adatia, M.McPartlin, M.J.Mays, M.J.Morris and P.R.Raithby, J. Chem. Soc., Dalton Trans., 1989, 1555.
101 H.-J.Müller, K.Polborn, M.Steimann and W.Beck, Chem. Ber., 1989, 122, 1901.
102 S.Ernst, S.Kohlmann and W.Kaim, J. Organometal. Chem., 1988, 354, 177.
103 S.N.Misra and K.Venkatasubramanian, Bull. Chem. Soc. Jpn., 1988, 61, 4482.
104 D.Coucouvanis, S.Al-Ahmad, A.Salifoglou, R.W.Dunham and R.H.Sands, Angew. Chem., Int. Ed. Engl., 1988, 27, 1353.
105 A.Ricalton and M.W.Whiteley, J. Organometal. Chem., 1989, 361, 101.
106 A.Varshney and G.M.Gray, Inorg. Chim. Acta, 1988, 148, 215.
107 T.S.A.Hor and H.S.O.Chan, Inorg. Chim. Acta, 1989, 160, 53.
108 M.J.Winter and S.Woodward, J. Organometal. Chem., 1989, 361, C18.
109 F.-W.Grevels, J.Jacke, P.Betz, C.Krüger and Y.H.Tsay, Organometallics, 1989, 8, 293.
110 M.Herberhold and J.Ott, Z. Naturforsch., B, 1988, 43b, 682.
111 Y.Ishikawa, P.A.Hackett and D.M.Rayner, J. Phys. Chem., 1989, 93, 652.
112 M.Ichikawa, P.E.Hoffmann and A.Fukuoka, J. Chem. Soc., Chem. Comm., 1989, 1395.
113 R.R.Andréa, W.G.J.de Lange, D.J.Stufkens and A.Oskam, Inorg. Chem., 1989, 28, 318.
114 A.Strube, G.Huttner and L.Zsolnai, Angew. Chem., Int. Ed. Engl., 1988, 27, 1529.
115 G.W.Harris and N.J.Coville, J. Crystallogr. Spectrosc. Res., 1989, 19, 451.
116 R.H.Hill and B.J.Palmer, Organometallics, 1989, 8, 1651.
117 T.M.Miller, K.J.Ahmed and M.S.Wrighton, Inorg. Chem., 1989, 28, 2347.
118 G.F.P.Warnock, L.C.Moodie and J.E.Ellis, J. Am. Chem. Soc., 1989, 111, 2131.
119 M.S.Corraine and J.D.Atwood, Inorg. Chem., 1989, 28, 3781.
120 W.E.Klotzbücher, J. Mol. Struct., 1988, 174, 5.
121 S.Dobos, A.Beck and L.Guczi, J. Mol. Struct., 1988, 174, 331.
122 T.J.Henly and J.R.Shapley, Organometallics, 1989, 8, 2729.
123 P.Ormos, D.Braunstein, H.Frauenfelder, M.K.Hong, S.L.Lin, T.B.Sauke and R.D.Young, Proc. Natl. Acad. Sci. U.S.A., 1988, 85, 8492.
124 T.Ogura and T.Kitagawa, Rev. Sci. Instrum., 1988, 59, 1316.
125 A.J.Dixon, P.Glyn, M.A.Healy, P.M.Hodges, T.Jenkins, M.Poliakoff and J.J.Turner, Spectrochim. Acta, Part A, 1988, 44A, 1309.
126 D.Braunstein, A.Ansari, J.Berendzen, B.R.Cowen, K.D.Egeberg, H.Frauenfelder, M.K.Hong, P.Ormos, T.B.Sauke, R.Scholl, A.Schulte, S.G.Sligar, B.A.Springer, P.J.Steinbach and R.D.Young, Proc. Natl. Acad. Sci. U.S.A., 1988, 85, 8497.
127 S.Han, J.F.Madden, L.M.Siegel and T.G.Spiro, Biochem., 1989, 28, 5477.
128 N.S.Schroeder, R.Funchess, R.A.Jacobson and R.J.Angelici, Organometallics, 1989, 8, 521.
129 J.Takács, L.Markó and L.Párkányi, J. Organometal. Chem., 1989, 361, 109.

130 A.Shaver, I.S.Butler and J.P.Gao, Organometallics, 1989, 8, 2079.
131 N.C.Norman and P.M.Webster, Z. Naturforsch., B, 1989, 44b, 91.
132 J.N.Moore, P.A.Hansen and R.M.Hochstrasser, J. Am. Chem. Soc., 1989, 111, 4563.
133 S.Dobos, S.Nunziate-Cesaro and B.V.Lokshin, J. Mol. Struct., 1988, 174, 325.
134 J.E.Ellis and Y.S.Chen, Organometallics, 1989, 8, 1350.
135 C.Bowers and P.K.Dutta, J. Phys. Chem., 1989, 93, 2596.
136 K.M.Rao, G.Spoto, E.Guglielminotti and A.Zecchina, Inorg. Chem., 1989, 28, 243.
137 R.T.Baker, J.C.Calabrese, P.J.Krusic, M.J.Therien and W.C.Trogler, J. Am. Chem. Soc., 1988, 110, 8392.
138 E.Solymosi and J.Raskó, J. Catal., 1989, 115, 107.
139 P.L.Bogdan and E.Weitz, J. Am. Chem. Soc., 1989, 111, 3163.
140 J.L.Robbins, J. Catal., 1989, 115, 120.
141 S.A.R.Knox, K.A.Macpherson, A.G.Orpen and M.C.Rendle, J. Chem. Soc., Dalton Trans., 1989, 1807.
142 J.J.Venter and M.A.Vannice, Inorg. Chem., 1989, 28, 1634.
143 J.J.Venter and M.A.Vannice, J. Am. Chem. Soc., 1989, 111, 2377.
144 D.J.Szalda, E.Fujita and C.Creutz, Inorg. Chem., 1989, 28, 1446.
145 T.T.Chiong and A.A.Davydov, Zh. Fiz. Khim., 1989, 63, 271.
146 T.M.Bockman and J.K.Kochi, J. Am. Chem. Soc., 1989, 111, 4669.
147 R.Tannenbaum, U.K.Dietler and G.Bor, Inorg. Chim. Acta, 1988, 154, 109.
148 L.W.H.Leung, S.C.Chang and M.J.Weaver, J. Chem. Phys., 1989, 90, 7426.
149 R.Dictor and S.Roberts, J. Phys. Chem., 1989, 93, 5846.
150 B.H.Weiller, E.P.Wasserman, R.G.Bergman, C.B.Moore and G.C.Pimentel, J. Am. Chem. Soc., 1989, 111, 8288.
151 A.Rubailo, V.P.Selina and Yu.S.Varshavskii, Koord. Khim., 1988, 14, 957.
152 S.T.Belt, F.-W.Grevels, W.E.Klotzbücher, A.McCamley and R.N.Perutz, J. Am. Chem. Soc., 1989, 111, 8373.
153 J.H.B.Chenier, M.Histed, J.A.Howard, H.A.Joly, H.Morris and B.Mile, Inorg. Chem., 1989, 28, 4114.
154 P.Dufour, L.Huang, A.Choplin, R.Sanchez-Delgado, A.Théolier and J.-M.Basset, J. Organometal. Chem., 1988, 354, 243.
155 A.Fumagalli, S.Martinengo, D.Galli, A.Albinati and F.Ganazzoli, Inorg. Chem., 1989, 28, 2476.
156 A.Fukuoka, L.F.Rao and M.Ichikawa, Nippon Kagaku Kaishi, 1989, 561.
157 P.Gelin, A.Auroux, Y.Ben Taarit and P.C.Gravelle, Appl. Catal., 1989, 46, 227.
158 D.M.Vandenberg, T.Chin-Choy and P.C.Ford. J. Organometal. Chem., 1989, 366, 257.
159 P.C.Servaas, D.J.Stufkens and A.Oskam, Inorg. Chem., 1989, 28, 1780.
160 R.L.Sweany, M.A.Polito and A.Moroz, Organometallics, 1989, 8, 2305.
161 P.B.Davies, N.A.Martin, M.D.Nunes, D.A.Pape and D.K.Russell, Chem. Phys. Lett., 1989, 156, 553.
162 M.P.Keyes, L.U.Gron and K.L.Watters, Inorg. Chem., 1989, 28, 1236.
163 K.M.Rao, G.Spoto and A.Zecchina, Langmuir, 1989, 5, 319.
164 L.L.Sheu, H.Knözinger and W.M.H.Sachtler, J. Am. Chem. Soc., 1989, 111, 8125.
165 P.Braunstein, M.Ries, C. de M. de Bellefon, Y.Dusausoy and J.-P.Mangeot, J. Organometal. Chem., 1988, 355, 533.

166 L.W.H.Leung, A.Wieckowski and M.J.Weaver, J. Phys. Chem., 1988, 92, 6985.
167 S.C.Chang, L.W.H.Leung and M.J.Weaver, J. Phys. Chem., 1989, 93, 5341.
168 M.R.Mucalo and R.P.Cooney, J. Chem. Soc., Chem. Comm., 1989, 94.
169 M.A.Kohler, N.W.Cant, M.S.Wainwright and D.L.Trimm, J. Catal., 1989, 117, 188.
170 R.B.Dyer, O.Einarsdottir, P.M.Killogh, J.J.Lopez-Garriga and W.H.Woodruff, J. Am. Chem. Soc., 1989, 111, 7657.
171 J.H.B.Chenier, C.A.Hampson, J.A.Howard and B.Mile, J. Phys. Chem., 1989, 93, 114.
172 Y.Ikezawa, H.Saito, H.Matsubayashi and G.Toda, J. Electroanal. Chem. Interfacial Electrochem., 1988, 252, 395.
173 J.H.B.Chenier, J.A.Howard, H.A.Joly, B.Mile and M.Tomietto, Can. J. Chem., 1989, 67, 655.
174 G.V.Chertikhin, I.L.Rozhanskii, L.V.Serebrennikov and V.F.Shevel'kov, Zh. Fiz. Khim., 1988, 62, 2256.
175 W.G.Hatton, N.P.Hacker and P.H.Kasai, J. Phys. Chem., 1989, 93, 1328.
176 R.S.Grev and H.F.Schaeffer, J. Am. Chem. Soc., 1989, 111, 5687.
177 A.P.Borisov and V.D.Makhaev, Russ. J. Inorg. Chem., 1988, 33, 1746.
178 A.Kurbonbekov, T.Kh.Alikhanova and U.Mirsaidov, Russ. J. Inorg. Chem., 1989, 34, 347.
179 J.A.Jensen and G.S.Girolami, Inorg. Chem., 1989, 28, 2107.
180 J.A.Jensen and G.S.Girolami, Inorg. Chem., 1989, 28, 2114.
181 C.Ting and L.Messerle, J. Am. Chem. Soc., 1989, 111, 3449.
182 S.Bettonville and J.Goffart, J. Organometal. Chem., 1988, 356, 297.
183 D.Baudry, E.Bulot, P.Carpin, M.Ephritikhine, M.Lance, M.Nierlich and J.Vigner, J. Organometal. Chem., 1989, 371, 155.
184 H.Plenio and H.W.Roesky, Z. Naturforsch., B, 1988, 43b, 1575.
185 J.F.Leboeuf, J.C.Leblanc and C.Moïse, J. Organometal. Chem., 1989, 364, C22.
186 M.N.Ackermann, C.R.Barton, C.J.Deodene, E.M.Specht, S.C.Keill, W.E.Schreiber and H.Kim, Inorg. Chem., 1989, 28, 397.
187 T.A.George, R.K.Hayes, M.Y.Mohammed and C.J.Pickett, Inorg. Chem., 1989, 28, 3269.
188 H.Kang and J.Zubieta, J. Chem. Soc., Chem. Comm., 1988, 1192.
189 T.Yoshida, T.Adachi, T.Ueda, M.Kaminaka, N.Sasaki, T.Higuchi, T.Aoshima, I.Mega, Y.Mizobe and M.Hidai, Angew. Chem., Int. Ed. Engl., 1989, 28, 1040.
190 D.Fenske, K.Völp and K.Dehnicke, Z. Naturforsch., B, 1988, 43b, 1125.
191 A.Frankenau and K.Dehnicke, Z. Naturforsch., B, 1989, 44b, 493.
192 A.Görge, K.Dehnicke and D.Fenske, Z. Naturforsch., B, 1989, 44b, 117.
193 A.Görge, K.Dehnicke and D.Fenske, Z. Naturforsch., B, 1988, 43b, 677.
194 K.Beran, G.Steinke and R.Mews, Chem. Ber., 1989, 122, 1613.
195 A.H.Klahn and D.Sutton, Organometallics, 1989, 8, 198.
196 S.G.Fox, V.M.Browne and P.Hollins, J. Chem. Soc., Chem. Comm., 1989, 697.
197 W.Hiller, K.Hösler and K.Dehnicke, Z. Anorg. Allg. Chem., 1989, 574, 7.
198 J.E.Pate, P.K.Ross, T.J.Thamann, C.A.Reed, K.D.Karlin, T.N.Sorrell and E.I.Solomon, J. Am. Chem. Soc., 1989, 111, 5198.
199 I.I.Grinval'd, B.V.Lokshin, N.K.Rudnevskii, A.D.Zorin, V.P.Mar'in and E.N.Karataev, Zh. Obshch. Khim., 1988, 58, 937.
200 S.E.Miller and D.A.House, Inorg. Chim. Acta, 1989, 157, 29.

201 D.C.Bradley, M.B.Hursthouse and H.R.Powell, J. Chem. Soc., Dalton Trans., 1989, 1537.
202 Z.Wang and A.J.Lees, J. Organometal. Chem., 1989, 363, 335.
203 S.Yurdakul, S.Akyüz and J.E.D.Davies, J. Mol. Struct., 1988, 174, 435.
204 H.K.van Dijk, J.van der Haar, D.J.Stufkens and A.Oskam, Inorg. Chem., 1989, 28, 75.
205 S.Yurdakul, S.Akyüz and J.E.D.Davies, J. Mol. Struct., 1989, 197, 355.
206 T.Yabe, D.R.Anderson, L.K.Orman, Y.J.Chang and J.B.Hopkins, J. Phys. Chem., 1989, 93, 2302.
207 D.R.Anderson and J.B.Hopkins, A.I.P.Conf. Proc., 1987, 172, 715.
208 Y.J.Chang, S.Laneman and J.B.Hopkins, Chem. Phys. Lett., 1989, 156, 421.
209 A.M.Pyle, J.P.Rehmann, R.Meshoyrer, C.V.Kumar, N.J.Turro and J.K.Barton, J. Am. Chem. Soc., 1989, 111, 3051.
210 R.Hage, J.G.Haasnoot, D.J.Stufkens, T.L.Snoeck, J.G.Vos and J.Reedijk, Inorg. Chem., 1989, 28, 1413.
211 T.C.Strekas, H.D.Gafney, S.A.Tysoe, R.P.Thummel and F.Lefoulon, Inorg. Chem., 1989, 28, 2964.
212 A.Kirsch-Deresmaeker, L.Jacquet, A.Masschelein, F.Vanhecke and K.Heremans, Inorg. Chem., 1989, 28, 2465.
213 J.B.Cooper and D.W.Wertz, Inorg. Chem., 1989, 28, 3108.
214 H.E.Toma, P.S.Santos and C.Cipriano, Spectrosc. Lett., 1988, 21, 909.
215 H.Feilchenfeld, X.Gao and M.J.Weaver, Chem. Phys. Lett., 1989, 161, 321.
216 G.Krüger, A.Roodt, J.G.Leipoldt and R.van Eldik, Inorg. Chem., 1989, 28, 3073.
217 D.W.Ball, R.H.Hauge and J.L.Margrave, High Temp. Sci., 1988, 25, 95.
218 D.P.Dowling and W.K.Glass, Spectrochim. Acta, Part A, 1988, 44A, 1351.
219 N.Farrell and Y.Qu, Inorg. Chem., 1989, 28, 3416.
220 D.W.Ball, R.H.Hauge and J.L.Margrave, Inorg. Chem., 1989, 28, 1599.
221 A.M.A.Bennett, G.A.Foulds and D.A.Thornton, Spectrochim. Acta, Part A, 1989, 45A, 219.
222 G.Ciani, M.Moret, A.Sironi, S.Bruni, F.Cariati, A.Pozzi, T.Manfredini, L.Menabue and G.C.Pellacani, Inorg. Chim. Acta, 1989, 158, 9.
223 J.G.Contreras and H.Cortés, Spectrochim. Acta, Part A, 1989, 45A, 1025.
224 A.V.Sakharov, V.O.Gel'mbol'dt, V.F.Sukhoverkov and A.A.Ennan, Russ. J. Inorg. Chem., 1989, 34, 302.
225 A.S.Hinman, B.J.Pavelich and K.McGarty, Can. J. Chem., 1988, 66, 1589.
226 R.S.Czernuszewicz, K.A.Macor, X.Y.Li, J.R.Kincaid and T.G.Spiro, J. Am. Chem. Soc., 1989, 111, 3860.
227 G.Terzian, B.Moubaraki, M.Mossoyan-Deneux and D.Benlian, Spectrochim. Acta, Part A, 1989, 45A, 675.
228 P.Klinzing, A.El-Kholi, U.Muller, K.Dehnicke and K.Findeisen, Z. Anorg. Allg. Chem., 1989, 569, 83.
229 R.R.Andréa, W.G.J.de Lange, D.J.Stufkens and A.Oskam, Inorg. Chim. Acta, 1988, 149, 77.
230 O.V.Quinzani, A.H.Jubert and P.J.Aymonino, J. Raman Spectrosc., 1989, 20, 141.
231 H.E.Toma, E.Stadler and P.S.Santos, Spectrochim. Acta, Part A, 1988, 44A, 1365.
232 C.Su, Y.D.Park, G.Y.Liu and T.G.Spiro, J. Am. Chem. Soc., 1989, 111, 3457.

233 W.A.Oertling, A.Salehi, C.K.Chang and G.T.Babcock, J. Phys. Chem., 1989, 93, 1311.
234 U.Bobinger, R.Schweitzer-Stenner and W.Dreybrodt, J. Raman Spectrosc., 1989, 20, 191.
235 D.L.Rousseau, M.Sassaroli, Y.C.Chang and S.Dasgupta, Ann. N.Y. Acad. Sci., 1988, 550, 223.
236 W.J.Chuang, S.Johnson and H.E.van Wart, J. Inorg. Biochem., 1988, 34, 201.
237 J.T.Sage, D.Morikis and P.M.Champion, J. Chem. Phys., 1989, 90, 3015.
238 D.L.Rousseau, Y.C.Ching, M.Brunori and G.M.Giacometti, J. Biol. Chem., 1989, 264, 7878.
239 D.J.Davies, M.K.Frame and D.A.Johnson, Biochim. Biophys. Acta, 1988, 936, 61.
240 V.Palaniappan and J.Terner, J. Biol. Chem., 1989, 264, 16046.
241 K.Gersonde, N.T.Yu, S.H.Lin, K.M.Smith and D.W.Parish, Biochem., 1989, 28, 3960.
242 R.G.Alden, J.D.Satterlee, J.Mintorovich, I.Constantinidis, M.R.Ondrias and B.I.Swanson, J. Biol. Chem., 1989, 264, 1933.
243 B.A.Crawford and M.R.Ondrias, J. Phys. Chem., 1989, 93, 5055.
244 C.D.Tait, J.M.Garner, J.P.Collman, A.P.Sattelberger and W.H.Woodruff, J. Am. Chem. Soc., 1989, 111, 7806.
245 D.J.Szalda, C.L.Schwarz, J.F.Endicott, E.Fujita and C.Creutz, Inorg. Chem., 1989, 28, 3214.
246 Y.Ozaki, K.Aoyagi, K.Iriyama, H.Ogoshi and T.Kitagawa, J. Phys. Chem., 1989, 93, 3842.
247 C.R.Bartholomew, A.A.McConnell and W.E.Smith, J. Raman Spectrosc., 1989, 20, 595.
248 Y.A.Sarma, Spectrochim. Acta, Part A, 1989, 45A, 649.
249 A.Salehi, W.A.Oertling, H.N.Fonda, G.T.Babcock and C.K.Chang, Photochem. Photobiol., 1988, 48, 525.
250 J.H.Schneider, J.Odo and K.Nakamoto, Nucleic Acids Res., 1988, 16, 10323.
251 L.Hi and T.Jiang, Guangpuxue Yu Guangfu Fenxi, 1989, 9, 6 (Chem. Abs., 1989, 111, 66877).
252 P.C.Servaas, D.J.Stufkens and A.Oskam, Inorg. Chem., 1989, 28, 1774.
253 S.P.Greiner, J.Winzenburg, B.van Maltzan, C.J.Winscom and K.Mobius, Chem. Phys. Lett, 1989, 155, 93.
254 A.K.Shiemke, J.A.Shelnutt and R.A.Scott, J. Biol. Chem., 1989, 264, 11236.
255 P.A.Apanasevich, V.V.Kvach and V.A.Orlovich, J. Raman Spectrosc., 1989, 20, 125.
256 R.S.Czernuszewicz, X.Y.Li and T.G.Spiro, J. Am. Chem. Soc., 1989, 111, 7024.
257 X.Y.Li, R.S.Czernuszewicz, J.R.Kincaid and T.G.Spiro, J. Am. Chem. Soc., 1989, 111, 7012.
258 O.K.Song, M.J.Yoon, J.R.Chang and D.Kim, Bull. Korean Chem. Soc., 1989, 10, 39.
259 J.A.Shelnutt, J. Phys. Chem., 1989, 93, 6283.
260 R.J.Donohoe, M.Atamian and D.F.Bocian, J. Phys. Chem., 1989, 93, 2244.
261 M.Atamian, R.J.Donohoe, J.S.Lindsay and D.F.Bocian, J. Phys. Chem., 1989, 93, 2236.
262 V.A.Walters, J.C.de Paola, G.T.Babcock and G.E.Leroi, J. Am. Chem. Soc., 1989, 111, 8300.

263 R.Aroca, R.E.Clavijo, C.A.Jennings, G.S.Kovacs, J.M.Duff and R.O.Loutfy, Spectrochim. Acta, Part A, 1989, 45A, 957.
264 R.G.Keil, D.W.Johnson, M.A.Fryling and J.F.O'Brien Inorg. Chem., 1989, 28, 2764.
265 T.Klapötke, Polyhedron, 1989, 8, 311.
266 D.Fenske, E.Hartmann and K.Dehnicke, Z. Naturforsch., B, 1988, 43b, 1611.
267 J.P.Bullock and K.R.Mann, Inorg. Chem., 1989, 28, 4006.
268 D.I.Zubritskaya, D.I.Semenishin, O.Ya.Borovaya and I.S.Komarinskaya, Russ. J. Inorg. Chem., 1989, 34, 134.
269 A.Khabou, K.Dehnicke, K.Findeisen and D.Fenske, Z. Naturforsch., B, 1988, 43b, 1616.
270 A.C.Filippou and E.O.Fischer, J. Organometal. Chem., 1989, 365, 317.
271 C.Ergezinger, A.El-Kholi, U.Müller and K.Dehnicke, Z. Anorg. Allg. Chem., 1989, 568, 55.
272 A.C.Filippou and W.Grünleitner, Z. Naturforsch., B, 1989, 44b, 666.
273 A.C.Filippou, E.O.Fischer and H.G.Alt, Z. Naturforsch., B, 1988, 43b, 654.
274 M.F.N.N.Carvalho, A.J.L.Pombeiro, E.G.Bakalbassis and C.A.Tsipis, J. Organometal. Chem., 1989, 371, C26.
275 D.Lentz and R.Marshall, Chem. Ber., 1989, 122, 1223.
276 S.Warner and S.J.Lippard, Inorg. Chem., 1989, 28, 3008.
277 M.M.Vergara, E.L.Varetti, G.Rigotti and A.Navaza, J. Phys. Chem. Solids, 1989, 50, 951.
278 C.O.Della Vedova, J. Raman Spectrosc., 1988, 19, 453.
279 A.Vogler and H.Kunkely, Inorg. Chim. Acta, 1988, 150, 1.
280 K.J.Brewer, W.R.Murphy and J.D.Petersen, Inorg. Chim. Acta, 1989, 159, 93.
281 S.K.Doorn and J.T.Hupp, J. Am. Chem. Soc., 1989, 111, 1142.
282 G.F.Warnock and N.J.Cooper, Organometallics, 1989, 8, 1826.
283 K.R.Dunbar, J. Am. Chem. Soc., 1988, 110, 8247.
284 J.Wu, P.E.Fanwick and C.P.Kubiak, J. Am. Chem. Soc., 1989, 111, 7812.
285 M.Hvastijová, J.Kohout, H.Kühler and G.Ondrejovič, Z. Anorg. Allg. Chem., 1988, 566, 111.
286 K.Ashley, M.Lazaga, M.G.Samant, H.Seki and M.R.Philpott, Surf. Sci., 1989, 219, L590.
287 S.Maier, W.Hiller, J.Strähle, C.Ergezinger and K.Dehnicke, Z. Naturforsch., B, 1988, 43b, 1628.
288 J.C.Dougal, P.Gans, J.B.Gill and L.H.Johnson, Pure Appl. Chem., 1988, 60, 1731.
289 J.Meyer, J.Willnecker and U.Schubert, Chem. Ber., 1989, 122, 223.
290 H.W.Roesky, J.Schimkowiak, M.Noltemeyer and G.M.Sheldrick, Z. Naturforsch., B, 1988, 43b, 949.
291 H.Aslan, K.Yünlü, R.D.Fischer, G.Bombieri and F.Benetollo, J. Organometal. Chem., 1988, 354, 63.
292 C.Ergezinger, F.Weller and K.Dehnicke, Z. Naturforsch., B, 1988, 43b, 1621.
293 R.Lechler, H.-D.Hausen and J.Weidlein, J. Organometal. Chem., 1989, 359, 1.
294 L.Jäger, H.Köhler, A.I.Brusilovets and V.V.Skopenko, Z. Anorg. Allg. Chem., 1988, 564, 85.
295 B.Lösking and H.Willner, Angew. Chem., Int. Ed. Engl., 1989, 28, 1255.
296 J.S.Thrasher and K.V.Madappat, Angew. Chem., Int. Ed. Engl., 1989, 28,

1256.
297 E.Carmona, E.Gutiérrez-Puebla, A.Monge, P.J.Pérez and L.J.Sánchez, Inorg. Chem., 1989, 28, 2120.
298 P.Hoffmann and R.Mattes, Inorg. Chem., 1989, 28, 2092.
299 N.Al-Obaidi, A.J.Edwards, C.J.Jones, J.A.McCleverty, B.D.Neaves, F.E.Mabbs and D.Collison, J. Chem. Soc., Dalton Trans., 1989, 127.
300 J.Y.Chen, K.R.Grundy and K.N.Robertson, Can. J. Chem., 1989, 67, 1187.
301 R.P.Sidebotham, P.D.Beer, T.A.Hamor, C.J.Jones and J.A.McCleverty, J. Organometal. Chem., 1989, 371, C31.
302 J.Müller, I.Sonn and T.Akhnoukh, J. Organometal. Chem., 1989, 367, 133.
303 C.Glidewell, R.J.Lambert, M.B.Hursthouse amd M.Motevalli, J. Chem. Soc., Dalton Trans., 1989, 1989, 2061.
304 M.Zhou, S.Cao, S.Ye and D.Chen, Gaodeng Xuexao Huaxue Xuebao, 1989, 10, 209 (Chem. Abs., 1989, 111, 160942).
305 N.M.Sinitsyn, V.N.Kokunova and A.A.Svetlov, Russ. J. Inorg. Chem., 1988, 33, 1336.
306 E.Yu.Bobkova, N.L.Rogalevich, N.B.Borkovskii, G.G.Novitskii, V.N.Kokunova and N.M.Sinitsyn, Russ. J. Inorg. Chem., 1989, 34, 524.
307 N.M.Sinitsyn, V.N.Kokunova, G.G.Novitskii and E.Yu.Bobkova, Russ. J. Inorg. Chem., 1988, 33, 1172
308 E.Yu.Bobkova, N.L.Rogalevich, G.G.Novitskii, A.A.Svetlov, N.M.Sinitsyn and K.V.Kradenov, Russ. J. Inorg. Chem., 1989, 34, 539.
309 W.Erley, Surf. Sci., 1988, 205, L771.
310 D.Ouafi, F.Mauge and J.C.Lavalley, Bull. Soc. Chim. France, 1989, 363.
311 T.Chivers, K.S.Dhathathreyan, C.Lensink, A.Meetsma, J.C.van de Grampel and J.L.de Boer, Inorg. Chem., 1989, 28, 4150.
312 G.L.Wood, E.N.Duesler, R.T.Paine and H.Nöth, Phosphorus Sulfur Silicon Relat. Elem., 1989, 41, 267.
313 J.Ellermann, M.Moll and N.Will. Z. Anorg. Allg. Chem., 1989, 574, 41.
314 R.T.Carlin and R.E.McCarley, Inorg. Chem., 1989, 28, 280.
315 W.Krampe, T.Kruck and K.F.Tebbe, Chem. Ber, 1989, 122, 1477.
316 M.Herberhold and K.Schamel, Z. Naturforsch., B, 1988, 43b, 1274.
317 D.R.Derringer, P.E.Fanwick, J.Moran and R.A.Walton, Inorg. Chem., 1989, 28, 1384.
318 R.-M.Catalá, D.Cruz-Garritz, P.Sosa, P.Terreros, H.Torrens, A.Hills, D.L.Hughes and R.L.Richards, J. Organometal. Chem., 1989, 359, 219.
319 X.Tang, H.Xu, Y.Huang and W.Zhai, Fenxi Cuihua, 1988, 2, 95 (chem. Abs., 1989, 110, 45653).
320 V.K.Agrawal and M.Trenary, J. Vac. Sci. Technol., A, 1989, 7, 2235.
321 J.Ellermann and N.Will. Z. Naturforsch., B, 1989, 44b, 127.
322 H.G.M.Edwards and D.W.Farwell, J. Mol. Struct., 1989, 197, 203.
323 M.Baudler and C.Block, Z. Anorg. Allg. Chem., 1988, 567, 7.
324 T.Shido, K.Asakura and Y.Iwasawa, J. Chem. Soc., Farad. Trans., 1, 1989, 85, 441.
325 C.Li, K.Domen, K.Maruya and T.Onishi, J. Chem. Soc., Chem. Comm., 1988, 1541.
326 C.Li, K.Domen, K.Maruya and T.Onishi, J. Am. Chem. Soc., 1989, 111, 7683.
327 M.K.Chaudhuri, B.Das and R.A.Lal, J. Mol. Catal., 1989, 51, 59.
328 M.T.H.Tarafder, M.B.H.Howlader, B.Nath, R.Khan and A.A.M.A.Islam, Polyhedron, 1989, 8, 977.
329 M.T.H.Tarafder and A.A.M.A.Islam, Polyhedron, 1989, 8, 109.

330 I.I.Grinval'd, B.V.Lokshin, N.K.Rudnevskii and V.M.Fomin, Izv. Akad. Nauk SSSR, Ser. Khim., 1988, 58.
331 E.M.Nour, A.B.El-Sayed and S.A.Sadeek, Bull. Chem. Soc. Jpn., 1989, 62, 317.
332 C.Djordjevic, M.Lee and E.Sinn, Inorg. Chem., 1989, 28, 719.
333 A.Antiñolo, J.M.de Ilarduya, A.Otero, P.Royo, A.M.Lanfredi and A.Tiripicchio, J. Chem. Soc., Dalton Trans., 1988, 2685.
334 N.J.Campbell, A.C.Dengel, C.J.Edwards and W.P.Griffith, J. Chem. Soc., Dalton Trans., 1989, 1989, 1203.
335 J.W.Faller and Y.Ma, J. Organometal. Chem., 1989, 368, 45.
336 J.T.Lin, S.K.Yeh, G.H.Lee and Y.Wang, J. Organometal. Chem., 1989, 361, 89.
337 R.Bhattacharyya, S.Biswas, J.Armstrong and E.M.Holt, Inorg. Chem., 1989, 28, 4297.
338 H.D.Burkett and S.D.Worley, J. Am. Chem. Soc., 1989, 111, 5992.
339 M.Perrée-Fauvet, A.Gaudemer, J.J.Girerd, C.Boucly-Goester and P.Boucly, Inorg. Chem., 1989, 28, 3533.
340 L.L.Zaitseva, A.A.Kruplov, A.V.Romanov, V.E.Samsonov and E.G.Teterin, Russ. J. Inorg. Chem., 1988, 33, 1452.
341 W.A.Herrmann, R.A.Fischer, W.Amslinger and E.Herdtweck, J. Organometal. Chem., 1989, 362, 333.
342 T.K.G.Erikson and J.M.Mayer, Angew. Chem., Int. Ed. Engl., 1988, 27, 1527.
343 U.L.Kettler, P.H.Barrett and R.G.Pearson, N.A.T.O., S.I.Ser. B, 1987, 158, 813.
344 W.Micklitz, S.G.Bott, J.G.Bentsen and S.J.Lippard, J. Am. Chem. Soc., 1989, 111, 372.
345 D.T.Sawyer, M.S.McDowell, L.Spencer and P.K.S.Tsang, Inorg. Chem., 1989, 28, 1166.
346 L.M.Proniewicz, T.Isobe and K.Nakamoto, Inorg. Chim. Acta, 1989, 155, 91.
347 K.Nakamoto, I.R.Paeng, T.Kuroi, T.Isobe and H.Oshio, J. Mol. Struct., 1988, 189, 293.
348 R.Holze, Electrochim. Acta, 1988, 33, 1619.
349 S.Kaminaka, T.Ogura, K.Kitagishi, T.Yonetani and T.Kitagawa, J. Am. Chem. Soc., 1989, 111, 3787.
350 M.A.Esteruelas, F.Sola, L.A.Oro, U.Mayer and H.Werner, Angew. Chem., Int. Ed. Engl., 1988, 27, 1563.
351 L.M.Proniewicz, J.Odo, J.Góral, C.K.Chang and K.Nakamoto, J. Am. Chem. Soc., 1989, 111, 2105.
352 L.M.Proniewicz, A.Bruha, K.Nakamoto, E.Kyuno and J.R.Kincaid, J. Am. Chem. Soc., 1989, 111, 7050.
353 H.J.Lawson and J.D.Atwood, J. Am. Chem. Soc., 1989, 111, 6223.
354 E.Carmona, J.M.Marín, P.Palma, M.Paneque and M.L.Poveda, Inorg. Chem., 1989, 28, 1895.
355 G.Thakur and A.L.Verma, Spectrochim. Acta, Part A, 1989, 45A, 615.
356 C.R.Bhattacharjee, M.K.Chaudhuri and R.N.D.Purkayastha, Inorg. Chim. Acta, 1989, 160, 147.
357 M.Yalpani, R.Köster and R.Boese, Chem. Ber., 1989, 122, 19.
358 J.Flannagan, W.P.Griffith, R.D.Powell and A.R.West, J. Chem. Soc., Dalton Trans., 1989, 1651.
359 M.G.Ivanov, S.D.Vashchenko, T.A.Denisova and I.I.Kalinichenko, Russ. J. Inorg. Chem., 1989, 34, 184.

360 M.Jaber, F.Bertin and G.Thomas-David, Bull. Soc. Chim. Fr., 1988, 470.
361 P.N.Kapoor, A.K.Bhagi, H.K.Sharma and R.N.Kapoor, J. Organometal. Chem., 1989, 369, 281.
362 R.Kurpiel-Gorgol. Pol. J. Chem., 1988, 62, 41.
363 L.G.Tejuca, A.T.Bell and V.C.Corbertin, Appl. Surf. Sci., 1989, 37, 353.
364 D.E.Morris and D.E.Hobart, J. Raman Spectrosc., 1988, 19, 231.
365 M.A.Vaidya and S.C.Jain, Ind. J. Chem., Sect. A, 1989, 28A, 135.
366 C.Li, Y.Sakata, T.Arai, K.Domen, K.Maruya and T.Onishi, J. Chem. Soc., Farad. Trans., 1, 1989, 85, 929.
367 W.Hertl, Langmuir, 1989, 5, 96.
368 U.Thewalt and T.Güthner, J. Organometal. Chem., 1989, 361, 309.
369 D.Rehder, W.Priebsch and M.von Oeynhausen, Angew. Chem., Int. Ed. Engl., 1989, 28, 1220.
370 P.O.Ikekwere and A.A.Adenyi, Synth. React. Inorg. Met.-Org. Chem., 1989, 19, 87.
371 J.E.Tackett, Appl. Spectrosc., 1989, 43, 490.
372 F.Lloret, M.Julve, M.Hollar, I.Castro, J.Latorre, J.Faus, X.Solans and I.Morgenstern-Badarau, J. Chem. Soc., Dalton Trans., 1989, 729.
373 M.Sarkar and S.Sen, Z. Anorg. Allg. Chem., 1989, 570, 159.
374 M.J.Chetcuti, J.C.Gordon, K.A.Green, P.E.Fanwick and D.Morgenstern, Organometallics, 1989, 8, 1790.
375 R.Kapoor, B.K.Bahi, P.Kaur and P.Kapoor, Polyhedron, 1988, 7, 2175.
376 I.B.Baranovich, A.N.Zhilyaev and L.M.Dikareva, Russ. J. Inorg. Chem., 1988, 33, 1802.
377 J.Ribas, M.Monfort, C.Diaz and X.Solans, An. Quim., Ser. B, 1988, 84, 186.
378 S.K.Shakhatreh, E.G.Bakalbassis, C.A.Tsipis and J.Mrozinski, Z. Anorg. Allg. Chem., 1989, 574, 199.
379 W.Clegg, P.A.Hunt, B.P.Straughan and M.A.Mendiola, J. Chem. Soc., Dalton Trans., 1989, 1127.
380 D.G.Berniyazova, E.M.Shvarts and V.G.Kalacheva, Russ. J. Inorg. Chem., 1989, 34, 335.
381 M.K.Wang, K.K.Lim and K.H.Houng, Chung-kuo Nung Yeh Hua Hueh Hui Chih, 1988, 26, 245 (Chem. Abs., 1989, 110, 38238).
382 P.G.Harrison and A.Guest, J. Chem. Soc., Farad. Trans, 1, 1989, 85, 1897.
383 A.Glowacki, F.Huber and H.Preut, Rec. Trav. Chim. Pays-Bas, 1988, 107, 278.
384 M.Domagala, F.Huber and H.Preut, Z. Anorg. Allg. Chem., 1989, 574, 130.
385 I.K.Kireeva, V.A.Trofimov, N.B.Generalova and A.Yu. Tsivadze, Koord. Khim., 1989, 15, 463.
386 V.Zhelyazkov, G.Georgiev, Zh.Nickolov and M.Miteva, Spectrochim. Acta, Part A, 1989, 45A, 625.
387 D.W.James, R.E.Mayes, W.H.Leong, I.M.Jamie and G.Zhen, Farad. Discuss. Chem. Soc., 1988, 85, 269.
388 M.G.B.Drew, K.C.Lee and K.F.Mok, Inorg. Chim. Acta, 1989, 155, 39.
389 P.Sobota, T.Płuziński, J.Utko and T.Lis, Inorg. Chem., 1989, 28, 2217.
390 O.A.Raevskii, V.E.Zubareva, I.I.Bulgak and D.G.Batyr, Koord. Khim., 1988, 14, 1193.
391 F.H.Elsner, T.D.Tilley, A.L.Rheingold and S.J.Geib, J. Organometal. Chem., 1988, 358, 169.

392 U.Patt-Siebel, H.Thullen, U.Müller and K.Dehnicke, Z. Naturforsch., B, 1988, 43b, 795.

393 N.M.Titov, L.N.Baikova, I.I.Kalinichenko and L.V.Krupnova, Russ. J. Inorg. Chem., 1989, 34, 200.

394 A.M.El-Hendawy, W.P.Griffith, C.A.O'Mahoney and D.J.Williams, Polyhedron, 1989, 8, 519.

395 K.Jacob and K.H.Thiele, Z. Anorg. Allg. Chem., 1988, 564, 81.

396 V.V.Kravchenko, G.G.Sadikov, N.S.Rukk, E.V.Savinkina, M.G.Zaitseva, L.A.Butman, L.Yu.Alikberova and B.D.Stepin, Russ. J. Inorg. Chem., 1989, 34, 843.

397 Yu.Ya.Kharitonov, E.I.Tabidze and G.V.Tsintsadze, Russ. J. Inorg. Chem., 1988, 33, 1312.

398 W.R.Carper, D.B.Coffin and J.R.Addis, Spectrochim. Acta, Part A, 1989, 45A, 391.

399 A.N.Yunuskhodzaev, Yu.Ya.Kharitonov and A.F.Dusmatov, Russ. J. Inorg. Chem., 1989, 34, 696.

400 N.N.Khod'ko, T.A.Kolevich, T.A.Dik and D.S.Umreiko, Russ. J. Inorg. Chem., 1988, 33, 1025.

401 A.M.Orlova, K.A.Solntsev, N.T.Kuznetsov, Z.M.Alikhanova and L.V.Goeva, Russ. J. Inorg. Chem., 1988, 33, 1023.

402 P.C.Vyas, C.K.Ojha, S.Mittal and G.K.Joshi, J. Ind. Chem. Soc., 1988, 65, 611.

403 I.Pavlik, K.H.Thiele, K.Jacob and M.Glanz, Z. Anorg. Allg. Chem., 1989, 569, 106

404 Z.K.Odinets, N.A.Sal'nikova, A.K.Molodkin and T.M.Ivanova, Russ. J. Inorg. Chem., 1989, 34, 467.

405 P.P.K.Claire, G.R.Willey and M.G.B.Drew, J. Chem. Soc., Dalton Trans., 1989, 57.

406 P.Paninić and H.Melder, Polyhedron, 1989, 8, 627.

407 M.Kersting, C.Friebel, K.Dehnicke, M.Krestel and R.Allmann, Z. Anorg. Allg. Chem., 1988, 563, 70.

408 K.L.Leighton, K.R.Grundy and K.N.Robertson, J. Organometal. Chem., 1989, 371, 321.

409 N.D.Mitrofanova, B.Kh.Kushikbaeva, L.I.Martynenko and S.A.Il'ichev, Russ. J. Inorg. Chem., 1989, 34, 72.

410 S.Wimmer, P.Castan, F.L.Wimmer and N.P.Johnson, J. Chem. Soc., Dalton Trans., 1989, 403.

411 A.Jouini and M.Dabbabi, J. Soc. Chim. Tunis., 1987, 2, 29.

412 K.V.Ramana and S.Singh, J. Raman Spectrosc., 1989, 20, 169.

413 V.N.Krasil'nikov and M.R.Glazyrin, Russ. J. Inorg. Chem., 1989, 34, 191.

414 M.N.Bhattacharjee, M.K.Chaudhuri and N.S.Islam, Inorg. Chem., 1989, 28, 2420

415 P.J.Kellett, M.J.Pawlik, L.F.Taylor, R.G.Thompson, M.A.Levstik, O.P.Anderson and S.H.Strauss, Inorg. Chem., 1989, 28, 440.

416 T.A.Fomina, A.N.Zhilyaev, I.V.Kuz'menko, N.N.Chalisova, A.N.Sotsolev, M.A.Porai-Koshits and I.B.Baranovskii, Russ. J. Inorg. Chem., 1989, 34, 879.

417 M.N.Bhattacharjee, M.K.Chaudhuri and M.Devi, Polyhedron, 1989, 8, 457.

418 L.V.Kocets, N.G.Ben' and D.S.Umreiko, Russ. J. Inorg. Chem., 1988, 33, 1020.

419 K.D.S.Mudher, K.Krishnan, D.M.Chakraburtty and N.C.Jayadevan, J. Less-Common Met., 1988, 143, 173.

420 P.A.Mosier-Boss, R.D.Bos, C.J.Gabriel, S.Szpak, J.J.Smith and R.J.Nowak, J. Chem. Soc., Farad. Trans., 1, 1989, 85, 11.
421 R.Kapoor, P.Wadwawan, P.Kapoor and J.F.Sawyer, Can. J. Chem., 1988, 66, 2367.
422 L.A.Aslanov, T.V.Galuzina, O.N.Gilyarov, A.L.Kamshanyi, B.N.Kuilkovskii, T.L.Novoderezhkina, V.A.Trofimov and A.Yu.Tsivadze, Russ. J. Inorg. Chem., 1988, 33, 1197.
423 V.P.Babaeva and V.Ya.Rosolovskii, Russ. J. Inorg. Chem.,1989, 34, 193.
424 Z.K.Nikitina, N.V.Krivtsov, Yu.V.Chuprakov and V.Ya.Rosolovskii, Russ. J. Inorg. Chem., 1988, 33, 1143.
425 C.M.Mikulski, M.K.Kurlan, M.Bayne, M.Gaul and N.M.Karayannis, Wuji Huaxue, 1988, 4, 83 (Chem. Abs., 1989, 111, 32715).
426 J.L.Pascal and M.Fourati, Compt. Rend. Acad. Sci., Ser. 2, 1989, 308, 281.
427 R.Usón, A.Laguna, M.Laguna and M.Concepción-Gimeno, J. Chem. Soc., Dalton Trans., 1989, 1883.
428 V.Krug, G.Koellner and U.Müller, Z. Naturforsch., B, 1988, 43b, 1501.
429 W.A.Schenk and G.H.J.Hilpert, Chem. Ber., 1989, 122, 1623.
430 T.Nojo, G.Matsubayashi and T.Tanaka, J. Organometal. Chem., 1989, 159, 49.
431 R.Lozano, J.Roman, A.Moragues and F.de Jesus, An. R. Acad. Farm., 1988, 54, 688.
432 A.I.Hadjikyriacou and D.Coucouvanis, Inorg. Chem., 1989, 28, 2169.
433 M.L.Ziegler, K.Blechschmitt, H.Bock, E.Guggolz and R.P.Korswagen, Z. Naturforsch., B, 1988, 43b, 590.
434 W.P.Ehrenreich, M.Herberhold and A.F.Hill. J. Organometal. Chem., 1989, 371, 303.
435 U.Schubert, J.Kron and H.Hörnig, J. Organometal. Chem., 1988, 355, 243.
436 P.C.Servaas, D.J.Stufkens, A.Oskam, P.Vernooijs, E.J.Barends, D.J.A.de Ridder and C.H.Stam, Inorg. Chem., 1989, 28, 4104.
437 M.H.Desbois and D.Astruc, Angew. Chem., Int. Ed. Engl., 1989, 28, 460.
438 A.K.Singh and J.K.Basumatary, J. Organometal. Chem., 1989, 364, 73.
439 K.K.Pandey, D.T.Nehete, S.K.Tewari, S.Rewari and R.Bhardwaj, Polyhedron, 1988, 7, 709.
440 M.G.Gribanovskata, O.V.Rudnitskaya, V.N.Pichkov, I.V.Lipnitskii and G.G.Novitskii, Koord. Khim., 1989, 15, 991.
441 A.A.M.Aly and A.I.El-Said, Z. Naturforsch., B, 1989, 44b, 323.
442 V.I.Pekn'no, A.V.Grafov, N.G.Aleksandrova and S.V.Volkov, Ukr. Khim. Zh., (Russ. Ed.), 1989, 55, 3.
443 C.Keller, D.Walther, J.Reinhold and E.Hoyer, Z. Chem., 1988, 28, 410.
444 J.J.Criado, A.Carrasco, B.Macias, J.M.Salas, M.Medarde and M.Castillo, J. Organometal. Chem., 1989, 160, 37.
445 R.Usón, J.Forniés, M.Tomás and I.Ara, J. Chem. Soc., Dalton Trans., 1989, 1011.
446 Kh.Kh.Khakimov, T.Kh.Taksanova and V.K.Khakimova, Koord. Khim., 1988, 14, 1507.
447 R.M.Olk, W.Dietzsch, K.Köhler, R.Kirmse, J.Reinhold, E.Hoyer, L.Golic and B.Olk, Z. Anorg. Allg. Chem., 1988, 567, 131.
448 P.Geboes and H.O.Desseyn, Spectrochim. Acta, Part A, 1988, 44A, 963.
449 J.Adel, F.Weller and K.Dehnicke, Z. Naturforsch., B, 1988, 43b, 1094.
450 R.M.H.Banda, I.G.Dance, T.D.Bailey, D.C.Craig and M.L.Scudder, Inorg. Chem., 1989, 28, 1862.

451 K.K.Yusuff and E.J.Mathew, Synth. React. Inorg. Met.-Org. Chem., 1989, 19, 15.
452 R.K.Chadha, J.E.Drake and A.B.Sarkar, J. Organometal. Chem., 1988, 143, 31.
453 C.Silvestru and I.Haiduc, J. Organometal. Chem., 1989, 365, 83.
454 D.L.A.de Faria, N.S.Gonçalves and P.S.Santos, Spectrochim. Acta, Part A, 1989, 45A, 643.
455 H.Plenio and H.W.Roesky, Z. Naturforsch., B, 1989, 44b, 94.
456 A.M.Mityukov, V.V.Zelentsov, N.A.Subbotina, P.E.Kazin, A.I.Zhirov, M.G.Felin and B.V.Zhadanov,, Russ. J. Inorg. Chem., 1989, 34, 47.
457 W.Kelm and W.Preetz, Z. Anorg. Allg. Chem., 1988, 565, 7.
458 W.Kelm and W.Preetz, Z. Anorg. Allg. Chem., 1989, 568, 106.
459 R.H.Herber, G.Nan, J.A.Potenza, H.J.Schugar and A.Bino, Inorg. Chem., 1989, 28, 938.
460 C.Sartori and W.Preetz, Z. Anorg. Allg. Chem., 1988, 565, 22.
461 C.Sartori and W.Preetz, Z. Anorg. Allg. Chem., 1989, 572, 151.
462 K.Bütje and W.Preetz, Z. Naturforsch., B, 1988, 43b, 574.
463 D.K.Paul, M.L.McKee, S.D.Worley, N.W.Hoffman, D.H.Ash and J.Gautney, J. Phys. Chem., 1989, 93, 4598.
464 J.L.Mesa, T.Rojo, M.I.Arriortua, G.Villeneuve, J.V.Folgado, A.Beltrán-Porter and D.Beltrán-Porter, J. Chem. Soc., Dalton Trans., 1989, 53.
465 Y.Huang and G.Wu, Spectrochim. Acta, Part A, 1989, 45A, 123.
466 S.Ishiguro, T.Takamuku and H.Ohtaki, Bull. Chem. Soc. Jpn., 1988, 61, 3901.
467 E.C.Alyea, K.J.Fisher and S.Johnson, Can. J. Chem., 1989, 67, 1319.
468 B.Korpar-Čolig, Z.Popović, M.Bruvo and I.Vicković, Inorg. Chim. Acta, 1988, 150, 113.
469 A.Heinrich and W.Preetz, Z. Naturforsch., B, 1988, 43b, 1327.
470 W.Heininger, K.Polborn and G.Nagorsen, Z. Naturforsch., B, 1988, 43b, 857.
471 F.Birkeneder, R.W.Berg, H.A.Hjuler and N.J.Bjerrum, Z. Anorg. Allg. Chem., 1989, 573, 170.
472 M.S.Sastry, R.Ghose and A.K.Ghose, Transition Met. Chem., 1988, 13, 447.
473 Z.Rzączyńska and W.Brzyska, Monatsh., 1989, 120, 231.
474 Y.Lu, R.Pan, D.Ren, H.Jiang, C.Yin and J.Ma, Gaodeng Xuexiao Huaxue Xuebao, 1988, 9, 1068 (Chem. Abs., 1989, 110, 146566).
475 E.N.Lebedeva, E.M.Proskurina, V.K.Guseva, O.V.Galaktionova, S.S.Korovin and M.G.Zaitseva, Russ. J. Inorg. Chem., 1988, 33, 1279.
476 H.Plenio, H.W.Roesky, F.T.Edelmann and M.Noltemeyer, J. Chem. Soc., Dalton Trans., 1989, 1815.
477 R.D.Lampeka, V.V.Skopenko, T.Yu.Sliva and H.Hennig, Ukr. Khim. Zh. (Russ. Ed.), 1988, 54, 675.
478 L.A.de Learie, R.C.Haltiwanger and C.G.Pierpont, Inorg. Chem., 1989, 28, 644.
479 F.Bottomley, P.S.White, M.Mukaida, K.Shimura and H.Kakihaha, J. Chem. Soc., Dalton Trans., 1988, 2965.
480 A.M.El-Hendawy, W.P.Griffith, F.I.Taha and M.N.Moussa, J. Chem. Soc., Dalton Trans., 1989, 901.
481 M.Herberhold and A.F.Hill, J. Organometal. Chem., 1989, 363, 371.
482 A.M.Heyns and D.de Waal, Spectrochim. Acta, Part A, 1989, 45A, 905.
483 K.C.Satpathy, A.K.Panda, R.Mishra and I.Panda, Acta Chim. Hung., 1988,

125, 527.
484 A.E.Bukanova, T.P.Sidorova, A.V.Chuvaev, G.A.Razgonyaeva and L.K.Shubochkin, Russ. J. Inorg. Chem., 1989, 34, 254.
485 M.Ghedini, M.Longeri, F.Neve, A.M.M.Lanfredi and A.Tiripicchio, J. Chem. Soc., Dalton Trans., 1989, 1217.
486 F.J.Quaeyhaegens, S.P.Perlepes and H.O.Desseyn, Spectrochim. Acta, Part A, 1989, 45A, 809.
487 B.Walther, H.Hartung, M.Maschmeier, U.Baumeister and B.Messbauer, Z. Anorg. Allg. Chem., 1988, 566, 121.
488 S.Al-Ahmad, H.Hudali and M.H.Zaghal, Inorg. Chim. Acta, 1988, 147, 1.
489 D.P.Singh and K.Singh, J. Inst. Chem. (India), 1988, 60, 76.
490 J.A.Winter, F.T.Lin and R.E.Shepherd, Inorg. Chim. Acta, 1989, 155, 155.
491 S.Jaworski, H.Schöllhorn, P.Eisenmann, U.Thewalt and B.Lippert, Inorg. Chim. Acta, 1988, 153, 31.
492 H.Schöllhorn, U.Thewalt and B.Lippert, J. Am. Chem. Soc., 1989, 111, 7213.
493 V.Benham, I.S.Butler, K.Okamoto and T.Theophanides, Inorg. Chim. Acta, 1988, 153, 145.
494 A.M.El-Roudi and E.M.Soliman, Synth. React. Inorg. Met.-Org. Chem., 1988, 18, 797.
495 E.A.Bezzubets, E.K.D'yachenko and V.E.Ivanov, Zh. Obshch. Khim., 1989, 59, 435.
496 F.Zalaru, A.Meghea, D.Ionescu and F.Cornea, Rev. Roum. Chim., 1988, 33, 969.
497 K.K.Narang, R.A.Lal, R.K.Thapa and V.K.Gautam, Transition Met. Chem., 1988, 13,212.
498 A.B.Akbarov, Russ. J. Inorg. Chem.,1988, 33, 1782.
499 A.Bury, A.E.Underhill, M.B.Fleet, P.J.Keymer, A.Stevens and P.S.Gomm, Inorg. Chim. Acta, 1989, 158, 181.
500 F.Lloret, M.Julve, J.Faus, Y.Journaux, M.Philoche-Levisalles and Y.Jeannin, Inorg. Chem., 1989, 28, 3702.
501 J.G.H.du Preez and B.J.A.M.van Brecht, J. Chem. Soc., Dalton Trans., 1989, 253.
502 P.Braunstein, T.M.G.Carneiro, D.Matt, F.Balegroune and D.Grandjean, J. Organometal. Chem., 1989, 367, 117.
503 M.I.L.Palomino, B.E.Zaitsev, R.K.Gridasova and T.M.Ivanova, Russ. J. Inorg. Chem., 1988, 33, 1340
504 A.N.Yunuskhodzhaev, Yu.Ya.Kharitonov, A.F.Dusmatov and S.D.Nasirdinov, Russ. J. Inorg. Chem., 1989, 34, 486.
505 R.A.Lal, H.Askari and S.S.Bhattacharjee, Synth. React. Inorg. Met.-Org. Chem., 1988, 18, 717.
506 I.S.Ahuja, G.L.Yadava and S.Tripathi, Ind. J. Chem., Sect. A, 1989, 28A, 167.
507 M.R.Bermejo, A.Fernandez, M.Gayoso, A.Castiñeiras, W.Hiller and J.Strähle, Polyhedron, 1988, 7, 2561.
508 R.A.Slavinskaya, T.A.Kovaleva, V.Zh.Ushanov and S.M.Kim, Izv. Akad. Nauk KazSSR, Ser. Khim., 1989, 25 (Chem. Abs., 1989, 111, 166136).
509 R.A.Slavinskaya, T.A.Kovaleva, M.K.Nasirdinova, A.A.Erzhanov and L.V.Li, Koord. Khim., 1989, 15, 333.
510 B.K.Dwivedi, H.N.Mehrotra and K.Bhatnagar, Proc. Natl. Acad. Sci. India, Sect. A, 1988, 58, 341.
511 L.P.Battaglia, A.B.Corradi, G.Pelosi, P.Tarasconi and C.Pelizzi,

J. Chem. Soc., Dalton Trans., 1989, 671.
512 A.Macias, A.Rodriguez, R.Bastida, H.de Blas, A.Sousa and D.Rodriguez, An. Quim., Ser. B, 1988, 84, 279.
513 A.Kumar and L.K.Mishra, J. Ind. Chem. Soc., 1989, 66, 268.
514 E.Lindner and V.Käss, Chem. Ber., 1989, 122, 2269.
515 N.J.Bryson, D.Brenner, J.Lister-James, A.G.Jones, J.C.Dewan and A.Davison, Inorg. Chem., 1989, 28, 3825.
516 S.Pohl, U.Bierbach and W.Saak, Angew. Chem., Int. Ed. Engl., 1989, 28, 776.
517 M.Herberhold and A.F.Hill. J. Organometal. Chem., 1989, 368, 111.
518 L.N.Lomakina, V.M.Ivanov and V.K.Belyaeva, Koord. Khim., 1988, 14, 1098.
519 N.K.Singh and R.Tripathi, Transition Met. Chem., 1988, 13, 346.
520 N.Saha, A.K.Adak and A.Misra, Synth. React. Inorg. Met.-Org. Chem., 1988, 18, 989.
521 N.V.Mel'nikova, Russ. J. Inorg. Chem., 1988, 33, 1785.
522 M.T.Pinillos, M.P.Jarauta, D.Carmona, L.A.Oro, M.C.Apreda, C.Foces-Foces and F.H.Cano, J. Chem. Soc., Dalton Trans., 1989, 1987.
523 C.A.O'Mahoney, I.P.Parkin, D.J.Williams and J.D.Woollins, J. Chem. Soc., Dalton Trans., 1989, 1179.
524 I.Zahn, K.Polborn and W.Beck, Chem. Ber., 1989, 122, 253.
525 E.S.Raper, J.R.Creighton, J.D.Wilson, W.Clegg and A.Milne, Inorg. Chim. Acta, 1989, 155, 77.
526 K.S.Satpathy, A.K.Panda, R.Mishra and A.Mohapatra, Synth. React. Inorg. Met.-Org. Chem., 1989, 19, 23.
527 R.Usón, A.Laguna, M.Laguna, M.N.Fraile, P.G.Jones and C.F.Erdbrügger, J. Chem. Soc., Dalton Trans., 1989, 73.
528 S.K.Singh, Y.Singh, A.K.Rai and R.C.Mehrotra, Polyhedron, 1989, 8, 633.
529 T.Chivers, J.Fait and K.J.Schmidt, Inorg. Chem., 1989, 28, 3018.
530 R.Cohen, J.Kimel, F.Elster, E.Peled and S.Efrima, J. Power Sources, 1989, 26, 5.
531 R.K.Agarwal, L.Singh and P.Kumar, Natl. Acad. Sci. Lett. (India), 1988, 11, 249.
532 S.C.O'Neal and J.W.Kolis, Inorg. Chem., 1989, 28, 2780.
533 C.Natarajan and P.R.Athappan, Ind. J. Chem., Sect. A, 1988, 27A, 691.
534 M.A.El-Hinnawi, A.M.Al-Ajlouni, J.N.Abunasser, A.K.Powell and H.Vahrenkamp, J. Organometal. Chem., 1989, 359, 79.
535 U.C.Sarma, K.P.Sarma and R.K.Poddar, Polyhedron, 1988, 7, 1727.
536 P.A.Lay, R.H.Magnuson and H.Taube, Inorg. Chem., 1989, 28, 3001.
537 A.L.Balch, B.J.Davis and M.M.Olmstead, Inorg. Chem., 1989, 28, 3148.
538 T.Hoffmann, H.Hoffmann, M.Karabit, P.Zdunneck, E.Wenschuh, J.Reinhold and M.Schüler, Z. Anorg. Allg. Chem., 1988, 656, 91.
539 E.Wenschuh and R.Zimmering, Z. Anorg. Allg. Chem., 1989, 570, 93.
540 G.Annibale, L.Cattalini, V.Bertolasi, V.Ferretti, G.Gilli and M.L.Tobe, J. Chem. Soc., Dalton Trans., 1989, 1265.
541 F.R.Hartley, S.G.Murray and A.Wilkinson, Inorg. Chem., 1989, 28, 549.
542 J.Vicente, A.Arcas, M.V.Borrachero, E.Molins and C.Miravitlles, J. Organometal. Chem., 1989, 359, 127.
543 C.Bellitto, M.Bonamico, V.Fares, P.Imperatori and S.Patrizio, J. Chem. Soc., Dalton Trans., 1989, 719.
544 N.W.Alcock, A.W.G.Platt, H.H.Powell and P.G.Pringle, J. Organometal. Chem., 1989, 361, 409.

7
Mössbauer Spectroscopy

BY S. J. CLARK, J. D. DONALDSON, S. M. GRIMES, AND M. J. K. THOMAS

1. INTRODUCTION

The format of this year's report is the same as that used last year. This introductory section provides details of books and review articles published during the review period. It is followed by sections that deal with the theoretical aspects of Moessbauer spectroscopy and with advances in instrumentation and methodology. Sections 4, 5 and 6 contain detailed reviews on iron-57, tin-119, and other Moessbauer effects respectively, while Section 7 reviews papers published on conversion electron and back scattering gamma-resonance spectroscopy. In view of the current intense interest in low T_c superconducting oxides, Moessbauer data relating to these materials are examined in detail. The isotopes (energies in keV in parenthesis) that have been mentioned, some of them in review articles, during the review period include: Fe-57(14.412 and 136.32), Ni-61(67.4), Zn-67(93.26),Ge-73(13.3), Kr-83(9.40), Ru-99(89.36), Sn-119(23.875), Sb-121(37.15), Te-125(35.46), I-127 (57.6), I-129(27.72), Cs-133 (81.0), Pr-141(145.4), Nd-145(67.3), Sm-149(45.8), Eu-151(21.64), Gd-155(86.54), Dy-161(80.7), Er-166 (80.56), Er-167(79.32), Tm-169(8.4), Yb-170(84.26), Yb-174(66.7), Ta-181(6.24), W-182(100.1), W-183(46.5 and 99.1) Ir-193(73.1), Au-197(77.36), and Np-237(59.54).

The twelfth volume of the literature service "Moessbauer Effect Data and Reference Journal" was published in 1989.[1]

Books and Reviews. The 3rd Seeheim Workshop on Moessbauer spectroscopy was held at the Lufthansa Training Centre, Seeheim, West Germany in May 1988 and the proceedings of the workshop can be found in reference 2.

An historical perspective to gamma-resonance spectroscopy has been presented by Moessbauer.[3] In the review he discusses nuclear resonance fluorescence and recoilless nuclear resonance absorption.

A number of general reviews on Moessbauer spectroscopy have appeared in this review year,[4-9] and a general discussion of the applications of conversion electron Moessbauer spectroscopy (CEMS) has been presented by Nasu.[10]

The application of Moessbauer spectroscopy to the study of tellurium in GaAs and $Al_xGa_{1-x}As$ has been reviewed and the author discusses the information that can be obtained from such studies on the DX-centre, which is known to affect the electronic and opto-electronic behaviour of donor atoms like Te.[11] Other reviews have described the application of gamma-resonance spectroscopy to the study of interstitial mobility,[12] the determination of physical parameters in 4*f*-3*d* intermetallics,[13,14] the kinetics of ordering in metallic alloys,[15] and the study of metals.[16]

The application of Moessbauer spectroscopy to structural studies of superionic conducting borate and semiconducting vanadate glasses[17] and metallic glasses[18,19] have been reviewed.

A large number of authors have reviewed the use of the Moessbauer effect in the study of magnetic materials.[20-29] Areas covered by the reviewers include: the study of magnetic phase transitions and phase transitions in magnetic materials;[22] magnetism in superconducting systems;[23] low-dimensional magnetic solids;[24] microcrystalline materials;[25] $R_2Fe_{14}B$ and related compounds,[28] and a comparison of the measurement of spin fluctuations in magnetic materials by Moessbauer spectroscopy and muon SR spectroscopy.[29]

In the field of metallurgy, applications of Moessbauer spectroscopy discussed during this review period include: the role of light interstitial elements in changing the physical properties of alloys;[30] the quantitative determination of the content of various phases and phase transformations;[31] structural elucidation in amorphous metals,[32] and corrosion studies.[33,34]

Two groups have independently discussed the application of gamma-resonance spectroscopy to solid state physics.[35,36] Ion implantation,[37] the study of defects,[38] and the characterization of semiconductors[39] have also been reviewed.

Frauenfelder has emphasised the use of the Debye-Waller factor in relation to protein crystal structure and dynamics.[40] The application of Rayleigh scattering of Moessbauer radiation to the study of protein dynamics[41] and studies of ferritin and iron porphyrins[42-44] have been reviewed.

Reviews on the uses of gamma-resonance spectroscopy in inorganic chemistry have covered a very wide range of topics. Areas that have been discussed include: inorganic photochemical reactions at low temperature;[45] spin equilibria in iron(II) complexes;[46,47] transition metal oxides;[48] high temperature superconductors;[49] triangular metal cluster complexes;[50] mixed-valence compounds;[51] the properties and reactions of substituted pentacyanoferrate(II) complexes;[52] organotin complexes;[53] organotellurium compounds,[54] and the actinides.[55]

Other fields where the applications of Moessbauer spectroscopy have been examined in review papers are: catalysis;[56-58] soil science and mineralogy;[59,60] relativity;[61] charge density waves in molybdenum blue bronzes;[62] spin dynamics;[63] optical cooling of trapped ions;[64] the evolution of the solar system;[65] and the examination of electrode-solution interfaces by in situ methods.[66]

The latest developments in data analysis for Moessbauer spectroscopy[67] and the experimental status of Moessbauer linewidths have also been the subjects of reviews.[68]

2. THEORETICAL

Improvements to the calibration scheme for Moessbauer isomer shifts have lead to new best values for the relative changes of the nuclear charge radius of 1.8×10^{-4}, 8.0×10^{-5}, -3.4×10^{-4}, and 4.3×10^{-4} for Sn-119, Te-125, I-127, and I-129, respectively.[69] Uncertainties are estimated to be less than 10%.

A branch matrix algorithm for the fitting of experimental spectra with Lorentzian lineshapes has been suggested.[70] It was derived from the analysis of the modified interrelations among the peak parameters of the Lorentz curve in the iterative process for the fitting of the Moessbauer spectrum and the treatment of the matrix branching to the algebraic equations of the Newton-Gauss algorithm. The branch matrix algorithm gave faster convergence, shorter iteration time and lower computer memory requirements. A Newton-Raphson algorithm implemented on an accelerator based on a network of INMOS transputers, interfaced to a personal computer has been shown to give significant improvements in the time taken to fit a spectrum to a number of Lorentzian-shaped lines.[71] A model for numerical analysis of Moessbauer spectra has been established and the regression technique developed for using the iteration method to evaluate the optimum parameters in the nonlinear model.[72] This model was then applied to the analysis of iron catalysts. A method has been derived for the maximum entropy evaluation of the distribution of hyperfine parameters from overlapped Moessbauer spectra for crystalline or noncrystalline alloy solid solutions, which was subjected to weak constraint conditions.[73] The method proved to be more reliable for fitting overlapping Moessbauer spectra. Improved discrimination of noisy spectra has been achieved in correlation based spectral analysis.[74] Gaussian, pseudo-Lorentzian, and Pearson type VII distribution profiles and convolutions of a Gaussian and a Lorentzian have been applied to fit complex gamma-resonance spectra of metallic glasses.[75] Experimental and simulated Moessbauer lines were analyzed by a DISPA-plot method.

A new model for the description of dynamic properties of macromolecules, especially globular proteins has been proposed.[76] The model proposed the existence of time-relaxation spectra determining the time characteristics of biomacromolecular dynamics. Numerical modelling of experimental gamma-resonance spectra has been used for the investigation of time dependence of mean square displacement of Moessbauer nuclei for their limited and unlimited diffusion.[77]

A theoretical analysis has been proposed to explain the sagging of the Moessbauer effect probability which occurs in different materials at phase-transition points and the reduction of this probability resulting from the irradiation of solids.[78] These observations were explained in terms of the heterogeneity of the matter under these conditions. Several different properties of a Moessbauer fractal have been calculated: the recoil free fraction, the Raman scattering intensity, and the spin-lattice relaxation time.[79] The recoil free fraction of a Moessbauer nucleus in a fractal was calculated by considering the displacement of an atom in a fractal which exhibited a scale invariance for a noninteger fraction dimensionality and was in agreement with the measurements in deoxygenated myoglobin.

Moessbauer absorber thickness has been optimized by considering Nowik's model of bound diffusion.[80] The optimum thickness is dependent on the ratio of the harmonic force constant to damping force constant and the variation between the optimum thickness and this ratio was obtained for various values of absorption cross section. A theoretical expression for an appropriate Moessbauer scatterer thickness for which the scattering line shape assumes the unusual absorption character has been derived.[81] The dependence of this thickness on nuclear resonance absorption, conversion electron and total electronic attenuation was found analytically. This analysis will be useful for studying interference effects for poly-crystalline scatterers.

Three papers have dealt with the development of the gamma-ray laser. Hoy and Taylor suggest that this will depend on the use of the Moessbauer effect[82] but Cohen has reexamined the possibility of a gamma-ray laser allowing recoil.[83] Baldwin believes that the central problem to developing the laser, that pumping can inhibit or destroy the Moessbauer and Borrmann effects, can be mitigated by a superlattice geometry.[84]

Several approximating expressions have been proposed for the theoretical lineshape of Moessbauer absorption spectra.[85] The theory of the Moessbauer lineshape in fluctuating environments has been applied to the case of thin films of a diluted ferromagnetic alloy, and the distribution of the spin autocorrelation time, characterizing the lineshape in the vicinity of a phase transition point across the sample, calculated.[86] The stochastic theory of Moessbauer lineshapes has been formulated in such a way to allow the evaluation of the spectral shapes for absorbers of arbitrary thickness, texture, and an anisotropic Lamb-Moessbauer factor.[87] Intensification of the radiation channel of a nuclear reaction during the interaction of gamma-rays with nuclei in a fluctuating absorber has been observed.[88] The time behaviour of Moessbauer gamma-radiation when passed through a rf-vibrating resonant medium and then absorbed at definite energy shifts revealed two mechanisms of radiation transfer.[89] The first is independent of the presence of a vibrating filter between the source and the detector and the second is rf-induced, due exclusively to the interactions of gamma quanta with the rf phonons in the vibrating crystal.

A model has been developed for the internal motion in macromolecules which allows for low-frequency acoustic thermal oscillations of constituent particles, shifts and turns of molecular groups and fragments.[90] The collective excitations were treated in terms of dynamic variables that describe the deviations of the shape of the macromolecule from that in equilibrium. The calculated temperature dependences of the spectral parameters were compared with experimental data.

Gamma-resonance spectroscopy has been used to determine the interatomic force constants of metallic Fe, Cr, Co, Ni, Cu, Rh, Pd, and Pt at room temperature.[91] The results were in good agreement with the constants determined from phonon spectra of the metals. The values obtained by using Gupta and Lal's formula were consistently lower than those calculated by other methods, due to the neglect of the temperature dependence of the chemical shifts. A transient pulse of the Moessbauer radiation in a thick resonant absorber, in the case of an isomer shift, has been studied theoretically and experimentally.[92] On transmission the incident pulse decomposed into a fast pulse, an analogue of the Sommerfeld-Brillouin precursive pulse in optics, and a slow pulse connected with the motion of a nuclear polariton packet.

The important influences on the Moessbauer lineshape of non-interacting superparamagnetic particles have been discussed and the superparamagnetic relaxation rate estimated.[93] The theoretical variation law for the saturation-free intensity ratio of a paramagnetic doublet has been written in a way which allows one to check easily whether experimental data which are intended to be fitted by a theoretical spectrum or subspectrum follow this law.[94] The law was simplified even further by taking the absorber symmetries into account.

A number of authors have reported on studies of nuclear reflection of Moessbauer radiation under a range of conditions.[95-99]

3. INSTRUMENTATION AND METHODOLOGY.

A microprocessor controlled data acquisition system designed for Moessbauer spectroscopy has been described, which operates in both pulse height analysis and multichannel scaling with high counting rates ($>$ 20 MHz) and resolution ($\leq$ 8192 channels per spectrum).[100] The system makes use of two counters and of a double memory block. The accumulated data can be transferred to a microcomputer for storage and analysis from one of the two blocks while the other continues the acquisition process. An ultrahigh-vacuum spherical electrostatic spectrometer for surface Moessbauer studies has been constructed and it has been used to show that the recoil-free fraction at room temperature of very thin oxide layers on the surface of an iron foil is low.[101] A spectrometer, developed for time-differential Moessbauer spectroscopy, has a high count rate capability, good time resolution, and flexible data handling.[102]

In order to improve the resolution of diffraction Moessbauer spectra, which entail directing resonant photons at a crystal at the Bragg angle and measuring the diffracted photons to obtain a spectrum, the crystal was stratified along its thickness into a surface layer and a bulk layer having distinct electronic or magnetic characteristics.[103] The crystal was then oriented so that interactions between the resonant photons and Moessbauer nuclei in the bulk region were minimized. When the crystal was magnetically ordered with an artificially formed domain structure comprising two parallel layers with distinct magnetic directions, the crystal was oriented so that the magnetic field direction of the sublattice in the bulk region was perpendicular to the plane of the scattered photons.

The epitaxial growth of paramagnetic garnet films of composition $\{Y_{2.9}La_{0.1}\}[Sc_{1.5}Fe_{0.5}](Fe_3)O_{12}$ have been reported.[104] These films, in principle, present noticeable advantages over ferrimagnetic ones when used as Moessbauer monochromators of synchrotron radiation.

The principles affecting the precision characteristics of motion of gamma-resonance spectrometers based on an electrodynamic control device have been examined and a number of concrete measures proposed to allow decrease of the integral and differential nonlinearity of the velocity scale.[105]

A new cryostat for Moessbauer spectroscopic measurements has been developed around a cryogenerator of the Gifford McMahon type operating with He gas in a closed circuit.[106] Both source and absorber can be cooled to about 30 K and the unwanted vibrations caused by the He compression/expansion cycle were limited to about 0.05-0.06 $mm.s^{-1}$. An in situ cell for Moessbauer resonance studies allows measurements up to 1100 K with a gas pressure ranging from (1-2) x 10^5 Pa.[107]

It has been shown that different designs of counter structure and filling gas are necessary to operate gas-filled proportional counters in different temperature regions.[108] The operation of a helium-filled proportional counter has been investigated at temperatures below 4.2 K.[109] The gas gain as a function of the anode voltage was not appreciably changed by the temperature of the filling gas, but continuous electric discharges were induced at lower anode voltages as the gas temperature was decreased. This type of counter provides a new method to detect nuclear radiation at low temperatures. A typical application of this counter is for resonance-electron Moessbauer spectroscopy at low temperatures.

A frequency modulation spectrometer has been used to measure the isomer shifts of ferromagnetic absorbers without recourse to mechanical tuning.[110] The velocity scale has also been calibrated by recording an interference video pulse, which is the result of a corresponding conversion from the movement of the rod of the vibrator in a Moessbauer spectrometer and, simultaneously with the recording, the time interval is measured between the moments of the recording of the interference video pulse.[111]

A new method for obtaining Moessbauer spectra was described.[112] The method comprises recording two spectra of one sample and determining their intensities at equal Doppler rates. To increase the accuracy of measurements by compensation of parabolic nonlinearity, the first spectrum is recorded at the energy of the Moessbauer radiation and the second at the energy of one of the non-resonance radiations, while the Moessbauer spectrum is determined according to the ratio of intensities of the first and second spectra.

The dependence of the Moessbauer spectral parameters on the magnitude of the counting losses of electronic counting apparatus have been examined for transmission and conversion gamma-resonance spectra at an arbitrary background level.[113] Theoretically and experimentally the counting losses change all the parameters of the spectra. The effect of gamma-ray collimation in Moessbauer absorption spectra when thin absorbers are used has been modelled to study the shift and broadening of spectral lines.[114] A method of gamma-resonance spectroscopy with synchronous movement of source, filter, and sample has been patented.[115]

A regression model for Moessbauer resonance spectra that uses a linear optimization method for background and intensities, allows half-widths and positions to be fitted by means of the Davidson-Fletcher-Powell method.[116] Improvements were reached in global convergence properties and speed of execution in relation to the simple Newton method. A method to analyze velocity noise has been presented.[117] A simple method for the estimation of the average hyperfine field has been proposed.[118] The intensity ratios in the Moessbauer spectra of the sample perpendicular to the gamma-ray propagation direction and tilted by 45° in the longitudinal and transverse planes yield both angle coordinates of the hyperfine field direction without supposing the direction vector normalization.

The direct measurement of the background in the discriminator window is possible if the diffraction of Moessbauer radiation on crystals not containing resonant nuclei is used.[119] Narrow intense lines with an increase in transmission, positioned symmetrically relative to zero velocity appear in the Moessbauer spectra of iron foils when the amplitude of the excitation pulses is increased to 20 V at 78 K, as a result of coherent scattering of gamma-rays at predetermined frequencies of elastic vibrations.[120]

A number of publications in this review period have considered the methodology for obtaining information from biopolymers using Moessbauer spectroscopy. Methods discussed include: time-resolved analysis of gamma-ray diffuse scattering;[121] Fourier transform gamma-resonance spectroscopy;[122] Rayleigh scattering;[123] and angular-dependent Rayleigh scattering.[124]

A technique for generating very high pressures by means of a diamond anvil cell for use in Moessbauer spectroscopy has been described.[125]

The Debye-Waller factor of Na has been measured as a function of temperature from 80-295 K using Moessbauer gamma-ray scattering.[126] The high energy resolution provided by this technique allowed separation of the elastic scattering from the inelastic thermal diffuse scattering.

The excitation of coherent transients in Moessbauer spectra was previously limited by the high powers required to modulate the nuclear phases. An orders-of-magnitude increase has been reported in the efficiency through which such phenomena can be produced.[127] Magnetic modulation of the quantum phases of ^{57}Fe nuclei in paramagnetic media was produced by employing large amplitude spin waves transported from ferromagnetic sources.

A gravitational null red-shift experiment was carried out using the Moessbauer effect to compare the red-shift between two states of Fe-57 in different chemical environments.[128] An upper bound of 3.5×10^{-5} was found for the relative difference in red-shift, in agreement with the principle of equivalence.

The possible existence of neutral bosons which are feebly bound to nuclear matter, and have a mass smaller than a few tenths of a keV, has been considered. To search for them, the possibility has been discussed of shielding a relatively large Moessbauer source against its conventional activity, and looking for new types of penetrating monochromatic radiation that would resonantly excite the same Moessbauer transition in an outside detector.[129]

4. IRON-57

General Topics.- Fundamental Studies and Metallic Iron The ^{57}Fe isomer shifts and magnetic hyperfine fields in a number of iron compounds have been calculated using self-consistent spin-polarized electronic structure calculations.[130] Experimental trends were explained and a value for the mean radius difference involved in ^{57}Fe isomeric transition of $\Delta\langle r^2\rangle = -20 \times 10^{-3}$ fm^2 was obtained. A new systematic relation for the isomer shifts of ^{57}Fe in different elemental hosts has been derived from the isomer shifts of iron impurities, amorphous alloys and crystalline intermetallics.[131] Unlike previous systems, no dramatic fall in the isomer shift in metalloid hosts can be justified. The full potential linearized APW method has been used in band structure calculations for *bcc*-Fe, *hcp*-Fe, FeAl, FeTi, and FeCo.[132] While the main contribution to the isomer shift was due to the valence 4*s* electrons, a small contribution from core and semicore 3*s* electronic states was identified.

The Moessbauer effect has been used to study changes in discontinuous, 1.3-3.0 nm thick, films of iron as functions of temperature and time.[133] The spectra were fitted using the Van der Woude and Dekker lineshape for each nuclear hyperfine transition and allowing for both uniaxial and transverse spin relaxation. Two distinct magnetic phases, with different long-range order parameters and isomer shifts, were identified and identified with "core" and "surface" sites. *In situ* Moessbauer transmission experiments have been carried out on the passive film formed on iron in a dilute borate buffer.[134] The passive film showed a decreasing isomer shift and increasing quadrupole splitting as the anode potential was increased. Moessbauer, Rutherford backscattering and X-ray methods have been used in a study of a foil of *alpha*-Fe following implantation of Fe^{4+} and carbon.[135] The temperature dependences of the Moessbauer linewidth showed that the field of elastic stress on the iron matrix was caused by the amorphous precipitation of iron carbide.

The Moessbauer spectra of highly disperse, electroprecipitated, iron powders have been recorded at 4.2, 80 and 300 K.[136] The particle surfaces were found to consist of 3-7 nm thick layers of Fe_3O_4, which possibly contain Fe_2O_3, coated with high-spin ferric species. Other studies have reported Moessbauer data for iron produced by the decomposition of $Fe(CO)_5$,[137] and for 26 nm iron particles coated with cobalt ferrite.[138]

Iron-57 Impurity Studies, Polymers, and Ion-Exchange. A series of papers have reported the Moessbauer spectra of Fe_2O_3-doped $LiNbO_3$. The composition dependence of the isomer shift showed two steps: at 6 mol.% Fe_2O_3 (the turning point of the hexagonal axis parameter) and at 11 mol.% (where phase separation occurs).[139] Above 9% only Fe^{3+} was taken into the host lattice. The ferroelectric phase transition in $LiNbO_3$ was associated with a strong anisotropy, reflecting large vibrations of the ferric ions along the c-axis and the strong coupling of Fe^{2+} to the Li-sublattice.[140] The different effects of preparation by water quenching and by double-roller quenching were also reported.[141]

Iron impurity atoms in alpha- and gamma-Al_2O_3 will enter the cation sublattice as Fe(III).[142] The phase transition from alpha to delta is associated with a change from octahedral to tetrahedral symmetry of Fe^{3+}. The short-range-order structure of zinc in vitreous and crystalline zinc borates has been investigated with the ^{57}Fe Moessbauer effect.[143] The Fe^{2+} site showed a transition from tetrahedral in the crystalline form to octahedral in the vitreous phase. Phase interactions in the Gd_2O_3-HfO_2 system have been probed by doping with Fe_2O_3.[144] The interaction kinetics were studied at 900-1300°C, and a mixture of superparamagnetic iron compounds identified.

The Moessbauer spectrum of ^{57}Fe doped CdS has been studied at 4.2-273 K.[145] A hyperfine splitting of 88 kOe was observed at 4.2 K, a small quadrupole splitting at 77 K and at 273 K a single absorption line. Iron impurity centres in CaF_2 were studied.[146] The quadrupole splitting data showed that the introduction of Fe^{2+} into CaF_2 is accompanied by distortions of the local environment, and the data for Fe^{3+} are consistent with tetragonal Fe-F centres in which Fe^{3+} and F^- are separated by 0.23 nm. The Moessbauer hyperfine spectrum of isolated $Fe(acac)_3$ molecules adsorbed on silica gel has also been reported.[147]

A Moessbauer effect investigation carried out on a series of conducting polymers, including the tetrachloroferrates of polyacetylene, poly-*p*-phenylene, poly(*p*-thiophenylene), polypyrrole, poly-(*N*-methylpyrrole), and poly(3-butylthiophene), has been described.[148,149] While all the isomer shifts were very similar, the quadrupole splitting of the polypyrrole complex was significantly different from those of the other systems. The data suggested that, in the polypyrrole, the dopant tetrahedra are more distorted due to hydrogen bonding with the NH groups. When $FeSO_4$ or $FeCl_3$ is used as a stabilizer in polymethacrylates, the isomer shift of the polymer is the same as that of the pure iron compound.[150,151] The quadrupole splitting values of the polymers were, however, very different. A Moessbauer study of divinylbenzene-styrene-$FeCl_3$ complexes has identified two different types of benzene ring coordinated to $FeCl_3$.[152] The iron was found to be more closely bound to the divinylbenzene unit than to the styrene unit.

Data on the submicroporosity of perfluorinated ion-exchange sulphocationic membranes has been obtained from the Moessbauer spectra of ^{57}Fe nuclei precipitated into the membrane pores.[153] Different methods of membrane hydrolysis, the passage of current through them and the precipitation process were all found to change the characteristic size of the micropores.

Matrix Isolation, Liquid Crystals, and Emission Studies. Magnetic hyperfine interactions of Fe in solid ammonia have been studied.[154] The hyperfine parameters of the paramagnetic $FeNH_3$ suggest an iron electronic configuration of $[Ar]3d^7 4s$, while the field dependence of the spectra indicates a strong axial magnetic anisotropy. An *ab initio* study of the electronic properties of matrix-isolated and free $FeCl_2$ and $FeBr_2$ was described.[155] The relative isomer shifts of noble-gas-matrix-isolated $FeCl_2$ and Fe^0 was analysed on the basis of the calculated electron densities.

Moessbauer measurements have been made on a 0.8 wt.% solution of 1,1'-diacetylferrocene in the liquid-crystal 4-*n*-octyloxybenzylidene-4'-ethylaniline at 100-290 K in the crystalline solid phase and at 100-150 K in the cold smectic-*B* phase.[156] Linear Debye behaviour was seen for the solid phase over the entire temperature range, the cold smectic phase exhibited Debye behaviour, and the smectic glass gave a Moessbauer-Debye temperature about 12 K lower than for the crystalline phase.

Emission Moessbauer spectroscopy has been used to study ^{57}Co electron capture after-effects in the isomorphous compounds $KCoF_3$ and $KNiF_3$:^{57}Co.[157] In $KNiF_3$ there was preferential stabilization of the aliovalent Fe(III) state, as well as the presence of less-abundant species. The latter were attributed to the reluctance of the Co^{2+} dopants to substitute at Ni^{2+} sites. The K X-ray-gated emission Moessbauer spectra of CoF_2 and $CoF_2.2H_2O$ were measured at room temperature using the coincidence technique.[158] A difference was found in the relative intensity of Fe(II)/Fe(III) between the X-ray-gated and non-gated emission spectra. The results were interpreted in terms of local radiolytic effects of water crystallization and the chemical after-effects associated with electron-capture decay. The gamma-X-ray coincidence technique has also been used to study $FeSO_4.7H_2O$:^{57}Co.[159] The formation of anomalous charge states was explained in terms of radiolysis of the water and sulphate coordination shells around the decaying atom. Other Moessbauer emission studies have examined an incommensurate phase in oxygen-deficient $SrMn_{1-x}Co_xO_{3-y}$,[160] and antiferromagnetism in ^{57}Co-doped La_2CuO_{4-y}.[161]

__Compounds of Iron.__- __High-spin Iron(II) Compounds.__ The Moessbauer spectrum of the high temperature polymorph of Li_2FeCl_4 has been found to be consistent with a disordered distribution of cations among the octahedral sites of an inverse spinel structure.[162] In the room temperature modification there was a 1:1 ordering of Li^+ and Fe^{2+} ions on the octahedral sites. The crystal structure, magnetic properties and Moessbauer spectra of $[Me_3NH]FeCl_3.2H_2O$ (FeTAC), a one-dimensional Ising ferromagnet, have been studied.[163] The structure is isomorphous with that of the Co and Ni analogues, and the system becomes antiferromagnetic near 3.12 K. Analysis of the polycrystalline and single crystal Moessbauer spectrum showed that V_{zz} is positive, corresponding to an orbital single ^5B ground term based on the real $\underline{d}_{xy}$ orbital, and the easy axis of magnetization lies in the __ab__ plane. Non-linear excitations in the Ising-type antiferromagnetic[164] and ferromagnetic systems[165] have been investigated in temperature and field dependent Moessbauer relaxation studies. Data were obtained which allowed the determination of relaxation rates over up to 8 decades in frequency. Gamma-resonance spectra were also reported for $FeF_2.4H_2O$,[166] and the disordered complex $FeNbF_6$.[167]

The layered compound $(NH_4)FePO_4.H_2O$ has been the subject of a magnetic and Moessbauer study.[168] The temperature dependence of the hyperfine field (H_{eff} = 22.3 at 6 K) showed a first order transition at 26 K. Above 300 K the quadrupole splitting was strongly temperature dependent, with the hyperfine field at 12° to the *b*-axis. Ferrous gluconate has been investigated with the Moessbauer effect at 80-423 K.[169] Two Fe(II) sites were identified in the hydrated phase and a single site in a heated sample. No Goldanskii-Karyagin effect was detected. Semi-empirical MO-LCAO-INDO calculations have been used to study the iron(II) complexes FePcLL' (L,L' = H_2O, CO, $HCONH_2$, py; Pc = phthalocyanine).[170] The calculations were in general agreement with experimental data. The ΔE_Q parameter was found to be sensitive to the geometry of the first coordination shell and to those ligand effects which modify the iron *d*-orbital occupancies.

Moessbauer effect and electronic spectral data have been collected for a series of bis(dimethylglyoximato)iron(II) complexes which contain imidazole, pyridine, and pyrazine derivatives in the axial position.[171] The isomer shifts increase with the pK_a of the heterocyclic ligands, and show an inverse correlation with the energies of the metal-to-equatorial charge-transfer transitions. The diphosphine complexes $FeX_2(1,2-C_6H_4(PPh)_2)$, (where X = Cl, Br, or I) have been prepared.[172] While the Cl-complex has a single high-spin *trans*-octahedral form, the Br and I complexes each have two isomers. One of the iodide complexes is a low-spin five-coordinate compound.

Other Moessbauer studies on iron(II) compounds reported during the review period include: the structural and valence state of Fe^{2+} in the K-Mg-Fe-CO_3 system;[173] complexes of $FeSO_4$ with nicotinic hydrazide;[174] bimetallic complexes containing 1-oxy-2,6-di[(N,N-biscarboxymethyl)aminomethyl]-4-chlorobenzol;[175] the squaric acid complex $[FeL(OH)]_2.BuOH$;[176] and the electronic structure of $[Fe(p\text{-}Bz\text{-}Py)_4(NCS)_2$.[177]

<u>High-spin Iron(III) Compounds.</u> Two isomers of $FeF_3.3H_2O$ have been studied with magnetic susceptibility and Moessbauer spectroscopy measurements.[178] Both phases are low-dimensional antiferromagnets at higher temperatures and become 2-dimensionally ordered below 6.5 and 14.7 K respectively. Amorphous $FeF_3.0.4HF$ has been studied in

the magnetic transition range centred at 39 K.[179] Analysis of the distributions of the quadrupole splitting and hyperfine field show that the spectra can be interpreted by a static broadened field distribution only up to 36 K. Two papers have investigated the Moessbauer spectra of quasi-one-dimensional K_2FeF_5. The effects on spin-fluctuations of substituting iron by non-magnetic gallium were studied at 1.3 K,[180] and the appearance of 3-dimensional magnetic ordering in the hydrate below 1 K was described.[181] Non-linear excitations in $Rb_2Mn_{1-x}Fe_xF_5.H_2O$ (x = 0-0.5) were studied near the Neel temperature.[182] Definite relaxation effects were attributed to short-range order with temperature-dependent relaxation times. Changes in the gamma-resonance spectra of the quasi-1-, 2-, and 3-dimensional antiferromagnets K_2FeF_5, $KFeF_4$, and $Cs_2FeCl_5.H_2O$ have been studied as the magnetic ordering temperature was approached from below.[183] Changes in lineshape, which were characteristic of a slowing of spin relaxation rates, were observed. The isomorphous salts $M_2FeF_5.H_2O$ (A = K or NH_4) have been studied.[184] While the K salt has a constant quadrupole splitting, that of the NH_4 salt is strongly temperature dependent, due to thermal reorientation of the NH_4^+ ion.

The ferrimagnetic silver fluorides $Ag_5M_3F_{14}$ (M = V, Cr, Fe) have been characterized using X-ray, gamma-resonance, and magnetic techniques.[185] The Moessbauer spectra of the iron compounds were recorded at 4.2-300 K and discussed in terms of localization of the ferric cations in the lattice. Moessbauer data were also reported for Na_2NiFeF_7;[186] for FeF_3-MnF_2-PbF_2 glasses;[187] and for $MFeCl_4$ (M = Li, Na, K, Rb, Cs).[188]

Magnetic ordering of monoclinic $FeNaP_2O_7$ has been studied with the Moessbauer effect in a zero applied field.[189] Spontaneous magnetic ordering occurred below 29.2 K and the hyperfine fields of the two antiferromagnetic sublattices showed coincident Brillouin-type temperature dependences. Room temperature Moessbauer and IR data have been reported for a series of Fe(III) complexes of nine dicarboxylic acids (**Table 1**).[190] Tris complexes are formed with acids up to pimelic, while the C8-C10 acids formed bis complexes. A regular trend can be seen in the quadrupole splitting values and magnetic moment data are consistent with high-spin iron(III) in an octahedral site. Other investigations have examined phase transitions in $Na_3Fe_2(PO_4)_3$;[191] identified multiple iron sites in $M_2Fe_3(OH)_5(CrO_4)_3.3H_2O$ (M = Li, Na);[192] and iron(III) chloride and sulphate complexes with the hydrazides of nicotinic and picolinic acids.[174]

Table 1 Room temperature Moessbauer data for iron(III) complexes of dicarboxylic acids.

Complex	Isomer shift ($mm.s^{-1}$)	Quadrupole splitting ($mm.s^{-1}$)	μ_{eff} (B.M.)
$Na_3[Fe(COOCH:CHCOO)_2]5H_2O$	0.62	0.74	5.25
$Na_3[Fe(COO(CH_2)COO)_2]3H_2O$	0.69	0.58	5.64
$Na_3[Fe(COO(CH_2)_2COO)_2]6H_2O$	0.67	0.68	6.02
$Na_3[Fe(COO(CH_2)_3COO)_2]3H_2O$	0.65	0.53	5.82
$Na_3[Fe(COO(CH_2)_4COO)_2]2H_2O$	0.63	0.50	5.77
$Na_3[Fe(COO(CH_2)_5COO)_2]2H_2O$	0.68	0.55	5.90
$Na[Fe(COO(CH_2)_6COO)_2]2H_2O$	0.70	0.63	6.01
$Na[Fe(COO(CH_2)_7COO)_2]2H_2O$	0.64	0.65	5.40
$Na[Fe(COO(CH_2)_8COO)_2.2H_2O]H_2O$	0.72	0.68	5.83

Magnetic susceptibility and Moessbauer effect measurements have been made on $Fe(S_2PPh_2)_3$.[193] The Moessbauer isomer shift data suggested the presence of a greater ligand-to-iron π-bonding contribution from diphosphinate than with either dithiocarbamate or xanthate ligands. The benzoxazole complex Fe(2-methylbenzoxazole) has been studied at 85 K.[194] The presence of magnetic order was attributed to a superexchange effect mediated by the π-electrons of the ligand rings.

Other iron(III) complexes for which Moessbauer spectra have been recorded during the review period include: [Fe(MeOH)Q(NCS)], [FeQCl].EtOH, and $[(FeQ)_2O].2H_2O$, (where H_2Q = an asymmetric N^1,N^4-disubstituted S-methylisothiocarbazide;[195] ferric complexes with Schiff bases derived from sulphonamide drugs;[196] iron complexes of a series of benzoxazole derivatives;[197] a radical-cation salt, $BBDTA[FeCl_4].MeCN$ (where BBDTA = benzo[1,2-<u>d</u>:4,5-<u>d'</u>]bis[1,3,2] dithiazole;[198] the complex sulphides $[NH_2Et_2]_6[Fe_8S_{15}]$ and $[NH_2(CHMe_2)_2]_{10}[Fe_{16}S_{29}]$;[199] complexes with nitrogen-oxygen donor ligands;[200] iron nitrate complexes with <u>o</u>-$HOC_6H_4CH:N(CH_2)_n]_2NR$;[201] and polynuclear Cr(III) and Fe(III) complexes with phosphonic acids.[202]

Intercalation Compounds Containing Iron. Moessbauer spectroscopy has been used to study the new graphite intercalation compounds of the type C_n-$FeCl_4$-$(MeNO_2)_y$, for different oxidation levels of graphite during intercalation and de-intercalation.[203] While the $FeCl_4^-$ ion retains its integrity during intercalation, the de-intercalation proceeds concomitantly with an Fe(III)-Fe(II) reduction process within the graphite. The mixed intercalation compound $FeCl_3$-$NiCl_2$-graphite, which contains both long-range magnetic order and (super)paramagnetism, has been characterized by a range of techniques, including Moessbauer spectroscopy.[204] A charge transfer of 0.2 electrons per iron atom in the $FeCl_3$ layers was found at 4.2 K, which is the same as in singly intercalated $FeCl_3$-graphite compounds. Gamma-resonance measurements were also made on an phase transitions and the thermal decomposition of stage I and stage II graphite-iron chloride compounds.[205,206] During pyrolysis, the ratio between $FeCl_3$ and $FeCl_2$ changed as iron was lost.

Mixed-Valence Compounds and Unusual Electronic States. A spatial localization-delocalization model has been proposed to explain the temperature transitions in Moessbauer spectra of mixed-valence molecular crystals.[207] At low temperatures, a strong intercluster dipolar interaction causes the electron delocalization in the Fe^{3+}-Fe^{2+} cluster and the doublet structure of the spectrum. At higher temperatures, a decreased molecular field leads to delocalization and an averaged spectrum.

Moessbauer and ESR spectroscopies have been used to study the complex $[Fe_2(salmp)_2]^{n-}$ (where n = 0-2 and H_3salmp = bis(salicylideneamino)-2-methylphenol).[208] Ferromagnetic coupling between the two high-spin iron was found with an S = 9/2 ground state. At 100 K a fast electronic relaxation results in a spectrum of two quadrupole doublets consistent with localized Fe(II) and Fe(III) sites. As the temperature is raised a new doublet, corresponding to a delocalized state, grows in intensity until it accounts for 60% of the total area at 297 K. A similar coexistence of trapped- and average-valence states was found in 1,1"'-bis(o-methylbenzyl)-biferrocenium$^+$ $TCNE^-$.[209] In this case an abrupt increase in the relative intensity occurs above 180 K suggesting a change in an intermolecular effect on the mixed-valence state.

Other studies on mixed-valence iron(II,III) compounds reported during the review period include: $LiFe_2F_6$;[210] the bridged dimer (py)PcFeCNPcFe(NH_3);[211] binuclear complexes with 2,6-bis[bis(2-benzimidazolylmethyl)aminomethyl]-4-methylphenolate(1-).[212] A novel mixed-valence compound, ($Fe_2L_2(\mu$-OH$)_3$]($ClO_4)_2$.2MeOH.2H_2O, where L = N,N',N"-trimethyl-1,4,7-triazacyclononane, was prepared and characterized using EXAFS, ESR, UV/visible, and Moessbauer spectroscopic methods.[213] At 4.2 K, a doublet was found at 0.74 mm.s^{-1}, corresponding to two equivalent iron sites with an average valency of +2.5. The data are consistent with a class III, Fe(II/III) species having an S = 9/2 ground state.

A series of Fe(IV) phthalocyanine 5- and 6-coordinated μ-carbido dimers $(FePc)_2C$ and $[Fe(L)Pc]_2C$ (L = acetone, pyridine, 4-methylpyridine, 1-methylimidazole, and piperidine) have been prepared and characterized by techniques including the Moessbauer effect.[214] A new iron(VI) compound $K_2Sr(FeO_4)_2$ has been prepared.[215] It is antiferromagnetic with a Neel temperature of 3 K and a magentic hyperfine field of 87 kOe at 2 K. The Moessbauer isomer shift was -0.91 mm.s^{-1} (relative to iron metal) at room temperature.

Spin-crossover Systems and Unusual Spin States. The structural, magnetic and spectral properties of iron(II) complexes with N^1-(pyridin-2-yl)-3,5-dimethylpyrazole, $[FeL_3]X_2.nH_2O$ (X = BF_4, ClO_4, I; n= 1, 2, 2 respectively), were described.[216] These salts undergo a reversible temperature-induced spin transition both in the solid state and in solution. The Moessbauer data for the $[FeL_3][BF_4]_2$ salt showed the presence of two low-spin ($^1A_1$) species at low temperature and two high-spin ($^5T_2$) species at elevated temperatures. A combined crystallographic phase change/low-spin to high-spin crossover was observed in iron(II) pyrazolylborate, $Fe[HB(pz)_3]_2$ at about 400 K.[217] Both the magnetic and Moessbauer measurements indicated that the ground state is populated with a temperature-dependent mixture of two spin states between 295 and 450 K.

Steric influences on the ground state of iron(II) in $[FeL_3]^{2+}$ (where L = 3,3'-dimethyl-2,2'-bipyridine) have been studied.[218] The anomalous temperature-dependent magnetic and spectroscopic properties were attributed to a temperature-induced singlet($^1A_1$)-quintet($^5T_2$) spin-transition. Lattice dynamics in $Fe(bptn)(NCS)_2$ (bptn = bis(2-pyridylmethyl)trimethylenediamine) and $Fe(tpa)(NCS)_2$

(tpa = tris(2-pyridylmethyl)amine) were also investigated.[219] Both complexes show a gradual high-spin/low-spin transition over the temperature range 4.2-300 K.

The temperature and pressure dependence of the quadrupole splitting of the Moessbauer transition in the spin crossover compound $[Fe_xZn_{1-x}(2\text{-pic})]Cl_2$.EtOH and its deuterated analogue (2-pic = 2-picolylamine) have been studied.[220] Changes in the quadrupole splitting were a linear function of the lattice deformation. The dependence of ΔE_Q on temperature, pressure, and on the fraction of molecules in the high-spin state in the x = 1 compounds was interpreted as a result of the change of the lattice contribution to the electric field gradient at the nucleus only. The intrinsic electric field gradients of the molecules remained unchanged under the cooperative interaction between the spin crossover complex molecules. This is consistent with a simple elastic interaction between the complexes rather than a Jahn-Teller-type interaction. Pressure dependent Moessbauer spectra were also used in a study of the spin equilibria in $[Fe(2\text{-pic})_3]PF_6$ (2-pic = 2-aminomethylpyridine) and $[Fe(6\text{-mepy})_3tren]PF_6$ ($(6\text{-mepy})_3$tren = tris[4-[(6-methyl)-2-pyridyl]-3-aza-3-butenyl]amine).[221] The rate constants for the high-spin to low-spin intersystem crossing were determined by a line-shape analysis of the Moessbauer spectra, and the essential features of the quadrupole-quadrupole relaxation spectra discussed. Thermodynamic analysis of the high-spin fraction allowed the determination of the intermolecular interactions constants and the volume changes of the lattice.

Magnetic and Moessbauer effect data for $[Fe(2,2'\text{bithiazole})_3]X_2$ (X = ClO_4, BF_4) show anomalous temperature dependences arising from a temperature induced spin crossover.[222] The perchlorate has two modifications with different behaviours. In the first, the relatively small fraction of the quintet state in the low-spin lattice increases only slightly from 89-343 K. The second modification, however, is essentially high-spin at high temperature and low-spin at low temperature. A pressure induced high-spin to low-spin crossover was observed in $Fe(dppen)_2X_2$, where X = Cl or Br; and dppen = cis-1,2-bis(diphenyl-phosphino)ethylene.[223] While the Cl-complex undergoes an abrupt first order phase transition at 7-9 kbar, the Br-complex undergoes a more gradual transition above about 60 kbar. In both compounds about 20% of the molecules remain in the high-spin state up to the highest pressures studied. The high-spin to low-spin transition in bis(selenocyanato)bis(2,2'-bi-2- thiazoline) iron(II) has also been described.[224]

A continuous spin-crossover has been observed in the iron(III) solvates $[Fe(L)_2]ClO_4.S$ (where L = the monoanionic Schiff base derived from 3-ethoxysalicylaldehyde and N-(3-aminopropyl)aziridine and S = PhF, PhCl, PhBr, PhI, or o-$C_6H_4Cl_2$).[225] The Moessbauer spectra of all the solvates show a single quadrupole split doublet over the temperature range 110-300 K. The rate of spin interconversion is greater than the Moessbauer time scale (10^{-7} s) but was slower than the EPR time scale (10^{-10} s).

A new discontinuous ferric spin-crossover complex $[Fe(HL)_2]Cl$, where H_2L = pyridoxal thiosemicarbazone has been characterized.[226] Variable temperature Moessbauer and ESR data show the presence of a first order phase transition, and magnetic data a hysteresis with the transition temperature at 245 K (cooling) and 256 K (heating). The cooperative nature of this transition was attributed to the extended coupling of the iron(III) complexes though intermolecular hydrogen-bonding. High-spin to low-spin ($^6A_1$ to $^2T_2$) transitions have also been studied in the FeQ(NCS)(py) complexes, in which H_2Q = 4-HOC_6H_4CH:NN:C(SeMe)N:CHC_6H_3OH-2-R-5 (R = H, NO_2),[227] and 2-HOC_6H_4CH:NN:C(SMe)N:CHC_6H_3OH-2-NO_2-5,[228] and in tris(carbodithioato)iron(III) complexes.[229]

A novel diiron complex containing low-spin and high-spin centres in close proximity has been characterized by UV-visible, IR and Moessbauer spectroscopies.[230] The complex, $[(H_2O)_4Fe(bpm)Fe(CN)_4]$ (where bpm = 2,2'-bipyrimidine), gave a Moessbauer spectrum similar to the sum of the spectra of high-spin $(H_2O)_4Fe(bpm)^{2+}$ and low-spin $Fe(CN)_4(bpm)^{2-}$.

Low-spin and Covalent Compounds. Prussian blue and its thermal decomposition products have been studied using Moessbauer and infra-red spectroscopy, and X-ray and electrical conductivity measurements.[231] The Moessbauer data show that the coordination environment of the Prussian blue is unchanged below 200°C, while flipping of the CN ligands occurs above 250°. An asymmetry of the relative line intensities in the Moessbauer spectra of powdered $Na_3[Fe(CN)_5NH_3].3H_2O$ has been studied.[232] The effect is due to a preferred orientation of the electric field gradient which is assumed to have a conical distribution around the normal of the absorber plane.

A comparison between bulk $Fe(CO)_5$ and its films on graphite has shown that dynamical anomalies occurring between 110 and 180 K affect the Moessbauer parameters.[233,234] In a 2.3 layer film, anomalous clustering of the film results in a coexistence of bulk material with a compressed monolayer, and with novel effects on the quadrupole splitting. Gamma-resonance spectroscopy was also used in a study of $Fe(CO)_3[(CF_3)_2C_2S_2)]$ and its derivatives.[235] The electronic properties of a series of tri- and tetranuclear organo-iron clusters have been studied by the comparison of Fenske-Hall SCF MO calculations with Moessbauer isomer shift data.[236] A linear correlation was found between the calculated effective nuclear charge and the isomer shift. While the 4s orbital populations are generally similar, in the cases of cis- and trans-$[CpFe(CO)_2]_2$ the 4s orbital populations are significantly lower. Moessbauer spectra have been recorded at 80 K for a series of 13 trinuclear cluster compounds prepared by the reaction of $(\mu\text{-RSXMgS})Fe_2(CO)_6$ with $(\eta^5\text{-}C_5H_5)Fe(CO)_2I$.[237] All the spectra contain two quadrupole doublets in a 2:1 area ratio, corresponding to two equivalent six-coordinate low-spin Fe(II) sites, and one low-spin Fe(III) site. Two metalladiboranes, $[Fe(CO)_4(\eta^2\text{-}B_2H_5)]^-$ and $(\eta^5\text{-}C_5H_5)Fe(CO)_2\text{-}(\eta^2\text{-}B_2H_5)]$, have been studied.[238] Both X-ray and Moessbauer data support a three-centre, two-electron bond between iron and the diborane moiety.

During the review period an unusually large number of publications on ferrocene and ferrocene-like compounds were published. A gamma-resonance study has obtained a Moessbauer lattice temperature of 100 K for the inclusion complex of poly(beta-cyclodextrin) and ferrocene.[239] This temperature is significantly lower than that of polycrystalline ferrocene (114 K) and is consistent with weak host-guest interactions which allow the ferrocene molecule to have a significant mobility within the host lattice. The products of the reaction of $Fe(II)(\eta^5\text{-}C_5Me_4H)_2$ with cyano acceptors were identified as iron(III) 1:1 linear charge-transfer salts with the composition $[Fe(C_5Me_4H)_2]^{\cdot+}$ $[anion]^{\cdot-}$.[240] The Moessbauer spectra of the TCNQ and TCNE salts were consistent with complete charge transfer and showed no sign of any strong intramolecular magentic coupling.

The synthesis and the ^{57}Fe and ^{197}Au Moessbauer spectra of the bimetallic complex [μ-1,1'-bis(diphenylphosphino)ferrocene]bis-(chlorogold) have been reported.[241] The iron spectrum gave the parameters IS = 0.50 $mm.s^{-1}$ (relative to Fe), and QS = 2.33 $mm.s^{-1}$, similar to those of ferrocene. The effects of heating ferricenium μ-oxo-bis[trichloroferrate(III)] and its pyridinium analogue at 18-200°C were studied with the Moessbauer effect.[242] The ferricinium complex was reduced to a mixture of Fe(II) compounds at 200°C. Unstable adducts of ferrocene and methylferrocene, and stable adducts of biferrocene, diferrocenylmethane, 1,2-diferroceylethane, and [2]- and [3]ferrocenophane with HgX_2 salts have been characterized.[243] The large quadrupole splittings (2.93-3.29 $mm.s^{-1}$) and the large low-field shifts in the ^{13}C-CP-MAS NMR spectra of the adducts suggest that direct chemical bonds are formed between Fe and Hg in these compounds.

Two papers have reported the Moessbauer data of mixed valence biferrocenium halide salts.[244,245] The valence states and temperature dependences of biferrocenium and 1',1"'-dihalobiferrocenium salts with $FeCl_4$ and $CuBr_2$ were found to be quite different from those of the corresponding triiodide salts,[244] and the spectra of I_3 salts of bihaloferrocenes containing two different halogen substituents were given.[245] The Moessbauer effect was also used to characterize the side-by-side fused biferrocene N,N-dimethyl-2,5"-(2-azoniapropane-1,3-diyl)biferrocene, which contains a fulvalene ligand fused into a cis conformation, and its I^- and I_5^- salts.[246]

A study of the iron(I) organometallic complexes $CpFe(C_6Et_6)$ and $(C_5Me_5)Fe(C_6HEt_5)$ was described.[247] Analysis of the temperature dependence of the quadrupole splittings suggested the existence of a vibronic reduction of the spin-orbit constant, similar to that found in other 19-electron systems, and indicated several phase transitions. Binuclear organoiron sandwiches in their 36-, 37-, and 38-electron states have been studied with the Moessbauer effect in external magnetic fields.[248] The couplings between the 19 e^- subunits were correlated to the electron transfer to the bridge in the mixed valence state. The Moessbauer spectra of the 37-electron complexes $[Fe_2Fv(arene)_2]^+$ PF_6^- (I), where Fv = fulvalene, show that these are delocalized mixed valence Fe(I)Fe(II) complexes with 40% spin density of each iron and no Jahn-Teller activity.[249] Two

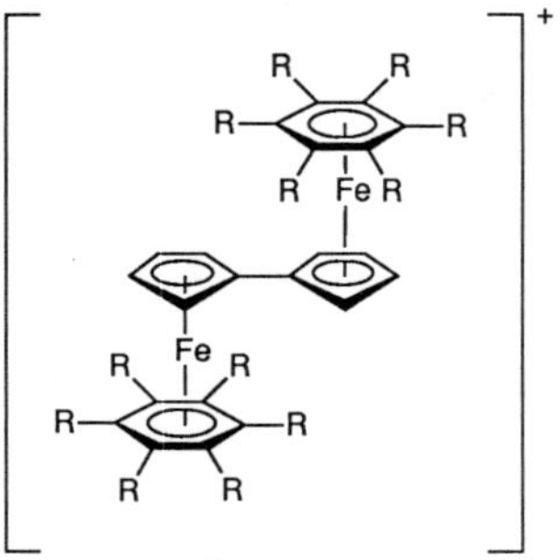

electron reduction gave the corresponding 38e$^-$ biradicals, which had a room temperature Jahn-Teller activity similar to that of the monomeric Fe(I) complexes. When the arene was C_6Me_6 a ferromagnetic transition was found at 37 K. Gamma-resonance data for pentamethylcyclopentadienyl(dithiocarbamato)iron complexes with 17 to 19 electrons were also reported.[250]

The fluoroborate salts of [1.1]ferrocenylruthenocenophane and [1.1]ferrocenylruthenocenophane-1,2-dione have been prepared and characterized.[251] Iron-57 Moessbauer spectroscopy studies suggest the presence of an intramolecular electron transfer between the ruthenium and iron atoms in these materials. The same authors also used Moessbauer spectroscopy and ^{13}C-CP-MAS to study adducts formed between ferrocenylruthenocene and its derivatives and the Lewis acids $HgCl_2$, $SnCl_4$, and I_2.[252-254] The data were consistent with the presence of direct chemical bonds between the central metal atoms and the Lewis acids.

Biological Systems and Related Compounds. An *ab initio* LCAO-SCF-MO study of the effects of different Fe-N_{py} bond distances on the Moessbauer spectral parameters of Fe(porphyrin)$(py)_2$.[255] Both the quadrupole splitting and the isomer shift increase with the Fe-N distance and, at the equilibrium distance, were in agreement with the experimental values for low-spin Fe(III) and Fe(II). A series of FeLCl and (μ-O)$(FeL)_2$ (where H_2L = a *meso*-tetraphenyl porphyrin with 1-4 *p*-Me groups) were studied.[256] The Moessbauer data were compared with data for similar high-spin Fe(III) porphyrins, and a good correlation was found between the Hammett σ_R° factor and the quadrupole splitting. The Moessbauer spectrum of the intermediate spin (S = 1) four-coordinate heme, (protoporphyrinato IX)iron(II), dispersed in a 5% CTAB solution, has been recorded.[257] The quadrupole splitting values follow the sigma-donating ability of the porphyrin ligands. The same porphyrins were also studied in aqueous and non-aqueous solvents in the presence of excess cyanide ions.[258]

The admixed intermediate-spin porphyinate $[Fe(OEP)L_2]ClO_4$ (H_2OEP = octaethylporphyrin; L = 3,5-dichloropyridine) has been characterized using crystallography and magnetic and Moessbauer measurements.[259] The solid state magnetic and Moessbauer results were fitted using common parameters according to the theory of Maltempo. A nitride-bridged iron octaphenyltetraazaporphyrin has been studied.[260] The Moessbauer parameters, IS = -0.042 $mm.s^{-1}$ and QS = 1.769 $mm.s^{-1}$, are consistent with a predominantly Fe(IV) character. Gamma-resonance spectroscopy has also been used in studies on five-coordinate complexes of tetraphenyl-21-thiaporphyrin,[261] and a ferromagnetically coupled iron(III)-porphyrin-dicopper(II) system.[262]

A semiempirical INDO/S-RHF open-shell procedure has been used to calculate the Moessbauer quadrupole splittings for the low-lying spin states of a series of 18 high-, intermediate-, and low-spin ferric heme complexes.[263] The results produced insights into the effects of geometry and axial ligands on the relative energies and observable properties of various spin states of such ferric heme complexes. The same method was also used to study the resting state active sites of the closely related heme proteins cytochrome c peroxidase, metmyoglobin, and catalase.[264] A theoretical and Moessbauer study of hydrazine-hemes was reported.[265] The quadrupole splitting and contributions from valence electrons and the lattice were related to the separation between the Fe and axial N-ligands, and the ligand basicity.

A range of 32 hemin derivatives of tetra(p-tolyl)porphyrin were prepared and their IR, visible and room temperature Moessbauer spectra measured.[266] The type of Moessbauer spectrum obtained, usually depended on the sigma-basicity of the axial ligand. The Moessbauer spectra of a number of model compounds were described. The high-spin iron(II) state in the catalytic cycle of cytochrome P450 was studied.[267,268] A Moessbauer spectrum similar to that of aquo met-myoglobin was obtained from $[(protoporphyrin\ IX)Fe(H_2O)_2]^+$ dispersed in aqueous frozen solutions of detergent micelles.[269]

Gamma-resonance spectroscopy was also used to study mesoverdohemochrome IX dimethyl ester and related porphyrins;[270] and to compare iron(III) pheothytins and iron(III) porphyrins.[271]

Diferric transferrin samples labelled with ^{57}Fe at the N- or the C-terminal bonding sites were compared using zero field Moessbauer spectroscopy at 15 K.[272] While the samples gave similar spectra, some of the line positions differed, suggesting a possible difference between the structures of the two binding sites. Other workers have carried out a similar study at 4.2 K in a range of magnetic fields.[273] This study, however, found that the line positions and relative intensities of the spectra to be identical, indicating that the ligand environments are the same, a suggestion supported by the finding that the strong-field spectra of the two monoferric transferrins are identical.

Ferritin stores iron within a hollow protein shell as an Fe(III) hydrous oxide core. Depending on the amount of iron in the core of horse spleen ferritin, the 4.1 K Moessbauer spectrum may be paramagnetic, a quadrupole doublet or superparamagnetic. A plant ferritin has been obtained which has the same hyperfine parameters as mammalian ferritin but larger iron aggregates.[274] Due to the larger particle size, a magnetic ordering temperature of 240 K was obtained, thus identifying the iron compound as ferrihydrite. A systematic study of the early stages of the core formation was carried out.[275] In samples with up to 40 iron atoms per molecule, a mixture of isolated iron(III) atoms, oxo-bridged ferric dimers, and small and large iron(III) clusters was found.

In vivo Moessbauer spectroscopy has been used to obtain information about the cell distribution pattern of iron in *Escherichia coli*.[276,277] The only species detected in the cell spectra were novel high-spin six-coordinated ferrous and ferric complexes which are not cytochromes, Fe-S proteins or bacterioferritin. Further experiments found similar proteins in other bacterial and fungal systems. The binuclear iron centre in the ribonucleotide reductase subunit B2 from *E. coli* has been characterized in its native ferric state and in a new ferrous form.[278] Based on data from a sample enriched in ^{57}Fe, the native B2 subunit was shown to contain two binuclear Fe clusters each associated to its own tyrosyl radical. The Moessbauer and ESR parameters of the reduced form resemble those of methane monooxygenase in its diferrous form. In other Moessbauer effect experiments, various states of iron-dioxygen interaction in iron-substituted horse liver alcohol dehydrogenase have been investigated;[279,280] the Moessbauer spectra of the spin-coupled centre in porcine purple acid phosphatase were analysed;[281] and the acid-induced reduction and spin-crossover in iron-bleomycin was followed.[282]

Redox intermediates of the hydrogenase of Desulphovibrio gigas, which contains one [3Fe-4S] and two [4Fe-4S] centres, have been generated by redox-titration of the natural substrate prior to study by EPR and Moessbauer spectroscopy.[283] After reduction the [4Fe-4S] clusters could be distinguished, and both showed atypical magnetic hyperfine coupling constants, suggesting differences from the clusters of bacterial ferredoxins. An investigation of the low-potential Fe-S cluster F_x in the photosystem I core protein of Synechococcus 6301 was described.[284] The Moessbauer data for the oxidized and reduced forms support the identification of F_x as a [4Fe-4S] cluster. The oxidized and semi-reduced states of the Fe-Mo cofactor of Azobactor vinelandii nitrogenase have been investigated.[285] The Moessbauer spectra indicated an Fe:Mo ratio of 5:1 with similar environments for all iron atoms. Other cubane-type cluster compounds for which Moessbauer spectra have been reported during the review period include: $[Fe_2S_2(MoS_4)_2]^{2+}$;[286] a series of compounds containing the $[Mo_2Fe_7S_8L_{12}]$ (where HL = HSPh, p-, or m-toluenethiol);[287,288] and double-cubane clusters with the $[ReFe_3S_4]$ core.[289]

Moessbauer spectroscopy has been used to study human adult and foetal haemoglobin (Hb) with different subunit structures in its deoxy- and CO- forms.[290] The spectra were interpreted assuming that the iron(II) environments in the different subunits were non-equivalent and differences of 0.21-0.25 $mm.s^{-1}$ were found between the quadrupole splittings of the different sub-units. Moessbauer effect, electron microscopy and electron diffraction studies have been used to study iron cores in various human and animal haemosiderins.[291,292] Iron-cores of haemosiderins produced normally and by various pathological conditions showed significantly different magnetic properties. Haemosiderin isolated from patients with secondary (transfusional) overload had a goethite structure which showed anomalous behaviour in its temperature-dependent Moessbauer spectra, and haemosiderin cores from cases of primary haemochromatosis, which have an amorphous iron(III) oxide structure, gave gamma-resonance spectra characteristic of a magnetically disordered material. The Moessbauer effect can be used to measure haemoglobin saturation.[293] While red cells showed a normal oxygen saturation below 60%, cells from patients with Hodgkin's lymphoma or leukaemia were 90% saturated.

Iron has been detected in three kinds of livexes (a type of blood product) using Moessbauer and ESR spectroscopy.[294] Both high- and low-spin ferric ions were identified. The Moessbauer parameters of a sample from human amyloidotic liver tissues were recorded.[295] A change of ferritin into haemosiderin was observed together with a simultaneous decrease in iron content. The iron Moessbauer effect has been used in the treatment of cancer.[296] The effect of ^{57}Fe(III)bleomycin on human breast cancer (*in vitro*) and tumour-bearing mice (*in vivo*) was studied alone and together with resonant Moessbauer irradiation. The iron bleomycin reduced cell proliferation *in vitro* and also markedly reduced cell proliferation *in vivo* when used with about 10^{-5} Gy of Moessbauer radiation. Some medicines containing dried ferrous sulphate and ferrous fumarate were studied, and the amounts of rehydration and oxidation were determined.[297]

Time-resolved Moessbauer spectroscopy has been used to study light-induced changes in photosynthetic model complexes of iron quinones on a millisecond time-scale.[298] Part of the high- spin iron(II) present was converted into iron(III) by exposure to near-IR radiation within 6 ms. The chemical nature of iron in the tomato plant was also studied.[299] The seeds, leaves, stems, and roots contained only high-spin iron(III) and, while most the of plant gave similar hyperfine parameters, the seeds gave a spectrum with smaller quadrupole splitting than the other parts.

Oxide and Chalcogenide Compounds Containing Iron. - Hydroxides. The dehydration of $Fe(H_2O)_6^{3+}$ exchanged onto a cation exchange membrane has been investigated.[300] After drying, the samples were quenched in liquid nitrogen, before Moessbauer spectra were recorded at 4.2 K. The following dehydration sequence was identified:

$$Fe(H_2O)_6^{3+} + Fe(H_3O_2)Fe^{5+} \longrightarrow FeOFe^{4+} \longrightarrow Fe(OH)_2Fe^{4+} \longrightarrow Fe^{2+}$$

The effects of temperature on the corrosion process was simulated by the oxidation of $Fe(OH)_2$ in sulphated alkaline media.[301] Goethite was always found while magnetite and lepidocrocite were never found together. A similar study has followed the anodic oxidation of Fe^{2+} at different values of pH and potential.[302] The formation of theta-FeOOH, $Fe(OH)_3$ and gamma-Fe_2O_3 was shown to be pH dependent. A high resolution six-line Moessbauer spectrum of ferrihydrite has been studied.[303] The results suggest that the fitted doublets represent a continuum of environments surrounding Fe^{3+} rather than discrete sites. The data could not positively differentiate between tetrahedral and octahedral ferric sites, nor could it definitely show the presence of tetrahedral Fe^{3+}.

The effects of substituting elements on the Moessbauer spectra have been used to examine the substitution of Ni, Co, Mn, and Al for Fe in synthetic goethite.[304] The approximate distribution function of the effective magnetic field in the Fe nucleus and its integral width was used as a measure of substitution in alpha-FeOOH. Products obtained by heating $(La_{0.1}Fe_{0.9})OOH$ at 100-1250°C were characterized by Moessbauer and magnetic measurements.[305] The thermal behaviour of the ferric gel was significantly modified by the presence of lanthanum.

Wuestite and Related Oxides. Samples of the non-stoichiometric magnesia wuestite, $(Fe_xMg_{1-x})_yO$, were investigated using the ^{57}Fe Moessbauer effect.[306] The spectra were fitted by a model involving interstitial tetrahedral Fe(III) doublet, an octahedral Fe(III) singlet, a defect associated octahedral Fe(II) doublet, and a bulk octahedral Fe(II) doublet. Iron has been found in two forms in the quenched metastable Fe-O-S system.[307] In Fe-O clusters the iron state was similar to that in FeO, *viz.* high-spin Fe(II) with a small admixture of Fe(III), while the Fe-O-S clusters contained low-spin Fe(II).

Haematite and Related Oxides. The problem of line area measurements in the Moessbauer analysis of spin directions in magnetic recording media has been addressed.[308] Asymmetric and broadened line shapes observed in the room temperature spectra of gamma-Fe_2O_3 mean that results based on line intensity analyses are incorrect, and care has to be taken when using area data.

Optical, magnetic and Moessbauer studies of Martian haematite have been reported.[309] Features in the remotely sensed spectral data are consistent with the presence of iron(III) in both super-paramagnetic and larger particles of haematite. The data were compared with measurements made on haematite dispersed in silica. Magnetic-field-induced texture in gamma-Fe_2O_3 has been measured with the Moessbauer effect.[310] The results were compared with those expected for a system of non-interacting particles with uniaxial anisotropy. Clustering and magnetic coupling between gamma-Fe_2O_3 colloids was investigated.[311] The magnetic anisotropy showed a minimum with the aggregation state. Gamma-resonance data have also been presented for gamma-Fe_2O_3 synthesized from ferrous fumarate hemihydrate.[312]

The effect of Ga substitution on the properties of alpha-Fe_2O_3 have been studied using X-ray diffraction and Moessbauer spectroscopy.[313] The magnetic hyperfine field decreased with increasing Ga substitution, collapsing completely at about 50% Ga. Chromium doping of Al_2O_3-$^{57}Fe_2O_3$ was studied at room temperature.[314] A new kind of cross-relaxation process involving two Fe^{3+} ions and one Cr^{3+}, which destroys the paramagnetic hyperfine structure of Fe^{3+}, was proposed. In another paper, Moessbauer spectroscopy was used to follow the kinetics of the reaction of Fe_2O_3 with ZrO_2-Sc_3O_3 mixtures at 850-1300°C over a period of 100 hours.[315]

<u>Magnetite and Spinel-type Oxides.</u> Moessbauer spectroscopy has been used to study powdered magnetite, with a controlled Fe:O ratio.[316] A sudden change in the spectra was found near the Verwey transition for closely stoichiometric samples, and a gradual change in highly cation-deficient samples. Extraterrestrial magnetite, containing about 11% maghemite, was identified in a carbonaceous chondrite meteorite, from its room temperature Moessbauer spectrum.[317] The effects of additives (dextran and sucrose) on ferrite films ($Fe_{3-x}O_4$) prepared by the ferrite plating technique have been studied.[318] The Moessbauer data showed changes in magnetic properties and the Fe(II)/Fe(III) ratio in the films. Changes in the Verwey transition temperature in two series of Li- and Mg-substituted magnetites have been observed.[319] The Moessbauer data also provided information on the electronic exchange and electron charge distribution.

The effect of cooling rate on the cation distribution in $CoFe_2O_4$ has been determined by Moessbauer spectroscopy in a magnetic field of 80 kOe.[320] With increasing cooling rate, the quenched structure becomes increasingly disordered, but becomes saturated just short of a random distribution. X-ray, magnetic and Moessbauer studies have been carried out on $Ca_xCo_{1-x}Fe_2O_4$ ferrite.[321] A solid solution is formed only when $x < 0.2$, then the system becomes a mixture of paramagnetic and ferrimagnetic phases. Ultrafine (5-20 nm) particles of $NiFe_2O_4$, $Ni_{0.5}Zn_{0.5}Fe_2O_4$, and $CoFe_2O_4$ have been prepared under hydrothermal conditions.[322-324] Moessbauer measurements show that these particles are superparamagnetic with hyperfine parameters slightly different from those reported for multidomain bulk materials. Gamma-resonance spectroscopy was used in the quality control of epitaxial $NiFe_2O_4$ films.[325] The magnetic anisotropy was found to depend on the growth conditions, the magnetic field at the Fe^{3+} increasing with the growth temperature. Cation distributions have been obtained in the spinel $InMFeO_4$ (M = Ni, Mn, Co, Mg).[326] While Ni and Mg were found almost exclusively on the octahedral sites the preference for the tetrahedral site decreases according to the series: $Mn^{2+} > In^{3+} > Fe^{3+} > Co^{2+}$.

A Moessbauer and magnetic susceptibility study has found that $BaFe_2O_4$ shows antiferromagnetic behaviour at room temperature, and ferrimagnetic behaviour at 4.2K and that the magnetization reversal mechanism of $BaFe_2O_4$ doped with Ti and Co changes from mostly coherent to fully coherent.[327] Moessbauer evidence has found very few ferric ions in the bipyramidal 2b sites of the substituted materials.[328] This confirms that the change in reversal mechanism is due to an actual physical interruption of the superexchange interactions between Fe^{3+}. The magnetic properties of $Zn_xMg_{1-x}Fe_2O_4$ were also studied.[329] A change from ferrimagnetic to paramagnetic behaviour was seen as x increased.

The ferrimagnetic spinel $CuFe_{0.7}Cr_{1.3}O_4$ has been studied with the Moessbauer effect at 4.2-347 K.[330,331] Iron was present as Fe^{3+} in tetrahedral sites. Line broadening affects and changes in the quadrupole splitting around the Neel temperature were observed. Magnetic and Moessbauer data were obtained for $MgFe_{2-x}Cr_xO_4$.[332] The ligand-cation 3d electron transfer parameters and the cation-cation exchange interaction integrals were obtained by using the superexchange bond model. Moessbauer spectroscopy has also been used to study the magnetic structure in single crystal $TbFe_2O_4$, an antiferromagnetic Ising spin system in a triangular lattice.[333] Three kinds of Fe^{3+} were studied in detail.

Other studies on spinel-related systems with trivalent species replacing iron(III) include: magnetic properties in $FeGa_2O_4$ and $FeAl_2O_4$;[334] single crystals of $(Mg,Fe)(Al,Fe)_2O_4$;[335] and magnetic order in ^{57}Fe-doped $Zn(Cr,Ga)O_4$.[336] A number of systems containing tetravalent ions were also studied. The materials investigated include: Ti^{4+}-substituted lithium-zinc ferrite,[337] lithium-chromium ferrite,[338] and copper ferrite;[339] Sn^{4+}-doped nickel ferrite;[340] Ge^{4+}-substituted zinc ferrite;[341] and magnesium ferrite containing Si, Ge, Ti and Sb.[342]

Superconducting Ceramics Containing Iron. The interest in high critical temperature superconductors remains high, and a fairly consistent picture is emerging from the literature. Although the bulk of the papers describe Moessbauer effect studies on the iron-doped 1-2-3 ceramic, $YBa_2(Cu_{1-x}Fe_x)_3O_{7-y}$, some papers described work on other superconducting ceramics.

In $YBa_2(Cu_{1-x}Fe_x)_3O_{7-y}$ the iron preferentially substitutes into the Cu(1) lattice sites.[343-347] The Moessbauer signals for these atoms correspond to iron in tetragonal and orthorhombic environments, and the relative abundances of these absorptances have been used to determine the oxygen-deficiency parameter.[345] Over 90% of the Fe was found to be substituted into the Cu(1) sites in samples with x up to 0.1.[347] A point charge model was used with Moessbauer spectra collected at 78-273 K to determine the location of iron doped into $YBa_2Cu_3O_{7-y}$.[348] Changes in the hyperfine parameters were related to a lattice softening near 110 K which was, in turn, related to the Peierls instability of the one-dimensional Cu(1)-O chains. Magnetic ordering of iron in the Cu(1) sites was observed in samples at low temperatures or high iron concentrations.[345-347] Antiferromagnetic ordering was reported at about 18 K and 1.5 K respectively for samples in which x = 0.06, y = -0.2 and x = 0.01, y = 0.[347] An antiferromagnetic or spin-glass-like ordering was also found in another study.[349] Magnetically split spectra in $YBa_2(Cu_{1-x}Fe_x)_3O_{7-y}$ with x = 0.01-0.04, recorded at 4.2 K and below, showed broadening which was fitted to the relaxation model of Blume and Tjon.[350] It was considered that such magnetic behaviour was due to iron atoms substituting on Cu(1) sites at twinning boundaries.

Different thermal treatments have been used to prepare samples of $YBa_2(Cu_{0.96}Fe_{0.04})_3O_{7-y}$ with iron occupying both the Cu(1) and Cu(2) sites.[351] The substitution of Fe for Cu in the Cu(2) sites had little effect on the superconducting properties. Neutron diffraction and Moessbauer data have shown that the iron on the Cu(2) site is antiferromagnetically ordered.[352] The temperature dependence of the hyperfine field indicated a 3-dimensional spin wave excitation. Other workers have obtained a Neel temperature of 415 K for Fe in the Cu(2) site.[353] Moessbauer spectra were obtained on a ceramic with x = 0.015 at low temperature in an external field of 5.5 T.[354] The spectra were fitted by considering a diamagnetic spin relaxation shielding of the component spectra. From these results was proposed that the planes containing Cu(2) are mainly superconducting, and that isolated magnetic ions coexist with superconductivity.

In addition to the iron entering either of two copper sites, the oxygen deficiency of the lattice produces sites with different Fe-O coordination numbers. This can produce Moessbauer spectra of three doublets attributed to iron with 4, 5 and 6,[355] or 2, 4, and 6 oxygen neighbours.[356] Changes in the oxygen deficiency can also lead to changes in the spectra. Anomalies in the temperature variation of the Moessbauer spectra of iron in the Cu(1) and Cu(2) sites have been related to changes in the populations of the O(4) and O(5) sites.[357] Samples of $YBa_2Cu_3O_{7-y}$ and $PrBa_2Cu_3O_{7-y}$ doped with 3% Fe were found to develop magnetic order at room temperature when the oxygen content was reduced.[358] The tendency to bond with oxygen atoms was presumed to be responsible for the discreteness of the Moessbauer spectra as a function of oxygen depletion. In the same way, the hyperfine parameters of iron in the Cu(1) positions have been described by changes in the oxygen coordination about iron.[359,360] Neutron and Moessbauer effect data have been recorded for $YBa_2(Cu_{1-x}Fe_x)_3O_{7-y}$ in the orthorhombic (x = 0.01, 0.02) and tetragonal ($x \geq 0.05$) phases.[361] Local ordering of oxygen about Fe encourages the placement of a second Fe atom in an adjacent site leading to clustering and the formation of chains of Fe atoms and low temperature magnetic ordering.

In contrast to macroscopically doped ^{57}Fe, carrier-free ^{57}Co has been found to be a useful microprobe atom for the Cu(1) sites in $YBa_2Cu_3O_{7-y}$.[362] While the oxygen-rich superconductor was found to have a square planar Cu(1) site as the dominant species, the main species in the Cu(1) site of the oxygen deficient ceramic had two axial oxyanions and showed an internal magnetic field at room temperatures. Moessbauer emission spectroscopy was also carried out on $YBa_2Cu_3O_{7-y}$:^{57}Co oxides over the temperature range 77-300 K.[363] From an analysis of the isomer shift and quadrupole splitting values, as well as the relative locations of Fe impurity levels in the matrix, it was argued that: (i) although most of the Co atoms enter the lattice in the Cu(1) sites, some of them have higher coordination numbers; (ii) the daughter ^{57}Fe exists as localized Fe^{4+} ions; (iii) the parent ^{57}Co enters as a localized Co^{3+} state. The results were consistent with the observed increase in oxygen number on doping with Fe or Co, and the existence of localized magnetic moments. A combined absorption and emission has confirmed the identification of high-spin Fe^{4+} in three of the four Moessbauer sites, with Fe^{3+} in the other.[364] In this work, however, the Co was thought to be present mostly as Co^{4+}.

Each iron site in the 1:2:3 superconducting lattice produces a quadrupole split spectrum. measurements on textured and powdered samples have found a negative sign to the electric field gradient at both Cu(1) and Cu(2) sites.[356] The sign directions, and the asymmetry parameters of the four quadrupole components have been determined.[365] Other authors have used Moessbauer and NQR/NMR measurements to show that the sign of the e.f.g. for ^{57}Fe on the Cu(2) sites is different for the superconducting and semiconducting states of $YBa_2Cu_3O_{7-y}$.[366] When y = 6.9 (superconducting), the ratio QS(2)/QS(1) was about 1.39, with both V_{zz} values negative and the asymmetry parameters close to 0 and 1 for the Cu(2) and Cu(1) sites. When y = 6.0 (semiconducting), QS(2) was much smaller and $V_{zz}(2)$ was positive.

The presence of Fe^{4+} in several 1:2:3 superconductors has been noted,[363,364,367] and various combinations of valency and spin-state for the iron dopants have been identified. Most workers have found only Fe^{3+} or Fe^{4+}, although there have been a Soviet reports of high-spin Fe^{2+},[368] and low-spin Fe^{4+}.[369] In samples where Fe has entered both sites, Fe^{3+} was found to be substituted for Cu with pyramidal coordination, and a III/IV equilibrium was found for Fe replacing Cu in square planar environments.[370] In a study of a series of ceramics with $x < 0.02$ three Moessbauer doublets were obtained at room temperature.[371] These were assigned to high-spin Fe(IV) in the Cu(2) site, intermediate spin Fe(III) in the Cu(1) site, and a tetrahedral high-spin Fe(III) site. A concentration dependent spin-crossover was found in the 4.2 K Moessbauer spectra of $YBa_2Cu_{3-x}Fe_xO_{7-y}$.[372] The ferric ions changed from a low-spin ($S = 3/2$) configuration when $x = 0.1$ to a high-spin ($S = 5/2$) state at $x = 0.4$.

Other ^{57}Fe Moessbauer studies on Y-Ba-Cu-O superconducting ceramics include: comparison with ^{119}Sn Moessbauer from the replacement of Cu by Fe or Sn;[373-375] the possibility of substituting iodine for oxygen;[376] and of replacing Ba by K.[377] The effect on the Moessbauer spectrum of changing the rare-earth atom was also studied.[378,379] In $RBa_2(Cu_{0.97}Fe_{0.03})_3O_{7-y}$ (R = Y, Pr, Er) a magnetic spectrum was obtained with the most oxygen-deficient Y sample, for all the Pr samples, and for none of the Er samples.[378] The Moessbauer spectra of $RBa_2Cu_3O_{7-y}$ (R = Gd, Dy, Eu) were studied using the resonances of ^{155}Gd, ^{161}Dy, ^{151}Eu, and ^{57}Fe.[379] Local properties of the Fe sites in the Fe-doped Gd compound were investigated.[379,380] Iron shows the same general behaviour as in the Y compound.[380] The Moessbauer spectra of $EuBa_2Cu_3O_{7-y}$:Fe has been fitted to four doublets in the orthorhombic phase and two in its tetragonal phase.[381] Iron showed a strong preference for entering the Cu(1) site as Fe^{4+}. Gamma-resonance data were also obtained for $HoBa_2(Cu_{1-x}Fe_x)_3O_{7-y}$ from room temperature down to 4.2 K.[382] Three quadrupole doublets were seen until the onset of paramagnetic relaxation at 10 K. No magnetic ordering was found.

Other superconducting ceramics studied with the ^{57}Fe Moessbauer effect in the review period are: $Tl_2Ca_{1.5}Ba_{1.7}(Cu_{1-x}Fe_x)_3O_9$[383] and $TlBaCaCuO_{4.5-y}$:Fe;[384] $Bi_2Sr_2CaCu_2O_y$, a 2:2:1:2 superconductor in which iron occupies environments similar to those found in 1:2:3 ceramics,[355] and the related oxide $Bi_2Sr_{3-x}Ca_xFe_2O_9$;[385] antiferromagnetic ordering was found in the 2:1:4 superconducting oxides La_2CuO_{4-y} and $La_{1.85}Sr_{0.15}CuO_{4-y}$;[386] and phase changes have been followed in Fe-doped $La_{2-x}Sr_xCuO_{4-y}$.[387]

Other Oxides. The Moessbauer spectra of the perovskite system $Sr(Fe_{1-x}Ti_x)O_{3-y}$ has been studied between room temperature and 78 K.[388] The oxygen vacancy concentration was calculated and the relationship between Fe^{4+} content and conductivity was found. Gibb and his co-workers have published a series of papers on nonstoichiometric perovskites of the type $Sr_2LaFe_3O_{8+y}$.[389-392] Moessbauer data obtained over a wide temperature range showed that cubic $Sr_2RFe_3O_{8+y}$, with y = 0.6-1.0 and R = La, contains a mixture of a high-temperature paramagnetic averaged valence, Fe(IV) state and a low-temperature antiferromagnetic mixed valence Fe(III)/Fe(V) state.[389] When $y < 0.6$, however, the system contains a mixture of two ordered vacancy phases.[390] Moessbauer data were reported for the same system when R = La, Ce, Pr, Nd, Sm, Eu, Gd, Dy, Er, Yb and Y[390] and for $Ba_{1-x}La_xFeO_{3-y}$.[391] While the behaviour of the Ba compound was similar to that of the Sr phases, ordered-vacancy structures were found at high values of y. Other Moessbauer effect measurements reported on perovskite oxides include: a study of the magnetic properties of $BiFeO_3$-$BaTiO_3$ and $BiFeO_3$-$Pb(Ti,Zr)O_3$ glassy films;[392] an examination of re-entrant spin-glass behaviour in $Fe_xMn_{1-x}TiO_3$;[393] and a report of spin-glass behaviour in the double perovskites Sr_2FeRuO_6 and $BaLaNiRuO_6$.[394]

The preparation of M-type $BaFe_{12}O_{19}$ particles by glass crystallization was followed with the Moessbauer effect.[395] An increase in the Fe^{3+} content on octahedral sites was increased to 70-90% by annealing above 780°C. An industrially produced barium hexaferrite, $BaFe_{12-2x}Co_xTi_xO_{19}$ (x = 0.72-0.84) was studied by magnetic methods and Moessbauer spectroscopy.[396] The Moessbauer data was able to exclude the presence of superparamagnetic particles as the origin of the observed scattering in the magnetization values. The five components of the Moessbauer spectrum of a CoMo-doped barium ferrite were studied as functions of iron content, heat-treatment, and sample temperature.[397] Substitution occurred mainly on the 12k site, with a resultant reduction of the moment on the 2a site. Other investigations on hexaferrites include: the substitution of Ba by Pr in $BaFe_{11.2}O_{17.8}$;[398] the fate of Fe^{2+} in $FeLaAl_{11}O_{19}$;[399] and the characterization of $SrFe_2Fe_{16}O_{27}$.[400]

Other iron oxide systems studied with the Moessbauer effect during the review period include: orthorhombic beta-$NaFe_{1-x}Al_xO_2$ solid solutions;[401] the orthoferrites $MFeO_2$ (M = Li, Na, K, Rb,

Cs);[402] $LiAl_{3.6}Fe_{1.4}O_8$;[403] the solid solutions $LuMn_{1-x}Fe_xO_3$ and the metastable discrete phase $Lu_2Mn_2Fe_2O_7$;[404] $Bi_2Fe_4O_9$ irradiated with Xe;[405] Mo and Xe irradiated yttrium iron garnet;[406] and $A_2Fe_2BO_7$ (where A = Ba, Sr; B = Ge, Si).[407]

Inorganic Oxide Glasses Containing Iron. The system $x(Fe_2O_3)$-$y(Al_2O_3)$-$(10-x-y)SiO$ has been investigated using Moessbauer and X-ray methods. When x = 3.33-5 and y = 2-4, the material has a spinel-type structure which shows either broad quadrupole doublets or complex Zeeman patterns, depending on the values of x and y.[408] The magnetic hyperfine fields and Curie temperature increase with Al_2O_3 content. For $5Fe_2O_3$-$4Al_2O_3$-SiO T_c was 900 K and hyperfine fields, extrapolated to absolute zero, of 540 and 485 kOe were obtained for the tetrahedral and octahedral sites, respectively.[409] The valence state of iron in borosilicate glasses and its influence on the crystallization of the glass matrix was studied.[410] The iron valency was dependent on the glass composition and the Fe_2O_3 content. Gamma-resonance spectra of the quenched and heat-treated glasses $10NiO$-$10Fe_2O_3$-$30Na_2O$-$50SiO_2$ were studied.[411] After heat-treatment above 710° magnetic splittings typical of precipitated $NiFe_2O_4$ were produced. Iron-doped soda-lime glasses, of close to standard composition were also investigated.[412]

Liquid shear viscosities were measured along the alkali and alkali-earth silicate-ferrite joins, SiO_2-$M_nFe_nO_{2n-y}$ (where X = Li, Na, K, Rb, Cs, Mg, Ca, Sr).[413] The oxidation states of all liquids that could be quenched to glasses were determined by Moessbauer spectroscopy. The Fe^{2+} contents of the quenched glasses increased with increasing temperature and are higher for the Li and alkali-earth ferrosilicates than for the other alkali metals. Redox ratios of Fe and the structural positions of Fe^{3+} and Fe^{2+} in melts in the system Na-Al-Si-Fe-O were determined with the Moessbauer effect. In most cases the value of $\log(Fe^{2+}/Fe^{3+})$ was linearly correlated with the oxygen fugacity and 1/T. The results showed that in natural magmatic liquids, the proportion of tetrahedral iron(III) increases systematically as the liquids become more felsic.

The environment around iron in Bi_2O_3-Fe_2O_3 has been studied by ESR and Moessbauer measurements.[415] A monotonic increase in isomer shift with Fe_2O_3 content was attributed to a reducing covalency in the Fe-O bonds. The average value of the electric field gradient also increased with the iron content. The magnetic properties of the related ternary system $Li_2O-Bi_2O_3-Fe_2O_3$ have been studied.[416] Moessbauer results identified magnetic frustration between two magnetic phases, a short-range antiferromagnetic phase and a ferromagnetic phase, co-existing in $25(Li_2O)-25(Bi_2O_3)-50(Fe_2O_3)$. Iron was used as a Moessbauer probe element to study the structure of sodium oxide-tellurium oxide glasses.[417] The results suggest the Fe^{3+} ions were present at substitutional sites of Te^{4+} ions in distorted TeO_4 trigonal bipyramids. Semiconducting and superionic conducting silver vanadate glasses containing 40 mol.% AgI and 1 mol.% Fe_2O_3 were studied.[418] Changes in the structure of the vanadate network were deduced from the composition dependence of the Fe^+ Moessbauer spectrum. The Moessbauer spectra of calcium phosphate glasses with different Fe_2O_3 contents were also reported.[419]

The Fe^{2+}/Fe^{3+} ratios of 47 simulated nuclear waste glass samples were determined by wet chemical and Moessbauer analyses.[420] While the results were comparable, it was suggested that the wet chemical analysis is preferable for highly radioactive glasses until such glasses were shown to give satisfactory Moessbauer spectra.

<u>Chalcogenides.</u> Spin-polarized multiple-scattering calculations have been made for $(FeS_6^{2-})_n$ (n = 8-10) clusters.[421] The calculated total spin and charge densities were used with the 3*d* and 4*s* atomic populations to interpret the Moessbauer hyperfine parameters of the spinel greigite, Fe_3S_4. Diffusional broadening of the Moessbauer absorption in nonstoichiometric $Fe_{1-x}S$ has been studied over the temperature range 600-1073 K.[422] The difference in diffusion coefficients obtained from Moessbauer and tracer experiments for x = 0.003 was explained by considering a change in diffusion mechanism from a two-dimensional to a one-dimensional pathway and a small correlation factor. The transformation of iron compounds, mostly FeAsS and FeS_2, during the roasting of an arsenical gold ore concentrate has been studied.[423] As the FeAsS decomposed, bound gold was converted into metal.

Three papers have studied thallium iron sulphides. The magnetic field on the Fe nuclei were found to make an angle of 32° with the [001] axis of the $TlFeS_2$ crystal.[424] Both monoclinic $TlFeS_2$ and orthorhombic $Tl_3Fe_2S_4$ show 3-dimensional antiferromagnetic order with Neel temperatures of 190 and 90 K respectively.[425] An intermediate valence state was found in the orthorhombic compound. A similar mixed-valence state was deduced from the isomer shift, 0.37 $mm.s^{-1}$, of $Cu_3Tl_2FeS_4$.[426]

Redox processes of Li_xFeS_2 in lithium batteries were studied.[427] A new electrochemically active ferric sulphide, formed by the de-intercalation of Li_2FeS_2, had a structure different to both pyrite and marcasite. A Moessbauer/electrochemical cell for the in situ monitoring of FeS_2 and Li_2FeS_2 cathodes has been described and used to identify recharge products in lithium batteries.[428] Moessbauer spectroscopy was used to determine magnetic properties of $KFeS_2$;[429] and to investigate $Ta_{0.95}Fe_{0.05}S_3$.[430] Moessbauer effect studies on the thiospinels $(Fe,Co)Cr_2S_4$,[431,432] and $Cu_{0.15}Fe_{0.3}Co_{0.07}Cr_2S_4$,[433] and on the mixed chalcogenide spinel system $Cd_{0.5}Fe_{0.1}Cr_2(S,Se)_4$[434] have been described. Iron was also used as a dopant to obtain gamma-resonance spectra of the semiconductors $Ge_{28.5}Pb_{15}S_{56.5}$,[435] Ga_2S_3, Ga_2Se_3,[436] and semimagnetic HgSe,[437] and to study the magnetic properties of $CuRh_2Se_4$.[438]

Applications of Iron-57 Moessbauer Spectroscopy.- Catalysts The Moessbauer effect has been used to characterize the nature and dispersion of the iron phases present in iron sulphide catalysts supported on various carbons, on Al_2O_3, and on SiO_2 which were used for thiophene hydrodesulphurization (HDS).[439] The Moessbauer Debye temperatures of the sulphide catalysts were compared with those of their oxidic precursors and used to give a qualitative description of the interaction between the iron-containing phase and the carbon supports. In-situ Moessbauer spectroscopy was used to study the sulphidation of C-supported iron oxide catalysts at temperatures down to 4.2 K.[440] The conversion of Fe(III) oxide to iron sulphide proceeded through oxidic high-spin Fe^{2+} and FeS_2. Moessbauer data have been used to compare the active species in P-promoted Fe and Fe-Mo HDS catalysts.[441] While the oxidic phases on both C and Al_2O_3 showed improved dispersion in the presence of P, the formation of an iron phosphate phase caused reduced activity on the C-supported catalysts, without affecting the alumina supported species.

Several papers have reported the Moessbauer spectra of Fischer-Tropsch catalysts of supported iron metal. The effects of sample preparation on the ratio of magnetic to superparamagnetic Fe_2O_3 on iron supported on alumina and silica was studied,[442] and changes in the chemical state of iron during the calcination, reduction and carburization processes were examined.[443] The effects of the preparation variables and of exposure to CO on Fe, dispersed on carbon black, alumina and graphite, were studied.[444,445] The Moessbauer data showed that samples of iron highly dispersed on carbon black, prepared from $Fe_3(CO)_{12}$, may be more spherical and possibly more widely separated than conventionally prepared particles,[444] these catalysts also adsorbed large amounts of CO, to give Moessbauer spectra characteristic of $Fe(CO)_5$.[445] Moerup and his co-workers have used in situ Moessbauer spectroscopy to follow the formation of small Fe particles on high-surface area SiO_2.[446] The particle size distribution, determined from the magnetic field dependence of the Moessbauer spectra, was critically dependent on the preparation conditions. Extended reduction reduced Fe^{2+} ions to very small, 6 nm, particles of alpha-Fe, which are superparamagnetic at room temperature and difficult to detect in the absence of a magnetic field. A similar study has examined the reduction behaviour of Fe/Al_2O_3 and Fe/SiO_2 catalysts.[447] While reduction of the alumina-supported Fe proceeds via a stabilized $Fe_{1-x}O$ phase, and on silica via an iron(II) silicate phase. Amorphous catalysts have been prepared by calcination at 120°C.[448] Phases formed by dehydration and by treatment with CO, H_2, and CO/H_2 mixtures were identified.

Gamma-resonance spectroscopy has been employed in studies of bimetallic iron-containing catalysts used for CO hydrogenation. Silica-supported Fe-Ir catalysts were examined in situ using the ^{57}Fe and ^{193}Ir Moessbauer effects.[449] Iron spectra recorded for used catalysts showed the formation of particles of ε'-$Fe_{2.2}C$ on reduced carbon under the high pressures of the catalytic reaction. The same authors followed the reduction of Fe-Ce oxide catalysts in different gaseous reducing agents.[450] Other systems investigated were: Fe-Re, Fe-K, and Fe-Re-K on SiO_2;[451] Fe-Re and Fe-Ir on silica;[452] unsupported $Fe-ZrO_2$;[453] Fe-Mn and K-Fe-Mn on carbon;[454] and Fe-Ni alloys.[455] Changes in Pd- and Fe-exchanged NaX and NaZSM-5 zeolite catalysts as a result of activation with hydrogen and during the $CO-H_2$ reaction were reported,[456] and PtFe/NaY was characterized after pretreatment in O_2 and H_2.[457]

A $Pb-FePO_4$ catalyst for the oxidative dehydrogenation of isobutyric acid to meathcrylic acid has been characterized by ESR and

Moessbauer spectroscopy.[458] While two non-equivalent Fe^{3+} sites were identified, only one of these was catalytically active. The same reaction can be catalysed by $FePO_4$, $Fe_2P_2O_7$ and $Fe_7(PO_4)_6$. Moessbauer studies of these phosphates before and after reaction showed that the starting phases were transformed into a mixed valence Fe(II,III) phosphate.[459] Changes in the iron oxide phases in supported Fe-Zn catalysts used for butene dehydrogenation have been described,[460] and the interactions between Fe_2O_3 and the promoters in Fe_2O_3-K_2O-CaO catalysts used for ethylbenzene dehydrogenation were studied.[461] Two Soviet studies on ammonia synthesis catalysts,[462,463] and data on Mo-Fe and Mo-Fe-Cr catalysts for methanol oxidation[464] were also reported.

Minerals and Coal. The Moessbauer effect has been combined with AAS to identify the minerals talc, ilmenite, and chlorite in talc rocks and schists.[465] The recoil free fractions of chlorite and ilmenite were determined. The use of Moessbauer spectroscopy for evaluating changes in redox conditions during ore mineralization, and hence for prospecting has been examined.[466] A survey of clinopyroxenes in prospecting for uranium ores in albitized granite was used as an example. Pyrite samples from a Chinese Au-Ag ore deposit have been used to determine the relationship between isomorphous impurities in pyrite (e.g. Co, As, Sb) and the ^{57}Fe Moessbauer parameters at 298 K.[467] The geological significance of the pyrite qudrupole splitting was discussed. In India, the Fe^{2+}/Fe^{3+} ratio in weathered samples from agricultural and forest lands were used to show the need for fertilizers in some parts of Sikkim,[468] and iron phases in sediments from a well in the Thar desert were identified.[469]

The Moessbauer effect has been used to determine iron compounds in Polish anthracites and lignites.[470] Pyrite, ankerite, jarosite and Fe(II) in illite minerals were detected. The coking similarity of North American coals has been studied.[471] While the iron-based mineral matter in the coals were very different, the alpha-Fe_2O_3 size distributions in the ashes were very similar. Iron-phases in ashes from Indian Railways coal engines,[472] and in agglomerated ash from the gasification of Chinese coal have been characterized.[473]

Corrosion Studies. The rusting condition of the cables of suspension bridges are usually measured by self-induction measurements, which assume that rusts of the same chemical composition always have the same magnetic properties. Where this assumption is questionable Moessbauer spectra were used to provide additional data on some detected defects.[474] The composition of the rust layer formed on weathering steel and pure iron after several wet-dry cycles in a SO_2-polluted atmosphere was investigated.[475] Poorly crystalline ferrihydrite, $FeSO_3.3H_2O$, and $FeSO_4.4H_2O$ were the only corrosion products. When mild steel was corroded in $NaNO_3$ and Na_2SO_4, the Moessbauer spectra showed that both solutions contained only gamma-FeOOH and $Fe(OH)_3$.[476] Other Moessbauer spectroscopy studies on corrosion products have examined carbon steel in a tropical marine atmosphere,[477] the surface equipment of an oil production well,[478] carbon steel in the cooling circuit of a nuclear power plant,[479] and the coating layer of a steel welding wire.[480]

Bacterial corrosion products from the surface of mild steel samples were analysed by Moessbauer spectroscopy, and X-ray and electron diffractions and green rust 2 was found to be the major species.[481] Moessbauer spectra of the green rust 2, run in vacuum at 78-460 K, showed changes in the abundances of 2 of the 3 Fe(II) sites, with the ferric sites due to electron exchange.[482] After further oxidation in air, the third iron(II) site was not oxidized, indicating that a $FeSO_4$ layer in the green rust remained unchanged.

The rust transforming abilities of 9 commercial converters have been studied by Moessbauer spectroscopy.[483] Only one compound was shown to actually change the chemical nature of rusted mild steel, and accelerated salt spray tests showed product performances which ranged from harmful to beneficial. Chemical changes produced on an iron substrate by by a water-thinned anticorrosive primer were also reported.[484] Products from reactions between iron and specific primer components were reported.

<u>Other Applications.</u> The Moessbauer effect has been used to examine the different oxidation and spin states of iron samples of street dust.[485] The Fe(II)/Fe(III) ratio and the relative area of the magnetic phases provided indications of the origin of the sample. The effect of alkali ions on the strength development of cement paste has been monitored by Moessbauer spectroscopy.[486] The accuracy of the Moessbauer method for the determination of the cement strength was demonstrated. The effect of the dehydration of

cement pastes on the states of Fe has been related to the decrease in the compressive strength of Portland cement.[487] On the basis of these results Moessbauer spectroscopy can be used to estimate the degree of fire in concrete buildings.

Moessbauer spectroscopy has been used to study clays whose industrial uses depend on the Fe content.[488] Reactions occurring during clay firing were monitored. Following dehydroxylation of the clay minerals, the quadrupole splitting of 8-coordinate iron(III) increased abruptly before reverting to lower values at higher temperatures. It was suggested that iron oxides account for a higher proportion of the total iron content than is commonly recognized, and should be taken into account for the correct interpretation of the Moessbauer spectra of clays. A Chinese study has obtained room temperature Moessbauer data on 13 samples of pottery from different sites, produced by different methods.[489] The results show that the paramagnetic ferrous quadrupole splittings of 11 samples increase linearly with the age of the pottery, and that the magnetic component almost vanishes in very ancient pottery. A firing study of the terracotta army of the Qin dynasty was also described.[490] An Indian study of pottery from Tamilnadu has attempted to correlate the Moessbauer parameters with the oxidation states of iron, the colour, the temperature of firing and the provenance.[491] The black colour in early (ca. 2000 B.C.) Cypriote and Nubian pottery was also studied and attributed to the presence of both Fe^{2+} and carbon, while the red colour is mainly due to Fe^{3+} ions.[491a]

5. TIN-119

General Topics. Electronic structure calculations of compounds containing tin as a component, as well as of substitutional tin impurities in elemental metals and semiconductors, have been presented.[492] The linear muffin-tin-orbital band-structure and Green-function methods were used and the calculations were self-consistent and scalar-relativistic using the local-density-approximation for exchange and correlation. As an isolated impurity in metals, the Moessbauer isomer shift shows a minimum in the middle of the transition series due to increased hybridization with the metal *d*-bands. The isomer shift is largest for the sp metals to the right of the transition metals. The strong covalency of the group-IV semiconductors significantly reduces the isomer shifts of substitutional tin atoms as compared with the sp metals. Moessbauer spectroscopy has been used to study the hyperfine field at ^{119}Sn impurities within rare earth metals such as Dy and Tb.[493] In microregions around these impurities the magnetic and physical properties differ from those of the rare earth matrix.

Scattering Moessbauer spectroscopy has been used for the non-destructive testing for tin in the surface layers of twelve ancient Chinese bronze mirrors.[494] All of the tin existed as either SnO_2 or as Cu-Sn alloy.

The Moessbauer effect probability has been measured in single crystals of tin oriented in the (100) and (001) planes under hydrostatic pressures of up to 40 kbar.[495] The anisotropy of the probability is pressure dependent. Myoren and Osaka have carried out a theoretical study of the ^{119}Sn Moessbauer spectral shape resulting from the bond-angle distribution of a tin atom in a Si matrix.[496] The application of this theory to experiments on ^{119}Sn gamma-resonance spectroscopy of sputtered amorphous-Si:H films showed that the generation of fewer overconstrained regions due to SiH_2 bonding configurations appreciably decreases the standard deviation of the bond angle deviation for the tin atom.

The influence of the type of cation and deviation from stoichiometry of a series of tin-doped transition metal antimonides on the Moessbauer parameters for tin introduced into the anion sublattice has been investigated.[497] Excess occupation of the trigonal bipyramidal, interstitial NiAs structure was observed. The vacancy-tin interaction has been studied in a dilute Al-Sn alloy by gamma-resonance spectroscopy and positron annihilation.[498] A simple model was developed for the interpretation of the results, and included monovacancy-, divacancy-, and tetravacancy-Sn complexes, and more complicated Sn-vacancy agglomerates as important participants in the recovery process. Changes observed in the Moessbauer spectra were explained by the gradual transformation of the complexes to simpler ones through transitional configurations. Only two lines were observed in the Moessbauer spectra and suggest that a single vacancy has a minor effect on the isomer shift of the line corresponding to tin in substitutional vacancy sites.

The temperature dependencies of the electronic conductivity and Moessbauer spectra of tin impurities in beta-In_2S_3 were studied.[499] Motiejunas and his co-workers have shown that the isomer shift of the isostructural tin(II) chalcogenides increases with increasing ligand electronegativity.[500] The observed correlation between the Moessbauer isomer shift and the hole concentration in SnTe has been explained on the peculiarities of the band structure.[501] Several structural transitions, as well as ferroelectric transitions, have been observed in $Sn_{1-x}Ge_xTe$.[502] The electron exchange between ionic tin centres in $Pb_{96}Sn_2NaTl(S,Se)_{100}$ solid solutions has been studied by gamma-resonance spectroscopy.[503] The Sn(II)/Sn(IV) ratio decreases as the amount of sulphur increases.

Systematic low-temperature in-situ Moessbauer effect studies on vapour-quenched amorphous Mn_xSn_{1-x} alloys showed that the magnetic behaviour of the system was correctly displayed by the transferred magnetic hyperfine interactions detected at the ^{119}Sn site.[504] The reaction of Cu with highly dispersed Sn under high pressure during plastic flow gives Cu_3Sn, which has high thermal and oxidation stability.[505] Aged and fresh Ag-Sn alloys and Cu-Ag-Sn dental alloys have been examined by X-ray diffraction and Moessbauer spectroscopy.[506] Although the X-ray traces of the aged Ag-Sn showed only gamma-Ag-Sn and free Ag, the Moessbauer spectra showed Sn(IV) oxide also. A low Debye temperature showed that the oxide was in intimate dynamical contact with the metallic matrix. Temperature dependent Moessbauer effect measurements on $AuSn_4$, $PdSn_4$, and PtSn4 in the temperature range 10-295 K showed the presence of a tin(IV) site.[507] Variation of the thermal shift and the Debye-Waller factor showed that the mean-squared displacement and mean-squared velocity of the tin nuclei were characterized by Debye temperatures of 238, 255, and 304 K for $AuSn_4$, $PdSn_4$, and $PtSn_4$, respectively.

Moessbauer spectra of Sn-119 at a concentration of 0.2% in alloys of Fe with Si, Al, and Rh have been presented.[508] For Rh concentrations above 3%, an intermetallic compound was formed in which Sn-119 has a positive hyperfine field of 57 kOe and an isomer shift 0.1 $mm.s^{-1}$ less than is found at tin in iron. Comparison of ^{57}Fe and ^{119}Sn hyperfine fields indicates that in the canted phase of Au-19%Fe-2%Sn, the transverse components of the magnetic moments are locally parallel.[509]

The role of conduction electrons in the formation of the hyperfine magnetic field at tin nuclei has been examined using ^{119}Sn introduced as impurities into $TbCu_{1-z}Zn_x$.[510] In Tb-Cu a chaotic distribution of antiferromagnetic domains was observed. Magnetization and ^{119}Sn Moessbauer effect studies have been performed on the pseudoternary systems, $U_xTh_{1-x}CoSn$ and $UCoAl_{1-x}Sn_x$.[511] The magnetic hyperfine fields at the tin nuclei are due to local surrounding of U atoms which carry magnetic moments. Local surrounding effects are indicated by multicomponent spectra obtained for the two pseudoternary series.

Gamma-resonance spectroscopy has been used to characterize epitaxial layers of $Ga_{1-x}al_xAs$ (x = 0-1) doped with ^{119}Sn-enriched tin.[512] The GaAs and AlAs samples gave Moessbauer spectra that were interpreted in terms of two tin sites corresponding mainly to substitutional shallow donors and clustered species. Samples with x = 0.3-0.4 yield significantly altered spectra with a new type of dominating site. This was interpreted as due to the DX-centre with an electronic structure altered because of electron localization and a local distortion of cubic symmetry.

Inorganic Tin(II) and Tin(IV) Compounds. Minerals of the stannite group have been studied by Moessbauer spectroscopy, with special reference to the question of stannite-kesterite miscibility.[513] Changes in the Moessbauer linewidth as a function of mineral composition indicate that all the samples have a single phase structure; which would preclude the existence of a stannite-kesterite miscibility gap. Several tin minerals were also studied by Chinese workers in order to determine the oxidation states of the tin.[514] A weak tin(II) doublet observed in stannite and franckeite was attributed to an impurity phase. Tin in cassiterite and andradite occurred consistently as tin(IV). The electronegativities for oxygen and sulphur were calculated on the basis of the isomer shift values and the results are quite close to those given by Pauling.

A Moessbauer investigation of $CsSnCl_3$ over the temperature range 85-300 K showed linear dependences of both the isomer shift and the quadrupole splitting, indicating that this compound does undergo no phase changes over the temperature range studied.[515] Successive phase transitions have been observed for the isomorphous compounds $CsSnBr_3$ and $MeNH_3SnBr_3$.[516] For the latter compound, the Moessbauer spectra changed markedly with decreasing temperature. These findings suggested a large distortion of the $SnBr_6$ octahedron and were interpreted on the basis of the 3c-4e bond model.

The products formed in the reaction between $[Pd(bpy)_2]Cl_2$ (bpy = 2,2'-bipyridyl) and $SnCl_2$ were characterized by Moessbauer spectroscopy.[517] Other reactions, reported this review year, for which the Moessbauer effect was used for characterization of products were: the complexing of Pd by Sn(II) in aqueous solutions of citric and tartaric acids;[518] the complexing of Rh(III) with Sn(II) in aqueous tartaric acid solutions;[519] complexation of Pd(II) and Sn(II) heteronuclear chlorides and sulphates,[520] and the complexation of Rh, Ir, Ru, and Os with tin(II) in aqueous oxalic acid.[521]

The reaction between SnF_2 and MCl (M = Na, K and NH_4) in aqueous solution gave MSn_2F_4Cl for M = K and NH_4.[522] Tin-119 Moessbauer results for the K salt indicate that the tin(II) lone pair is stereochemically active and has considerable $\underline{p}_z$ character. The larger isomer shift and smaller quadrupole splitting of the NH_4 compound was attributed to p-electron withdrawal from the tin due to hydrogen bonding.

The geometrical and chemical bonding structure of isolated molecules consisting of a number of interstitial gas atoms bound to a substitutional 5sp element atom in a noble metal matrix has been studied for Sn-119 and Sb-121.[523] The structural information obtained was discussed in relation to the structure of conventional molecules and to the constraints introduced by the metal matrix.

The all-electron, first-principles, discrete-variational method has been employed to study the electronic structures of SnF_4^-, $SnCl_4$, $SnBr_4$, and SnI_4.[524] Values of the electron density at the tin nuclei were derived and related to the Sn-119 isomer shifts. Differences in the electron density were discussed in terms of the chemical bonding between tin and the halogen atoms.

Thin films of SnO_x (x = 1-2) have been deposited by electron beam evaporation of SnO_2 and studied by X-ray, electron-diffraction and Moessbauer spectroscopy.[525] Similar films deposited by reactive sputtering[526] and pulsed laser evaporation[527] were also studied.

The sensitivity to humidity of the semiconducting ceramic material Zn_2SnO_4 has been studied and the contribution of the tin atom to the conduction mechanism in the compound was shown.[528] Moessbauer spectroscopy is a good method for humidity sensitivity studies. The computer-fitted Moessbauer spectral parameters for cadmium stannate films prepared by dip-coating, showed that there was apparently no tin(II) present in the films, in agreement with previous work.[529] The Moessbauer effect in $Eu_2Sn_2O_7$ has been used to study changes in the effective charges on atoms occupying coordination sites in the compound which are different to those in the original oxides used to synthesize this compound.[530] Changes in the effective charges on the metal atoms, determined in this way, can be considered as criteria of the donor-acceptor interaction in mixed oxide systems.

Superconducting Ceramics. Studies on ceramic high temperature superconductors by ^{119}Sn gamma-resonance spectroscopy continue to be of interest. Very low temperature measurements on Sn impurities in $La_{1.83}Sr_{0.15}CuO_4$ showed that a magnetic hyperfine field at the tin nucleus appears below about 7 K.[531] The origin of the magnetic hyperfine interaction was ascribed to collective spin fluctuations in the copper oxide. Tin has been shown to replace copper in $TlBaCaCuO_{4.5+y}$, with unusual localization of the tin sites.[375]

Anomalies in the temperature dependencies of the probability of the Moessbauer effect on tin impurity nuclei in superconducting $YBa_2Cu_3O_{7-y}$ have been observed near the phase-transition points, T_c = 83 K and T_N = 190 K.[532] These anomalies were interpreted in terms of softening of the phonon spectrum in those temperature regions. Tin exists in $YBa_2Cu_{3-x}Sn_xO_{7-y}$ (x = 0.05, 0.1, 0.2, 0.3, and 0.4) as Sn(IV) and it occupies the Cu(1) and Cu(2) sites.[533] A similar study for x = 0.2 and 0.3 revealed two components, both with isomer shifts corresponding to tin(IV).[534] The temperature dependence of the area of only one component showed an anomaly near the superconducting temperature for both samples, suggesting anisotropic behaviour of the lattice rigidity. Tin(IV) was also found in a sample with x = 0.05.[373] The separation of Cu charge states in $YBa_2Cu_3O_{7-y}$ doped with ^{119}Sn and ^{57}Fe has been supported by studies on samples prepared under different conditions of heat and gas treatment.[374]

Anomalies in the temperature dependence of the Moessbauer effect were also observed near the T_c for $EuBa_2(Cu_{1-y}Sn_y)_3O_{7-x}$ (where y = 0.006).[535] The anomaly was again explained by softening of the phonon mode corresponding to oscillations of the copper atoms. A more detailed study of this system detected more than two different tin states by decomposition of the Moessbauer spectrum.[536] These tin sites were associated with non-equivalent copper sites replaced by Sn atoms in the perovskite-type lattice. Anomalous changes were observed in the isomer shifts and area fractions in the spectra, as measured at room temperature and at 77 K, and were interpreted by assuming a low temperature phase transformation and phonon softening. Time dependent changes were found in the tin spectra recorded at 77 K in the case of the highest tin concentration.

Evidence for phonon softening near 215 K was observed for $DyBa_2Cu_{2.85}Sn_{0.15}O_{7-x}$ by Moessbauer spectroscopy.[537] There was also evidence of strong correlation between superconductivity and lattice anharmonicity. The fluctuations of the isomer shift indicated that, as far as the electron density at tin sites is concerned, the sample has a high degree of instability. A transferred hyperfine interaction at temperatures below 60 K has been observed for tin in $GdBa_2(Cu_{0.99}Sn_{0.1})_3O_7$.[538] The onset of the magnetic interaction at the nonparamagnetic Sn-119 probe was explained in terms of local depletion of holes due to Sn(IV) doping. These experimental findings provide strong evidence for spin-spin correlations originating from the Cu^{2+} moments and present in the superconducting state.

Organotin Compounds. The Sn-Cl distances in octahedral $SnCl_4L_2$ compounds have been found to be linearly correlated to Moessbauer partial quadrupole splitting (pqs) data.[539] The correlation can be used to predict Sn-Cl distances, which are expected to agree with experimental values within 2 pm, although there is no theoretical explanation for the linear relationship. The Sn-Cl distance can be expected to increase for negative pqs values, since the pqs for a ligand becomes more negative as the donor strength of the ligand increases, and the Sn-Cl bond becomes weaker as the Sn-L bond becomes stronger. Adducts of ferrocenylruthenocene, ferrocenylruthenocenylmethane and biruthenocene with $SnCl_4$ have been studied with the ^{119}Sn Moessbauer effect.[254] The presence of covalently bonded tin(IV) species, together with data from ^{13}NMR and ^{57}Fe

Moessbauer measurements suggest the presence of a direct chemical bond between tin and iron.

Moessbauer data have been obtained for a series of organotin anions, e.g., $Ph_3Sn^- M^+$ (M = Li, Na, K).[540] The spectra of these anions in solution show the presence of two species, a singlet in the tin(IV) region and a doublet in the tin(II) region (**Table 2**). The doublet arises from the naked Ph_3Sn^- ion and its counter ion may or may not be solvated. The single line arises from the contact ion pair Ph_3SnM in solution, or from the precipitated salt.

Table 2 Frozen Solution Sn-119 Moessbauer Data at 77 K of some Organotin Anions.

Compound	IS mm/s	QS mm/s	% area
Ph_3SnLi (THF)	1.30	0.0	38
	2.15	1.87	62
Ph_3SnLi (NH_3)	2.03	1.83	100
Ph_3SnNa (THF)	2.14	1.86	100
Ph_3SnNa (NH_3)	2.09	1.86	100
Ph_3SnK (THF)	1.42	0.0	40
	2.16	1.82	60
Ph_3SnK (NH_3)	2.05	1.88	100

The correlation occurring between the tin-119 Moessbauer isomer shift and the partial atomic charges on tin for R_3SnX (R = Me, Et, Pr, Bu, cyclohexyl; X = F, Cl, Br, I) has been investigated by Silvestri.[541] The results of this correlation suggest that Ph_3SnF and Ph_3SnCl are five-coordinate polymers while Ph_3SnBr and Ph_3SnI are monomeric species.

The configuration of the tin environment in binding sites of the microcrystalline SnR_3-rat Hb complexes (where R = Me, Et, Bu; and Hb = haemoglobin), as well as in model systems consisting of SnR_3 moieties in aqueous media and in organic solutions was investigated by gamma-resonance spectroscopy. Using the point-charge model rationalization of the experimental quadrupole splitting measured at 77 K and also through Moessbauer-Zeeman spectra recorded at 4.2 K in a transverse external magnetic field of 6 T [542] a distorted trigonal bipyramidal structure was inferred for the Hb complexes, where the ligand axial atoms come from a cysteine and histidine side chain.

Moessbauer spectral data for $[(Bu_2SnO_2CCH{:}CHC_6H_4OMe\text{-}4)_2O]_2$ suggest that this compound adopts a ladder structure containing five-coordinate tin atoms, while dibutyltin bis(cinnamate) ester has a distorted six-coordinate <u>trans</u>-R_2SnO_4 structure.[543] The tributyltin ester is a five-coordinated polymer in the solid state and is a four-coordinate monomer in solution.

The Moessbauer spectra of a series of organotin molybdates, R_2SnMoO_4 (R = Me, Et, Bu, octyl), $(R_3Sn)_2MoO_4$ (R = Bu, Ph, c-C_6H_{11}, $PhMe_2CCH_2$), and $(R_3Sn)(R'_3Sn)MoO_4$ (R = Bu, R' = Ph, c-C_6H_{11}; R = Ph, R' = c-C_6H_{11}) at 80 K are in accord with the tin atoms in the dialkyltin molybdates having a <u>trans</u>-R_2SnX_4 octahedral geometry, and those in the bis(triorganotin) molybdates adopting a <u>trans</u>-R_3SnX_2 trigonal bipyramidal stereochemistry, due to intermolecular bridging by the MoO_4 ligands.[544] The synthesis and characterization of $SnEt_2$(dapt), SnBuCl(dapt), and $SnBu_2$(daps) [where H_2dapt = 2,6-diacetylpyridine bis(2-thenoylhydrazone); and H_2daps = 2,6-diacetylpyridine bis(salicyloylhydrazone)] has been reported.[545] For complexes containing an SnR_2 moiety the quadrupole splitting values fall in the range 3.12-4.29 $mm.s^{-1}$ and are consistent with five-, six-, or seven-coordinate organotin compounds. There is poor discrimination between six- and seven-coordinate tin using partial quadrupole splitting values. Moessbauer spectroscopy has shown that in the structure of $[(SnBu_2(O_2CC_6H_4NH_2\text{-}o))_2O]_2$ one molecule in the dimer has one six- and one five-coordinate tin atom, while the other contains two six-coordinate tin atoms.[546] Moessbauer and electronic spectral data are consistent with distorted octahedral structures for $RSnCl_3L$ and R'_2SnCl_2L (R = Ph, Bu; R' = Ph, Me; L = 4-(2-pyridylazo)-N,N-dimethylaniline) in the solid state.[547]

Moessbauer spectroscopy has been used to show that degradation of bis(tributyl)tin oxide within a neoprene rubber matrix leads to tin(IV) oxide, by progressive hydrolysis and de-alkylation.[548]

A large number of authors have used gamma-resonance spectroscopy as one method of characterizing organotin compounds.[549-568]

6. Other Elements

This section reviews the published data for elements other than iron and tin. In each of the three main subsections (main-group elements, transition-metal elements and lanthanide and actinide elements) the isotopes are discussed in order of increasing atomic number.

Main-Group Elements. - Krypton (Kr-83) The 9.4 keV Moessbauer gamma rays in Kr-83 at 4.2 K have been used to investigate the lattice dynamics and interactions of solid krypton bubbles in aluminum.[569] The spectra show the presence of krypton in two environments (i) krypton atoms within the bubble which give rise to a simple line spectrum and (ii) krypton atoms at the Kr-Al interface which give rise to a quadrupole split resonance line. The increasing bubble size (from 1.5 to 3.7 mm) resulting from annealing the Kr-implanted aluminum sample at 700 K is reflected both in decrease in isomer shift and a decrease in the characteristic Moessbauer temperature from 89 to 64K. The changes in the Moessbauer parameters of the two components measured at 4.2K at different annealing temperatures are shown in **Table 3**

Table 3 Moessbauer parameters of Kr in aluminum annealed at different temperatures. (IS relative to ^{83}RbCl).[569]

	Component		
Temp.	bulk	Kr at Kr-Al interface	
K	IS (mm/s)	IS (mm/s)	QS (mm/s)
293	0.029	0.17	1.83
553	0.014	-0.06	1.94
573	0.008	-0.06	1.90
603	0.006	-0.04	2.21
628	-0.13	-0.06	2.21
648	-0.18	-0.04	1.94
700	-0.21	-0.07	2.27

Germanium (Ge-73). The 13.3 keV Ge-73 Moessbauer transition has been calibrated as a result of a study of the electronic structures and isomer shifts of Ge, Sn and Sb impurities within the diamond-structure Si, Ge and Sn lattices as hosts.[570] A value of 1.7×10^{-3} is calculated for $\Delta R/R$ for Ge-73.

Antimony (Sb-121). The effects of the substitution of Sb for Bi in the superconducting oxides $BiCaCu_2O_y$ and $Bi_2Sr_2CaCu_2O_y$ have been studied.[571] The Moessbauer parameters for the antimony sites in these oxides are similar to those for Sb_2O_5 showing that the antimony is present in the 5+ oxidation state and that it has an oxygen environment similar to that in Sb_2O_5. Similar results have been found for antimony substitution in $Pb_{0.3}Bi_{1.8}Sr_{1.9}Ca_2Cu_3O_y$.[572] Gamma-resonance spectroscopy has also been used to determine Sb(III)/Sb(V) concentration ratios in silicate glasses.[573]

Antimony-121 Moessbauer spectra for chalcogenide glasses in the Ge-Sb-Se system have been interpreted in terms of $SbSe_{3/2}$ structural units being important over the whole glass forming region.[574] Although in the extended glass forming region, with Sb contents greater than 50 atom%, the presence of additional structural units is suggested.

The Moessbauer isomer shift and quadrupole coupling constants for $K_2SeO_4SbF_3$ have been reported as -4.65 $mm.s^{-1}$ relative to InSb and 14.3 $mm.s^{-1}$ respectively.[575] These parameters are similar to those found for $K_2SO_4SbF_3$ and are consistent with the lone-pair distorted antimony environment found in the crystal structure of $K_2SeO_4SbF_3$.

Two papers published in the review period have dealt with the Sb-121 data for antimony halides intercalated into graphite.[576,577] The Moessbauer parameters of stage-1 and stage-2 intercalation compound with $SbCl_5$ and SbF_5 revealed the presence of both Sb(III) and Sb(V) species with relative abundances that were dependent of the method of preparation of the intercalate. In the $SbCl_5^-$ graphites the Sb(III) spectra have been interpreted in terms of $SbCl_3$ intercalate molecules with their three fold axis orientated approximately parallel to the graphite layer planes. In under-saturated stage-2 materials Sb(V):Sb(III) ratios of 1.4:2.2 were found with the Moessbauer data being interpreted in terms of $SbCl_6^-$ species. In saturated stage-1 materials the predominant species were ~~now~~ found to be Sb(V) moieties.[576] Data for the intercalates $C_nSbCl_5.MCl_3$ (M = Al,Fe, Sb or As) show that the Sb is present mainly in the 5+ oxidation state,[577] but that both $SbCl_5$ and $SbCl_6$-type moieties coexist and it has been suggested that the

MCl_3 species act as links between the Sb species. For the SbF_5-graphites, the Sb(V):Sb(III) ratio is particularly dependent upon the method of preparation although the Sb(V) species are always the major component. The data further suggest that any Sb(III) species are anions such as SbF_4^- rather than SbF_3.

Reaction of tricyclopentadienyl antimony with SbX_3 (X= Cl,Br,I) results in the formation of Cp_2SbX and $CpSbX_2$. The Sb-121 data for the organometallic products have been obtained and the isomer shift data interpreted in terms of the ligand's electronegativity and π -donating abilities.[578]

Tellurium (Te-125). The Te-125 and I-129 Moessbauer spectra have been obtained for both the tetragonal and orthorhombic forms of tellurium(IV) oxide. The Te-125 spectra were obtained in the emission mode using CuI(I-125) as source. The shift and splitting parameters of the orthorhombic and tetragonal oxide are similar with the hyperfine parameters: IS = 0.73 $mm.s^{-1}$ (relative to ZnTe), QS = 6.48 $mm.s^{-1}$; and IS = 0.70 $mm.s^{-1}$ (from ZnTe) and QS = 6.80 $mm.s^{-1}$ respectively.[579]

Tellurium-125 and I-129 Moessbauer data have also been combined in a study of Te-Se mixed chains.[580] The quadrupole splittings obtained from ^{125}Te experiments are larger for isolated chains than the trigonal phase, suggesting that the anisotropy of Te bonding is increased by reducing the inter-chain coupling between adjacent chains. Both the Te-125 and I-129 data suggest that the molecular character of the amorphous phase is much stronger than in the trigonal phase. The Te-125 effect has also been used to study the annealing behaviour of amorphous Te_xC_{1-x} films prepared by radio-frequency sputtering.[581] The results suggest that the interaction between Te chains in the amorphous clusters is weak compared to that in crystalline tellurium and that it decreases with decreasing tellurium content.

A correlation has been observed between the lack of electronic activation of tellurium-doped GaAs at high doping levels and the development of a quadrupole splitting in the Te-125 gamma resonance spectrum.[582] The authors tentatively suggest that the quadrupole split spectrum arises from the DX-centre.

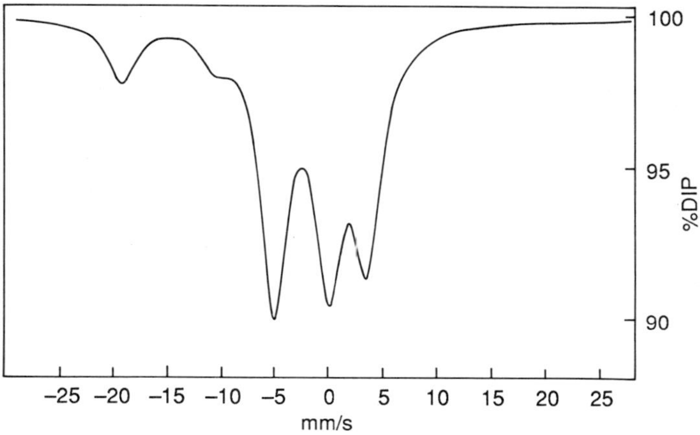

Figure 1 Moessbauer spectrum of $I(OOCCF_3)_3$, at 4.2 K relative to $K^{127}I$. (Based on data from reference 583).

Iodine (I-127, and I-129). The I-127 Moessbauer spectra have been recorded at 4.2 K for a series of tris(carboxylato)iodine(III) derivatives using a ^{127m}Te-containing Mg_3TeO_6 source.[583] The I-127 spectrum of $I(CF_3COO)_3$ (**Figure 1**) shows a well resolved line at high negative velocities that arise from the 5/2 ground state to 7/2 excited state transition. The spectra of the other compounds in the series are similar indicating that they all have a positive value for the quadrupole coupling constant as would be expected for an iodine atom surrounded by three equatorial carboxylate groups, and two axial non-bonding electron pairs. The quadrupole splitting parameters are sensitive to the number of carboxylate groups bonded to iodine. For example, when two carboxylate groups in $I(CH_3COO)_3$ are replaced by two organic groups with I-C bonds the value of the quadrupole coupling constant drops from about +3400 to about +1900 MHz. The values of the chemical isomer shifts from the tris(carboxylato) iodine(III) compounds lie in the range -1.60 to 11.83 $mm.s^{-1}$ from KI and are the lowest yet recorded for iodine(III) compounds. This suggests that the I-O bonding involves using predominantly p-electrons from the iodine with very little or no s-electron participation.

I-129 Moessbauer data have been used in conjunction with Te-125 data in studies of TeO_4,[579] Te-Se chain compounds,[580] and Te-C alloys.[581] The I-129 emission spectra for the orthorhombic and tetragonal forms of TeO_2 are similar suggesting, like the Te-125 data, that the Te sites are very similar in both lattices.

The structural characteristics of crystalline and vitreous phases in the MgO-TeO_2 system have also been studied by I-129 emission Moessbauer spectroscopy.[584] The spectra consist of eight clearly resolved lines from the quadrupole splitting and both the isomer shifts and the quadrupole coupling constants decrease with increasing MgO content. This is interpreted as an increasing transfer of p-electron density to the tellurium atom.

Caesium (Cs-133). The isomer shifts data from the Cs-133 daughter isotope from Xe-133 implanted in a number of metal lattices and in diamond and graphite have been measured.[585] Some of the shifts had unusually large values which were attributed to very high local pressure effects at the impurity atom sites. The formation of impurity-vacancy complexes was, however, thought to result in a decrease in the isomer shift.

Transition-Metal Elements. - Nickel (Ni-61). The unusually large magnetic hyperfine fields found at octahedral Ni^{2+} sites in some Ni-Mn spinels in Ni-61 Moessbauer studies have attributed this to supertransferred hyperfine fields.[586]

Zinc (Zn-67). A shift in the zinc-67 Moessbauer resonance line has been observed when a single crystal of ZnO is rotated in the gamma ray beam. Three possible explanations of their observation have been considered: (i) the presence of an electric dipole moment of the photon; (ii) interference effects; and (iii) differences between the quadrupole splitting parameter of the source and the ZnO absorber.[587] A phase modulation theory has also been used to explain the influence of alternating magnetic fields on the gamma resonance of Zn-61 in ZnO.[588]

A high pressure spectrometer has been designed for higher resolution of the 93.3 keV Zn-67 resonance and the spectrometer has been used to study the effects of high pressure (up to 5.8GPa) on zinc metal.[589,590] Major changes in the spectra occur with changes in pressure. The recoil-free fraction increases linearly by a factor of three at 5.8 GPa while the shift decreases linearly with applied pressure. The quadrupole interaction also varies with pressure because the reduction in the c/a ratio of the zinc unit cell with pressure leads to a decrease in the electric field gradient at the zinc nucleus which is partly compensated for by an increase caused by the reduction in volume.

The sharp line of Zn-67 is very sensitive to small disturbances and the observation of highly anisotropic recoilless fractions along the a-axis and the ab plane in a zinc crystal at 4.2, 20.8 on 47 K have been explained using a dynamical model. The experimental data for the temperature dependent second order Doppler shift and specific heat are found to be in good agreement with the values calculated from the model.[591]

Time domain data for the nuclear Zeeman effect have been obtained from Zn-67 experiments.[592] The results are said to be in agreement with a theory for magnetic phase modulation. Experimental data from Zn-67 studies are also included in a review of the temporal aspects of Moessbauer resonance.[593]

Ruthenium (Ru-99). The electronic structures of the ligand bridged binuclear ions $[(NH_3)_5Ru\text{-pyz-}Ru(NH_3)_5]^{n+}$ (n = 4-6; pyz = pyrazine) and $[(NH_3)_5Ru\text{-bqd-}Ru(NH_3)_5]a^{m+}$ (M = 4,5; bqd = benzoquinone dimine) have been calculated from self consistent charge extended Hueckel calculations. The Ru-99 shift and quadrupole splitting parameters for the ions were shown to be in good agreement with the values calculated from the model.[594]

Ruthenium-99 and iron-57 Moessbauer measurements have been used to clarify the magnetic properties of the Heusler alloy $RuFe_2Si$. The Ru-99 spectra suggested that although Ru atoms in the alloy were substituted by Fe[B] preferentially they were also substituted by Fe[A,C].[595]

Tungsten (W-183). The ^{183}W Moessbauer gamma radius was used in scattering studies of the Debye-Waller factor for sodium over the temperature range 80-295K.[596]

Iridium (Ir-193). Three papers published during the review year are concerned with studies on supported iridium catalysts. Silica-supported Fe-Ir catalysts have, for example, been studied with both the ^{57}Fe and ^{193}Ir effects following hydrogen pretreatment.[449] Conventional Ir-193 absorption experiments have been carried out on Ir and Pt in supported catalysts.[597,598] Use of Moessbauer studies along with EXAFS data showed that there is a preference of Ir for reduction in the catalysts.[598]

Gold (Au-197). Au-197 emission studies with the 77 keV gamma-rays as source have been used to study the platinum in silica supported catalysts for automotive catalytic converters.[597]

The Au(I) and Au(III) sites in $Cs_2Au_2X_6$ (X = Cl, Br, I) were studied with the ^{197}Au Moessbauer effect. The differences between the electronic environments of the two Au sites decrease as charge transfer interactions increase from the chloride to the iodide.[599]

The composition of gold species absorbed on to activated carbon was studied by Moessbauer spectroscopy.[600] The species $Au(CN)_2^-$ was identified on carbon contacted with typical Au/Pb solutions. Comparison of gold cyanide species on activated carbon with the spectra of the solid crystalline phases $M[Au(CN)]_2$ (M = Na, K), $Ca[Au(CN)_2]_2$ and $Gd[Au(CN)_2]_3$ showed that the adsorbed species had increased isomer shifts and quadrupole splittings and more symmetric spectra than the crystalline phases.[601]

The transformation of chemically bound gold into metallic gold during the roasting of an arsenical gold ore has been studied quantitatively.[294] The bound gold was converted into metal in parallel with the decomposition of FeAsS and an increase in cyanide leachability, with the refractory nature of the ore caused by gold chemically bonded to FeAsS. The ratio of the Lamb-Moessbauer *f*-factors of gold bound in the FeAsS component were determined to be f(Au:FeAsS)/f(Au) = 1.48.

The interaction of secondary phosphines with a variety of Au(I) compounds has been studied by a number of techniques, including ^{197}Au Moessbauer spectroscopy. In the presence of bases or polar solvents gold(I) phosphide, $[Au(\mu\text{-}PR_2)]_n$, compounds are formed. The Moessbauer parameters (**Table 4**) are consistent with the formation of P-Au-P coordinations.[602] The data for complexes of the type $[AuX(PHR_6)]$ (where X = Cl or Br; R = Ph or p-tolyl) similarly were found to be consistent with P-Au-X bonding and to have isomer shift and quadrupole splitting values of about 3.5 and 7.00 $mm.s^{-1}$ respectively. The X-ray structure and Moessbauer spectrum of $Fdpp(AuCl)_2$ (Fdpp = bis(diphenylphosphino)ferrocene) has been measured.[241] The Moessbauer data showed an asymmetric doublet of QS = 6.93 $mm.s^{-1}$ at IS = 3.81 $mm.s^{-1}$ (relative to Au metal) for a nearly linear P-Au-Cl linkage with typical bond distances.

Table 4 Au-197 Moessbauer data for $[Au(\mu\text{-}PR_2)]_n$ compounds.[602]

R-	IS (mm/s)	QS (mm/s)
Ph-	4.83	8.50
$p\text{-}MeC_6H_4\text{-}$	4.93	8.55
$p\text{-}(t\text{-}Bu)C_6H_4\text{-}$	4.95	8.61
Et-	5.23	9.02
n-Oct-	5.33	9.59

A number of papers published in the past year have described gamma resonance studies on gold-containing clusters.[603,604] Gold-197 and specific heat measurements were carried out on the polynuclear cluster compounds $[Au_9(PPh_3)_8](NO3)_3$, $[Au_{11}(PPh_3)_7](SCN)_2$ and $Au_{55}(PPh_3)_{12}Cl_6$. The Moessbauer data were used to provide information on the vibrational motions of the distinguishable gold sites in these clusters.

The Au-197 data for the $AuPt_3$ and Au_2Pt_2 in $[AuPt_3(\mu\text{-}CO)_3L_4]^+$, $[Au_2Pt_2L_4(CNR)_4]^{2+}$ species (where L = $P(C_6H_{11})$ or $P(p\text{-}FC_6H_4)_3$ and R = xylyl) have been obtained.[605] For the $AuPt_3$ clusters, the Moessbauer parameters fall within the band of values associated with linear gold(I) complexes, suggesting that the gold is <u>sp</u>-hybridized and that it is contributing only a single valence electron to the $AuPt_3$ cluster. Although the gold atoms in Au_2Pt_2 clusters are structurally not equivalent, there is no evidence for this in their Moessbauer spectra. Separate resonance peaks have, however, been observed for the non-equivalent gold atoms in the cluster in $[Au_6(P(C_6H_{11})Ph_2)_6]^{2+}$ with shift and splitting values of 3.58 and 8.39 $mm.s^{-1}$ and 3.23 and 6.20 $mm.s^{-1}$ respectively.[605] The nitrate of the centred crown cluster ion $[Au_9L_9]^{3+}$ (where L is

$P(C_6H_4OMe\text{-}p)_2)$ has a shift of 3.52 $mm.s^{-1}$ (from gold metal) and a splitting of 6.95 $mm.s^{-1}$, values that are slightly greater than its isohedral fragment analogues and consistent with the observed shortening of Au-Au distances. No separate signal is, however, resolved for the central gold atom. P-31 NMR and Au-197 Moessbauer data show that the $PtAu_8$ cluster in $PtAu_8(PPh_3)_8(NO_3)_2$ is a central crown with Pt in the centre and the eight Au atoms in the crown.[606] The Moessbauer data are also consistent with Pt being the central atom in the cluster $[Pt(H)(PPh_3)(AuPPh_3)_7]^{2+}$.[607]

The Moessbauer spectra of ^{197}Au in CsAu at 4.2K and at pressure of up to 40 kbar show that a phase transition from the CsCl-type lattice to a structure of lower symmetry occurs. The spectral changes at the transition indicate a decrease in electron density on Au, the presence of electric field gradient in the high pressure phase and an increase in the Debye temperature.[608]

Lanthanide and Actinide Elements.- During the review period papers have been published on Moessbauer effect studies using isotopes of neodymium, europium, gadolinium, dysprosium, erbium, thulium, ytterbium, and neptunium.

Neodymium (Nd-145). The 72.5 keV Neodymium-145 Moessbauer resonance has been used to study the paramagnetic relaxation of Nd^{3+} in glasses.[609] The spectra were strongly dependent on temperature and and the Nd^{3+} concentration.

Europium (Eu-151). A value of -87 for the lattice Sternheimer factor for Eu-151 in $Eu(OH)_3$ was obtained from Moessbauer spectra obtained over the temperature range 4.2-300K.[610] The Moessbauer parameters for $EuPO_4$ have been obtained and the contributions of the valance electrons and the crystal field to the electric field gradient calculated.[611] The mixed metal oxides $LnEuO_3$ (Ln = La, Ce and Pr) have been prepared for the first time. The Eu-151 shifts increase with increasing atomic number of Ln and with decreasing lanthanide contraction unit cell volume and the spectra are fitted to twelve transitions.[612,613] The relationship between Eu-151 shifts and unit cell volume and of the compounds $LnEuO_3$ (Ln = Gd, Tb, Dy, Ho, Er, Tm, Yb and Lu) has also been discussed.[614]

A number of papers published during the review year have dealt with the ^{151}Eu effect in superconducting europium barium copper oxide ceramic phases.[615-620] The temperature dependence of the isomer shift and quadrupole splitting parameters has been used to show that the contribution of the excited electronic state of Eu^{3+} to the parameter is greater in the superconducting phase than in the semiconducting phase of oxygen deficient $EuBa_2Cu_3O_{7-x}$.[615]

The gamma-resonance parameters for the phases of composition $Eu(Ba_{2-x}Eu_x)Cu_3O_{7-y}$, with values of x = 0-0.5, show that all of the Eu atoms are present in the trivalent non-magnetic state and that there is very little difference between the electronic structures of Eu^{3+} ions in trivalent metal ion sites and Eu^{3+} substituting at Ba^{2+} sites. The decrease in T_c with increasing values of x has been attributed to structural and electronic effects rather than Eu^{3+} magnetism.[616] Other authors have also shown that the Eu-151 spectra for the superconducting $Ba_2EuCu_3O_y$ phases at temperatures between 4.2 and 350 K are in the range expected for Eu^{3+} with no magnetic splitting.[617] In a study of the electrical-quadrupole interactions at the europium site in $EuBa_2Cu_3O_{7-y}$ Dunlap et al found a small negative coupling constant at both 4.2 and 300 K. This observation is at variance with previous theoretical predictions and the reasons for the failure of these point charge models applied to high T_c superconductors is discussed.[618] A negative value (-4.46 $mm.s^{-1}$) has also been found for the electric field gradient from a gamma resonance study on magnetically aligned samples of $EuBa_2Cu_3O_{7-x}$. The main axis of the e.f.g. tensor was found to be parallel to the c-axis of the crystal.[619]

The temperature dependence of the Moessbauer effect probability of $EuBa_2Cu_3O_7$ has been used to show that there are no anomalies in the vibrations of the Eu^{3+} ions in the lattice over the temperature range 10-300 K.[620]

The ^{119}Sn and ^{151}Eu Moessbauer spectra of the tin-doped phases $EuBa_2(Cu_{1-x}Sn_x)O_{7-y}$ (x = 0.05, 0.10 and 0.20) show the presence of at least two Sn(IV) sites but that all the Eu was present as Eu^{3+}. Some observed temperature dependences observed in both the Eu-151 and Sn-119 parameters were found to be associated with the presence of different Sn sites and with a low temperature phase change.[621]

Cubic and rhombohedral phases have been found in the oxide ion conductor system $(Bi_2O_3)_{1-x}(Eu_2O_3)_x$ for x = 0.3 and x = 0.2 respectively. The cubic phase transforms to the rhombohedral phase with applied pressure of 4 GPa at 873 K, but at 4 GPa and 1073 K the final phase is monoclinic. The cubic, rhombohedral and monoclinic phases have Eu-151 Moessbauer shifts of 0.86, 0.76, and 0.74 $mm.s^{-1}$ with linewidths of 2.57, 2.52 and 2.97 $mm.s^{-1}$ respectively. These linewidths are narrower than that for Eu_2O_3 (3.24 $mm.s^{-1}$) suggesting that the Eu atoms in the europium bismuth oxide phases have a single well defined local environment.[622].

Europium-zirconium selenide intercalates ($Eu_xZrSe_{1.95}$) have been prepared and characterized.[623] The Eu is located in octahedral sites in the Van der Waals's gap and its Moessbauer parameters show that it is less oxidized than in similar compounds with TiS_2. Moessbauer spectroscopy has also been used to show that about 40% of the Eu^{3+} ions in a europium oxide-silica supported Pt catalyst can be reduced to Eu^{2+} at 773 K.[624]

Moessbauer data along with magnetic and L_m edge measurements on amorphous Eu_xRb_{1-x} alloys show that an in homogeneous valence admixture occurs in the disordered phase. For x = 0.64 the isomer shift is characteristic of Eu(II) but for lower values of x, a second peak due to Eu(III) appears in the spectra and when x drops to 0.33 nearly all of the Eu is present in the trivalent state.[625] The Eu-151 spectra of EuPtP are complex with several temperature-dependent resonance lines indicating a complex europium valency behaviour.[626] Valence changes for Eu in $[Eu(Ir_{1-x}Pd_x)_2Si_2]$ and $[EuCo_2Si_{2-x}Ge_x]$ have also been detected by the Eu-151 effect. For the former, the Moessbauer data suggest a homogenous mixed-valence system with the room temperature europium valency decreasing as x increases from 0-0.125. At lower temperatures, 48, 54, and 78 K, a sharp continuous valence transition from Eu^{2+} to Eu^{3+} is detected for x = 0.75, 0.81 and 0.94 respectively.[627] The Moessbauer data for europium in $EuCo_2Si_2$ and $EuCo_2Ge_2$ are consistent with Eu(III) in the silicide and Eu(II) in the germanide. In the mixed system $EuCo_2Si_{2-x}Ge_x$ separate narrow Moessbauer lines are found for the Eu(III) and Eu(II) sites. The relative areas of the two lines for $EuCoSi_{1.5}$ are found to be strongly temperature and pressure dependent giving values of the Moessbauer Debye temperature of 175 K for the Eu(II) site and 250 K for the Eu(III) site.[628] Moessbauer shift data also suggest a change in valence of Eu in $Eu_2Ni_3Si_5$ from 2.6 at 300 K, to 2.7 at 4.2 K.[629]

A number of gamma resonance studies have been carried out on various phases doped with europium. Conversion electron ^{151}Eu data suggest that Eu doped Si has the Eu impurity in proper lattice sites.[630] The valence state of Eu-doped Ga_2S_3 and Ga_2Se_3 is 2+ according to the Moessbauer spectra of the doped species,[631] while the reduction of Eu(III) in beta -Al_2O_3 has been studied at 295 K by the Moessbauer effect.[632] Both Eu-151 Moessbauer spectroscopy and luminescence measurements show the presence of both Eu^{2+} and Eu^{3+} in the phosphors $SrAl_{12}O_{19}$:Eu.[633] The linewidth of the Eu^{3+} resonance decreases with increasing Eu content of the phosphor and this has been explained in terms of spin-spin relaxation, while the maximum in the Eu^{3+} line width at 25% Eu content is accounted for by the presence of non-equivalent Eu^{3+} sites.

Gadolinium (Gd-155). The temperature dependence of the hyperfine interactions at the Gd nuclei in $Gd_2Fe_{14}B$ have been studied using the 86.5 keV Moessbauer resonance.[634] There are two Gd sites in this boride and the results suggest that the Fe-Gd exchange interactions at the two sites may be different. The mixed borides $(Gd_{0.8}M_{0.2})_{1-x}Fe_4B_4$, where M = Gd, Nd and Sm, were also studied. Their spectra indicate a large quadrupole interaction but at 4.2 K they are very different because of differences in the direction of the preferred moment for Gd imposed by lanthanide substitution.[635] The Gd Moessbauer spectra of $GdFe_{10.8}M_{1.2}$ (M = Ti and W) have been obtained at 4.2 K and the contributions of the core and conduction electrons to the hyperfine field calculated.[636] Crystal field effects in the phases GdM_2Si_2, where M is a transition metal, have been studied by measurement of Gd-155 quadrupole splittings.[637]

The local magnetic and structural properties of $LnBa_2Cu_3O_{7-y}$ (Ln = Gd, Dy, Eu) superconductors have been studied with the Moessbauer resonances of ^{155}Gd, ^{161}Dy and ^{151}Eu, and ^{57}Fe.[379] In the Gd compound different magnetic ordering behaviours of the Gd-sublattice were found in the orthorhombic (superconducting) and tetragonal (non-superconducting) phases. The electric quadrupole splittings observed for the Gd and Dy systems were discussed in connection with the lattice e.f.g.'s derived for the Gd system.

Dysprosium (Dy-161). Superconducting and non-superconducting phases of $DyBa_2Cu_3O_{7-x}$ have been examined by ^{161}Dy Moessbauer spectroscopy at temperatures above 0.05 K.[638] No evidence is found for any coupling of the two Dy sites present by a conduction band. In the same compound, the magnetic moments of the crystal electric field ground states in the orthorhombic and tetragonal phases were derived from paramagnetic hyperfine splittings at 1.4 K.

A complete modulated spin structure has been suggested for DyM_2Si_2 (M = Ru, Rh) compounds on the basis of their Moessbauer spectra.[639,640] The quadrupole interactions found in the spectra of $DyRu_2X_2$ (X = Si, Ge) are said to be in agreement with e.f.g calculations.[641] Dy-161 data have also been used in studies of: (i) magnetic and crystal field interactions in dysprosium iron carbides,[642] and $DyNi_5$;[643] (ii) hyperfine interactions in Dy-Fe intermetallics.[644]

Erbium (Er-166). Erbium-166 Moessbauer spectra have been reported for $ErNi_5$,[643] for ErM_2Si_2 (M = Ru, Rh)[639] and Er_2Fe_{14}.[642] Erbium Moessbauer effect data have been collected for the high T_c superconductor $ErBa_2Cu_3O_{7-x}$ phase and for its oxygen-deficient non-superconducting tetragonal form down to 0.05 K.[645] The spectra provide evidence that magnetic correlations are still present in the low-temperature paramagnetic Er^{3+} region.

Thulium (Tm-169). The temperature dependence of the quadrupole splittings in $TmBa_2Cu_3O_x$ (x = 6.6 and 6.9) has been interpreted in terms of crystal field effects rather than to changes occurring at T_c.[646] The nature of the ^{169}Tm Moessbauer spectrum of thulium pyrogermanate has been predicted from calculations from hyperfine field splitting data obtained from magnetic susceptibility measurements.[647] The Tm-169 spectra of an amorphous 40% Tm-Ag alloy, collected over the temperature range 1.2 K to ambient, consist of two components viz. a magnetically split resonance line and a non-magnetic central doublet that are consistent with a short-range-order model.[648] At 1.4 K the thulium Moessbauer spectra for Tm_2Al consists of two six-line, magnetically split, resonances with differing magnetic and quadrupole hyperfine parameters. As the temperature increases the spectrum changes until at 4.2 K the weaker of the two sub-spectra has collapsed to a single central peak. At even higher temperatures the intensity of the remaining six-line component weakens but retains a nearly temperature independent magnetic splitting. These observations are consistent with the presence of two Tm sites that do not interact to any extent.[649] Tm-169 data have also been reported for $Tm_2Fe_{14}C_{75}$ and TmM_5 (M = Ni and Co).[643]

Ytterbium (Yb-170 and 174). Comparisons have been made between calculations of the temperature dependence of the 4*f* electron quadrupole moment and the ytterbium Moessbauer data for $YbCu_2Si_2$. The results were found to lead to an identification of characteristic parameters for the material,[650] and the data for the material have been explained in terms of an inter-configuration fluctuation model.[651]

Ytterbium-170 data have been reported for YbSb,[652] YbX (X = P, As, Sb),[653] and $LaBe_{13}$.[654] Two low temperature phase transitions were detected in YbSb, a phase which orders antiferromagnetically below T_N = 0.32 K, and a second, uncharacterized, phase existing between 0.32 and 5 K.

Values for crystal-field parameters and ytterbium-exchange energies have been derived by obtaining a simultaneous fit of the Yb-174 spectra of $Yb_2Fe_{14}B$ to five sub-spectra.[655] The two ytterbium sites in Yb_2Fe_{17} were also studied by Yb-174 Moessbauer spectroscopy.[656]

Neptunium (Np-237). The X-ray absorption near-edge structure of neptunium compounds with the formal valences III+ to VI+ were studied.[657] For non-metallic Np compounds the weighted mean values of the L_{III}-XANES shifts show a clear correlation with the isomer shifts of the ^{237}NP Moessbauer resonance. The Moessbauer emission spectra of $^{241}AmO_2$ dissolved in PuO_2 have been recorded at 77-296 K.[658] Both Np(IV) and Np(V) species were identified. Covalent effects in neptunium compounds have been correlated with the ^{237}Np isomer shifts.[659] For neptunates, the 5*f*-orbital populations decrease, and the 6*d* populations increase with increasing valency. The 5*f*-orbital populations are higher in neptunyl compounds than in neptunates.

A Moessbauer study of the molecular structure of the organometallic Np compounds Cp_4Np, CpNpX, $Cp_3Np^+X_2$, $Cp_3Np.CC.NpCp_3$, $(COT)_2Np$ (COT = 1,5-cyclooctatetraene) and $Cp_3Np.Z$ (Z = THF, Et_2O) has been described.[660,661] The spectra are strongly affected by paramagnetic relaxation effects, and the data was fitted using an anisotropic paramagnetic relaxation model. In general, distortions from axially symmetric hyperfine parameters were observed, no magnetic ordering was found above 1.5 K, and anisotropic relaxation frequencies were needed to fit the spectra.

High pressure Moessbauer spectroscopy was used to study the magnetic properties of $NpOs_2$, $NpAl_2$, $NpCo_2Si_2$, NpAs, and $NpSn_3$.[662] Behaviour ranging from band-like electron behaviour to highly localized 5*f* electron magnetism was seen. In the heavy-fermion intermetallics $NpBe_{12}$, $NpRu_2$, $NpIr_2$, and $NpOs_2$ the bulk and local magnetic susceptibilities were equal and proportional to the Moessbauer isomer shift.[663] Quasirelativistic MO calculations and gamma-resonance data were used to investigate the two neptunium sites in $NpPu_xPu_{4-x}$.[664,665] In alloys with 50-90 atom% Pu, the quadrupole splittings, 30 and 35 $mm.s^{-1}$ with asymmetry parameters 0.22 and 0.29 respectively, were almost independent of the Pu concentration. In the boride $Np_{1-x}Pu_xB_2$ the isomer shift was almost constant at -14.5 $mm.s^{-1}$ (relative to $NpAl_2$), consistent with Np(IV).[666] A large quadrupole interaction, observed at high temperatures, decreases slowly with the Np concentration, while magnetic patterns appear at low temperatures for $x \leq 0.5$.

7. Backscatter and Conversion Electron Moessbauer Spectroscopy.

As in previous years, this section presents a summary of the developments in methodology and instrumentation followed by descriptions of the recent applications of the technique in studies of chemical compounds.

A spectrometer with an electrostatic cylindrical mirror analyser for depth selective conversion electron Moessbauer spectroscopy was constructed.[667] The use of the spectrometer for depth selective investigations of materials containing ^{57}Fe in natural abundance is described. A number of publications have been concerned with developments in the counting of conversion electrons describing various high-temperature proportional counters,[668,669] gas-flow proportional and gas-filled cascade detectors,[670] and an optimized scintillation counter.[671] The use of ^{57}Fe surface enrichment to improve the sensitivity of conversion X-ray Moessbauer spectroscopy (CXMS) has also been described.[672]

The intensity of backscattered Moessbauer gamma rays has been calculated in total geometry and the contribution of resonance, non-resonance, elastic and inelastic scattering to the spectra discussed.[673] The calculations show that problems involved in the geometry of total back scattering can be solved when proportional counters recorded only gamma radiation. Simulations of electronic relaxation following nuclear decay of ^{57}Fe have been carried out to determine the mechanisms of the generation of electrons of energy less than 15 eV that have been detected in CEMS.[674,675] The results indicate that low energy Auger transitions and shake-off events account for about 20% of the total resonant CEMS signal.[674] These low energy (less than 15 eV) electrons can be used to probe the top 5 nm of a surface while the high energy (7.3 keV) electrons generated can probe regions as deep as 300 nm and together they can provide a valuable method for non-destructive depth profiling.[675] In ultra-high vacuum applications of depth selective CEMS or integral CEMS optimization of experimental performance is of the utmost importance and attempts have been made to characterize the statistical quality of the experiments.[676] The use of in situ CEMS with a parallel plate avalanche counter for determining the layer thicknesses of thin films has been described,[677] as has combined backscatter conversion electron and backscatter photon Moessbauer spectroscopy for depth selective studies on a wide range of thin films.[678]

Iron-57. - Implantation Studies and Films. The properties of the surface of single crystals of Fe_3BO_6 at different temperatures have been studied by conversion electron methods.[669] The effects of ion-beam-induced atomic mixing and of subsequent thermal transformations at the Fe:SiO_2 interface were studied with CEMS and small-angle X-ray diffraction measurements. The ion-beam sample in the as-mixed state was shown to contain Fe_7SiO_{10} and some Fe-Si-O complexes but annealing the sample for several hours at 450° leads to an increase in the Fe(II):Fe(III) ratio.[679] CEMS has also been used to study the effects of iron implantation into ZrO_2.[680] At 0.1 atom% levels the iron atoms are present in substituted Zr^{4+} sites while at 1-10% levels the iron atoms form Fe^{2+}-(O-vacancy) complexes.

The ^{57}Fe conversion electron technique was used to determine the relative concentrations of epsilon-Fe_2N, epsilon-Fe_3N and gamma'-Fe_4N in iron samples subjected to N_2^+ implantation. The results were correlated with the nitrogen diffusion process deduced from the nuclear profile taken on the same samples.[681] In another study,[682] it was shown that alpha-martensite is formed at low N implantation doses but that both epsilon-Fe_3N and epsilon-$Fe_{2+n}N$ are formed at higher doses. In further investigations, the technique was used to study both N and P implantation,[683] and the N and Ar implantation,[684] of iron. CEMS was also been used to study iron implanted alloys, for example, it has been used in studies of the iron nitrides formed in N implanted austenitic steel,[685] and low carbon steel.[686]

CEMS and a superconducting quantum interference device have been used to measure the magnetic hyperfine field and the spontaneous magnetization of ultrathin Fe(110) films grown epitaxially on Ag(111) substrate films.[687] The ground state hyperfine field at the Fe/Ag interface is enhanced relative to the bulk.

Conversion electron studies on amorphous Metglas films of the type $Fe_{78}B_{11}Si_9C_2$[667] and $Fe_{78}B_{13}Si_9$/Si[688] have been reported. The results of work on the thin films: Fe-Al,[689-691] $Fe_{60}Mg_{40}$,[692] $Fe_{60}Co_{40}$,[693] Fe-Co,[694] Fe-Tb,[695] Fe on Ni,[696,697] and Fe on Ag[698] were also described.

Corrosion Products and Alloys. The use of the conversion electron technique to study the corrosion of iron foil was the subject of four publications in the review period.[678,699-701] In one of the publications,[701] the technique was used to study the corrosion processes on iron and steel and the method of inhibition of the corrosion. The combined use of CEMS with other surface sensitive techniques has shown: (i) that the corrosion resistance of the film formed on some Si-containing stainless steels after nitric acid treatment contains no iron oxides but only SiO_2; (ii) that Fe-3%Si transformer sheet steel after thermal pretreatment is covered with a layer of iron oxides containing large amounts of Fe_2SiO_4 and alpha-Fe containing SiO_2 just under the metal surface; and (iii) that the passivation layers arising from sulphate or phosphate treatments rely for their effectiveness on a very thin layer of a highly disordered structure which provides the transition from the cubic to orthorhombic phases. The corrosion of a mild steel has been followed after surface enrichment with ^{57}Fe and coating with protective compounds.[672]

Conversion X-ray Moessbauer and back scattering techniques have been used to examine the carbide phases present in the surface layers of various grades of steel: including tool steels,[702,703] thermomechanically treated spring steels,[704] high manganese steels,[705] and Ti-implanted high-speed steels.[706]

CEMS has been used to show that there is no hydrogen-rich phase at the surface of the cubic alpha-hydride phase, $(Fe_{0.8}Ni_{0.2})TiH_x$, and that the hydrogen atoms are uniformly distributed through the bulk material.[707] Other hydride phases studied by conversion electron Moessbauer methods include Fe-Ti hydrides and Pd-coated Fe-Ti hydrides.[678] Surface studies have also been carried out on a number of bulk alloys including Fe-29.5%Ni,[708] amorphous $Fe_{89}Zr_{11}$,[709] and $Fe_{40}Ni_{40}Pl_4B_6$ alloys.[710]

Compounds of Iron. A number of publications during the review year have been concerned with studies of various iron oxide phases on a number of substrates. Reactions induced by a field-generated plasma at liquid/solid interfaces have been used to synthesize a metastable form of iron oxide which was characterized by electron microscopy and CEMS.[711] Films of a number of iron oxides including FeO, Fe_3O_4 and alpha-Fe_2O_3 have been deposited on the (0001) face of a sapphire crystal by vapour deposition and characterized by X-ray diffraction and CEMS.[712] The Moessbauer experiments showed that the as-grown films had good stoichiometry and the Verwey transition was clearly detected by both CEMS and resistivity measurements down to 77 K. To study the lattice mismatch between Fe_3O_4 and the substrate, which resulted in columnar oxide growth, the surface was enriched with Fe-57 at the deepest layers. CEMS was also used to study iron oxide and ferrite layers deposited on

Al_2O_3 by pulsed ruby laser evaporation and it was found that the stoichiometry and microstructure of the films depended upon the deposition conditions and particularly on the partial oxygen pressure.[713]

Conversion electron Moessbauer spectroscopy was used to study the surface reaction between pyrites and diethyldithiocarbamate as a model for floatation of non-ferrous sulphide minerals. The interaction with the dithiocarbamate was found to occur on about 20% of the total pyrite surface via Fe(III) bonding.[714] The effects of iron overlayers on MoS_2 have also been studied.[678] CEMS has been used to follow the interaction between HF and iron oxides on an iron surface in the presence and absence of oxygen and water.[715] During the initial stages of the reaction with moist 40% HF, the predominant product is $FeF_3.3H_2O$ which is partly reduced by the substrate to $FeF_2.4H_2O$. Both Fe_3O_4 and alpha-Fe_2O_3 are reduced directly to $FeF_2.H_2O$ by reaction with the HF. During the reaction with dry HF vapours, the reduction by the substrate does not occur to the same extent. Back-scatter Moessbauer spectroscopy has also been used, along with other techniques, to study (2-formyl-1-chlorovinyl)ferrocene film electrodes prepared by electrochemical polymerization.[716]

<u>Other Elements.</u> As in previous years, very few results have been reported on conversion electron studies using isotopes other than Fe-57. There are reports on the use of Sn-119, Au-197 and Eu-151.

The Sn-119 Moessbauer effect was used to study the effects of Xe^+ irradiation on Fe-Sn bilayers[717] and on Si-Sn multilayers.[718] Both Au L_3 EXAFS and Au-197 CEMS data have been used to carry out depth profile studies on gold films on Si(111) single crystal faces.[719] CEMS with Eu-151 has also been used to show that the impurity atoms in Eu-doped SiO_2 are in normal substitutional sites.[630]

REFERENCES

1. "Moessbauer Effect Reference and Data Journal", ed. J.G. Stevens, V.E. Stevens, R.W. White, and J.L. Gibson, Moessbauer Effect Data Center, Univ. N. Carolina, Ashville, N.C. 28814, 1989, Vol.12.
2. Hyperfine Interact., 1989, 47-48.
3. R. L. Moessbauer, in "George de Hevesy, 1885-1966: Festschr.", G. Marx (ed.), Akad. Kiado, Budapest, 1988.
4. J. J. Bara, Postepy Fiz., 1989, 40, 261.
5. Y. Xia, Hejishu, 1988, 11, 1.
6. I. Czako-Nagy and A. Vertes, Trends Anal. Chem. (Pers. Ed.), 1988, 7, 305.
7. R. W. Gomez Gonzalez, Rev. Mex. Fis., 1988, 34, 442.
8. V. I. Goldanskii, in "George de Hevesy, 1985-1966: Festschr.", G. Marx (ed.), Akad. Kiado, Budapest, 1988.
9. C. Pietzsch, Isotopenpraxis, 1988, 24, 402.
10. S. Nasu, Nippon Oyo Jiki Gakkaishi, 1989, 13, 38.
11. G. Langouche, Hyperfine Interact., 1989, 47-48, 85.
12. F. Aubertin and U. Gonser, Z. Metallovd., 1989, 80, 398.
13. P. C. M. Gubbens, A. M. Van der Kraan, and K. H. J. Buschow, Hyperfine Interact., 1988, 50, 685.
14. K. Krop, Physica B, 1989, 159, 66.
15. B. Fultz and H. H. Hamdeh, Hyperfine Interact., 1988, 45, 55.
16. F. E. Fujita, Mater. Trans., JIM, 1989, 30, 87.
17. T. Nishida, J. Non-Cryst. Solids, 1989, 108, 87.
18. G. Longworth in "Moessbauer Spectrosc. Appl. Inorg. Chem., Volume 2", G. J. Long (ed.), Plenum, New York, 1987.
19. J. Hesse, Hyperfine Interact., 1989, 47-48, 357.
20. A.H. Morrish, Mater. Sci. Eng., 1989, B3, 383.
21. D. Niarchos, Hyperfine Interact., 1988, 49, 367.
22. C.E. Johnson, Hyperfine Interact., 1988, 49, 19.
23. B.D. Dunlap and C.W. Kimball, Hyperfine Interact., 1988, 49, 187.
24. M.F. Thomas in "Moessbauer Spectrosc. Appl. Inorg. Chem., Volume 2", G. J. Long (ed.), Plenum, New York, 1987.
25. S. Moerup, in "Moessbauer Spectrosc. Appl. Inorg. Chem., Volume 2", G. J. Long (ed.), Plenum, New York, 1987.
26. R. Waeppling, Hyperfine Interact., 1989, 47-48, 251.
27. Z.M. Stadnik and G. Stroink, Hyperfine Interact., 1989, 47-48, 275.
28. M. Rosenberg, P. Deppe, and T. Sinnemann, Hyperfine Interact., 1988, 45, 3.
29. L. Asch, J. Chappert, and G.M. Kalvius, Hyperfine Interact., 1988, 42, 1137.
30. F. Aubertin, S. Abel, G. Wagner, P. Schaaf, and U. Gonser, Hyperfine Interact., 1989, 47-48, 379.
31. G. Principi and F. Gauzzi, Metall. Ital., 1988, 80, 511.
32. U. Gonser, C.T. Limbach, and F. Aubertin, J. Non-Cryst. Solids, 1988, 106, 395.
33. A. Vertes, Hyperfine Interact., 1988, 45, 105.
34. A. Vertes and I. Czako-Nagy, Electrochim. Acta, 1989, 34, 721.
35. M. Huang and Y. Chen, Harbin Kexue Jishu Daxue Xuebao, 1988, 72.
36. Y. Zheng, Wuli, 1988, 17, 646.
37. G. Langouche, Hyperfine Interact., 1988, 45, 199.
38. L.A. Nylandsted, J.W. Petersen, and G. Weyer, Mater. Sci. Forum, 1989, 38-41, 1137.
39. G. Weyer, NATO ASI Ser., Ser. E, 1988, 144, 167.
40. H. Frauenfelder, Int. J. Quantum Chem., 1989, 35, 711.

41. V.I. Goldanskii and Yu.F. Krupyanskii, Hyperfine Interact., 1988, 39, 341;Q. Rev. Biophys., 1989, 22, 39.
42. R.B. Frankel, G.C. Papaefthymiou, and G.D. Watt in "Moessbauer Spectrosc. Appl. Inorg. Chem., Volume 2", G. J. Long (ed.), Plenum, New York, 1987.
43. P.G. Debrunner, Phys. Bioinorg. Chem. Ser., 1989, 4, 137.
44. V.I. Khleskov, B.N. Burykin, A.B. Smirnov, Zh. Strukt. Khim., 1989, 30, 148.
45. T. Tominaga, Kokagaku, 1989, 13, 17.
46. H. Toftlund, Coord. Chem. Rev., 1989, 94, 67.
47. P. Adler, A. Hauser, A. Vef, H. Spiering, and P. Guetlich, Hyperfine Interact., 1989, 47-48, 343.
48. D.A. Scherson, D.A. Corrigan, C. Fierro, and R. Carbonio, ACS Symp. Ser., 1988, 378, 257.
49. T. Shinjo and S. Nasu, Springer Ser. Mater. Sci., 1989, 11, 166.
50. R.D. Cannon and R.P. White, Prog. Inorg. Chem., 1988, 36, 195.
51. Y. Masuda and H. Sano, Kikan Kagaku Sosetsu, 1988, 157.
52. D.H. Macartney, Rev. Inorg. Chem., 1988, 9, 101.
53. A. Kaur and G. K. Sandhu, J. Chem. Sci., 1986, 12, 1.
54. F.J. Berry, in "Chem. Org. Selenium, Tellurium Compd., Volume 2", S. Patai (ed.), Wiley, Chichester, 1987.
55. G.M. Kalvius, J. Nucl. Mater., 1989, 166, 5.
56. J.W. Niemantsverdriet, Hyperfine Interact., 1989, 47-48, 219.
57. A. Pattek-Janczyk, Wiss. Z. Friedrich-Schiller-Univ. Jena: Naturwiss. Reihe, 1988, 37, 835.
58. B. S. Clausen and H. Topsoe, Hyperfine Interact., 1989, 47-48, 203.
59. S. Li and L. Cha, Turang (Nanjing), 1988, 20, 218.
60. E. Murad, Hyperfine Interact., 1989, 47-48, 33.
61. B. Lin, Hebei Daxue Xuebao, Ziran Kexueban, 1987, 69.
62. C. Schlenker, NATO ASI Ser., Ser. B, 1987, 155, 477.
63. S. Dattagupta, Hyperfine Interact., 1988, 49, 253.
64. P.E. Toschek, GSI-Rep., GSI-89-10, Proc. Workshop Cryst. Ion Beams, 1989, 259.
65. J. M. Knudsen, Hyperfine Interact., 1989, 47-48, 3.
66. M. Fleischmann, Proc. Robert A. Welch Found. Conf. Chem. Res., 1986, 30, 90.
67. X. Sun and S. Li, Wuli, 1989, 18, 211.
68. P. Boolchand, J. Quant. Spectrosc. Radiat. Transfer, 1988, 40, 777.
69. E. Hartmann and W. Winkler, Isotopenpraxis, 1988, 24, 411.
70. J. Huang, J. Chen, and G. Wang, Yingyong Kexue Xuebao, 1988, 6, 329.
71. F.W.D. Woodhams, J. Phys. E, 1988, 21, 1067.
72. J.F. Lee, M.D. Lee, and P.K. Tseng, Hua Hsueh, 1987, 45, 50.
73. Z. Xu and R. Ma, Wuli Xuebao, 1988, 37, 1949.
74. D.P. Hoffmann, A. Proctor, and D.M. Hercules, Anal. Chem., 1989, 61, 898.
75. M. Miglierini, Nucl. Instrum. Methods Phys. Res., 1989, B36, 475.
76. I.V. Uporov, I.N. Nikolaev, A.R. Panchenko, and K.V. Shaitan, Biol. Nauki, 1989, 25.
77. A.R. Mkrtchyan, A.Z. Atsagortsyan, T.N. Nadzharyan, and N.M. Gasparyan, Izv. Akad. Nauk Arm. SSR. Fiz., 1988, 23, 274.
78. V.I. Yukalov, Solid State Commun., 1989, 69, 393.
79. K.N. Shrivastava, Phys. Status Solidi B, 1989, 153, 547.
80. A. Kumar, M.R. Singh, P.R. Sarma, and K.C. Tripathi, J. Phys. D, 1989, 22, 465.
81. A. Kumar, M.R. Sinh, and K.C. Tripathi, Nucl. Instrum. Methods Phys. Res., 1989, B36, 189.

82. G.R. Hoy and R.D. Taylor, J. Quant. Spectrosc. Radiat. Transfer, 1988, 40, 763.
83. L. Cohen, J. Quant. Spectrosc. Radiat. Transfer, 1988, 40, 735.
84. G.C. Baldwin, J. Quant. Spectrosc. Radiat. Transfer, 1988, 40, 685.
85. Yu.F. Babikova, E.S. Bakanova, N.S. Kolpakov, and K.E. Nilov, Zavod. Lab, 1988, 54, 45.
86. A. Urbaniak-Kucharczyk and Z. Osetek, Phys. Status Solidi B, 1989, 152, 267.
87. P. Adler, P. Poganiuch, and H. Spiering, Hyperfine Interact., 1989, 52, 47.
88. S.L. Popov, G.V. Smirnov, and Yu.V. Shvyd'ko, Pis'ma Zh. Eksp. Teor. Fiz., 1989, 49, 651.
89. L. Tsankov, Phys. Status Solidi A, 1989, 114, 341.
90. A.V. Zatovskii and V. Klishko, Mol. Biol. (Moscow), 1989, 23, 1091.
91. T.A. Orazbaev, D.K. Kaipov, U.Kh. Asylbaev, and E.F. Galyutina, Izv. Akad. Nauk Kaz. SSR, Ser. Fiz.-Mat., 1988, 18.
92. M. Haas, V. Khiahnyakov, E. Realo, and J. Jogi, Phys. Status Solidi B, 1988, 149, 283.
93. D.H. Jones, Hyperfine Interact., 1989, 47-48, 289.
94. O. Ballet, Phys. Chem. Miner., 1989, 16, 331.
95. D.P. Siddons, J.B. Hastings, G. Faigel, J.R. Grover, P.E. Haustein, and L.E. Berman, Rev. Sci. Instrum., 1989, 60, 1649.
96. U. Van Buerck, G.V. Smirnov, and R.L. Moessbauer, J. Phys. C, 1988, 21, 5843.
97. V.N. Gavrilov, E.V. Zolotoyabko, and E.M. Iolin, Izv. Akad. Nauk SSSR, Ser. Fiz., 1988, 52, 1762.
98. G.V. Smirnov, U. Van Buerck, and R.L. Moessbauer, J. Phys. C, 1989, 21, 5835.
99. U. Van Buerck, Hyperfine Interact., 1989, 47-48, 127.
100. F. Cavatorta, A. Deriu, and C. Schiaretti, Hyperfine Interact., 1988, 46, 715.
101. Z.M. Stadnik, H.R. Borsje, A.E.M. Swolfs, W.H.A. Leenders, and J.C. Fuggle, Rev. Sci. Instrum., 1989, 60, 708.
102. M. Alflen, C. Hennen, F. Tuczek, H. Spiering, P. Guetlich, and Z. Kajcsos, Hyperfine Interact., 1989, 47-48, 115.
103. V.G. Labushkin, E.R. Sarkisov, and I. G. Tolpekin, U.S.S.R. Pat., SU 1444657.
104. G. Balestrino, E. Gerdau, M. Grove, R. Hollatz, E. Milani, A. Paoletti, P. Paroli, R. Rueffer, H. D. Ruether, and W. Sturhahn, J. Phys. Colloq., 1988, C8, 2031.
105. V.A. Evdokimov, Yu.G. Kononov, A.S. Lobko, and A.A. Fedorov, Prib. Tekh. Eksp., 1989, 53.
106. G. Alonzo, M. Consiglio, N. Gionfriddo, and G. Bardi, Hyperfine Interact., 1988, 46, 703.
107. L. Fournes, Y. Potin, J.C. Grenier, and P. Hagenmuller, Rev. Phys. Appl., 1989, 24, 463.
108. Y. Isozumi, T. Fujii, R. Katano, and S. Kishimoto, Natl. Lab. High Energy Phys., KEK (Jpn.) KEK-88-5, 1988, 103.
109. Y Isozumi, S. Ito, T. Fujii, and R. Katano, Rev. Sci. Instrum., 1989, 60, 3262; R. Katano, T. Fujii, T. Kobayashi, K. Fukumura, and Y. Isozumi, Nucl. Instrum. Methods Phys. Res., 1988, A280, 285.
110. T.W. Sinor, P.W. Reittinger, and C.B. Collins, Rev. Sci. Instrum., 1989, 60, 3258.
111. M.E. Vakhonin, R.N. Gall, O.A. Gordeev, V.A. Gotlib, S.M. Irkaev, and V. V. Kuptiyanov, CA111:87063e.
112. Yu.A. Kurinyi, V.P. Beskachko, and G.P. Vyatkin, Otkrytiya Izobret., 1989, 186.

113. S.D. Karakishev and Yu.V. Baldokhin, Prib. Tekh. Eksp., 1988, 50.
114. H. Flores-Llamas and H. Jimenez-Dominguez, Rev. Mex. Fis., 1989, 35, 33.
115. S.M. Irkaev, V.V. Kupriyanov, V.A. Semenkin, and M.M. Sokolov, Brit. Pat., GB 2204395.
116. A. Alamino Bouzo, O. Hernandez Gonzalez, D. Izada Rodriquez, and J. Matutes Aguino, Rev. Cubana Fis., 1988, 8, 3.
117. A. Baldini, F. Del Giallo, F. Pieralli, and G. Spina, Nucl. Instrum. Methods Phys. Res., 1989, B42, 393.
118. T. Zemcik, Isotopenpraxis, 1988, 24, 408.
119. D. Cher and V.A. Sarkisyan, Pis'ma Zh. Tekh. Fiz., 1989, 15, 44.
120. Ya.P. Glazkova, Izv. Vyssh. Uchebn. Zaved., Fiz., 1988, 31, 109; 111.
121. G.U. Nienhaus, J. Heinzl, E. Huenges, and F. Parak, Nature, 1989, 338, 665.
122. S.K. Basovets, I.V. Uporov, K.V. Shaitan, Yu.F. Krupyanskii, I.V. Kurinov, I.P. Suzdalev, A.B. Rubin, and V.I. Goldanskii, Hyperfine Interact., 1988, 39, 369.
123. V.I. Goldanskii, T.V. Zhuravleva, Yu.F. Krupianskii, I.V. Kurinov, and I.P. Suzdalev, Fiz. Mnogochastichnykh Sist., 1988, 85.
124. G.U. Nienhaus, H. Hartmann, F. Parak, J. Heinzl, and E. Huenges, Hyperfine Interact., 1989, 47-48, 299.
125. M. Pasternak and R.D. Taylor, Hyperfine Interact., 1989, 47-48, 415.
126. M.L. Crow, G. Schupp, W.B. Yelon, J.G. Mullen, and A. Djedid, Phys. Rev. B, 1989, 39, 909.
127. T.W. Sinor, P.W. Reittinger, and C.B. Collins, Phys. Rev. Lett., 1989, 62, 2547.
128. H. Vucetich, R.C. Mercader, G. Lozano, G. Mindlin, A.R. Lopez Garcia, and J. Desimoni, Phys. Rev. D, 1988, 38, 2930.
129. A. De Rujula and K. Zioutas, Phys. Lett. B, 1989, 217, 354.
130. O. Eriksson and A. Svane, J. Phys.: Condensed Matter, 1989, 1, 1589.
131. I. Dezsi, U. Gonser, and G. Langouche, Phys. Rev. Lett., 1989, 62, 1659.
132. P. Blaha and K. Schwartz, J. Phys. Colloq., 1988, C8, 101.
133. A.A. Hirsch, T. Massalha, and G. Galeczki, J. Magn. Magn. Mater, 1988, 75, 209.
134. J. Eldridge and R.W. Hoffman, J. Electrochem. Soc., 1989, 136, 955.
135. V.V. Igrushin, V.G. Kirichenko, Yu.G. Mashkarov, A.I. Snurnikova, V.V. Chekin, and E.A. Reznichenko, Vopr. At. Nauki Tekh. Ser.: Fiz. Rad. Povrezh. Rad. Material., 1988, 28.
136. O.M. Mikhailik, N.I. Shklovskaya, I.P. Suzdalev, V.K. Imshennik, and A.A. Chuiko, Zh. Prikl. Khim., 1988, 61, 2414.
137. J. Tanaka, K. Nomura, Y. Ujihara, T. Takahashi, and Y. Uchida, Sogo Shikensho Nenpo, 1988, 47, 209.
138. H.X. Lu, X.Y. Mao, Y.W. Du, W. Yu, W.F. Chen, J. Phys. Colloq, 1988, C8, 1839.
139. F. Mouahid, H. Engelmann, I. Dezsi, U. Gonser, and B. Molnar, Phase Transitions, 1988, 13, 73.
140. T. Tomov, H. Engelmann, I. Dezsi, and U. Gonser, Solid State Commun., 1989, 69, 41.

141. H. Engelmann, F. Mouahid, I. Dezsi, B. Molnar, U. Gonser, D. Seibert, J. Dahlem, and F. Tuczek, Appl. Phys. A, 1989, A48, 211.
142. R.I. Zakharchenya and P.P. Seregin, Fiz. Tverd. Tela, 1989, 31, 260.
143. V.D. Khalilev, P.P. Seregin, A.N. Krasnov, and A.A. Agzamov, Fiz. Khim. Stekla, 1988, 14, 774.
144. G.A. Teterin, E.M. Menchuk, F.K. Egorov, E.G. Semin, and T.M. Shkol'nikova, Ukr. Khim. Zh., 1989, 55, 229.
145. R.P. Vardapetyan, S.I. Reiman, Yu.N. Ivanov, E.M. Arutyunyan, and T.V. Sarkisyan, Izv. Akad. Nauk Arm. SSR, Fiz., 1989, 24, 47.
146. V.I. Aleshin, P.B. Fabrichnyi, V.V. Karelin, Izv. Akad. Nauk SSSR, Neorg. Mater., 1988, 24, 1894.
147. Y. Sakai, A. Kageyama, H. Nishioji, and T. Tominaga, Radiochim. Acta, 1989, 46, 73.
148. I. Kulszewicz-Bajer, A. Pron, J. Suwalski, Z. Kucharski, and S. Lefrant, Synth. Met., 1989, 28, C225.
149. I. Kulszewicz-Bajer, A. Pawlicka, J. Plenkiewicz, A. Pron, and S. Lefrant, Synth. Met., 1989, 28, 335.
150. A.S. Brar and G.S Kapur, Hyperfine Interact, 1989, 45, 323.
151. G.S. Kapur and A.S. Brar, J. Radioanal. Nucl. Chem., 1989, 136, 169.
152. D. Feng and B. Liu, Xinan Shifan Daxue Xuebao, Ziran Kexueban, 1988, 58.
153. K. Kh. Chiribova and S.F. Timashov, Makromol. Chem., Macromol. Symp., 1989, 27, 311.
154. E.M. Baggo-Saitovitch, J. Terra, and F.J. Litterst, Phys. Rev. B, 1989, 39, 6403.
155. E.L. Bominaar, J. Guillin, A. Sawaryn, and A.X. Trautwein, Phys. Rev. B, 1989, 39, 72.
156. R.P. Marande, Phys. Rev. B, 1989, 39, 46.
157. M. Devillers, J. Ladriere, and D. Apers, Radiochim. Acta, 1989, 46, 197.
158. Y. Watanabe, M. Nakada, K. Endo, H. Nakahara, and H. Sano, J. Radioanal. Nucl. Chem., 1989, 136, 257.
159. T. Kobayashi, Bull. Chem. Soc. Jpn., 1989, 62, 576.
160. P.D. Battle, T.C. Gibb, and R. Strange, J. Solid State Chem., 1989, 81, 217.
161. S. Jha, C. Mitros, A. Lahamer, S. Yehia, G.M. Julian, R.A. Dunlap, G. Stroink, and Z.M. Stadnik, Hyperfine Interact, 1989, 50, 607.
162. H.D. Lutz, A. Pfitzner, W. Schmidt, E. Riedel, and D. Prick, Z. Naturforsch., A, 1989, 44, 756.
163. R.E. Greeney, C.P. Landee, J.H. Zhang, and W.M. Reiff, Phys. Rev. B, 1989, 39, 2200.
164. H.H.A. Smit, H.J.M. De Groot, M. Elmassalami, R.C. Thiel, and L.J. De Jongh, Physica B, 1989, 154, 237.
165. M. Elmassalami, H.H.A. Smit, R.C. Thiel, and L.J. De Jongh, Physica B, 1989, 154, 267.
166. J. Swanepool, J. Phys. Chem. Solids, 1989, 50, 935.
167. V. Delobbe, J. Chassaing, D. Bizot, M. Quarton, P. Lacorre, Y. Calage, M. Leblanc, and, G. Ferey, J. Magn. Magn. Mater., 1988, 74, 165.
168. J.E. Greedan, K. Reubenbauer, T. Birchall, M. Ehlert, D.R. Corbin, M.A. Subramanian, J. Soild State Chem., 1988, 77, 376.
169. D.S. Kulgawczuk, K. Reubenbauer, and B. Sepiol, J. Radioanal. Nucl. Chem., 1989, 131, 43.

170. V. Valenti, P. Fantucci, F. Cariati, G. Micera, M. Petrera, and N. Burriesci, Inorg. Chim. Acta, 1988, 148, 191.
171. H.E. Tomas, L.A. Morino, and H.R. Rechemberg, Spectrosc. Lett., 1989, 22, 795.
172. J.E. Barclay, G.J. Leigh, A. Houlton, and J. Silver, J. Chem. Soc., Dalton Trans., 1988, 2765.
173. A.M. Bychkov and A.V. Polosin, Dokl. Akad. Nauk SSSR, 1988, 303, 368.
174. L.A. Aliev. R.M. Bagirov, A.Pl. Narimanidze, R.A. Stukan, G.V. Tsintsadze, R.Sh. Kurtanidze, and M.A. Mdivani, Koord. Khim., 1989, 15, 389.
175. G. Favero, U. Russo, M. Vidali, and B. Zarli, Polyhedron, 1988, 7, 2703.
176. L.A. Hall, Polyhedron, 1989, 8, 1335.
177. M. Andruh, G. Filoti, and M. Brezeanu, Rev. Roum. Chim., 1988, 33, 673.
178. I. Dezsi, S.G. Sankar, L.N. Mulay, J.F. Houlihan, and T. Pannaparayil, J. Phys. Colloq., 1988, C8, 1463.
179. J. Teillet, A. Hauet, and J.M. Greneche, J. Phys. Colloq., 1988, C8, 1095.
180. D.H. Jones. J. Chadwick, and M.F. Thomas, J. Phys.: Condensed Matter, 1989, 1, 147.
181. L. Takas and W.M. Reiff, J. Phys. Chem. Solids, 1989, 50, 33.
182. J. Pebler, Inorg. Chem., 1989, 28, 1038.
183. J. Chadwick, M.F. Thomas, C.E. Johnson, and D.H. Jones, J. Phys. C, 1988, 21, 6159.
184. C.S.M. Partiti, H.R. Rechenberg, and J.P. Sanchez, J. Phys. C, 1988, 21, 5825.
185. J. Grannec, A. Yacoubi, A. Tressaud, L. Fournes, and P. Hagenmuller, Mater. Res. Bull, 1988, 23, 1611.
186. Y. Laligant, Y. Calage, G. Heger, J. Pannetier, and G. Ferey, J. Solid State Chem., 1989, 78, 66.
187. Y. Kawamoto and D. Koumyoji, J. Electrochem. Soc., 1989, 136, 1816.
188. J. Cerisier, C. Guillot, P. Palvadeau, and J. Rouxel, Eur. J. Soild State Inorg. Chem., 1988, 25, 35.
189. L. Terminiello and R.C. Mercador, Hyperfine Interact, 1989, 50, 651.
190. R.B. Lanjewar, A.M. Waditwar, and A.N. Garg, J. Radioanal. Nucl. Chem., 1988, 125, 75.
191. I.S. Lyubutin, O.K. Mel'nikov, S.E. Sigarev, and V.G. Terziev, Solid State Ionics, 1988, 31, 197.
192. P.J. Mineely, Thermochim. Acta, 1989, 137, 197.
193. B. Leon and D.K. Straub, Inorg. Chim. Acta, 1989, 156, 17.
194. R. Pinna, G. Ponticelli, F. Aramu, A. Delunas, and V. Maxia, Trans. Met. Chem., 1988, 13, 217.
195. N.V. Gerbeleu, M.A. Yampol'skaya, S.G. Shova, M.S. Byrke, Yu.A. Simonov, K.D. Turte, F.A. Spatar, V.B. Arion, and A.A. Dvorkin, Zh. Neorg. Khim., 1989, 34, 877.
196. A. Garg and J.P. Tandon, J. Prakt. Chem., 1989, 33, 157.
197. R. Pinna, G. Ponticelli, F. Aramu, A. Delunas, and V. Maxia, Trans. Met. Chem., 1988, 13, 426.
198. G. Wolmershaeuser, G. Wortmann, and M. Schnauber, J. Chem. Res., Synop., 1988, 358.
199. D.B. Dell'Amico, F. Calderazzo, L. Labella, G. Albanese, and A. Deriu, Gazz. Chim. Ital., 1988, 118, 741.
200. S.K. Agarwal and V.K. Tyagi, Orient. J. Chem., 1989, 5, 98.
201. X. Wang, M.E. Kotun, W.T. pennington, and J.C. Fanning, Inorg. Chim. Acta, 1988, 154, 189.

202. B.V. Roy, N. Palta, D.M. Puri, and N.D. Sharma, Trans. Met. Chem., 1988, 13, 281.
203. D. Billaud, A. El Haouri, and R. Gerardin, Synth. Met., 1989, 29, F241.
204. D.G. Rancourt, B. Hun, and S. Flandrois, Can. J. Chem., 1988, 66, 776.
205. I.A. Bulgak, A.S. Skoropanov, A.A. Vecher, B.A. Abdikamalov, A.G. Knizknik, Zh. Fiz. Khim., 1989, 63, 2280.
206. A.G. Khiznik, R.A. Stukan, T.A. Kizina, A.S. Skoropanov, A.A. Vecher, L.D. Kvacheva, and Yu.N. Novikov, Zh. Neorg. Khim., 1989, 34, 1688.
207. B.S. Tsukerblat and S.I. Klokishner, Dokl. Akad. Nauk. SSSR, 1989, 305, 144.
208. K.K. Surerus, E. Munck, B.S. Snyder, and R.H. Holm, J. Am. Chem. Soc., 1989, 111, 5501.
209. S. Nakashima, Y. Masuda, and H. Sano, Chem. Lett., 1988, 1851.
210. J.L. Fourquet, E. Le Samedi, and Y. Calage, J. Solid State Chem., 1988, 77, 84.
211. M. Hanack and A. Hirsch, Synth. Met., 1989, 29, F9.
212. M. Suzuki, H. Oshio, A. Uehara, K. Endo, M. Yanaga, S. Kida, and K. Saito, Bull. Chem. Soc. Jpn., 1988, 61, 3907.
213. S. Dureeke, P. Chaudhuri, K.Pohl, K. Weighardt, X.Q. Ding, E. Bill, A. Swaryn, A.X. Trautwein, H. Winkler, S.J. Gurman, J. Chem. Soc., Chem. Commun., 1989, 59.
214. C. Ercolani, M. Gardini, V.L. Goedken, G. Pennesi, G. Rossi, U. Russo, and P. Zanonato, Inorg. Chem., 1989, 28, 3097.
215. S. Ogasawara, M. Takano, and Y. Bando, Bull. Inst. Chem. Res., Kyoto Univ., 1988, 66, 64.
216. A.T. Baker, N.J. Ferguson, H.A. Goodwin, and A.D. Rae, Aust. J. Chem., 1989, 42, 623.
217. F. Grandjean, G.J. Long, B.B. Hutchinson, L. Ohlhausen, P. Neill, and J.D. Holcomb, Inorg. Chem., 1989, 28, 4406.
218. D.C. Craig, H.A. Goodwin, and D. Onggo, Aust. J. Chem., 1989, 41, 1157.
219. A.A. Yousif, H. Winkler, H. Toftlund, A.X. Trautwein, and R.H. Herber, J. Phys.: Condensed Matter, 1989, 1, 7103.
220. H. Koeppen, E. Meissner, L. Wiehl, H. Spiering, and P. Guetlich, Hyperfine Interact, 1989, 52, 29.
221. P. Adler, H. Spiering, and P. Guetlich, J. Phys. Chem. Solids, 1989, 50, 587.
222. D.C. Craig, H.A. Goodwin, D. Onggo, and A.D. Rae, Aust. J. Chem., 1989, 41, 1625.
223. J.K. McCusker, M. Zvagulis, H.G. Drickamer, and D.N. Hendrickson, Inorg. Chem., 1989, 28, 1380.
224. E. Koenig, G. Ritter, J. Dengler, and J. Nelson, Inorg. Chem., 1989, 28, 611.
225. W. Yang and C. Xie, Huaxue Xuebao, 1989, 47, 333.
226. M. Mohan, P.H. Madhuranath, A. Kumar, M. Kumar, N.K. Jha, Inorg. Chem., 1989, 28, 96.
227. M.S. Byrke, M.A. Yampol'skaya, N.V. Gerbeleu, V.V. Zelentsov, K.I. Turte, K.M. Indrichan, B.Ya. Kuyavskaya, and M.D. Revenko, Zh. Neorg. Khim, 1988, 33, 2559.
228. M.A. Yampol'skaya, Yu.A. Simmonov, K.I. Turte, M.Yu.. Antipin, V.K. Bel'skii, Yu.T. Struchkov, N.V. Gerbeleu, and M.S. Byrke, Zh. Neorg. Khim, 1988, 33, 3100.
229. B.S. Manhas and S. Bala, Polyhedron, 1988, 7, 2465.
230. K.J. Brewer, W.R. Murphy, and J.D. Peterson, Inorg. Chim. Acta, 1989, 159, 93.
231. H. Inoue, T. Nakazawa, T. Mitsuhashi, T. Sharai, and E. Fluck, Hyperfine Interact., 1989, 46, 725.

232. J.H. De Araujo, M.A.B. De Araujo, and F.F.T. De Araujo, J. Radioanal. Nucl. Chem., 1989, 136, 313.
233. R. Brener and H. Shechter, Phys. Rev. B, 1989, 39, 2803.
234. R. Brener and H. Shechter, Z. Naturforsch, 1988, 43, 855.
235. F.J. Berry, C.J. Jones, J.A. McCleverty, J. Sharpe, N.A. Bailey, and M. Deadman, J. Organomet. Chem., 1988, 353, 209.
236. G.J. Long and J.F. O'Brien, Hyperfine Interact., 1988, 40, 101.
237. J. Zhang, L. Song, Z. Zhang, R. Liu, Z. Cheng, and J. Wang, Hyperfine Interact., 1988, 40, 363.
238. T.J. Coffy, G. Medford, J. Plotkin, G.J. Long, J.C. Huffman, and S.G. Shore, Organometallics, 1989, 8, 2404.
239. D. Koradecki, M. Lukasiak, A. Pron, W. Kutner, and J. Suwalski, Polym. Commun., 1989, 30, 61.
240. J.S. Miller, D.T. Glatzhofer, D.M. O'Hare, W.M. Reiff, A. Chakraborty, and A.J. Epstein, Inorg. Chem., 1989, 28, 2930.
241. D.T. Hill, G.R. Girard, F.L. McCabe, R.K. Johnson, P.D. Stupik, J.H. Zhang, W.M. Reiff, and D.S. Eggleston, Inorg. Chem., 1989, 28, 3529.
242. B.J. Howlin and J. Silver, Inorg. Chim. Acta, 1988, 141, 227.
243. M. Watanabe, Y. Masuda, I. Motoyama, and H. Sano, Bull. Soc. Chem. Jpn., 1988, 61, 3479.
244. M. Kai, M. Katada, and H. Sano, Chem. Lett., 1988, 1523.
245. M. Kai, I. Motoyama, K. Izumi, M. Motomi, Y. Masuda, and H. Sano, Chem. Lett., 1988, 1037.
246. W. Zhang, S.R. Wilson, and D.N. Hendrickson, Inorg. Chem., 1989, 28, 4160.
247. F. Varret, J.P. Mariot, J.R. Hamon, and D. Astruc, Hyperfine Interact, 1988, 39, 67.
248. J. Guillin, M.H. Desbois, M. Lacoste, D. Astruc, and F. Varret, J.Phys. Colloq., 1988, C8, C8-837.
249. M.H. Desbois, D. Astruc, J. Guillin, F. Varret, A.X. Trautwein, and G. Villeneuve, J. Am. Chem. Soc., 1989, 111, 5800.
250. M.H. Desbois, and D. Astruc, Angew. Chem. (Int. Ed.), 1989, 28, 460.
251. M. Watanabe and H. Sano, Chem. Lett., 1988, 1457.
252. M. Watanabe, Y. Masuda, I. Motoyama, and H. Sano, Bull. Chem. Soc. Jpn., 1988, 61, 827.
253. M. Watanabe, Y. Masuda, I. Motoyama, and H. Sano, Hyperfine Interact, 1988, 40, 355.
254. M. Watanabe, Y. Masuda, I. Motoyama, and H. Sano, J. Coord. Chem., 1988, 18, 141.
255. M. Saito and H. Kashiwagi, Chem. Phys. Lett., 1989, 155, 557.
256. H. Abu-Soud and J. Silver, Inorg. Chim. Acta, 1988, 153, 139.
257. O.K. Medhi and J. Silver, J. Chem. Soc., Chem. Commun., 1989, 1199.
258. J. Silver and J.A. Taies, Inorg. Chim. Acta, 1989, 159, 231.
259. W.S. Scheidt, S.R. Osvath, J.J. Lee, C.A. Reed, B. Shaevitz, and G.P. Gupta, Inorg. Chem., 1989, 28, 1591.
260. P.A. Stuzhin, L. Latos-Grazinski, anard A. Jezierski, Trans. Met. Chem., 1989, 14, 341.
261. L. Latos-Grazinski, J. LIsowski, M.M. Olmstead, and A.L. Balch, Inorg. Chem., 1989, 28, 1183.
262. C.A. Koch, C.A. Reed, G.A. Brewer, N.P. Rath, W.S. Scheidt, G. Gupta, and G. Lang, J. Am. Chem. Soc., 1989, 111, 5800.
263. F.U. Axe, C. Flowers, G.H. Loew, and A. Waleh, J. Am. Chem. Soc., 1989, 111, 7333.
264. F.U. Axe, A. Waleh, L. Chantranupong, and G.H. Loew, Int. J. Quantum Chem, 1989, 35, 181.

265. V.I. Khleskov, A.B. Smirnov, and R.E. Garibov, Koord. Khim., 1989, 15, 308.
266. C.D. Shaffer and D.K. Straub, Inorg. Chim. Acta, 1989, 158, 167.
267. J. Silver and J.A. Taies, Inorg. Chim. Acta, 1988, 153, 235.
268. M. Schappacher, L. Ricard, J. Fischer, R. Weiss, R. Monteil-Montoya, E. Bill, and A.X. Trautwein, Inorg. Chim., 1989, 28, 4639.
269. O.K. Medhi, A. Houlton, and J. Silver, Inorg. Chim. Acta, 1989, 161, 213.
270. S. Saito, S. Sumita, K. Iwai, and H. Sano, Bull. Chem. Soc. Jpn., 1988, 61, 3539.
271. H. Hori, K. Kadono, H. Inoue, T. Shirai, and E. Fluck, J. Radioanal. Nucl. Chem., 1989, 136, 159.
272. F. Kilar, A. Gustavsson-Seidel, and L. Haeggstroem, Biochem. Biophys. Res. Commun, 1989, 158, 755.
273. S.A. Kretchmar, M. Teixeira, B.H. Huynh, and K.N. Raymond, Biol. Met., 1989, 1, 26.
274. E.R. Bauminger and I. Nowik, Hyperfine Interact., 1989, 50, 489.
275. E.R. Bauminger, P.M. Harrison, I. Nowik, and A. Treffry, Biochemistry, 1989, 28, 5466.
276. B.F. Matzanke, G.I. Mueller, A.X. Trautwein, and E. Bill, Eur. J. Biochem., 1989, 183, 371.
277. B.F. Matzanke, E. Bill, G.I. Mueller, G. Winkelmann, and A.X. Trautwein, Hyperfine Interact., 1989, 47-48, 311.
278. J.B. Lynch, C. Juarez-Garcia, E. Muenck, and L. Que, J. Biol. Chem., 1989, 264, 8091.
279. E. Bill, X.Q. Ding. H. Winkler, A.X. Trautwein, and U. Gonser, Hyperfine Interact., 1989, 47-48, 329.
280. E. Bill, C. Haas, X.Q. Ding, W. Maret, H. Winkler, A.X. Trautwein, and M. Zeppezauer, Eur. J. Biochem., 1989, 180, 111.
281. J.T. Sage, Y.M. Xia, P.G. Debrunner, D.T. Keogh, J. De Jersey, and B. Zerner, J. Am. Chem. Soc., 1989, 111, 7239.
282. A. Levy, P.T. Manoharan, J.M. Rifkind, J.C. Walker, F.C. Haberle, N.G. Kumar, J. D. Glickson, and G.A. Elgavish, Biochem. Biophys. Acta, 1989, 991, 97.
283. M. Teixeira, I. Moura, A.V. Xavier, J.J.G. Moura, J.G. Jose, and J. LeGall, J. Biol. Chem., 1989, 264, 16435.
284. V. Petrouleas, J.J. Brand, K.G. Parrett, and J.H. Golbeck, Biochemistry, 1989, 28, 8980.
285. W.E. Newton, S.F. Gheller, R.H. Sands, and W.R. Dunham, Biochem. Biophys. Res. Commun., 1989, 162, 882.
286. A. Mueller, A. Hildebrand, M. Penk, H. Boegge, E. Bill, and A. Trautwein, Inorg. Chim. Acta, 1988, 148, 11.
287. H. Liu, B. Kang, J. Cai, L. Huang, D. Wu, F. Wang, Z. Guo, A. Cong, and J. Lu, Jiegou Huaxue, 1988, 7, 171.
288. B. Kang, H. Liu, F. Wang, Z. Guo, W. Cheng, C. Lu, X. Wang, and J. Lu, Yingyong Huaxue, 1988, 5, 67.
289. S. Ciurli, M.J. Carney, R.H. Holm, and G.C. Papaethymiou, Inorg. Chem., 1989, 28, 2696.
290. M.I. Ostrakh, V.A. Semenkin, B.N. Burykin, and V.I. Khleskov, Mol. Phys., 1989, 66, 531.
291. D.P.E. Dickson, N.M.K. Reid, S. Mann, V.J. Wade, R.J. Ward, and T.J. Peters, Biochem. Biophys. Acta, 1989, 957, 81.
292. D.P.E. Dickson, N.M.K. Reid, S. Mann, V.J. Wade, R.J. Ward, and T.J. Peters, Hyperfine Interact., 1989, 45, 225.

293. I. Ortalli, G. Pedrazzi, L. Mangoni, and V. Rizzoli, Phys. Scr., 1988, 38, 639.
294. J.L. Bara, S.N. Szubinska, and L. Misiak, J. Biochem. Biophys. Methods, 1989, 19, 241.
295. M. Chorazy, R. Meczynski, and P.J. Panek, J. Radioanal. Nucl. Chem., 1988, 128, 201.
296. R.L. Mills, C.W. Walter, L. Venkataraman, K. Pangm and J.F. Farrell, Nature, 1988, 336, 787.
297. M. Sud, N. Bala, V. Liveleen, and S.P. Puri, Ind. J. Pure Appl. Chem., 1988, 26, 701.
298. K.R.P.M. Rao, S. Padhye, and P. Garge, Phys. Lett. A, 1989, 137, 232.
299. S. Ambe, Appl. Radiat. Isot., 1989, 40, 671.
300. A. Meagher and B. Rodmacq, Nouv. J. Chem., 1988, 12, 961.
301. A.A. Olowe and J.M.R. Genin, Hyperfine Interact., 1989, 46, 445.
302. Yu.Yu. Esipenko, A.M. Sukhotin, N.V. Rasskazova, and G.B. Alyushina, Elektrokhimiya, 1988, 24, 1554.
303. C.A. Cardile, Clays Clay Miner., 1988, 36, 537.
304. C. Diaz, N.R. Furet, V.I. Nikolaev, V.S. Rusakov, and M.C. Cordeiro, Hyperfine Interact., 1989, 46, 689.
305. I.P. Saraswat, G.S. Rao, C.L. Sharma, V.K. Garg, and A.C. Vajpei, J. Mater. Sci., 1989, 24, 2132.
306. K. Nomura, Y. Hara, K. Koumoto, Y. Kunihito, H. Yanigida, and Y. Ujihara, Sogo Shikensho Nenpo (Toyko Daigaku Kogakubu), 1988, 47, 205.
307. A.G. Bobkovskii, L.Sh. Tsemekhman, V.P. Mintsis, A.G. Kim, and G.P. Ermakov, Rasplavy, 1989, 122.
308. R.J. Pollard, J. Appl. Phys., 1989, 66, 3392.
309. R.V. Morris, D.G. Agresti, H.V. Lauer, J.A. Newcomb, T,D, Shelfer, and A.V. Murali, J. Geophys. Res., Solid Earth Planets, 1989, 94, 2760.
310. A. Meagher, S.W. Charles, and S. Wells, J. Phys. Colloq., 1988, C8, 1845.
311. E. Tronc and J.P. Jolivet, J. Phys. Colloq., 1988, C8, 1823.
312. A. Venkataraman and A.J. Mukhedkar, J. Mater. Sci., 1988, 23, 3505.
313. S. Music, S. Popovic, and M. Ristic, J. Mater. Sci., 1989, 24, 2722.
314. L. Cianchi, F. Gulisano, M. Mancini, and G. Spina, Hyperfine Interact., 1989, 52, 7.
315. G.A. Teterin. E.M. Menchuk, F.K. Egorov, T.M. Shkol'nikova, and E.G. Semin, Ukr. Khim. Zh., 1988, 54, 799.
316. G. Haley, J.G. Mullen, and J.M. Honig, Solid State Commun., 1989, 69, 285.
317. M.B. Madsen, S. Moerup, and J.M. Knudsen, Hyperfine Interact., 1989, 50, 659.
318. T. Itoh, M. Abe, and T. Tamaura, J. Phys. Colloq., 1988, C8, 2015.
319. C. Djega-Mariadassou, T. Merceron, J.L. Dormann. P. Porte, and P. Renaudin, Phys. Status Solidi A, 1989, 112, 601.
320. M.R. De Guire, R.C. O'Handley, and G. Kalonji, J. Appl. Phys., 1989, 65, 3167.
321. G.J. Baldha, R.V. Upadhyay, and R.G. Kulkarni, J. Mater. Sci., 1988, 23, 3357.
322. T. Pannaparayil, R. Marande, S. Komarneni, and S.G. Sankar, J. Appl. Phys., 1988, 64, 5641.
323. T. Pannaparayil and S. Komarneni, IEEE Trans. Magn., 1989, 25, 4233.

324. T. Pannaparayil, R. Marande, and S. Komarneni, J. Phys. Colloq., 1988, C8, 947.
325. S.I. Yushchuk, S.A. Yur'ev, P.I. Shevchuk, P.S. Kostyuk, and S.I. Khomin, Izv. Akad. Nauk. SSSR, Neorg. Mater., 1989, 25, 119.
326. R. Gerardin, A. Alebouyeh, J.F. Brice, O. Evrard, and J.P. Sanchez, J. Solid State Chem., 1988, 76, 398.
327. K. Melzer, M. Lukasiak, B. Lippold, J. Suwalski, and A. Martin, Cryst. Res. Technol., 1989, 24, K9.
328. A.R. Corradi, D.E. Speliotis, A.H. Mossish, Q.A. Pankhurst, X.Z. Zhou, G. Bottoni, D. Candolfo, A. Cecchetti, and F. Masoli, IEEE Trans. Magn., 1988, 22, 2862.
329. C.Y. Hong, S.D. Baek, W.C. Lee, and K.H. Lee, Sae Mulli, 1989, 29, 65.
330. H.N. Ok, K.S. Baek, E.J. Choi, and Y.K. Kim, Sae Mulli, 1988, 28, 364.
331. H.N. Ok, K.S. Baek, and E.J. Choi, Phys. Rev. B, 1989, 40, 84.
332. A.D. Balaev, O.A. Bayukov, and A.F. Savitskii, Phys. Status Solidi A, 1989, 152, 639.
333. M. Tanaka, H. Iwasaki, K. Siratori, and I. Shindo, J. Phys. Soc. Jpn., 1989, 58, 1433.
334. J.L. Soubeyroux, D. Fiorani, E. Agostinelli, S.C. Bhargava, and J.L. Dormann, J. Phys. Colloq., 1988, C8, 1117.
335. H. Haneda, M. Isobe, A. Watanabe, S. Shirasaki, J. Cryst. Growth, 1988, 92, 155.
336. J.L. Dormann, S.C. Bhargava, J. Jove, and D. Fiorani, Hyperfine Interact., 1989, 50, 625.
337. S. Phanjoubam, D. Kothari, J.S. Baijal, and P. Kishan, Solid State Commun., 1988, 68, 549.
338. D. Kothari, S. Phanjoubam, J.S. Baijal, P. Kishan, and C.Prakash, J. Phys. C, 1988, 21, 6169.
339. R.V. Upanhyay and R.G. Kulkarni, J. Magn. Magn. Mater., 1988, 74, 327.
340. J.S. Baijal, D. Kothari, S. Phanjoubam, and C. Prakash, Solid State Commun., 1989, 69, 277.
341. C.S. Lee and W.G. Kim, J. Korean Phys. Soc., 1989, 22, 57.
342. M. Fayek, M. Elnimr, N. Nada, H. Saleh, and Y. Eid, Ann. Phys., 1989, 46, 247.
343. V.A. Tsurin, A.M. Sorkin, N.P. Filippova, and V.L. Kozhevnikov, Pis'ma Zh. Eksp. Teor. Fiz., 1988, 48, 444.
344. V. Sedykh, S. Nasu, and F.E. Fujita, Solid State Commun., 1988, 67, 1063.
345. A. Simopoulos and D. Niarchos, Phys. Rev. B, 1988, 28, 8931.
346. T. Shinjo, S. Nasu, T. Kohara, T. Takabatake, and M. Ishikawa, J. Phys. Colloq., 1988, C8, 2207.
347. P. Dalmas de Reotier, E. Gil, P. Vulliet, A. Yaouanc, P. Chaudouet, J.P. Senateur, and F. Weiss, J. Phys. Colloq., 1988, C8, 2203.
348. S.K. Xia, Z.B. Zhao, C.R. Wang, R.Z. Ma, G.H. Cao, and J.Y. Ping, Solid State Commun., 1989, 70, 141.
349. E. Baggio-Saitovitch, F.J. Litterst, I.S. Azevedo, and R.B. Scorzelli, Hyperfine Interact., 1989, 50, 521; 529.
350. S. Suharan, J. Chadwick, D.B. Hannon, D.H. Jones, and M.F. Thomas,, Solid State Commun., 1989, 70, 817.
351. S. Katsuyama, Y. Ueda, and K. Kosuge, Funtai oyobi Funmatsu Yakin, 1988, 35, 896.
352. Z.Q. Qiu, Y.W. Du, H. Tang, and J.C. Walker, J. Magn. Magn. Mater., 1989, 78, 359.
353. E.R. Bauminger, I. Felner, M. Kowitt, and I. Nowik, J. Phys. Colloq., 1988, C8, 2211.

354. S.C. Bhargava, J.L. Dormann, J. Jove, O. Gorochov, C. Djega-Mariadassou, H. Pankowska, and R. Suryanayanan, J. Phys. C, 1988, 21, L905.
355. O. Bremert, C. Michaelson, and H.U. Krebs, J. Appl. Phys., 1989, 65, 1018.
356. V.D. Checheskii, V.V. Eremenko, V.A. Finkel, and N.E. Kaner, Fiz. Nizk. Temp. (Kiev), 1989, 15, 92.
357. V.D. Checheskii, V.V. Eremenko, V.A. Finkel, and N.E. Kaner, Fiz. Nizk. Temp. (Kiev), 1988, 14, 1099.
358. L.J. Swartzendruber, L.H. Bennett, M.Z. Harford, and M. Rubenstein, J. Supercond., 1988, 1, 219.
359. A.M. Khasanov, S.S. Tsarevskii, and A.B. Liberman, Fiz. Tverd. Tela, 1988, 30, 2688.
360. Y. Nishihara, H. Obara, T. Katayama, H. Oyangi, K. Murata, Y. Kimura, and K. Kajimura, J.Phys. Soc. Jpn., 1989, 58, 2925.
361. B.D. Dunlap, J.D. Jorgensen, C. Segre, A.E. Dwight, J.L. Matykiewicz, H. Lee, W. Peng, and C.W. Kimball, Physica C, 1989, 158, 397.
362. A. Nath, S. Nagy, M.W. Barsoum, S.D. Tyagi, and Y. Wei, Solid State Commun., 1988, 68, 181.
363. J. Fontcuberta, X. Obradors, M. Valler, and J.M. Gonzalez-Calbet, Z. Phys. B, 1988, 73, 143.
364. L. Bottyan, B. Mohar, D.L. Nagy, I.S. Szucs, J. Toth, J. Dengler, G. Ritter, and J. Schober, Phys. Rev. B, 1988, 38, 11373.
365. P. Boolchand, C. Blue, K. Elgaid, I. Zitovsky, D. McDaniel, W. Huff, G. Goodman, G. Lemon, D.E. Farrell, and B.S. Chandrasakhar, Phys. Rev. B, 1988, 38, 11313.
366. R.A. Brand, C. Sauer, H. Luetgemeier, B. Rupp, and W. Zinn, Physica C, 1988, 156, 539.
367. Z.Q. Qiu, Y.W. Du, H. Tang, J.C. Walker, J.S. Morgan, and W.A. Bryden, J. Appl. Phys., 1988, 64, 5947.
368. V.A. Varnek, L.V. Chernikova, I.P. Asanov, V.N. Ikorskii, V.A. Nadolinnyl, S.A. Gromilov, P.P. Samoilov, V.E. Fedoriv, and L.N. Mazalov, Zh. Strukt. Khim., 1989, 30, 190.
369. Yu.T. Pavlyukhin, N.G. Khainovskii, Ya.Yu. Medikov, and A.I. Rykov, Izv. Sib. Otd. Akad. Nauk SSSR, Ser. Khim. Nauk, 1988, 107.
370. S.N. ShringiO.M. Prakash, R.V. Vadnere, S. Prasad, and N. Sharma, Pramana, 1989, 32, L699.
371. T. Nishida, H. Ide, Y. Maeda, H. Nasu, T. Yagi, A. Sakai, and Y. Takashima, Bull. Chem. Soc. Jpn., 1989, 62, 61.
372. M. Eibshutz, M.E. Lines, and J.M. Tarascon, Phys. Rev. B, 1988, 38, 8858.
373. A.M. Khasanov, S.S. Tsarevskii, and A.B, Liberman, Fiz. Tverd. Tela, 1988, 30, 2836.
374. Yu.T. Pavlyukhin, N.G. Khainovskii, Ya.Ya. Medikov, and A.I. Rykov, Pramana, 1988, 31, L445.
375. E. Kuzmann, Z. Homonnay, A. Vertes, I. Halasz, J. Bankuti, and I. Kirschner, J. Radioanal. Nucl. Chem., 1989, 135, 373.
376. Yu.A. Osipan, O.V. Zharikov, G.V. Novikov, N.S. Sidorov, V.I. Kulakov, L.V. Sypavina, R.K. Nikolaev, and A.M. Gromov, Physica C, 1989, 159, 137.
377. I. Nowik, M. Kowitt, I. Felner, and E.R. Bauminger, Phys. Rev. B, 1988, 38, 6677.
378. M. Rubenstein, M.Z. Harford, L.J. Swartzendruber, and L.H. Bennett, J. Phys. Colloq., 1988, C8, 2209.
379. G. Wortmann, A. Kolodziejczyk, M. Bergold, G. Stadermann, C.T. Simmons, and G. Kaindl, Hyperfine Interact., 1989, 50, 555.

380. S.T. Lin, C.L. Lin, K.C. Lin, C. Tien, and I.M. Jiang, Physica C, 1989, 159, 188.
381. K. Donnelly, J.M.D. Coey, S. Tomlinson, and J.M. Greneche, Physica C, 1988, 156, 579.
382. C.E. Violet, P.A. Hahn. T. Datta, C. Almasan, and J. Estrada, J. Appl. Phys., 1988, 64, 5911.
383. Y. Zeng, X. Yan, M. Lin, Y. Ren, X. Meng, Y. Lu, Q. Tu, X. Wu, and L. Sang, Phys. Rev. B, 1989, 40, 727.
384. E. Kuzmann, Z. Homonnay, A. Vertes, I. Halasz, J. Bankuti, and I. Kirschner, J. Radioanal. Nucl. Chem., 1989, 136, 121.
385. M. Hervieu, C. Michel, N. Nguyen, R. Retoux, and B. Raveau, Eur. J. Solid State Inorg. Chem., 1988, 25, 375.
386. H. Tang, Y.W. Du, Z.Q. Qiu, J.C. Walker, J.S. Morgan, and W.A. Bryden, J. Appl. Phys., 1988, 64, 5950.
387. P. Imbert, G. Jehanno, and J.A. Hodges, Hyperfine Interact., 1989, 50, 599.
388. Y. Zhuang, Y. Lin, D. Zhu, Y. Zheng, and Z. Yu, J. Am. Ceram. Soc., 1989, 72, 1444.
389. P.D. Battle, T.C. Gibb, and S. Nixon, J. Solid State Chem., 1988, 77, 124.
390. P.D. Battle, T.C. Gibb, and S. Nixon, J. Solid State Chem., 1989, 79, 75; 86.
391. T.C. Gibb and M. Matsuo, J. Solid State Chem., 1989, 81, 83.
392. T. Fujii, S. Jinzenji, Y. Asahara, A. Kajima, and T. Shinjo, J. Appl. Phys., 1988, 64, 5434.
393. A. Ito, H. Aruga, S. Morimoto, and H. Yoshizawa, J. Phys. Colloq., 1988, C8, 1129.
394. P.D. Battle, T.C. Gibb, C.W. Jones, and F. Studer, J. Solid State Chem., 1989, 78, 281.
395. K. Melzer, O. Knauf, and R. Russbuelt, Isotenopraxis, 1988, 24, 423.
396. A. Deriu, E. Calabrase, A. Paoluzi, and G. Turilli, Hyperfine Interact., 1988, 45, 271.
397. D.G. Agresti, T.D. Shelfer, Y.K. Hong, and Y.J. Paig, IEEE Trans. Magn., 1989, 25, 4069.
398. P.P. Kiruchok, N.B. Voronina, and M.M. Kindrat, Izv. Akad. Nauk. SSSR, Neorg. Mater., 1989, 25, 1182.
399. E. Tronc, F. Laville, M. Gasperin, A.M. Lejun, and D. Vivien, J. Solid State Chem., 1989, 81, 192.
400. N. Suarez-Almodovar, J.L. Sanchez-Llamazares, F. Leccabue, R. Panizzieri, and R. Xue, Mater. Lett., 1989, 8, 127.
401. J.J. Bara, B.F. Bogancz, T. Jarowska, J. Leciejewicz, M. Styczynska, and A. Szytula, J. Magn. Magn. Mater, 1988, 75, 73.
402. T.N. Sudziloskaya, N.N. Rzhevskaya, E.G. Stepanov, V.V. Yun, and G.R. Kotel'nikov, Koord.Khim., 1989, 15, 96.
403. S.C. Bhargava, J. Phys. Colloq., 1988, C8, 1051.
404. R. Gerardin, H. Aqachmar, I. Nodari, J.F. Brice, and O. Evrard, J. Phys. Chem. Solids, 1989, 50, 43.
405. D. Groult, M. Hervieu, N. Nguyen, and B. Raveau, J. Solid State Chem., 1988, 76, 248.
406. M. Toulemonde, Philos. Mag. A, 1988, 58, 799.
407. D. Barb, S. Constantinescu, D. Tarina, I.S. Lyubutin, B.V. Mill, V.G. Terziev, T.D. Dmitreva, and A.V. Butashin, Hyperfine Interact., 1989, 50, 645.
408. S.H. Lee and K.P. Chae, Solid State Commun, 1988, 68, 971.
409. S.H. Lee, K.P. Chae, W.T. Kim, D.C. Chung, and J.D. Chung, Sae Mulli, 1988, 28, 511.
410. S. Music, J. Mater. Sci. Lett., 1989, 8, 537.

411. C.Y. Hong, S.D. Baek, C.H. Moon, and K.H. Lee, Sae Mulli, 1988, 29, 60.
412. M. Bizi, B. Hannoyer, and M. Lenglet, Riv. Stn. Sper. Vetro, 1989, 19, 75.
413. D.B. Dingwell, Am. Mineral., 1989, 74, 1038.
414. B.O. Mysen and D. Virgo, Am. Mineral., 1989, 74, 58.
415. K. Tanaka, K. Kamiya, T. Yoko, S. Tanabe, K. Hirao, and N. Soga, J. Non-Cryst. Solids, 1989, 109, 289.
416. S. Nakamura and N. Ichinose, Jpn. J. Appl. Phys., Part 1, 1989, 28, 984.
417. T. Nishida, S. Saruwatari, and Y. Takashima, Bull. Chem. Soc. Jpn., 1988, 61, 4093.
418. T. Nishida, H. Ide, Y. Takashima, T. Yagi, S. Tomarigushi, T. Ishizuka, and A. Sakai, J. Mater. Sci., 1989, 24, 1687.
419. A.M. Sanad, I. Kashif, M.A. Khaled, S.A. Aly, and H. Farouk, Phys. Chem. Glasses, 1989, 30, 27.
420. R.T. Hunter, M. Edge, A. Kalivretenos, K.M. Brewer, N.A. Brock, A.E. Hawkes, and J.C. Fanning, J. Am. Ceram. Soc., 1989, 72, 943.
421. M. Braga, S.K. Lie, C.A. Taft, and W.A. Lester, Phys. Rev. B, 1988, 38, 10837.
422. T. Tsuji, H. Tsuchiya, and K. Naito, Solid State Ionics, 1989, 35, 331.
423. F.E. Wagner, P.M. Swash, P. Marion, and P. Marion, Hyperfine Interact., 1988, 46, 681.
424. G.D. Sultanov, V.D. Chercherskii, R.M. Mirzababaev, S.G. Ibragimov, and G.D. Guseinov, Fiz. Tverd. Tela., 1988, 30, 875.
425. D. Welz, P. Deppe, W. Schaefer, H. Sabrowsky, and M. Rosenberg, J. Phys. Chem. Solids, 1989, 50, 297.
426. K. Forcher, W. Lottermoser, and G. Amthauer, Neues Jahrb. Mineral., Abh., 1989, 160, 25; 63.
427. R. Brec, E. Prouzet, and G. Ouvrard, J. Power Sources, 1989, 26, 325.
428. R. Fong, C.H.W. Jones, and J.R. Dahn, J. Power Sources, 1989, 26, 333.
429. J. Zink and K. Nagorny, J. Phys. Chem. Solids, 1988, 49, 1429.
430. D. Dai, A.H. Morrish, X.Z. Zhou, and Z. Shan, J. Phys. Colloq., 1988, C8, 155.
431. R.K. Gubaidullin and Sh.R. Safin, Izv. Vyssh Uchebn. Zaved., Fiz, 1988, 331, 119.
432. H.N. Ok, K.S. Baek, H.S. Lee, Y.K. Kim, abd J.C. Sur, J. Korean Phys. Soc., 1989, 22, 367.
433. K.S. Baek, H.N. Ok, and J.C. Sur, Phys. Rev. B, 1988, 39, 2800.
434. E. Riedel and H. Anik, Z. Anorg. Allg. Chem., 1988, 567, 173.
435. G.A. Bordovskii, N.A. Savinova, and L.N. Seregina, Phys. Status Solidi A, 1989, 109, K51.
436. I.A. Askerov, G.K. Aslanov, F.S. Nasredinov, and B.G. Tagiev, Fiz. Tekh. Poluprovodn., 1989, 23, 1083.
437. I. Nowik, E.R. Bauminger, A. Mycielski, and H. Szymczak, Physica B, 1988, 153, 215.
438. C.S. Kim, J. Childress, and C.L. Chien, J. Appl. Phys., 1988, 64, 5886.
439. C.K. Groot, P.J.G.D. Van de Gender, W.S. Niedzwiedz, A.M. Van der Kraan, V.H.J. De Beer, and R. Prins, Bull. Soc. Chim. Belg., 1988, 97, 167.
440. W.L.T.M. Ramselaar, R.H. Hadders, E. Gerkema, V.H.J. De Beer, E.M. Van Oers, and A.M. Van der Kraan, Appl. Catal., 1989, 51, 263.

441. W.L.T.M. Ramselaar, S.M.A.M. Bouwens, V.H.J. De Beer, and A.M. Van der Kraan, Hyperfine Interact., 1988, 46, 599.
442. S. Zhang, C. Wang, L. Pan, H. Cai, D. Liang, and L. Lin, Ranlio Huaxue Xuebao, 1988, 16, 289.
443. Z. Huang and Z.Li, Ranlio Huaxue Xuebao, 1988, 16, 313.
444. S.A. Stevenson, S.A. Goddard, M. Arai, and J.A. Dumesic, J. Phys. Chem., 1989, 93, 2058.
445. S.A. Stebenson and J.A. Dumesic, J. Mol. Catal., 1989, 52, 387.
446. B.S. Clausen, H. Topsoe, and S. Moerup, Appl. Catal., 1989, 48, 327.
447. A.F.H. Wielers, A.J.H.M. Kock, C.E.C.A. Hop, J.W. Geus, and A.M. Van der Kraan, J. Catal., 1989, 117, 1.
448. F.J. Berry and M.R. Smith, J. Chem. Soc., Faraday Trans. I, 1989, 85, 467.
449. F.J. Berry and S. Jobson, Hyperfine Interact., 1988, 46, 559.
450. F.J. Berry, S. Jobson, and S.L. Jones, Hyperfine Interact., 1988, 46, 613.
451. Z. Schay, K. Lazar, I. Bogyay, and L. Guczi, Appl. Catal., 1989, 51, 33.
452. Z. Schay, K. Lazar, K. Matusek, I. Bogyay, and L. Guczi, Appl. Catal., 1989, 51, 49.
453. F.J. Berry, S. Jobsen, and M.R. Smith, Hyperfine Interact., 1988, 46, 607.
454. A.A. Chen, J. Phillips, J.J. Venter, and M.A. Vannice, J. Catal., 1989, 118, 443.
455. T. Shigematsu, M. Ohtsuki, K. Watanabe, Y. Bando, and N. Nakanishi, Funtai oyobi Funmatsu Yakin, 1989, 36, 56.
456. K. Lazar, B.M. Choudary, and L. Guczi, Hyperfine Interact., 1988, 46, 591.
457. V.R. Balse, W.M.H. Sachtler, and J.A. Dumesic, Catal. Lett., 1988, 1, 275.
458. X. Cui, M. Jin, Y. Ji, and Y. Jheng, Cuihua Xuebao, 1988, 9, 241.
459. J.M. Millet, C. Virely, M. Forissier, P. Bussiere, and J.C. Vedrine, Hyperfine Interact., 1988, 46, 619.
460. A.V. Golub'ev, Yu.V. Maksimov, A.I. Matveev, I.P. Suzdalev, E.N. Amirbekov, F.V. Aliev, and A.D. Kuliev, Kinet. Katal., 1988, 29, 1508.
461. R. Chen and Y. Zhu, Xiamen Daxue Xuebao, Ziran Kexueban, 1988, 27, 633.
462. P.Ya. Kolotyrkin, Yu.V. Maksimov, Yu.V. Baldokin, and A.V. Krylova, Dokl. Akad. Nauk. SSSR, 1988, 302, 1416.
463. G.I. Kileimikov, Yu.V. Maksimov, V.V. Dudoladov, I.P. Suzdalev, I.G. Brodskaya, and L.M. Dmitrenko, Kinet. Katal., 1989, 30, 927.
464. D. Klisurski, I. Mitov, K. CHeshkova, S. Angelov, K. Petrov, and S. Georgiev, Kinet. Katal., 1989, 30, 403.
465. F. Aramu, V. Maxia, and A. Delunas, Nuovo Cimento Soc. Ital. Fis., 1989, 11 D, 891.
466. L. Izoret, J.N. Rimbert, F. Dumas, and P. Maruejol, Hyperfine Interact., 1988, 46, 659.
467. P.Liu, Yanshi Kuangwuxue Zashi, 1988, 7, 167.
468. S.C. Das, S.K. Sengupta, and N. Chaudhuri, Indian J. Pure Appl. Chem., 1988, 26, 680.
469. A.N. Nigam, R.P. Tripathi, H.S. Singh, R.S. Gambhir, N.G. Lukose, Fuel, 1989, 68, 209.
470. J.L. Komraus and E.S. Popiel, Koks, Smola, Gaz, 1988, 33, 203.
471. A. Newman and P. Bussiere, Hyperfine Interact., 1988, 46, 645.

472. K.S. Harchand, S.P. Taneja, D. Raj, and P. Sharma, Fuel Process. Technol., 1989, 21, 19.
473. Z. Li and Z. Zhang, Hejishu, 1988, 11, 17.
474. B. Kounde, A. Raharinaivo, A.A. Olowe, D. Rezel, P. Bauer, and J.M.R. Genin, Hyperfine Interact., 1988, 46, 421.
475. J.F. Marco, J. Davalos, J.R. Gancedo, and M. Garcia, Hyperfine Interact., 1988, 46, 453.
476. K. Choudhury, D. Das, S. Mukherjee, and C.K. Majumdar, J. Mater. Sci. Lett., 1988, 7, 1331.
477. J.A. Jaen and B. Fernandez, Electrochim. Acta, 1989, 34, 885.
478. T. Peev, A. Alaka, and V. Rusanov, J. Radioanal. Nucl. Chem., 1988, 128, 207.
479. A.V. Dolenko, V.P. Korneev, V.A. Mamet, V.F. Tyapkov, and V.A. Yurmanov, At. Energ., 1988, 65, 257.
480. E. De Grave, R.M. Persoons, R.E. Vanderberghe, and D.G. Chambaere, Hyperfine Interact., 1988, 45, 353.
481. A.A. Olowe, P. Bauer, J.M.R. Genin, and J. Guezennec, Corrosion (Houston), 1989, 45, 229.
482. A.A. Olowe, J.M.R. Genin, and P. Bauer, Hyperfine Interact., 1988, 46, 437.
483. J.R. Gancedo, M. Garcia, W. Francisco, M. Morcillo, and S. Feliu, Hyperfine Interact., 1988, 46, 461.
484. L.Ya. Buziner, L.N. Vasil'ev, L.N. Aristova, V.Ya. Grigor'ev, and A.I. Plyusnina, Zh. Prikl. Khim., 1989, 62, 455.
485. S.S. Ismail, G. Weisinger, P. Bottig, and F. Grass, J. Trace Microprobe Tech., 1988, 6, 535.
486. M.Y. Hassaan, M.A. Moutawie, and N.A. Eissa, Hyperfine Interact., 1988, 46, 739.
487. M.Y. Hassaan, S.H. Salah, and N.A. Eissa, Hyperfine Interact., 1988, 46, 733.
488. E. Murad and U. Wagner, Hyperfine Interact., 1988, 45, 161.
489. X. Pan, G. Jin, and Z. Sun, Hejishu, 1989, 12, 10.
490. G. Qin, X. Pan, and S. Li, Archeometry, 1989, 31, 3.
491. K. Ramaswamy, M. Dheenathayalu, and S. Barathan, Phys. Earth Planet. Inter., 1989, 56, 230.
491a. I.N. Makundi, A. Waern-Sperber, and T. Ericsson, Archeometry, 1989, 31, 54.
492. A. Svane, J. Chim. Phys. Phys.-Chim. Biol., 1988, 86, 911.
493. S. K. Godovikov, Fiz. Tverd. Tela, 1989, 31, 97.
494. Y. Wu, Wusan Jiance, 1989, 11, 169.
495. A.A. Opalenko and A.A. Kornilova, Phys. Status Solidi B, 1988, 150, 25; Fiz. Tverd. Tela, 1989, 31, 128.
496. H. Myoren and Y. Osaka, Jpn. J. Appl. Phys., Part 1, 1989, 28, 1165.
497. E.A. Vasil'ev, V.A. Virchenko, T.M. Tkachenko, and A.V. Shablovskii, Dokl. Akad. Nauk BSSR, 1988, 32, 692.
498. Z. Homonnay, K. Suvegh, C. Szeles, and A. Vertes, Hyperfine Interact., 1988, 45, 389.
499. Kh.B. Rakhmatullaev, M.A. Sagatov, F.S. Nasredinov, N.A. Savinova, and P.P. Sergin, Izv. Akad. Nauk SSSR, Neorg. Mater., 1989, 25, 333.
500. S. Motiejunas, D. Baltrunas, V.I. Garasim, and P.M. Starik, Phys. Status Solidi B, 1988, 148, K161.
501. S. Motiejunas, D. Baltunas, N. Geciauskaite, and K. Makariunas, Phys. Status Solidi B, 1989, 154, 341.
502. I.A. Dobryakov, Kristallografiya, 1988, 33, 1297.
503. P.V. Nistiryuk, A.A. Agzamov, I.V. Nistiryuk, F.S. Nasredinov, and P.P. Seregin, Izv. Akad. Nauk SSSR, Ser. Fiz., 1988, 52, 1716.

504. V. Drago, E. Baggio Saitovitch, and M.M. Abd-Elmeguid, Phys. Rev. B, 1988, 38, 8992; Hyperfine Interact., 1988, 51, 1061.
505. L.B. Kukushina, V.A. Zhorin, and V.Ya. Rochev, Zh. Fiz. Khim., 1989, 63, 1921.
506. W.J. Allen, R.J. Pollard, and J.D. Cashion, Hyperfine Interact., 1988, 45, 381.
507. P.P. Vaishnava, A.E. Dwight, W.R. Dunham, and W.W. Zhao, Hyperfine Interact., 1988, 51, 1011.
508. T.E. Cranshaw, J. Phys: Condens. Matter, 1989, 1, 829.
509. A. Ait-Bahammou, C. Meyer, F. Hartmann-Boutron, Y. Gros, and I.A. Campbell, J. Phys. Colloq., 1988, C8, 1157.
510. S.I. Reiman and F.D. Khamdamov, Izv. Akad. Nauk SSSR, Ser. Fiz., 1988, 52, 1730.
511. V. Sechovsky, L. Havela, V.I. Krylov, and A.V. Andreev, Hyperfine Interact., 1988, 51, 943.
512. P. Gibart, D.L. Williamson, B. El Jani, and P. Basmaji, Inst. Phys. Conf. Ser., 1988, 91, 379.
513. A. Baldini, G.P. Bernardini, L. Cianchi. F. Gulisano, M. Mancini, and G. Spina, Hyperfine Interact., 1988, 46, 695.
514. J. Jiang, W. Huang, Z. Cai, and Y. Liu, Yanshi Kuangwuxue Zashi, 1988, 7, 71.
515. E.L. Saw, M.H. Kuok, C.T. Yap, and J.B. Ward, Phys. Status Solidi A, 1989, 114, K223.
516. K. Yamada, S. Nose, T. Umehara, O. Tsutomu, and S. Ichiba, Bull. Chem. Soc. Jpn., 1988, 61, 4265.
517. I.E. Kolosova, V.I. Kuzuetsov, N.B. Shitova, and E.N. Yurchenko, Koord. Khim., 1988, 14, 1672.
518. P.G. Antonov, I.A. Agapov, and M.N. Maksimov, Zh. Prikl. Khim., 1989, 62, 494.
519. P.G. Antonov and E.N. Emel'yanov, Zh. Prikl. Khim., 1989, 62, 498.
520. A.N. Glebov, I.G. Nagaitseva, Yu.I. Sal'nikov, and A.S. Khramov, Zh. Neorg. Khim., 1988, 33, 2859.
521. P.G. Antonov, E.N. Emel'yanova, and I.A. Agapov, Izv. Vyssh. Uchebn. Zaved., Khim. Khim. Tekhnol., 1988, 31, 23.
522. G. Denes and K. Tam, Can. Ceram. Q., 1988, 57, 39.
523. H. Andreasen, D.O. Boerma, L. Niesen, W. Segeth, and H. De Waard, Hyperfine Interact., 1989, 47-48, 69.
524. J. Terra and D. Guenzburger, Phys. Rev. B, 1989, 39, 50.
525. H.G. Neumann, P. Zeggel, and K. Melzer, J. Non-Cryst. Solids, 1989, 108, 128.
526. F.C. Stedile, B.A.S. De Barros, C.V.B. Leite, F.L. Freire, I.J.R. Baumvol, and F.L. Schreiner, Thin Solid Films, 1989, 170, 285.
527. R.D. Vispute, V.P. Godbole, S.M. Chaudhari, S.M. Kanetkar, and S.B. Ogale, J. Mater. Res., 1988, 3, 1180.
528. N. Gao, H. Yu, X. Zhang, Y. Liu, Z. Zhang, and Z. Zhou, Wuji Cailiao Xuebao, 1989, 4, 125.
529. C.M. Cardile, A.J. Koplick, R. McPherson, and B.O. West, J. Mater. Sci. Lett., 1989, 8, 370; C.M. Cardile, J. Phys. Chem. Solids, 1989, 50, 221.
530. G.A. Teterin, V.F. Zinchenko, and Yu.V. Vorob'ev, Ukr. Khim. Zh., 1988, 54, 1244.
531. T. Shinjo, T. Mizutani, N. Hosoito, T. Kusuda, T. Takabatake, K. Matsukuma, and H. Fujii, Physica C, 1989, 159, 869.
532. V.M. Cherepanov, M.A. Chuev, S.S. Yakimov, V.Ya. Goncharov, S.A. Smirnov, and A.A. Bush, Pis'ma Zh. Eksp. Teor. Fiz., 1989, 49, 378.
533. H. Zhang, G.M. Wang, and Q.R. Zhang, Phys. Lett. A, 1989, 138, 517.

534. K.G. Prasad, R.P. Sharma, B. Jayaram, S.K. Agarwal, A. Gupta, A.V. Narlikar, and R. Srinivasan, Hyperfine Interact., 1988, 50, 543.
535. V.A. Andrianov, O.L. Anisimova, M.G. Kozin, A.Yu. Pentin, V.S. Shpinel, O.M. Ivanenko, and K.V. Mitsen, Pis'ma Zh. Eksp. Teor. Fiz., 1989, 49, 80.
536. E. Kuzmann, Z. Homonnay, A. Vertes, I. Kirschner, M. Gal, K. Torkos, B. Csakvari, K. Solymos, G. Horvath, et al, Hyperfine Interact., 1988, 46, 747.
537. C. Wang, L.T. Chen, M.L. Chu, H.L. Chang, F.W. Hsu, Y.S. Gou, T.J. Yang, and T.M. Uen, Chin. J. Phys. (Taipei), 1988, 26, 262.
538. M. Pasternak and R.D. Taylor, Solid State Commun., 1989, 69, 1157.
539. D. Tudela, M.A. Khan, and J.J. Zuckerman, J. Chem. Soc., Chem. Commun., 1989, 558.
540. T. Birchall and J. Vetrone, Hyperfine Interact., 1988, 40, 291; J. Chem. Soc., Chem. Commun., 1988, 877.
541. A. Silvestri, Inorg. Chim. Acta, 1988, 149, 5.
542. R. Barbieri, A. Silvestri, M.T. Lo Giudice, G. Ruisi, and M.T. Musmeci, J. Chem. Soc., Dalton Trans., 1989, 519.
543. H.K. Sharma, S. Lata, K.K. Sharma, K.C. Molloy, and P. C. Waterfield, J. Organomet. Chem., 1988, 353, 9.
544. P.J. Smith and B.N. Patel, Recl. Trav. Chim. Pays-Bas, 1988, 107, 167.
545. C. Carini, G. Pelizzi, P. Tarasconi, C. Pelizzi, K.C. Molloy, and P.C. Waterfield, J. Chem. Soc., Dalton Trans., 1989, 289.
546. S.P. Narula, S.K. Bharadwaj, H.K. Sharma, G. Mairesse, P. Barbier, and G. Nowogrocki, J. Chem. Soc., Dalton Trans., 1988, 1719.
547. A.K. Kumar, T.K. Chattopadhyay, and E. Rivarola, Indian J. Chem., 1988, 27A, 912.
548. D.W. Allen, S. Bailey, J.S. Brooks, and B.F. Taylor, Chem. Ind., 1988, 762.
549. R. Barbieri, A. Silvestri, and F. Huber, Appl. Organomet. Chem., 1988, 2, 525.
550. D.K. Srivastava, V.D. Gupta, H. Noeth, and W. Rattay, J. Chem. Soc., Dalton Trans., 1988, 1533.
551. A.S. Kothiwal, A. Singh, A.K. Rai, and R.C. Mehrotra, Indian J. Chem., 1988, 27A, 507.
552. T.T. Bamgboye and O.A. Bamgboye, Inorg. Chim. Acta, 1988, 144, 249.
553. T.T. Bamgboye, Inorg. Chim. Acta, 1988, 145, 243.
554. C. Silvestru and I. Haiduc, J. Organomet. Chem., 1989, 365, 83.
555. M. Gielen, C. Vanbellinghen, J. Gelan, and R. Willem, Bull. Soc. Chim. Belg., 1988, 97, 873.
556. M. Gielen, R. Willem, A. Delmotte, E. Joosen, A. Meriem, M. Melotte, C. Vanbellinghen, B. Mahieu, P. Lelieveld, Main Group Met. Chem., 1989, 12, 55.
557. R. Barbieri, A. Silvestru, and F. Huber, Appl. Organomet. Chem., 1988, 2, 457.
558. A.K. Varshney and S. Varshney, Synth. React. Inorg. Met.-Org. Chem., 1988, 18, 505.
559. G. Pelizzi, P. Tarasconi, C. Pelizzi, K.C. Molloy, and P. Waterfield, Main Group Met. Chem., 1987, 10, 353.
560. A.G. Davies, S.D. Slater, D.C. Povey, and G.W. Smith, J. Organomet. Chem., 1988, 352, 283.
561. S.J. Blunden, B.N. Patel, P.J. Smith, and B. Sugavanam, Appl. Organomet. Chem., 1987, 1, 241.

562. K.C. Molloy, S.J. Blunden, and R. Hill, J. Chem. Soc., Dalton Trans., 1988, 1259.
563. K.C. Molloy, T.G. Purcell, D. Cunningham, P. McCardle, and T. Higgins, Appl. Organomet. Chem., 1987, 1, 119.
564. W.K. Ng, W. Chen, V.G. Kumar Das, and R.J. Butcher, J. Organomet. Chem., 1989, 361, 53.
565. M.F. Mahon, K.C. Molloy, and P.C. Waterfield, J. Organomet. Chem., 1989, 361, C5.
566. M.J. Heeg, R.H. Herber, C. Janiak, J.J. Zuckerman, H. Schumann, and W.F. Manders, J. Organomet. Chem., 1988, 346, 321.
567. M. Dargatz, H. Hartung, E. Kleinpeter, B. Rensch, D. Schollmeyer, and H. Weichmann, J. Organomet. Chem., 1989, 361, 43.
568. T. Chivers, G.Y. Denes, S.W. Liblong, and J.F. Richardson, Inorg. Chem., 1989, 28, 3683.
569. G.L. Zhang and L. Niesen, J. Phys., 1989, 1, 1145.
570. A. Svane, J. Phys. C, 1988, 21, 569.
571. W. Peng, R.H. Hannon, H. Lee, A.P. Genis, V.J. Melim, G.W. Kimball, B. Darowski, and D.G. Hinks, Phys. Lett. A, 1989, 139, 91.
572. B. Dabrowski, D.R. Richards, D.G. Hinks, R.H. Hannon, W. Peng, H. Lee, A.P. Genis, V.J. Melim, and C.W. Kimball, Physica C, 1989, 160, 281.
573. B. Stahlberg, B.D. Mosel, W. Mueller-Warmuth, and F.G.K. Baucken., Glastech Ber., 1988, 61, 335.
574. H. Cremers, B.D. Mosel, W. Mueller-Warmuth, G.H. Frischat, and V. Braetsch., Phys. Chem. Glasses, 1989, 30, 79.
575. D. Mascherpa-Corral, B. Ducourant, and S. Alberola, J. Solid State Chem., 1988, 76, 276.
576. G. Wortmann, F. Godler, B. Perscheid, G. Kainal, and R. Schloegl, Synth. Met., 1988, 26, 109.
577. H. Mehner and H. Preiss, Isotopenpraxis, 1988, 24, 416.
578. T.F. Berlitz, J. Pebler, and J. Lorbeth, J. Organomet. Chem., 1988, 348, 175.
579. H. Sakai and Y. Maeda, Bull. Chem. Soc. Jap., 1989, 62, 33.
580. H. Sakai and M. Yao, M. Inui, K. Maruyama, K. Tamura, K. Takimoto, and H. Endo, J. Phys. Soc. Jpn., 1988, 57, 3587.
581. K. Tsunetomo, T. Sugishima, T. Imura, Y. Osaka, and H. Sakai, J. Appl. Phys., Part 1, 1989, 28, 671.
582. G. Langouche, H. Bemelmans, J. Odeurs, G. Borghs, M. De Potter, W. Deraedt, and M. Van Rossum, Mater. Sci. Forum, 1988, 38-41, 1245.
583. T. Birchall, C.S. Frampton, and P. Kapoor, Inorg. Chem., 1989, 28, 636.
584. H. Kageoka, K. Yamada, S. Ichiba, and H. Sakai, Kyoto Daigaku Genshiro Jikkensho Gakujutsu Koenkai Koen Yoshishu, 1989, 23, 59.
585. I. Dezsi, H. Pattyn, E. Verbiest, and M. Van Rossum, Phys. Rev. B, 1989, 39, 6321.
586. D. Gryffroy, R.E. Vandenberghe, E. De Grave, and R.M. Persoons, Hyperfine Interact., 1988, 50, 631.
587. T. Ruskov, R. Ruskov, I. Spirov, A. Kirov, T. Tomov, and S. Asenov, Nucl. Instrum. Methods Phys. Res., 1989, B36, 82.
588. E. Ikonen, J. Hietaniemi, and T. Katila, Gov. Rep. Announce. Index, 1988, 88, 834,781.
589. W. Potzel, W. Adlassnig, J. Moser, C. Schaefer, M. Steiner, and G.M. Kalvius, Phys. Rev. B, 1989, 39, 8236.
590. W. Adlassnig, W. Potzel, J. Moser, C. Schaefer, M. Steiner, and G.M. Kalvius, Nucl. Instrum. Methods Phys. Res., 1989, A277, 485.

591. S.P. Tewari and P. Silotia, J. Phys., 1989, 1, 5165.
592. E. Ikonen, P. Helistoe, J. Hietaniemi, and T. Katila, Gov. Rep. Announce. Index, 1988, 88, 834,780.
593. E. Ikonen, J. Hietaniemi, T. Katila, J. Linden and I. Tittonen, Hyperfine Interact., 1989, 47-48, 139.
594. S. Joss, K.M. Hasselbach, H.B. Buergi, R. Wordel, F.E. Wagner and A. Ludi, Inorg. Chem., 1989, 28, 1815.
595. Y. Kobayashi, M. Katada, H. Sano, T. Okada, K. Asai, and F. Ambe, J. Radioanal. Nucl. Chem., 1989, 136, 387.
596. Phys. Rev. B, 1988, 39, 909.
597. H. Von Brandis, F.E. Wagner, J.A. Sawicki and J.H. Rolston575.
598. J.A. Sawicki, T. Tyliszczak, F.E. Wagner, J.H. Rolston, A.P. Hitchcock, and D.E. Clegg, Physica B, 1989, 158, 203.
599. N. Kojima, H. Kitagawa, and H. Sakai, Kyoto Daigaku Genshiro Jikkensho Gakujutsu Koenkai Koen Yoshishu, 1989, 23, 181.
600. J.D. Cashion, A.C. McGrath, P. Volz, and J.S. Hall, Trans. -Inst. Min. Metall., Sect. C, 1988, 97, 129.
601. A.C. McGrath, J.S. Hall, and J.D. Cashion, Hyperfine Interact., 1988, 46, 673.
602. D.B. Dyson, R.V. Parish, C.A. McAuliffe, R.G. Pritchard, R. Fields, and B. Beagley, J. Chem. Soc., Dalton Trans., 1989, 907.
603. H.H.A. Smit, R.C. Thiel, and L.J. De Jongh, Z. Phys. D, 1989, 12, 193.
604. H.H.A. Smit, P.R. Nugteren, R.C. Thiel, and L.J. De Jongh, Physica B, 1988, 153, 33.
605. R.V. Parish, L.S. Moore, A.J.J. Dens, D.M.P. Mingos, and D.J. Sherman, J. Chem. Soc., Dalton Trans., 1989, 539.
606. J.J. Bour, R.P.F. Kanters, P.P.J. Schlebos and J.J. Steggerda, Recl. Trav. Chim. Pays-Bas, 1988, 107, 211.
607. R.P.F. Kanters, J.J. Bour, P.P.J. Schlebos, W.P. Bosman, H. Behm, and J. Steggerda, Inorg. Chem., 1989, 28, 2591.
608. S.S. Hafner and J. Stanek, Veroeff. Forschungsbereichs Goe-Kosmoswiss., 1988, 17, 102.
609. M. Winterer and E. Moersen, Hyperfine Interact., 1988, 50, 807.
610. S. Bhattacharyya, and D. Ghosh, J. Magn. Magn. Mater., 1989, 80, 276.
611. Y.Q. Jia and M.Z. Jin, J. Less-Common Met., 1989, 152, 25.
612. W.H. Shu, X. Zhao, W.M. Xu, X.X. Liu, M.Z. Jin, and D.M. Wu, J. Less-Common Met., 1988, 149, 101.
613. W. Su, W. Xu, X. Zhao, X. Liu, D. Wu, and M. Jin, Phys. Rev. B, 1989, 40, 102.
614. X.Liu, W. Su, M.Jin, W. Xu, D. Wu, and M. Liu, Kexue Tongbao (For. Lang. Ed.), 1988, 33, 1507.
615. P. Chaudouet, J.P. Senateur, F. Wiess, P. Dalmas de Reotier, P. Vulliet, and A. Yaouanc, J. Phys. Colloq., 1988, C8, 2161.
616. C.V. Tomy, R. Nagarajan, S.K. Malik, D.T. Adroja, N.C. Soni, and R. Prasad, Solid State Commun., 1988, 68, 531.
617. M. Taniwaki and H. Sasaki, Kogakubu Kenkyu Hokoku (Hokkaido Diagaku), 1988, 53.
618. Z.M. Stadnik, G. Stroink, and R.A. Dunlap, Phys. Rev. B, 1989, 39, 9108.
619. M. Lippmaa, E. Realo, and K. Realo, Phys. Lett. A, 1989, 139, 353.
620. V.A. Adrianov, O.L. Anisimova, M.G. Kozin, A.Yu. Pentin, K.V. Mitsen, and O.M. Ivanenko, Fiz. Tverd Tela, 1988, 30, 2859.
621. E. Kuzmann, Z. Homonnay, A. Vertes, M. Gal, K. Torkos, B. Csakvari, G.K. Solymos, G. Horvath, and J. Bankuti, Phys. Rev. B, 1989, 39, 328.
622. P.D. Battle, T.C. Gibb, P. Lightfoot, and D.C. Munro, J. Solid State Chem., 1989, 83, 31.

623. P. Le Bail, P. Colombet, and J. Rouxel, Solid State Ionics, 1989, 34, 127.
624. B.A. Milligan and J.A. Dumesic, J. Catal., 1989, 115, 180.
625. D. Malterre, A. Menny, P. Delcroix, J. Durand, G. Krill, and G. Marchal, Phys. Rev. B, 1989, 39, 10655.
626. N. Lossau, H. Kierspel, J. Langen, W. Schlabitz, D. Wohlleben, A. Mewis, and C. Sauer, Z. Phys. B, 1989, 74, 227.
627. L. Fournes, B. Chevalier, B. Lloret, and J. Etourneau, Z. Phys. B, 1989, 75, 501.
628. B. Perscheid, I. Nowik, G. Wortmann, G. Schmeister, G. Kaindl, and I. Felner, Z. Phys. B, 1989, 73, 511.
629. C. Godart, L.C. Gupta, C.V. Tomy, S. Patil, R. Nagarajan, E. Beaurepaire, R. Vijayaraghavan, and J.V. Yakhmi, Mater. Res. Bull., 1988, 23, 1781.
630. A. Bhagawat, M.D. Kurup, k.G. Prasad, and R.P. Sharma, Difus. Defect Data Pt. B, 1988, Bl-2, 33.
631. I.M. Askerov, V.F. Kobelev, v.F. Masterov, O.B. Tagiev, K.F. Shtel'makh, and L.F. Likholit, Fiz. Tekh. Poluprov., 1989, 23, 1307.
632. M.A. Salzberg, J.O. Thomas, and R. Waeppling, Solid State Ionics, 1987, 28-30, 1563.
633. H.T. Hintzen, C.J.M. Denissen, and H.M. Van Noort, Mater. Res. Bull, 1989, 24, 247.
634. G. Czjzek, H.J. Bornemann, and H.S. Li, J. Magn. Magn. Mater., 1989, 80, 23.
635. J.L. Oddou, M. Boge, C. Jeandy, and H.R. Rechenberg, J. Phys., Colloq., 1988, C8, 429.
636. M.W. Dirken, R.C. Thiel, and K.H.J. Buschow, J. Less-Common Met., 1989, 146, L15.
637. M.W. Dirken, R.C. Thiel, and K.H.J. Buschow, J. Less-Common Met., 1989, 147, 97.
638. J.A. Hodges, P. Imbert, J.B. Marimon de Cuhna, and J.P. Sanchez, Hyperfine Interact., 1988, 50, 569.
639. J.P. Sanchez, K. Tomola, and R. Kmiec, J. Phys. Colloq., 1988, C8, 435.
640. K. Tomala, J.P. sanchez, and R. Kmiec, J. Phys., 1989, 1, 4415.
641. M. Boge, C. Jeandey, J.L. Oudou, and Y.K. Yakinthos, J. Phys., Colloq., 1988, C8, 443.
642. P.C.M. Gubbens, A.M. Van der Kraan, and K.H.J. Buschow, J. Phys., Colloq., 1988, C8, 591.
643. P.C.M. Gubbens, A.M. Van der Kraan, and K.H.J. Buschow, Phys. Rev. B, 1989, 39, 12548.
644. J. Pszczola, Zesz. Nauk. Akad. Gorn.-Hutn. im. Stanislawa Staszica, Fiz., 1987, 1150, 5.
645. J.A. Hodges, P. Imbert, J.B. Marimon da Cuhna, and J.P. Sanchez, Physica C, 1989, 160, 49.
646. P.C.M. Gubbens, J.J. Van Loef, A.M. Van der Kraan, and D.M. De Leeuw, J. Magn. Magn. Mater., 1988, 76-77, 615.
647. A. Sengupta, S. Bhattacharyya, and D. Ghosh, Phys. Lett. A, 1989, 140, 261.
648. L.S. Fritz and N.S. Dixon, Hyperfine Interact., 1989, 52, 1.
649. G.J. Bowden, J.M. Cadogan, R.K. Day, and J.B. Dunlop, Hyperfine Interact., 1988, 39, 359.
650. V. Zevin, G. Zwicknagl, and P. Fulde, J. Magn. Magn. Mater., 1988, 76-77, 475.
651. R. Currat, R.G. Lloyd, P.W. Mitchell, A.P. Murani, and J.W. Ross, Physica B, 1989, 156-157, 812.

652. P. Bonville, J.M. Broto, A. Fert, F. Gonzalez-Jiminez, A. Hamzic, F. Hulliger, P. Imbert, G. Jehanno, and J.B. Marimon da Cuhna, J. Phys. Colloq., 1988, C8, 727.
653. P. Bonville, J.A. Hodges, F. Hulliger, P. Imbert, G. Jehanno, J.B. Marimon da Cuhna, and H.R. Ott, J. Magn. Magn. Mater., 1988, 76-77, 473.
654. P. Bonville, P. Imbert, G. Jehanno, and F. Gonzalez-Jiminez, J. Phys., Colloq., 1988, C8, 729.
655. C. Meyer, J.P. Gavigan, G. Czjzek, and H.J. Bornemann, Solid State Commun., 1989, 69, 83.
656. K.H.J. Buschow, G. Czjzek, H.J. Bornemann, and R. Kmiec, Solid State Commun., 1989, 71, 759.
657. S. Bertram, G. Kaindl, J. Jove, and M. Pages, Physica B, 1989, 158, 508.
658. V.M. Filin, V.F. Gorbunov, and S.A. Ulanov, Radiokhimiya, 1989, 31, 13.
659. G.V. Ivonova and J. Jove, Radiokhimiya, 1989, 31, 31.
660. G. Adrian, Kernforschungszent. Karlsruhe, [Ber.], KfK, KfK 4356.
661. G. Adrian, H. Appel, R. Bohlander, H. Haffner, and B. Kanellakopoulos, Hyperfine Interact., 1988, 40, 275.
662. U. Potzel, J. Moser, W. Potzel, S. Zwirner, W. Schiessl, F.J. Litterst, G.M. Kalvius, J. Gal, and S. Fredo, Hyperfine Interact., 1989, 47-48, 399.
663. J. Gal, S. Fredo, S. Tapuchi, J. Jove, M. Pages, A. Cousson, W. Potzel, F.J. Litterst, J. Moser, and G.M. Kalvius J. Magn. Magn. Mater., 1988, 7-77, 650.
664. V.S. Nefedov and V.M. Filin, Radiokhimiya, 1989, 31(1), 8.
665. V.M. Filin, L.F. Imofeeva, and N.I. Nogin, Radiokhimiya, 1989, 31(2), 8.
666. R. Chipaux, D. Bonniaseau, M. Boge, and J. Larroque, J. Magn. Magn. Mater., 1988, 74, 67.
667. C. Sauer, A. Holzwarth, Z. Kajcsos, and W. Zinn, Nucl. Instrum. Methods Phys. Res., 1988, B34, 377.
668. V.Ya. Goncharov, M.A. Polikarpov, I.V. Trushin, and S.S. Yakimov, Prib. Tekh. Eksp., 1989, 232.
669. A.S. Kasmin, and V.P. Rusakov, Prib. Tekh. Eksp., 1988, 56.
670. Yu.V. Petrikin, Yu.V. Baldokhin, A.A. Kiselev, and A.P. Kuprin, Prib. Tekh. Eksp., 1989, 91.
671. Z. Kajcsos, C. Sauer, R.A. Brand, W. Zinn, R. Kurz, M.A.C. Ligtenberg, and G. Van Aller, Hyperfine Interact., 1988, 46, 707.
672. J.S. Brooks and S. Thorpe, Hyperfine Interact., 1989, 47-48, 159.
673. G.N. Belozerskii, A.A. Dozorov, and B.C. Pavlov, Izv. Vyssh. Uchebn. Zaved., Fiz., 1989, 32, 23.
674. J.S. Zabinski and B.J. Tatarchuk, Nucl. Instrum. Methods Phys. Res., 1989, B42, 379.
675. J.S. Zabinski and B.J. Tatarchuk, Thin Solid Films, 1988, 166, 213.
676. Z. Kajcsos, C. Sauer, and W. Zinn, Hyperfine Interact., 1989, 51, 1147.
677. L. Sbriz, M. Villagran, and G. Valconi, J. Radioanal. Nucl. Chem., 1988, 128, 387.
678. B.J. Tatarchuk, Gov. Rep. Announce Index (U.S.), 1989, 89, Abs. No. 915314.
679. S. Bhagwat, S.N. Yedave, D.M. Phase, S.M. Chaudhari, S.M. Kanetkar, and S.B. Ogale, Phys. Rev. B, 1989, 40, 700.
680. J.A. Sawicki, G. Marest, and B. Cox, NATO ASI Ser., Ser. E, 1989, 170, 209.

681. G. Terwagne, M. Piette, P. Bertrand, and F. Bodart, Mater. Sci. Eng., 1989, B2, 195.
682. G. Terwagne, G. Van Tendeloo, and S.E. Donnelly, J. Appl. Phys., 1989, 65, 4225.
683. H. Reuther, Isotenopraxis , 1988, 24, 419.
684. D.L. Williamson, Y. Qu, R. Wei, W.S. Sampath, and P.J. Wilbur, Mater. Res. Soc. Symp. Proc., 1989, 128, 409.
685. G. Wagner, T. Louis, R. Leutenecker, and U. Gonser, Hyperfine Interact., 1988, 46, 501.
686. S.M.M. Ramos, L. Amaral, M. Behar, G. Marest, A. Vasquez, and F.C. Zawislak, Hyperfine Interact., 1988, 46, 483.
687. G. Lugert and G. Bayreuther, Phys. Rev. B, 1988, 38, 11068.
688. L. Mei, B. Weidong, and S. Bi, J. Phys. Colloq., 1988, C8, 1759.
689. M.A.Z. Vasconcellos, S.R. Teixeira, P.H. Dionisio, W.H. Schreiner, I.J.R. Baumvol, Nucl. Instrum. Methods Phys. Res., 1988, A280, 557.
690. S.R. Teixeira, F.L. Freire, and I.J.R. Baumvol, Appl. Phys. A, 1989, 48, 481.
691. A. Fnidiki and J.P. Eymery, J. Phys. Colloq., 1988, C8, 1735.
692. A. Fnidiki, J.P. Eymery, and M.F. Denanot, Hyperfine Interact., 1988, 45, 295.
693. J.P. Riviere, P. Boullaud, and J.P. Eymery, Phys. Lett. A, 1989, 138, 223.
694. F.W.A. Dirne and C.J.M. Denissen, J. Magn. Magn. Mater, 1989, 78, 122.
695. J.P. Eymery, A. Fnidiki, R. Krishnan, M. Tessier, and J.P. Vitton, Phys. Rev. B, 1988, 38, 11931.
696. G.B. Fratucello, E. Colombo, O. Donzelli, and F. Ronconi, Hyperfine Interact., 1988, 45, 255.
697. G.B. Fratucello, E. Colombo, O. Donzelli, and F. Ronconi, J. Phys. Colloq., 1988, C8, 1698.
698. F.A. Volkening, B.T. Jonker, J.J. Krebs, G.A. Prinz, and N.C. Koon J. Phys. Colloq., 1988, C8, 1699.
699. T. Kobayashi, K. Fukumura, Y. Isozumi, and R. Katano, Hyperfine Interact., 1988, 39, 419.
700. Yu.V. Maksimov, P.Ya. Kolotyrkina, Yu.V. Baldokhin, and A.V. Krylova, Izv. Akad. Nauk. SSSR, Ser. Khim., 1989, 1470.
701. W. Meisel, Hyperfine Interact., 1988, 45, 73.
702. P. Schaaf, P. Bauer, U. Gonser, Hyperfine Interact., 1988, 46, 541.
703. P. Schaaf, P. Bauer, U. Gonser, Z. Metallkd., 1989, 80, 77.
704. C. Pietzsch and U. Thieme, Isotenopraxis , 1988, 24, 405.
705. H. Tanimoto, S. Nasu, F.E. Fujita, Y. Shima, and K. Nakaoka, Nippon Kinzoku Gakkaishi, 1989, 52, 909.
706. M.A. El Khakani, H. Jaffrezic, G. Marest, N. Moncoffre, and J. Tousset, Mater. Sci. Eng., 1989, A115, 37.
707. P. Raj, P. Suryanatana, S. Sathyamoorthy, and A.M. Iyer, Mater. Res. Bull., 1989, 24, 717.
708. V.A. Tsurin, A.M. Sorkin, V.A. Pavlov, and A.P. Stepanov, Poverkhnost, 1988, 131.
709. E. Kuzmann, A. Vertes, Y. Ujihara, M. Fujinami, P. Kovacs, T. Ando, and T. Matsumoto, Hyperfine Interact., 1988, 45, 279.
710. A.A. Novakova, and G.V. Sidorova, Hyperfine Interact., 1988, 45, 337.
711. S.V. Ghaisas, N.R. Bulakh, S.A. Kulkarni, S.M. Kanetkar, S.B. Ogale, S.K. Date, and A. Mitra, J. Phys. D, 1989, 22, 364.
712. T. Fujii, M. Takano, R. Katano, Y. Bando, and Y. Isozumi, J. Appl. Phys., 1989, 66, 3168.

713. S. Joshi, R. Nawathey, V.N. Koinkar, V.P. Godbole, S.M. Chaudhari, S.B. Ogale, and S.K. Date, J. Appl. Phys., 1989, 64, 5647.
714. P.M. Solozhenkin, G.Yu. Pulatov, and Yu. Boldokhin, Dokl. Akad. Nauk. Tadzh.SSR, 1988, 31, 668.
715. J. Phys. Chem. Solids, 1989, 50, 369.
716. X.You, Z. Ding, X. Peng, and G. Xue, Electrochim. Acta, 1989, 34, 249.
717. I. Czako-Nagy, A. Vertes, L.M. Gratton, C. Tosello, and G. Principi, J. Radioanal. Nucl. Chem., 1989, 136, 275.
718. G. Massouras, J.A. Roger, A. Perez, G. Fuchs, and L. Romana, Hyperfine Interact., 1988, 46, 509.
719. J.A. Sawicki, A.P. Hichcock, and T. Tyliszczak, Physica B, 1989, 158, 681.

8
Gas-phase Molecular Structures Determined by Electron Diffraction

BY D. W. H. RANKIN AND H. E. ROBERTSON

1 Introduction

In this chapter we report on structural information determined for gas-phase molecules by electron diffraction and published in 1989. One notable feature is the high proportion of work reported this year emanating from the USSR. It is not clear whether this will be a permanent feature, or it is just an abnormal perturbation of the usual pattern. It certainly makes our task as reviewers more difficult, because many of the papers are published only in Russian, and the English translations of others are not available at the time this review is written. We apologise for errors and omissions, and acknowledge the assistance of Professor V. Spirodonov and Dr. V. Mastryukov in interpreting some of the papers.

The compounds included in this report are listed below, with the sections in which they can be found.

Section 2, Main Groups I, II and III: $(NaF)_2$, $(TlF)_2$, $(KOH)_2$, $(CsOH)_2$, InI, InI_3, In_2I_4, In_2I_6, B_4Cl_4, $(1,2\text{-}C_2B_{10}H_{11})_2$, $Hg(1,2\text{-}C_2B_{10}H_{11})_2$.

Section 3, Main Group IV: NC-C≡C-CN, ClCONCO, $(C_6F_5)_2$, perfluorodecalin, SiH_3CN, Me_2SiCl_2, Bu^tMe_2SiCl, $Si(Pr^i)_4$, $(Me_3Si)_3CSiCl_3$, $Si(C_5Me_5)_2$, $NMe(SiH_3)_2$, $NEt(SiH_3)_2$, $NPr^i(SiH_3)_2$, $NBu^t(SiH_3)_2$, GeI_4, GeI_2, $Bu^t{}_3SnBr$.

Section 4, Main Group V: Me_2NNH_2, CF_2=N-NCF_2, perfluoroquinolizidine, $N(n\text{-}C_3F_7)_3$, 2-Me-5-vinyltetrazole, o-$C_6H_4Br(NO_2)$, m-$C_6H_4Br(NO_2)$, 2,6-$C_6H_3Br(NO_2)_2$, 1,3,5-$C_6Cl_3(NO_2)_3$, PF_4H, PF_3H_2, $PO(OPF_2)_3$, 2-Br-1,3,2-dioxarsenane, $C_6H_5AsBr_2$, o-$C_6H_4(AsBr_2)Br$, p-$C_6H_4(AsBr_2)Br$, o-$C_6H_4(AsCl_2)Cl$, Me_3SbCl_2.

Section 5, Main Group VI: 3-Br-furan, FSSF, S=SF_2, S_2F_{10}, CH_3SF_3, CF_3SF_3, $Me_3SiNSNSiMe_3$, p-$C_6H_4(SH)_2$, 1,3,5-trithiane, tetrahydrothiophene-1-oxide, *cis*-2-Me-tetrahydrothiophene-1-

oxide, perfluorotetrahydrothiophene, *trans*-methyl (chlorovinyl)-sulphone.

Section 6, Transition Metals: ZnF_2, $CdBr_2$, CuF_2, CrF_3, $LaCl_3$, $TiCl_3$, TiI_3, $MeTiCl_3$, $W(OMe)_6$, $Cr(OBu^t)_4$, $Cu(S_2CNMe_2)_2$, $Zn(S_2CNMe_2)_2$, $Fe(C_5Me_5)(P_5)$.

As usual we give distances and errors (quoted in parentheses after values of refined parameters) as they are reported in the original papers, and we indicate whether the papers are r_a, r_g, r_α etc.

2 Compounds of Elements in Main Groups I, II and III

There have been several studies of structures of alkali-metal halides in recent years, and this year one further paper has been added to the collection.[1] In this new work attention has been concentrated on the dimer of sodium fluoride, which accounts for $21^{+6}_{-4}\%$ of the gas at 1380 K. Refinements were based on observed vibrational frequencies and electron diffraction intensities, and yielded the equilibrium Na-F distance 2.072(7) Å, and FNaF angle, 96(3)°, and all F matrix elements except those for the B_{2u} and B_{3u} vibrations. The standard enthalpy of dissociation for the dimer at 298 K was given as 256.1(44) kJ mol^{-1} and the enthalpy of dissociation as 141.7 J mol^{-1} K^{-1}.

The percentage of the dimer of thallium fluoride at 700 K was found to be 51^{+7}_{-12} in a new electron diffraction study.[2] The structure of the dimer was taken from published spectroscopic work, and the dimer was then shown to be planar, with D_{2h} symmetry. The r_e Tl-F distance, obtained in a refinement based on diffraction data and vibrational frequencies, was 2.264(10) Å, while the FTlF angle was 73(1)°, and the F matrix elements for the A_{1g} and B_{1g} vibrational modes were also refined. Values for ΔH^o and ΔS^o for dissociation of the dimer at 298 K were given as 134.8(34) kJ mol^{-1} and 134.7 J mol^{-1} K^{-1} respectively.

Potassium and caesium hydroxides, studied in the gas phase at 990 and 803 K respectively,[3] also consist of mixtures of monomers and dimers. For the caesium compound, the percentage dimer is given as 31^{+18}_{-6}, and r_g parameters are 2.62(1) Å for Cs-O, 3.95 Å for Cs···Cs and 2.87(58) Å for O···O, assuming a

D_{2h} structure. In refinements based on the four heaviest atoms alone, equilibrium parameters could be determined, with r(Cs-O) 2.58(1) and <OCsO 81.3(10)°. Three F matrix elements were also refined. ΔH_o for dissociation of dimer at 298 K is 146^{+14}_{-11} kJ mol^{-1} and ΔS_o is 129.3 J mol^{-1} K^{-1}. Potassium hydroxide is 37(3)% dimer, with r_e(K-O) 2.374(9) Å and <OKO 83.0(10)°, ΔH^o $181.6^{+7.5}_{-4.9}$ kJ mol^{-1} and ΔS^o 137.9 J mol^{-1} K^{-1}. In this case four F matrix elements were refined.

The iodides of indium are difficult to study, as there are four species known in the gas phase (InI, InI_3, In_2I_6 and a compound best described as $InInI_4$), and vapours above solid indium iodides may consist of several of these.[4] The simplest compound, InI, has an r_e bond length of 2.757(3) Å, while in InI_3, which has D_{3h} symmetry, the r_α bond length is 2.617(5) Å. In_2I_6 has D_{2h} symmetry, with two bridging iodine atoms, r_a(In-I) 2.614(5) Å for terminal bonds and 2.80(3) Å for the bridges. The terminal IInI angle is 130(7)° and the corresponding bridge angle is 80(3)°. In_2I_4 is similar to In_2I_6, with two terminal iodine atoms removed. The bond lengths are 2.640(6) Å for the terminal atoms, 2.740 (6) Å for the bridges from the four-coordinated indium, and 3.53(9) Å for the bridges from the two-coordinated indium atoms. The terminal IInI angle is 125(5)° and the corresponding bridge angle is 97(4)°. Standard enthalpies for dissociation of In_2I_6 and In_2I_4 (giving InI + InI_3) at 298 K are 100(4) and 111(8) kJ mol^{-1} and entropies for the same reactions are 161(10) and 142(8) J mol^{-1} k^{-1}.

The structure of tetraboron tetrachloride in the gas phase has been determined using a sample of less than 20 mg.[5] The boron cage is a regular tetrahedron with r_a(B-B) 1.703(11) Å, and the chlorine atoms are bound, one to each boron atom, with r(B-Cl) 1.693(Å). Although the two bond-lengths are very similar, they can both be determined reliably from the non-bonded B···Cl and Cl···Cl distances.

In 1,1′-bis(1,12-dicarbadodecaborane) there is an unusual C-C bond, between two six-coordinate carbon atoms. However, it is not easy to measure its length directly by electron diffraction, because there are so many other similar B-C and B-B distances, and so the reported distance (r_a) of 1.553(26) Å is not as precise as one would hope for.[6] It is longer than bonds between four-coordinate carbon atoms, but considerably shorter than C-C bonds within an icosahedral framework, as in 1,2-dicarbadodecaborane. Distances within with the carbaborane core, for which D_{5d} symmetry was assumed, were B(2)-B(3) 1.795(13), B(2)-B(7) 1.760(17) and B-C 1.736(11) Å. The experimental data gave no significant information on the conformation adopted, and a staggered arrangement was assumed in the final refinement.

Local D_{5d} symmetry with $B_{10}C_2$ cages was also assumed in a study of a related compound, $Hg(1,2\text{-}C_2B_{10}H_{11})_2$, in which the two carbon atoms, one from each carbaborane group, are linked through a mercury atom.[7] In this case r_α parameters were given, and within the core the distances were: B(2)-B(3) 1.786(4), B(2)-B(7) 1.774(10) and B-C 1.701(7) Å. The Hg-C systems refined to 2.084(13) Å, and linear co-ordination at the mercury atom was assumed. Structures of 1,2-dicarbadecaborane itself, and with various substituent atoms at carbon, including carbon, lithium, phosphorus, germanium, tin, iodine, platinum and mercury, are known, and there seems to be no significant variations in the cage structure, despite the wide range of substituents used.

3 Compound Elements in Main Group IV

The structure of dicyanoactylene in the vapour phase has been determined for the first time.[8] Electron diffraction data were used with vibrational corrections calculated from a harmonic force field to give an r_α^{o} structure, with r(C≡N) 1.161(5), r(C-C) 1.367(3) and r(C≡C) 1.198(11) Å. Using this structure and harmonic corrections the ground-state rotation constant, B_o, was calculated to be 0.044891(86) cm^{-1}. The pure rotational Raman spectrum did not give B_o directly, because the spectrum consisted mainly of unresolved contributions from excited states of two low-frequency vibrational modes. Analysis gave a mean value, B_{av}, of 0.044867(19) cm^{-1}. This very close agreement with B_o is not expected, as B for long, linear molecules usually increases with bending excitation. So either the vibrational corrections are for some reason inadequate, or the molecule undergoes some unusual structural change when it bends.

A study of chlorocarbonylisocyanate at room temperature has shown that the gas consists of a mixture of *syn* and *anti* conformers, the *anti* form predominating.[9] About 75% is in this form, and this corresponds to a free energy difference of 3.0(13) kJ mol^{-1}. Differences between geometrical parameters for the two conformers are fixed at values calculated *ab initio*. These calculations reproduced the experimental *anti* geometry well but predicted the *syn* form to be the more stable. Refined r_α parameters included C=O(isocyanate) 1.139(16), C=O(carbonyl) 1.201(16), N=C 1.218(23), N-C 1.384(6), C-Cl 1.757(5) Å <CNC 127.1(16), <N-C=O 124.8(15), <NCCl 115.8(8) and <N=C=O 173.4(23)°.

The structure of perfluorobiphenyl has been reinvestigated as part of a study of the effects of substituents on internal rotation in biphenyls.[10] A large-amplitude torsional motion about the central C-C bond was modelled in several different ways, and in the most satisfactory one the minimum in the potential function corresponded to a dihedral angle between ring planes of 64.4(7)°. Other r_a parameters included distances C-F(mean) 1.3275(9), C(1)-C(2) 1.404(4), C(2)-C(3) 1.391(7), C(3)-C(4) 1.384(10), C(1)-C(1′) 1.492(5) Å and <C(2)C(1)C(6) 117.9(4)°.

Perfluorodecalin exists as a mixture of *cis* and *trans* isomers in almost equal amounts. To study the structures of these two isomers by electron diffraction necessitates making many assumptions, so that the total number of refined parameters becomes small enough to cope with.[11] Only chair conformations were considered, and each ring was assumed to have D_{3h} symmetry, with all C-C distances and all CCC angles equal. Each local C_2CF_2 fragment was assumed to have C_{2v} symmetry. Under these conditions the major parameters (r_g) were as follows, the values for the *trans* isomer preceding that for the *cis* isomer in each case: *r*(C-F) 1.338(2), 1.339(2); *r*(C-C) 1.556(3), 1.553(3) Å <CCC 111.9(2), 120.0(2); <FCF 109.2(3), 107.6(4)°. Space-filling models showed that both the structures appear to be ellipsoids, with their surfaces uniformly covered with fluorine atoms.

Molecular structure of silylcyanide has been determined in the gas phase by electron diffraction and from rotation constants, in solution in a liquid crystal by n.m.r.

spectroscopy, and in the crystalline phase at 100 K by X-ray crystallography.[12] The combined analysis of all of the gas-phase data gave r_{av}(Si-C) 1.8496(4), r(C≡N) 1.1563(5), r(Si-H) 1.4873(11) and <HSiC 107.25(4)°. In the crystal N···Si interactions between molecules lead to lengthening of the Si-C distance to 1.8996(18) and narrowing of the HSiC angles to 100.4(7)°. In the liquid crystal solution the structure is mid-way between those of gaseous and crystalline phases, and it is believed that there is weak co-ordination of cyano groups in the solvent to the silicon atoms.

New diffraction data and old rotation constants have been analysed together to give an improved structure for Me_2SiCl_2.[13] Two different structures fit the electron diffraction data, one with the angle CSiC at 108° and one with it at 114°. The former is not consistent with the rotational data. The refined angles (r_α basis) obtained in the combined analysis are 114.2(2)° for CSiC and 107.5(1)° for ClSiCl, and r_g distances are 1.850(4) for Si-C and 2.054(2) Å for Si-Cl.

The t-butyl group in t-butyldimethylsilylchloride has no apparent steric effect other than a small tilt of the three-fold local symmetry axis of the group [1.6(17)°] away from the Si-methyl groups.[14] The fully staggered conformation is adopted, and other paramaters are exactly as would be expected in uncrowded molecules: r_a(Si-C) 1.875(1), r(C-C) 1.539(2), r(Si-Cl) 2.086(2) Å, <ClSiC(Bu) 106.3(11), <C(Me)SiC(Me) 111.1(26) and <CCC 109.2(3)°.

Tetra(iso-propyl)silane is rather more crowded, but by adopting a conformation which minimises interactions between the isopropyl groups, most of the geometrical parameters are able to remain fairly close to their normal values.[15] The overall symmetry is S_4, with the propyl groups each twisted 14.8(7)° away from positions in which the molecular symmetry was D_{2d}, and models with D_2 symmetry were rejected. The Si-C and C-C bond lengths (r_a), 1.919(2) and 1.549(1) Å respectively, were both a little greater than normal, and the SiCC angles, 113.3(3)° also reflected the steric crowding. The CCC angles in the propyl groups were found to be 108.5(5)°. With S_4 symmetry the angles at the central silicon atom were not all equal: two of them were 115.7(7)°, and the other four were therefore 106.4(4)°.

Tris(trimethylsilyl)(trichlorosilyl)methane is another sterically crowded compound, whose structure is very similar to that of tetrakis(trimethylsilyl)methane.[16] The angles at the central carbon atom are close to tetrahedral, being 108.1(6)° between the $SiMe_3$ groups and 110.9(6)° between $SiMe_3$ and $SiCl_3$ groups. The average bond lengths (r_a) to this central atom is 1.909(8)°, but the difference between the two types of bond could not be determined. The Si-C distance in the $SiMe_3$ groups is 1.878(6) Å, so the crowding is seen in the lengthening of the inner Si-C bonds. It is also apparent in the C(Me)SiC(Me) and ClSiCl angles, which are 107.0(11) and 104.4(11)° respectively. All four silyl groups are twisted by 23(3)° from the fully eclipsed configuration, in which the central $C(SiX_3)_4$ skeleton approximates to T_d symmetry, reducing this to approximately T symmetry.

Bis(pentamethylcyclopentadienyl)silicon, the first stable derivative of silicon in oxidation state(II), is also a remarkable compound so far as its structure is concerned. In the crystalline phase, some of the molecules have a ferrocene-like structure, with the silicon atom sandwiched between two parallel planes, but in other molecules the ring planes are not parallel, although the silicon atom lies close to the point at which the ring C_5 axes intersect. The latter structure is the one found in the gas phase.[17] Many models were considered, but only those with two η^5 rings fitted the data at all well. In the best one the angle between the rings was 22.3(12)° and the rings were staggered with respect to one another, so that the overall molecular symmetry was C_s. This thermal average (r_a) structure could, of course, be significantly different from the equilibrium structure, and the possibility that the rings are parallel cannot be ruled out. The co-existence of two arrangements in the crystal indicates that little energy is involved in bending at silicon, and it is very likely that there is a large amplitude bending vibration in the gas phase. Refined parameters included 2.129(12) Å for the distance from the silicon atom to the ring centres and 0.021(11) Å for the distance from the silicon atom to the point where the ring C_5 axes intersected, giving Si-C distances in the range 2.34 - 2.55 Å.

Four disilylamines with different alkyl substituents on

nitrogen (methyl,[18] ethyl,[18] isopropyl,[19] and t-butyl[19]) have been studied. The parameters listed in Table 1 show clear effects attributable to the size of the alkyl groups.

Table 1 - Geometrical parameters (r_a) in some silyl amines

	r(Si-N)/Å	r(C-N)/Å	<SiNSi/°	<SiNC/°
$NMe(SiH_3)_2$	1.723(1)	1.483(3)	125.2(2)	117.5(1)
$NEt(SiH_3)_2$	1.726(1)	1.461(5)	123.0(2)	118.5(1)
$NPr^i(SiH_3)_2$	1.727(1)	1.487(5)	121.6(2)	{121.9(5)
				{116.5(5)
$NBu^t(SiH_3)_2$	1.735(1)	1.479(8)	118.4(4)	120.8(2)

In all the compounds the bonds to nitrogen are coplanar, but as the alkyl group increases in size the SiNSi angle gets narrower, from 127.7(1)° in $NH(SiH_3)_2$ to 118.4(4)° in $NBu^t(SiH_3)_2$. In the ethyl amine the C-C bond lies almost in a plane perpendicular to that of the $NCSi_2$ skeleton, so the overall symmetry is close to C_s. The C-C distance in this compound is 1.558(14) Å, and the CCN angle is 112.7(7)°.

In the isopropyl amine the alkyl group is twisted so that the unique C-H bond nearly eclipses one of the Si-N bonds, again giving C_s symmetry. But in this case the mirror plane bisects the plane of the $NCSi_2$ group, leading to non-equal SiNC angles. On the more crowded side the angle of 121.9° is close to those in the N-butyl amine, whereas the angle on the less crowded side is similar to that in the N-methyl compound. Apart from the large change in angles at nitrogen, the parameters are remarkably invariant. The lengthening of the Si-N bond in the butyl amine by about 0.01 Å is the only indication of steric strain.

In a study of the iodides of germanium, the structure of GeI_4 was first determined with a sample temperature of

350 K.[20] The Ge-I distance (r_g) was 2.515(5) Å, and the data were fully consistent with the expected tetrahedral structure. Using vibrational frequencies as experimental data and refining F matrix elements as well as geometrical parameters gave $r_e{}^h$ 2.507 Å. Germanium diiodide was generated by heating solid germanium with GeI_4. The resulting gases included from 70 to 83% GeI_2, with more or less equal amounts of GeI_4 and I_2 accounting for the rest of the mixture. The Ge-I distance ($r_e{}^h$) refined 2.540 Å (no error is quoted) and the IGeI angle refined to 102.1°. In other germanium dihalides the bonds are longer than in the tetrahalides by 0.06 to 0.08 Å, and angles are 97, 100 and 101° in the difluoride, dichloride and dibromide respectively.

Tri(t-butyl)tin bromide shows some evidence of steric crowding,[21] most notably in the wide CSnC angle of 113.7(8)°. (Angles are r_α distances r_g in this work.) The t-butyl groups avoid much of the strain by twisting 17(7)° away from the fully staggered arrangement, so that the molecule has overall C_3 symmetry. Torsion angles of this magnitude are very frequently found in compounds with three or four t-butyl or tris(trimethylsilyl) groups. Other major parameters are r(Sn-C) 2.201(17) Å, 0.04 Å longer than in di(t-butyl)tin dichloride and 0.09 Å longer than in dimethyltin dichloride, r(Sn-Br) 2.533(5) Å, r(C-C) 1.532(5) Å and <SnCC 110.4(4)°.

4 Compounds of Elements in Main Group V

The conformation of 1,1-dimethylhydrazine has been studied by electron diffraction and by MNDO and *ab initio* calculations at the 3-21G level.[22] The angle between the planes bisecting the NC_2 and NH_2 groups was found experimentally to be 97(8)°, while the *ab initio* calculations gave an energy minimum at 90°. The MNDO calculations indicated that there were two minima, one at 90 and one at 180°, the latter being very slightly lower in energy (<1 kJ mol^{-1}) but the barrier between the two was only 1 kJ mol^{-1}. Other experimental parameters (r_a) were r(N-C) 1.470(2), r(N-N) 1.435(2) Å, <NNC 108.3(3) and <CNC 111.9(5)°.

Perfluoro-2,3-diaza-1,3-butadiene, CF_2=N-N=CF_2, was shown by electron diffraction to have a planar *anti* structure, and there was no evidence for the presence of a second conformer.[23] High level *ab initio* calculations, including allowance for electron correlation, indicated that a *gauche* conformation, 7.6 kJ mol^{-1} higher in energy, should also be present, but at no more than 10% abundance. There is therefore a marked contrast with perfluorobutadiene, for which the *gauche* conformation is the most stable, presumably because this reduces non-bonded F···F interactions. In the diaza compound r_a bond lengths are 1.421(12) for N-N, 1.264(7) for C=N and 1.304(4) Å for C-F, while the NNC and FCF angles are 112.7(9) and 110.8(9)° respectively. The CF_2 groups are tilted 5.0(8)° towards the nitrogen lone pair of electrons, so one NCF angle in just below 120° and the other is nearly 130°. The contrast with the small NNC angle is remarkable.

In Section 3 the structures of the *cis* and *trans* isomers of perfluorodecalin were reported. Perfluoroquinolizidine is the corresponding compound with one bridging CF group replaced by nitrogen, and in this case only the *trans* isomer, in which the nitrogen lone pair is *trans* to the remaining bridging fluorine atom, was observed.[24] Again it was necessary to make many assumptions about the structure, to reduce the problem to manageable proportions. Parameters (r_g) reported include distances C-C (all equal) 1.554(4), N-C 1.452(11) and C-F 1.343(2) Å, <C(2)NC(6) 116.8(14) [C(6) is the bridging carbon atom] and <FCF (all equal) 107.9(5)°.

Perfluorotri-n-propylamine is an even more difficult molecule for study by electron diffraction, because so many conformations are possible. The best fit to the experimental data was obtained with C_3 symmetry and with the three N-C bonds coplanar.[24] The N-C and mean C-C distances (r_g) of 1.459(8) and 1.570(5) Å respectively are longer than those in $N(CF_3)$ and C_2F_6, and indicate that there are strong steric interactions, which are also shown by wide angles NNC 118.1(12) and CCC 113.7(7)°. The difference between the inner and outer C-C bond lengths was 0.02(1) Å, giving individual distances of 1.580(8) and 1.560(8) Å. The FCF angles were 110.3(4)° in the CF_3 groups and 105.4(4)° in the CF_2 groups.

The structure of 2-methyl-5-vinyltetrazole, illustrated below, has so many similar bond lengths that it is impossible to determine it by electron diffraction alone.[25]

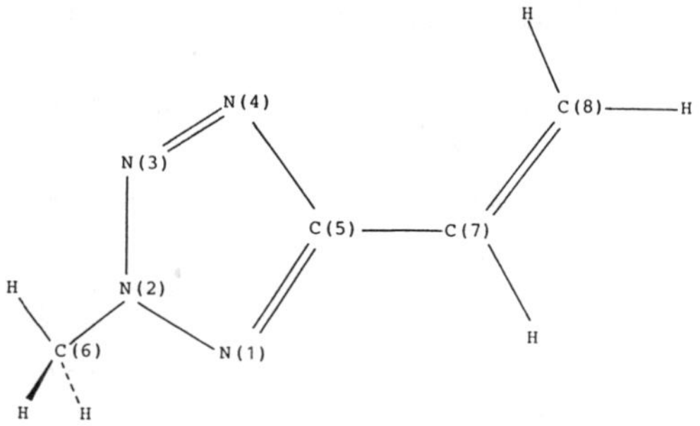

It was therefore necessary to fix differences between most of the bond lengths and the only ones refined were *r*[C(5)-N(1)] 1.335(2) Å, tied to all the other C-N, N-N and C=C distances, and *r*(C-C) 1.459(9) Å (both such bonds assumed to be of equal length). Four bond angles were also refined: N(4)C(5)C(7) 126.5(37), N(1)C(5)N(4) 113.2(47), N(2)N(1)C(5) 102.0(9) and CCC 127.2(30)°. The angle between the planes of the ring and the vinyl group was 10(10)° so the whole molecule is effectively planar.

Four structures of benzene derivatives with both nitro groups and halogen atoms as a substituents have been reported recently. The *ortho* and *meta* isomers of bromonitrobenzene and 2,6-dinitrobromobenzene have been studied,[26] and in each case the NO_2 group is twisted out of plane of the ring, by 43(3), 25(5) and 58(2)° respectively. These are thermal average parameters, and further analysis showed that the equilibrium angles were much smaller, being 43° for o-$C_6H_4Br(NO_2)$, while the equilibrium conformation for m-$C_6H_4Br(NO_2)$ is planar. In the dinitro compound the two nitro groups were twisted so that the overall molecular symmetry was C_2. Other important parameters (r_g) are listed in Table 2. The most notable features are the shortening of the C-Br bond in m-$C_6H_4Br(NO_2)$ and the apparent variability in the ONO angles. Some CCN angles are also listed, but it is not stated which carbon atoms are involved.

In 1,3,5-trinitro-2,4,6-trichlorobenzene,[27] the three nitro groups are all twisted out of the ring plane in the same sense, by 79.2(4)°, so that the overall molecular symmetry is D_3 in the thermal average structure, although it is quite possible that the equilibrium structure has D_{3h} symmetry,

Table 2 - Geometrical parameters (r_g) in some bromonitrobenzenes

	p-$C_6H_4Br(NO_2)$	m-$C_6H_4Br(NO)_2$	2,6-$C_6H_3Br(NO_2)_2$
r(N=O)/Å	1.218(3)	1.238(3)	1.229(3)
r(C-N)/Å	1.494(14)	1.448(14)	1.469(9)
r(C-C)/Å	1.388(3)	1.394(3)	1.393(3)
r(C-Br)/Å	1.894(6)	1.865(8)	1.899(9)
<ONO/°	128.6(15)	128.8(14)	125.6(9)
<CCC(at NO_2 group/°	120.4(24)	121.4(15)	122.7(9)
<CCC (at Br)/°	119.6(12)	121.4(10)	117.1(12)

with the groups twisted 90°. In this compound the ONO angle is very wide, 131.4(3)°. The bond lengths (r_g) are 1.399(3) for C-C, 1.464(6) for C-N, 1.217(2) for N-O and 1.699(4) for C-Cl. There is no significant deviation of the ring from regular hexagonal symmetry, with the angle at carbon atoms carrying NO_2 substituents being 120.6(5)°.

Both tetrafluorohydridophosphorane, PF_4H and trifluorohydridophosphorane, PF_3H_2, adopt trigonal bipyramidal structures with hydrogen atoms occupying equatorial sites, and C_{2v} molecular symmetry.[28] In PF_4H the equatorial and axial P-F bond lengths are 1.537(4) and 1.591(4) Å respectively, and the corresponding distances in PF_3H_2 are 1.544(4) and 1.613(2) Å. The axial P-F bonds are lengthened on replacement of equatorial fluorine by hydrogen as the axial P-F distance in PF_5 is 1.580 Å. The lengthening of the equatorial P-F bonds is much less significant. The hydrogen atom positions are not well determined, but the HPF_{ax} and HPF_{eq} angles in PF_4H are known once the heavy atoms have been located. These angles are 87.8(2) and 123.0(9)°, indicating that the axial fluorine atoms are bent towards and the equatorial fluorine atoms are bent away from the hydrogen atom. In PF_3H_2 the HPF_{eq} angle is poorly determined [119.6(44)°] but the $F_{ax}PF_{eq}$ angle, 91.7(2)°, again indicates bending of the axial fluorine atoms towards the hydrogen atoms.

The difluorophosphino ester of phosphoric acid, $PO(OPF_2)_3$, has been studied in both gaseous and crystalline phases.[29] In the gas phase the molecule has C_3 symmetry. Relative to the position in which the outer P-O bonds eclipses the P=O bond and the FPF bisectors eclipse the inner P-O bonds, the torsion angles about the P-O bonds are both 127°, with standard deviations of 1-2°. The r_a bond lengths were P=O 1.401(10), P-O(mean) 1.583(11), with an insignificant difference in length between the inner and outer bonds, and P-F 1.581(10)°. The angles were O-P-O 102.8(7), POP 132.0(16), OPF 98.3(6) and FPF 96.4(14)°. On cooling two crystalline phases were formed, the higher-temperature form converting to a second form without destruction of the crystal at 77 K. In the lower-temperature form the molecules are aligned along the 3-fold axis, with short contacts between all four phosphorus atoms in one molecule and the unique oxgygen atom of its neighbour (3 contacts of 2.88 Å and one of 3.14 Å). These lead to lengthening of the P=O bond by 0.08 Å to 1.478 Å, widening of the O-P-O angles by 5° to 107.6°, and changes in the conformation of the OPF_2 groups. The inner P-O bonds are significantly shorter in the crystal than in the vapour, while the reverse is true of the outer P-O bonds.

The six-membered heterocycle, 2-bromo-1,3,2-dioxarsolane, illustrated below, exists in the chair form with the bromine atom in an axial position in the gas phase at 360 K. The

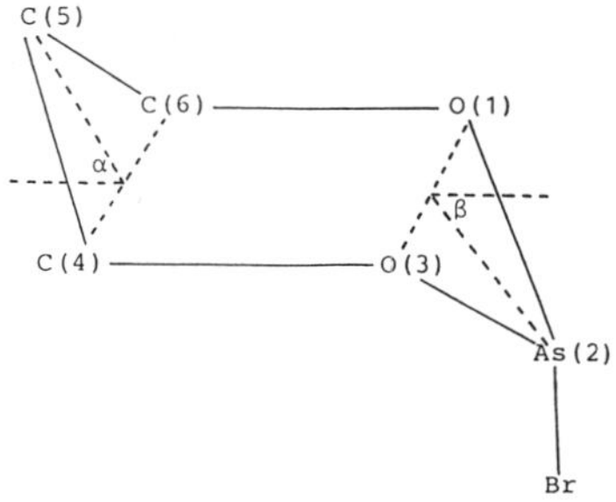

ring structure is far from being regular, with the angles α and β shown in the figure 49.5(43) and 32.4(18)° respectively, the angles including CCC 110.5(27) and OAsO 100.9(16), and bond lengths (r_a) C-C 1.539(15), C-O 1.444(6) and As-O 1.733(3) Å. The angle OAsBr refined to 97.8(8)°, so all the ring atoms adopt their normal configurations and they together lead to the observed ring structure. The As-O bonds are 0.026 Å shorter than in the corresponding five-membered ring compound, 2-bromo-1,3,2-dioxarsolane, and several ring angles are substantially wider than in that more highly strained system.

Structures of four dihalo(phenyl)arsines, three of them also having halogen substituents in the benzene ring, have been reported. For $C_6H_5AsBr_2$ the data could be fitted equally well with the $AsBr_2$ group twisted to 0-15 or 75-90° from the position in which the bisector of the $AsBr_2$ group lay in the ring plane.[31] The conformation made little difference to other refining parameters, and those given in Table 3 are averages of values obtained in the refinements.

Table 3 - Geometrical parameters (r_a) for some aryldihaloarsines

	$C_6H_5AsBr_2$	p-C_6H_4 ($AsBr_2$)Br	o-C_6H_4 ($AsBr_2$)Br	o-C_6H_4 ($AsCl_2$)Cl
r(C-C)/Å	1.398(4)	1.396(4)	1.400(4)	1.396(4)
r(C-X)/Å		1.866(14)	1.886(17)	1.735(7)
r(As-C)/Å	1.952(11)	1.967(14)	1.954(17)	1.955(9)
r(As-X)/Å	2.359(2)	2.361(2)	2.368(3)	2.183(3)
<CAsX/°	100.7(8)	98.7(9)	97.1(9)	98.8(8)
<XAsX/°	98.7(5)	99.3(6)	99.1(6)	96.7(7)

There was similar uncertainty about the conformation of p-$C_6H_4(AsBr_2)Br$, with twist angles of 15 and 60-75° fitting well.[31] For o-$C_6H_4(AsBr_2)Br$ the zero twist position was defined as that in which the $AsBr_2$ bisector lay in the ring plane, on the side away from the ring bromine substituent.[32] The observed angle, 154(3)° thus indicates that the three bromine atoms are well away from one another, although not quite the maximum separation, which would correspond to a twist angle of 180°. The twist angle observed for the analogous chlorine compound, o-$C_6H_4(AsCl_2)Cl$,[33] was 145(5)°. Other parameters for these four compounds are listed in Table 3. Results of normal cooridinate analyses and CNDO calculations are given in the original papers.

It is well known that in general the more electronegative substituents in phosphoranes tends to occupy the axial positions. There are relatively few five-coordinate derivatives of arsenic and antimony, and the structure of trimethylantimony(V) dichloride has been studied[34] to see whether the same rules apply to this compound. They do. The structure is trigonal bipyramidal, with axial chlorines, although it was not possible to rule out small distortions from the ideal D_{3h} symmetry for the framework. There was no evidence at all for any isomer with one or two chlorines in equatorial positions. With such a regular sturcture, there are just two parameters to find the heavy atom structure. These were r_g(Sb-C) 2.107(6) and r(Sb-Cl) 2.460(6) Å.

5 Compounds of Elements in Main Group VI

Combined analysis of existing electron diffraction and microwave data has given a structure of 3-bromofuran which is more precise than that obtained using diffraction data alone.[35] However, it was still necessary to make some structural assumptions, and C_{2v} symmetry was imposed on the ring. This is inevitably somewhat unsatisfactory, particularly as halogen substiuents are known to change distances and angles between neighbouring atoms. The major parameters (r_a and $<_\alpha$) are: C-Br 1.862(5), C-C 1.466(10), C=C 1.351(12), C-O 1.364(9) Å, <COC 110.5(6), <CCO 108.0(7) and <C-CBr 126.9(9)°. From these parameters the remaining ring angle, CCC, can be deduced to be 106.8°, and <C=CBr is 126.3°.

There are two isomers of disulphur difluoride, F-S-S-F, which has a structure related to that of hydrogen peroxide, and $S{=}SF_2$, with a pyramidal central sulphur atom. Both have been studied by combined analysis of new electron diffraction data and existing rotation constants.[36] The experimental r_{av}

parameters for FSSF are r(S–S) 1.890(2), r(S-F) 1.635(2) Å, <SSF 108.3(2)° and τ(FSSF) 87.7(4)°. The extremely short S-S bond is of particular interest, and *ab initio* calculations were performed to try to understand the reasons for this shortness. Self-consistent field calculations did not account completely for the short bond, but once allowance was made for electron correlation at the MP2 level with a double-zeta basis, the structure was reproduced reasonably well. It appears that there is a substantial degree of (p-p)π bonding between the sulphur atoms, which is lost if the molecule twists into either the *syn* or the *anti* planar configuration, so both barriers to rotation and the torsional vibration frequency are very high. The calculations show that in both planar forms the S-S bond is much longer than in the equilibrium skew form. These calculations also predict correctly that FSSF is less stable than SSF_2, although the lower level calculations led to the reverse conclusion. Experimental parameters for SSF_2 (r_{av}) were r(S=S) 1.856(2), r(S-F) 1.608(2), <SSF 108.1(2) and <FSF 91.7(3)°. It can be seen that the double sulphur-sulphur bond is not much shorter than the nominally single bond in the other isomer.

An unusually long S-S bond is found in S_2F_{10},[37] in which the r_a distance is 2.274(5) Å. It seems that there is a systematic lengthening of S-S bonds as the oxidation state of the sulphur atoms is increased. The two sulphur atoms are very nearly exactly octahedral [<$S_{ax}SF_{eq}$ 89.8(1)°] although the equatorial S-F bonds are 0.027(6) Å longer than the axial bonds. The mean S-F distance in this compound is 1.569(2) Å, so the axial bonds, at 1.547(6) Å, are shorter than in SF_6. The two SF_5 groups are staggered, and if strict D_{4d} symmetry was relaxed the effective torsion angle was 7.7(5)°. Assuming a cosine potential function and a classical distribution, a barrier to internal rotation of 10.2 kJ mol^{-1} was deduced.

Both trifluoro(methyl)sulphurane, CH_3SF_3, and trifluoro-(trifluoromethyl)sulphurane, CF_3SF_3, have been studied by gas-phase electron diffraction and *ab initio* calculations, including allowance for electron correlation at the MP2 level.[38] In each molecule the alkyl group is in an equatorial site of a trigonal bipyramidal structure and the equatorial S-F bond eclipses one of the C-H or C-F bonds, giving molecular C_s symmetry. The eclipsed conformation is rather unexpected, particularly for the trifluoromethyl compound, but it is predicted by the *ab initio* calculations. The most significant parameters (r_a) for CH_3SF_3 are r(S-F_{ax}) 1.689(1), r(R-F_{eq}) 1.575(5), r(S-C) 1.790(13) Å, <$S_{ax}SF_{ax}$ 174.6(8) and <CSF_{eq} 102.9(8)°. For CF_3SF_3, parameters include r(S-F_{ax}) 1.679(4), r(S-F_{eq}) 1.596(11), r(S-C) 1.911(7), r(C-F) (mean) 1.324(2) Å, <$F_{ax}SF_{ax}$ 165.2(25), <CSF_{eq} 100.7(16), <SCF 107.5(6)°. In both cases the axial fluorines are bent away from the lone pair of electrons on the sulphur atom, and the equatorial atoms are also bent away from this lone pair, as indicated by the CFS_{eq} angles, which are much less than 120°. These angles were reproduced well by the *ab initio* calculations, as were the distances, with the exception of the S-C bond length in CF_3SF_3. Even with the highest level calculations this was underestimated by 0.04 Å, a feature which has been noted before for bonds between two second-row atoms in highly fluorinated compounds.

Sulphur di-imides, RNSNR, with *syn,syn*, *syn,anti* and *anti,anti* configurations are known, so it was of interest to know what conformation was adopted by $Me_3SiNSNSiMe_3$ in the gas phase.[39] In fact it has a very distorted *syn,syn* arrangement, with the $SiMe_3$ groups each twisted 42(1)° out of the NSN plane by rotation about the S=N bonds, giving molecular C_2 symmetry. Parameters (r_a) include r(S=N) 1.536(3), r(Si-N) 1.738(3) and r(Si-C) 1.869(1) Å, <NSN 129.5(10) <SNSi 132.9(7)°. *Ab initio* calculations on the model compound $H_3SiNSNSiH_3$ predict a much wider angle at nitrogen, but these calculations take no account of electron correlation, nor of the substantial electronic and steric effects of the six methyl groups on the silicon atoms.

Para-benzene dithiol has C_{2h} symmetry, and it was assumed that the $C_6H_4S_2$ core had D_{2h} symmetry.[40] The mean r_g C-C distance was 1.399(3) Å, and the difference between the lengths of the four bonds adjacent to the substituted atoms and the other two was 0.005(5) Å, not very significant. Similarly, the CCC angle at the substituted atoms was 120.1(2)°, so the SH group clearly has little effect on the ring geometry. The C-S bond length is reported to be 1.775(4) Å, r(S-H) is 1.359(11) Å and the angle CSH 96.5(20)°. The data are consistent with the large torsional motions about the C-S bonds, with mean twist angles of up to 30°.

A chair conformation with C_{3v} symmetry has been observed for 1,3,5-trithiane, which is considerably more puckered than cyclohexane, presumable a consequence of the narrow angles at the sulphur atoms.[41] The angles in the r_g structure are 115.8(1)° at carbon and 99.1(4)° at sulphur, and the S-C bond length is 1.812(4)°. The SCSC dihedral angle is 65.2(5)°.

The structures of tetrahydrothiophene-1-oxide and its *cis*-2-methyl derivative have both been studied by electron diffraction.[42] Both are said to have asymmetric ring conformations, which can be represented as points on the pseudorotational pathways which start with half-chair rings with C_2 symmetry. However, the description of molecules with such low symmetry and with very large amplitude motions in terms of single rigid structures is not always helpful, and there must be doubts about whether the reported data really represent structures at or near the potential energy minima. In each molecule the S=O bond occupies an axial position and the methyl group in the second compound is therefore in an equatorial position. The following parameters were reported, with the value for the unsubstituted compound given first in each case: r(S-C) 1.828(4), 1.834(4); r(S=O) 1.484(3), 1.485(3) Å, <CSC 92.0(3), 91.7(4) and <CSO 110.0(4), 105.3(3)°. The C-C distances and angles at carbon are generally as expected, although one SCC angle in the parent compound is as small as 102.2(3)°.

Geometrical parameters obtained for perfluorotetrahydrothiophene, its 1-oxide and its 1,1-dioxide[43] are given in Table 4. The large variations in S-C distances are consistent with those observed previously in $S(CF_3)_2$ and its oxidised derivatives. The rings are all strongly puckered with half-chair conformations and C_2 symmetry. The degree of puckering can be represented by the twist angle of the C(3)-C(4) bond about the C_2 axis. This angle was 25.2(5), 24.6(5) and 23.2(6)° in the three compounds, and it is noted that the effect of perfluorination is to decrease the degree of puckering in the sulphide and sulphoxide, but to increase it in the sulphone.

Table 4 - Geometrical parameters for perfluorotetrahydrothiophene and its derivatives

	C_4F_8S	C_4F_8SO	$C_4F_8SO_2$
r(C–C)(mean)/Å	1.548(4)	1.553(5)	1.556(6)
r(S-C)/Å	1.822(4)	1.903(4)	1.882(7)
r(C-F)(mean)/Å	1.338(2)	1.337(2)	1.327(3)
r(S=O)/Å		1.464(5)	1.427(4)
<OSO/°			125.9(24)
<CSO/°		104.3(11)	108.2(8)
<CSC/°	94.5(3)	89.4(4)	93.2(10)
<SCC/°	107.4(3)	110.2(4)	107.7(10)
<CCC/°	106.4(4)	106.5(6)	107.9(12)

In a study of *trans*-methyl(chlorovinyl)sulphone reported in 1987 it was necessary to constrain many parameters, because the number of independent distances and angles was far greater than could be determined from electron diffraction data. Now the analysis has been repeated using constraints obtained from *ab initio* calculations at the 4-21G* level. As there are two conformers present, it was necessary to fix all the difference parameters, both between conformers and between related paramaters within each molecule, e.g. between S-C(methyl) and S-C(vinyl). Differences between S-C and C-Cl, and between S=O and C-C distances were also fixed. This led to unrealistically short S-C and C=C bonds, so some of the differences were changed, somewhat arbitrarily, and then acceptable geometry was obtained. The precise values of distances and angles are clearly not very reliable in these circumstances. One point of interest is that in one of the conformers the C=C bond eclipses one of the S=O bonds. One hydrogen atom of the phenyl group is thus close (2.42 Å) to an oxygen atom, and it is probable that intramolecular hydrogen-bonding stabilises this form.

6 Compounds of Transition Metals

Studies of metal halides at high temperatures continue to be reported, and particularly attention has been paid to investigating whether dihalides have linear or bent equilibrium structures. Data for zinc difluoride, volatalised at 1320 K,[45] and cadmium dibromide, for which 750 K was adequate,[46] have been published this year. The thermally averaged Zn-F distance (r_g) in ZnF_2 is 1.742(4) Å, and the F···F distance shows a shrinkage of 0.031(20) Å (i.e. it is this much less than twice the bond length). Analysis based on diffraction data and experimental vibration frequencies, refining four *F*-matrix elements and the bond length, assuming a curvilinear bending coordinate, (r_e^{ch}) gave 1.743(4) Å for the Zn-F distance. The force field was consistent with a bending frequency of 160 cm^{-1}. For $CdBr_2$ the r_g Cd-Br distance was 2.394(2) Å, with a Br···Br shrinkage of 0.045(8) Å, and the r_e^{ch} distance was 2.395(3) Å. The equilibrium distance, r_e, was estimated to be 2.386(5) Å, and the bending frequency to be 57(13) cm^{-1}.

Copper(II) fluoride on vapourisation at 1220 K yielded a mixture of CuF_2 and CuF.[47] The best fit to the electron diffraction data was obtained assuming equal amounts of both compounds to be present, but nothing in the range of 35 to 85% of CuF_2 could be excluded. A bond length in CuF obtained spectroscopically (1.7449 Å) was assumed and the refined distance in the CuF_2 depended on the proportions of the two compounds assumed to be present. Allowing for this uncertainty gave r_g 1.713(12), and the F···F shrinkage was 0.023(47) Å, so the molecule is linear within experimental error. The bond length is the shortest for the first transition series difluorides. The bonds shorten regularly from MnF_2 to CuF_2, and there is then an increase to ZnF_2.

Four transition metal trihalides have also been studied. In studies of chromium trifluoride and lanthanum trichloride[48] analysis was based on electron diffraction data and vibrational frequencies, and yielded the bond lengths, harmonic force constants, and an angle describing the position of the potential minimum for the out-of-plane bending vibration. For CrF_3 the latter term was 0.2(4)° and the Cr-F bond distance was 1.720(1) Å. For $LaCl_3$ the La-Cl distance was 2.560(2) Å, and the potential minimum corresponded to displacement of the La-Cl bonds by 10.9(5)° from their positions in a planar structure. The barrier to inversion was about 240 cm^{-1}, with five vibrational levels below the barrier, so this compound can truely be said to be pyramidal. Two other trihalides, $TiCl_3$ and TiI_3 were generated by heating the corresponding tetrahalides with metallic titanium.[49] Mass spectrometry showed that at 1000 K the gas in both systems consisted predominantly of the trihalides. Analysis of the electron diffraction data showed that 96% purity had been obtained for $TiCl_3$, and 88% for TiI_3. Assuming trigonal planar structures, r_g(Ti-Cl) was found to be 2.203(5) Å and r(Ti-I) 2.568(6) Å with X···X shrinkages of 0.085(24) and 0.084(21) Å respectively. The latter agreed well with the value calculated from spectroscopic data, but the calculated value of the former was only 0.036 Å.

The structure of methyltitaniumtrichloride has been reinvestigated by electron diffraction,[50] using samples of CH_3TiCl_3 and CD_3TiCl_3. The Ti-Cl and Ti-C bond lengths (r_a) are the same as had been reported earlier, at 2.185(3) and 2.047(6) Å, and the ClTiC angle is also unchanged, at 105.6(2)°. However, the C-H bond length reported is 1.096(6) Å, much shorter than was earlier believed to be the case, and the TiCH angle is wider, at 109.0(17)°. The methyl group is thus perfectly normal, and there is no sign of deformation of the hydrogen atoms towards the titanium atom.

Tungsten hexamethoxide has an octahedral WO_6 core but because the ligands are bent at oxygen the overall symmetry is lower. Models with D_{3d}, S_6 and D_3 symmetry were tried in analyses of electron diffraction data.[51] The D_3 model gave the best fit, but it is possible that some S_6 conformer is also present. The D_{3d} form was definitely ruled out. The r_a bond lengths W-O and O-C were 190.2(3) and 140.1(3) Å respectively, <WOC was 132.4(4)° and the OWOC dihedral angle 61(2)°.

A similar problem arises with chromium tetra-tert-butoxide.[52] The angles at the chromium atoms are tetrahedral within experimental error, but the bent CrOC arrangement [142(2)°] lowers the overall symmetry, to S_4. The other major parameters are r_a(Cr-O) 175.1(7), r(O-C) 141.1(12), r(C-C) 151.7(7) Å and <OCC 107.8(4)°. With this geometry there is little crowding of the bulky tert-butoxide groups, but the four groups together shield the metal atom efficiently, so that the molecule appears to be almost spherical, with hydrogen atoms on the outside. This accounts for the low reactivity of the compound towards Lewis bases and for its high volatility.

The bis(dimethyldithiocarbamato) complexes of copper and zinc differ in a striking way: the copper atom has almost square-planar coordination (D_{2h} symmetry), whereas the zinc compound has a configuration with D_{2d} symmetry, close to tetrahedral.[53] It is believed that the crystal-field stabilisation energy in the copper compound is insufficient to overcome steric interactions. This observation shows the importance of studying such compounds in the gas phase, where there are no perturbations caused by the crystal lattices. The most important parameters (r_g, $<_\alpha$, with the parameter for the copper compound given before that for the zinc compound) are: r(M-S) 2.284(9), 2.384(8); r(C-S) 1.716(10), 1.727(10); r(C=N) 1.334(18), 1.351(17); r(C-N) 1.476(18), 1.479(17) Å, <C=N-C) 124.9(13), 122.5(12), <SMS(chelate) 78.8(7), 79.7(6)°.

A ferrocene-like compound, undreamed of just a few years ago, completes this years' report and reminds us that new compounds with unpredictable structures are continually being prepared.[54] (Cyclopentaphosphorus)(pentamethylcyclopentadienyl) iron(II) contains an iron atom sandwiched between P_5 and C_5Me_5 rings, which are coplanar and staggered. The molecule has C_{5v} symmetry. The P-P, Fe-C and Fe-P distances (r_a) are 2.117(4), 2.135(11) and 2.377(5) Å respectively. The P-P and Fe-C distances are reversed from those found in crystalline Fe(C_5Me_4Et)(P_5), and care was taken to check that the reversal is real, and not an artefact of the experiment. The P-P distance is unusually short, and it would be interesting to study related compounds. Can anyone make Fe(P_5)$_2$?

References

1. S.B. Lapshina, G.V. Girichev and S.A. Shlykov, Zh. Strukt. Khim., 1989, 30(3), 49.

2. G.V. Girichev, S.B. Lapshina and N.I. Giricheva, Zh. Strukt. Khim., 1989, 30(1), 42.

3. G.V. Girichev and S.B. Lapshina, Zh. Strukt. Khim., 1989, 30(3), 55.

4. N.I. Giricheva, G.V. Girichev, V.M. Petrov, V.A. Titov and T.P. Chusova, Zh. Strukt. Khim., 1988, 29(5), 46.

5. P.T. Brain, A.J. Downs, M. Fanfarillo, M.J. Goode, A.G. Massey, D.W.H. Rankin and H.E Robertson, J. Mol. Struct., 1989, 192, 163.

6. O.V. Dorofeeva, V.S. Mastryukov, L.V. Vilkov, A.V. Golubinskii and A.I. Kovredov, Zh. Strukt. Khim., 1988, 29(6), 137.

7. V.S. Mastryukov, A.V. Polyakov, A.V. Golubinskii, L.V. Vilkov, V.Z. Kampel and V.I. Bregadze, Metalloorganich Khimiya, 1989, 2, 684.

8. K.W. Brown, J.W. Nibler, K. Hedberg and L. Hedberg, J. Phys. Chem., 1989, 93, 5679.

9. H.G. Mack, H. Oberhammer and C.O. Della Vedova, J. Mol. Struct. (Theochem)., 1989, 200, 277.

10. O. Bastiansen, S. Gundersen and S. Samdal, Acta. Chem. Scand., 1989, 43, 6.

11. H.-G. Mack and H. Oberhammer, J. Mol. Struct., 1989, 197, 321.

12. P.D. Blair, A.J. Blake, R.W. Cockman, S. Cradock, E.A.V. Ebsworth and D.W.H. Rankin, J. Mol. Struct., 1989, 93, 279.

13. A.V. Belyakov, V.S. Zavgorodnii and V.S. Mastryukov, Zh. Strukt. Khim., 1989, 30(1), 34.

14. D.G. Anderson, D.W.H. Rankin and H.E. Robertson, J. Mol. Struct., 1989, 195, 253.

15. D.G. Anderson, D.W.H. Rankin, H.E. Robertson, C.M.F. Frazao and H. Schmidbaur, Chem. Ber., 1989, 122, 2213.

16. A.H. Cowley, M. Pakulski, D.G. Anderson, D.W.H. Rankin and H.E. Robertson, J. Mol. Struct., 1989, 196, 21.

17. P. Jutzi, U. Holtmann, D. Kanne, C. Krüger, R. Blom, R. Gleiter and I. Hyla-Kryspin, Chem. Ber., 1989, 122, 1629.

18. D.G. Anderson and D.W.H. Rankin, J. Mol. Struct., 1989, 195, 261.

19. D.G. Anderson and D.W.H. Rankin, J. Chem. Soc., Dalton Trans., 1989, 779.

20. N.I. Giricheva, G.V. Girichev, S.A. Shlykov, V.A. Titov and T.P. Chusova, Struktura i Svoestva Molekul, Ivanovo, 1988, 105.

21. A.V. Belyakov, A.V. Golubinskii, L.V. Vilkov, N.V. Girbasova, E.T. Bogoradovskii and V.S. Zavgorodnii, Zh. Strukt. Khim., 1989, 30(4), 53.

22. O.A. Litvinov, L.V. Ermolaeva, V.V. Zverev and V.A. Naumov, Zh. Strukt. Khim., 1989, 30(2) 64.

23. H. Oberhammer, C.W. Bauknight and D.D. DesMarteau, Inorg. Chem., 1989, 28, 4340.

24. H.-G. Mack and H. Oberhammer, J. Mol. Struct., 1989, 197, 329.

25. N.I. Sadova, A.A. Ivanov, M.V. Popik, I.F. Shishkov, L.V. Vilkov and Yu.A. Pankrushev, Zh. Strukt. Khim., 1989, 30(2), 56.

26. O.G. Batyukhnova, N.I. Sadova, Yu.M. Syshchikov, L.V. Vilkov and Yu.A. Pankrushev, Zh. Strukt. Khim., 1988, 29(4), 53.

27. E.I. Kulikova, N.I. Sadova, M.V. Popik, A.A. Ivanov, L.V. Vilkov, N.F. Pyatakov and Yu. A. Pankrushev, Vest. Mosk. Univ. Ser. 2 (Khim.), 1989, 30(4), 338.

28. A.J. Downs, G.S. McGrady, E.A. Barnfield, D.W.H. Rankin and H.E. Robertson, J. Chem. Soc., Dalton Trans., 1989, 545.

29. A.J. Blake, M.J. Davis, E.A.V. Ebsworth, D.W.H. Rankin and A.J. Welch, J. Chem. Soc., Dalton Trans., 1989, 223.

30. N.M. Zaripov, R.G. Khusnutdinov and V.A. Naumov, Zh. Strukt. Khim., 1989, 30(5), 55.

31. N.M. Zaripov, R.G. Khusnutdinov, I.A. Litvinov, L.V. Ermolaeva and V.A. Naumov, Zh. Strukt. Khim., 1989, 30(5), 61.

32. N.M. Zaripov, R.G. Khusnutdinov, V.A. Naumov and R.R. Shagidullin, Zh. Strukt. Khim., 1989, 30(3), 61.

33. N.M. Zaripov, R.G. Khusnutdinov, V.A. Naumov, R.R. Shagidullin and G.I. Kokarev, Zh. Strukt. Khim., 1988, 29(4), 64.

34. Q. Shen and R.T. Hemmings, J. Mol. Struct., 1989, 197, 349.

35. A.V. Belyakov, G.A. Sherbak and L.V. Vilkov, Zh. Strukt. Khim., 1989, 30(4), 47.

36. C.J. Marsden, H. Oberhammer, O. Lösking and H. Willner, J. Mol. Struct., 1989, 193, 233.

37. H. Oberhammer, J. Mol. Struct., 1989, 192, 171.

38. A.J. Downs, G.S. McGrady, E.A. Barnfield, D.W.H. Rankin, H.E. Robertson, J.E. Boggs and K.D. Dobbs, Inorg. Chem., 1989, 28, 3286.

39. D.G. Anderson, D.W.H. Rankin, H.E. Robertson and J.D. Woollins, J. Chem. Soc., Dalton Trans., 1989, 859.

40. G. Portalone, A. Domenicano, Gy. Schultz and I. Hargittai, J. Mol. Struct. (Theochem)., 1989, 186, 185.

41. Z. Bencze, A. Kucsman, G. Schultz and I. Hargittai, Acta Chem. Scand. 1989, 43, 953.

42. Gy. Forgács, Gy. Schultz, I. Hargittai, I. Jalsovszky and A. Kucsman, J. Chem. Soc., Faraday II., 1989, 85, 303.

43. H.-G. Mack, H. Oberhammer, J.M. Shreeve and X.-B. Xia, J. Mol. Struct., 1989, 196, 57.

44. E. Vajda, P. Friedman, I. Hargittai, D. Hnyk, L. Schäfer and K. Siam, J. Mol. Struct., 1989, 213, 309.

45. G.V. Girichev, A.G. Gershikov and N.Yu. Subbotina, Zh. Strukt. Khim., 1989, 29(6), 139.

46. A.G. Gershikov, Zh. Strukt. Khim., 1989, 30(6), 169.

47. N.Yu. Subbotina, G.V. Girichev and V.V. Ostropikov, Zh. Strukt. Khim., 1989, 30(4), 42.

48. V.P. Spiridonov, A.G. Gershikov and V.S. Lyutsarev, Vest. Mosk. Univ., Ser. 2 (Khim.), 1989, 30(1), 23.

49. G.V. Girichev, S.A. Shlykov, V.N. Petrova, N.Yu. Subbotina, S.B. Lapshina and T.G. Danilova, Izv. Viz., Khim. Khim. Technol., 1988, 31(8), 46.

50. P. Briant, J. Green, A. Haaland, H. Møllendal, K. Rypdal and J. Tremmel, J. Am. Chem. Soc., 1989, 111, 3434.

51. A. Haaland, K. Rypdal, H.V. Volden, E. Jacob and J. Weidlein, Acta Chem. Scand., 1989, 43, 911.

52. E.G. Thaler, K. Rypdal, A. Haaland and K.G. Caulton, Inorg. Chem., 1989, 28, 2431.

53. K. Hagen, C.J. Holwill and D.A. Rice, Inorg. Chem., 1989, 28, 3239.

54. R. Blom, T. Brück and O.J. Scherer, Acta Chem. Scand., 1989, 43, 458.